INTEGRABLE HAMILTONIAN SYSTEMS

SYSTEMS

Geometry, Topology, Classification

CRC Press
Taylor & Francis Group
Boca Raton London New York

CRC Press is an imprint of the
Taylor & Francis Group, an **informa** business

A CHAPMAN & HALL BOOK

INTEGRABLE HAMILTONIAN SYSTEMS

SYSTEMS

Geometry, Topology, Classification

A. V. Bolsinov and A. T. Fomenko

CRC Press
Taylor & Francis Group
6000 Broken Sound Parkway NW, Suite 300
Boca Raton, FL 33487-2742

First issued in paperback 2019

© 2004 by Chapman & Hall/CRC
CRC Press is an imprint of Taylor & Francis Group, an Informa business

No claim to original U.S. Government works

ISBN-13: 978-0-415-29805-6 (hbk)
ISBN-13: 978-0-367-39450-9 (pbk)

Library of Congress Cataloging-in-Publication Data

Bolsinov, A. V. (Aleksei Viktorovich)
[Integriruemye gamil'tonovy sistemy. English]
Integrable Hamiltonian systems : geometry, topology, classification / by A.V. Bolsinov, A.T. Fomenko.
p. cm.
Includes bibliographical references and index.
ISBN 0-415-29805-9 (alk. paper)
1. Hamiltonian systems. 2. Geodesic flows. 3. Geodesics (Mathematics) I. Fomenko, A. T. II. Title.

QA614.83.B6413 2004
515'.39—dc22   2003067457

Visit the Taylor & Francis Web site at
http://www.taylorandfrancis.com

and the CRC Press Web site at
http://www.crcpress.com

Contents

Preface

The aim of this book is to discuss new qualitative methods in the study of integrable Hamiltonian systems.

It is well known that many systems of differential equations that appear in physics, geometry, and mechanics and describe quite different phenomena, turn out nevertheless to be closely connected (in some sense, similar). Studying such links (in other words, diffeomorphisms of various kinds between different systems) has been a subject of many papers since Maupertuis, Euler, Jacobi, and Minkowski. Nowadays this question (in connection with integrability problems) has been discussed in papers of S. Smale, J. Marsden, J. Moser, M. Adler, H. Knörrer, L. Gavrilov, V. V. Kozlov, S. P. Novikov, A. P. Veselov, A. I. Bobenko, and others.

What kind of isomorphisms do we mean? Depending on the statement of a problem they can be quite different. In the present book we shall mainly discuss the following three types of equivalence relations among dynamical systems: conjugacy, orbital equivalence (topological and smooth), and Liouville equivalence (in the case of integrable systems). We recall that two smooth dynamical systems σ^t and σ'^t are called topologically (smoothly) conjugate if there exists a homeomorphism (diffeomorphism) ξ between the manifolds on which the systems are given transforming the systems to each other, i.e., $\sigma'^t = \xi \sigma^t \xi^{-1}$. In other words, the conjugacy means that the systems under consideration are actually identical. Using another terminology, one can simply say that the systems are transformed into each other by means of a certain change of variables (without changing the time).

The second equivalence relation, namely orbital equivalence, is somewhat weaker. It is supposed that the homeomorphism (diffeomorphism) ξ maps the trajectories of the first system to those of the second one (in general, without preserving the parameter t (time) on these trajectories). It is clear that conjugate systems are orbitally equivalent, but not vice versa. Nevertheless, the orbital equivalence relation is rather strong. In particular, all qualitative properties of dynamical systems (such as stability, the structure of limit sets, types of closed trajectories, etc.) are preserved under orbital isomorphisms.

The third equivalence relation, the so-called Liouville equivalence, appears in the case of Hamiltonian dynamical systems that are integrable in Liouville sense. Two such systems are said to be Liouville equivalent if their phase spaces are foliated in the same way into Liouville tori.

The question on the classification of dynamical systems with respect to these equivalence relations is classical. Among many well-known results in this direction (i.e., in the classification theory for dynamical systems), it is worth mentioning:

a) local theory in a neighborhood of an equilibrium point or a closed trajectory (H. Poincaré, H. Dulac, G. D. Birkhoff, K. T. Chen, and others; for the present state of this theory, see reviews [9], [16]);

b) global classification of Morse–Smale flows (E. A. Leontovich, A. G. Maier [210], [211], M. M. Peixoto [286], Y. A. Umanskiĭ [350]) and flows of a special type on two-dimensional surfaces (S. Kh. Aranson, V. Z. Grines [14]).

c) study of the topology of integral manifolds and Liouville foliations of integrable systems (S. Smale [316], A. A. Oshemkov [277], M. P. Kharlamov [178], [179], T. I. Pogosyan [293], [294], [295], Ya. V. Tatarinov [334], Nguyen Tien Zung [260], [261], R. Cushman and L. Bates [89], R. Cushman and H. Knörrer [90], M. Audin [22], [23], L. Gavrilov [141], [142], [143]).

It is clear that the solution of the classification problem for smooth dynamical systems of general type is hardly possible. That is why it would be natural to confine oneself to consideration of some special class of systems with similar properties.

In this book we present the solution to the Liouville and orbital classification problems for one of the most important classes of dynamical systems, namely, non-degenerate integrable Hamiltonian systems with two degrees of freedom.

The basis of this classification is a new approach in the qualitative theory of integrable Hamiltonian systems proposed by A. T. Fomenko [117], [118], [123], and the theory of topological classification for such systems developed then jointly with H. Zieschang, S. V. Matveev, A. V. Bolsinov, and A. V. Brailov in a series of papers [46], [47], [50], [53], [65], [73], [134], [135].

A. T. Fomenko proposed to assign to each integrable Hamiltonian system a certain graph W as a topological invariant of the system, the so-called *molecule*. By means of this invariant, it is possible to describe completely the structure of the foliation of the isoenergy surface into invariant Liouville tori and to classify, as a result, such systems up to Liouville equivalence. As a final invariant, A. T. Fomenko and H. Zieschang introduced the so-called marked molecule W^*.

This molecule W^* can be naturally considered as a portrait of the integrable Hamiltonian system, which contains much very useful information on it. However, in the case of orbital classification we have to solve a more delicate problem. Namely, we need to describe the foliation of an isoenergy surface into integral trajectories (but not only into Liouville tori) and to classify the systems up to orbital equivalence. It is clear that for this we must make the portrait of the system more detailed by completing it (i.e., the marked molecule) with new information (i.e., orbital invariants). In other words, the problem is to find and to describe a complete set of additional orbital invariants (both in the topological and smooth cases). This problem has been solved by the authors, and its solution is presented as part of this book.

At present, many interesting results have been obtained in the study on the topology of integrable systems. In particular, the topological portraits have been described and the Liouville classification has been given for many specific physical and mechanical systems with two degrees of freedom. We mention here papers by E. V. Anoshkina, V. V. Kalashnikov (Jr.), B. S. Kruglikov, V. S. Matveev, Nguyen Tien Zung, O. E. Orel, A. A. Oshemkov, L. S. Polyakova, E. N. Selivanova, P. Topalov, V. V. Trofimov, P. Richter, H. Dullin, A. Wittek, L. Gavrilov, M. Audin, R. Cushman, L. Bates, and H. Knörrer.

The present book is an introduction to the problem of the classification of integrable systems. One of its features is a systematic character of the presentation of a large material which was previously available only in journal papers. We aimed to talk about all of these studies in a simple and understandable way, counting also upon students in physics, mathematics, and mechanics.

In the first part of our book (Chapters 1–9), we present the foundation of the classification theory and related topics in the topology of integrable systems.

The second part (Chapters 10–16) presents various applications of the classification theory in physics, mechanics, and geometry. We discuss here general topological methods for the analysis of specific dynamical systems, without touching upon the algebro-geometric approach which is also a powerful tool in studying the qualitative properties of algebraically integrable systems, their Liouville foliations, bifurcation of tori, etc. This is a very interesting and branched subject, which is presented, in particular, in the book [22] by M. Audin. We concentrate our attention on more general methods which work both in the cases of algebraic integrability and when there is no algebraic background.

Two classes of integrable systems are considered in detail. These are integrable cases in rigid body dynamics and integrable geodesic flows of Riemannian metrics on two-dimensional surfaces.

Of course, many applications of the theory of topological invariants of integrable systems have remained out of this book. Moreover, this direction is still actively developing. For example, just recently new interesting results have been obtained by Yu. A. Brailov, N. V. Korovina, E. A. Kudryavtseva, V. V. Korneev, V. O. Manturov, and E. Ya. Tatarinova.

In conclusion we would like to express our deep gratitude to V. V. Kozlov, S. V. Matveev, H. Zieschang, V. V. Sharko, I. K. Babenko, Ya. V. Tatarinov, I. A. Taimanov, A. M. Stepin, Yu. N. Fedorov, N. N. Nekhorochev, J. Marsden, L. Gavrilov, P. Richter, H. Dullin, and A. Wittek for extremely useful discussions, as well as to our students; permanent contacts with them have been very important for us. Special thanks are due to Andrey Oshemkov for preparing the manuscript for publication, his remarks and comments.

The work on this book was supported by the Russian Foundation for Basic Research (grants 02-01-00998 and 01-01-00583).

At present, many interesting results have been obtained in the study on the topology of integrable systems. In particular, the topological portraits have been described and the Liouville classification has been given for many specific physical and mechanical systems with two degrees of freedom. We mention here papers by E. V. Anoshkina, V. V. Kalashnikov (Jr.), B. S. Kruglikov, V. S. Matveev, Nguyen Tien Zung, O. E. Orel, A. A. Oshemkov, L. S. Polyakova, E. N. Selivanova, P. Topalov, V. V. Trofimov, P. Richter, H. Dullin, A. Wittek, L. Gavrilov, M. Audin, R. Cushman, L. Bates, and H. Knörrer.

The present book is an introduction to the problem of the classification of integrable systems. One of its features is a systematic character of the presentation of a large material which was previously available only in journal papers. We aimed to talk about all of these studies in a simple and understandable way, counting also upon students in physics, mathematics, and mechanics.

In the first part of our book (Chapters 1–9), we present the foundation of the classification theory and related topics in the topology of integrable systems.

The second part (Chapters 10–16) presents various applications of the classification theory in physics, mechanics, and geometry. We discuss here general topological methods for the analysis of specific dynamical systems, without touching upon the algebro-geometric approach which is also a powerful tool in studying the qualitative properties of algebraically integrable systems, their Liouville foliations, bifurcation of tori, etc. This is a very interesting and branched subject, which is presented, in particular, in the book [22] by M. Audin. We concentrate our attention on more general methods which work both in the cases of algebraic integrability and when there is no algebraic background.

Two classes of integrable systems are considered in detail. These are integrable cases in rigid body dynamics and integrable geodesic flows of Riemannian metrics on two-dimensional surfaces.

Of course, many applications of the theory of topological invariants of integrable systems have remained out of this book. Moreover, this direction is still actively developing. For example, just recently new interesting results have been obtained by Yu. A. Brailov, N. V. Korovina, E. A. Kudryavtseva, V. V. Korneev, V. O. Manturov, and E. Ya. Tatarinova.

In conclusion we would like to express our deep gratitude to V. V. Kozlov, S. V. Matveev, H. Zieschang, V. V. Sharko, I. K. Babenko, Ya. V. Tatarinov, I. A. Taimanov, A. M. Stepin, Yu. N. Fedorov, N. N. Nekhorochev, J. Marsden, L. Gavrilov, P. Richter, H. Dullin, and A. Wittek for extremely useful discussions, as well as to our students; permanent contacts with them have been very important for us. Special thanks are due to Andrey Oshemkov for preparing the manuscript for publication, his remarks and comments.

The work on this book was supported by the Russian Foundation for Basic Research (grants 02-01-00998 and 01-01-00583).

Chapter 1

Basic Notions

1.1. LINEAR SYMPLECTIC GEOMETRY

Definition 1.1. A *symplectic space* is defined as a vector (real or complex) space V endowed with a non-degenerate bilinear skew-symmetric form

$$\omega(a,b) = \sum \omega_{ij} a^i b^j .$$

This form is called a *symplectic structure* on V.

If we fix a basis $e_1, \ldots, e_m$ in V, then ω is uniquely defined by its matrix $\Omega = (\omega_{ij})$, where $\omega_{ij} = \omega(e_i, e_j)$. This matrix is skew-symmetric and non-degenerate. This implies immediately that the dimension of the symplectic space V is even, since

$$\det \Omega = \det \Omega^\top = \det(-\Omega) = (-1)^m \det \Omega ,$$

where $m = \dim V$.

Two symplectic spaces V and V' of the same dimension are *isomorphic*, that is, there exists a linear isomorphism $h : V \to V'$ such that $\omega(a,b) = \omega'(ha, hb)$ for any vectors a and b. This follows from the so-called *linear Darboux theorem*:

Proposition 1.1. *In a symplectic space V of dimension $2n$, there exists a basis $e_1, \ldots, e_n, f_1, \ldots, f_n$ in which the matrix Ω has the form*

$$J = \begin{pmatrix} 0 & E \\ -E & 0 \end{pmatrix} ,$$

where $E = E_n$ is the identity $(n \times n)$-matrix.

Such a basis is called *canonical* or *symplectic*.

Definition 1.2. A linear subspace L in V is called *isotropic* if the form ω vanishes on L, that is, $\omega(a,b) = 0$ for any $a, b \in L$. A maximal isotropic subspace is called a *Lagrangian subspace*.

It is easy to verify that an isotropic subspace L is Lagrangian if and only if its dimension is equal to n. As an example of Lagrangian subspaces, one can consider n-dimensional planes spanned on the vectors $e_1, \ldots, e_n$ or on the vectors $f_1, \ldots, f_n$ of the canonical basis.

Definition 1.3. A linear transformation $g: V \to V$ is called *symplectic* if it preserves the symplectic structure, i.e., $\omega(a, b) = \omega(ga, gb)$ for any vectors $a, b \in V$.

Definition 1.4. The set of all symplectic transformations $g: V \to V$ forms a group, which is called the *symplectic group* and is denoted by $\mathrm{Sp}(2n, \mathbb{R})$ (or $\mathrm{Sp}(2n, \mathbb{C})$ in the complex case), where $2n = \dim V$.

Proposition 1.2.

a) *Symplectic transformations are unimodular, i.e.,* $\det g = 1$ *for any* $g \in \mathrm{Sp}(2n, \mathbb{R})$ *(or* $g \in \mathrm{Sp}(2n, \mathbb{C})$*).*

b) *The characteristic polynomial* $P(\lambda) = \det(g - \lambda E)$ *of a symplectic transformation* g *satisfies the property*

$$P(\lambda) = \lambda^{2n} P\left(\frac{1}{\lambda}\right).$$

In particular, if λ *is an eigenvalue of* g*, then* λ^{-1} *is an eigenvalue of the same multiplicity.*

Proof. a) To prove the first statement, it suffices to consider the $2n$-form $\tau = \underbrace{\omega \wedge \omega \wedge \ldots \wedge \omega}_{n \text{ times}}$. Since ω is non-degenerate, τ is a non-zero form of the maximal degree on V. Therefore τ can be interpreted as an oriented volume form. A symplectic transformation g preserves any power of ω and, in particular, the volume form $\tau = \omega^{(n)}$. Hence, $\det g = 1$.

b) It follows from the definition of a symplectic transformation that $g^{\top} \Omega g = \Omega$. Let us rewrite this relation as $g = \Omega^{-1} g^{-1^{\top}} \Omega$. Hence

$$P(\lambda) = \det(g - \lambda E) = \det(\Omega^{-1} g^{-1^{\top}} \Omega - \lambda E)$$
$$= \det \Omega^{-1}(g^{-1} - \lambda E)^{\top} \Omega = \det(g^{-1} - \lambda E) = \det g^{-1} \det(E - \lambda g).$$

Since $\det g = \det g^{\top} = 1$, we finally have

$$P(\lambda) = \det(E - \lambda g) = \det \lambda\left(\frac{1}{\lambda} E - g\right) = \lambda^{2n} \det\left(g - \frac{1}{\lambda} E\right) = \lambda^{2n} P\left(\frac{1}{\lambda}\right). \quad \square$$

The following statement describes the properties of the real symplectic group.

Proposition 1.3.

a) $\mathrm{Sp}(2n, \mathbb{R})$ *is a non-compact real Lie group of dimension* $n(2n + 1)$.

b) *The Lie algebra* $\mathrm{sp}(2n, \mathbb{R})$ *of this group consists of the matrices* A *satisfying the relation* $A^{\top} \Omega + \Omega A = 0$. *If the basis is canonical, i.e.,* $\Omega = J$*, then*

$$A = \begin{pmatrix} A_1 & A_2 \\ A_3 & -A_1^{\top} \end{pmatrix},$$

where A_1 *is an arbitrary* $(n \times n)$*-matrix, and the matrices* A_2 *and* A_3 *are symmetric.*

c) *From the topological point of view the symplectic group* $\mathrm{Sp}(2n, \mathbb{R})$ *is diffeomorphic to the direct product of the unitary group* $\mathrm{U}(n)$ *and the Euclidean vector space* $\mathbb{R}^{n(n+1)}$.

d) $\mathrm{Sp}(2n, \mathbb{R})$ *is (arcwise) connected.*

e) $\mathrm{Sp}(2n, \mathbb{R})$ *is not simply connected and its fundamental group is* $\mathbb{Z}$.

Proof. a) and b) Without loss of generality we shall assume $\Omega = J$ (by using a canonical basis). Then $\mathrm{Sp}(2n, \mathbb{R})$ can be presented as a subgroup in the group $\mathrm{GL}(2n, \mathbb{R})$ given by the matrix equation $g^{\top} J g = J$, which can be considered as a system of polynomial (namely, quadratic) equations. In other words, $\mathrm{Sp}(2n, \mathbb{R})$ is a linear algebraic group and, consequently, as all such groups, is a Lie group [156]. Its non-compactness follows, for example, from the fact that the matrices $\mathrm{diag}(\lambda, \ldots, \lambda, \lambda^{-1}, \ldots, \lambda^{-1})$ are symplectic for any $\lambda \in \mathbb{R}$.

To compute the dimension of the symplectic group, let us find its tangent space at the unit $e \in \mathrm{Sp}(2n, \mathbb{R})$. In other words, let us describe the corresponding Lie algebra $\mathrm{sp}(2n, \mathbb{R})$. Suppose A is an arbitrary element from the tangent space $T_e \mathrm{Sp}(2n, \mathbb{R}) = \mathrm{sp}(2n, \mathbb{R})$. Then there exists a smooth curve $g(t) \subset \mathrm{Sp}(2n, \mathbb{R})$ such that $g(0) = E$ and $\dfrac{dg}{dt}(0) = A$. By differentiating the relation $g^{\top}(t) J g(t) = J$ for $t = 0$, we obtain $\dfrac{dg^{\top}}{dt}(0) J + J \dfrac{dg}{dt}(0) = 0$, or $A^{\top} J + J A = 0$.

Conversely, let a matrix A satisfy the relation $A^{\top} J + J A = 0$. Consider the smooth curve $g(t) = \exp(tA)$ and show that it entirely belongs to the group $\mathrm{Sp}(2n, \mathbb{R})$. Indeed, let us differentiate the expression $g^{\top} J g(t)$, taking into account that $\dfrac{d}{dt} \exp(tA) = A \exp(tA)$. We get

$$\frac{d}{dt}(g^{\top} J g(t)) = \left(\frac{d}{dt} \exp(tA) \right)^{\top} J \exp(tA) + (\exp(tA))^{\top} J \frac{d}{dt} \exp(tA)$$

$$= (\exp(tA))^{\top} (A^{\top} J + J A) \exp(tA) = 0.$$

Thus, $g^{\top} J g(t)$ is a constant matrix. But for $t = 0$, we have $g(0) = E$; that is why, in fact, $g^{\top} J g(t) \equiv J$ for any t. Hence $g(t)$ entirely belongs to $\mathrm{Sp}(2n, \mathbb{R})$, and its tangent vector $\dfrac{dg}{dt}(0) = A$ belongs to the Lie algebra $\mathrm{sp}(2n, \mathbb{R})$.

Note that the relation $A^{\top} J + J A = 0$ means exactly that the matrix $J A$ is symmetric, and the mapping $A \to J A$ defines a linear isomorphism of the symplectic Lie algebra $\mathrm{sp}(2n, \mathbb{R})$ to the space of symmetric matrices. Therefore, the dimension of $\mathrm{Sp}(2n, \mathbb{R})$ is equal to that of the space of symmetric $(2n \times 2n)$-matrices, i.e., $\dim \mathrm{Sp}(2n, \mathbb{R}) = n(2n + 1)$. This proves (a) and (b).

c) Let us identify the symplectic space $\mathbb{R}^{2n}$ with the n-dimensional complex space $\mathbb{C}^n$. Consider the Hermitian form $(a, b) = a_1 \bar{b}_1 + \ldots + a_n \bar{b}_n$ in $\mathbb{C}^n$ and suppose that the symplectic structure in $\mathbb{R}^{2n} = \mathbb{C}^n$ coincides with the imaginary part of the Hermitian form: $\omega(a, b) = \mathrm{Im}(a, b)$.

On the other hand, in the same space one has the Euclidean structure $\langle a, b \rangle = \mathrm{Re}(a, b)$. The complex structure operator I (i.e., multiplication by the imaginary unit i) is uniquely defined by the relation $\langle Ia, b \rangle = \omega(a, b)$. In particular, in the canonical basis $e_1, \ldots, e_n, i e_1, \ldots, i e_n$, the matrices of the symplectic and complex structures coincide.

Recall that the transformations that preserve the Hermitian structure are called *unitary*. It is clear that the group of unitary transformations $U(n)$ is a subgroup of the symplectic group $\mathrm{Sp}(2n, \mathbb{R})$ (after identification of $\mathbb{R}^{2n}$ with $\mathbb{C}^n$). Moreover,

$$U(n) = \mathrm{Sp}(2n, \mathbb{R}) \cap O(2n, \mathbb{R}),$$

where $O(2n, \mathbb{R})$ is the group of transformations preserving the Euclidean structure $\langle \cdot, \cdot \rangle$.

Let $L \subset \mathrm{sp}(2n, \mathbb{R})$ be the subspace consisting of symmetric matrices. We now show that every symplectic matrix $g \in \mathrm{Sp}(2n, \mathbb{R})$ can uniquely be presented in the form $g = U \exp S$, where U is a unitary matrix, $S \in L$. And vice versa, each matrix $U \exp S$, where $U \in U(n)$ and $S \in L$, is symplectic.

To this end, consider the matrix $g^\top g$. It is symmetric and positively defined. Therefore, there exists a single positively defined symmetric matrix R such that $R^2 = g^\top g$. Next, for this matrix R there exists a single symmetric matrix S such that $R = \exp S$. Now we set $U = gR^{-1}$ and show that $g = UR = U \exp S$ is the desired decomposition.

First, let us check that U and R are symplectic. Since g, $g^\top$ are symplectic matrices and $R^2 = g^\top g$, we conclude that $R^2 = \exp 2S$ is symplectic. This implies immediately that all matrices of the form $\exp tS$ are symplectic.

Next, since $U = gR^{-1} = g \exp(-S)$, the matrix U is symplectic too. Besides, this matrix is orthogonal. Indeed, $U^\top U = (R^\top)^{-1} g^\top g R^{-1} = R^{-1} R^2 R^{-1} = E$. This means that the corresponding linear transformation preserves both the symplectic structure $\mathrm{Im}(\cdot, \cdot)$ and the Euclidean structure $\mathrm{Re}(\cdot, \cdot)$. In other words, this means that $U \in \mathrm{Sp}(2n, \mathbb{R}) \cap O(n, \mathbb{R}) = U(n)$.

Finally, being the tangent vector of the smooth curve $\exp(tS) \subset \mathrm{Sp}(2n, \mathbb{R})$, the symmetric matrix S belongs to the Lie algebra $\mathrm{sp}(2n, \mathbb{R})$.

The (polar) decomposition $g = U \exp S$ defines a diffeomorphism between the symplectic group $\mathrm{Sp}(2n, \mathbb{R})$ and the direct product $U(n) \times L$, where L is the subspace in the Lie algebra $\mathrm{sp}(2n, \mathbb{R})$ that consists of symmetric matrices. A straightforward calculation shows that $\dim L = n(n+1)$.

d) and e). The connectedness of the symplectic group follows the connectedness of $U(n)$ and L. Besides, $\pi_1(\mathrm{Sp}(2n, \mathbb{R})) = \pi_1(U(n)) = \mathbb{Z}$. This completes the proof. $\square$

1.2. SYMPLECTIC AND POISSON MANIFOLDS

Definition 1.5. A *symplectic structure* on a smooth manifold M is a differential 2-form ω satisfying the following two properties:

1) ω is closed, i.e., $d\omega = 0$;

2) ω is *non-degenerate* at each point of the manifold, i.e., in local coordinates, $\det \Omega(x) \neq 0$, where $\Omega(x) = (\omega_{ij}(x))$ is the matrix of this form.

The manifold endowed with a symplectic structure is called *symplectic*.

Is it possible to endow an arbitrary manifold with a symplectic structure? The answer is negative. The manifold should satisfy at least several natural restrictions.

Proposition 1.4. *A symplectic manifold is even-dimensional.*

Proof. This follows immediately from the fact that ω defines the structure of a symplectic space on each tangent space of the manifold. Since the form ω is non-degenerate, the tangent space has to be even-dimensional. □

Proposition 1.5. *A symplectic manifold is orientable.*

Proof. A manifold is orientable if on each tangent space one can naturally define an orientation which depends continuously on the point. In the case of a symplectic manifold M it can be done in the following way. Consider a differential form $\tau = \underbrace{\omega \wedge \omega \wedge \ldots \wedge \omega}_{n \text{ times}}$. It is clear that τ nowhere vanishes. Consider an arbitrary basis $e_1, e_2, \ldots, e_{2n}$ in an arbitrary tangent space $T_x M$ and assume its orientation to be positive by definition if $\tau(e_1, e_2, \ldots, e_{2n}) > 0$, and negative otherwise. In other words, a manifold is orientable if there exists a differential form of the maximal rank on it which nowhere vanishes. In our case, such a form exists; this is $\tau = \omega^{(n)}$. Sometimes, τ is called a symplectic volume form. □

Proposition 1.6. *If a symplectic manifold is compact, then its form ω represents a non-zero two-dimensional cohomology class. In particular, $H^2(M, \mathbb{R}) \neq 0$.*

Proof. Assume the contrary. Let the symplectic structure ω be exact and $\omega = d\alpha$. Consider then the $(2n - 1)$-form

$$\varkappa = \alpha \wedge \underbrace{\omega \wedge \ldots \wedge \omega}_{n-1 \text{ times}} .$$

It is easy to see that $d\varkappa = \omega \wedge \ldots \wedge \omega$ is the symplectic volume form. But then, taking into account the Stokes formula, we come to a contradiction:

$$\text{vol}(M) = \int_M \omega \wedge \ldots \wedge \omega = \int_{\partial M = \varnothing} \varkappa = 0.$$

□

The simplest example of a symplectic manifold is a two-dimensional orientable surface (the sphere with handles). The symplectic structure on it is simply the area form.

Another example is the symplectic space $\mathbb{R}^{2n}$ with the standard symplectic structure $\omega = dp_1 \wedge dq_1 + \ldots + dp_n \wedge dq_n$.

Besides these simplest examples, consider three more classes of symplectic manifolds: cotangent bundles, Kähler manifolds, and orbits of the coadjoint representation.

1.2.1. Cotangent Bundles

Let M be a smooth manifold (not necessary symplectic), and let T^*M be its cotangent bundle. First, consider the so-called *action 1-form* α on T^*M. Recall that a 1-form on a manifold is a function that assigns a number to every tangent vector. Let ξ be a tangent vector to the cotangent bundle at a point $(x, p) \in T^*M$. By definition, we set

$$\alpha(\xi) = p(\pi_*(\xi)),$$

where $\pi_*: T(T^*M) \to TM$ is the natural projection generated by the projection $\pi: T^*M \to M$. It is easy to see that, in local coordinates, α has the form

$$\alpha = p_1 dq_1 + \ldots + p_n dq_n,$$

where $q_1, \ldots, q_n$ are local coordinates on the manifold M and $p_1, \ldots, p_n$ are the corresponding coordinates on the cotangent space. As the symplectic structure on T^*M, we take the form $\omega = d\alpha$. Obviously, it satisfies all necessary conditions.

1.2.2. The Complex Space $\mathbb{C}^n$ and Its Complex Submanifolds. Kähler Manifolds

Consider the standard Hermitian form $(z, w) = \sum z_i \overline{w}_i$ in $\mathbb{C}^n$. It is easy to see that its imaginary part is a symplectic structure on $\mathbb{C}^n$. Consider an arbitrary complex submanifold in $\mathbb{C}^n$, for example, given by a system of polynomial equations. By restricting the imaginary part of the Hermitian structure to this submanifold, we obtain a closed differential 2-form. It is automatically non-degenerate, since the restriction of the Hermitian structure to a complex submanifold is evidently an Hermitian structure again, and its imaginary part is always non-degenerate.

Recall that a *Kähler structure* on a complex manifold is defined to be an Hermitian structure whose imaginary part is a closed 2-form.

It is easy to see that a Kähler manifold and each of its complex submanifold are symplectic (with respect to the imaginary part of the Hermitian structure).

Examples of Kähler manifolds are the complex projective space $\mathbb{C}P^n$ and any of its complex projective submanifolds.

Let us describe the symplectic structure ω on $\mathbb{C}P^n$.

Let $(z_0 : z_1 : \ldots : z_n)$ be homogeneous coordinates in $\mathbb{C}P^n$. Consider one of the corresponding charts U_0 and define the complex coordinates in the usual way:

$$w_1 = \frac{z_1}{z_0}, \ldots, w_n = \frac{z_n}{z_0}, \qquad (z_0 \neq 0).$$

In this chart, ω is defined by the following explicit formula:

$$\omega = \frac{i}{2\pi} \left(\frac{\sum dw_k \wedge d\overline{w}_k}{1 + \sum |w_k|^2} - \frac{\left(\sum \overline{w}_k dw_k\right) \wedge \left(\sum w_k d\overline{w}_k\right)}{\left(1 + \sum |w_k|^2\right)^2} \right).$$

It is easy to verify that in another chart U_j this formula takes a similar form. Thus, we obtain a well-defined non-degenerate closed form on the whole projective space $\mathbb{C}P^n$.

1.2.3. Orbits of Coadjoint Representation

Consider the Lie algebra G of an arbitrary Lie group $\mathfrak{G}$. Consider the dual space G^* and define the *coadjoint action* of this group on it. For simplicity, it is assumed that we deal with matrix Lie groups, so that the adjoint representation is the usual conjugation of matrices. Recall the main notation. Let

$$x, y, a \in G, \quad A \in \mathfrak{G}, \quad \xi \in G^*.$$

Recall that Ad and ad are linear operators on the Lie algebra, and Ad^* and ad^* are linear operators on the coalgebra. By definition,

$$\mathrm{Ad}_A x = A^{-1} x A, \qquad \mathrm{ad}_a x = [a, x] = ax - xa.$$

The operator $\mathrm{Ad}_A^* \xi : G^* \to G^*$ is defined by the equation (for every y)

$$\mathrm{Ad}_A^* \xi(y) = \xi(\mathrm{Ad}_A^{-1} y),$$

and similarly

$$\mathrm{ad}_a^* \xi(y) = \xi(- \mathrm{ad}_a y) = \xi([y, a]).$$

Consider now a covector $\xi \in G^*$ and its orbit under the coadjoint action of $\mathfrak{G}$

$$O(\xi) = \{\eta = \mathrm{Ad}_A^* \xi \mid A \text{ runs over } \mathfrak{G}\}.$$

This is a smooth manifold. Define a symplectic structure ω on it. Recall that a differential 2-form is uniquely defined if we define a skew-symmetric bilinear form on every tangent space. Consider the tangent space $T_\xi O(\xi)$ to the orbit at a point ξ (this point does not differ from other ones). It is easy to verify that

$$T_\xi O(\xi) = \{\eta = \mathrm{ad}_a^* \xi \mid a \text{ runs over the Lie algebra } G\}.$$

Now take two arbitrary tangent vectors of the form

$$\eta_1 = \mathrm{ad}_{a_1}^* \xi \quad \text{and} \quad \eta_2 = \mathrm{ad}_{a_2}^* \xi$$

and by definition set

$$\omega(\eta_1, \eta_2) = \xi([a_1, a_2]).$$

Of course, one has, first of all, to check the correctness of the above definition. The point is that η_i can be written as $\mathrm{ad}_{a_i}^* \xi$ in different ways. Suppose, for instance, that

$$\eta_1 = \mathrm{ad}_{a_1}^* \xi = \mathrm{ad}_{a_1 + b}^* \xi.$$

Then we have $\mathrm{ad}_b^* \xi = \mathrm{ad}_{a_1 + b}^* \xi - \mathrm{ad}_{a_1}^* \xi = 0$ and, consequently,

$$\xi([a_1 + b, a_2]) = \xi([b, a_2]) + \xi([a_1, a_2]) = -\mathrm{ad}_b^* \xi(a_2) + \xi([a_1, a_2]) = \xi([a_1, a_2]),$$

which means the correctness of the definition.

It remains to check that ω is non-degenerate and closed. The non-degeneracy is easily checked. Indeed, suppose that there exists a vector η_1 such that $\omega(\eta_1, \eta_2) = 0$ for any tangent vector η_2. This is equivalent to the fact that for any $a_2 \in G$ we have

$$\omega(\eta_1, \eta_2) = \xi([a_1, a_2]) = -\operatorname{ad}^*_{a_1} \xi(a_2) = -\eta_1(a_2) = 0 .$$

Since a_2 is arbitrary, $\eta_1 = 0$. This means exactly that ω is non-degenerate. The closedness of ω in fact follows from Jacobi's identity in the Lie algebra G, but the standard proof requires some additional facts about Poisson brackets which we shall discuss below.

We now study some basic local properties of symplectic manifolds.

Let H be a smooth function on a symplectic manifold M. We define the vector of *skew-symmetric gradient* sgrad H for this function by using the following identity:

$$\omega(v, \operatorname{sgrad} H) = v(H) ,$$

where v is an arbitrary tangent vector v.

In local coordinates $x^1, \ldots, x^{2n}$, we obtain the following expression:

$$(\operatorname{sgrad} H)^i = \sum \omega^{ij} \frac{\partial H}{\partial x^j} ,$$

where ω^{ij} are components of the inverse matrix to the matrix Ω.

Definition 1.6. The vector field sgrad H is called a *Hamiltonian vector field*. The function H is called the *Hamiltonian* of the vector field sgrad H.

One of the main properties of Hamiltonian vector fields is that they preserve the symplectic structure ω.

Proposition 1.7. *Let g_t be the one-parameter group of diffeomorphisms (the Hamiltonian flow) corresponding to the Hamiltonian field $v = \operatorname{sgrad} f$. Then $g_t^*(\omega) = \omega$ for any $t \in \mathbb{R}$.*

Proof. It is sufficient to show that the Lie derivative of ω with respect to v is identically zero. Since ω is closed, we have

$$L_v \omega = d(v \lrcorner \, \omega) ,$$

where $v \lrcorner \omega$ denotes a 1-form obtained by substituting v into the 2-form ω, that is, $v \lrcorner \omega(\xi) = \omega(v, \xi) = \omega(\operatorname{sgrad} f, \xi) = -df(\xi)$ for any tangent vector ξ. Thus,

$$L_v(\omega) = d(df) = 0 ,$$

which was to be proved. $\square$

The above argument also implies the converse statement. If a vector field v preserves the symplectic structure ω, then the form $v \lrcorner \omega$ is closed and, therefore, at least locally there exists a function f such that $v \lrcorner \omega = df$ or, equivalently, $v = \operatorname{sgrad} f$. The vector fields satisfying this property are called *locally Hamiltonian*.

Definition 1.7. Let f and g be two smooth functions on a symplectic manifold M. By definition, we set $\{f, g\} = \omega(\operatorname{sgrad} f, \operatorname{sgrad} g) = (\operatorname{sgrad} f)(g)$.

This operation $\{\cdot, \cdot\}: C^\infty \times C^\infty \to C^\infty$ on the space of smooth functions on M is called the *Poisson bracket*.

Proposition 1.8 (Properties of the Poisson bracket). *The Poisson bracket satisfies the following conditions:*

1) *bilinearity on* $\mathbb{R}$;
2) *skew-symmetry*

$$\{f,g\} = -\{g,f\}\,;$$

3) *Jacobi's identity*

$$\{g,\{f,h\}\} + \{h,\{g,f\}\} + \{f,\{h,g\}\} = 0\,;$$

4) *the Leibniz rule*

$$\{fg,h\} = f\{g,h\} + g\{f,h\}\,;$$

5) *the operator* sgrad *defines a homeomorphism between the Lie algebra of smooth functions on the manifold* M *and that of smooth vector fields, in other words, the identity*

$$\mathrm{sgrad}\{f,g\} = [\mathrm{sgrad}\,f, \mathrm{sgrad}\,g]$$

holds (in particular, Hamiltonian vector fields form a subalgebra);

6) *a function* f *is a first integral of the Hamiltonian vector field* $v = \mathrm{sgrad}\,H$ *if and only if* $\{f,H\} = 0$ *(in particular, the Hamiltonian* H *is always a first integral of the field* $\mathrm{sgrad}\,H$ *).*

Proof. The bilinearity and skew-symmetry of the Poisson bracket are evident. Let us prove Jacobi's identity. Recall the following well-known Cartan formula:

$$d\omega(\xi,\eta,\zeta) = \xi\omega(\eta,\zeta) - \omega([\xi,\eta],\zeta) + (\text{cyclic permutation})\,,$$

where ω is an arbitrary 2-form, and ξ, η and ζ are vector fields. We apply it in the case when ω is a symplectic structure and $\xi = \mathrm{sgrad}\,f$, $\eta = \mathrm{sgrad}\,g$, and $\zeta = \mathrm{sgrad}\,h$. Since ω is closed, we have

$$\mathrm{sgrad}\,f(\omega(\mathrm{sgrad}\,g, \mathrm{sgrad}\,h)) - \omega([\mathrm{sgrad}\,f, \mathrm{sgrad}\,g], \mathrm{sgrad}\,h)$$
$$+ (\text{cyclic permutation}) = 0\,.$$

Rewrite this expression in a slightly different way:

$$\mathrm{sgrad}\,f(\{g,h\}) - [\mathrm{sgrad}\,f, \mathrm{sgrad}\,g](h) + (\text{cyclic permutation}) = 0\,.$$

By rewriting once more, we obtain Jacobi's identity:

$$\{f,\{g,h\}\} - \mathrm{sgrad}\,f(\mathrm{sgrad}\,g(h)) + \mathrm{sgrad}\,g(\mathrm{sgrad}\,f(h)) + (\text{cyclic permutation})$$
$$= \{g,\{f,h\}\} + (\text{cyclic permutation}) = 0\,.$$

This proof implies the following useful observation: Jacobi's identity for the Poisson bracket is in fact equivalent to the closedness of the symplectic structure ω.

The Leibniz rule easily follows from the similar rule for the skew-symmetric gradient:

$$\mathrm{sgrad}(fg) = f\,\mathrm{sgrad}\,g + g\,\mathrm{sgrad}\,f\,.$$

Let us prove (5). By differentiating a function h along the vector field $\mathrm{sgrad}\{f, g\}$, we have

$$
\begin{aligned}
\mathrm{sgrad}\{f, g\}(h) = \{\{f, g\}, h\} = & \quad \text{(by virtue of Jacobi's identity)} \\
= & \{f, \{g, h\}\} - \{g, \{f, h\}\} \\
= & \;\mathrm{sgrad}\,f(\mathrm{sgrad}\,g(h)) - \mathrm{sgrad}\,g(\mathrm{sgrad}\,f(h)) \\
= & \;[\mathrm{sgrad}\,f, \mathrm{sgrad}\,g](h)\,,
\end{aligned}
$$

which was to be proved.

Property (6) evidently follows from the definition of a Poisson bracket. □

Sometimes, when constructing Hamiltonian mechanics, instead of a symplectic structure on a manifold, one takes a Poisson bracket as the initial structure. In this case, the Poisson bracket is not necessary assumed to be non-degenerate.

Definition 1.8. A *Poisson manifold* is a smooth manifold endowed with a *Poisson bracket*, that is, a bilinear skew-symmetric operation $\{\cdot, \cdot\}$ on the space of smooth functions satisfying Jacobi's identity and the Leibniz rule.

It is easily verified that a Poisson structure on a manifold can be equivalently defined as a skew-symmetric tensor field $A^{ij}(x)$ satisfying the relation

$$A^{j\alpha}\frac{\partial A^{ki}}{\partial x^\alpha} + A^{i\alpha}\frac{\partial A^{jk}}{\partial x^\alpha} + A^{k\alpha}\frac{\partial A^{ij}}{\partial x^\alpha} = 0\,.$$

The relationship between the Poisson bracket and the tensor field A^{ij} is very simple and natural:

$$\{f, g\} = A^{ij}\frac{\partial f}{\partial x^i}\frac{\partial g}{\partial x^j}\,;$$

and the above restriction to the components of A^{ij} is exactly equivalent to Jacobi's identity.

If a Poisson bracket is non-degenerate, then the Poisson manifold is symplectic with respect to the symplectic form $\omega = A^{-1}$, which is uniquely defined by the identity $\omega_{ij}A^{jk} = \delta_i^k$.

As an example of a degenerate Poisson bracket, we indicate the *Lie–Poisson bracket* on the Lie coalgebra G^*:

$$\{f, g\} = \sum c_{jk}^i x_i \frac{\partial f}{\partial x_j}\frac{\partial g}{\partial x_k}\,,$$

where c_{jk}^i are structural constants of the Lie algebra G in some basis, and $x_1, \ldots x_s$ are coordinates in the dual space G^* with respect to the dual basis.

This bracket becomes non-degenerate on the orbits of the coadjoint representation. The symplectic structure defined by it on the orbits coincides with the structure, which was described above.

1.3. THE DARBOUX THEOREM

Theorem 1.1 (G. Darboux).   *For any point of a symplectic manifold M^{2n}, there exists an open neighborhood with local regular coordinates $p_1, \ldots, p_n, q_1, \ldots, q_n$ in which the symplectic structure ω has the canonical form $\omega = \sum dp_i \wedge dq_i$.*

The canonicity condition for the symplectic structure $\omega = \sum dp_i \wedge dq_i$ can equivalently be rewritten in terms of the Poisson bracket, i.e.,

$$\{p_i, p_j\} = 0, \quad \{p_i, q_j\} = \delta_{ij}, \quad \{q_i, q_j\} = 0 \qquad 1 \le i, j \le n.$$

Proof. We first prove the following lemma.

Lemma 1.1.   *Let $p_1, \ldots, p_n$ be n independent commuting functions on a symplectic manifold M^{2n} given in some neighborhood of a point $x \in M^{2n}$. Then there exist n independent functions $q_1, \ldots, q_n$ which complete the set $p_1, \ldots, p_n$ up to a canonical coordinate system, i.e., such that $\{p_i, p_j\} = 0$, $\{p_i, q_j\} = \delta_{ij}$, $\{q_i, q_j\} = 0$ for all $1 \le i, j \le n$.*

Proof. 1) Let us consider the linearly independent vector fields $v_i = \operatorname{sgrad} p_i$, corresponding to the functions $p_1, \ldots, p_n$. Since

$$[\operatorname{sgrad} p_i, \operatorname{sgrad} p_j] = \operatorname{sgrad}\{p_i, p_j\} = 0,$$

it follows that $v_1, \ldots, v_n$ commute.

2) According to the Frobenius theorem [328], for the commuting vector fields v_i there exists a local regular coordinate system $x_1, \ldots, x_n, y_1, \ldots, y_n$ such that $v_i = \operatorname{sgrad} p_i = \partial/\partial x_i$, $1 \le i \le n$.

3) Let us write p_i as functions of the new coordinate $x_1, \ldots, x_n, y_1, \ldots, y_n$, that is, $p_i = p_i(x, y)$. We state that, in fact, $p_i = p_i(y)$, i.e., they do not depend on x. Indeed,

$$\frac{\partial}{\partial x_j}(p_i) = \operatorname{sgrad} p_j(p_i) = \{p_j, p_i\} = 0.$$

4) Now, instead of (x, y), we consider (x, p) as local coordinates. It can be done because the functions $p_i = p_i(y)$ are, by assumption, independent.

5) Let us prove that the Poisson brackets of the functions x and p have the form: $\{x_i, x_j\} = \lambda_{ij}(p)$, $\{p_i, x_j\} = \delta_{ij}$, $\{p_i, p_j\} = 0$. Indeed,

$$\{p_i, x_j\} = \operatorname{sgrad} p_i(x_j) = \frac{\partial}{\partial x_i}(x_j) = \delta_{ij}.$$

Further, the brackets $\{x_i, x_j\}$ are represented as some functions $\lambda_{ij}(x, p)$. Let us prove that they do not depend on x. Indeed,

$$\frac{\partial}{\partial x_k}\{x_i, x_j\} = \operatorname{sgrad} p_k\{x_i, x_j\} = \{p_k, \{x_i, x_j\}\}$$

$$= \{x_j, \{x_i, p_k\}\} + \{x_i, \{p_k, x_j\}\} = 0,$$

since $\{x_s, p_t\} = \delta_{st} = \text{const}$. Thus, $\{x_i, x_j\} = \lambda_{ij}(p)$.

6) We now "improve" the functions x to obtain a canonical coordinate system. To this end, we shall try to find the coordinates q in the form $q_j = x_j - f_j(p)$. The functions f_j must satisfy the following conditions: $\{p_i, q_j\} = \delta_{ij}$, $\{q_i, q_j\} = 0$. Therefore, we have

$$\{p_i, q_j\} = \{p_i, x_j - f_j(p)\} = \{p_i, x_j\} - \{p_i, f_j(p)\} = \delta_{ij} + 0 = \delta_{ij}.$$

Thus, the first condition $\{p_i, q_j\} = \delta_{ij}$ is automatically fulfilled.

7) Next:

$$\{q_i, q_j\} = \{x_i - f_i(p), x_j - f_j(p)\}$$
$$= \{x_i, x_j\} - \{x_i, f_j(p)\} + \{x_j, f_i(p)\} = \lambda_{ij} - \frac{\partial f_i}{\partial p_j} + \frac{\partial f_j}{\partial p_i}.$$

Here we use the fact that

$$\{x_i, f_j(p)\} = \sum_k \frac{\partial f_j}{\partial p_k} \{x_i, p_k\} = - \sum_k \delta_{ik} \frac{\partial f_j}{\partial p_k} = - \frac{\partial f_j}{\partial p_i}.$$

This follows from the general identity

$$\{f, g(s_1, \ldots, s_m)\} = \sum_{k=1}^{m} \frac{\partial g}{\partial s_k} \{f, s_k\}.$$

Thus, in order for the condition $\{q_i, q_j\} = 0$ to be fulfilled, it is necessary and sufficient that $\lambda_{ij} - \frac{\partial f_i}{\partial p_j} + \frac{\partial f_j}{\partial p_i} = 0$. For such a system of equations to be solvable with respect to unknown functions f_i, it is necessary and sufficient that the following compatibility condition holds:

$$\frac{\partial \lambda_{\alpha\beta}}{\partial p_\gamma} + \frac{\partial \lambda_{\gamma\alpha}}{\partial p_\beta} + \frac{\partial \lambda_{\beta\gamma}}{\partial p_\alpha} = 0.$$

The latter identity is actually fulfilled. Indeed, $\{\{x_\alpha, x_\beta\}, x_\gamma\} = \{\lambda_{\alpha\beta}, x_\gamma\} = \frac{\partial \lambda_{\alpha\beta}}{\partial p_\gamma}$, and, therefore, the compatibility condition is equivalent to Jacobi's identity:

$$\{\{x_\alpha, x_\beta\}, x_\gamma\} + \{\{x_\gamma, x_\alpha\}, x_\beta\} + \{\{x_\beta, x_\gamma\}, x_\alpha\} = 0.$$

Thus, the functions q_i constructed above satisfy all necessary conditions. □

We now turn to the proof of the Darboux theorem. By virtue of lemma 1.1, it remains to show that in a neighborhood of any point $x \in M^{2n}$ there always exist independent functions $p_1, \ldots, p_n$ in involution. In fact, a stronger inductive statement holds.

Lemma 1.2. *Given k independent functions $p_1, \ldots, p_k$ in involution, where $k < n$, there always exists a function p_{k+1} independent of them and such that $\{p_{k+1}, p_i\} = 0$ for $1 \leq i \leq k$.*

Proof. Following the proof of lemma 1.1, we obtain that for $p_1, \ldots, p_k$ there exists a local regular coordinate system $x_1, \ldots, x_k, y_1, \ldots, y_{2n-k}$ such that $\operatorname{sgrad} p_i = \partial/\partial x_i$, $1 \leq i \leq k$. As above, it follows from this that

$$\frac{\partial p_i}{\partial x_j} = \operatorname{sgrad} p_i(p_j) = \{p_i, p_j\} = 0,$$

i.e., $p_i = p_i(y_1, \ldots, y_{2n-k})$ for $1 \leq i \leq k$.

Since $k < 2n - k$, the number of functions y's is greater than that of the functions p's depending on y's. Consequently, there exists a new function $p_{k+1}(y)$, independent of $p_1(y), \ldots, p_k(y)$. Next we have: $\{p_{k+1}, p_i\} = -\dfrac{\partial p_{k+1}}{\partial x_i} = 0$, i.e., the function p_{k+1} is in involution with all of $p_1, \ldots, p_k$. □

This completes the proof of the Darboux theorem. □

We also give another, more formal, proof of the Darboux theorem [150].

Consider the symplectic form ω in some fixed point $P \in M^{2n}$. By a linear coordinate change, we can always reduce the matrix Ω (at the point P) to the canonical form

$$\Omega_0 = \begin{pmatrix} 0 & E \\ -E & 0 \end{pmatrix}.$$

Such a reduction is possible, in general, at the single point P only. Now consider a new form ω_0 with the constant matrix Ω_0 (in the same neighborhood of P). Our aim is to find a diffeomorphism of the neighborhood $U(P)$ into itself which sends the form ω into ω_0. It is clear that the new coordinates defined by such a diffeomorphism will be canonical Darboux coordinates.

Let us find the family of diffeomorphisms φ_t such that

$$\varphi_t^* \omega = \omega_t = (1 - t)\omega + t\omega_0 .$$

For $t = 1$ we get the desired diffeomorphism φ_1 sending ω into ω_0. To find the family φ_t, we differentiate the above expression with respect to t and consider the obtained differential equation

$$L_{\xi_t} \omega_t = \omega_0 - \omega ,$$

where L_{ξ_t} denotes the Lie derivative along the vector field $\xi_t = \dfrac{d\varphi_t}{dt}$. Since ω_t is closed, the left-hand side of the equation can be written in the form

$$L_{\xi_t} \omega_t = d(\xi_t \lrcorner \, \omega_t) ,$$

where $\xi_t \lrcorner \, \omega_t$ denotes the result of substitution of the field ξ_t into ω_t, i.e., the differential 1-form defined by the identity $\xi_t \lrcorner \, \omega_t(v) = \omega_t(\xi_t, v)$, where v is an arbitrary tangent vector. On the other hand, $\omega_0 - \omega$ is a closed form and, therefore, it is locally exact (in some neighborhood of P) and can be represented as $\omega_0 - \omega = d\alpha$. Moreover, without loss of generality, we can assume the 1-form α to be zero at P. We now find the vector field ξ_t from the relation $\xi_t \lrcorner \, \omega_t = \alpha$.

Since ω_t is non-degenerate, it can be done and, besides, uniquely. As a result, we obtain a family of smooth vector fields ξ_t, $t \in [0,1]$. Moreover, $\xi_t(P) = 0$ for all t. Consider now the family of diffeomorphisms φ_t which satisfy the differential equation $\xi_t = \dfrac{d\varphi_t}{dt}$ with the initial condition $\varphi_0 = \mathrm{id}$. Then, by construction, $\varphi_1^* \omega = \omega_0$, as required. The Darboux theorem is proved.   □

Definition 1.9. Let (M^{2n}, ω) be a symplectic manifold. A smooth submanifold $N \subset M$ is called

1) *symplectic* if the restriction of ω onto N is non-degenerate;

2) *Lagrangian* if dim $N = n$ and the restriction of ω onto N vanishes identically.

Example. Consider a canonical coordinate system $p_1, \ldots, p_n, q_1, \ldots, q_n$ and an arbitrary smooth function $S = S(q_1, \ldots, q_n)$. Then the submanifold N given as a graph

$$ p_1 = \frac{\partial S}{\partial q_1}, \quad \ldots, \quad p_n = \frac{\partial S}{\partial q_n} $$

is Lagrangian. The converse statement is also true: if a Lagrangian submanifold N can be presented in canonical coordinates as a graph $p_i = P_i(q_1, \ldots, q_n)$, then (at least locally) there exists a function $S = S(q_1, \ldots, q_n)$ such that $P_i = \dfrac{\partial S}{\partial q_i}$.

Another example of Lagrangian submanifolds are Liouville tori of integrable Hamiltonian systems, which we shall discuss in the next section.

1.4. LIOUVILLE INTEGRABLE HAMILTONIAN SYSTEMS. THE LIOUVILLE THEOREM

Let M^{2n} be a smooth symplectic manifold, and let $v = \operatorname{sgrad} H$ be a Hamiltonian dynamical system with a smooth Hamiltonian H.

Definition 1.10. The Hamiltonian system is called *Liouville integrable* if there exists a set of smooth functions $f_1, \ldots, f_n$ such that

1) $f_1, \ldots, f_n$ are integrals of v,

2) they are functionally independent on M, i.e., their gradients are linearly independent on M almost everywhere,

3) $\{f_i, f_j\} = 0$ for any i and j,

4) the vector fields $\operatorname{sgrad} f_i$ are complete, i.e., the natural parameter on their integral trajectories is defined on the whole real axis.

Definition 1.11. The decomposition of the manifold M^{2n} into connected components of common level surfaces of the integrals $f_1, \ldots, f_n$ is called the *Liouville foliation* corresponding to the integrable system $v = \operatorname{sgrad} H$.

Since $f_1, \ldots, f_n$ are preserved by the flow v, every leaf of the Liouville foliation is an invariant surface.

The Liouville foliation consists of regular leaves (filling M almost in the whole) and singular ones (filling a set of zero measure). The Liouville theorem formulated below describes the structure of the Liouville foliation near regular leaves.

Consider a common regular level T_ξ for the functions $f_1, \ldots, f_n$, that is, $T_\xi = \{x \in M \mid f_i(x) = \xi_i, \ i = 1, \ldots, n\}$. The regularity means that all 1-forms df_i are linearly independent on T_ξ.

Theorem 1.2 (J. Liouville). *Let* $v = \mathrm{sgrad}\, H$ *be a Liouville integrable Hamiltonian system on* M^{2n}, *and let* T_ξ *be a regular level surface of the integrals* $f_1, \ldots, f_n$. *Then*

1) T_ξ *is a smooth Lagrangian submanifold that is invariant with respect to the flow* $v = \mathrm{sgrad}\, H$ *and* $\mathrm{sgrad}\, f_1, \ldots, \mathrm{sgrad}\, f_n$;

2) *if* T_ξ *is connected and compact, then* T_ξ *is diffeomorphic to the n-dimensional torus* T^n *(this torus is called the Liouville torus)*;

3) *the Liouville foliation is trivial in some neighborhood of the Liouville torus, that is, a neighborhood* U *of the torus* T_ξ *is the direct product of the torus* T^n *and the disc* D^n;

4) *in the neighborhood* $U = T^n \times D^n$ *there exists a coordinate system* $s_1, \ldots, s_n, \varphi_1, \ldots, \varphi_n$ *(which is called the action-angle variables), where* $s_1, \ldots, s_n$ *are coordinates on the disc* D^n *and* $\varphi_1, \ldots, \varphi_n$ *are standard angle coordinates on the torus, such that*

   a) $\omega = \sum d\varphi_i \wedge ds_i$,

   b) *the action variables* s_i *are functions of the integrals* $f_1, \ldots, f_n$,

   c) *in the action-angle variables* $s_1, \ldots, s_n, \varphi_1, \ldots, \varphi_n$, *the Hamiltonian flow* v *is straightened on each of the Liouville tori in the neighborhood* U, *that is,* $\dot{s}_i = 0$, $\dot{\varphi}_i = q_i(s_1, \ldots, s_n)$ *for* $i = 1, 2, \ldots, n$ *(this means that the flow* v *determines the conditionally periodic motion that generates a rational or irrational rectilinear winding on each of the tori)*.

Proof.

1) Since the functions $f_1, \ldots, f_n$ commute with each other, they are first integrals not only for $v = \mathrm{sgrad}\, H$, but also for each of the flows $\mathrm{sgrad}\, f_i$. Therefore, their common level surface T_ξ is invariant under these flows, and moreover, being linearly independent, the vector fields $\mathrm{sgrad}\, f_1, \ldots, \mathrm{sgrad}\, f_n$ form a basis in every tangent plane of T_ξ. The fact that T_ξ is a Lagrangian submanifold is now implied by the formula $\omega(\mathrm{sgrad}\, f_i, \mathrm{sgrad}\, f_j) = \{f_i, f_j\} = 0$.

2) The flows $\mathrm{sgrad}\, f_1, \ldots, \mathrm{sgrad}\, f_n$ pairwise commute and are complete. This allows us to define the action Φ of the Abelian group $\mathbb{R}^n$ on the manifold M^{2n} generated by shifts along the flows $\mathrm{sgrad}\, f_1, \ldots, \mathrm{sgrad}\, f_n$. This action can be defined by an explicit formula. Let g_i^t be the diffeomorphism shifting all the points of M^{2n} along the integral trajectories of the field $\mathrm{sgrad}\, f_i$ by time t. Let $(t_1, \ldots, t_n)$ be an element of $\mathbb{R}^n$. Then

$$\Phi(t_1, \ldots, t_n) = g_1^{t_1} g_2^{t_2} \ldots g_n^{t_n}.$$

Lemma 1.3. *If the submanifold* T_ξ *is connected, then it is an orbit of the* $\mathbb{R}^n$*-action.*

Proof. Consider the image of the group $\mathbb{R}^n$ in M under the following mapping

$$A_x : (t_1, \ldots, t_n) \to \Phi(t_1, \ldots, t_n)(x),$$

where x is a certain point in T_ξ. Since the fields $\mathrm{sgrad}\, f_i$ are independent, this mapping is an immersion, i.e., is a local diffeomorphism onto the image. Thus, the image of $\mathbb{R}^n$ (that is, the orbit of x) is open in T_ξ. If we assume that the submanifold T_ξ is not a single orbit of the group $\mathbb{R}^n$, then it is the union of at least two orbits. Since each of them is open, T_ξ turns out to be disconnected. But this contradicts our assumption. $\quad\square$

Lemma 1.4. *An orbit $O(x)$ of maximal dimension of the action of the group $\mathbb{R}^n$ is the quotient space of $\mathbb{R}^n$ with respect to some lattice $\mathbb{Z}^k$. If $O(x)$ is compact, then $k = n$, and $O(x)$ is diffeomorphic to the n-dimensional torus.*

Proof. Every orbit $O(x)$ of a smooth action of $\mathbb{R}^n$ is a quotient space (homogeneous space) of $\mathbb{R}^n$ with respect to the stationary subgroup H_x of the point x. It is clear that the subgroup H_x is discrete, since the mapping A_x is a local diffeomorphism. Recall that a discrete subgroup has no accumulation points. In particular, inside any bounded set, there is only a finite number of elements of this subgroup.

Let us prove by induction that H_x is a lattice $\mathbb{Z}^k$.

Suppose $n = 1$. Take a non-zero e_1 of H_x on the line $\mathbb{R}^1$ which is nearest to zero. Then all the other elements of H_x have to be multiples of e_1. Indeed, if an element e is not multiple to e_1, then for some k we have

$$ke_1 < e < (k+1)e_1 .$$

But then the element $e - ke_1$ is closer to zero than e_1. We come to a contradiction and, consequently, $H(x)$ is the lattice generated by e_1.

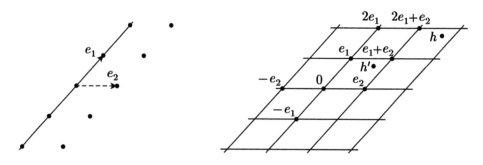

Figure 1.1   Figure 1.2

Suppose $n = 2$. As e_1 we choose a non-zero element nearest to zero on the plane $\mathbb{R}^2$ and consider the straight line $l(e_1)$ generated by it (Fig. 1.1). All the elements of H_x lying on $l(e_1)$ are multiple to e_1. Then two possibilities appear. It may happen that all elements of H_x lie on $l(e_1)$. Then the proof is complete. The other possibility is that there exist elements of H_x which do not belong to $l(e_1)$. Then, as e_2, we take a non-zero element nearest to the line $l(e_1)$. It is easy to see that such an element exists. We now wish to prove that all elements of H_x are linear combinations of e_1 and e_2 with integer coefficients. Assume the contrary, and take an element $h \in H_x$ which cannot be decomposed into e_1 and e_2 with integer coefficients. Then, we divide the plane into parallelograms generated by e_1 and e_2 (Fig. 1.2). The element h turns out to be in one of them and, moreover, does not coincide with any vertex of parallelograms.

It is clear that, moving h by an appropriate integer-valued combination of e_1 and e_2, we shall find an element h' closer to $l(e_1)$ than e_2. This contradiction shows that $H(x)$ is generated by e_1 and e_2.

Continuing this argument by induction, we obtain that there exists a basis $e_1, \ldots, e_k$ in the subgroup H_x such that each of its elements is a unique linear combination of the basis vectors with integer coefficients. In other words, $H(x)$ is the lattice generated by $e_1, \ldots, e_k$.

If $k < n$, then the quotient space $\mathbb{R}^n / \mathbb{Z}^k$ is a cylinder, i.e., the direct product $T^k \times \mathbb{R}^{n-k}$, where T^k is a k-dimensional torus. In particular, the orbit is compact for $n = k$ only, and, in this case, $O(x)$ is diffeomorphic to the torus T^n. $\square$

This proves item (2).

3) We now prove that a neighborhood U of the torus T_ξ is a direct product of T^n by a disc D^n.

This fact follows from the following more general and well-known theorem. Suppose $f: M \to N$ is a smooth mapping of smooth manifolds, and $y \in N$ is a regular value of f, that is, at each point of the preimage $f^{-1}(y)$, the rank of df is equal to the dimension of N. In particular, $\dim M \geq \dim N$. Suppose, in addition, that $f^{-1}(y)$ is compact. Then there exists a neighborhood D of a point y in N such that its preimage $f^{-1}(D)$ is diffeomorphic to the direct product $D \times f^{-1}(y)$. Moreover, the direct product structure is compatible with the mapping f in the sense that $f: D \times f^{-1}(y) \to D$ is just the natural projection. It follows from this, in particular, that each set $f^{-1}(z)$, where $z \in D$, is diffeomorphic to $f^{-1}(y)$.

4) The construction of the action-angle variables. Consider a neighborhood $U(T_\xi) = T_\xi \times D^n$ of the Liouville torus T_ξ. Choose a certain point x on each of the tori T depending smoothly on the torus. Consider T as the quotient space $\mathbb{R}^n / H_x$ and fix a basis $e_1, \ldots, e_n$ in the lattice H_x. Notice that the basis will smoothly depend on x. Indeed, the coordinates of the basis vector $e_i = (t_1, \ldots, t_n)$ are the solutions of the equation $\Phi(t_1, \ldots, t_n)x = x$, where x is regarded as a parameter. According to the implicit function theorem, the solutions of this equation depend on x smoothly. Note that the assumptions of this theorem holds, since $\dfrac{\partial}{\partial t_j} \Phi(t)x = \operatorname{sgrad} f_j(\Phi(t)x)$, and the vector fields $\operatorname{sgrad} f_j$ are linearly independent.

Let us now define certain angle coordinates $(\psi_1, \ldots, \psi_n)$ on the torus T_ξ in the following way. If $y = \Phi(a)x$, where $a = a_1 e_1 + \ldots + a_n e_n \in \mathbb{R}^n$, then $\psi_1(y) = 2\pi a_1 \pmod{2\pi}, \ldots, \psi_n(y) = 2\pi a_n \pmod{2\pi}$.

Such a coordinate system satisfies the following evident property: the vector fields $\partial/\partial \psi_1, \ldots, \partial/\partial \psi_n$ and $\operatorname{sgrad} f_1, \ldots, \operatorname{sgrad} f_n$ are connected by a linear change with constant coefficients, i.e., $\partial/\partial \psi_i = \sum c_{ik} \operatorname{sgrad} f_k$.

Let us write again the form ω in coordinates $(f_1, \ldots, f_n, \psi_1, \ldots, \psi_n)$:

$$\omega = \sum_{i,j} \tilde{c}_{ij}\, df_i \wedge d\psi_j + \sum_{i,j} b_{ij}\, df_i \wedge df_j .$$

In this decomposition, the terms of the form $a_{ij}\, d\psi_i \wedge d\psi_j$ are absent, since the Liouville tori are Lagrangian. We state that the coefficients $\tilde{c}_{ij}$ of the symplectic form exactly coincide with the coefficients c_{ij} and, in particular, do not depend on $\psi_1, \dots, \psi_n$. Indeed,

$$\tilde{c}_{ij} = \omega\left(\frac{\partial}{\partial f_i}, \frac{\partial}{\partial \psi_j}\right) = \omega\left(\frac{\partial}{\partial f_i}, \sum c_{kj}\,\mathrm{sgrad}\,f_k\right)$$

$$= \sum c_{kj}\,\omega\left(\frac{\partial}{\partial f_j}, \mathrm{sgrad}\,f_k\right) = \sum c_{kj}\frac{\partial f_k}{\partial f_j} = c_{ij} = c_{ij}(f_1, \dots, f_n).$$

We now show that the functions b_{ij} do not depend on $(\psi_1, \dots, \psi_n)$ either. Since ω is closed, we get

$$\frac{\partial b_{ij}}{\partial \psi_k} = \frac{\partial c_{kj}}{\partial f_i} - \frac{\partial c_{ki}}{\partial f_j}.$$

The function b_{ij} is 2π-periodic on ψ_k (as a function on the torus), but, as we see, its derivative $\dfrac{\partial b_{ij}}{\partial \psi_k}$ does not depend on ψ_k. It follows from this that the function b_{ij} itself does not depend on ψ_k.

This observation implies another important corollary. Let us write the form ω in the following way: $\omega = \left(\sum c_{ij}df_j\right) \wedge d\psi_i + \sum b_{ij}df_i \wedge df_j = \sum \omega_i \wedge d\psi_i + \beta$, where $\omega_i = \sum c_{ij}df_j$ and $\beta = \sum b_{ij}df_i \wedge df_j$ are forms on the disc D^n (which do not depend on $(\psi_1, \dots, \psi_n)$). As ω is closed, so are ω_i and β.

Lemma 1.5. *In the neighborhood $U(T_\xi)$, the form ω is exact, i.e., there exists a 1-form α such that $d\alpha = \omega$.*

Proof. This claim is a particular case of the following general statement. Let Y be a submanifold in X, and there exists a mapping $f\colon X \to Y \subset X$ homotopic to the identity mapping $\mathrm{id}\colon X \to X$. Then a closed differential form $\varkappa$ is exact on X if and only if its restriction $\varkappa|_Y$ onto Y is exact. In our case, when X is a neighborhood of a Liouville torus, and Y is this Liouville torus itself, we even have the stronger condition $\omega|_{T_\xi} = 0$, because T_ξ is Lagrangian. Therefore, ω is exact.

The same, however, can be proved by a straightforward calculation. Since ω_i and β are closed on the disk, they are exact and, therefore, there exist functions s_i and a 1-form $\varkappa$ on the disk D^n such that $ds_i = \omega_i$ and $d\varkappa = \beta$. Let $\alpha = \sum s_i\,d\psi_i + \varkappa$. Then $d\alpha = \sum ds_i \wedge d\psi_i + d\varkappa = \omega_i \wedge d\psi_i + \beta = \omega$. $\square$

Consider the functions $s_1 = s_1(f_1, \dots, f_n)$, $\dots$, $s_n = s_n(f_1, \dots, f_n)$ and show that they are independent. Indeed, from the formula $\omega = \sum ds_i \wedge d\psi_i + \beta$, it follows that the matrix Ω of the symplectic structure ω has the form

$$\Omega = \left(\begin{array}{c|c} 0 & c_{ij} \\ \hline -c_{ij} & b_{ij} \end{array}\right),$$

where $c_{ij} = \dfrac{\partial s_i}{\partial f_j}$. Therefore, $\det \Omega = (\det C)^2$ and $\det C \neq 0$, where C is the Jacobi matrix of the transformation $s_1 = s_1(f_1, \dots, f_n), \dots, s_n = s_n(f_1, \dots, f_n)$, and we can consider a new system of independent coordinates $(s_1, \dots, s_n, \psi_1, \dots, \psi_n)$.

Next, we represent $\varkappa$ in the form $\varkappa = g_i ds_i$ and make one more change $\varphi_i = \psi_i - g_i(s_1, \ldots, s_n)$. Geometrically, this means that we change the initial points of reference for the angle coordinates on the Liouville tori. The level lines and even basis vector fields are not changed.

Finally, let us show that the constructed system of action-angle variables $(s_1, \ldots, s_n, \varphi_1, \ldots, \varphi_n)$ is canonical. We have

$$
\begin{aligned}
\sum ds_i \wedge d\varphi_i &= \sum ds_i \wedge d(\psi_i - g_i(s_1, \ldots, s_n)) \\
&= \sum ds_i \wedge d\psi_i + \sum dg_i(s_1, \ldots, s_n) \wedge ds_i \\
&= \sum ds_i \wedge d\psi_i + d\varkappa = \sum ds_i \wedge d\psi_i + \beta = \omega .
\end{aligned}
$$

Thus, the action-angle variables have been constructed.

It remains to prove that the flow $v = \operatorname{sgrad} H$ straightens on Liouville tori in coordinates $(s_1, \ldots, s_n, \varphi_1, \ldots, \varphi_n)$.

Indeed, $\operatorname{sgrad} s_i = \partial/\partial\varphi_i$; hence $\dfrac{\partial H}{\partial \varphi_i} = \operatorname{sgrad} s_i(H) = \{s_i(f_1, \ldots, f_n), H\} = 0$, i.e., H is a function of $s_1, \ldots, s_n$ only. Consequently,

$$
v = \operatorname{sgrad} H = \sum_i \frac{\partial H}{\partial s_i} \operatorname{sgrad} s_i = \sum_i \frac{\partial H}{\partial s_i} \partial/\partial\varphi_i ,
$$

moreover, the coefficients $\dfrac{\partial H}{\partial s_i}$ depend only on the action variables $s_1, \ldots, s_n$, i.e., are constant on Liouville tori.

This completes the proof of the Liouville theorem. $\square$

COMMENT. Note that the action variables $s_1, \ldots, s_n$ can be defined by an explicit formula. Let $U(T_\xi) = D^n \times T^n$ be the neighborhood of the Liouville torus. By fixing a certain basis $e_1, \ldots, e_n$ in the lattice related to the torus T_ξ, we uniquely define the set of basis cycles $\gamma_1, \ldots, \gamma_n$ in the fundamental group $\pi_1(T^n) = \mathbb{Z}^n$. By continuity, these cycles can be extended to all Liouville tori from the neighborhood $U(T_\xi)$.

To every Liouville torus, we now assign the set of real numbers $s_1, \ldots, s_n$ by the following formula:

$$
s_i = \frac{1}{2\pi} \oint_{\gamma_i} \alpha ,
$$

where α is a differential 1-form in $U(T_\xi)$ such that $d\alpha = \omega$ (α is usually called the *action form*). As a result, there appears the set of smooth functions on $U(T_\xi)$

$$
s_1 = s_1(f_1, \ldots, f_n), \quad \ldots, \quad s_n = s_n(f_1, \ldots, f_n) ,
$$

which coincide (up to a constant) with the action variables. To verify this, it suffices to consider the form $\sum s_i \, d\varphi_i$ as α.

Let us make some general remarks on the action variables, related to a more general situation, when we consider an arbitrary smooth family of Lagrangian submanifolds.

Let $\{L_f\}$ be a smooth family of compact Lagrangian submanifolds in a symplectic manifold M^{2n}, where f is a parameter of the family taking values in some simply connected region $C \subset \mathbb{R}^k$ (or, for simplicity, in a disc $C = D^k$). In particular, in the case of an integrable Hamiltonian system, these are Liouville tori parameterized by the values of the first integrals $f_1, \ldots, f_n$.

Let $[\gamma]$ be an element of the fundamental group $\pi_1(L_f)$. Since the parameter of this family varies in a simply connected region, we may assume that $[\gamma]$ is naturally fixed in the fundamental group of each of the Lagrangian submanifold L_f, where $f \in C$.

Suppose first that ω is exact and choose α such that $d\alpha = \omega$. Then, to each submanifold L_f we can assign a real number $s(f)$ (the so-called action related to the fixed element $[\gamma] \in \pi_1(L_f)$) by the formula

$$s_\gamma(f) = \frac{1}{2\pi} \oint\limits_{\gamma_f} \alpha \,,$$

where the integration is taken over a smooth cycle γ_f which lies on L_f and represents $[\gamma]$. As a result, on the space of parameters (that is, on the set of Lagrangian submanifolds under consideration) there appears a function $s: C \to \mathbb{R}$ called the *action* (related to $[\gamma]$).

If the symplectic structure is not exact, then the analogous construction can be carried out as follows. Consider a family of submanifolds $\{L_f\}$ as a family of embedding, i.e., as a mapping $F: L \times C \to M^{2n}$, where $F|_{L \times \{f\}}: L \times \{f\} \to M^{2n}$ is an embedding whose image is L_f. Then, in spite of the fact that ω is not exact on M^{2n}, its pull-back $F^*\omega$ is exact on $L \times C$, and we can send the construction onto $L \times C$ by setting again

$$s_\gamma(f) = \frac{1}{2\pi} \oint\limits_{\gamma_f} \alpha \,,$$

where α is a 1-form on $L \times C$ such that $d\alpha = F^*\omega$, and γ_f is a cycle on $L \times \{f\}$.

Finally, we list some general properties of the actions $s_i(f)$.

First, the value of the action on the Liouville torus does not depend on the choice of γ_i in its homotopy class. This immediately follows from the fact that the form α, being restricted to the Lagrangian submanifold, is exact, since $d(\alpha|_{T_\xi}) = \omega|_{T_\xi} = 0$.

Second, the action s_i is defined up to an additive constant. This is related to the ambiguity of the choice of α. Indeed, let $d\alpha = d\alpha' = \omega$. This means that $\alpha' = \alpha + \beta$, where β is a certain closed form. Then

$$s'_\gamma(f) = \frac{1}{2\pi} \oint\limits_{\gamma_f} \alpha' = \frac{1}{2\pi} \oint\limits_{\gamma_f} \alpha + \frac{1}{2\pi} \oint\limits_{\gamma_f} \beta = s_\gamma(f) + \frac{1}{2\pi} \oint\limits_{\gamma_f} \beta \,.$$

If L_{f_1} and L_{f_2} are two different Lagrangian submanifolds from our family, then

$$\oint\limits_{\gamma_{f_1}} \beta = \oint\limits_{\gamma_{f_2}} \beta = \text{const} \,,$$

since the cycles γ_{f_1} and γ_{f_2} are homologous in M^{2n}. Thus, the actions s_γ and s'_γ differ by a constant which depends on the closed form β, but does not depend on the Lagrangian submanifold from the family.

Third, let us observe one important property of the action in the case of the Liouville foliation of an integrable Hamiltonian system.

Proposition 1.9. *Let $U(T_\xi) = D^n \times T^n$ be the neighborhood of the Liouville torus; we fix a certain non-trivial cycle γ on each of the Liouville tori which depends continuously on the torus. Consider the corresponding action function*

$$s_\gamma(f_1, \ldots, f_n) = \frac{1}{2\pi} \oint_\gamma \alpha,$$

where the integral is taken over the cycle γ lying on the torus that corresponds to given values $f_1, \ldots, f_n$ of the first integrals. Then all the trajectories of the Hamiltonian vector field sgrad s_γ *are closed with the same period 2π and are homologous to the cycle γ.*

Proof. Without loss of generality, we may assume that γ coincides with the cycle γ_1 corresponding to the first angle coordinate φ_1 in action-angle variables $(s_1, \ldots, s_n, \varphi_1, \ldots, \varphi_n)$. In particular, $s_\gamma = s_1$. So, we see that sgrad $s_\gamma = \partial/\partial\varphi_1$, and our statement becomes evident. $\square$

1.5. NON-RESONANT AND RESONANT SYSTEMS

Consider a Liouville torus T of an integrable system v. According to the Liouville theorem, in the action-angle variables, the vector field v on T has the form $\dot{\varphi}_1 = c_1, \ldots, \dot{\varphi}_n = c_n$, where c_i are some constants, which are called *frequencies*. When T varies, these frequencies also vary in general.

Definition 1.12. A Liouville torus T is called *resonant* if there exists a non-trivial vanishing linear combination of frequencies with integer coefficients, i.e.,

$$\sum k_i c_i = 0,$$

where $k_i \in \mathbb{Z}$ and $\sum k_i^2 \neq 0$. Otherwise, the torus is called *non-resonant*.

A Liouville torus is non-resonant if and only if the closure of every integral trajectory lying on it coincide with the whole torus. Conversely, in the resonant case, the closure of any trajectory is a torus of a strictly less dimension.

Definition 1.13. An integrable system is called *non-resonant* on M^{2n} (or on some invariant subset) if almost all its Liouville tori are non-resonant. A system is called *resonant* if all its Liouville tori are resonant.

REMARK. In the smooth case, an integrable system need not to belong to one of these classes (i.e., to be either resonant or non-resonant). We do not consider such systems, since the systems that appear in practice are analytic and always of a certain type: they are either resonant or non-resonant.

1.6. ROTATION NUMBER

Consider an integrable Hamiltonian system $v = \operatorname{sgrad} H$ with two degrees of freedom on a symplectic manifold M^4; let T be an arbitrary Liouville torus. Consider the angle variables φ_1 and φ_2 on this torus (constructed in the Liouville theorem). Then v is written as follows:

$$\dot{\varphi}_1 = c_1, \qquad \dot{\varphi}_2 = c_2.$$

Definition 1.14. The *rotation number* ρ of the integrable system v on the given Liouville torus T is the ratio

$$\rho = \frac{c_1}{c_2}.$$

The value of ρ depends on the Liouville torus T and, therefore, is a function of the action variables s_1 and s_2 (or of the first independent integrals f_1 and f_2, since the action variables and these integrals can be expressed through each other).

Proposition 1.10. *Consider the Hamiltonian H as a function of s_1, s_2. Then*

$$\rho = \frac{\partial H/\partial s_1}{\partial H/\partial s_2}.$$

Proof. Since the form ω takes the canonical form in action-angle variables, we get

$$\operatorname{sgrad} H = \frac{\partial H}{\partial s_1}\partial/\partial\varphi_1 + \frac{\partial H}{\partial s_2}\partial/\partial\varphi_2.$$

Now the desired formula follows directly from the definition of ρ.   □

In fact, the notion of a rotation number is topological. It can be seen, for example, from the following proposition.

If T is resonant, then all integral trajectories of v are closed and homologous to each other (and even isotopic on the torus). Let λ and μ be a basis in the fundamental group of the torus, i.e., two independent cycles that are homologous to the coordinate lines of the angle variables φ_1 and φ_2. Then the closed trajectory γ of the field v can be decomposed with respect to this basis as follows:

$$\gamma = p\lambda + q\mu.$$

Proposition 1.11. *The rotation number on a resonant torus is equal to p/q.*

In other words, in this case, the rotation number ρ defines (and is itself defined by) the topological type of γ uniquely up to its orientation.

Proof. Since the torus is resonant, $\rho = \dfrac{c_1}{c_2}$ is rational and, therefore, $\dfrac{c_1}{c_2} = \dfrac{p}{q}$ for some integers p and q. Clearly, two vector fields

$$v = c_1\,\partial/\partial\varphi_1 + c_2\,\partial/\partial\varphi_2 \quad\text{and}\quad p\,\partial/\partial\varphi_1 + q\,\partial/\partial\varphi_2$$

have the same integral trajectories. But the trajectories of the second field are evidently closed and of the type $p\lambda + q\mu$.   □

REMARK. In some cases, for these reasons, the resonant tori are called *rational*, and non-resonant tori are called *irrational*.

To compute the rotation number ρ, it is not necessary to use action-angle variables. Usually, it is quite a non-trivial problem to find them. Instead of action-angle variables, one can use an arbitrary system of periodic coordinates

$$(x \bmod 2\pi, y \bmod 2\pi)$$

on the torus whose coordinate lines are homotopic to level lines of the canonical angle variables φ_1 and φ_2. It is convenient to consider x and y as coordinates on the 2-plane that covers the torus T. Let $x(t)$ and $y(t)$ be the coordinates of a point of an arbitrary integral trajectory of v (after its lifting to the covering plane).

Proposition 1.12. *The following formula holds:*

$$\rho = \lim_{t \to \infty} \frac{x(t)}{y(t)}.$$

Proof. By choosing new periodic coordinates on the torus, we only change the shape of the fundamental domain (Fig. 1.3). Moreover, we may suppose that the vertices of the lattice remain the same.

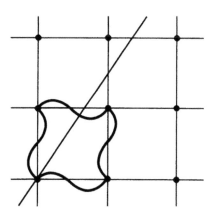

Figure 1.3

Hence, for some constant C we have:

$$|x(t) - \varphi_1(t)| \le C, \qquad |y(t) - \varphi_2(t)| \le C.$$

But, since $\varphi_1(t) = c_1 t + \text{const}$ and $\varphi_2(t) = c_2 t + \text{const}$, we obtain

$$\lim_{t \to \infty} \frac{x(t)}{y(t)} = \frac{c_1}{c_2},$$

as was to be proved. □

The above formula can be taken as the definition of the rotation number. Note that here we do not use the fact that the system in question is Hamiltonian. That is why the rotation number is defined for a larger class of dynamical systems on the torus. See, for example, [9], [14], and [16].

Note that the rotation number depends on the choice of a basis on the torus. Choosing the action-angle variables, we indicate, as a result, a certain basis of cycles on the corresponding Liouville torus. These cycles are just level lines of the angle variables on the torus. Conversely, having chosen a basis of cycles on the torus, we can construct those action-angle variables for which the basis cycles are (homologous to) the level lines of the angle variables.

It is useful to understand how the rotation function transforms when the pair of basis cycles λ, μ is replaced by the cycles λ', μ'. It is well known that for every pair of bases, there exists an integer matrix such that

$$\begin{pmatrix} \lambda \\ \mu \end{pmatrix} = \begin{pmatrix} a_1 & a_2 \\ a_3 & a_4 \end{pmatrix} \begin{pmatrix} \lambda' \\ \mu' \end{pmatrix}.$$

Proposition 1.13. *Let ρ be the rotation number for the pair of cycles λ and μ, and let ρ' be that for the other pair of cycles λ' and μ'. Then ρ and ρ' are connected by the relation*

$$\rho' = \frac{\rho a_1 + a_3}{\rho a_2 + a_4}.$$

Proof. This relation easily follows from the standard formulas for the transformation of the coordinates of a vector under a change of basis.   □

This formula allows us to define the rotation number ρ' in the case when λ' and μ' do not form a basis on the torus but are linearly independent. In this case, the transition matrix is, generally speaking, rational (but not integer).

An analog of the rotation number can also be defined in the case of integrable systems with many degrees of freedom. Let us fix a basis on a Liouville torus; we choose the corresponding angle coordinates $\varphi_1, \ldots, \varphi_n$ in which a Hamiltonian vector field v is straightened and takes the form

$$\dot{\varphi}_1 = c_1, \quad \ldots, \quad \dot{\varphi}_n = c_n.$$

As an analog of the rotation number, it is natural to consider the set of frequencies up to proportionality, i.e.,

$$(c_1 : c_2 : \ldots : c_n).$$

1.7. THE MOMENTUM MAPPING OF AN INTEGRABLE SYSTEM AND ITS BIFURCATION DIAGRAM

Let M^{2n} be a symplectic manifold with an integrable Hamiltonian system $v = \operatorname{sgrad} H$, and $f_1, \ldots, f_n$ be its independent integrals in involution. Let us define the smooth mapping

$$\mathcal{F}: M^{2n} \to \mathbb{R}^n, \qquad \text{where} \quad \mathcal{F}(x) = (f_1(x), \ldots, f_n(x)).$$

Definition 1.15. The mapping $\mathcal{F}$ is called the *momentum mapping*.

Definition 1.16. A point $x \in M$ is called a *critical* (or *singular*) point of the momentum mapping $\mathcal{F}$ if $\operatorname{rank} d\mathcal{F}(x) < n$. Its image $\mathcal{F}(x)$ in $\mathbb{R}^n$ is called a *critical value*.

Let $K \subset M$ be the set of all critical points of the momentum mapping.

Definition 1.17. The image of K under the momentum mapping, i.e., the set $\Sigma = \mathcal{F}(K) \subset \mathbb{R}^n$, is called the *bifurcation diagram*.

Thus, the bifurcation diagram is the set of all critical values of $\mathcal{F}$. According to the Sard theorem, the set Σ has zero measure in $\mathbb{R}^n$. In most examples of integrable systems appearing in physics and mechanics, the bifurcation diagram Σ is a manifold with singularities. In other words, it consists of several strata (pieces) Σ^i each of which is a smooth i-dimensional surface in $\mathbb{R}^n$. We can write this

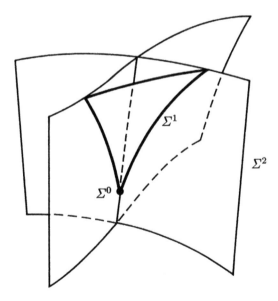

Figure 1.4

as $\Sigma = \Sigma^0 + \Sigma^1 + \ldots + \Sigma^{n-1}$, where different strata do not intersect with each other and the union of them gives the whole of Σ. The boundary of each strata Σ^i is contained in the union of strata of strictly less dimension (Fig. 1.4). In this case Σ is called a stratified manifold. Some of Σ^i's can be empty.

In the generic case, the complement of Σ, i.e., $\mathbb{R}^n \setminus \Sigma$ is open and everywhere dense in $\mathbb{R}^n$. The set $\mathbb{R}^n \setminus \Sigma$ can consist of several connected components. In some cases, they will be called *chambers*.

The momentum mapping and its bifurcation diagrams are closely connected with the Liouville foliation on M^{2n}.

First, a leaf of the Liouville foliation is a connected component of the inverse image of a point under the momentum mapping. In what follows, we shall always assume that all the leaves of the Liouville foliation are compact. This condition certainly holds if either the manifold M^{2n} itself or the level surface of the Hamiltonian H is compact.

Second, Σ is the image of the singular leaves of the Liouville foliation.

Third, the Liouville foliation is locally trivial over each chamber. In particular, the inverse images of any two points of a certain chamber are diffeomorphic to the disjoint union of k Liouville tori, where k is the same for both these points.

The bifurcation diagram allows us to trace the bifurcations of Liouville tori when changing the values of first integrals $f_1, \ldots, f_n$. For example, let points a and b be connected by a smooth segment γ that intersects the bifurcation diagram Σ at a point c. A certain number of Liouville tori hang over the point a, and a certain (possibly different) number of Liouville tori hang over b. Since $\gamma(t)$ goes along the segment, the Liouville tori move smoothly in M^{2n}, and over the point c they can undergo a topological bifurcation (see Fig. 1.5). For example, one torus can split into two new tori.

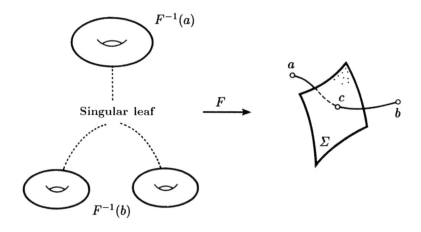

Figure 1.5

If $n = 2$ (i.e., if we consider a system with two degrees of freedom), then Σ usually consists of some segments of smooth curves on the 2-plane and, perhaps, individual isolated points. Here the chambers are two-dimensional open domains on the plane.

1.8. NON-DEGENERATE CRITICAL POINTS OF THE MOMENTUM MAPPING

We wish to emphasize here one important idea. Below we shall deal with typical singularities of the momentum mapping of integrable systems. It turns out that, despite their complexity, they admit, in principle, a reasonable and quite visual description. At the same time, from the viewpoint of the general theory of singularities of smooth mappings, those of the momentum mapping are neither typical nor stable. The point is that in the generic case the dimension of the set of critical points for a smooth mapping from M^{2n} into $\mathbb{R}^n$ is equal to $n-1$, whereas the typical singularities of the momentum mapping have doubled dimension, namely $2n-2$. It is explained by the fact that on M^{2n} there is the Poisson action of $\mathbb{R}^n$ which spreads singular points of the momentum mapping (if a point turns out to be singular, then all points obtained from it by the action of $\mathbb{R}^n$ are also critical). That is why the generic character of singularities of the momentum mapping should be considered not in the abstract, broad sense, but only in the class of mapping generated by the action of the Abelian group $\mathbb{R}^n$. Note that after this the problem of the classification of such singularities does not become simpler or more complicated. It becomes just different.

1.8.1. *The Case of Two Degrees of Freedom*

Consider a Hamiltonian system $v = \operatorname{sgrad} H$ on a four-dimensional symplectic manifold M^4.

Definition 1.18. The set of points given by the equation $H(x) = \text{const}$ is called an *isoenergy surface.*

If $H(x) = h$, then we denote the corresponding isoenergy surface by Q_h. As we already know, it is always invariant with respect to the Hamiltonian field v.

Consider an isoenergy three-dimensional surface $Q = Q^3$. In what follows, we shall assume it to be a smooth compact submanifold in M^4. In particular, we shall consider only those 3-surfaces on which $dH \neq 0$.

In the case of two degrees of freedom, for integrability of the system v it is sufficient to have just one additional integral f functionally independent of the energy integral H. This integral f, being restricted to Q, is a smooth function which always has some critical points (by virtue of the compactness of Q). Clearly, these critical points of f on Q coincide with those of the momentum mapping $\mathcal{F} = (H, f)$ which belong to Q. That is why it is natural to study the singularities of the momentum mapping in terms of the restriction of f onto Q. We shall denote this restriction by the same letter f as before.

Lemma 1.6. *The integral f cannot have isolated critical points on Q.*

Proof. According to our assumption, $dH|_Q \neq 0$. Hence, the vector field $v = \operatorname{sgrad} H$ differs from zero at each point of Q. Consider the integral curve γ of the field v passing through a critical point x of the integral f. The Hamiltonian flow v preserves the function f and, in particular, sends critical points to critical ones. Therefore, γ entirely consists of critical points.   □

Thus, f cannot ever be a Morse function on a non-singular isoenergy surface. To what class of functions does it actually belong? It turns out that in real problems of physics and mechanics, the situation described in the following definition is typical.

Definition 1.19. We shall say that f is a *Bott* function on Q if all of its critical points are organized into non-degenerate critical submanifolds. This means that the set of critical points is the disjoint union of some smooth submanifolds each of which is non-degenerate in the following sense. The second differential d^2f is non-degenerate on the subspace transversal to the submanifold (at each point). In other words, the restriction of f on the transversal to the submanifolds is a Morse function.

The appearance of such functions in the theory of integrable systems is quite natural. Roughly speaking, they play here the same role as Morse functions in the usual theory of functions on smooth manifolds, i.e., they are the most non-degenerate from the point of view of Poisson actions.

Consider the integrable system v, and suppose that f is a Bott function on a certain regular compact isoenergy surface Q.

Proposition 1.14. *Connected critical submanifolds of the integral f on Q are diffeomorphic either to a circle, or to the torus, or to the Klein bottle.*

Proof. Since Q is three-dimensional, the critical submanifolds of f can be either one-dimensional or two-dimensional. In the one-dimensional case, each connected component of such a manifold is a circle (because of the compactness argument). In the two-dimensional case, on such a manifold there is a smooth nowhere vanishing vector field $v = \operatorname{sgrad} H$ (since we assume that $dH \neq 0$ on Q). Such a vector field can exist only on the torus and Klein bottle.   □

Thus, the critical submanifolds have a rather simple structure. In our book (for two degrees of freedom) we shall mainly consider those integrable systems which have neither critical tori nor Klein bottles on the three-dimensional isoenergy surface in question. In other words, we shall study the systems whose critical submanifolds can be only circles.

This assumption is motivated by the following two reasons.

1. Just such systems usually occur in real problems in mechanics and physics.

2. It is possible to show (see [170], [171]) that, by a small perturbation, one can turn an integrable Hamiltonian system with a Bott integral into an integrable Hamiltonian system of the above type (i.e., turn critical tori and Klein bottles into a collection of non-degenerate critical circles).

Below we shall need an analog of the Morse lemma for the case of Bott functions. We now formulate and prove this Morse–Bott lemma in the case of arbitrary dimension.

Let M be an orientable smooth manifold, and f be a smooth function on it. Let N^{n-k} be a connected smooth compact submanifold in M^n whose normal bundle is trivial. Suppose N is a non-degenerate critical submanifold of f, that is, $df(x) = 0$ for any point $x \in N$, and the Hessian d^2f is non-degenerate on the transversal subspace to N (at each point $x \in N$). Let the index of d^2f be equal to λ. It is easy to see that, under these assumptions, λ does not depend on the choice of a point x on N.

Further, consider two subbundles E_- and E_+ in the normal bundle $E(N)$ which are orthogonal with respect to the form d^2f and such that d^2f is negative definite on E_- and positive definite on E_+. These submanifolds can be constructed in the following way. Suppose we are given a Riemannian metric on M. Then, at each point $x \in N$, the Hessian d^2f (regarded as a form on $E(N)$) can be reduced to the diagonal form in some orthonormal basis. Although such a basis is not uniquely defined, the two subspaces $E_+(x)$ and $E_-(x)$ spanned on the vectors corresponding to positive and negative eigenvalues of the Hessian d^2f, are well-defined. It is clear that the topological type of the subbundles E_+ and E_- does not depend on the choice of the Riemannian metric. Note that the dimension of $E_-(x)$ is λ.

Lemma 1.7 (Generalized Morse–Bott lemma). *Suppose both the subbundles E_+ and E_- are trivial. Then in some neighborhood $U(N)$ of the submanifold N there exist smooth independent functions $x_1, \ldots, x_k$ such that $x_i|_N = 0$ for all $i = 1, 2, \ldots, k$ and*

$$ f = c - x_1^2 - \ldots - x_\lambda^2 + x_{\lambda+1}^2 + \ldots + x_k^2 , $$

where $c = f|_N$.

Proof. Consider orthonormal (with respect to d^2f) bases $e_1(x), \ldots, e_\lambda(x)$ and $e_{\lambda+1}(x), \ldots, e_k(x)$ in fibers $E_+(x)$ and $E_-(x)$ of the bundles E_+ and E_-. Suppose that these bases smoothly depend on $x \in N$. Such bases evidently exist because E_+ and E_- are trivial. Clearly, the coordinates $y_1, \ldots, y_\lambda, y_{\lambda+1}, \ldots, y_k$ (relative to the chosen basis) can be considered as smooth independent functions in some neighborhood of N. This neighborhood can locally be identified with a neighborhood of the zero section in the normal bundle.

This collection of functions (coordinates) $y_1, \ldots, y_\lambda, y_{\lambda+1}, \ldots, y_k$ can be completed by functions $y_{k+1}, \ldots, y_n$ up to a coordinate system defined in a neighborhood of a given point $x \in N$. Here the functions $y_1, \ldots, y_\lambda, y_{\lambda+1}, \ldots, y_k$ are defined on the whole neighborhood of $U(N)$, whereas the functions $y_{k+1}, \ldots, y_n$ are defined only in some neighborhood of the point x. Without loss of generality, we can consider $y_{k+1}, \ldots, y_n$ as local coordinates on N, which are constant on the fibers of $E(N)$. It will be convenient to define the position of a point y in the following way: $(y_1, \ldots, y_k, \alpha)$, where α denote the point on the base N that is the projection of y. Then the quadratic form d^2f takes the following form at points of the critical submanifold:

$$ d^2f = -dy_1^2 - \ldots - dy_\lambda^2 + dy_{\lambda+1}^2 + \ldots + dy_k^2 . $$

After this, we can in fact literally repeat the proof of the classical Morse lemma for the function $f(y_1, \ldots, y_k, \alpha)$ thinking of α as a multidimensional parameter. Following the standard proof of the Morse lemma [241], we represent $f(y)$ in the form

$$ f(y_1, \ldots, y_k, \alpha) = \sum_{1 \le i,j \le k} y_i y_j h_{ij}(y_1, \ldots, y_k, \alpha + c) . $$

Here

$$ h_{ij}(y_1, \ldots, y_k, \alpha) = \int_0^1 \int_0^1 t \frac{\partial^2 f}{\partial y_i \partial y_j}(tuy_1, \ldots, tuy_k, \alpha) \, dt \, du . $$

Besides,

$$h_{ij}(0) = \frac{1}{2} \cdot \frac{\partial^2 f}{\partial y_i \partial y_j}(0).$$

Consider then the expression $\sum_{1\le i,j\le k} y_i y_j h_{ij}(y_1,\ldots,y_k,\alpha)$ as a quadratic form with coefficients $h_{ij}(y_1,\ldots,y_k,\alpha)$, and reduce it to the normal form by standard changes of coordinates.

It is important that this procedure can be carried out simultaneously for all α. As a result, the quadratic form $\sum_{1\le i,j\le k} y_i y_j h_{ij}(y_1,\ldots,y_k,\alpha)$ can be reduced to the diagonal form $-x_1^2 - \ldots - x_\lambda^2 + x_{\lambda+1}^2 + \ldots + x_k^2$, where the new functions $x_1,\ldots,x_k$ are related to the previous variables $y_1,\ldots,y_k,\alpha$ by the "linear formulas":

$$x_i = \sum_{1\le j\le k} a_{ij}(y,\alpha)y_j.$$

Moreover, the matrix $(a_{ij}(y,\alpha))$ is upper triangular with positive entries on the diagonal. Such a matrix exists, is uniquely defined, and smoothly depends on y and α. In fact, the proof is completed. However, in conclusion we explain where the triviality of the foliations $E_-(N)$ and $E_+(N)$ has been used. Note that not every quadratic form in y with coefficients depending on a parameter α can be reduced to the diagonal form simultaneously for all α by means of a transformation which smoothly depends on α. But in our case, if $y = 0$, then the required transformation exists for all α, because for $y = 0$ the form $\sum h_{ij}y_i y_j$ is already written in the canonical way. In other words, $a_{ij}(0,\alpha) = \delta_{ij}$. But in such a case, in a small neighborhood of zero (with respect to variables $y_1,\ldots,y_k$), the desired matrix $(a_{ij}(y,\alpha))$ also exists and, therefore, it is uniquely defined and depends smoothly on y and α. $\quad\square$

1.8.2. Bott Integrals from the Viewpoint of the Four-Dimensional Symplectic Manifold

We now consider the critical points of the momentum mapping $\mathcal{F}$ not on Q, but from the viewpoint of the ambient 4-manifold M. The closed set K of all critical points can be stratified by the rank of the momentum mapping, i.e., can be represented as the union

$$K = K_1 + K_0,$$

where $K_1 = \{x \mid \operatorname{rank} d\mathcal{F}(x) = 1\}$ and $K_0 = \{x \mid \operatorname{rank} d\mathcal{F}(x) = 0\}$.

Proposition 1.15. *The set K_1 is the union of all one-dimensional orbits of the Poisson action of the group $\mathbb{R}^2$ on M^4. The set K_0 consists of all fixed points of this action.*

Proof. This statement holds for an arbitrary number of degrees of freedom. Indeed, the dimension of any orbit of the group $\mathbb{R}^n$ is equal to the rank of the system of vectors $\{\operatorname{sgrad} f_i\}$. This rank, in turn, is equal to the rank of the system $\{df_i\}$. $\quad\square$

Let us ask ourselves which points of K_1 and K_0 can naturally be called non-degenerate. We begin with K_1.

Consider a point $x \in M^4$ such that rank $d\mathcal{F}(x) = 1$. The orbit of this point is one-dimensional and diffeomorphic to either the real line or the circle.

Suppose, for definiteness, that $dH(x) \neq 0$. Then, according to the Darboux theorem, there exists a canonical coordinate system (p_1, q_1, p_2, q_2) in a neighborhood of x such that $H = p_1$. As f and H commute, the function f does not depend on q_1, i.e., $f = f(p_1, p_2, q_2)$.

Since $x \in K_1$ is a critical point of $\mathcal{F}$, we have

$$\frac{\partial f}{\partial p_2}(x) = \frac{\partial f}{\partial q_2}(x) = 0.$$

Definition 1.20. The point $x \in K_1$ is called *non-degenerate* for the momentum mapping $\mathcal{F}$ if the matrix

$$\begin{pmatrix} \dfrac{\partial^2 f}{\partial p_2^2} & \dfrac{\partial^2 f}{\partial p_2 \partial q_2} \\[2ex] \dfrac{\partial^2 f}{\partial p_2 \partial q_2} & \dfrac{\partial^2 f}{\partial q_2^2} \end{pmatrix}$$

is non-degenerate at x. We denote the set of all such points by K_1^*.

The definition is correct, that is, does not depend on the choice of a canonical coordinate system. This follows from the fact that the above matrix exactly coincides with the Hessian of the function f restricted to the two-dimensional transversal to the one-dimensional orbit $Q = \{p_1 = \mathrm{const}\}$, which entirely consists of critical points of f.

In other words, f is a Bott function on the three-dimensional level $\{H = \mathrm{const}\}$ (locally in a neighborhood of x). In particular, the above non-degeneracy condition forbids the existence of critical tori and Klein bottles on the level $\{H = \mathrm{const}\}$.

The previous definition of non-degeneracy has been formulated in terms of local coordinates. It is possible to give another equivalent definition which does not use any coordinates. It will be useful to verify the non-degeneracy condition, because usually it is not so easy to find the canonical coordinate system used above.

Let the rank of $d\mathcal{F}(x)$ be equal to 1. Then the differentials df and dH are dependent at this point, i.e., there exist λ and μ such that

$$\lambda\, df(x) + \mu\, dH(x) = 0.$$

Here λ and μ are defined uniquely up to proportionality. Let L be a tangent line to the one-dimensional orbit of the action of $\mathbb{R}^2$. It is a one-dimensional subspace (in the tangent space to M^4) generated by the linearly dependent vectors sgrad f and sgrad H. Let L' be the three-dimensional subspace orthogonal to L in the sense of the symplectic form.

Definition 1.21. The critical point x is called *non-degenerate* if the rank of the symmetric 2-form $\lambda\, d^2f(x) + \mu\, d^2H(x)$ on the subspace L' is equal to 2.

This 2-form is well-defined, because $\lambda\, df(x) + \mu\, dH(x) = 0$.

Note that the rank of $\lambda\, d^2 f(x) + \mu\, d^2 H(x)$ cannot be equal to 3 on L', since the one-dimensional subspace $L \subset L'$ belongs to the kernel of this form. Let us prove this. Let, for definiteness, L be generated by $v = \operatorname{sgrad} f$ (recall that $\operatorname{sgrad} f(x)$ and $\operatorname{sgrad} H(x)$ are linearly dependent). Compute the value of the form on the pair of vectors v and ξ, where ξ is arbitrary. Since f and H commute, we get

$$(\lambda\, d^2 f + \mu\, d^2 H)(v,\xi) = \xi(\operatorname{sgrad} f(\lambda f + \mu H)) = \xi(\{f, \lambda f + \mu H\}) = 0.$$

Definitions 1.20 and 1.21 are equivalent. Indeed, let us rewrite Definition 1.21 in the special system of Darboux coordinates (p_1, q_1, p_2, q_2) from Definition 1.20. Then L is generated by the vector $\partial/\partial q_1$, and L' is generated by the vectors $\partial/\partial q_1, \partial/\partial p_2, \partial/\partial q_2$. Since the function f does not depend on the variable q_1, and $H = p_1$, it follows that the matrix of the form $\lambda\, d^2 f(x) + \mu\, d^2 H(x)$ has the form

$$\lambda \begin{pmatrix} 0 & 0 & 0 \\ 0 & \dfrac{\partial^2 f}{\partial p_2^2} & \dfrac{\partial^2 f}{\partial p_2 \partial q_2} \\ 0 & \dfrac{\partial^2 f}{\partial p_2 \partial q_2} & \dfrac{\partial^2 f}{\partial q_2^2} \end{pmatrix}.$$

Thus, in local coordinates Definitions 1.20 and 1.21 coincide.

What is the topological structure of the subset K_1^* of non-degenerate critical points of the momentum mapping?

Proposition 1.16.

a) K_1^* is a two-dimensional smooth symplectic submanifold in M^4.

b) For each point x of K_1^* there exists a neighborhood U in K_1^* which is diffeomorphic to a 2-disc and such that its image $\mathcal{F}(U)$ in $\mathbb{R}^2$ is a regular curve δ without self-intersections.

c) Let $\dot{\delta} = (a,b)$ be a tangent vector to the curve δ at the point $\mathcal{F}(x)$; then

$$b \operatorname{sgrad} H(x) - a \operatorname{sgrad} f(x) = 0.$$

Proof. Let us use Definition 1.20 of the non-degeneracy of critical points. Choose a canonical coordinate system (p_1, q_1, p_2, q_2) in which $H = p_1$ and $f = f(p_1, p_2, q_2)$. Then, in a neighborhood of x, the set of critical points of the momentum mapping $\mathcal{F}$ is given by the two equations: $\dfrac{\partial f}{\partial p_2} = 0$ and $\dfrac{\partial f}{\partial q_2} = 0$. By virtue of non-degeneracy of the matrix

$$\begin{pmatrix} \dfrac{\partial^2 f}{\partial p_2^2} & \dfrac{\partial^2 f}{\partial p_2 \partial q_2} \\ \dfrac{\partial^2 f}{\partial p_2 \partial q_2} & \dfrac{\partial^2 f}{\partial q_2^2} \end{pmatrix},$$

these two equations define (locally) a smooth two-dimensional submanifold in M. Since the above matrix remains non-degenerate in some neighborhood of x, all critical points from this neighborhood also turn out to be non-degenerate. Note that this matrix is a part of the Jacobi matrix for the system of two equations $\dfrac{\partial f}{\partial p_2} = 0$, $\dfrac{\partial f}{\partial q_2} = 0$; therefore, according to the implicit function theorem, we obtain that the variables p_2 and q_2 can locally be expressed as single-valued functions of p_1 and q_1, i.e., the set K_1^* is locally represented as a smooth graph.

Let us show that K_1^* is symplectic. To this end, we need to verify that the restriction of ω on it is non-degenerate.

Indeed, by substituting the functions $p_2 = p_2(p_1, q_1)$ and $q_2 = q_2(p_1, q_1)$ into the form $\omega = dp_1 \wedge dq_1 + dp_2 \wedge dq_2$ and recalling that they do not actually depend on q_1 (i.e., $p_2 = p_2(p_1)$ and $q_2 = q_2(p_1)$), we see that the restriction of ω onto K_1^* coincides with $dp_1 \wedge dq_1$ and, consequently, is non-degenerate.

We now prove the second statement of Proposition 1.16.

Write the momentum mapping in coordinates (p_1, q_1, p_2, q_2) and restrict it to K_1^*. Since one can take p_1 and q_1 as local coordinates on K_1^*, and since $p_2 = p_2(p_1)$ and $q_2 = q_2(p_1)$, the momentum mapping takes the form:

$$H = p_1, \qquad f = f(p_1, p_2(p_1), q_2(p_1)) = \widetilde{f}(p_1).$$

Thus, the image of a small neighborhood of $x \in K_1^*$ under the momentum mapping is a smooth curve given as the graph $f = \widetilde{f}(p_1) = \widetilde{f}(H)$.

It remains to prove (c).

In the chosen coordinate system we have

$$\operatorname{sgrad} f(x) = \frac{\partial \widetilde{f}}{\partial p_1} \operatorname{sgrad} H(x).$$

On the other hand, taking p_1 as a local parameter on the curve δ, we get

$$\dot{\delta} = \left(1, \frac{\partial \widetilde{f}}{\partial p_1}\right).$$

This completes the proof of Proposition 1.16. $\square$

Corollary. *Suppose* $x \in K_1^*$ *and the straight line* $H = H(x)$ *intersects the bifurcation diagram* Σ *transversally at the point* $\mathcal{F}(x)$. *Then* x *is a regular point for the function* H *restricted to* K_1^*.

Proof. Under above assumptions, as we have seen before, one can suppose that H has the form p_1, whereas p_1 and q_1 are regular local coordinates on the submanifold K_1^* in a neighborhood of x. Therefore, the function $H|_{K_1^*}$ has no singularity at x. $\square$

The submanifold K_1^* consists of one-dimensional orbits of the Poisson action of the group $\mathbb{R}^2$. Let us remove all non-compact (i.e., homeomorphic to $\mathbb{R}$) orbits of this group from K_1^*. The remaining part K_1^{**} of K_1^* is foliated into circles.

Divide K_1^{**} into connected components.

Corollary. *The image (under* $\mathcal{F}$*) of each connected component of* K_1^{**} *in* $\mathbb{R}^2$ *is either an immersed line, or an immersed circle.*

Consider the set $K \setminus K_1^{**}$. It consists of one-dimensional non-compact orbits of the action of $\mathbb{R}^2$, one-dimensional degenerated orbits, and zero-dimensional orbits (fixed points of the action). Then consider its image $\mathcal{F}(K \setminus K_1^{**})$ as the subset of the bifurcation diagram Σ and call it the set of *singular points* of Σ. Thus, Σ consists of regular (perhaps, intersecting) curves and singular points, among which there may be single isolated points of Σ, cusp points, tangent points, etc. (Fig. 1.6).

Figure 1.6

We now can give a visual explanation of the condition that the integral f is a Bott function on the isoenergy surface $Q = \{H = \text{const}\}$. Take the line $H = \text{const}$ in the two-dimensional plane $\mathbb{R}^2\,(H, f)$. Suppose that this line does not pass through singular points of Σ and intersects the smooth pieces of Σ transversally. Then f will be a Bott function on Q (notice that Q is the preimage of this line in M^4). Moreover, all its critical submanifolds will be circles, i.e., among them there will be neither tori nor Klein bottles.

Now let $x \in K_0$ and $dH(x) = df(x) = 0$, i.e., x is a fixed point of the Poisson action of $\mathbb{R}^2$. Then the group $\mathbb{R}^n$ acts naturally on the tangent space $T_x M$ to the manifold M at the point x. Since the group preserves the form ω, it induces linear symplectic transformations on $T_x M$. Therefore, we obtain some Abelian subgroup $G(H, f)$ in the group $\text{Sp}(4, \mathbb{R})$ of symplectic transformations of $T_x M$. Denote its Lie algebra by $K(H, f)$. Clearly, $K(H, f)$ is a commutative subalgebra in the Lie algebra $\text{sp}(4, \mathbb{R})$ of the group $\text{Sp}(4, \mathbb{R})$. It is easy to see that $K(H, f)$ is generated by the linear parts of the vector fields $\text{sgrad}\, H$ and $\text{sgrad}\, f$.

Definition 1.22. We say that $x \in K_0$ is a *non-degenerate singular point* of the momentum mapping $\mathcal{F}$ if $K(H, f)$ is a Cartan subalgebra in $\text{sp}(4, \mathbb{R})$. In particular, this requirement implies that the commutative subalgebra $K(H, f)$ must be two-dimensional.

How to check the condition that $K(H, f)$ is a Cartan subalgebra? To answer this question, first of all we describe this subalgebra in terms of the functions H and f. To this end, consider their quadratic parts, i.e., the Hessians d^2H and d^2f. They generate the linear symplectic operators $A_H = \Omega^{-1}d^2H$ and $A_f = \Omega^{-1}d^2f$ which coincide with the linearizations of the Hamiltonian vector fields $\text{sgrad}\, H$ and $\text{sgrad}\, f$ at the singular point $x \in K_0$. Indeed, the linearization of the field $w = \text{sgrad}\, f$ has the form

$$\frac{\partial w^i}{\partial x^j} = \frac{\partial}{\partial x^j}\left(\omega^{ik}\frac{\partial f}{\partial x^k}\right) = \omega^{ik}\frac{\partial^2 f}{\partial x^j \partial x^k} = (\Omega^{-1}d^2f)^i_j\,.$$

Thus, we see that the subalgebra $K(H, f)$ is generated by the linear operators $A_H = \Omega^{-1}d^2H$ and $A_f = \Omega^{-1}d^2f$.

A commutative subalgebra in $\mathrm{sp}(4, \mathbb{R})$ is a Cartan subalgebra if and only if it is two-dimensional and contains an element (a linear operator) whose eigenvalues are all different.

Thus, first we need to make sure that the forms d^2f and d^2H are independent. Then it should be checked that some linear combination $\lambda A_f + \mu A_H$ has different eigenvalues.

The real symplectic Lie algebra $\mathrm{sp}(4, \mathbb{R})$ contains precisely four types of different, i.e., pairwise non-conjugate, Cartan subalgebras. Let us list them. Let the symplectic structure ω be given (at the point x) by the canonical matrix

$$\Omega = \begin{pmatrix} 0 & -E \\ E & 0 \end{pmatrix}.$$

Then the Lie algebra $\mathrm{sp}(4, \mathbb{R})$ is represented (in the standard matrix representation) by the following 4×4-matrices (see Proposition 1.3):

$$\begin{pmatrix} P & Q \\ R & -P^\top \end{pmatrix},$$

where P, Q, and R are 2×2-matrices, Q and R are symmetric, and P is arbitrary.

Theorem 1.3. *Let K be a Cartan subalgebra in the Lie algebra $\mathrm{sp}(4, \mathbb{R})$. Then it is conjugate to one of the four Cartan subalgebras listed below:*

$$\begin{pmatrix} 0 & 0 & -A & 0 \\ 0 & 0 & 0 & -B \\ A & 0 & 0 & 0 \\ 0 & B & 0 & 0 \end{pmatrix}, \quad \begin{pmatrix} -A & 0 & 0 & 0 \\ 0 & 0 & 0 & -B \\ 0 & 0 & A & 0 \\ 0 & B & 0 & 0 \end{pmatrix}, \quad \begin{pmatrix} -A & 0 & 0 & 0 \\ 0 & -B & 0 & 0 \\ 0 & 0 & A & 0 \\ 0 & 0 & 0 & B \end{pmatrix}, \quad \begin{pmatrix} -A & -B & 0 & 0 \\ B & -A & 0 & 0 \\ 0 & 0 & A & -B \\ 0 & 0 & B & A \end{pmatrix},$$

$$\text{Type 1} \qquad\qquad \text{Type 2} \qquad\qquad \text{Type 3} \qquad\qquad \text{Type 4}$$

where A and B are arbitrary real numbers.

The listed subalgebras are not pairwise conjugate.

COMMENT. Cartan subalgebras of the four above types are not conjugate. This follows from the fact that their elements have eigenvalues of different types.

Type 1: four imaginary eigenvalues iA, $-iA$, iB, $-iB$.

Type 2: two real and two imaginary eigenvalues $-A$, A, iB, $-iB$.

Type 3: four real eigenvalues $-A$, A, $-B$, B.

Type 4: four complex eigenvalues $A - iB$, $A + iB$, $-A + iB$, $-A - iB$.

Consider a Cartan subalgebra in $\mathrm{sp}(4, \mathbb{R})$ and choose an element in it whose eigenvalues are all different (such an element is called regular). Proposition 1.2 implies that the eigenvalues of an operator from $\mathrm{sp}(4, \mathbb{R})$ separate into pairs $\lambda, -\lambda$. Hence, the spectrum of the regular element has one of the four above types and, therefore, the element itself is conjugate to one of the matrices indicated in Theorem 1.3. Moreover, the conjugation can be carried out by a symplectic transformation (this is equivalent to the fact that an operator with simple spectrum from the Lie algebra $\mathrm{sp}(4, \mathbb{R})$ can be reduced to the canonical form in a symplectic basis). Since two regular elements turn out to be conjugate, the Cartan subalgebras that contain them are also conjugate (by means of a symplectic transformation).

It follows easily from the fact that Cartan subalgebras are centralizers of their regular elements.

Note that over $\mathbb{C}$, i.e., in the complex Lie algebra $\mathrm{sp}(4,\mathbb{C})$, any two Cartan subalgebras are conjugate.

Theorem 1.3 can be reformulated as an answer to the question about the canonical form of the Hessians of two commuting functions H and f at a non-degenerate singular point.

Theorem 1.4. *In a neighborhood of a non-degenerate singular point $x \in K_0$, there exist canonical coordinates (p_1, q_1, p_2, q_2) in which the Hessians of H and f are simultaneously reduced to one of the following forms.*

1) *Center–center case:*
$$d^2 H = A_1(dp_1^2 + dq_1^2) + B_1(dp_2^2 + dq_2^2),$$
$$d^2 f = A_2(dp_1^2 + dq_1^2) + B_2(dp_2^2 + dq_2^2).$$

2) *Center–saddle case:*
$$d^2 H = A_1 dp_1 dq_1 + B_1(dp_2^2 + dq_2^2),$$
$$d^2 f = A_2 dp_1 dq_1 + B_2(dp_2^2 + dq_2^2).$$

3) *Saddle–saddle case:*
$$d^2 H = A_1 dp_1 dq_1 + B_1 dp_2 dq_2,$$
$$d^2 f = A_2 dp_1 dq_1 + B_2 dp_2 dq_2.$$

4) *Focus–focus case:*
$$d^2 H = A_1(dp_1 dq_1 + dp_2 dq_2) + B_1(dp_1 dq_2 - dp_2 dq_1),$$
$$d^2 f = A_2(dp_1 dq_1 + dp_2 dq_2) + B_2(dp_1 dq_2 - dp_2 dq_1).$$

Proof. The proof is implied by the above algebraic classification of Cartan subalgebras in $\mathrm{sp}(4,\mathbb{R})$. □

In fact, in a neighborhood of a non-degenerate singular point, a stronger theorem holds: not only the Hessian, but also the functions H and f themselves can simultaneously be reduced to some canonical form in appropriate symplectic coordinates. More precisely, the following important statement takes place.

Theorem 1.5. *Let the manifold M^4, the symplectic structure ω, and the functions H and f be real analytic. Then in a neighborhood of a non-degenerate singular point $x \in K_0$, there exist canonical coordinates (p_1, q_1, p_2, q_2) in which H and f are simultaneously reduced to one of the following forms.*

1) *Center–center case:*
$$H = H(p_1^2 + q_1^2, \, p_2^2 + q_2^2),$$
$$f = f(p_1^2 + q_1^2, \, p_2^2 + q_2^2).$$

2) *Center–saddle case:*
$$H = H(p_1 q_1, \, p_2^2 + q_2^2),$$
$$f = f(p_1 q_1, \, p_2^2 + q_2^2).$$

3) *Saddle–saddle case:*
$$H = H(p_1 q_1, \, p_2 q_2),$$
$$f = f(p_1 q_1, \, p_2 q_2).$$

4) *Focus–focus case:*
$$H = H(p_1 q_1 + p_2 q_2, \, p_1 q_2 - p_2 q_1),$$
$$f = f(p_1 q_1 + p_2 q_2, \, p_1 q_2 - p_2 q_1).$$

REMARK. Theorems 1.3, 1.4, and 1.5 have natural multidimensional general-izations (see Theorems 1.6 and 1.7 below). Theorem 1.3 and its multidimensional version (Theorem 1.6) is due to Williamson [362], [363]. Theorem 1.5 was first proved by Russmann [306], and was then generalized to the multidimensional case by Vey [355] and Ito [166].

1.8.3. Non-degenerate Singularities in the Case of Many Degrees of Freedom

We now give the definition of a non-degenerate singular point of rank i for the momentum mapping $\mathcal{F}$ in the multidimensional case.

Let $f_1, \ldots, f_n$ be smooth commuting functions on M^{2n} and

$$\mathcal{F}: M^{2n} \to \mathbb{R}^n, \qquad \text{where} \quad \mathcal{F}(x) = (f_1(x), \ldots, f_n(x)),$$

be the corresponding momentum mapping.

Let $K = K_0 + K_1 + \ldots K_{n-1}$ be the set of its critical points. Here K_i is the set of points $x \in K$, where the rank of $d\mathcal{F}(x)$ is equal to i (such points are called *singular points of rank i*).

For any $x \in K_i$, it is always possible to find a regular linear change of functions $f_1, \ldots, f_n$ so that the new functions, which we denote by $g_1, \ldots, g_n$, satisfy the following properties:

1) the first $n - i$ functions $g_1, \ldots, g_{n-i}$ have singularity at the point x, i.e., $dg_1(x) = \ldots = dg_{n-i}(x) = 0$;

2) the gradients of the functions $g_{n-i+1}, \ldots, g_n$ are linearly independent at x.

Since $dg_1(x) = \ldots = dg_{n-i}(x) = 0$, the corresponding Hamiltonian vector fields $\operatorname{sgrad} g_1, \ldots, \operatorname{sgrad} g_{n-i}$ vanish at the point x. Therefore, their linear parts $A_1, \ldots, A_{n-i}$ can be considered as linear operators from the Lie algebra of the symplectic group $\operatorname{Sp}(2n, \mathbb{R})$ which acts on the tangent space $T_x M$. Since the vector fields $\operatorname{sgrad} g_1, \ldots, \operatorname{sgrad} g_{n-i}$ commute pairwise, so do the corresponding operators $A_1, \ldots, A_{n-i}$.

Now consider the i-dimensional subspace L in $T_x M$ generated by the vectors $\operatorname{sgrad} g_{n-i+1}, \ldots, \operatorname{sgrad} g_n$. Note that L is the tangent space to the orbit of the Poisson action passing through the point x.

Let L' be the skew-orthogonal complement (in the sense of the symplectic form Ω) to the subspace L in $T_x M$. It contains L, since the functions $g_{n-i+1}, \ldots, g_n$ are in involution and the subspace L is, therefore, isotropic.

Lemma 1.8. *The subspace L belongs to the kernel of every operator $A_1, \ldots, A_{n-i}$. The images of operators $A_1, \ldots, A_{n-i}$ are contained in L'.*

Proof. Recall that if A_v is the linearization of the field v at its singular point x, and ξ is an arbitrary vector field, then

$$A_v(\xi) = [\xi, v].$$

In our case, we have: $A_j(\operatorname{sgrad} g_k) = [\operatorname{sgrad} g_k, \operatorname{sgrad} g_j] = 0$, because all the functions g_s are pairwise in involution (i.e., the corresponding fields $\operatorname{sgrad} g_s$ commute).

Therefore, L belongs to the kernel of A_j, $1 \leq j \leq n - i$.

Let us prove the second statement. The properties of the symplectic form Ω and an operator $A \in \mathrm{sp}(2n, \mathbb{R})$ imply that for any two vectors ξ, η the following identity holds (Proposition 1.3):

$$\Omega(A\xi, \eta) = -\Omega(\xi, A\eta).$$

Hence, im A belongs to the skew-orthogonal complement to ker A. In particular, since $L \subset \ker A_j$, it follows that im $A_j \subset L'$, as was to be proved. $\square$

Therefore, we can define a natural action of the operators $A_1, \ldots, A_{n-i}$ on the quotient space L'/L. We use the same notation for the induced operators.

Lemma 1.9. *There is a natural symplectic structure $\widetilde{\Omega}$ on the quotient space L'/L such that the operators $A_1, \ldots, A_{n-i}$ acting on L'/L are elements of the Lie algebra of the symplectic group $\mathrm{Sp}(\widetilde{\Omega}, L'/L) \simeq \mathrm{Sp}(2(n-i), \mathbb{R})$.*

Proof. It is clear that L is the kernel of the restriction of Ω to L'. That is why we can define a non-degenerate skew-symmetric form $\widetilde{\Omega}$ on L'/L:

$$\widetilde{\Omega}(\xi L, \eta L) = \Omega(\xi, \eta).$$

Then, for each operator A_s we need to verify the identity

$$\widetilde{\Omega}(A_s(\xi L), \eta L) = -\widetilde{\Omega}(\xi L, A_s(\eta L)),$$

which is equivalent, by Proposition 1.3, to the condition $A_s \in \mathrm{sp}(2(n - i), \mathbb{R})$.

We have

$$\widetilde{\Omega}(A_s(\xi L), \eta L) = \widetilde{\Omega}(A_s(\xi)L, \eta L) = \Omega(A_s(\xi), \eta)$$
$$= -\Omega(\xi, A_s(\eta)) = -\widetilde{\Omega}(\xi L, A_s(\eta)L) = -\widetilde{\Omega}(\xi L, A_s(\eta L)),$$

as required. $\square$

We can now consider the commutative subalgebra $K(x, \mathcal{F})$ in a real symplectic Lie algebra $\mathrm{sp}(2(n - i), \mathbb{R})$, generated by the operators $A_1, \ldots, A_{n-i}$.

Definition 1.23. A critical point $x \in K_i$ of the momentum mapping $\mathcal{F}$ is called *non-degenerate* if $K(x, \mathcal{F})$ is a Cartan subalgebra in $\mathrm{sp}(2(n - i), \mathbb{R})$.

Note that the subalgebra $K(x, \mathcal{F})$ does not depend on the linear change of the initial functions $f_1, \ldots, f_n$ to the new function $g_1, \ldots, g_n$, which we made above. In fact, $K(x, \mathcal{F})$ is completely determined by the Poisson action of the group $\mathbb{R}^n$ in a neighborhood of x.

Note that the above described procedure is equivalent to the local reduction with respect to the Poisson action of the group $\mathbb{R}^i$ generated by the functions $g_{n-i+1}, \ldots, g_n$ independent at the point x: we first make such a reduction that x becomes fixed, and then, for a fixed point of rank zero, we repeat the definition given above for two degrees of freedom.

How to verify that $K(x, \mathcal{F})$ is a Cartan subalgebra in $\mathrm{sp}(2(n - i), \mathbb{R})$? The criterion is quite simple. The subalgebra $K(x, \mathcal{F})$ must have dimension $n - i$ and contain an element with different eigenvalues. This allows us to reformulate the definition of a non-degenerate point in the following way.

Definition 1.24. A critical point x of rank i is said to be *non-degenerate* if the Hessians $d^2 g_1(x), \ldots, d^2 g_{n-i}(x)$ are linearly independent on L' and, for some $\lambda_1, \ldots, \lambda_{n-i}$, the characteristic polynomial

$$P(\mu) = \det\left(\lambda_1\, d^2 g_1(x) + \ldots + \lambda_{n-i}\, d^2 g_{n-i}(x) - \mu \Omega\right)\big|_{L'}$$

has $2(n - i)$ different non-zero roots.

This definition can be also reformulated as follows.

Let us consider the subspace linearly generated by the functions $f_1, \ldots, f_n$ as a Lie algebra. Let K_x be the stationary subalgebra of the point x, i.e., the subalgebra consisting of the functions f such that $df(x) = 0$. As before, let L denote the tangent space to the orbit of x, i.e., the subspace in $T_x M$ generated by the vectors $\mathrm{sgrad}\, f_1, \ldots, \mathrm{sgrad}\, f_n$, and L' denote the skew-orthogonal complement to L.

Definition 1.25. The critical point x of rank i is called *non-degenerate* if the two conditions are fulfilled:

1) for each function $f \in K_x$, different from zero, the quadratic form $d^2 f(x)$ is not identically zero on the subspace L';

2) there exists a function $f \in K_x$ such that the polynomial

$$P(\mu) = \det\left(d^2 f(x) - \mu \Omega\right)\big|_{L'}$$

has $2(n - i)$ different non-zero roots.

Let us clarify the connection between the latter definition and the two previous ones. It is easy to see that the subalgebra $K(x, \mathcal{F})$ is the image of the stationary subalgebra K_x of the point x under the natural mapping $f \to d^2 f(x)$. The first condition in Definition 1.25 means that this mapping is an isomorphism and, consequently, is equivalent to the fact that $\dim K(x, \mathcal{F}) = n - i$. The second condition means that $K(x, \mathcal{F})$ contains a regular element (i.e., an operator whose eigenvalues are all different). Also note that the subspace L', to which we restrict all the Hessians, has dimension $2n - i$. That is why the polynomial $P(\mu)$ has $2n - i$ roots. However, i of these roots are always equal to zero, since the i-dimensional subspace $L \subset L'$ lies in the kernel of all Hessians. The second requirement is that all the other roots are different.

Denote by K_i^* the set of non-degenerate critical points of rank i. It is possible to prove that this set is a smooth symplectic submanifold in M^{2n} of dimension $2i$. Besides, it is smoothly foliated into i-dimensional orbits of the Poisson action of the group $\mathbb{R}^n$.

It is worth distinguishing the class of integrable systems all of whose critical points (of all ranks) are non-degenerate. Such systems actually exist and we give some examples below.

1.8.4. Types of Non-degenerate Singularities in the Multidimensional Case

According to Definition 1.24, a critical point of the momentum mapping $\mathcal{F}$ of rank i is non-degenerate if and only if the subalgebra $K(x, \mathcal{F})$ is a Cartan subalgebra in the real symplectic Lie algebra $\mathrm{sp}(2(n - i), \mathbb{R})$. Recall that, unlike the complex symplectic Lie algebra $\mathrm{sp}(n - i, \mathbb{C})$, in the real Lie algebra $\mathrm{sp}(2(n - i), \mathbb{R})$ there exist Cartan subalgebras that are not conjugate. They have different types and these types in fact classify non-degenerate critical points of the momentum mapping in the real case. Such a classification was obtained by J. Williamson [362]. To formulate his theorem, it is convenient to consider the real Lie algebra $\mathrm{sp}(2m, \mathbb{R})$ as the space of homogeneous quadratic polynomials on the symplectic space $\mathbb{R}^{2m}$ with the canonical 2-form ω (we denote its matrix by Ω). The commutator in this algebra is the usual Poisson bracket of polynomials (regarded as usual functions).

COMMENT. The connection between this model and the construction from the previous section is explained as follows. Consider a homogeneous quadratic polynomial f and the corresponding vector field $\mathrm{sgrad}\, f$. This field has a singular point at zero, and we can, therefore, consider its linearization A_f as a linear operator. It is easily verified that the mapping $f \to A_f$ is an isomorphism of the algebra of quadratic polynomials onto the algebra $\mathrm{sp}(2m, \mathbb{R})$.

Theorem 1.6 (J. Williamson). *Let $K \subset \mathrm{sp}(2m, \mathbb{R})$ be a Cartan subalgebra. Then there exist a symplectic coordinate system $x_1, \ldots, x_m, y_1, \ldots, y_m$ in $\mathbb{R}^{2m}$ and a basis $e_1, \ldots, e_m$ in K such that each of the quadratic polynomials e_i takes one of the following forms:*
1) $e_i = x_i^2 + y_i^2$ *(elliptic type)*,
2) $e_i = x_i y_i$ *(hyperbolic type)*,
3) $e_i = x_i y_{i+1} - x_{i+1} y_i$, $e_{i+1} = x_i y_i + x_{i+1} y_{i+1}$ *(focus–focus type)*.

It is seen from this theorem that the type of the Cartan subalgebra K is completely determined by the collection of three integers (m_1, m_2, m_3), where $m_1 + m_2 + 2m_3 = m$, and m_1 is the number of basic elements of elliptic type, m_2 is that of hyperbolic type, m_3 is the number of pairs of basic elements e_i, e_{i+1} of the focus–focus type.

COMMENT. The type (m_1, m_2, m_3) of the Cartan subalgebra K in $\mathrm{sp}(2m, \mathbb{R})$ can be described in a slightly different way. Consider a regular element in K. It can be represented as a quadratic polynomial f given by some symmetric matrix which we denote again by f. Consider the characteristic equation with respect to the symplectic form ω:

$$\det(f - \mu\Omega) = 0.$$

The roots of this equation can be divided into three groups:
1) pairs of imaginary roots $i\alpha$, $-i\alpha$;
2) pairs of real roots β, $-\beta$;
3) quadruples of complex conjugate roots $\alpha + i\beta$, $\alpha - i\beta$, $-\alpha + i\beta$, $-\alpha - i\beta$.
Denote the number of elements in each group by m_1, m_2, m_3 respectively. These are just the same numbers as m_1, m_2, m_3, which we used for the classification of Cartan subalgebras.

Therefore, every non-degenerate singularity of the momentum mapping $\mathcal{F}$ is associated with a certain Cartan subalgebra which is classified by a triple of integers (m_1, m_2, m_3). That is why it is natural to call this triple (m_1, m_2, m_3) the *type of the given singularity*.

Finally we obtain that a non-degenerate singularity of the momentum mapping $\mathcal{F}$ is characterized by four integer numbers:

$$\left(i = \text{rank of the singularity}, \ (m_1, m_2, m_3) = \text{its type}\right).$$

Here $m_1 + m_2 + 2m_3 + i = n$, where n is the number of degrees of freedom of the given integrable system.

Our next aim is to describe the structure of the Liouville foliation in a neighborhood of a non-degenerate singularity in M^{2n}. First, we define some model Liouville foliation in $\mathbb{R}^{2n}$. Consider a collection of n functions of the following form:

$$F_j = p_j^2 + q_j^2,$$
$$F_k = p_k q_k,$$
$$F_l = p_l q_{l+1} - q_l p_{l+1}, \quad F_{l+1} = p_l q_l + p_{l+1} q_{l+1},$$
$$F_s = p_s,$$

where

$$j = 1, \ldots, m_1, \qquad k = m_1+1, \ldots, m_1+m_2,$$
$$l = m_1+m_2+1, \ m_1+m_2+3, \ \ldots, \ m_1+m_2+2m_3-1,$$
$$s = m_1+m_2+2m_3+1, \ldots, n.$$

It is clear that all these functions (F_j, F_k, F_l, F_s) commute pairwise with respect to the Poisson bracket and are functionally independent. Therefore, they define some model (canonical) Liouville foliation $\mathcal{L}_{\text{can}}$ in a neighborhood of the point 0, which is evidently a non-degenerate singular point of the momentum mapping $\mathcal{F}_{\text{can}} \colon \mathbb{R}^{2n} \to \mathbb{R}^n$, where $\mathcal{F}_{\text{can}}(x) = (F_1(x), \ldots, F_n(x))$. It is easy to see that the rank of this singularity is equal to i, and its type is (m_1, m_2, m_3).

The statement formulated below is a direct corollary of Russmann–Vey theorem on the reduction of commuting functions to the canonical (normal) form in a neighborhood of a non-degenerate singular point.

Theorem 1.7 (Local linearization theorem). *Given a real-analytic integrable Hamiltonian system with n degrees of freedom on a real-analytic symplectic manifold M^{2n}, the Liouville foliation in a neighborhood of a non-degenerate singular point of rank i and of type (m_1, m_2, m_3) is locally symplectomorphic to the model Liouville foliation $\mathcal{L}_{\text{can}}$ with the same parameters. In particular, any two Liouville foliations with the same parameters (i, m_1, m_2, m_3) are locally symplectomorphic.*

Theorem 1.7 can be reformulated in a different way. Let $f_1, \ldots, f_n$ be arbitrary commuting functions. Consider their Taylor expansions at the non-degenerate singular point of the mapping $\mathcal{F}$ in a canonical coordinate system. Let us remove all terms of these expansions except for linear and quadratic. It is easy to see that the functions obtained remain commuting and will define some Liouville foliation $\mathcal{L}_0$, which can be naturally considered as the linearization of the initial foliation $\mathcal{L}$ given by the functions $f_1, \ldots, f_n$. Theorem 1.7 states that locally the foliation $\mathcal{L}$ is symplectomorphic to its linearization $\mathcal{L}_0$.

Another important corollary of Theorem 1.7 is that locally a non-degenerate singularity has the type of the direct product whose components represent elementary singularities and the number of components is $m = i + m_1 + m_2 + m_3$. More precisely, we mean the following. The symplectic manifold M^{2n} can be represented as the direct product $M^{2n} = M_1 \times \ldots \times M_m$ of symplectic manifolds each of which carries the momentum mapping with an elementary singularity. This mapping is given by one of model functions F_j, F_k, F_s or by the pair F_l, F_{l+1}.

This decomposition allows us to understand how the bifurcation diagram looks like in a neighborhood of a non-degenerate critical point. It will be diffeomorphic to the bifurcation diagram of the model momentum mapping $\mathcal{F}_{\text{can}}$. In what follows, we shall call it the *canonical bifurcation diagram* and denote it by Σ_{can}. Let us describe its structure in greater detail.

First, let us see how the image of the momentum mapping and its bifurcation diagram look like in the case of elementary non-degenerate singularity. As we know, there are three such singularities: center, saddle, and focus. Besides, the decomposition can have components without singularities (given by the fourth group of functions of the form $F_s = p_s$). These four possibilities are listed in Fig. 1.7:

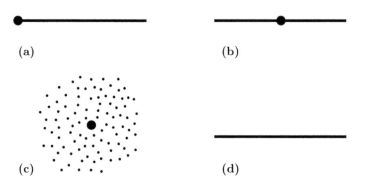

Figure 1.7

(a) $M = \mathbb{R}^2$, $F = p^2 + q^2$ (center), the image of the momentum mapping is a ray, and the bifurcation diagram Σ is its origin;

(b) $M = \mathbb{R}^2$, $F = pq$ (saddle), the image of the momentum mapping is a straight line, and Σ is a point on it;

(c) $M = \mathbb{R}^4$, $F_1 = p_1 q_2 - q_1 p_2$, $F_2 = p_1 q_1 + p_2 q_2$ (focus–focus), the image of the momentum mapping is a plane, and Σ is an isolated point on it;

(d) $M = \mathbb{R}^2$, $F = p$ (regular case), the image of the momentum mapping is a straight line, and Σ is empty.

Since the model singularity has the type of the direct product with components of the four types listed above, it follows that the image U_{can} of the momentum mapping $\mathcal{F}_{\text{can}}$ and its bifurcation diagram Σ_{can} are also decomposed into direct product in the following natural sense. Consider the images U_i of the momentum mapping for each of components M_i together with the corresponding bifurcation diagram $\Sigma_i \subset U_i$. Then

$$U_{\text{can}} = U_1 \times \ldots \times U_m,$$
$$\Sigma_{\text{can}} = U_{\text{can}} \setminus ((U_1 \setminus \Sigma_1) \times (U_2 \setminus \Sigma_2) \times \ldots \times (U_s \setminus \Sigma_s)).$$

Since all possible components, i.e., pairs (U_i, Σ_i), are known and listed in Fig. 1.7, it is not difficult to describe explicitly the local bifurcation diagram Σ for each type of a non-degenerate singularity. Of course, we mean the description up to a diffeomorphism.

Let us emphasize that each type of singularity (i, m_1, m_2, m_3) defines its bifurcation diagram uniquely. Moreover, bifurcation diagrams corresponding to singularities of different types are all different themselves, i.e., there exists a one-to-one correspondence between types of singularities and their (local) bifurcation diagrams.

As an example, we list all bifurcation diagrams for non-degenerate singularities in the case of three degrees of freedom. Here the quadruples (i, m_1, m_2, m_3) are as follows:

1) $(3, 0, 0, 0)$ (regular point),
2) $(2, 1, 0, 0)$ (elliptic singular point of rank 2; center),
3) $(2, 0, 1, 0)$ (hyperbolic singular point of rank 2; saddle),
4) $(1, 0, 0, 1)$ (focus–focus of rank 1),
5) $(1, 2, 0, 0)$ (center–center of rank 1),
6) $(1, 1, 1, 0)$ (center–saddle of rank 1),
7) $(1, 0, 2, 0)$ (saddle–saddle of rank 1),
8) $(0, 3, 0, 0)$ (center–center–center of rank 0),
9) $(0, 2, 1, 0)$ (center–center–saddle of rank 0),
10) $(0, 1, 2, 0)$ (center–saddle–saddle of rank 0),
11) $(0, 0, 3, 0)$ (saddle–saddle–saddle of rank 0),
12) $(0, 1, 0, 1)$ (center–focus–focus of rank 0),
13) $(0, 0, 1, 1)$ (saddle–focus–focus of rank 0).

The corresponding images of the momentum mapping and the bifurcation diagrams Σ_{can} in $\mathbb{R}^3$ are illustrated in Fig. 1.8. The shaded regions show the three-dimensional domains filled by the image of the momentum mapping $\mathcal{F}_{\text{can}}$. The bifurcation diagrams Σ_{can} shown in Fig. 1.8 correspond to the canonical models. These diagrams are composed of pieces of two-dimensional planes and straight lines. If, instead of this, we consider an arbitrary analytic integrable system, then the indicated pictures undergo some diffeomorphism.

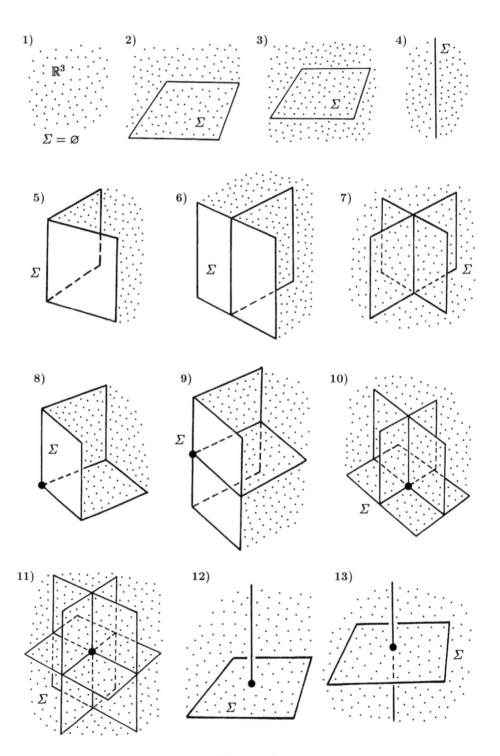

Figure 1.8

If an integrable system is not analytic, but only smooth, then the bifurcation diagram may split as shown in Fig. 1.9. Some curve (or surface) of Σ may split at the singular point into two tangent curves (or two tangent surfaces). Let us note that the tangency that appears at the moment of splitting has to have infinite order. In the analytic case, such splitting, of course, cannot appear. The smooth splitting is connected to the fact that, in the saddle case, the level line $\{pq = \varepsilon\}$ is not connected, but consists of two components. Each of these components can be mapped independently without "knowing anything" about the other one.

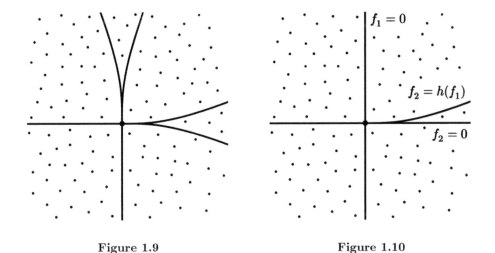

Figure 1.9   Figure 1.10

As an example, we consider the following pair of commuting functions:

$$f_1 = p_1 q_1 \,,$$
$$f_2 = p_2 q_2 + \lambda(p_1, q_1) \,,$$

where λ is a C^∞-function (but not analytic) given by the formula

$$\lambda(p_1, q_1) = \begin{cases} h(p_1 q_1) & \text{for } p_1 > 0, \ q_1 > 0, \\ 0 & \text{in the other cases}, \end{cases}$$

where the function $h(x)$ has zero of infinite order at $x = 0$. Here the singular points fill two surfaces

$$\{p_1 = 0, \ q_1 = 0\} \quad \text{and} \quad \{p_2 = 0, \ q_2 = 0\} \,.$$

The bifurcation diagram on the plane with coordinates f_1, f_2 consists of two straight lines $f_1 = 0$, $f_2 = 0$ and the curve $f_2 = h(f_1)$ (see Fig. 1.10). It is seen that the smooth splitting of the axis f_1 occurs at zero.

1.9. MAIN TYPES OF EQUIVALENCE OF DYNAMICAL SYSTEMS

We shall consider smooth dynamical systems and the following six types of equivalence relations:
1) the topological conjugacy of dynamical systems;
2) the smooth conjugacy of dynamical systems;
3) the orbital topological equivalence of dynamical systems;
4) the orbital smooth equivalence of dynamical systems;
5) the Liouville equivalence of integrable Hamiltonian systems;
6) the rough Liouville equivalence of integrable Hamiltonian systems.

Definition 1.26. Two smooth dynamical systems v_1 and v_2 on manifolds Q_1 and Q_2 are called *topologically* (resp. *smoothly*) *conjugate* if there exists a homeomorphism (resp. diffeomorphism) $\tau: Q_1 \to Q_2$ mapping the flow σ_1^t (corresponding to v_1) into the flow σ_2^t (corresponding to v_2), i.e., $\tau \circ \sigma_1^t = \sigma_2^t \circ \tau$.

In the case of smooth systems, the smooth conjugacy means that there exists a diffeomorphism τ that transforms v_1 into v_2, i.e., $v_2 = d\tau(v_1)$. In other words, smoothly conjugate systems can be obtained from each other by a "change of coordinates". In fact, they represent the same system but are written in different coordinates. The conjugacy is the strongest equivalence relation for dynamical systems, which means, in essence, an isomorphism.

Definition 1.27. Let v_1 and v_2 be two smooth dynamical systems on manifolds Q_1 and Q_2. They are called *topologically* (resp. *smoothly*) *orbitally equivalent* if there exists a homeomorphism (resp. diffeomorphism) $\xi: Q_1 \to Q_2$ mapping the oriented trajectories of the first system into those of the second one. Here we do not require that the parameter (time) on the trajectories is preserved. In other words, a trajectory is considered here as a curve without parameterization but with the orientation given by the flow.

Let us comment on this definition. Consider two systems of differential equations $\dfrac{dx}{dt} = f(x)$ and $\dfrac{dy}{d\tau} = g(y)$. What does their (smooth) orbital equivalence mean? It means that it is possible to find a regular change of variables and time

$$x = x(y), \qquad dt = \lambda(y)d\tau \quad (\text{here } \lambda(y) > 0)$$

that transforms the first system into the second one.

Let (M^4, ω) be a smooth symplectic manifold with an integrable Hamiltonian system $v = \operatorname{sgrad} H$. Consider the Liouville foliation on (M^4, ω) generated by v. As above, we assume that its leaves are compact. Having fixed an energy level, we obtain the Liouville foliation on the isoenergy surface $Q_h^3 = \{H = h\}$.

Definition 1.28. Two integrable Hamiltonian systems v_1 and v_2 on symplectic manifolds M_1^4 and M_2^4 (resp. on isoenergy surfaces Q_1^3 and Q_2^3) are called *Liouville equivalent* if there exists a diffeomorphism $M_1^4 \to M_2^4$ (resp. $Q_1^3 \to Q_2^3$) transforming the Liouville foliation of the first system to that of the second one.

In other words, two systems are Liouville equivalent if they have equal Liouville foliations.

Definition 1.29. Let v be a Liouville integrable Hamiltonian system on M^4. Consider the corresponding Liouville foliation on M^4. The topological space of its leaves (with the standard quotient topology) is called the *base of the Liouville foliation*. In other words, it is a topological space whose elements are, by definition, the leaves of the Liouville foliation (each leaf is replaced by a point). The base of the Liouville foliation on Q^3 is defined analogously.

In typical cases, the base of a Liouville foliation is not only a Hausdorff space, but also a CW-complex. For example, the base of the Liouville foliation on Q_h^3 is a one-dimensional graph W_h.

Definition 1.30. Two integrable Hamiltonian systems v_1 and v_2 on symplectic manifolds M_1^4 and M_2^4 are called *roughly Liouville equivalent* if there exists a homeomorphism between the bases of the corresponding Liouville foliations that can locally be lifted up to a fiber homeomorphism of the Liouville foliations (i.e., in a neighborhood of each point of the base). The rough Liouville equivalence for integrable systems restricted to their isoenergy 3-surfaces Q_1^3 and Q_2^3 is defined analogously.

In the case of three-dimensional isoenergy surfaces, one can define the rough Liouville equivalence in another way. Let us introduce the operation of twisting of a Liouville foliation on Q. First, cut Q along some regular Liouville torus T. As a result, we obtain a 3-manifold whose boundary consists of two tori. Let us glue them together by applying an arbitrary orientation preserving diffeomorphism. We obtain, generally speaking, a new 3-manifold with the structure of a Liouville foliation. We say that it is obtained from Q by *twisting* along the Liouville torus T. Then two integrable systems are roughly Liouville equivalent (on isoenergy surfaces) if and only if their Liouville foliations can be obtained one from the other as the result of several twisting operations. It is clear that Liouville equivalent systems are roughly Liouville equivalent (but not vice versa).

Definitions 1.28 and 1.30, of course, have a sense also for higher-dimensional integrable systems. The definitions of Liouville and rough Liouville equivalencies were introduced in [135], [65], and [123].

The Liouville equivalence of integrable systems can also be considered for any invariant (with respect to the Poisson $\mathbb{R}^n$-action) subsets in M^{2n}. For example, for saturated neighborhoods of singular leaves of Liouville foliations, which is studied in the theory of Liouville foliation singularities.

Note that two integrable Hamiltonian systems of the same dimension are, of course, Liouville equivalent in neighborhoods of their compact non-singular leaves (i.e., Liouville tori). This follows immediately from the Liouville theorem.

Chapter 2

The Topology of Foliations on Two-Dimensional Surfaces Generated by Morse Functions

2.1. SIMPLE MORSE FUNCTIONS

Consider the space of smooth functions on a smooth manifold. What is the structure of typical smooth functions that are in some sense generic? How do they differ from "exotic" functions? It is clear that the properties of a function are determined mostly by the character of its singularities, i.e., those points where the differential of the function vanishes. That is why we can reduce the above question to the question about functions with typical singularities.

Consider a smooth function $f(x)$ on a smooth manifold X^n, and let $x_1, \ldots, x_n$ be smooth regular coordinates in a neighborhood of a point $p \in X^n$. The point p is called *critical* (for the function f) if the differential

$$df = \sum \frac{\partial f}{\partial x_i} dx_i$$

vanishes at the point p. The critical point is called *non-degenerate* if the second differential

$$d^2 f = \sum \frac{\partial^2 f}{\partial x_i \partial x_j} dx_i \, dx_j$$

is non-degenerate at this point. This is equivalent to the fact that the determinant of the second derivative matrix is not zero.

By the well-known Morse lemma, in a neighborhood of a non-degenerate critical point, one can choose local coordinates in which f is written as a quadratic form:

$$f(x) = -x_1^2 - x_2^2 - \ldots - x_\lambda^2 + x_{\lambda+1}^2 + \ldots + x_n^2 .$$

The number λ is uniquely defined for any critical point and called its *index*.

There are three possible types of non-degenerate critical points for functions on two-dimensional surfaces: minimum, maximum, and saddle (see Fig. 2.1).

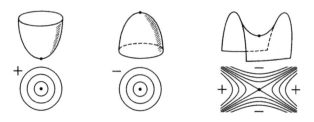

Figure 2.1

In appropriate coordinates, the function can be written as follows (if $f(p) = 0$):
1) $f = \quad x^2 + y^2$ (minimum, $\lambda = 0$),
2) $f = -x^2 - y^2$ (maximum, $\lambda = 2$),
3) $f = -x^2 + y^2$ (saddle, $\lambda = 1$).

Definition 2.1. A smooth function is called a *Morse function* if all its critical points are non-degenerate.

The following important statement holds [242]: the Morse functions are everywhere dense in the space of all smooth functions on a smooth manifold. In other words, any smooth function can be turned into a Morse function by arbitrarily small perturbation. Degenerate critical points split, as a result, into several Morse-type (i.e., non-degenerate) singularities.

It is also known that, if X^n is a closed manifold, then the Morse functions form an open everywhere dense subset in C^2-topology in the space of smooth functions on X^n.

In what follows, by $f^{-1}(r)$ we shall denote the preimage of a value r of f; besides, a will denote regular values of the function, i.e., those whose preimage contains no critical points. In this case, $f^{-1}(a)$ is a smooth submanifold in X^n according to the implicit function theorem.

By c we shall denote critical values of f, i.e., those in whose preimage there is at least one critical point. By arbitrarily small perturbation, one can do so that, on every critical level c (i.e., on the set of x's for which $f(x) = c$), there is exactly

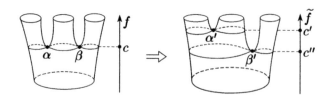

Figure 2.2

one critical point. In other words, the critical points which occur in the same level can be moved to close but different levels (Fig. 2.2). If each critical level $f^{-1}(c)$ contains exactly one critical point, then f is called a *simple* Morse function.

2.2. REEB GRAPH OF A MORSE FUNCTION

Let f be a Morse function on a compact smooth manifold X^n. For any $a \in \mathbb{R}$ consider the level surface $f^{-1}(a)$ and its connected components, which will be called *fibers*. As a result, on the manifold there appears the structure of a foliation with singularities. By declaring each fiber to be a point and introducing the natural quotient topology in the space Γ of fibers, we obtain some quotient space. It can be considered as the base of the foliation. For a Morse function, the space Γ is a finite graph.

Definition 2.2 [304], [152]. The graph Γ is called the *Reeb graph* of the Morse function f on the manifold X^n. A vertex of the Reeb graph is the point corresponding to a singular fiber of the function f, i.e., a connected component of the level that contains a critical point of f. A vertex of the Reeb graph is called an *end-vertex*, if it is the end of one edge only. Otherwise, the vertex is called *interior*.

Consider, for instance, the two-dimensional torus in $\mathbb{R}^3$ embedded as shown in Fig. 2.3, and take the natural height function to be a Morse function on the torus.

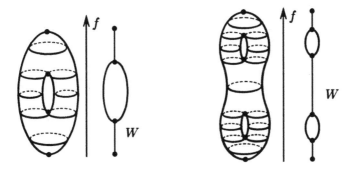

Figure 2.3

Then its Reeb graph has the form shown in Fig. 2.3. In Fig. 2.3 one can see another example of a Morse function on the sphere with two handles (again a height function). Here the Reeb graph is little more complicated.

Lemma 2.1. *The end-vertices of the Reeb graph correspond in a one-to-one manner to local minima and maxima of the function. The interior vertices of the Reeb graph correspond in a one-to-one manner to the singular fibers that contain saddle critical points.*

The proof is evident.   □

If it is known in advance whether the surface X is orientable or not, then the Reeb graph of a simple Morse function allows one to reconstruct the topology of the surface.

Theorem 2.1. *Let f be a simple Morse function on a closed two-dimensional orientable (or, respectively, non-orientable) surface X. Then its Reeb graph determines this surface uniquely up to a diffeomorphism.*

Proof. For a simple Morse function f, there is a natural one-to-one correspondence between its critical points and vertices of the Reeb graph. Indeed, each vertex of the Reeb graph corresponds to a certain critical fiber. Since f is simple, this critical fiber contains exactly one critical point. According to Lemma 2.1, we can uniquely divide the vertices of the Reeb graph into two classes: end-vertices (which denote local minima and maxima) and interior ones (which denote saddle critical points). It is well known that the Euler characteristic of a two-dimensional closed manifold is equal to the difference between the number of local minima and local maxima (i.e., the number of end-vertices) and the number of saddles (i.e., the number of interior vertices). Thus, using the Reeb graph only, we can compute the Euler characteristic of the surface. Since we know in advance whether the surface is orientable or not, this completes the proof, because the Euler characteristic is the complete topological invariant of a surface both in the orientable and non-orientable cases. □

If the function is complicated (not simple), then the analog of Theorem 2.1 does not hold. The point is that, in this case, there can be several critical points on a singular fiber. That is why, unlike the case of a simple Morse function, the information about the number and types of vertices of the Reeb graph does not allow one to compute the Euler characteristic of the surface.

2.3. NOTION OF AN ATOM

Let f be a Morse function on a surface X^2 (orientable or non-orientable). Let g be another Morse function on another surface Y^2. A natural question appears: when can these functions be regarded as equivalent ones?

Consider the two pairs (X^2, f) and (Y^2, g).

Every Morse function determines a foliation on the surface. By definition, its fibers are the connected components of levels of the function. In the neighborhood of each regular fiber, this foliation is trivial, being just the direct product of a circle by an interval. On the contrary, in the neighborhood of a singular fiber the foliation may have a rather complicated structure.

Definition 2.3. The Morse functions f and g on surfaces X^2 and Y^2 are called *fiberwise equivalent* if there exists a diffeomorphism

$$\lambda \colon X^2 \to Y^2 \,,$$

which transforms the connected components of level lines of the function f into those of the function g. Sometimes we shall say that the pair (X^2, f) is fiberwise equivalent to the pair (Y^2, g).

REMARK. Under the above fiber equivalence, two connected components of some level line of the function f may be mapped into connected components lying on different levels of the function g. In other words, connected components, which initially belong to the same level of one function, can move onto different levels of the other function.

It is a natural problem to give the classification of Morse functions on two-dimensional surfaces up to the fiber equivalence. To solve it, at first we need to study the local question, namely, to describe the local topological structure of singular fibers.

We begin with an informal definition. An *atom* is defined to be the topological type of a two-dimensional Morse singularity. In other words, this is the topological type of a singular fiber of the foliation defined on a two-dimensional surface by a Morse function. More precisely, we can reformulate this as follows.

Definition 2.4. An *atom* is a neighborhood P^2 of a critical fiber (which is defined by the inequality $c - \varepsilon \leq f \leq c + \varepsilon$ for sufficiently small ε), foliated into level lines of f and considered up to the fiber equivalence. In other words, an atom is the germ of the foliation on a singular fiber.

The atom P^2 is called *simple*, if the Morse function f in the pair (P^2, f) is simple. The other atoms are called *complicated*. The atom is called *orientable* (*oriented*) or *non-orientable* depending on whether the surface P^2 is orientable (oriented) or non-orientable.

REMARK. We are not interested yet in the orientation of the surface and the direction of increasing the function f.

REMARK. In some of our early papers on this subject, the notion of orientability and non-orientability of an atom had another sense and was related to the orientability or non-orientability of separatrix diagrams for hyperbolic periodic orbits of a Hamiltonian vector field.

For our purposes, it will be useful to introduce an important notion of an f-atom, or framed atom, which takes into account the direction of increasing the function f.

Let c be a critical value of f on X^2 and c' be a critical value of g on Y^2. Consider their singular fibers

$$f^{-1}(c) \quad \text{and} \quad g^{-1}(c')$$

and suppose that they are connected.

Definition 2.5. Morse functions f and g are called *fiberwise frame equivalent* in the neighborhoods of singular fibers $f^{-1}(c)$ and $g^{-1}(c')$ if there exist two positive numbers ε and ε' and a diffeomorphism

$$\lambda : f^{-1}(c - \varepsilon, c + \varepsilon) \to g^{-1}(c' - \varepsilon', c' + \varepsilon')$$

that maps level lines of f into those of g and preserves the direction of the growth of functions, i.e., λ maps $\{f > c\}$ into $\{g > c'\}$.

Denote the surface with boundary $f^{-1}(c - \varepsilon, c + \varepsilon)$ by P_c^2. Sometimes we shall omit the index c in this notation.

Definition 2.6. Consider a pair (P^2, f), where P^2 is a connected compact surface with boundary ∂P^2, and f is a Morse function on it with a single critical value c such that $f^{-1}(c - \varepsilon) \cup f^{-1}(c + \varepsilon) = \partial P^2$. The fiber frame equivalence class of this pair (P^2, f) is called an f-*atom* (or *framed atom*).

REMARK. Each atom corresponds to two f-atoms. They are obtained one from the other by changing the sign of the function given on the atom. Sometimes, these two atoms turn out to be equivalent (i.e., simply coincide).

2.4. SIMPLE ATOMS

What do level lines of a Morse function on a two-dimensional surface look like?

If a is a regular value of the function, then the corresponding level line consists of several non-intersecting smooth circles. Let us now look what happens to them when they cross a singular level. We begin with the case of an orientable surface.

2.4.1. *The Case of Minimum and Maximum. The Atom* A

First consider a non-singular level line which is close to a local maximum point. This line is a circle. As the regular value tends to the local maximum, the circle shrinks into a point (Fig. 2.4). Let us represent this evolution and the bifurcation in the following conventional, but quite visual manner. Every regular level line (a circle) we represent as one point which is located on the level a (Fig. 2.4). As a changes, this point moves running through a segment. At the moment, when the value of the function becomes critical (equal to c), a circle has shrunk into a point. Denote this event by the letter A with a segment going out of it. This segment is directed downwards.

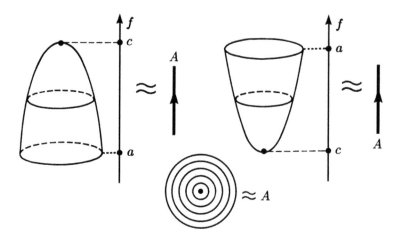

Figure 2.4

In the case of a local minimum, we proceed in the similar way (Fig. 2.4). Here the segment descends from above and ends with the letter A.

We shall also suppose that the letter A denotes a disc with the marked center foliated into concentric circles. We have obtained the example of the simple atom A.

There are two different f-atoms corresponding to the atom A. One of them corresponds to the maximum of a function, the other corresponds to its minimum. By convention, we shall distinguish them by putting the arrow on the edge showing the direction of increasing the function (Fig. 2.4).

2.4.2. The Case of an Orientable Saddle. The Atom B

If c is a critical saddle value, then the singular level line looks like a figure eight curve. As a tends to c, two circles are getting closer and, finally, touch at a point.

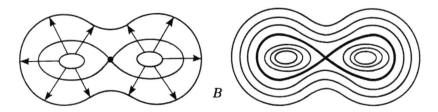

Figure 2.5

After this, the level line bifurcation happens and, instead of two, we obtain just one circle. This process is also shown in Fig. 2.5. By changing the direction, one can speak, conversely, about the splitting of one circle into two circles. The initial circle constricts in the middle; then two points of the circle glue together; and after that, the figure eight curve obtained divides into two circles. Proceeding in the same way as in the previous case, i.e., representing each regular circle by a point and looking after their evolution, we obtain the graph shown in Fig. 2.6. We denote this atom by B.

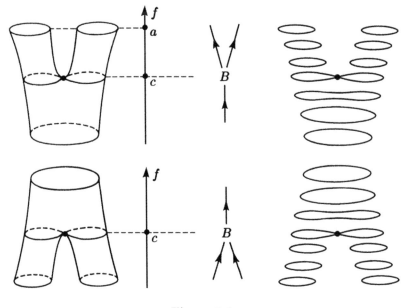

Figure 2.6

As in the previous case, there are two different f-atoms corresponding to B. We distinguish them by putting arrows on the three edges incident to the letter B (see Fig. 2.6). The corresponding graph with oriented edges describes either the splitting of one circle into two circles or the inverse bifurcation of two circles into one.

The atom B can be imaged in a slightly different way (Fig. 2.5), namely, as a flat disc with two holes foliated into level lines of the Morse function.

2.4.3. The Case of a Non-orientable Saddle. The Atom $\widetilde{B}$

Now let us give up the orientability assumption on X^2. The bifurcations of type A are similar in both orientable and non-orientable cases. A difference appears in the case of a saddle. First, let us see how a saddle bifurcation happens

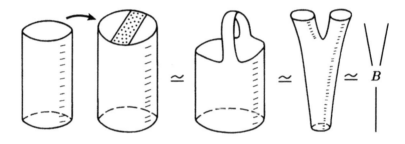

Figure 2.7

in the orientable case (Fig. 2.7). To a boundary circle (presenting the boundary of the manifold $\{f(x) \leq c - \varepsilon\}$, where ε is small enough) one glues a narrow strip (a rectangle). Moreover, we do this gluing operation in such a way that the surface obtained remains orientable. As a result, the boundary becomes homeomorphic to two circles. After replacing f by $-f$, the direction of the bifurcation changes: two boundary circles transform into one.

Consider now the case, when bifurcations happen inside a non-orientable surface. Some of them can be similar to those in the orientable case. However, there is

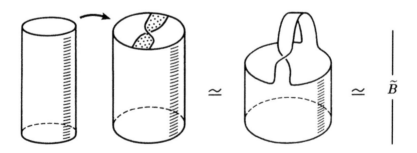

Figure 2.8

at least one bifurcation, which is in essence arranged in a different way. It is shown in Fig. 2.8. Here a twisted (by $180°$) strip is glued to the same boundary circle of the surface $\{f \leq c - \varepsilon\}$. As a result, a new Möbius strip appears inside the surface $\{f \leq c+\varepsilon\}$. It is clear that, in this case, after crossing the critical level c one circle $\{f = c - \varepsilon\}$ turns again into one circle $\{f = c + \varepsilon\}$. Using the above symbolics, i.e., presenting each regular level line (= circle) by a point, we have to draw the above described evolution as shown in Fig. 2.8: the letter $\tilde{B}$ which is put in the middle of a segment. This letter conventionally denotes the simplest non-orientable bifurcation.

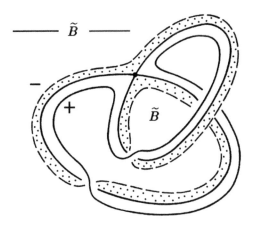

Figure 2.9

The surface $P^2 = f^{-1}(c - \varepsilon, c + \varepsilon)$ for the non-orientable atom $\tilde{B}$ is presented in Fig. 2.9. It is obtained by gluing two Möbius strips together.

2.4.4. The Classification of Simple Atoms

It is easy to see that any simple atom coincides with one of the atoms $A, B, \tilde{B}$ described above.

Theorem 2.2. *A simple atom has either the type A, or B, or $\tilde{B}$. These three atoms correspond to five f-atoms: two for the atom A, two for the atom B, and one for the atom $\tilde{B}$.*

Proof. It follows from the Morse lemma that any bifurcation of the surface $\{f \leq c-\varepsilon\}$, where f is a simple Morse function, under transition through the critical level c is reduced to gluing either a 2-disc or a rectangle to the boundary of the set $\{f \leq c-\varepsilon\}$. Gluing a 2-disc gives the atom A. Gluing a rectangle leads either to B or to $\tilde{B}$. $\square$

It is possible to give another proof of Theorem 2.2, which is useful for understanding the topology of atoms.

Consider a small disc around a critical saddle point of a Morse function f and take the region $\{c - \varepsilon \leq f \leq c + \varepsilon\}$ inside this disc. Then mark the subregions

where $f - c$ is positive and negative. We obtain an object which will be called
a *cross* (Fig. 2.10). The four oriented segments $\alpha, \beta, \gamma, \delta$ shown in Fig. 2.10

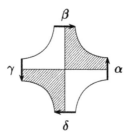

Figure 2.10

will be called its *ends*. The orientation on each of them indicates the direction
of increasing f, i.e., that of its gradient.

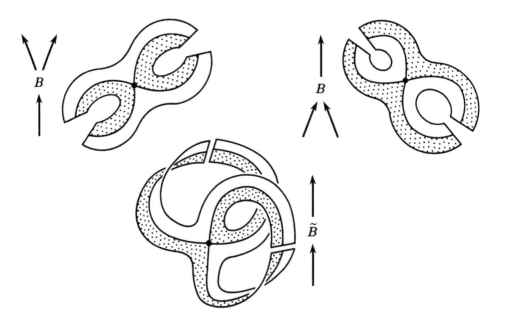

Figure 2.11

The whole surface $P^2 = \{c - \varepsilon \le f \le c + \varepsilon\}$ is obtained from this cross
by means of a simple procedure. One needs to glue pairwise the ends of the cross,
taking into account their orientation. It is clear that there are three possibilities
for gluing. They are shown in Fig. 2.11. As a result, we obtain two different
orientable f-atoms corresponding to the atom B, and one non-orientable f-atom
corresponding to the atom $\tilde{B}$. Thus, in the case of a saddle atom, there are no
other possibilities. The theorem is proved. □

2.5. SIMPLE MOLECULES

2.5.1. *Notion of a Simple Molecule*

Let f be a simple Morse function on a compact closed surface X^2 (orientable or non-orientable). Consider its Reeb graph Γ. The vertices of Γ correspond to critical fibers of f. Let us replace these vertices by corresponding atoms (either A, or B, or $\widetilde{B}$). We assume here that, for each atom, we have fixed its canonical model described above. Each edge of the Reeb graph incident to a certain vertex is assigned to one of the boundary circles of this model, and this correspondence is assumed to be fixed. This remark is important only for the atom B (in the case of simple atoms), because its boundary circles (corresponding to the edges incident to a vertex of Γ of degree 3) are not equivalent.

Definition 2.7. The graph obtained is called a *simple molecule W*.

In fact, the notion of the simple molecule does not differ yet from that of the Reeb graph. However, for complicated Morse functions the molecule W will carry more information than the Reeb graph Γ.

Endow the edges of W with the orientation corresponding to the direction of increasing f.

Definition 2.8. The directed graph obtained is called a *simple f-molecule*.

2.5.2. *Realization Theorem*

Consider a connected finite graph with vertices of degree 1, 2 or 3. At each vertex of degree 1 we put the atom A. At each vertex of degree 2 we put the atom $\widetilde{B}$, and at each vertex of degree 3 we put the atom B. Then we endow each edge of the graph with an orientation so that every saddle atom obtains at least one incoming edge and at least one outgoing edge.

Assume now that this graph admits an immersion into the 2-plane under which the edges will be all oriented upwards.

Theorem 2.3. *Any such graph is a simple molecule, i.e., there exists a two-dimensional surface and a simple Morse function on it such that its molecule coincides with this graph.*

The proof is evident. □

Note that, if the 2-surface X^2 is orientable, the molecule W has no atoms of type $\widetilde{B}$.

Theorem 2.4. *Let $W(X^2, f)$ and $W(Y^2, g)$ be simple molecules corresponding to Morse functions $f: X^2 \to \mathbb{R}$ and $g: Y^2 \to \mathbb{R}$ on orientable surfaces X and Y. If the molecules coincide, then X and Y are diffeomorphic, and the functions f and g are fiberwise equivalent.*

Proof. The molecule tells us from what pieces we should glue the surface and which of their boundary components should be glued together. Besides, we must glue the pieces taking into account their orientations. To do this, we need first to define some orientation on each atom. It can be done in two different ways.

But for the simple atoms A and B there always exists a homeomorphism of the atom into itself which changes the orientation but preserves the foliation. Therefore, the result of gluing does not depend on the choice of orientation on each atom. This proves the theorem. ◻

2.5.3. Examples of Simple Morse Functions and Simple Molecules

Example 1. Consider the standard height function on the torus embedded in $\mathbb{R}^3$, as shown in Fig. 2.12. It is clear that this function is a simple Morse function and its molecule has the form illustrated in Fig. 2.12. It is easy to see that this Morse function on the torus is minimal, i.e., has the minimal possible number of non-degenerate critical points.

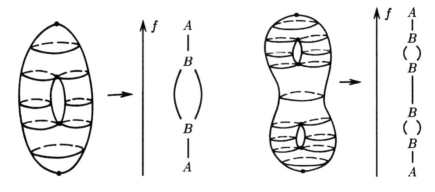

Figure 2.12   Figure 2.13

Example 2. A minimal simple Morse function on the pretzel, i.e., on the sphere with two handles, is realized as the height function on the embedding of the pretzel presented in Fig. 2.13. The corresponding simple molecule is also shown here.

Example 3. A minimal simple Morse function on the projective plane $\mathbb{R}P^2$ can be constructed in the following way. Recall that the projective plane can be presented as the result of gluing a square, as shown in Fig. 2.14. The foliation on $\mathbb{R}P^2$ into level lines of such a function and the corresponding molecule are presented in Fig. 2.15.

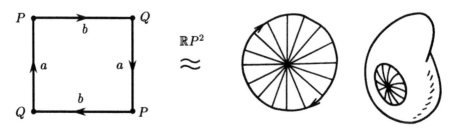

Figure 2.14

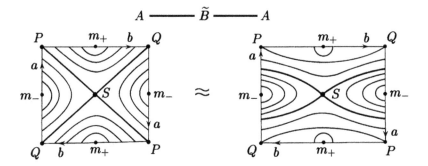

Figure 2.15

This function can be also written as follows.   Consider the homogeneous coordinates $(x : y : z)$ on $\mathbb{R}P^2$. Then the desired function has the form

$$f(x : y : z) = \frac{x^2 + 2y^2 + 3z^2}{x^2 + y^2 + z^2}.$$

Example 4. A minimal simple Morse function on the Klein bottle can be constructed in the following way. Define the Klein bottle as the result of gluing

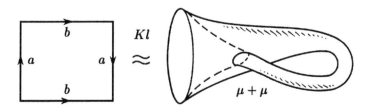

Figure 2.16

a square shown in Fig. 2.16. Then the level lines of the desired function and the corresponding simple molecule have the form shown in Fig. 2.17.

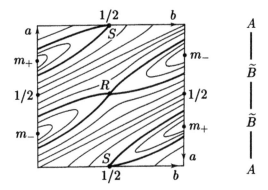

Figure 2.17

It is an interesting fact that this function can be realized as the height function on an appropriate immersion of the Klein bottle into $\mathbb{R}^3$. To this end, we should consider its standard immersion and then lay it down to the horizontal plane. We should also blow up the Klein bottle in order for the height function to have only one minimum and one maximum.

By the way, there exists another simple Morse function on the Klein bottle, whose molecule coincides with the simple molecule on the torus (Fig. 2.12). This function can be realized as the height function on the immersion of the Klein bottle

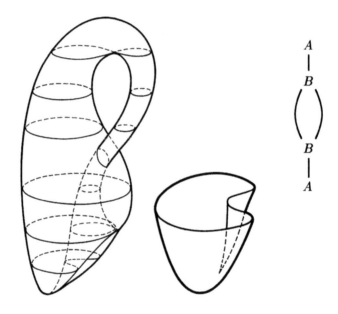

Figure 2.18

into $\mathbb{R}^3$ as shown in Fig. 2.18. We also presented the evolution of its level lines. Note that there are no atoms $\widetilde{B}$ in this molecule although the Klein bottle is non-orientable.

Example 5. Let us return once more to the simple Morse function on the projective plane that was constructed in Example 3. Its molecule is $A\!-\!\widetilde{B}\!-\!A$.

It turns out that it also can be realized as the height function on an appropriate immersion of $\mathbb{R}P^2$ into $\mathbb{R}^3$. Moreover, such an immersion is the well-known Boy surface. Recall its construction. Consider the standard immersion of the Klein bottle lying on the horizontal plane and cut it in halves as shown in Fig. 2.19. We obtain two immersed Möbius strips. Take only one of them, namely, the lower half (Fig. 2.19), and glue a disc along its boundary circle. The result is $\mathbb{R}P^2$. It is convenient to make such gluing in the following way. Lifting the horizontal plane up, we shall smoothly deform the boundary of the Möbius strip immersed into the plane, as illustrated in Fig. 2.20. When the curve is unfolded and transformed into the embedded circle, we glue it up by a disc. After this gluing, the Möbius strip turns into the projective plane.

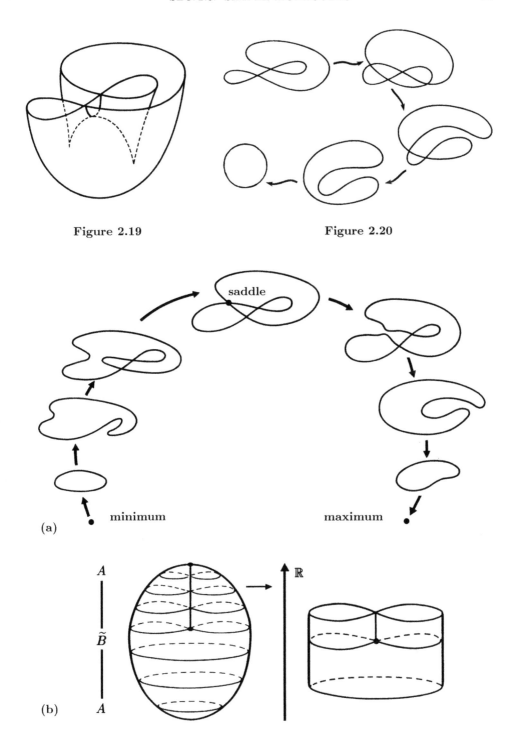

Figure 2.19

Figure 2.20

Figure 2.21

Thus, we have described the immersion of $\mathbb{R}P^2$ into $\mathbb{R}^3$, which is the Boy surface. Consider the height function on it (the projection to the vertical line). The evolution of its level lines is shown in Fig. 2.21(a). It is seen that this function has the only saddle point. Since $\mathbb{R}P^2$ is non-orientable, this point must correspond to the non-oriented saddle atom $\widetilde{B}$. As a result, we obtain the desired molecule A—$\widetilde{B}$—A.

It is possible to imagine this height function on $\mathbb{R}P^2$ in another, more visual way. Consider the well-known image of $\mathbb{R}P^2$ in $\mathbb{R}^3$, shown in Fig. 2.21(b). This is an algebraic surface $K \subset \mathbb{R}^3$ with singularities, which can be given by the following polynomial equation:

$$(k_1 x^2 + k_2 y^2)(x^2 + y^2 + z^2) - 2z(x^2 + y^2) = 0.$$

Now consider the height function $h: K \to \mathbb{R}$ on it. It is clear that it is possible to choose a smooth mapping $g: \mathbb{R}P^2 \to K$ such that the superposition $h \circ g: \mathbb{R}P^2 \to \mathbb{R}$ is a Morse function. It is seen then that $h \circ g$ has exactly three critical points, and its molecule has the form A—$\widetilde{B}$—A.

A model of the atom $\widetilde{B}$ with singularity in $\mathbb{R}^3$ is separately shown in Fig. 2.21(b). Here one of its boundary circles is "flat", the other is immersed into a plane.

Note that any simple Morse function on a two-dimensional surface (both orientable and non-orientable) can be realized as a height function on an appropriate immersion of the surface into $\mathbb{R}^3$.

2.5.4. The Classification of Minimal Simple Morse Functions on Surfaces of Low Genus

Using simple molecules, it is possible to give the classification of simple Morse functions on oriented surfaces of low genus up to fiber equivalence. In the class of simple Morse functions one can distinguish a subclass of simple minimal Morse functions. In this context, minimality means that the number of critical points of a function is minimal (for a given surface). As is well known, if f is a minimal Morse function on X^2, then it has exactly one minimum and one maximum, and the number of saddles is $2g$, where g is the genus of X^2 (number of handles). For example, on the sphere there is just one minimal Morse function (up to fiber equivalence). This is the height function on the standard embedding $S^2 \subset \mathbb{R}^3$.

To obtain this classification on the surface of genus g, it is sufficient to list all simple molecules that contain exactly two atoms A (one of which corresponds to a minimum, and the other corresponds to a maximum) and $2g$ atoms B (corresponding to $2g$ saddles).

Theorem 2.5. *The number of fiber equivalence classes of simple minimal Morse functions on a closed oriented surface of genus g is equal to*

   1   *for the sphere (i.e., $g = 0$),*
   1   *for the torus (i.e., $g = 1$),*
   3   *for the pretzel (i.e., $g = 2$),*
  16   *for the sphere with three handles (i.e., $g = 3$).*

The simple molecules corresponding to these classes are illustrated in Fig. 2.22.

The proof is a simple enumeration of simple molecules.   □

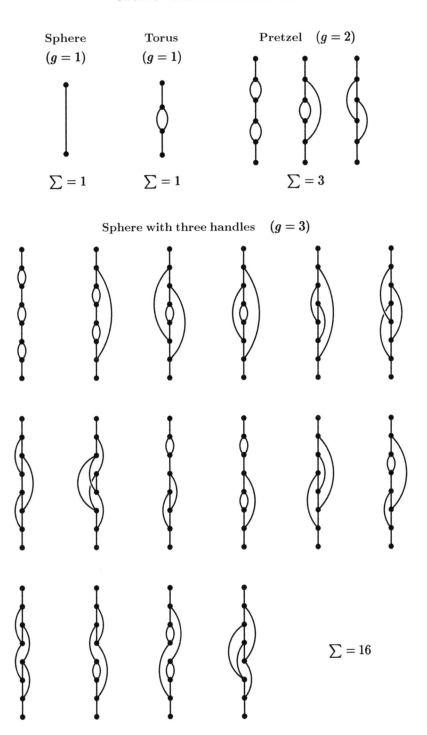

Sphere
$(g = 1)$

$\sum = 1$

Torus
$(g = 1)$

$\sum = 1$

Pretzel $(g = 2)$

$\sum = 3$

Sphere with three handles $(g = 3)$

$\sum = 16$

Figure 2.22

The fiber equivalence relation can be made a little stronger in the following way. Two Morse functions f and g on a surface M are called *topologically equivalent* if there exist diffeomorphisms $\xi\colon M \to M$ and $\eta\colon \mathbb{R} \to \mathbb{R}$ such that $f(\xi(x)) = \eta(g(x))$. In addition, we assume that the diffeomorphism η of the real line into itself preserves orientation.

The difference between the fiber equivalence and topological equivalence can be explained as follows. Let a certain level line of a Morse function consist of several connected components. Under fiber equivalence, these components are allowed to move to different levels of a Morse function, whereas under topological equivalence they have to remain on the same level (although the corresponding value of the function can change). In particular, when studying the topological equivalence, we can assume that the critical levels of a Morse function are naturally ordered (in ascending order). It is important that this order is not changed under topological equivalence of Morse functions. That is why the number of topological equivalence classes of Morse functions is greater than that for fiber equivalence. We formulate the theorem proved by E.V. Kulinich.

Theorem 2.6. *The number of topological equivalence classes of simple minimal Morse functions on a closed oriented surface of genus g is equal to*

1	*for the sphere (i.e., $g = 0$),*
1	*for the torus (i.e., $g = 1$),*
3	*for the pretzel (i.e., $g = 2$),*
31	*for the sphere with three handles (i.e., $g = 3$),*
778	*for the sphere with four handles (i.e., $g = 4$),*
37998	*for the sphere with five handles (i.e., $g = 5$),*
3171619	*for the sphere with six handles (i.e., $g = 6$).*

All the Reeb graphs corresponding to the Morse functions of the above type on the torus, pretzel, and sphere with three handles are presented in Fig. 2.23(a,b). In this figure, we take into account the mutual disposition of saddle critical levels of a Morse function. A critical point related to a greater value of the function is located higher than the critical points with smaller critical values. Those Reeb graphs which belong to the same fiber equivalence class are surrounded by the dotted line in Fig. 2.23. If we eliminate all duplicates from Fig. 2.23, we obtain Fig. 2.22.

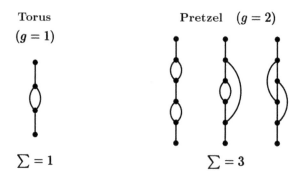

Figure 2.23(a)

Sphere with three handles $(g = 3)$

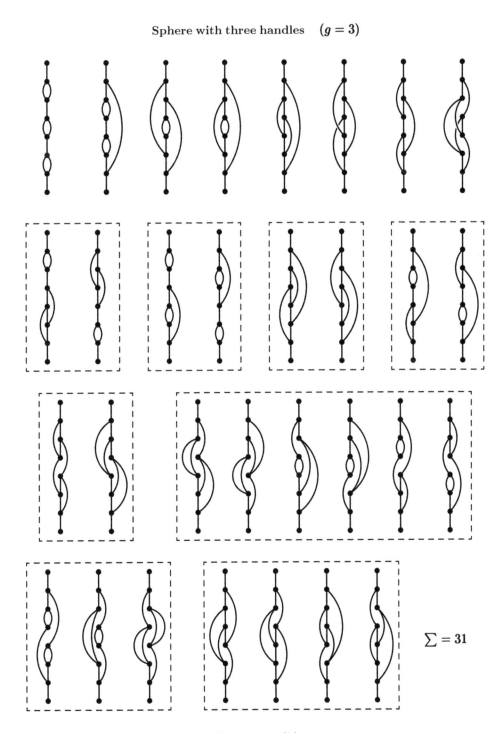

$\sum = 31$

Figure 2.23(b)

2.6. COMPLICATED ATOMS

An atom is called *complicated* if on the critical connected level of the function f there are several (more than one) critical points. Such objects naturally arise in many problems in geometry and physics. We give now a simple example. Suppose that a finite group G acts smoothly on a surface X^2, and let f be a G-invariant Morse function. Then, as a rule, such a function will be complicated. Indeed, if, for instance, the orbit of a critical point x belongs entirely to a connected component of the level line $\{f(x) = \text{const}\}$, then this level contains several critical points.

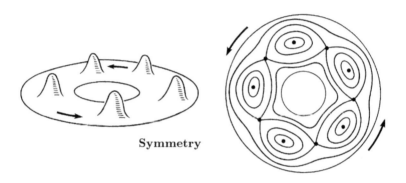

Symmetry

Figure 2.24

An example is shown in Fig. 2.24. Here the height function is invariant with respect to the group $\mathbb{Z}_5$. The connected critical level, containing five critical points, is also shown in Fig. 2.24. Of course, a small perturbation can make the function into a simple one by moving critical points into different levels.

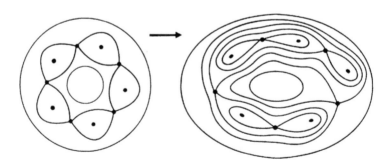

Figure 2.25

However, this destroys the $\mathbb{Z}_5$-symmetry, as is seen from Fig. 2.25. Thus, in the problems that require studying symmetries of different kinds, one has to investigate complicated Morse functions as an independent object.

As we show below, complicated Morse functions also appear naturally in the classification theory for Morse–Smale flows on two-dimensional manifolds.

The main representation of a complicated atom is its representation as a tubular neighborhood of the critical level of a Morse function.

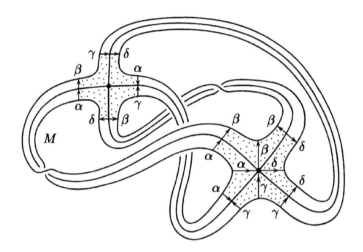

Figure 2.26

The atom is realized as a two-dimensional surface, which consists of several crosses and strips that connect them as shown in Fig. 2.26. Therefore, we can give another equivalent geometrical definition of an atom.

Definition 2.9. An *atom* is defined to be a pair (P^2, K), where P^2 is a connected compact two-dimensional surface with boundary (orientable or non-orientable), and $K \subset P^2$ is a connected graph satisfying the following conditions:

1) either K consists of a single point (i.e., isolated vertex of degree 0) or all the vertices of K have degree 4;

2) each connected component of the set $P^2 \setminus K$ is homeomorphic to an annulus $S^1 \times (0, 1]$, and the set of these annuli can be divided into two classes (positive annuli and negative ones) in such a way that, for each edge of K, there is exactly one positive annulus and exactly one negative annulus incident to the edge.

As before, every atom can be retracted (shrunk) onto its graph K.

We illustrate several pairs (P^2, K) that are not atoms in Fig. 2.27.

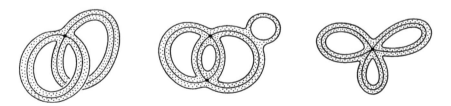

Figure 2.27

We shall consider atoms up to a natural equivalence: two atoms (P^2, K) and (P'^2, K') are equivalent if there exists a homeomorphism mapping P'^2 onto P^2, and K' onto K.

The partition of annuli into positive and negative ones described in Definition 2.9 can be done in two different ways (we can just replace negative annuli with positive ones). If this partition is fixed, then we come to the notion of an f-atom.

Definition 2.10. An f-*atom* is an atom from Definition 2.9 with a fixed partition of its annuli into positive and negative ones.

It is clear that one can define a Morse function f on an f-atom (P^2, K) in such a way that the graph K will be its critical level (for definiteness, $K = f^{-1}(0)$), and f will be positive on the positive annuli, and negative on the negative ones.

Definition 2.11. The vertices of the graph K, i.e., the critical points of f, are called the *vertices* of the atom. The number of vertices is called the *complexity* of the atom.

We usually denote the atom by some letter with several incoming and outgoing edges. Each edge represents a certain annulus of the atom (see above). Since every annulus has exactly one boundary circle, it is natural to say that the corresponding edge has the end-point which represents this boundary circle.

Definition 2.12. The ends of these edges are called the *ends* of the atom, and the number of the ends is called its *valency*. If we take into account the direction of increasing f, then we can naturally introduce the notion of *positive* and *negative* edges, corresponding to positive and negative annuli respectively. We shall assume the positive edges to be outgoing, and the negative edges to be incoming.

For convenience we endow the edges of the atom with the arrows indicating their orientation (Fig. 2.28).

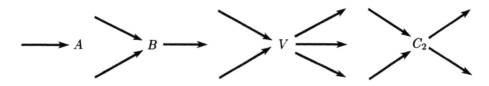

Figure 2.28

As before, the arrows on the edges of atoms show the direction of increasing the function f.

Consider boundary circles of the atom and glue up each of them by a 2-disc. We obtain a closed surface $\tilde{P}^2$ without boundary.

Definition 2.13. *Genus* g of the atom (P, K) is defined to be the genus of the surface $\tilde{P}^2$. If $\tilde{P}^2$ is orientable, then g is the number of handles, if $\tilde{P}^2$ is non-orientable, then g is the number of Möbius strips. The atom is called *planar* if the surface $\tilde{P}^2$ obtained is the sphere.

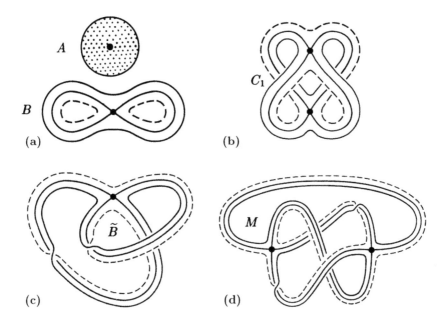

Figure 2.29

Some examples are presented in Fig. 2.29.

a) The orientable atoms A and B have genus zero, since $\widetilde{P}^2$ is the sphere.

b) The orientable atom C_1 has genus one, since $\widetilde{P}^2$ is the torus.

c) The non-orientable atom $\widetilde{B}$ has genus one, since $\widetilde{P}^2$ is the projective plane, i.e., the sphere with one Möbius strip.

d) The non-orientable atom M has genus two, since $\widetilde{P}^2$ is the Klein bottle, i.e., the sphere with two Möbius strips.

Given the atom, it is easy to find its genus. To this end, it suffices to compute the Euler characteristic of $\widetilde{P}^2$.

Proposition 2.1. *The Euler characteristic χ of the surface $\widetilde{P}^2$ is computed as follows*

$$\chi = V - E + R,$$

where V is the number of vertices, E is the number of edges of the graph K, and R is the number of annuli of the atom (P, K). Or, equivalently,

$$\chi = R - V$$

(since the number of vertices V is twice less than the number of edges E).

Proof. The graph K gives a cell decomposition of the surface $\widetilde{P}^2$. Therefore, the number $V - E + R$ coincides with the alternated sum of zero-dimensional, one-dimensional, and two-dimensional cells, which is evidently equal to the Euler characteristic. $\square$

2.7. CLASSIFICATION OF ATOMS

2.7.1. *Classification Problem*

There are two kinds of atoms: atoms of type A and saddle atoms. The atoms of type A are all isomorphic. If we consider an atom as an equivalence class, then it is better to say that there is just one atom A. That is why the classification problem in fact relates to saddle atoms.

In this section we shall speak about saddle atoms only, referring to them as simply atoms (unless otherwise specified).

As we show below, the atoms admit a rather natural classification. We shall present an algorithm of enumerating all atoms. Then we shall indicate an algorithm which compares two given atoms and answers the question of whether they are equivalent or not. Let us note that the classification of atoms is not a trivial problem. Indeed, an atom is a pair: a surface with a graph embedded into it. Consider first a more general problem when there are no restrictions to the embedded graph. Let (P^2, K) and (P'^2, K') be two such pairs. Does there exist a homeomorphism that maps P onto P' and K onto K'? This more general problem is algorithmically solvable. But the answer for certain pairs can be a rather unwieldy procedure, since the problem is related to deep properties of fundamental groups of two-dimensional surfaces. In our case, there is a facilitating circumstance that the graph K imbedded into P possesses some additional properties. In particular, we know that the complement of K in P is a disjoint union of annuli. This makes it possible not only to construct abstract algorithms for enumerating and recognition, but also to realize this algorithm on a computer.

2.7.2. *Algorithm of Enumeration of Atoms*

It is clear that we may restrict ourselves with atoms of fixed complexity. Thus, consider the set of all atoms with a fixed number of vertices m.

Take m colored crosses. A colored cross is shown in Fig. 2.10. The arrows on its edges are directed from "white" to "black". Mark the ends of all crosses by letters in such a way that each letter occurs exactly twice. In other words, we divide the collection of ends into pairs and mark each pair by a common letter. Then we glue the ends marked by the same letter taking into account their orientation (i.e., adjusting the arrows).

It is easily seen that, as a result of this operation, we obtain a certain atom. The black regions of crosses give positive annuli, and the white regions give negative annuli. By exhausting all possibilities for gluing, we construct a certain set of surfaces. We then select only connected surfaces among them. As a result, we evidently obtain the set of all atoms of fixed complexity m, but, perhaps, with duplicates. In other words, this list is excessive, because two different procedures of gluing can generate the same surface with a graph. Nevertheless, we obtain the complete list of atoms.

2.7.3. Algorithm of Recognition of Identical Atoms

It remains to solve the recognition problem for the finite list of atoms of complexity m, i.e., to answer the question whether two atoms from the list obtained are equivalent or not.

Consider two atoms that were obtained as a result of gluing crosses. It is convenient to reformulate the problem as follows. Consider a collection of crosses: (cross 1), (cross 2), ..., (cross m), and two codes giving the rules for gluing their ends (Fig. 2.30). One needs to find out whether the atoms obtained are homeomorphic as surfaces with embedded graphs.

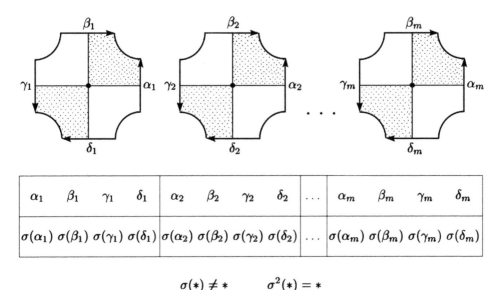

α_1	β_1	γ_1	δ_1	α_2	β_2	γ_2	δ_2	...	α_m	β_m	γ_m	δ_m
$\sigma(\alpha_1)$	$\sigma(\beta_1)$	$\sigma(\gamma_1)$	$\sigma(\delta_1)$	$\sigma(\alpha_2)$	$\sigma(\beta_2)$	$\sigma(\gamma_2)$	$\sigma(\delta_2)$	...	$\sigma(\alpha_m)$	$\sigma(\beta_m)$	$\sigma(\gamma_m)$	$\sigma(\delta_m)$

$$\sigma(*) \neq * \qquad \sigma^2(*) = *$$

Figure 2.30

Determining a certain rule of gluing is equivalent to fixing an element σ in the finite group of permutations S_{4m}. Recall that each cross has four ends, so the set of all ends consists of $4m$ elements. To determine the gluing operation, for each end x one has to indicate the other end $\sigma(x)$ which is glued with x. This correspondence $x \to \sigma(x)$ can be considered as a permutation of $4m$ elements. However this permutation σ is not arbitrary, but satisfies some natural conditions:

    1) since the ends are glued pairwise, the permutation σ is an involution.

    2) since each x cannot be glued with itself, then $\sigma(x) \neq x$.

In other words, σ is an involution without fixed points. Conversely, any such involution is realized as a certain gluing of crosses whose result is an atom.

Let us denote the subset of such involutions in S_{4m} by G_m. Now the question is what elements $\sigma \in G_m$ give the same atom?

Assume that two gluing operations give the same result. This means that there exists a homeomorphism from one atom onto the other. But each atom consists of crosses (with indicated gluing of their ends). The homeomorphism gives, consequently, some homeomorphism of the disjoint union of m crosses onto itself. This homeomorphism can be considered as a composition of two transformations. The first one is a permutation of crosses. The second is some symmetry of a cross onto itself (which takes into account the coloring, i.e., maps "white to white" and "black to black"). There are exactly four such symmetries corresponding to the elements of the group $\mathbb{Z}_2 \oplus \mathbb{Z}_2$.

We can reformulate this by saying that we have described the action of some subgroup $H \subset S_{4m}$ on G_m.

To clarify if two gluing operations are equivalent, it is sufficient to check whether the corresponding permutations belong to the same orbit of the H-action on G_m. Since H is finite, as well as G_m, the answer is given, for example, just by exhaustion.

Since, from the very beginning, we have fixed the coloring of the crosses, the above algorithm in fact recognizes equivalent f-atoms, but not just atoms. To solve the classification problem for atoms, one needs to consider one more transformation, namely the permutation of white and black colours on all the crosses simultaneously, i.e., on the whole atom. This extends H by means of the group $\mathbb{Z}_2$. The generator of this additional group $\mathbb{Z}_2$ acts as follows: on each cross we apply the symmetry with respect to the vertical line passing through its center (Fig. 2.30).

The above algorithm is, of course, rather simple and natural, but not sufficiently effective because it requires considering too many possible cases. That is why an interesting problem is to find another, more effective algorithm. We discuss it in the next section.

2.7.4. Atoms and f-Graphs

The useful reformulation of the notion of an atom, which we discuss below, was proposed by A. A. Oshemkov [279].

We begin with the definition of an abstract f-graph Γ, and then shall explain how f-graphs are connected with atoms.

Definition 2.14. A finite connected graph Γ is said to be an f-graph if the following conditions hold:

1) all vertices of Γ have degree 3;

2) some of the edges of Γ are oriented in such a way that each vertex of Γ is adjacent to two oriented edges, one of which enters this vertex and the other goes out of it (a vertex can be the beginning and the end of the same oriented edge if this edge is a loop);

3) each non-oriented edge of Γ is endowed with a number ± 1.

REMARK. Note that an f-graph Γ is not assumed to be embedded into any surface. This is a discrete object which can be completely defined as a list of oriented edges (i, j) and non-oriented edges (k, l, ε), where i, j, k, l are the numbers of the vertices of Γ, and $\varepsilon = \pm 1$ is the mark associated with a non-oriented edge.

It follows from condition (2) in the above definition of an f-graph that its oriented edges form non-intersecting oriented cycles. Besides, each vertex of such a cycle is incident to exactly one non-oriented edge. Using this observation, we can also describe an f-graph in the following way.

Consider a collection of non-intersecting oriented circles. Choose an even number of points on them in an arbitrary way. Then divide this set of points into pairs and connect the corresponding pairs by non-oriented segments (Fig. 2.31). The result is just an f-graph. This construction is similar to the so-called chord diagrams used in the knot theory (see [27]).

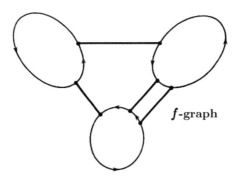

Figure 2.31

Definition 2.15. We call two f-graphs *equivalent* if one of them can be obtained from the other by a sequence of the following operations. It is allowed to change orientation on all the edges of a certain cycle and, at the same time, to change all marks on the non-oriented edges incident to this cycle (i.e., $\pm 1 \rightarrow \mp 1$). If both of the ends of a non-oriented edge belong to the same cycle, then the mark is not changed. The equivalence classes of f-graphs are called f-*invariants*.

It turns out that there exists a natural one-to-one correspondence between f-invariants and f-atoms introduced in Section 2.3. We now describe this correspondence explicitly.

Consider an arbitrary f-atom. It can be represented as a surface P^2 with boundary and a Morse function g on it with a single critical value equal to zero. Consider the separatrices of g going from the boundary of negative annuli into the critical points of g, i.e., vertices of the graph $K = g^{-1}(0)$ (Fig. 2.32). Each pair of separatrices coming into a vertex gives a non-oriented edge of an f-graph Γ. The vertices of Γ are the end-points of the separatrices lying on the boundaries of negative annuli. Fixing an orientation on each boundary circle of negative annuli of the f-atom, we obtain the oriented edges of the f-graph Γ. These oriented edges are simply the arcs of oriented boundary circles between the end-points of the separatrices.

To complete the construction of the f-graph, it remains to put marks on non-oriented edges. It can be done using the following rule. Consider a small neighborhood of a non-oriented edge in P^2. This is a rectangle two of whose opposite edges lie on the boundary circles of negative annuli and, consequently, are oriented. If these edges induce the same orientation on the boundary of the rectangle, then we put $\varepsilon = +1$. If these orientations are opposite, we put $\varepsilon = -1$.

Let us note that the above rectangle can be considered as a handle which is glued to negative circles of the atom after passing through the critical value of f, as it is usually done in the classical Morse theory.

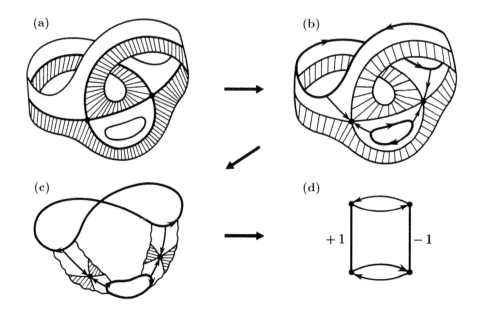

Figure 2.32

Let us give some comments to Fig. 2.32.

The initial f-atom is presented in Fig. 2.32(a) as a surface P^2 with an embedded graph K. The positive annuli of the f-atom are shaded.

In Fig. 2.32(b), we indicate the boundaries of negative annuli with the fixed orientation on them. We also show the separatrices coming into critical points.

Figure 2.32(c) shows the boundaries of negative annuli (= oriented edges of the f-graph), the separatrices (= non-oriented edges of the f-graph), and the neighborhoods of separatrices, which allow us to determine marks ε on non-oriented edges.

In Fig. 2.32(d), we show the result of our construction, i.e., the final f-graph. It is a representative of an f-invariant.

Thus, starting from the f-atom, we have constructed a certain f-graph. In this process we fixed orientations on the boundary circles of negative annuli. However, it is easy to see that, after changing the orientation, we obtain an equivalent f-graph. Therefore, there is a well-defined mapping β from the set of f-atoms into the set of f-invariants.

Theorem 2.7 (A. A. Oshemkov).   *The mapping β establishes a natural one-to-one correspondence between the set of f-atoms and the set of f-invariants.*

Proof.   To show that β is a one-to-one correspondence we shall explicitly construct the inverse mapping β^{-1} from the set of f-invariants to the set of f-atoms. First consider an f-atom represented by a pair (P^2, g). If we remove all incoming and outgoing separatrices out of the surface P^2, this surface will be divided into several hexagons of the following form: two opposite sides of the hexagon are segments of boundary circles of two annuli, one of which is negative and the other is positive; the diagonal parallel to them is an edge of the graph $K = g^{-1}(0)$; each of two remaining pairs of sides of the hexagon consists of two separatrices — incoming and outgoing (see Fig. 2.33).

It turns out that the f-graph contains complete information about how one should glue the hexagons obtained to obtain the initial f-atom. The reconstruction of the f-atom according to the rule of gluing given by the f-graph is just the desired mapping β^{-1}. An example of this procedure is shown in Fig. 2.34. An f-graph with numbered edges is shown on the left, the process of gluing the corresponding f-atom is presented on the right. We now give a formal description of this construction.

Consider an arbitrary f-graph and enumerate its oriented edges by integers from 1 to n. Take n hexagons of the above type. Then define an orientation on the boundary of each hexagon and denote its sides by $a_i^\pm, p_i^\pm, q_i^\pm$, where i is a number of an oriented edge of the f-graph (see Fig. 2.33(b)). The process of gluing the f-atom has two steps.

First, for each vertex of the f-graph, we glue the segment p_i^- with the segment q_j^- if the i-th edge comes into this vertex and the j-th edge leaves it. The orientation of p_i^- and q_j^- must be opposite. After this, we obtain a collection of annuli, one of whose boundary circles consists of segments a_i^- and the other has the form

$$\ldots a_i^+ q_i^+ p_j^+ a_j^+ q_j^+ p_k^+ a_k^+ \ldots,$$

where the boundary segments are oriented in the consistent way.

In the next step we glue these annuli along segments $q_i^+ p_j^+$. The rule of gluing is as follows.

1) The segment $q_i^+ p_j^+$ must be glued with the segment $q_k^+ p_m^+$, provided there exists a non-oriented edge of the f-graph connecting the vertex, which is the end of the j-th edge and the beginning of the i-th edge, with the vertex, which is the end of the m-th edge and the beginning of the k-th edge.

2) If the mark on this non-oriented edge is $+1$, then the orientation of the glued segments $q_i^+ p_j^+$ and $q_k^+ p_m^+$ must be opposite. If the mark is -1, then the orientation of $q_i^+ p_j^+$ and $q_k^+ p_m^+$ are consistent.

Thus, we have described the algorithm for reconstructing the f-atom from a given f-graph. It is easy to verify that, taking equivalent f-graphs, we obtain equivalent f-atoms. This gives a mapping from the set of f-invariants into the set of f-atoms. It is seen from our construction that this mapping is indeed inverse to β. This proves the theorem. $\square$

Thus, f-atoms can be coded by special graphs, which can be algorithmically enumerated (provided their complexity is bounded). Therefore, in the theory of atoms, a pair $(P, K)=$(surface, graph) can be replaced by one graph of special kind.

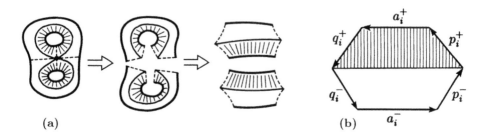

Figure 2.33

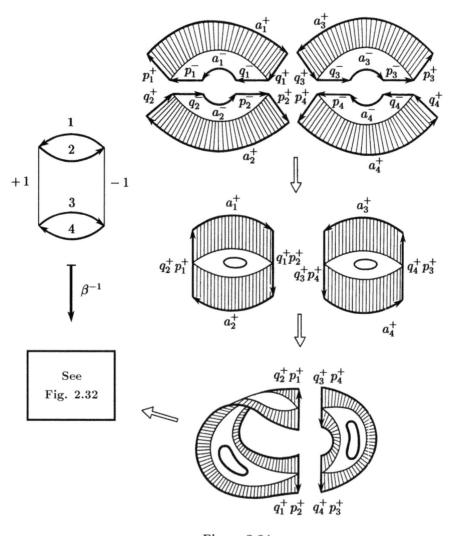

Figure 2.34

2.7.5. Oriented Atoms and Subgroups in the Group $\mathbb{Z} * \mathbb{Z}_2$

In this section, we consider only oriented atoms.

Definition 2.16. We call an f-atom *oriented* if the corresponding surface P is orientable, and the orientation on it is fixed.

COMMENT. Each f-atom is a frame equivalence class of pairs (P, f). For an oriented f-atom it is required that diffeomorphisms connecting equivalent pairs with each other preserve the fixed orientation of P.

In terms of f-invariants it means that the corresponding f-invariant can be represented by an f-graph all of whose marks are equal to $+1$. Thus, ignoring the marks (which are all $+1$), we see that the set of oriented f-atoms is exactly the set of f-graphs without marks.

We now assign to each f-graph (without marks) a certain subgroup of finite index in the free product $\mathbb{Z} * \mathbb{Z}_2$.

Consider an arbitrary f-graph Γ without marks and fix one of its vertices x. Consider all continuous paths on Γ beginning at the vertex x and ending at any other vertex of Γ. A path on the graph is considered in combinatorial sense as a sequence of edges.

Each path is uniquely divided into segments of three kinds.

1) Motion along a non-oriented edge. We denote such a segment by a.

2) Motion along an oriented edge in the direction given by the orientation on the edge. We denote such a segment by b.

3) Motion along an oriented edge in the direction which is opposite to the orientation on the edge. Such a segment is denoted by b^{-1}.

As a result, each path γ defines uniquely a word composed by the letters a, b, b^{-1}. And conversely, every such word uniquely defines a path γ on the f-graph.

We assume two words to be equivalent if one of them can be obtained from the other by removing the following pairs of neighboring letters $aa, bb^{-1}, b^{-1}b$ from the word, or by inserting such pairs. The set of such equivalence classes of words with the standard multiplication forms a group. Its unity is the equivalence class of the empty word. It is clear that this group is isomorphic to the free product $\mathbb{Z} * \mathbb{Z}_2$, since it is determined by two generators a, b and one relation $a^2 = e$.

Under this correspondence between words and paths on the f-graph with the beginning at x, equivalent words correspond to homotopic paths with fixed ends and vice versa. Thus, by fixing a vertex $x \in \Gamma$, we establish a bijection between homotopy classes of paths beginning at x and elements of the group $\mathbb{Z} * \mathbb{Z}_2$. Under this bijection, the set of closed paths on Γ corresponds to some subgroup H_Γ^x of $\mathbb{Z} * \mathbb{Z}_2$.

Having chosen another vertex $y \in \Gamma$, we obtain another subgroup H_Γ^y. It can be easily shown that the two subgroups H_Γ^x and H_Γ^y are conjugate. Indeed, as a conjugating element $g \in \mathbb{Z} * \mathbb{Z}_2$, i.e., such that $H_\Gamma^x = gH_\Gamma^y g^{-1}$, one can take an arbitrary element corresponding to the homotopy class of any path in Γ connecting x with y.

Let us denote by H_Γ the conjugacy class of the subgroups in $\mathbb{Z} * \mathbb{Z}_2$ corresponding to the set of closed paths in Γ. We consider here different sets of closed paths related to different points of the f-graph Γ. As a result, we have constructed a mapping δ from the set of f-graphs without marks into the set of conjugacy classes of the group $\mathbb{Z} * \mathbb{Z}_2$.

Theorem 2.8 (A. A. Oshemkov). *The mapping δ establishes a one-to-one correspondence between the set of f-graphs without marks and the set of conjugacy classes of subgroups in $\mathbb{Z} * \mathbb{Z}_2$ that have finite index and contain no elements of finite order.*

Proof. We first prove that any subgroup which is the image under the mapping δ has finite index and contains no elements of finite order. Consider right cosets by the subgroup H_Γ^x. Take the elements $hg \in \mathbb{Z} * \mathbb{Z}_2$, where $h \in H_\Gamma^x$, and g is a fixed element. Evidently, such elements correspond one-to-one to the paths in Γ that begin at the vertex x and end at some fixed vertex y (uniquely defined by the element g). Therefore, the index of H_Γ^x is equal to the number of vertices of the f-graph Γ and, consequently, is finite.

Next, any element of finite order in the group $\mathbb{Z}*\mathbb{Z}_2$ defined by two generators a, b and one relation $a^2 = e$ is conjugate to a and has order 2. This follows, for example, from Kurosh's theorem on subgroups in free products. Assume that a subgroup H_Γ^x contains such an element gag^{-1}. Then this element is associated with a closed path in Γ. Therefore, the path corresponding to ga ends at the same vertex y as the path corresponding to g does. But this means that the edge a is a loop (with the beginning and end at y), which is impossible by the definition of an f-graph.

Thus, we have shown that the images of f-graphs under the mapping δ are subgroups of finite index and without elements of finite order. To complete the proof, let us construct explicitly the mapping δ^{-1}.

Let H be a subgroup in $\mathbb{Z} * \mathbb{Z}_2$. Let us construct the graph of cosets by this subgroup. The vertices of this graph correspond to right cosets by H. Two of such vertices x and y are connected by a non-oriented edge if the corresponding conjugacy classes X and Y are connected by the relation $Xa = Y$, where a is the generator of $\mathbb{Z}_2$. Analogously, two vertices x and y are connected by an oriented edge if the corresponding conjugacy classes X and Y are connected by $Xb = Y$, where b is the generator of $\mathbb{Z}$. In this case, the edge is directed from x to y.

We now prove that, in the case when H is a subgroup of finite index without elements of finite order, the graph of cosets by H is an f-graph. Indeed, that graph is finite and does not contain non-oriented loops. It is also clear that every vertex x of this graph has degree 3. Moreover, x is incident to one non-oriented and two oriented edges (incoming and outgoing). These three edges connect x with the vertices that correspond to the conjugacy classes Xa, Xb^{-1}, Xb (we do not consider Xa^{-1}, because $a^{-1} = a$). The oriented edges can, however, coincide by forming a loop. It happens if $X = Xb$. It remains to show that the f-graph obtained does not depend on the choice of H in its conjugacy class. Indeed, if we replace the initial subgroup H with a conjugate subgroup $H' = gHg^{-1}$, then the above relations $Xa = Y$ and $Xb = Y$ will hold for right cosets $X' = gX$ and $Y' = gY$ with respect to H', since $H'g = gH$. This completes the proof. $\square$

COMMENT. We give another explanation of this construction in Section 2.8.

Corollary. *The mapping δ establishes a one-to-one correspondence between the set of f-graphs (without marks) and the set of conjugacy classes of free subgroups of finite index in $\mathbb{Z} * \mathbb{Z}_2$.*

2.7.6. Representation of Atoms as Immersions of Graphs into the Plane

Recall a classical theorem in two-dimensional topology.

Theorem 2.9.

a) *Every oriented atom* (P, K) *(considered as a surface P with a graph K in it, see above) admits a smooth orientation preserving immersion into the sphere (and, therefore, into the plane).*

b) *Two such immersions of the same atom can be transformed one to the other by means of a smooth isotopy and the loop-untying operation shown in Fig. 2.35.*

Figure 2.35

Proof. a) Every atom $V = (P^2, K)$ can be represented as a collection of crosses whose ends are connected by narrow strips. Each cross can be embedded into the 2-sphere separately (with preserving the orientation). After this, it remains to embed the strips connecting the crosses. Since P^2 is assumed to be orientable, it is evidently also possible.

b) Consider now two different immersions of the same atom $V = (P^2, K)$. To transform one of them into the other, consider first the crosses of the atom. It is clear that two embeddings of the collection of crosses can be matched by a smooth isotopy. Moreover, it can be done without self-intersections, since we can make the crosses sufficiently small. It remains to match the embeddings of the narrow strips connecting the crosses. Since the ends of these strips, being glued to the crosses, have been already matched, we can evidently transform these embeddings one to the other (separately for each strip), using the loop-untying operation (Fig. 2.35) if necessary. This proves the theorem. □

This theorem gives the possibility for a visual representation of the atom. Every immersion of the atom into the sphere is uniquely reconstructed from the immersion of the graph K. The point is that the immersion of the atom is the tubular neighborhood of the immersion of K (Fig. 2.36).

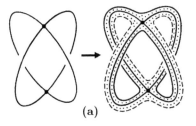
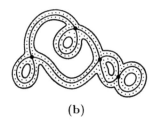

(a) (b)

Figure 2.36

However, we can confine ourselves to immersions of the graph into the plane. The point is that pulling an edge through infinity is equivalent to the appearance of two loops on the edge. But these loops, by definition, may be avoided (untied).

2.7.7. *Atoms as Cell Decompositions of Two-Dimensional Closed Surfaces*

It turns out that the classification of atoms is exactly equivalent to the classification of cell decompositions of closed surfaces. Recall that a cell decomposition of a surface is defined to be its representation as a union of some number of two-dimensional, one-dimensional, and zero-dimensional cells. Equivalently, we can assume that a cell decomposition is determined (uniquely up to a homeomorphism) by an embedding of a certain finite graph into the surface, which separates it into regions homeomorphic to open discs. Under this approach, the embedded graph is simply the one-dimensional skeleton of the cell decomposition, i.e., the union of zero- and one-dimensional cells.

How does an atom appear in this construction? The following statement holds: *there exists a one-to-one correspondence between f-atoms and cell decompositions of two-dimensional closed surfaces (or, equivalently, between f-atoms and embeddings of finite graphs into two-dimensional closed surfaces).* All the objects are considered here up to a homeomorphism.

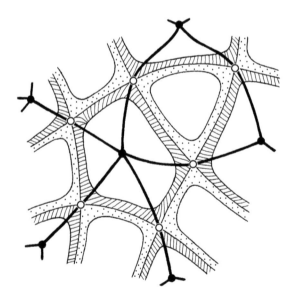

Figure 2.37

This statement is easy to prove. Indeed, take a cell decomposition of a surface and construct an atom from it. To this end, we connect the midpoints of a pair of neighboring edges of the given decomposition that meet at the same vertex. Doing this for each vertex (see Fig. 2.37), we obtain some graph K.

Evidently, all of its vertices (denoted by white points in Fig. 2.37) have degree 4. That is just the skeleton K of the atom we are constructing. It remains now to take a small tubular neighborhood of K in the surface. We obtain an atom. To make this operation well-defined, we need one additional remark. Two-dimensional annuli of the constructed atom can be naturally separated into two classes. The first class includes the annuli that are entirely contained inside the two-dimensional cells of the initial cell decomposition. For definiteness, we shall refer to them as positive annuli of the atom. The second class consists of the annuli surrounding the vertices of the initial decomposition. They should be called negative. As a result we get not just an atom, but an f-atom. Thus, every cell decomposition allows us to construct a certain f-atom.

Conversely, consider the f-atom corresponding to an atom (P, K) and the closed surface $\widetilde{P}$ obtained from P by gluing 2-discs to all of its boundary circles. The graph $K \subset \widetilde{P}$ determines some cell decomposition of $\widetilde{P}$. To guarantee the above one-to-one correspondence, we have to construct another cell decomposition. To this end, we mark the centers of negative two-dimensional discs and connect them by edges going through the vertices of the f-atom, as shown in Fig. 2.37. Note that each vertex of the f-atom gives exactly one such an edge. As a result, we obtain another embedded graph which determines a new cell decomposition of $\widetilde{P}$. Clearly, this operation will be inverse to that described above.

The statement is proved. $\square$

Every atom, as we know, is associated with two f-atoms. Each of them correspond to a cell decomposition of the surface $\widetilde{P}$. It is easy to see that these two decompositions are dual to each other.

In some sense, an f-atom can be considered as "semidual" object for a given cell decomposition of a closed surface. Starting with a cell decomposition, we construct the f-atom, and then, after "changing a sign" of the f-atom, we obtain the dual decomposition by using the inverse operation. Roughly speaking, the f-atom is located in the middle of the way from the initial cell decomposition to the dual one. That is why it is natural to consider an f-atom as a semidual object.

There are many problems in geometry and topology that can be reduced to the classification of cell decompositions of some surfaces. The above statement in fact shows that those problems are reduced to the description of f-atoms. As a good example, in Section 2.11 we discuss a new approach to the classification of Morse and Morse–Smale flows on two-dimensional surfaces in terms of f-atoms.

2.7.8. Table of Atoms of Low Complexity

It is convenient to denote every atom (P^2, K) by some letter with a number of incoming and outgoing edges. The end of each edge corresponds to a certain boundary circle of the surface P.

It is important to emphasize that, generally speaking, the ends of an atom (P, K) are not equivalent, because the boundary circles of the surface P are not equivalent in the sense that not every two of them can be matched by a homeomorphism of the pair (P, K) onto itself. We shall talk about the possible non-equivalence of the ends in greater detail later, when we discuss the notion of a molecule.

All atoms of low complexity (both orientable and non-orientable) are listed in Table 2.1 (see also [65], [123]). In the same table one can see the corresponding pairs of f-graphs, as well as the surfaces $\widetilde{P}$ obtained from P by gluing discs to all of its boundary circles.

As we already remarked, each atom is associated with two f-atoms. Sometimes they coincide, sometimes they don't. They coincide if and only if the atom has an additional symmetry, that is a homeomorphism which interchanges positive annuli with negative ones. Such a homeomorphism exists, for example, for the following atoms from Table 2.1:

$$C_1 , \quad C_2 , \quad D_2 .$$

The list of non-orientable atoms (of complexity ≤ 3) in Table 2.1 is completed by V. V. Korneev.

In Table 2.2 (see also, [65], [123]) we list all the graphs (spines) K for orientable atoms of complexity ≤ 5. For every such graph K we indicate the number of different orientable atoms for which K can be used as a spine. In other words, we indicate the number of admissible non-equivalent immersions of the graph K into the sphere.

The total number of oriented atoms of a given complexity (up to 5) is also shown in Table 2.2.

2.7.9. Mirror-like Atoms

Definition 2.17. An atom is called *mirror-like* if there exists a diffeomorphism of this atom onto itself that changes its orientation.

The mirror-likeness of an atom actually means that the atom possesses a special non-trivial symmetry. Let us consider some examples.

Proposition 2.2. *The atoms of complexity 1, 2, and 3 (see Table 2.1) are all mirror-like.*

Proof. It is seen from Table 2.1 that each of the listed atoms has a symmetry axis, i.e., admits a reflection changing the orientation. ☐

Not all atoms are mirror-like. An example of a non-mirror-like atom is presented in Fig. 2.36(b).

Two atoms obtained from each other by changing orientation are called *mirror symmetric*. If an atom is mirror-like, then it coincides with its mirror symmetric reflection.

Table 2.1. Atoms of complexity 1, 2, 3 and their f-graphs

No	ATOM	f-GRAPHS	CODE	GENUS
	Complexity 1, orientable			
1			$- A$	S^2
2			$- B<$	S^2
	Complexity 1, non-orientable			
$\tilde{1}$			$- \tilde{B} -$	$\mathbb{R}P^2$
	Complexity 2, orientable			
1			$- C_1 -$	T^2
2			$> C_2 <$	S^2
3			$- D_1 \lessgtr$	S^2

Table 2.1. Atoms of complexity $1,2,3$ and their f-graphs (continued)

No	ATOM	f-GRAPHS	CODE	GENUS
4			$\geq D_2 \leq$	S^2
	Complexity 2, non-orientable			
$\tilde{1}$			$-\tilde{C}_2-$	Kl
$\tilde{2}$			$-\tilde{C}_1\leq$	$\mathbb{R}P^2$
$\tilde{3}$			$-\tilde{D}_1\leq$	$\mathbb{R}P^2$
$\tilde{4}$			$-\tilde{D}_2-$	Kl

Table 2.1. Atoms of complexity $1, 2, 3$ and their f-graphs (continued)

No	ATOM	f-GRAPHS		CODE	GENUS
	Complexity 3, orientable				
1				$-E_1<$	T^2
2				$-E_2<$	T^2
3				$>E_3\leqq$	S^2
4				$-F_1<$	T^2
5				$>F_2\leqq$	S^2

Table 2.1. Atoms of complexity 1, 2, 3 and their f-graphs (continued)

No	ATOM	f-GRAPHS	CODE	GENUS
6			$-G_1 \lneqq$	S^2
7			$\gtrsim G_2 \lneqq$	S^2
8			$\gtrsim G_3 \lneqq$	S^2
9			$-H_1 \lneqq$	S^2
10			$\gtrsim H_2 \lneqq$	S^2

Table 2.1. Atoms of complexity 1, 2, 3 and their f-graphs (continued)

No	ATOM	f-GRAPHS	CODE	GENUS
	Complexity 3, non-orientable			
$\tilde{1}$			$-\tilde{E}_1-$	$S^2+3\mu$
$\tilde{2}$			$-\tilde{E}_2-$	$S^2+3\mu$
$\tilde{3}$			$-\tilde{E}_3<$	Kl
$\tilde{4}$			$-\tilde{E}_4<$	Kl
$\tilde{5}$			$-\tilde{E}_5<$	Kl

Table 2.1. Atoms of complexity $1, 2, 3$ and their f-graphs (continued)

No	ATOM	f-GRAPHS	CODE	GENUS
$\tilde{6}$			$> \tilde{E}_6 <$	$\mathbb{R}P^2$
$\tilde{7}$			$- \tilde{E}_7 \leqq$	$\mathbb{R}P^2$
$\tilde{8}$			$- \tilde{F}_1 -$	$S^2 + 3\mu$
$\tilde{9}$			$- \tilde{F}_2 -$	$S^2 + 3\mu$
$\widetilde{10}$			$- \tilde{F}_3 <$	Kl

Table 2.1. Atoms of complexity $1, 2, 3$ and their f-graphs (continued)

No	ATOM	f-GRAPHS	CODE	GENUS
$\widetilde{11}$			$-\widetilde{F}_4 <$	Kl
$\widetilde{12}$			$-\widetilde{F}_5 \lessgtr$	$\mathbb{R}P^2$
$\widetilde{13}$			$> \widetilde{F}_6 <$	$\mathbb{R}P^2$
$\widetilde{14}$			$> \widetilde{F}_7 <$	$\mathbb{R}P^2$
$\widetilde{15}$			$- \widetilde{G}_1 -$	$S^2 + 3\mu$

Table 2.1. Atoms of complexity $1, 2, 3$ and their f-graphs (continued)

No	ATOM	f-GRAPHS	CODE	GENUS
$\widetilde{16}$			$-\tilde{G}_2 <$	Kl
$\widetilde{17}$			$-\tilde{G}_3 <$	Kl
$\widetilde{18}$			$> \tilde{G}_4 <$	$\mathbb{R}P^2$
$\widetilde{19}$			$> \tilde{G}_5 <$	$\mathbb{R}P^2$
$\widetilde{20}$			$-\tilde{G}_6 \lessgtr$	$\mathbb{R}P^2$

Table 2.1. Atoms of complexity 1, 2, 3 and their f-graphs (continued)

No	ATOM	f-GRAPHS	CODE	GENUS
$\widetilde{21}$			$-\,\widetilde{G}_7\lesseqgtr$	$\mathbb{R}P^2$
$\widetilde{22}$			$-\,\widetilde{H}_1-$	$S^2+3\mu$
$\widetilde{23}$			$-\,\widetilde{H}_2<$	Kl
$\widetilde{24}$			$>\widetilde{H}_3<$	$\mathbb{R}P^2$
$\widetilde{25}$			$-\,\widetilde{H}_4\lesseqgtr$	$\mathbb{R}P^2$

Table 2.2. Spines of orientable atoms of complexity ≤ 5

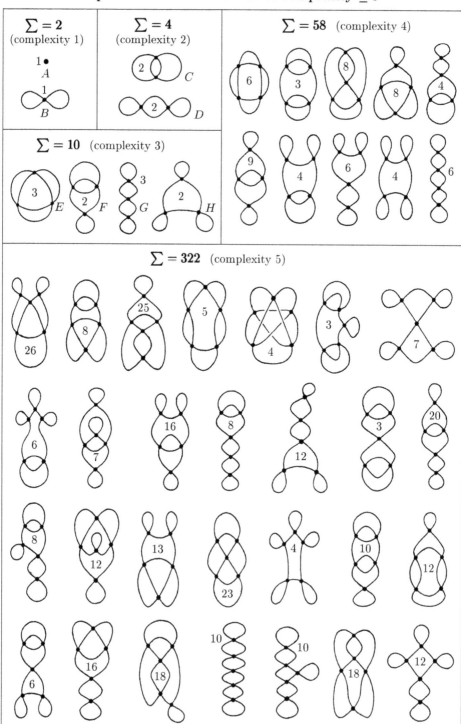

2.8. SYMMETRY GROUPS OF ORIENTED ATOMS AND THE UNIVERSAL COVERING TREE

2.8.1. *Symmetries of f-Graphs*

Studying symmetries of atoms we suppose that the orientation on the surface P^2 is not fixed, but the partition of the annuli into positive and negative ones is. Under this assumption, studying symmetries of an atom is equivalent to studying symmetries of any of two f-atoms corresponding to it.

Let us first consider orientable atoms only.

If we represent an atom V as a pair (P^2, K), then it is natural to define a symmetry of V as a homeomorphism of the pair (P^2, K) onto itself. However, the group of such homeomorphisms is too large. It is more interesting to consider the discrete group of symmetries of the atom which is obtained after factorization of the total group of homeomorphisms with respect to the subgroup of homeomorphisms isotopic to the identity mapping. That is why, speaking of symmetries of atoms, in what follows we mean that homeomorphisms of a pair (P^2, K) are considered up to isotopy, i.e., by definition, a symmetry of an atom is an equivalence class of isotopic homeomorphisms of the pair (P^2, K) onto itself.

By $\mathrm{Sym}(V)$ we denote the *group of proper symmetries* of an atom $V = (P^2, K)$, that is, the group of symmetries that preserve the orientation of the surface P^2 (and also preserve the partition of P^2 into positive and negative annuli). Note that this definition does not depend on the fact how exactly the orientation on the surface P^2 is chosen, and how P^2 is divided into positive and negative annuli.

By $\widehat{\mathrm{Sym}}(V)$ we denote the *total symmetry group* of an atom $V = (P^2, K)$ that includes both orientation preserving and orientation reversing symmetries. However, the decomposition of the surface P^2 into positive and negative annuli is still preserved.

Proposition 2.3. *If an oriented atom V is mirror-like, then the group of proper symmetries $\mathrm{Sym}(V)$ has index 2 in the total symmetry group $\widehat{\mathrm{Sym}}(V)$. Otherwise, these groups coincide: $\mathrm{Sym}(V) = \widehat{\mathrm{Sym}}(V)$. In particular, the order of $\widehat{\mathrm{Sym}}(V)$ is either equal to the order of $\mathrm{Sym}(V)$ or twice larger.*

The proof follows immediately from the definitions of the groups and mirror-like atoms.   □

We first concentrate on studying the group of proper symmetries $\mathrm{Sym}(V)$.

Consider now symmetries of f-graphs.

Definition 2.18. We shall say that an f-graph Γ_1 is *mapped* to an f-graph Γ_2 if there is a mapping between Γ_1 and Γ_2 as abstract graphs, i.e., vertices are mapped into vertices, and edges into edges. And, in addition, this mapping sends non-oriented edges to non-oriented, and oriented edges to oriented ones, but perhaps with simultaneous inversion of orientation on them (i.e., on all the oriented edges at the same time).

Definition 2.19. A *proper symmetry* of an f-graph Γ is an isomorphism of Γ onto itself that maps the oriented edges into oriented ones with preserving their orientation. We denote the group of proper symmetries of Γ by $\mathrm{Sym}(\Gamma)$.

An *improper symmetry* of Γ is an isomorphism of Γ onto itself that maps the oriented edges into oriented ones with simultaneous inversion of the orientation on all of them.

The set of all symmetries (both proper and improper) is called the *total symmetry group* of Γ. We denote it by $\widehat{\mathrm{Sym}}(\Gamma)$.

Proposition 2.4. *Let V be an oriented atom, and Γ the corresponding f-graph. Then the group $\mathrm{Sym}(V)$ is isomorphic to $\mathrm{Sym}(\Gamma)$, and the group $\widehat{\mathrm{Sym}}(V)$ is isomorphic to $\widehat{\mathrm{Sym}}(\Gamma)$.*

Proof. The proof immediately follows from the procedure of constructing the f-graph corresponding to the atom V (see Section 2.7.4). The point is that in the case of oriented atoms this procedure leads to an f-graph without marks ± 1 (see above). $\square$

Thus, we can study the symmetries of atoms V in terms of f-graphs. In other words, by computing the symmetry groups for f-graphs, we describe those for atoms V. The symmetry groups for atoms of low complexity are listed in Table 3.2 (Chapter 3).

2.8.2. The Universal Covering Tree over f-Graphs. An f-Graph as a Quotient Space of the Universal Tree

Consider an arbitrary f-graph Γ. If we think of it as a topological space, then we can consider various coverings over it. Moreover, it is easy to see that any covering space $\widetilde{\Gamma}$ has a natural structure of an f-graph. Indeed, one can uniquely put arrows on the covering space $\widetilde{\Gamma}$ in such a way that the projection $\widetilde{G} \to G$ is a mapping of f-graphs in the sense of Definition 2.18.

Theorem 2.10.

a) *For every f-graph Γ there exists a universal covering space D, which has a natural structure of an infinite f-graph. This universal f-graph is the same for all f-graphs. In other words, the universal covering spaces over any two f-graphs are isomorphic as infinite f-graphs.*

b) *The universal f-graph D is an infinite tree with vertices of degree 3 (Fig. 2.38). Each vertex is incident to one non-oriented edge, one incoming edge and one outgoing edge.*

c) *The group of proper symmetries $\mathrm{Sym}(D)$ of the universal f-graph D is isomorphic to the free product $\mathbb{Z} * \mathbb{Z}_2$. Its action on D is described as follows. Let x_0 be a fixed vertex of D. Then the generator b of infinite order in $\mathbb{Z} * \mathbb{Z}_2$ sends x_0 to the end of the oriented edge incident to x_0. The generator a of order 2 in $\mathbb{Z} * \mathbb{Z}_2$ sends x_0 to the other end of the non-oriented edge incident to x_0. After this, the action of the generators a and b on the other vertices of D is uniquely defined.*

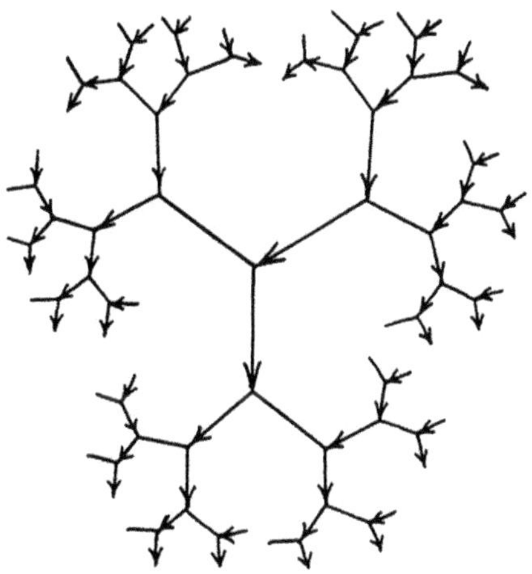

Figure 2.38

Proof. Let us prove (a) and (b) first. If we forget about the orientation on some edges of the f-graph Γ, then the universal covering D over Γ is clearly a tree, all of whose vertices have degree 3. This follows from the standard theory of covering spaces. Evidently, there is the only way to introduce orientation on some edges of this tree such that D becomes an infinite f-graph and the projection $D \to \Gamma$ is a mapping of f-graphs. It is clear that the structure of the space D is uniquely defined by the above properties. This proves items (a) and (b).

Let us prove (c). The structure of every f-graph, in particular, of the infinite f-graph D, is such that each of its proper symmetries is well-defined if the image of at least one vertex under this symmetry is given. Thus, if we fix some vertex of a given f-graph Γ, then the set of its proper symmetries is in one-to-one correspondence with the set of possible images of this vertex. Note that, for an arbitrary f-graph, the group of its proper symmetries, generally speaking, does not act transitively on the set of vertices.

Let us fix a vertex x_0 of the f-graph D. Evidently, x_0 can be moved to any other vertex x by an appropriate proper symmetry. Since D is a tree, there exists the only path on the f-graph D connecting x_0 with x. This path can be uniquely written as a word composed of letters a, b, and b^{-1}, where the letter a means the pass along a non-oriented edge of the tree, the letter b denotes the pass along the arrow on an oriented edge, and b^{-1} denotes the pass along an oriented edge, but in the direction opposite to the arrow on the edge (see Fig. 2.39). So, each vertex is associated with the word representing the path from the fixed vertex x_0 to it.

On the other hand, as was already shown, each vertex $x \in D$ is associated with the proper symmetry of D under which the fixed point x_0 is mapped into x. By comparing these two facts, we obtain a one-to-one correspondence between the proper symmetries of the universal tree D and the words composed from three letters a, b, b^{-1}. The letter a corresponds here to the generator of order 2 in $\mathbb{Z} * \mathbb{Z}_2$, and b corresponds to the generator of infinite order in $\mathbb{Z} * \mathbb{Z}_2$. It remains to check

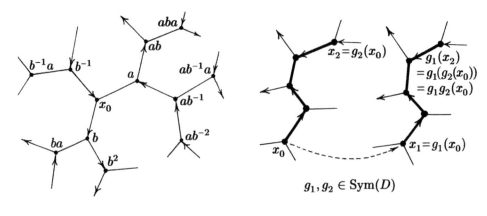

Figure 2.39   Figure 2.40

that the superposition of two proper symmetries corresponds to the standard multiplication of words. This fact easily follows from Fig. 2.40.

The statement (c) is proved. This completes the proof of Theorem 2.10.   □

Thus, there appears an action of $\mathbb{Z} * \mathbb{Z}_2$ on the universal tree D. Note that the quotient space of D with respect to this action, i.e., its orbit space, is the union of a circle and a segment (Fig. 2.41). Note that D is not a covering space over this union, because the action of $\mathbb{Z} * \mathbb{Z}_2$ on D is not free. The point is that, the transformations gag^{-1} (i.e., those conjugated to the generator a or, in other words, all the elements of order 2 in $\mathbb{Z} * \mathbb{Z}_2$) always have exactly one fixed point on the tree. This is the midpoint of a non-oriented edge. In general, it is easy to see that any involution on a tree has a fixed point. Note that the projection of D onto $D/\mathbb{Z} * \mathbb{Z}_2$, i.e., onto the union of a circle and a segment, is an infinite-sheeted covering everywhere except for the only point, namely the free end of the segment (Fig. 2.41).

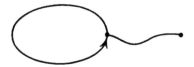

Figure 2.41

Thus, all f-graphs are located between the "maximal" f-graph, that is, the universal covering tree D, and the "minimal" graph $D/\mathbb{Z} * \mathbb{Z}_2$, that is, the union of a circle and a segment. We now want to describe all f-graphs in terms of quotient spaces of D.

Consider an arbitrary f-graph Γ. Since the universal covering space over Γ is D, we see that $\Gamma = D/G$, where G is a subgroup of $\mathbb{Z} * \mathbb{Z}_2$ which acts on D freely. This means that Γ is obtained from the tree D by factorization with respect to the free action of G.

Proposition 2.5.

a) *Let $\Gamma = D/G$ be an f-graph. Then the group G is naturally isomorphic to the fundamental group of the graph Γ.*

b) *The group G is naturally isomorphic to the fundamental group of the atom $V = (P, K)$ corresponding to Γ, i. e., $G = \pi_1(P)$.*

c) *The atom V and f-graph Γ are homotopically equivalent.*

The proof evidently follows from the definition of an f-graph. $\quad\square$

Proposition 2.6. *The subgroup $G \subset \mathbb{Z} * \mathbb{Z}_2 = \mathrm{Sym}(D)$ acts freely on the tree D if and only if it contains no elements of order 2, i.e., elements that are conjugate to the generator a.*

Proof. In one direction this statement has been proved: if the action of G is free, then G contains no elements of order 2.

Conversely, we need to verify that, if G has no elements of order 2, then the action is free. Assume, by contradiction, that some non-trivial symmetry has a fixed point. Then this point can only be the midpoint of a non-oriented edge. Indeed, if any other point is fixed, then the symmetry is the identity mapping, because we can easily find a fixed vertex.

Now consider the midpoint of a non-oriented edge e which is a fixed point of some symmetry h. Then h must have the following form: gag^{-1}, where g is the element of $\mathbb{Z} * \mathbb{Z}_2$ that sends x_0 into one of the ends of e. Indeed, the mapping gag^{-1} leaves the midpoint of e fixed. Then it must coincide with h on the ends of the edge e, and, consequently, it coincides with h everywhere. Hence $h = gag^{-1}$, and Proposition 2.6 is proved. $\quad\square$

2.8.3. The Correspondence between f-Graphs and Subgroups in $\mathbb{Z} * \mathbb{Z}_2$

Recall that we still consider only oriented atoms.

Theorem 2.11.

a) *Every free subgroup G of finite index k in $\mathbb{Z} * \mathbb{Z}_2$ corresponds to a finite f-graph $\Gamma = \Gamma(G)$ with k vertices which has the form $\Gamma = D/G$. Two such graphs $\Gamma(G_1)$ and $\Gamma(G_2)$ coincide if and only if the subgroups G_1 and G_2 are conjugate in $\mathbb{Z} * \mathbb{Z}_2$. In other words, there exists a natural one-to-one correspondence between f-graphs without marks and conjugacy classes of free subgroups of finite index in $\mathbb{Z} * \mathbb{Z}_2$.*

b) *The group $\mathrm{Sym}(\Gamma(G))$ of proper symmetries of $\Gamma(G)$ is isomorphic to the quotient group $N(G)/G$, where $N(G)$ is the normalizer of G in $\mathbb{Z} * \mathbb{Z}_2$ (i.e., $N(G) = \{h \in \mathbb{Z} * \mathbb{Z}_2 : hGh^{-1} = G\}$).*

c) *The order of $\mathrm{Sym}(\Gamma(G))$ is not greater than the number k of vertices of Γ (that is, the index of G in $\mathbb{Z} * \mathbb{Z}_2$). The order of the total symmetry group $\widehat{\mathrm{Sym}}(\Gamma(G))$ is not greater than $2k$.*

Proof. a) The first part of this statement is evident. It remains to prove that if $\Gamma(G_1) = \Gamma(G_2)$, then the subgroups G_1 and G_2 are conjugate. This follows from the fact that any homeomorphism of the base of the universal covering can always be lifted up to a homeomorphism of the universal covering onto itself. Therefore, under the mapping of the base into itself, the fiber over a point is mapped into the fiber over its image. Connecting the point with its image by some path on the base, we move the fiber along this path and obtain the required conjugation of the subgroups.

b) Each symmetry of an f-graph Γ, which is the base of the universal covering $D \to \Gamma = D/G$, can be lifted up to a mapping from D onto itself. Each such lifted mapping $h \in \mathbb{Z} * \mathbb{Z}_2$ must be consistent with the initial covering. This means that h satisfies the condition $hGh^{-1} \subset G$. Thus, the set of lifted mappings is exactly the normalizer $N(G)$ of the group G in $\mathbb{Z} * \mathbb{Z}_2$. It is easy to check that two elements of $N(G)$ correspond to the same symmetry of the f-graph $\Gamma = D/G$ if and only if they belong to the same coset by G in $N(G)$.

c) This statement in fact follows from (a) and (b). Indeed, (b) implies that the order of the symmetry group $\mathrm{Sym}(\Gamma(G))$ is the index of G in its normalizer $N(G)$. Since $N(G)$ is contained in $\mathbb{Z} * \mathbb{Z}_2$, the order of $\mathrm{Sym}(\Gamma(G))$ is not greater than the number of vertices of Γ, which is equal to the index of G in $\mathbb{Z} * \mathbb{Z}_2$. This completes the proof. $\square$

Let us summarize the results. If G is an arbitrary subgroup in $\mathbb{Z} * \mathbb{Z}_2$ which has finite index and contains no elements of finite order, then we can consider its action on the graph D. The quotient space is just that f-graph Γ which was constructed above as the f-graph corresponding to the conjugacy class of the given subgroup G (see Section 2.7.5).

Note that the corresponding projection $D \to \Gamma = D/G$ is the universal covering for the graph Γ, because G contains no elements of finite order. This fact guarantees that the action of G on D is free. In particular, this implies that G is the fundamental group of the graph Γ (and of the atom V). This group is obviously free as the fundamental group of a graph.

2.8.4. The Graph J of the Symmetry Group of an f-Graph. Totally Symmetric f-Graphs

Recall the definition of the Cayley graph of an abstract finitely generated group S. Let us choose and fix a system of generators $s_1, s_2, \ldots, s_n$ in S. This system is not assumed to be minimal. Some of its elements, for instance, can be expressed from the others.

We take all the elements of S as vertices of the graph $J = J_S$. Then we connect by an edge with mark s_i those pairs of elements $g, h \in S$ for which $g = hs_i$. The arrow on the edge is directed from h to g. Note that the graph J depends, of course, on the choice of generators. Different systems of generators lead, in general, to different graphs. Some examples of different graphs J for the cyclic group $S = \mathbb{Z}_6$ are shown in Fig. 2.42. In the first case, we take only one generator x (of order 6); in the second, we take two generators x^2 and x^3.

If a generator s_i is of order 2, then formally for each pair of elements transferring one to another under multiplication by s_i we should draw two edges connecting these elements, which are oriented in opposite directions. Instead of this, we shall

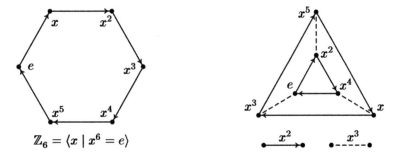

Figure 2.42

draw a single edge, but without any orientation. In particular, non-oriented edges of J in Fig. 2.42, related to the element x^3 of order 2, are shown by a dotted line.

Analogously, if we are given a subgroup $H \subset S$, then we can construct the graph $J(S, H)$ of cosets with respect to this subgroup. The construction is, of course, the same as above. The only difference is that the vertices of $J(S, H)$ correspond now to the cosets by H in G. The graph $J = J(S)$ of the whole group is obtained if $H = \{e\}$.

Proposition 2.7. *An f-graph $\Gamma(G)$ is isomorphic to the graph J of cosets of $\mathbb{Z} * \mathbb{Z}_2$ with respect to the subgroup G. The oriented edges of J correspond to the generator b of infinite order in $\mathbb{Z} * \mathbb{Z}_2$, and non-oriented edges of J correspond to the generator a of order 2.*

Proof. The graph J of the group $\mathbb{Z}(b) * \mathbb{Z}_2(a)$ was in fact illustrated in Fig. 2.39. As the unity of this group we can take the point denoted x_0 in Fig. 2.39. On the other hand, this graph coincides with the universal tree D. If we consider the tree as the group $\mathbb{Z} * \mathbb{Z}_2$, then symmetries of D are obtained by means of the free action of $\mathbb{Z}(b) * \mathbb{Z}_2(a)$ on itself under right multiplication. Then it becomes clear that the orbits of the action of the subgroup G on D (that is, vertices of the graph $\Gamma(G)$) correspond to cosets in $\mathbb{Z}(b) * \mathbb{Z}_2(a)$ with respect to G. Two such vertices x and y are connected by the arrow if and only if we have $Y = Xb$ for the corresponding cosets; and they are connected by a non-oriented edge if $Y = Xa$. The proposition is proved. □

We now introduce the useful notion of a totally symmetric f-graph. For simplicity, we confine ourselves with proper symmetries of f-graphs only.

Definition 2.20. An f-graph Γ is called *totally symmetric*, if the group of its proper symmetries acts transitively on the set of its vertices.

Let us clarify why we are speaking about the total symmetry. As was already explained, each proper symmetry of an f-graph is uniquely defined by the image of any single vertex of the f-graph. Consequently, the order of the group of proper symmetries is not greater than the number of vertices of the f-graph. Hence, it is natural to say that the f-graph is totally symmetric if and only if the number of its vertices is exactly equal to the order of the group of its proper symmetries.

Theorem 2.12.

a) *If an f-graph Γ is totally symmetric, then it is possible to choose the generators in the group $\mathrm{Sym}(\Gamma)$ of proper symmetries in such a way that the corresponding Cayley graph J of the group $\mathrm{Sym}(\Gamma)$ coincides with Γ.*

b) *If the order of $\mathrm{Sym}(\Gamma)$ for an f-graph Γ is equal to the number of its vertices, then Γ is totally symmetric.*

c) *Totally symmetric f-graphs correspond in one-to-one manner to normal free subgroups of finite index in $\mathbb{Z} * \mathbb{Z}_2$ (up to conjugation).*

Proof. a) Take an arbitrary vertex x_0 of the f-graph Γ and consider two symmetries $\tilde{b}$ and $\tilde{a}$ that move x_0 along oriented and non-oriented edges of G adjacent to the chosen vertex x_0.

The existence of such transformations follows from the assumption that the action of $\mathrm{Sym}(\Gamma)$ on Γ is transitive. It is easily verified that one can move the chosen vertex x_0 to any other vertex of Γ by means of the superpositions of $\tilde{b}$, $\tilde{b}^{-1}$, and $\tilde{a}$.

It is also clear that $\tilde{b}$ and $\tilde{a}$ generate the group $\mathrm{Sym}(\Gamma)$ of proper symmetries.

Let us fix $\tilde{b}$ and $\tilde{a}$ to be generators of $\mathrm{Sym}(\Gamma)$. Since the action of the symmetry group is transitive and free, there exists a one-to-one correspondence between vertices of the f-graph Γ and the elements of the group of proper symmetries $\mathrm{Sym}(\Gamma)$. Namely, the vertex $g(x_0)$ corresponds to the element $g \in \mathrm{Sym}(\Gamma)$. Moreover, the vertices $g_1(x_0)$ and $g_2(x_0)$ are connected by an oriented edge if and only if $g_1 = g_2\tilde{b}$. They are connected by a non-oriented edge if and only if $g_1 = g_2\tilde{a}$. This means exactly that Γ can be considered as the graph J of G. The first part is proved.

b) In fact, this follows from the definition of a totally symmetric f-graph.

c) This statement follows immediately from item (b) in Theorem 2.11. Indeed, the class of subgroups conjugated to a normal subgroup consists just of this subgroup itself. The order of the symmetry group is equal to the number of elements in the quotient group $(\mathbb{Z} * \mathbb{Z}_2)/G$, and this number, in turn, is equal to the number of vertices of $\Gamma(G)$. Conversely, if an atom is totally symmetric, then the order of the group $N(G)/G$ and that of $(\mathbb{Z} * \mathbb{Z}_2)/G$ coincide. Therefore, $N(G)$ coincides with the whole group $\mathbb{Z} * \mathbb{Z}_2$, i.e., G is a normal subgroup which was to be proved. $\square$

Let us now discuss the following natural question: for what finite groups is it possible to choose the generators in such a way that the corresponding graph J has the structure of an f-graph, i.e., describes a certain atom? The answer follows from Theorem 2.12.

Corollary.

a) *Let S be a finite group generated by two elements, one of which has order 2. Then the corresponding graph J is a totally symmetric f-graph.*

b) *Conversely, if Γ is a totally symmetric f-graph, then Γ is the graph J for some finite group S generated by two elements one of which has order 2. The group S is the group of proper symmetries of the f-graph Γ.*

REMARK. The system of two generators of S is not assumed here to be minimal. For example, the generator b can be the unit element of the group.

Recall that the description of all finite groups up to some fixed order is a very complicated problem. Nevertheless, for the groups of low order such a description exists.   For example, in the book [87] this description is given for all non-commutative finite groups whose order does not exceed 32. These groups are listed in the form of tables of their generators and relations. Since the order of a group is the number of vertices of the F-graph, this list allows us, in principle, to obtain the list of all totally symmetric f-graphs with at most 32 vertices.

To this end, we need to select from the list only those groups which admit a presentation with two generators one of which has order 2.

Note that there may exist several different presentations of this kind for the same group S. Therefore, we shall obtain not one, but several corresponding f-graphs. These f-graphs can be different, but all of them have the same symmetry group S.

Thus, we have described a natural relationship between totally symmetric f-atoms and the class of finite groups S given by the following presentation:

$$S = \langle a, b \mid a^2 = e, \ldots \rangle,$$

where "dots" denote other additional relations. They can be arbitrary, provided the generator a remains an element of order 2 (for example, relations $a = e$ or $a^3 = e$ are forbidden). The above corollary can be reformulated as follows.

Corollary. *There exists a one-to-one correspondence between the presentations of the form $S = \langle a, b \mid a^2 = e, \ldots \rangle$ that determine finite groups and totally symmetric oriented f-atoms. Moreover, S is the group of proper symmetries of the corresponding f-atom.*

It is an interesting question how to compute the genus of the totally symmetric atom given by a representation $S = \langle a, b \mid a^2 = e, \ldots \rangle$ of a finite group S. The answer is as follows.

Theorem 2.13 (Yu. A. Brailov). *Let $S = \langle a, b \mid a^2 = e, \ldots \rangle$ be a fixed presentation of a finite group S. Consider the atom $V = (P, K)$ corresponding to this presentation and the surface $\widetilde{P}$ obtained from P by gluing discs to all of its boundary circles. Then the Euler characteristic of $\widetilde{P}$ can be calculated as follows:*

$$\chi(\widetilde{P}) = |S| \left(\frac{1}{\text{order of } b} + \frac{1}{\text{order of } ab} - \frac{1}{2} \right).$$

In particular, the genus $g(V)$ of the atom V is $\dfrac{2 - \chi}{2}$.

In Table 2.3 we collect the complete list of all totally symmetric f-atoms with ≤ 6 vertices. Note that the number of vertices of an f-graph is always even. We see that, among the atoms of complexity ≤ 3, there are only 5 totally symmetric atoms: B, C_1, C_2, E_1, E_3.

Let us give some comments to Table 2.3. The standard notation for atoms is given in the first column.   The second column includes their f-graphs. The symmetry group S of an f-graph and its presentation are shown in the third column.   Finally, the two generators of S are described in the fourth column. The first of them is the element a of order 2.

It is seen from Table 2.3 that, to obtain the complete list of totally symmetric f-atoms, we can choose a pair of generating elements $a, b \in S$ in different ways. For example, we do so for the groups $\mathbb{Z}_2, \mathbb{Z}_6, D_3$. As a result, we obtain different totally symmetric f-graphs. In our case, each of these groups gives two f-graphs.

Table 2.3. Totally symmetric f-atoms with ≤ 6 vertices

TOTALLY SYMMETRIC ATOM	TOTALLY SYMMETRIC f-GRAPH	SYMMETRY GROUP	THE PAIR OF GENERATORS
$\begin{array}{c} \mid \\ B \\ /\backslash \end{array}$		$\mathbb{Z}_2 = \langle x \mid x^2 = e \rangle$	x , e
$\begin{array}{c} \backslash / \\ B \\ \mid \end{array}$		$\mathbb{Z}_2 = \langle x \mid x^2 = e \rangle$	x , x
$\begin{array}{c} \mid \\ C_1 \\ \mid \end{array}$		$\mathbb{Z}_4 = \langle x \mid x^4 = e \rangle$	x^2, x
$\begin{array}{c} \backslash / \\ C_2 \\ /\backslash \end{array}$		$\mathbb{Z}_2 \oplus \mathbb{Z}_2 = \langle x \mid x^2=e \rangle \oplus \langle y \mid y^2=e \rangle$	x , y
$\begin{array}{c} \mid \\ E_1 \\ /\backslash \end{array}$		$\mathbb{Z}_6 = \langle x \mid x^6 = e \rangle$	x^3, x^2
$\begin{array}{c} \backslash / \\ E_1 \\ \mid \end{array}$		$\mathbb{Z}_6 = \langle x \mid x^6 = e \rangle$	x^3, x
$\begin{array}{c} \backslash / \\ E_3 \\ /\mid\backslash \end{array}$		$D_3 = \langle x,y \mid x^2 = y^2 = (xy)^3 = e \rangle$	x , y
$\begin{array}{c} \backslash\mid/ \\ E_3 \\ /\backslash \end{array}$		$D_3 = \langle x,y \mid x^2 = y^2 = (xy)^3 = e \rangle$	x , xy

2.8.5. The List of Totally Symmetric Planar Atoms. Examples of Totally Symmetric Atoms of Genus $g > 0$

Recall that the genus of an atom $V = (P^2, K)$ is the genus of the two-dimensional closed surface $\widetilde{P}$ which is obtained from P^2 by gluing 2-discs to all of its boundary circles. We now discuss the question how many totally symmetric atoms of a fixed genus exist. In particular, the complete classification is obtained for planar atoms.

Theorem 2.14 (N. V. Korovina).

a) *For every integer $g > 1$, there exists only a finite number of totally symmetric atoms of genus g (see examples below).*

b) *The complete list of totally symmetric atoms of genus $g = 1$ consists of two infinite series described below.*

c) *The complete list of totally symmetric atoms of genus $g = 0$ (i.e., planar atoms) consists of one infinite series and three exceptional totally symmetric atoms described below.*

Proof. As we showed above (see Section 2.7.7 and Fig. 2.37), each atom (P, K) canonically corresponds to a certain cell decomposition of $\widetilde{P}$, i.e., its decomposition into polygons. It is easy to see that, if the atom is totally symmetric, then this decomposition is totally symmetric too. This means that, for each pair of edges of the decomposition, there exists a homeomorphism of this decomposition onto itself which maps the first edge to the second, and moreover, for each edge there exists a homeomorphism of this decomposition into itself which leaves this edge fixed, but interchanges its endpoints. We also assume that the homeomorphisms preserve orientation. It follows from this, in particular, that all polygons of the decomposition have the same number of edges, and all vertices have the same degree. Therefore, the problem is reduced to the description of totally symmetric decompositions of a closed oriented surface.

Thus, consider an arbitrary totally symmetric cell decomposition of a closed surface $\widetilde{P}$. Let n be the number of its edges, p the number of polygons, and q the number of vertices. Note that n coincides with the number of vertices of the corresponding atom (P, K). Denote the number of edges of a polygon by l, and the degree of a vertex by m. Since each edge connects two vertices and belongs to the boundary of two polygons (with multiplicity), the following relation holds:

$$pl = qm = 2n.$$

On the other hand, the Euler characteristic of $\widetilde{P}$ is expressed from n, p, q as follows:

$$\chi = p - n + q.$$

Hence we obtain a system of equations:

$$-n + 2n\left(\frac{1}{l} + \frac{1}{m}\right) = \chi, \quad pl = 2n, \quad qm = 2n.$$

We are interested in all integer positive solutions of it. Consider cases (a), (b), (c) step by step.

a) *Case of the sphere with g handles, where g > 1.* Here $\chi < 0$. The system of equations takes the following form:

$$n = \chi \cdot \frac{lm}{2(m+l) - lm}, \quad pl = 2n, \quad qm = 2n.$$

The solutions we are interested in (i.e., integer pairs (l, m)) are located in the region D between two hyperbolas (see Fig. 2.43). Indeed, the condition that $n = n(l, m) = \chi \cdot \dfrac{lm}{2(m+l) - lm}$ is positive implies $2(m+l) - lm < 0$, because $\chi lm < 0$. It is clear that the equation $2(m+l) - lm = 0$ defines a hyperbola on the plane (l, m) with asymptotes $l = 2$, $m = 2$; this is the lower hyperbola in Fig. 2.43.

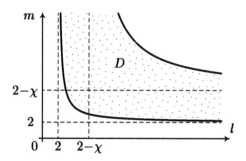

Figure 2.43

The other hyperbola is defined as follows. The inequality $p, q > 1$ implies that $\chi = -n + (p+q) \geq -n + 2$, i.e., $n \geq 2 - \chi$. Hence

$$\chi \cdot \frac{lm}{2(m+l) - lm} \geq 2 - \chi,$$

and, consequently, $lm \leq (2 - \chi)(l + m)$. The equation $lm = (2 - \chi)(l + m)$ defines the upper hyperbola.

Consider the level lines of the function $n(l, m)$. It is easy to see that all of them are hyperbolas. Moreover, for $n(l, m) = 2 - \chi$ we obtain the upper boundary of the region D, and then, as n increases, these hyperbolas move down tending to the lower hyperbola that corresponds to $n = \infty$.

We now prove that the number of the solutions we are interested in is finite. All integer pairs (l, m), being solutions, belong to the region D. Note that, if N is sufficiently large, then the hyperbola $n(l, m) = N$ becomes so close to the lower boundary $n(l, m) = +\infty$ that between these two hyperbolas there are no integer points at all. This means that possible values of n are bounded from above. On the other hand, n is the complexity of the corresponding atom, and, as is well known, there exists only a finite number of atoms of complexity $n \leq N$. This completes the proof of (a).

Let us consider some examples of totally symmetric atoms of genus $g > 1$. We shall describe two series of such examples. It is convenient to do this in terms of f-atoms. Two series of totally symmetric f-atoms X_n, Y_n, where $n \geq 4$, are presented in Fig. 2.44. The corresponding atoms are shown in Fig. 2.45. Their type evidently depends on whether the number of vertices n is even or odd. The genus of X_n is calculated as follows:

$$g = \begin{cases} \dfrac{n-1}{2} & \text{if } n \text{ is odd,} \\[2mm] \dfrac{n-2}{2} & \text{if } n \text{ is even.} \end{cases}$$

The genus of Y_n is calculated in a similar way:

$$g = \begin{cases} \dfrac{n-1}{2} & \text{if } n \text{ is odd,} \\[2mm] \dfrac{n}{2} & \text{if } n \text{ is even.} \end{cases}$$

The series of atoms are both totally symmetric. The symmetry group of the atoms of type X_n is isomorphic to $\mathbb{Z}_n \oplus \mathbb{Z}_2$, whereas that of Y_n is isomorphic to $\mathbb{Z}_{2n}$. In both cases, the order of the group is $2n$. It is worth, however, mentioning that, for odd n, the atoms X_n and Y_n are isomorphic. If n is even, X_{2p} and Y_{2p} are different (they have different genus and different symmetry groups). The isomorphism between X_{2p+1} and Y_{2p+1} can be obtained just by changing the sign of the function f on the f-atom X_{2p+1}.

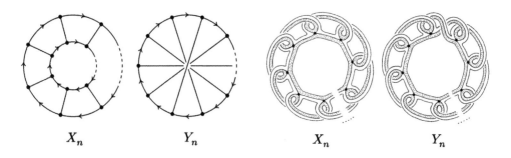

$$X_n \qquad\qquad Y_n \qquad\qquad\qquad X_n \qquad\qquad\qquad Y_n$$

Figure 2.44   Figure 2.45

It is useful to imagine both series of atoms on the standard fundamental polygon that represents the sphere with g handles. For the series X_n, one needs to take the fundamental $2n$-gon in the form

$$(a_1 a_n^{-1})(a_2 a_1^{-1})(a_3 a_2^{-1}) \ldots (a_i a_{i-1}^{-1}) \ldots (a_n a_{n-1}^{-1})$$

and to draw the skeleton K of the atom X_n on it as shown in Fig. 2.46. As usual, to reconstruct the sphere with g handles, one needs to glue the edges of the fundamental polygon marked by the same letters, taking into account their orientation.

For the series Y_n one needs to take the fundamental $2n$-gon in the form

$$(a_1 a_2 \ldots a_n)(a_1^{-1} a_2^{-1} \ldots a_n^{-1})$$

and to draw the skeleton K of the atom Y_n as shown in Fig. 2.47.

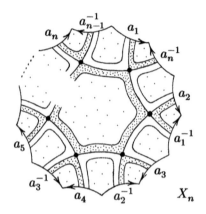

Figure 2.46   Figure 2.47

b) *Case of the torus* $(g = 1)$. The system of equations takes the form

$$lm = 2(l + m), \quad pl = 2n, \quad qm = 2n.$$

The complete list of all integer positive solutions of the system is:
1) $m = l = 4$, n is even;
2) $m = 3$, $l = 6$ (or, conversely, $m = 6$, $l = 3$), $n = 3k$, where $k \in \mathbb{N}$.

Let us construct two corresponding series of totally symmetric atoms.

Atoms of the first series $(m = l = 4$ and n is even$)$.
Consider the standard partition of the Euclidean space $\mathbb{R}^2$ into equal squares (see Fig. 2.48). Let (k_1, k_2) and $(k_2, -k_1)$ be a pair of orthogonal integer-valued vectors, where $k_1, k_2 \in \mathbb{Z}$. Consider the lattice on the plane, generated by these vectors. This is a square lattice in the sense that its fundamental region is the square spanned on the vectors (k_1, k_2) and $(k_2, -k_1)$. By taking the quotient space of $\mathbb{R}^2$ with respect to this lattice, we obtain a two-dimensional torus. This torus is divided into small squares coming from the initial partition of the plane. It is clear that this partition of the torus is totally symmetric in the above sense.

Now, in the same way as it was done in Section 2.7.7, we can reconstruct an atom from this partition. Let us mark the midpoints of the edges of all small squares and connect them pairwise, as shown in Fig. 2.49. We obtain a graph with vertices of degree 4. This is the graph K of the atom we are constructing. The atom itself appears as a tubular neighborhood of K (Fig. 2.49). The group of symmetries of this atom coincides with that of the partition of the torus into squares. The fact that the constructed atom is totally symmetric follows immediately from the same property for the initial partition. It is important that the lattice we used is invariant under the rotation through $\pi/2$.

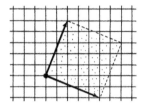

Figure 2.48   Figure 2.49

Atoms of the second series: $m = 3$, $l = 6$ (or, conversely, $m = 6$, $l = 3$), $n = 3k$. Now, instead of tiling by squares, we consider the partition of $\mathbb{R}^2$ into equilateral triangles (Fig. 2.50). Take an arbitrary vector of this triangle lattice and rotate it through $\pi/3$. Consider the new lattice generated by these two vectors. This lattice is, clearly, invariant under $\pi/3$-rotations. Now consider the torus that is the quotient space of $\mathbb{R}^2$ with respect to the chosen lattice. As a result, we obtain a symmetric partition of this torus into (initial) triangles. Then we repeat the above procedure of constructing an atom from the partition. Namely, we take the midpoints of the edges of triangles and connect them pairwise, as shown in Fig. 2.51. We obtain a totally symmetric atom all of whose positive cycles have the same length $l = 6$, and all negative ones have length $m = 3$. Or, conversely, $l = 3$, $m = 6$. It depends on the choice of the direction of increasing f on the atom.

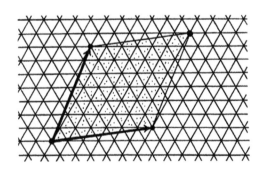

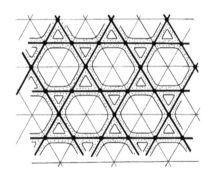

Figure 2.50   Figure 2.51

Let us prove now that these two series exhaust the set of totally symmetric atoms of genus $g = 1$, i.e., located on the torus.

As was remarked above, instead of totally symmetric atoms $V = (P, K)$, we may classify totally symmetric partitions of the closed surface $\widetilde{P}$ (in our case, the torus).

Since we are interested just in the torus, we can consider the standard universal covering of the torus by a plane. As a result, the atom V will be covered by an infinite atom $\widetilde{V}$ imbedded into the plane. Analogously, the totally symmetric partition of the torus will be covered by a *totally symmetric partition* of the plane. This follows easily from the fact that each symmetry can be lifted from the torus up to a symmetry on the covering plane.

We already know that the partition obtained must satisfy one of the following properties: either $m = l = 4$, or $m = 6$, $l = 3$, or, conversely, $m = 3$, $l = 6$. Hence, a symmetric decomposition can be either a partition into squares or triangles (we omit the case of hexagons, because this case is dual to that of triangles). It is easy to see that this decomposition coincides (up to a homeomorphism) with one of the standard partitions: either into squares or into triangles (see above).

The lattice we use to obtain the torus must be invariant under the symmetries of the atom. This means, in particular, that, in the case of the partition into squares, this lattice must be preserved under the rotation through $\pi/2$. Therefore, it has just that structure which we described above: one vector of its basis is arbitrary, the other is obtained from that by rotation through $\pi/2$. Thus, in the first case, we exhaust all possibility. The second case is very similar. The lattice must be invariant under rotations through $\pi/3$ and, consequently, the first basis vector of the lattice can be taken arbitrary, while the second one is obtained from the first by rotation through $\pi/3$. This completes the proof of (b).

c) *Case of the sphere* $(g = 0)$. The totally symmetric partitions of the sphere are well-known. We recall their description and construct the corresponding atoms.

If $g = 0$, then we have

$$n = \frac{2lm}{2(l + m) - lm}.$$

It is easily seen that all possible pairs of solutions (m, l) are as follows (see Fig. 2.52):

a) $m = l = 3$, $n = 6$;
b) $m = 3$, $l = 4$ (or $m = 4$, $l = 3$), $n = 12$;
c) $m = 3$, $l = 5$ (or $m = 5$, $l = 3$), $n = 30$;
d) $m = 2$, l is positive integer (or $l = 2$, m is positive integer), $n = l$.

The totally symmetric partitions related to these solutions are well known. These are the so-called Platonic solids and two more special partitions (see, for example, [157]). Recall that the Platonic solids are five classical polytopes illustrated in Fig. 2.53, namely, tetrahedron, cube, octahedron, dodecahedron and icosahedron. Two more special symmetric partitions are shown in Fig. 2.54. No other totally symmetric partitions on the two-dimensional sphere exist.

It remains to construct explicitly the atoms related to the Platonic solids and two special partitions. The skeletons K of the "Platonic" atoms are shown in Fig. 2.53 by a dotted line (see also the general rule for reconstructing an atom from a partition in Fig. 2.37). As is seen from Fig. 2.53, we in fact obtain not five, but only three different atoms.

Two special partitions of the two-dimensional sphere generate the same atom. Its skeleton K is shown in Fig. 2.54 by a dotted line.

In Fig. 2.55(a, b, c, d) the skeletons K of the totally symmetric atoms are shown as graphs on the plane. The totally symmetric atoms themselves are presented in Fig. 2.56(a, b, c, d). □

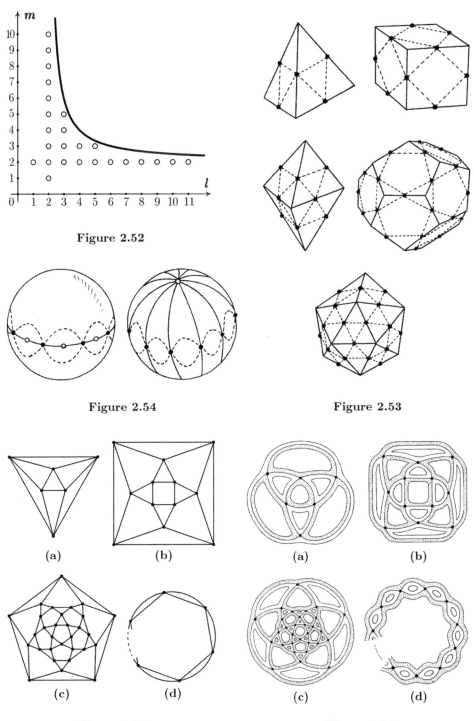

Figure 2.52

Figure 2.54

Figure 2.53

(a)

(b)

(c)

(d)

Figure 2.55

(a)

(b)

(c)

(d)

Figure 2.56

2.8.6. *Atoms as Surfaces of Constant Negative Curvature*

It turns out that there exists an inclusion of the group $\mathbb{Z} * \mathbb{Z}_2$ into the isometry group $\mathrm{Iso}(L^2)$ of the two-dimensional Lobachevskii plane L^2. Therefore, $\mathbb{Z} * \mathbb{Z}_2$ is represented by isometries of the Lobachevskii plane. To show this, we realize L^2 as the upper half-plane of the complex plane $\mathbb{C}^2$. Consider two isometries β, α on it given by the following linear-fractional transformations:

$$\alpha: z \to -\frac{1}{z}, \qquad \beta: z \to z + 2.$$

If we consider the (orientation preserving) isometries of L^2 as the quotient group $\mathrm{SL}(2, \mathbb{R})/\mathbb{Z}_2$, where $\mathbb{Z}_2 = \{\pm E\}$, then β and α can be defined by the matrices

$$\alpha' = \begin{pmatrix} 0 & 1 \\ -1 & 0 \end{pmatrix}, \qquad \beta' = \begin{pmatrix} 1 & 2 \\ 0 & 1 \end{pmatrix}.$$

Geometrically, the transformation β is the shift along the real axis and has infinite order, because its k-th degree has the form

$$(\beta')^k = \begin{pmatrix} 1 & 2k \\ 0 & 1 \end{pmatrix}.$$

The transformation α is the "rotation" through π around the point $i \in L^2 \subset \mathbb{C}$. This is, clearly, an involution. The square of α' is equal to $-E$, but it is the identity mapping in $\mathrm{Iso}_0(L^2) = \mathrm{SL}(2, \mathbb{R})/\mathbb{Z}_2$.

Lemma 2.2.   *The subgroup in $\mathrm{Iso}_0(L^2)$ generated by the transformations β and α is isomorphic to the free product $\mathbb{Z} * \mathbb{Z}_2$.*

The proof follows from the form of the matrices α' and β'.   □

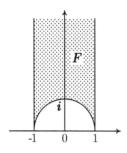

Figure 2.57

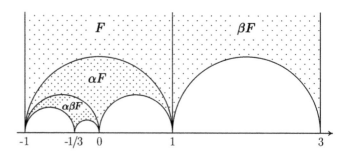

Figure 2.58

The fundamental domain $F(D)$ of the subgroup $\mathbb{Z} * \mathbb{Z}_2$ is presented in Fig. 2.57. It is a triangle all of whose vertices lie at infinity, i.e., on the absolute. Under the transformations β and α, this domain is reproduced as shown in Fig. 2.58. The transformation β shifts it to the right, and α "rotates" it around i through the angle π. It is easy to see that, applying the superpositions of β and α, we can, step by step, fill the whole upper half-plane by images of $F(D)$, as shown in Fig. 2.58.

We now can embed the tree D into the Lobachevskii plane in such a way that it will be mapped into itself under the above action of $\mathbb{Z} * \mathbb{Z}_2$. The simplest way to construct such an embedding is as follows. Let us take the "tripod" inside the fundamental domain $F(D)$, which consists of three geodesic segments starting from the center of $F(D)$ (Fig. 2.59). These geodesics, by construction, are orthogonal to the sides of the triangle $F(D)$.

Under the action of superpositions of β and α, the tripod generates the tree D (Fig. 2.59). To make this picture more visual, the same is shown on the Poincaré model, i.e., inside the unit circle (Fig. 2.60).

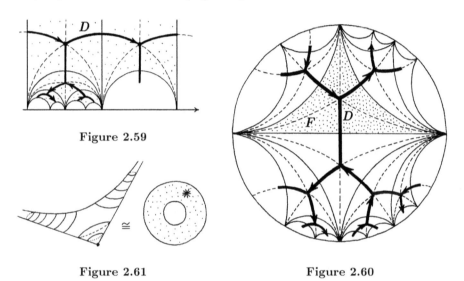

Figure 2.59

Figure 2.61   Figure 2.60

Consider now the quotient space $L^2/\mathbb{Z} * \mathbb{Z}_2$. From the topological point of view, we obtain an annulus with one marked point inside (Fig. 2.61) that is a fixed point of α. At this point, the smoothness is broken and we obtain a conic singularity. Such an annulus with a fixed point will soon appear as the atom A^*.

Now we "embed" the theory of f-atoms into hyperbolic geometry. It turns out that each f-atom can be presented as the quotient space of the Lobachevskii plane with respect to an appropriate subgroup of the isometry group.

More precisely, we can do this as follows. Take any subgroup $G \subset \mathbb{Z} * \mathbb{Z}_2$ of finite index and with no elements of finite order. The fundamental domain $F(G)$ of the group G is obtained as the union of several fundamental domains $F(D)$. Moreover, the number of such domains $F(D)$ is exactly equal to the index of G in $\mathbb{Z} * \mathbb{Z}_2$. In particular, the area of $F(G)$ is πk, where k is an index of G in $\mathbb{Z} * \mathbb{Z}_2$. The quotient space L^2/G is a complete (in the sense of the constant negative curvature metric on L^2/G) non-compact surface with parabolic ends.

Now let us consider a Morse function $\widetilde{f}$ which is invariant with respect to the action of $\mathbb{Z} * \mathbb{Z}_2$ (and, consequently, with respect to its subgroup G). Then $\widetilde{f}$ can be lowered down to the quotient space L^2/G. As a result, we obtain a Morse function $f: L^2/G \to \mathbb{R}$, which defines the structure of an atom on L^2/G.

Let us describe the "universal" function $\widetilde{f}$ on L^2. The level lines of the desired function $\widetilde{f}$ in the fundamental domain and its image under α are shown in Fig. 2.62. As a result, we obtain a quadrangle with the vertices at infinity. Its center is a saddle Morse point, and the geodesics connecting the midpoints of the opposite edges are critical level lines of $\widetilde{f}$. Moreover, the level lines meet the boundary of the fundamental domain orthogonally. This guarantees the smoothness of the function $\widetilde{f}$ after reproducing $F(D)$ by elements of $\mathbb{Z} * \mathbb{Z}_2$. As a result, we obtain the picture of level lines of $\widetilde{f}$ on the whole Lobachevskii plane. By comparing Fig. 2.62 and Fig. 2.63, one can easily see all the other (regular) level lines of $\widetilde{f}$.

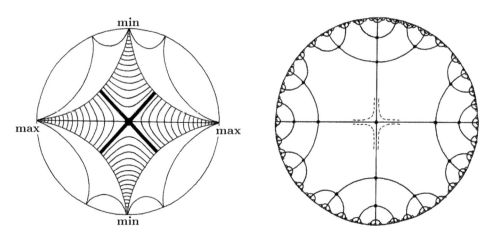

Figure 2.62   Figure 2.63

Lemma 2.3. *The function $\widetilde{f}: L^2 \to \mathbb{R}$ is invariant with respect to the action of $\mathbb{Z} * \mathbb{Z}_2$. All of its critical points are non-degenerate saddles.*

The proof is evident and follows directly from the above construction. $\square$

Let $G \subset \mathbb{Z} * \mathbb{Z}_2$ be an arbitrary subgroup of finite index and without elements of finite order.

Theorem 2.15. *The quotient space L^2/G of the Lobachevskii plane L^2 with respect to the group G is a non-compact two-dimensional surface P^2 with the complete constant negative curvature metric. The surface P^2 is naturally endowed with a Morse function f such that $\widetilde{f} = \pi \circ f$, where $\pi: L^2 \to P^2$ is a natural projection. The pair (P^2, f) represents a certain f-atom. And conversely, every f-atom can be obtained in this way.*

The proof follows easily from the construction of $\widetilde{f}$. $\square$

COMMENT. The surface P^2 has ends going to infinity. It is convenient to assume $\widetilde{f}$ to be equal zero on the critical level line. Then, by cutting the ends of P^2, i.e., considering the set $\{|f| \leq \varepsilon\}$, we obtain a compact two-dimensional surface with a Morse function on it, which defines a certain f-atom.

In Fig. 2.64, we illustrated the atom B as a complete constant negative curvature surface with three parabolic ends. The critical level of the Morse function, i.e., the figure eight curve, is realized as a closed minimal geodesic with one self-intersection point. The constant curvature model for the atom C_2 is presented in the same figure. In this case, the critical level consists of two closed geodesics intersecting each other at two points.

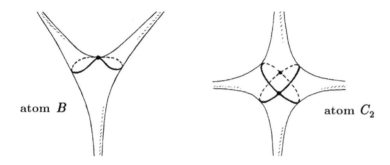

Figure 2.64

In Section 2.8.2, we described the covering of an f-graph by the universal tree D. Now we have realized this tree in the Lobachevskii plane (Fig. 2.60) as a graph composed from geodesic segments. Three such segments meet at each point of the graph. The angles between them are all equal to $2\pi/3$. Consequently, the tree D turns out to be a local minimal net (or, Steiner net). This object naturally occurs in several geometrical problems (see, for example, the book by A. O. Ivanov and A. A. Tuzhilin [165]). In our case, by taking the quotient space L^2/G, we obtain closed minimal nets on complete constant negative curvature surfaces. It is curious that all the edges of the constructed nets have equal length.

2.9. NOTION OF A MOLECULE

Let f be a Morse function on a surface X^2 (orientable or non-orientable). Its level lines define the structure of a foliation with singularities on X^2. We now wish to construct the invariant of this foliation. To that end, consider all critical values c_i of the function f and the corresponding critical levels $\{f = c_i\}$. Every such level is associated with a certain atom. The boundary circles of these atoms are connected by cylinders (tubes) which are one-parameter families of connected level lines of f (i.e., of circles). To the whole foliation we assign now a graph whose vertices are just the atoms (denoted by certain letters). More precisely, this means that each of its vertices is associated with a certain atom and, moreover, we indicate a one-to-one correspondence between the boundary circles of the atom and the edges of the graph that are incident to the vertex. Then we connect the ends of the vertices-atoms by the edges which correspond to the above one-parameter families of regular circles (i.e., by the tubes). An example is shown in Fig. 2.65. This construction is quite similar to the Reeb graph. The only difference is that now we distinguish the types of graph's vertices by assigning a certain atom to each of them.

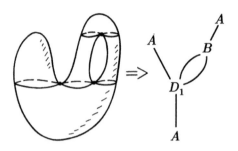

Figure 2.65

If the surface X^2 is orientable, then the vertices of the graph (more precisely, the corresponding atoms) are naturally assumed to be oriented. In other words, we should distinguish mirror symmetric atoms from each other (two atoms are called mirror symmetric if they are homeomorphic, but have opposite orientations). If the surface X^2 is non-orientable, then we consider atoms without taking into account their orientation.

Definition 2.21. The described graph is called the *molecule W* corresponding to the pair (X^2, f).

For convenience, each vertex of the graph W is replaced by the standard letter notation for the corresponding atom.

Which molecules are supposed to be identical (or equivalent)? The answer to this question is important if we want to compare different functions by means of molecules.

Definition 2.22. Two molecules W and W' are supposed to be *identical*, if there exists a homeomorphism $\xi: W \to W'$ which transforms edges to edges, atoms to atoms, and, moreover, this homeomorphism can be extended to the atoms themselves. This means the following. Let V and $V' = \xi(V)$ be two atoms considered as vertices of the molecules. Recall that these atoms correspond to some pairs (P, K) and (P', K'), where P and P' are two-dimensional surfaces with the embedded graphs K and K'. Since the edges of the molecules related to the atoms-vertices V and V' are in one-to-one correspondence with the boundary circles of P and P' respectively, ξ induces a natural mapping $\xi^*: \partial P \to \partial P'$. This mapping must be extendable up to a homeomorphism between the pairs (P, K) and (P', K').

COMMENT. For each atom, it is useful to consider its standard model as a surface P_{st} with a graph K_{st} (here "st" means "standard"). When defining a molecule W, we fix, in particular, some homeomorphism between the atom associated with a vertex and its standard model. We act similarly when the same atom occurs in some other molecule W'. Given a homeomorphism between the molecules W and W', it naturally induces a permutation of boundary circles of the standard model (for each pair of the corresponding vertices). In order for the molecules to be identical, one has to require this permutation to be generated by some homeomorphism of the standard model (P_{st}, K_{st}) onto itself. It should be emphasized that not every permutation of boundary circles of a surface P_{st} can be extended up to a homeomorphism of the atom (P_{st}, K_{st}) onto itself. This just means that, generally speaking, the ends of an atom are not equivalent. For example, for the atom D_1 the ends 1 and 3 are equivalent, but the ends 1 and 2 (as well as 2 and 3) are not (Fig. 2.66). In Fig. 2.66 one can see three molecules W_1, W_2, W_3, among which the molecules W_1 and W_2 are identical, and the molecules W_2 and W_3 differ.

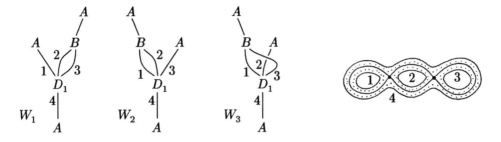

Figure 2.66

This problem cannot be avoided by assigning the same numbers to those boundary circles of the atom which can be transformed into each other by a homeomorphism of the atom onto itself. It can be seen, for example, from the atom V shown in Fig. 2.67. The molecules W_1 and W_2 are different in spite of the fact that all the ends $1, 2, 3, 4$ of the atom V are equivalent in the sense that each of them can be transformed into any other by an appropriate homomorphism $V \to V$.

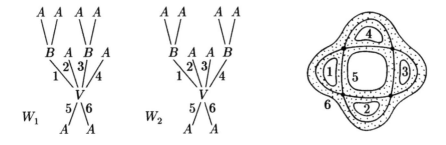

Figure 2.67

From the formal point of view, we would need to enumerate the ends of standard models of atoms, and then, when constructing molecules, to assign the corresponding number to each edge incident to the vertex associated to an atom. We shall not do this, since the atoms occurring in applications are, as a rule, quite simple: $A, B, \tilde{B}, C_1, C_2, D_1$. For these atoms, the correspondence between the edges associated with the letter notation of an atom and the boundary circles of this atom can be easily established. For example, the atom D_1, shown in Fig. 2.68, will be denoted by letter D_1 with one lower end and three upper ends. The lower end is related to the exterior boundary circle, and the three upper ones are related to the three other boundary circles. The middle upper edge corresponds to the central interior boundary circle.

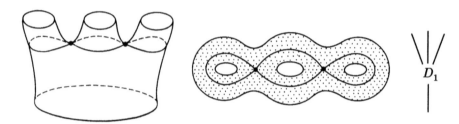

Figure 2.68

Theorem 2.16. *Let (X^2, f) and (X'^2, f') be two oriented surfaces with Morse functions, and W, W' the corresponding molecules. Then the pairs (X^2, f) and (X'^2, f') are fiberwise equivalent with preserving their orientation if and only if their molecules W and W' are identical.*

Proof. In one direction the statement is evident: if the pairs (X^2, f) and (X'^2, f') are fiberwise equivalent, then their molecules, of course, coincide.

Conversely, let W and W' be identical. Let us take a homeomorphism from W onto W'. It establishes a one-to-one correspondence between one-parameter families of regular level lines of the functions f and f', as well as between their critical fibers. A neighborhood of each critical level is an atom. It follows from the coincidence of the molecule that the corresponding atoms are equivalent. By definition, this means that f and f' are fiberwise equivalent in some neighborhoods of their critical levels. Then this equivalence must be extended to the remaining tubes foliated into regular level lines of f and f'. It can also be done, because the homeomorphism is already defined on the ends of the tubes. This proves Theorem 2.16. □

What happens to a molecule, if we change orientation on X^2 without changing the function f on it?

Proposition 2.8. *If we change orientation on X^2, then the atoms of the corresponding molecule are replaced with mirror symmetric ones.*

The proof is evident. □

One should not think that the molecule W is not changed under changing orientation on X if all the atoms are mirror-like. The point is that the ends of the atoms are not equivalent. That is why, changing orientation may lead to renumbering ends of atoms which is induced by the mirror symmetry of the atom onto itself.

We call an atom *strongly mirror-like* if it admits a mirror symmetry that does not permute its ends. Such are, for example, all the atoms of complexity 1 and 2, as is seen from Table 2.1.

Corollary. *Suppose all the vertices of the molecule W associated with a Morse function $f: X^2 \to \mathbb{R}$ are strongly mirror-like atoms. Then there exists an orientation reversing diffeomorphism $\xi: X^2 \to X^2$ such that $f(x) = f(\xi(x))$.*

Proof. It immediately follows from the fact that W is not changed if we change orientation on X^2. $\quad\square$

Thus, the molecule W is a quite powerful invariant, which allows one to answer many questions.

We now give the list of all simplest molecules, i.e., those composed of low complexity atoms.

Theorem 2.17. *The fiber equivalence classes for the Morse functions on the sphere and torus with at most 6 critical points are described by the molecules shown in Fig. 2.69. The listed molecules are all different. There are 8 such classes on the sphere and 14 classes on the torus.*

The proof is obtained by exhausting all possibilities. $\quad\square$

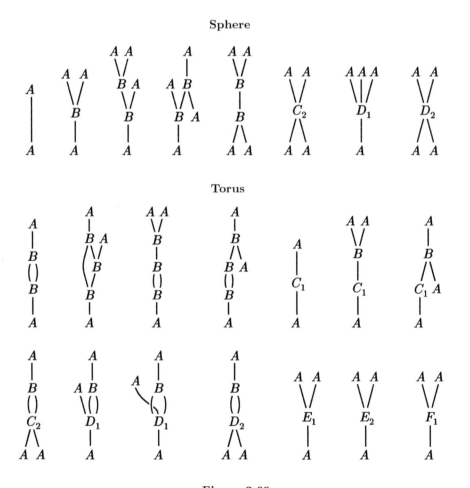

Figure 2.69

REMARK. In Theorem 2.6 the orientation is not important in the sense that the number of fiber equivalence classes with regard to orientation and that without regard to orientation are the same. In other words, each of the listed molecules does not change under the change of orientation on the surface. This follows from the properties of the atoms that occur in these molecules.

Consider an oriented surface X^2 as a symplectic manifold. Then every Morse function on it can be thought as the Hamiltonian of an integrable system with one degree of freedom. (Every Hamiltonian system with one degree of freedom is obviously integrable.) Its integral trajectories are determined by the level lines of the function. Therefore, the fiber equivalence of the pairs (X^2, f) and (X'^2, f') can in fact be considered as the orbital equivalence of the corresponding Hamiltonian systems sgrad f and sgrad f'. A little difference is that on every integral trajectory of the Hamiltonian field there is a natural orientation, namely, the direction of the flow. For the orbital equivalence of flows we have agreed to take into account this orientation (see above). To achieve this, we need to add additional information to the molecule by pointing out, for example, the direction of increasing f on each edge. It is natural to call such a molecule directed. Knowing the direction of increasing the function and the orientation on every atom, we can uniquely find the direction of the flow on each trajectory.

Corollary. *Let f and f' be Morse functions on oriented surfaces X^2 and X'^2. Then two Hamiltonian vector fields* sgrad f *and* sgrad f' *are orbitally equivalent (smoothly or topologically) if and only if the corresponding directed molecules W and W' are identical.*

Thus, the directed molecule is a complete orbital invariant for Hamiltonian systems with one degree of freedom.

2.10. APPROXIMATION OF COMPLICATED MOLECULES BY SIMPLE ONES

It is well known [242] that every Morse function (on a smooth manifold) can be approximated by a Morse function which has just one critical point on each of its critical levels. In other words, by an arbitrary small perturbation the critical points of the function can be moved into different levels. It is clear that such a perturbation of a Morse function makes its molecule into a simple one, i.e., into a molecule whose atoms are just A, B, and $\widetilde{B}$. The atom $\widetilde{B}$ may occur only in the case, when the surface X^2 is non-orientable.

In this sense, every complicated molecule is approximated by a simple one.

It should be noted that this approximation, generally speaking, is not uniquely defined. In other words, a complicated molecule may turn into different simple molecules under different perturbations. Moreover, a complicated atom may split in several different ways into a sum of simple atoms. Let us give an example. In Fig. 2.70 we show the complicated atom G_1 with three vertices, and two different ways of its splitting into a sum of simple atoms.

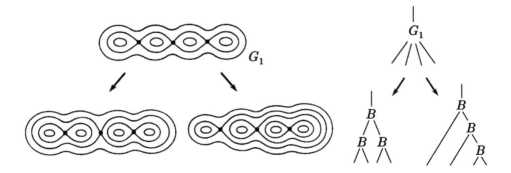

Figure 2.70

Note once more that the perturbation of a Morse function, making a complicated molecule into a simple one, may destroy an initial symmetry of the Morse function (symmetry usually makes a function complicated). By perturbing the function, we lose the information about this symmetry. Therefore, intending to study the symmetries of functions, one has to treat just complicated atoms and molecules, but not their perturbations.

We now want to discuss the following question. Take two Morse functions f and g on the same two-dimensional closed surface and try to deform them smoothly one into the other in the class of Morse functions. When is it possible? What conditions should be imposed to the functions in order for them to be connected by the above deformation?

One of the obvious conditions is that f and g must have the same number of local minima and local maxima. Then, the number of saddle critical points for f and that for g also coincide. The necessity of this condition follows from the fact that, during the deformation, critical points do not disappear and do not arise, because any such bifurcation means a transition through a degenerate singularity.

Therefore, we may reformulate this question as follows: is the space of Morse functions with a fixed number of local minima and maxima arcwise connected? For the sake of simplicity, in this section we consider only oriented surfaces. A natural idea is to study the deformations of Morse functions by using molecules. If we represent the Morse functions f and g by the corresponding molecules $W(f)$ and $W(g)$, then it is possible to construct a transformation of one molecule into the other by using some elementary bifurcations. On the first step, these molecules can be transformed into simple ones (that is, with atoms A and B only). Then we should consider the elementary bifurcations which interchange two neighboring saddle atoms in the molecule. These four bifurcations are shown in Fig. 2.71(a, b, c, d) and Fig. 2.72. Each of them is described by one of the atoms C_1, C_2, D_1, D_2. The corresponding evolution of level lines of the Morse function is illustrated in Fig. 2.71.

Theorem 2.18 (E. A. Kudryavtseva). *Let f be a simple Morse function on a two-dimensional closed surface, and $W(f)$ be its molecule. Then $W(f)$ can be reduced to the canonical form shown in Fig. 2.73 by means of the four elementary bifurcations listed above.*

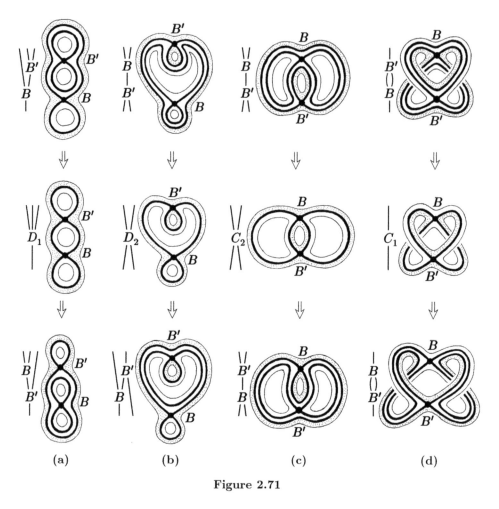

Figure 2.71

Figure 2.72

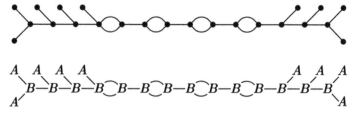

Figure 2.73

Since the molecule W determines the Morse function up to a fiber homeomorphism, this theorem implies the following result.

Corollary. *Given two Morse functions f and g on a closed two-dimensional surface X^2, they can be smoothly deformed one to the other up to a diffeomorphism of X^2 onto itself. In other words, there exist a diffeomorphism $\xi: X^2 \to X^2$ and a smooth deformation $\varphi_t: X^2 \to \mathbb{R}$ such that $\varphi_0 = f$ and $g = \varphi_1 \circ \xi$.*

In fact, a stronger statement is true. Namely: the space of Morse functions with a fixed number of local minima and maxima on a closed two-dimensional surface is arcwise connected. We could not find this result in the literature. The only proof we know has been recently obtained by S. V. Matveev. Note that this proof is rather non-trivial and uses deep low-dimensional topology techniques.

Consider a closed two-dimensional surface M, and let $F(M, p, q)$ denote the set of all Morse functions on M that have p local minima and q local maxima.

Theorem 2.19 (S. V. Matveev). *The space of Morse functions $F(M, p, q)$ is arcwise connected.*

It is useful to reformulate this theorem in terms of surfaces with boundary. Let P be a surface with boundary whose boundary components are divided into two classes $\partial_+ P$ and $\partial_- P$ (positive and negative circles). Let p be the number of negative circles and q be the number of positive ones. By $F(P)$ we denote the space of the Morse functions $f: P \to \mathbb{R}$ that satisfy the following properties:

a) f has only saddle critical points on P;
b) f has no critical points on the boundary ∂P;
c) $f|_{\partial_+ P} = +1$ and $f|_{\partial_- P} = -1$.

Theorem 2.20 (S. V. Matveev). *The space $F(P)$ is arcwise connected.*

COMMENT. Thus, the homotopy connecting two Morse functions $f, g \in F(M, p, q)$ can be chosen in such a way that the points of local minima and maxima remain fixed under this homotopy.

Developing the ideas of S. V. Matveev, E. A. Kudryavtseva proved several generalizations of this result.

Sometimes, when deforming a Morse function, it is useful to look after each of its critical points. It is also useful to take into account the behavior of separatrices. Namely, at each saddle critical point of a Morse function there are two incoming separatrices which form a smooth segment called a separatrix arc. During the deformation process, this arc is changed and undergoes some bifurcations by interacting with other analogous arcs. Let us endow each separatrix arc with some orientation. One more natural question is whether it is possible to deform one Morse function into the other in such a way that the orientations of separatrix arcs also coincide after the deformation. To answer these questions we introduce the following spaces of Morse functions "with framing".

Consider the space $\widetilde{F}(M, p, q)$ of Morse functions f on a closed two-dimensional surface M with the following properties:

1) f has p points of local minimum and q points of local maximum;
2) all local minima and maxima of f are assumed to be fixed on M;
3) the saddle critical points of f are enumerated, and this numeration is fixed. Such a function f is said to be a *Morse function with enumerated saddles*.

It is clear that $\widetilde{F}(M,p,q)$ can be considered as a covering space for $F(M,p,q)$. On $\widetilde{F}(M,p,q)$ we can define a natural action of the group S_r of permutations. Here r is the number of saddle critical points of f. By taking a quotient of $\widetilde{F}(M,p,q)$ with respect to this action, we obtain the space $F(M,p,q)$. The fiber of this covering is isomorphic to S_r.

Every homotopy f_t, $0 \le t \le 1$, in $F(M,p,q)$ defines (uniquely) some homotopy in $\widetilde{F}(M,p,q)$ (i.e., a homotopy preserving the numeration of saddle critical points).

Consider another space $F_+(M,p,q)$ of Morse functions f with properties (1), (2), and (4) on a closed two-dimensional surface M, where property (4) is defined as follows:

4) for each saddle critical point of a Morse function $f \in F(M,p,q)$, the orientation on the separatrix arc is chosen and fixed.

We shall say that the space $F_+(M,p,q)$ obtained is the space of *Morse functions with framed saddles*.

It is clear that $F_+(M,p,q)$ is a covering space for $F(M,p,q)$ with a fiber $(\mathbb{Z}_2)^r$.

Consider one more space $\widetilde{F}_+(M,p,q)$ of Morse functions with enumerated and framed saddles. An element $f \in \widetilde{F}(M,p,q)$ is a Morse function with enumerated saddles for each of which the orientation on the separatrix arc is fixed.

It is clear that $\widetilde{F}_+(M,p,q)$ is a covering space over $F(M,p,q)$ with a fiber isomorphic to the group $S_r \times (\mathbb{Z}_2)^r$.

Theorem 2.21 (E. A. Kudryavtseva [206]). *Let M be a closed connected two-dimensional surface. Then*

a) *the space $\widetilde{F}(M,p,q)$ of Morse functions with enumerated saddles is arcwise connected;*

b) *the space $F_+(M,p,q)$ of Morse functions with framed saddles is arcwise connected;*

c) *the space $\widetilde{F}_+(M,p,q)$ of Morse functions with framed and numbered saddles splits into two arcwise connected components.*

REMARK. According to Matveev's Theorem 2.20, we can choose a homotopy between two functions $f, g \in F(M,p,q)$ so that the points of local minima and maxima remain fixed. It is an interesting question whether we can do the same if we require, in addition, that the saddle points are also fixed.

2.11. CLASSIFICATION OF MORSE–SMALE FLOWS ON TWO-DIMENSIONAL SURFACES BY MEANS OF ATOMS AND MOLECULES

In this section, we show how the idea of atoms and molecules can be applied to the problem of orbital classification of Morse–Smale flows on closed two-dimensional surfaces. This construction appeared to be a result of the discussion on different applications of atoms and molecules by V. V. Sharko and the authors. Then A. A. Oshemkov developed this approach (see [67], [281] for details). Note that one can find in [281] a survey of different approaches to the classification of Morse–Smale flows suggested by M. M. Peixoto [286], G. Fleitas [115], and X. Wang [358].

Recall that vector fields v_1 and v_2 given on closed surfaces M_1 and M_2 are called *topologically orbitally equivalent* if there exists a homeomorphism $h: M_1 \to M_2$ that sends the trajectories of v_1 to those of v_2 while preserving their natural orientation.

Definition 2.23. A vector field v on a manifold M is called *structurally stable* if the topological behavior of its trajectories is not changed under small perturbations; i.e., after any sufficiently small perturbation $v \to \tilde{v}$, the field $\tilde{v}$ remains orbitally topologically equivalent to v.

According to Peixoto's theorem [286], [287], [288], structurally stable vector fields on two-dimensional surfaces are exactly Morse–Smale fields. In the case of two-dimensional surfaces, they can be defined in the following way.

Definition 2.24. A vector field v on a closed two-dimensional surface X^2 is called a *Morse–Smale field* if

1) v has a finite number of singular points and closed trajectories, and all of them are hyperbolic,

2) there are no trajectories going from a saddle to a saddle,

3) for each trajectory of v, its α- and ω-limit sets are either a singular point or a closed trajectory, i.e., a limit cycle.

For simplicity, we describe here the classification of Morse–Smale flows with no closed trajectories. Such flows are called *Morse flows*. In the case of general Morse–Smale flows on X^2, the approach we discuss below is realized in [281].

Morse flows have another natural description. They are exactly gradient-like flows without separatrices going from one saddle to another. A flow is called *gradient-like* if it is topologically orbitally equivalent to the flow $\operatorname{grad} f$ for some Morse function f and some Riemannian metric g_{ij} on the manifold.

It turns out that each Morse flow on a two-dimensional surface X^2 can be associated with some f-atom in such a way that the correspondence between f-atoms and topological orbital equivalence classes of Morse flows will be bijective. Let us describe this construction explicitly.

The singular points of a Morse flow can be divided into three types: sources, sinks, and saddles. Besides, the flow has separatrices that connect sources and sinks with saddles. Each saddle has two incoming and two outgoing separatrices.

Consider a small circle around each source which is transversal to the flow (Fig. 2.74(a)). Choose some orientation on it and mark the points of intersection with the separatrices. The marked points can be divided into pairs. Indeed, for each saddle there are two incoming separatrices. Then the curve consisting of these separatrices connects a pair of marked points (Fig. 2.74(a)).

Consider a graph whose vertices are the marked points and whose edges are of two types. The edges of the first type are the arcs of the circles around sources, the edges of the second type are the curves consisting of two separatrices connecting the pair of vertices. As a result, each vertex is incident to three edges, two of which (edges of the first type) are oriented, but the third (an edge of the second type) is not (see Fig. 2.74(b)). To obtain an f-graph, it remains to endow each non-oriented edge with mark $+1$ or -1. It can be done just in the same way as above (see Section 2.7.4). The rule is shown in Fig. 2.75. As a result, we obtain an f-graph, which was introduced for the classification of atoms.

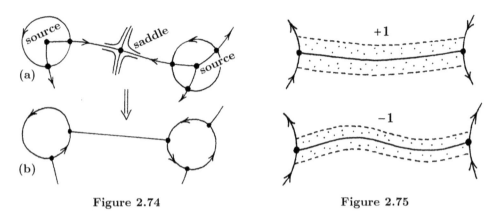

Figure 2.74   Figure 2.75

As we already know, the f-graphs corresponding to singularities of Morse functions are defined up to a natural equivalence relation (see Section 2.7.4). In the case of Morse flows, the situation is just the same. Namely, the marks on non-oriented edges depend on the choice of orientation on the circles around sources. If we change the orientation on one of them, then the marks are changed simultaneously on all non-oriented edges incident to this circle. This evidently leads to that equivalence relation between f-graphs which was already introduced in Section 2.7.4.

If the surface M is oriented, then all circles can be endowed with the canonical orientation so that the marks will be equal to $+1$, and as a result we can forget about them.

Thus, each Morse flow can be associated with some f-atom represented as an f-graph. This observation is the key point for the classification of Morse flows by means of atoms.

Note that in the above construction, one assumes the existence of at least one saddle singular point. But there is the simplest Morse flow which has no such point. This is the gradient flow of the height function on the two-dimensional sphere (Fig. 2.76). It flows from the south pole to the north one along the meridians. This flow does not have any natural f-graph. But we do not need this, because the flow with such a property is unique, i.e., uniquely defined up to orbital topological equivalence. In what follows, we shall assume that our flows are different from the simplest one.

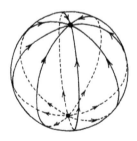

Figure 2.76

It is easy to see that the f-graph of a Morse flow (considered up to natural equivalence) is an orbital topological invariant of the flow. Moreover, this invariant is classifying, i.e., the following statement holds.

Theorem 2.22 (Classification of Morse flows on surfaces).

a) *There exists a natural one-to-one correspondence between f-invariants (or, equivalently, f-atoms) and Morse flows on closed surfaces (considered up to orbital topological equivalence).*

b) *Two Morse flows v_1 and v_2 on two-dimensional surfaces M_1 and M_2 are orbitally topologically equivalent if and only if the corresponding f-graphs are equivalent.*

In this form, the classification theorem has been obtained by V. V. Sharko and A. A. Oshemkov.

At the same time, it should be noted that the described relationship between Morse flows and f-atoms is in fact contained in the paper [240] written by K. R. Meyer in 1968, although he did not consider the classification problem for f-atoms. Let us describe this relationship explicitly. Consider a Morse flow v. Since v is gradient-like, this flow is orbitally equivalent to the gradient flow of some Morse function f. Can this function f be taken as an invariant of the flow? The answer is obviously negative. The point is that f is not well-defined. In particular, f depends on the choice of a Riemannian metric on the surface.

It turns out that this problem can be avoided in the following way. It suffices to choose a function f in such a way that all of its critical points are located on the same level. The fact that such a function exists can be seen from Theorem 2.22 about the correspondence between Morse flows and f-graphs. Indeed, we can take the f-graph corresponding to the Morse flow v, which is already embedded into the surface M. Then we construct the function on M corresponding to this f-graph. Evidently, all of its critical points are located on the same level. It is easy to see that the gradient flow of f is orbitally equivalent to v.

Theorem 2.23 (K. R. Meyer [240]). *Let M be a closed two-dimensional surface with a Riemannian metric on it.*

a) *Consider a Morse function $f: M \to \mathbb{R}$ all of whose critical points are located on the same level $f^{-1}(c)$, and consider its gradient flow with respect to the given Riemannian metric. The mapping $f \to \operatorname{grad} f$ establishes a natural one-to-one correspondence between fiber equivalence classes of such functions and orbital topological equivalence classes of Morse flows on M.*

b) *This one-to-one correspondence does not depend on the choice of a Riemannian metric on M.*

The above construction deals with Morse flows only. As we have just seen, their classification is equivalent to the classification of f-atoms. It turns out that Morse–Smale flows can be classified in a similar way, but instead of atoms we should consider the molecules similar to those which have been defined above (see [281] for details).

Chapter 3

Rough Liouville Equivalence of Integrable Systems with Two Degrees of Freedom

3.1. CLASSIFICATION OF NON-DEGENERATE CRITICAL SUBMANIFOLDS ON ISOENERGY 3-SURFACES

Consider a symplectic manifold M^4 with an integrable Hamiltonian system $v = \operatorname{sgrad} H$; let Q_h^3 be a non-singular compact connected isoenergy 3-surface in M^4. Let f be an additional integral of the system v that is independent of H. We denote its restriction to Q_h^3 by the same letter f. Recall that f is assumed to be a Bott function on Q_h^3. Our aim is to investigate the topology of the Liouville foliation on Q_h^3 defined by the given integrable system. Its non-singular leaves are Liouville tori, and the singular ones correspond to critical levels of the integral f on Q_h^3.

Proposition 3.1. *Let the system v be non-resonant on Q_h^3, and let the additional integral f be a Bott function. Consider the corresponding Liouville foliation on Q_h^3. Then this foliation is completely determined just by the Hamiltonian H and does not depend on the specific choice of the additional integral f.*

Proof. Since the system $v = \operatorname{sgrad} H$ is non-resonant, it follows that almost all Liouville tori are the closures of trajectories of v. Thus, almost all non-singular level surfaces of the integral f are uniquely defined by the Hamiltonian H itself. Clearly, if f and f' are two Bott integrals of the system v, then the coincidence of almost all non-singular level surfaces implies that the remaining level surfaces (both regular and singular) will coincide too. But this means that f and f' define the same Liouville foliation. $\square$

Nevertheless, it is convenient to use some fixed integral f in order to distinguish leaves of the Liouville foliation in Q_h^3.

Recall that, if $\operatorname{grad} H \neq 0$ on Q_h^3, then f does not have any isolated critical points on Q_h^3. Therefore, the critical points of the Bott integral f always form either one-dimensional or two-dimensional non-degenerate submanifolds in Q_h^3. Moreover, the connected critical submanifolds of f in Q_h^3 can be of three types only: circles, tori, and Klein bottles.

Proposition 3.2.

a) *Non-degenerate critical circles of the Bott integral f can be both manifolds of local minimum or maximum and saddle ones.*

b) *Non-degenerate critical tori and Klein bottles are submanifolds of local minimum or maximum.*

Proof. If S is a critical submanifold, and D is a normal disc to S, then the restriction of f onto D is a Morse function. If S is a circle, then the disc D is two-dimensional, and the Morse function can have either a local minimum, or a local

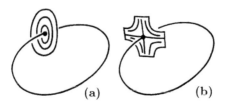

(a)   (b)

Figure 3.1

maximum, or a saddle at the center of the disc (Fig. 3.1). If the submanifold S is two-dimensional, then the normal disc D is one-dimensional and, therefore, f must have either a local maximum or a local minimum at the center of D. □

In fact, by taking a two-sheeted covering over the 3-surface Q_h^3, it is always possible to avoid the critical Klein bottles, namely, to unfold them into tori.

Proposition 3.3. *Let f be a Bott integral on $Q = Q_h^3$ having critical Klein bottles $K_1, \ldots, K_r$, and let $U(Q)$ be a sufficiently small open neighborhood of Q in M^4. Then there exists a two-sheeted covering*

$$\pi \colon (\tilde{U}(\tilde{Q}), \tilde{H}, \tilde{f}) \to (U(Q), H, f),$$

where $\tilde{U}(\tilde{Q})$ is a symplectic manifold with an integrable Hamiltonian system $\tilde{v} = \operatorname{sgrad} \tilde{H}$, and $\tilde{f}$ is its additional integral. Here $\tilde{H} = \pi^* H$, $\tilde{f} = \pi^* f$, $\tilde{v} = \pi^* v$ are the natural pull-backs of H, f, and v from $U(Q)$ onto $\tilde{U}(\tilde{Q})$. Under this covering, all the critical Klein bottles $K_1, \ldots, K_r$ unfold into two-dimensional critical tori $T_1, \ldots, T_r$ of the function $\tilde{f}$ on $\tilde{Q}$.

Proof. Let $K_1, \ldots, K_r$ be the critical Klein bottles. Consider their sufficiently small tubular neighborhoods $V(K_i) \subset Q$ invariant with respect to the flow v. Let us first show that the boundary of each neighborhood $V(K_i)$ is a torus. Indeed, without loss of generality, we can assume that the integral f has a local minimum on the critical Klein bottle K_i. Then, as an invariant tubular neighborhood, we can take the domain $V(K_i) = f^{-1}(c, c + \varepsilon)$, where $c = f(K_i)$. Its neighborhood $\partial V(K_i) = f^{-1}(c + \varepsilon)$ is a regular level surface of f and, therefore, consists of one or several Liouville tori. On the other hand, the boundary of each normal to the Klein bottle K_i segment consists of two points (Fig. 3.2). By projecting them down to K_i, we obtain a two-sheeted covering of the Klein bottle by the boundary of its normal tubular neighborhood. Hence, $\partial V(K_i)$ consists just of one torus. Otherwise, we would have come to the contradiction with the existence of the two-sheeted covering $\partial V(K_i) \to K_i$.

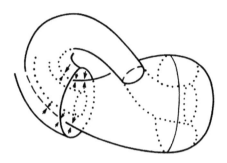

Figure 3.2

Now we cut the manifold Q along all Klein bottles $K_1, \ldots, K_r$. We obtain a 3-manifold W whose boundary consists of several tori $T_1, \ldots, T_r$. Take another copy W' of this 3-manifold and construct a new 3-manifold $\tilde{Q} = W + W'$ called the double of Q (Fig. 3.3) and obtained by the natural identification of the boundary tori: each boundary torus $T_i \subset \partial W$ is glued with its duplicate $T_i' \subset W'$.

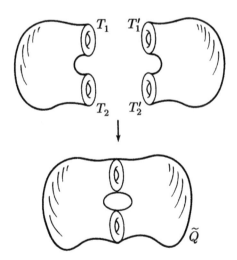

Figure 3.3

Then we define a natural projection of $\tilde{Q} = W + W'$ onto Q. To this end, we use the fact that W and W' are diffeomorphic to the manifold $Q \setminus (K_1 + \ldots + K_r)$. That is why the projection of W and W' onto $Q \setminus (K_1 + \ldots + K_r)$ is already defined in the natural way. Now, we map $Q \setminus (K_1 + \ldots + K_r)$ onto Q by projecting each torus T_i onto the Klein bottle K_i using the corresponding two-sheeted covering, i.e., by doing the gluing operation that is inverse to cutting along the Klein bottle.

REMARK. It is useful to illustrate the above cutting procedure and, then, the inverse gluing by the two-dimensional example of the Möbius strip (Fig. 3.4). By cutting the Möbius strip along its axis, we obtain an annulus. By doing the inverse operation, we map the annulus onto the Möbius strip. As a result, one of the boundary circles of the annulus covers twice the axis of the Möbius strip.

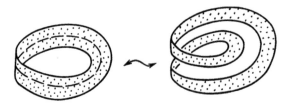

Figure 3.4

Thus, we have defined the projection $\pi \colon \widetilde{Q} \to Q$ such that the preimage of each Klein bottle K_i is the torus T_i. This projection can be viewed as a smooth mapping between smooth 3-manifolds. Since a neighborhood $U(Q)$ in M^4 is diffeomorphic to the direct product $Q \times I$, the projection π is naturally extended on the tubular neighborhood $U(\widetilde{Q}) = \widetilde{Q} \times I$ of $\widetilde{Q}$, where I is an interval. Finally, we lift all the necessary objects from the 4-manifold $U(Q)$ onto $U(\widetilde{Q})$: the form ω, the vector field v, the Hamiltonian H, the integral f. As a result, we have constructed the covering $\pi \colon \widetilde{U}(\widetilde{Q}) \to U(Q)$, which unfolds the critical Klein bottles K_i into the critical tori T_i, as required. □

In what follows, we shall mostly consider integrable systems that have no critical tori and Klein bottles on an isoenergy 3-surface. The following three reasons are an argument for this.

Reason 1.   As the analysis of specific systems in mathematical physics, mechanics, and geometry shows, in most cases, the critical tori and Klein bottles do not really appear. Those comparatively rare cases, where the critical tori and Klein bottle occur, will be discussed separately later on.

Reason 2.   From the topological point of view, the foliation into Liouville tori in a neighborhood of a critical torus is trivial. In other words, there are no singularities here. Moreover, by an appropriate changing of the integral f in a neighborhood of the critical torus (namely, by taking its square root), we can make this torus regular. Here we use the assumption that f is a Bott function, and, consequently, on a one-dimensional transversal, f is a quadratic function in one variable.

Reason 3.   Finally, it is possible to prove that any integrable system with a Bott integral can be turned into a system without critical tori and Klein bottles by an arbitrarily small perturbation in the class of integrable systems (that is, by a small perturbation of H and f).

Now let us describe the topological structure of a neighborhood of a critical circle of the integral f on Q^3.

Let S be a critical circle of the Bott integral, and let D be a transversal two-dimensional disc. Then, by definition, in appropriate coordinates, the function f on D can be written as follows:

$$f = \pm x^2 \pm y^2 .$$

The cases $f = x^2 + y^2$ and $f = -x^2 - y^2$ correspond to the minimum and maximum of f respectively (see Fig. 3.1(a)). The case $f = x^2 - y^2$ corresponds to a saddle (see Fig. 3.1(b)). On the disc D, there appears a foliation into level lines of f with one singular point at the center of D. Under the action of the flow v, the center of D moves along S, so does the disc itself (always being transversal to S). Since f is an integral, the foliation into level lines of f is preserved by this flow. Having made the complete revolution, the disc D returns to the initial position; this generates some diffeomorphism of the disc onto itself preserving the foliation on it. Thus, there appears a foliation of a tubular neighborhood of the critical circle S into two-dimensional leaves with a singularity along S.

As a result, the structure of the neighborhood of S is determined by a foliation preserving diffeomorphism of the disc onto itself. It is easy to see that only the following cases are possible.

a) If the critical circle S is minimal or maximal, then the foliation is trivial, being the direct product of the initial foliation on D by the circle S (Fig. 3.5).

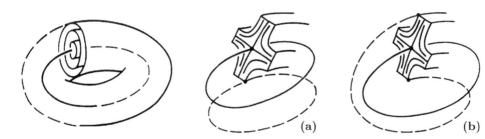

Figure 3.5   Figure 3.6

If the critical circle S is saddle, then the following two cases appear.

b-1) A diffeomorphism of D onto itself that preserves the function $f = x^2 - y^2$ is isotopic to the identity mapping. In this case, the foliation that appears on the neighborhood of S is trivial (Fig. 3.6(a)). It can be considered as the direct product of the cross shown in Fig. 3.6 and the circle S. In this case, we say that S has an *orientable separatrix diagram*.

b-2) A diffeomorphism of D onto itself that preserves the function $f = x^2 - y^2$ is isotopic to the central symmetry, i.e., the rotation through π (Fig. 3.6(b)). Here the foliation that appears on the neighborhood of S is no longer trivial. It is a skew product of the two-dimensional cross and the circle S. In particular, the skew product of the one-dimensional coordinate cross and the circle S is obtained as the union of two Möbius bands intersecting along their common axis S. In this case, we say that S has a *non-orientable separatrix diagram*.

COMMENT. Speaking of a separatrix diagram, we mean here the intersection of the set $\{f = 0\}$ with a tubular neighborhood of S. In the case when S is a hyperbolic trajectory of the Hamiltonian vector field v, this separatrix diagram coincides with the union of the stable and unstable submanifolds of the trajectory S. Notice that an orientable separatrix diagram represents the union of two annuli intersecting transversally along their common axis; and a non-orientable separatrix diagram is the analogous union of two Möbius strips (see Fig. 3.7).

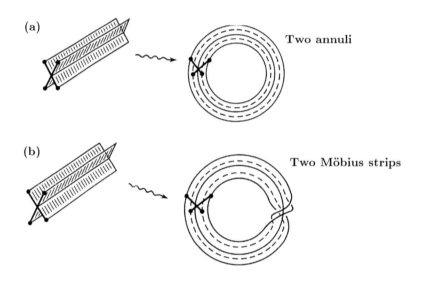

(a) Two annuli

(b) Two Möbius strips

Figure 3.7

In fact, one can prove the following analog of the Morse lemma.

Lemma 3.1. *In a sufficiently small tubular neighborhood of the critical circle* S, *one can choose coordinates* x, y, φ *(where* x, y *are coordinates on the disc* D, *and* φ *is a coordinate along* S*) such that the integral* f *will be written in these coordinates (in the whole neighborhood) as follows:*

a) $f = x^2 + y^2$ *(resp.* $f = -x^2 - y^2$*) in the case of the minimal (resp. maximal) circle (* f *does not depend on* φ*);*

b-1) $f = x^2 - y^2$ *in the case of a saddle circle with an orientable separatrix diagram (* f *does not depend on* φ*);*

b-2) $f = x^2 \cos \varphi - 2xy \sin \varphi - y^2 \cos \varphi$ *in the case of a saddle circle with a non-orientable separatrix diagram.*

Proof. The proof easily follows from the generalized Morse–Bott lemma (Proposition 1.15). □

Lemma 3.1 implies the following statement, which yields a classification of the Liouville foliations near non-degenerate critical circles.

Proposition 3.4. *Up to a diffeomorphism, there exist only three above types of Liouville foliations in a neighborhood of a critical circle. All of them are pairwise different.*

3.2. THE TOPOLOGICAL STRUCTURE OF A NEIGHBORHOOD OF A SINGULAR LEAF

To formulate the main result of this section, we introduce the notion of a Seifert fibration, which is of independent importance.

A Seifert fibration is a three-dimensional manifold represented as a union of disjoint simple closed curves, which are called fibers. See, for example, [237], [309], and [338]. In addition, these fibers should adjoin to each other in a good way. To explain what this means, we introduce the notion of a fibered solid torus.

A *solid torus* $D^2 \times S^1$ divided into fibers $\{*\} \times S^1$ is called a *trivially fibered solid torus*. To define a *non-trivially fibered solid torus*, we choose a pair of relatively prime numbers α, ν, where $\alpha > 1$. Consider the cylinder $D^2 \times I$ and glue its feet by rotation through the angle $2\pi\nu/\alpha$. As a result, we obtain a solid torus. The separation of the cylinder into the segments $\{*\} \times I$ determines a foliation of the solid torus into circles called *fibers*. The fiber obtained by gluing the ends of the segment $\{0\} \times I$ goes along the torus only once. It is called *singular*. Every other fiber goes along the torus exactly α times. The number α is called the *multiplicity of the singular fiber*. Two numbers (α, ν) are called the *parameters of the fibered solid torus* (or those of its singular fiber).

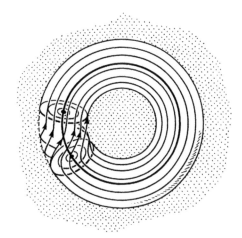

Figure 3.8   Figure 3.9

The fibered solid torus of type $(3, 2)$ is shown in Fig. 3.8, and the fibers of a solid torus of a general type are shown in Fig. 3.9.

Definition 3.1. A compact orientable three-dimensional manifold (with or without boundary) foliated into non-intersecting simple closed curves (fibers) is called a *Seifert manifold* if each of its fibers has a neighborhood consisting of the whole fibers and is homeomorphic to the fibered solid torus. A Seifert manifold with the given fiber structure is called a *Seifert fibration*.

It is easy to see that the fibered solid tori with parameters (α, ν) and $(\alpha, \nu + k\alpha)$ are fiberwise homeomorphic. Therefore, one can always assume that $0 < \nu < \alpha$. Moreover, it can easily be shown (see, for example, [237]) that the pair (α, ν), where $0 < \nu < \alpha$, is an invariant of a fibered solid torus.

Let Q be a Seifert manifold (with or without boundary). Introduce an equivalence relation on it by assuming two points to be equivalent if and only if they belong to the same fiber.

Definition 3.2. The quotient space of the manifold Q by this equivalence relation is denoted by P and is called the *base of the Seifert fibration*.

In other words, the space P is obtained from Q by shrinking its every fiber into a point. The images of the singular fibers are called the singular points of the base P.

Proposition 3.5. *The base P of every Seifert fibration is a compact two-dimensional surface (with or without boundary).*

Proof. See, for example, the book by A. T. Fomenko and S. V. Matveev [237]. □

In what follows, we shall consider only connected Seifert manifolds with boundary. The bases of the corresponding Seifert fibrations are two-dimensional connected surfaces with boundary. Let us point out all singular points on the base of a Seifert fibration and assign the type (α, ν) of the corresponding singular fiber to each of them.

Theorem 3.1. *Two Seifert fibrations Q and Q' with boundary are fiberwise homeomorphic with preservation of their orientation if and only if their bases are homeomorphic and the number and types of their singular points on the bases coincide.*

The proof can be found, for example, in [237] and [309]. □

We now return to an integrable system on an isoenergy 3-manifold Q^3. As before, we assume that the additional integral $f: Q^3 \to \mathbb{R}$ is a Bott function.

Consider a sufficiently small tubular three-dimensional neighborhood $U(L)$ of an arbitrary singular leaf L of the Liouville foliation on Q^3.

Theorem 3.2.

a) *In some four-dimensional neighborhood $V(L)$ of the 3-manifold $U(L)$ in M^4, there exists a smooth function F such that all its integral trajectories on $V(L)$ are closed. Moreover, one can assume that, for any such trajectory $\gamma(t)$, the relation $\gamma(0) = \gamma(2\pi)$ holds.*

b) *The function F commutes (i.e., is in involution) with the functions H (Hamiltonian) and f (integral). Their skew gradients are connected by the relation*

$$\operatorname{sgrad} F = \lambda \operatorname{sgrad} H + \mu \operatorname{sgrad} f$$

for some smooth functions λ and μ that are constant on each leaf of the Liouville foliation.

Such an integral F will be called *periodic*.

Proof.

1) Let us begin with the case when the integral f has a local minimum (or local maximum) on the critical circle S. Here S coincides with the singular leaf L. The topological structure of the three-dimensional neighborhood $U(L)$ has been already described in Lemma 3.1 and Fig. 3.5. This is a solid torus foliated into concentric tori, i.e., the direct product of a fibered disc D^2 by a circle S^1. Clearly, the four-dimensional neighborhood $V(L)$, in turn, can be considered as the direct product of $U(L)$ by a segment I. Thus, $V(L) = D^2 \times S^1 \times I$ is naturally foliated into circles of the form $\{a\} \times S^1 \times \{b\}$, where $a \in D^2$, $b \in I$. Moreover, each circle belongs to a certain Liouville torus. We now define the desired function F by the formula

$$F(x) = \frac{1}{2\pi} \int\limits_{\gamma(x)} \alpha \,,$$

where α is the differential 1-form such that $d\alpha = \omega$, and $\gamma(x)$ is the circle from the above foliation that passes through the point $x \in V(L)$. As we see, the definition of F is analogous to the construction of action variables (see the Liouville theorem above). Note that the function F is not changed under isotopy of γ. We have proved before that the integral trajectories of the vector field sgrad F lie on Liouville tori, are closed with period 2π, and are homologous to γ. Since F plays here the role of an action variable, it follows that, according to the Liouville theorem, F is a function of H and f. Therefore, sgrad F is a linear combination of the vector fields sgrad H and sgrad f. This completes the proof of (b) in the case of a local minimum or maximum.

Note that the structure of the direct product $D^2 \times S^1$ on the $U(L)$ is not uniquely defined. This leads to the ambiguity of the function F.

2) Now consider the saddle case. The scheme of the proof is the same, but one needs to make more precise what cycles γ should be used for the integration. As in the previous case, we take all the critical saddle circles $S_1, \ldots, S_k$ of the leaf L and their three-dimensional neighborhoods $U(S_1), \ldots, U(S_k)$ in Q. The structure of these neighborhoods is already known from Lemma 3.1. Namely, if a saddle circle S_i has the orientable separatrix diagram, then its neighborhood is the direct product of a two-dimensional cross by a circle. This neighborhood is foliated into circles each of which lies on a certain Liouville torus. In the non-orientable case, such a foliation into circles γ can also be defined. But, in this case, each circle γ different from S_i will turn twice along the axis S_i. This means that the neighborhood $U(S_i)$ has the structure of the fibered solid torus with parameters $(2, 1)$. In both cases, these foliations into circles are naturally extended onto a four-dimensional neighborhood $V(S_i)$.

Next, consider an arbitrary Liouville torus which is sufficiently close to the singular leaf L. This torus necessarily intersects one or several four-dimensional neighborhoods $V(S_i)$ (Fig. 3.10). Each of them allows us to choose and fix a circle on the torus (since each of the neighborhoods $V(S_i)$ is already foliated into circles). The circles obtained on the torus are mutually homologous non-trivial cycles (Fig. 3.10). This follows from the following lemma.

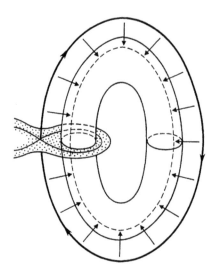

Figure 3.10

Lemma 3.2. *Let T_s be a smooth family of Liouville tori on each of which we choose a cycle (circle) γ_s which smoothly depends on s and, in addition, tends to a closed trajectory γ_0 of the Hamiltonian vector field $\operatorname{sgrad} H$ as $s \to 0$ (in C^1-metric). Then each cycle γ_s is non-trivial (i.e., not contractible) on the torus T_s.*

Proof. The continuity argument implies that all the cycles γ_s are simultaneously either trivial on the tori T_s or non-trivial. Assume the contrary, i.e., all of them are trivial. Then the smooth curve γ_s bounds a two-dimensional disc on the torus.

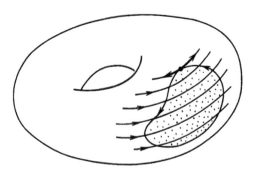

Figure 3.11

Clearly, there exists a point x_s on it at which the vectors $\operatorname{sgrad} H$ and $d\gamma_s/dt$ have opposite directions (see Fig. 3.11). The point is that, according to the Liouville theorem, the field $\operatorname{sgrad} H$ straightens after a suitable choice of coordinates on the torus. Now let $s \to 0$. Then the vector $d\gamma_s/dt$ tends to the vector $v = \operatorname{sgrad} H$ at each point. Clearly, these two facts contradict each other. □

We now turn to the proof of Theorem 3.2. On each Liouville torus close to L, we obtained a non-trivial circle γ (which is close to the periodic integral trajectory S_i of $v = \operatorname{sgrad} H$). If the Liouville torus passes by several periodic trajectories S_i, then we obtain several such circles on it. Since they do not intersect with each other, they are homologous. Note that this argument shows that they are homologous only up to orientation. Is it possible to orient them in such a way that they become homologous as *oriented* cycles? (We, of course, want these cycles γ to have the same orientation near each critical circle γ.) Generally speaking, it is not evident. Of course, on each single Liouville torus, all cycles γ located on it can be consistently oriented. But the problem is that the singular leaf L may be adjacent to several different families of Liouville tori. Consequently, we have to make consistent the orientations of cycles γ on different Liouville tori. Is this always possible? The answer is positive.

The question can be reformulated in the following way. Consider an arbitrary Liouville torus T lying in the neighborhood $V(L)$. There are several cycles γ on it which are homologous with each other up to orientation. Consider the "two-valued" function defined by the already known formula:

$$F(T) = \pm \frac{1}{2\pi} \int_{\gamma} \alpha \,.$$

Here, as before, α is the action form, i.e., $d\alpha = \omega$ (it is not difficult to show that such a form always exists in the neighborhood $V(L)$). Evidently, this function can be smoothly extended on the whole neighborhood $V(L)$. We need actually to show that this "two-valued" function splits into two single-valued functions each of which can be considered as the desired periodic integral.

To this end, it suffices to observe that F has the meaning of an action variable (see the Liouville theorem), since the cycles γ are non-trivial on the Liouville tori. Hence, the integral trajectories of ("two-valued") vector field $\operatorname{sgrad} F$ are all closed with period 2π. Besides, the following relation holds:

$$\operatorname{sgrad} F = \pm \alpha \operatorname{sgrad} H \pm \beta \operatorname{sgrad} f \,,$$

where α and β are constant on each leaf of the Liouville foliation. Consider this relation on the singular leaf L. Note that α and β cannot vanish simultaneously, (otherwise we would have obtained the contradiction with the 2π-periodicity of the trajectories). We now choose the signs of α and β in a certain way and fix our choice. As a result, we obtain a *single-valued* vector field on the singular leaf which vanishes nowhere. We can now extend it to the neighborhood $V(L)$, choosing its direction by continuity.

As a result, we obtain a smooth vector field in $V(L)$ all of whose trajectories are closed with period 2π and lie on leaves of the Liouville foliation. It remains to take its Hamiltonian as the desired periodic integral. $\square$

Corollary. *On $V(L)$, there exists a naturally defined Poisson S^1-action (namely, shifting along integral trajectories of the field $\operatorname{sgrad} F$ by the angle φ). The trajectories of $\operatorname{sgrad} F$ are just the orbits of this action.*

This result can be interpreted as follows. According to the Liouville theorem, one can define the action variables s_1, s_2 in a neighborhood of each non-singular leaf of the Liouville foliation. How do they behave in a neighborhood of a singular leaf L? Theorem 3.2 asserts that one of these action variables survives (provided the corresponding cycle is chosen in the right way), i.e., it is a smooth function in a neighborhood of L without singularities. Note that a similar result holds for non-degenerate systems in the multidimensional case (see [258], [259], [261]).

Theorem 3.2 implies the following statement which gives the description of the Liouville foliation in a three-dimensional invariant neighborhood $U(L) \subset Q^3$ of the singular leaf L. Namely, the periodic integral F allows us to define the structure of a Seifert fibration in a neighborhood of the singular leaf.

Theorem 3.3.

a) *The three-dimensional manifold $U(L)$ is a Seifert manifold whose singular fibers (if they exist) have the same type $(2, 1)$.*

b) *These singular fibers coincide exactly with the critical circles of the integral f possessing non-orientable separatrix diagrams.*

c) *If this Seifert fibration has no singular fibers, then the manifold $U(L)$ is the direct product of a two-dimensional surface $P(L)$ and a circle-fiber S^1.*

d) *The structure of the Seifert fibration on the manifold $U(L)$ and the structure of the Liouville foliation on $U(L)$ are consistent in the sense that each fiber of the Seifert fibration (a circle) lies on some leaf of the Liouville foliation. In particular, the integral f is constant on the fibers of the Seifert fibration.*

Proof. This statement is in fact a reformulation of the Corollary to Theorem 3.2. As oriented fibers of the Seifert fibration on $U(L)$, we simply take the oriented orbits of the S^1-action generated by the periodic integral F. Only part (b) needs a small comment. Due to orientability of fibers of the Seifert fibration and that of the neighborhood $U(L)$, the base $P = P(L)$ is an oriented two-dimensional surface with boundary. If there are no singular fibers in the Seifert fibration, then the fibration is locally trivial. Moreover, since the base $P(L)$ has a boundary, then no additional invariants (like an Euler number) exist, and, consequently, the Seifert fibration on $U(L)$ has the direct product type. □

In the case of a singular leaf L that contains critical circles with non-orientable separatrix diagrams, one can give another visual topological description of the neighborhood $U(L)$. It is easy to verify that, in this case, there exists a section $\widehat{P}$ of the Seifert fibration such that

1) $\widehat{P}$ is transversal to the fibers of the Seifert fibration,

2) every non-singular fiber intersects the surface $\widehat{P}$ twice, whereas the singular fibers (i.e., critical circles with a non-orientable separatrix diagram) intersect it only once.

Let us explain how this surface can be constructed. Since the leaf L is a deformation retract of its neighborhood $U(L)$, it suffices to construct a transversal section on the singular leaf only (and to extend it then to some neighborhood). We first construct such a section in small neighborhood of critical circles: in the case of the non-orientable separatrix diagram, we take an arbitrary transversal to the critical circle; and, in the orientable case, we take two non-intersecting transversals (which are obtained from each other by shifting by half-period π). It is easy to see that the singular leaf L is a union of two-dimensional orbits diffeomorphic to the annulus $S^1 \times D^1$, and critical circles. For each critical circle (and even for its neighborhood), the desired section $\widehat{P}$ has been constructed and we must extend it to each of the annuli (Fig. 3.12). This can evidently be done by connecting the initial pairs of points located on the opposite boundary circles of each annulus.

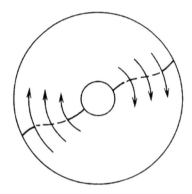

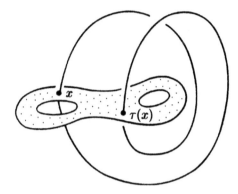

Figure 3.12   Figure 3.13

As a result, we obtain a section of the Seifert fibration on the singular leaf L which satisfies all required conditions. By extending it to the three-dimensional $U(L)$, we construct the desired surface $\widehat{P}$. It is easily seen that $\widehat{P}$ is connected provided the singular leaf L is connected.

On $\widehat{P}$, there exists a natural involution τ. Indeed, to every point $x \in \widehat{P}$, one can assign a point $\tau(x) \in \widehat{P}$ that is the second point of intersection of the fiber through x with $\widehat{P}$ (Fig. 3.13). The second point always exists and is different from x, unless x belongs to a singular fiber of the Seifert fibration. If x belongs to the singular fiber, then, obviously, $\tau(x) = x$.

Lemma 3.3.

a) *The mapping τ is an involution on $\widehat{P}$ whose fixed points are exactly the points of the intersection of $\widehat{P}$ with the singular fibers of the Seifert fibration.*

b) *The base P of the Seifert fibration on $U(L)$ is the quotient space of $\widehat{P}$ by the action of τ.*

The proof follows from the definition of τ. □

Note that the restriction of the integral F to $\widehat{P}$ can be regarded as a natural Morse function $f \colon \widehat{P} \to \mathbb{R}$. Clearly, the involution τ preserves the function f.

This construction allows us to represent $U(L)$ in the following form. Consider the cylinder $\widehat{P} \times [0, \pi]$ and glue its feet $\widehat{P} \times \{0\}$ and $\widehat{P} \times \{\pi\}$ by the action of τ, i.e., by identifying $(x, 0)$ with $(\tau(x), \pi)$ (Fig. 3.14). As a result, we obtain the desired 3-manifold $U(L)$, which is the skew product of $\widehat{P}$ by the circle.

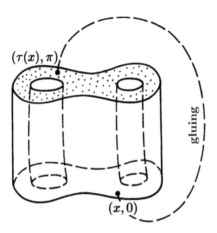

Figure 3.14

It would be useful to imagine how the 3-manifold $U(L)$ is foliated into two-dimensional Liouville tori. On $\widehat{P}$, there is a foliation into level lines of the function $\widehat{f}$. Therefore, the direct product $\widehat{P} \times [0, \pi]$ is foliated into 2-cylinders that are direct products of the level lines of $\widehat{f}$ by the segment $[0, \pi]$. Gluing the feet of the 3-cylinder $\widehat{P} \times [0, \pi]$ by involution τ, we see that these 2-cylinders are glued into two-dimensional Liouville tori foliating $U(L)$.

Thus, we see that the structure of the Liouville foliation on $U(L)$ is uniquely defined by a two-dimensional object, namely, a surface with a Morse function on it. Moreover, it is easy to see that such a pair (surface, function) defines a certain atom in the sense of Chapter 2. In fact, this observation means that there is a natural one-to-one correspondence between topological types of Liouville foliation singularities and two-dimensional atoms. In Sections 3.5 and 3.6, we discuss this relation in more detail and give the classification of generic three-dimensional singularities.

3.3. TOPOLOGICALLY STABLE HAMILTONIAN SYSTEMS

Definition 3.3 (see [65]). An integrable Hamiltonian system is called *topologically stable* on the isoenergy surface $Q_{h_0}^3 = \{H = h_0\}$ if, for sufficiently small variations of the energy level, the structure of the Liouville foliation of the system does not change. In other words, for a sufficiently small ε, the systems $(v, Q_{h_0}^3)$ and $(v, Q_{h_0 + \varepsilon}^3)$ are Liouville equivalent.

REMARK. What does the topological stability of a system mean? It is easy to see that the set of Liouville equivalence classes for integrable Hamiltonian systems (with Bott integrals) is discrete in a natural sense. Therefore, it is natural to expect that there exists just a finite number of bifurcation energy levels when the topology of the Liouville foliation is changed. Such energy levels can easily be recognized with the help of the bifurcation diagram of the momentum mapping $(H, f): M^4 \rightarrow \mathbb{R}^2$. If the straight line $\{H = h_0\}$ on the plane $\mathbb{R}^2$ intersects the bifurcation diagram transversely (Fig. 3.15) and does not pass through its singular points, then the system is topologically stable on $Q^3 = \{H = h_0\}$. Otherwise, the values h_0 are, as a rule, bifurcational.

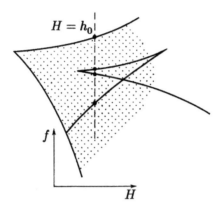

Figure 3.15

In the previous section, we have shown that the neighborhood of a singular leaf of the Liouville foliation has the structure of an orientable Seifert fibration. Is it possible to define the orientation on the fibers of this fibration in a canonical way? One of the possible ways is as follows. The critical circles of the integral f are, at the same time, the fibers of the Seifert fibration and closed trajectories of the Hamiltonian system. Therefore, on each of these circles, there exists the canonical orientation given by the flow. This orientation could be taken to be the canonical orientation on the fibers of the Seifert fibration. However, beforehand, one has to make sure that the orientations of all these critical circles are compatible among themselves, i.e., just coincide (we can compare them, since all the circles are fibers of the same connected orientable Seifert fibration). It turns out that the sufficient condition for the orientation compatibility is the topological stability of the system.

Let $Q^3 = Q^3_{h_0} = \{H = h_0\}$ be an isoenergy surface of an integrable Hamiltonian system $v = \operatorname{sgrad} H$ on a symplectic 4-manifold M^4. Let $f: Q^3 \rightarrow \mathbb{R}$ be a Bott integral of the system v, and let L be a singular leaf of the Liouville foliation on Q^3 given by the function f. Suppose $S_1, \ldots, S_k$ are critical circles of f lying on the singular leaf L and oriented by the flow v.

Proposition 3.6. *If the system v is topologically stable on Q^3, then all the circles $S_1, \ldots, S_k$ have the same orientation.*

Proof. Let us consider each of the critical circles $S_1, \ldots, S_k$ separately. Since S_i is non-degenerate, it follows that, from the viewpoint of the ambient manifold M^4, the circle S_i is contained in a one-parameter family $S_i(\varepsilon)$ of non-degenerate closed one-dimensional orbits of the Poisson action of $\mathbb{R}^2$ (generated by sgrad H and sgrad f). Consider the image of this family under the momentum mapping $\mathcal{F}\colon M^4 \to \mathbb{R}^2$. This will be a smooth curve δ_i, a piece of the bifurcation diagram (see Proposition 1.18).

Let us clarify what happens to the singular leaf L under small variation of the value h of the Hamiltonian H. By virtue of the topological stability of v, the structure of the singular leaf L is not changed. In particular, the critical circles $S_1(\varepsilon), \ldots, S_k(\varepsilon)$ remain on the same singular leaf $L(\varepsilon) \subset Q^3_{h_0+\varepsilon}$ for all sufficiently small ε (i.e., they do not move to different leaves). Hence all curves $\delta_1, \ldots, \delta_k$ coincide. Denote them simply by δ. According to Proposition 1.18, on the critical circles $S_1, \ldots, S_k$ the following relation holds:

$$b \, \text{sgrad} \, H - a \, \text{sgrad} \, f = 0 \,,$$

where a and b are the coordinates of the tangent vector to δ.

It is important that, in the case of a topologically stable system, the coefficients a and b are the same for all critical circles $S_1, \ldots, S_k$ lying on the singular leaf of the Liouville foliation.

To prove the consistency of orientations on $S_1, \ldots, S_k$, it is sufficient to compare the directions of the vector fields sgrad H and sgrad F, where F is the periodic integral (see Theorem 3.2), which defines the structure of an oriented Seifert fibration on $U(L)$. According to Theorem 3.2, there exist constants λ and μ such that the relation

$$\text{sgrad} \, F = \lambda \, \text{sgrad} \, H + \mu \, \text{sgrad} \, f$$

holds on the whole singular leaf L.

Using the additional relation

$$b \, \text{sgrad} \, H - a \, \text{sgrad} \, f = 0 \,,$$

we obtain the identity

$$\text{sgrad} \, F = \left(\lambda + \mu \frac{b}{a} \right) \text{sgrad} \, H$$

on each critical circle $S_1, \ldots, S_k$, where the coefficient $\lambda + \mu \dfrac{b}{a}$ does not depend on S_i.

Hence, the vectors sgrad F and sgrad H have either the same or opposite orientations simultaneously on all the critical circles. In any case, the orientations on $S_1, \ldots, S_k$ given by the flow sgrad H coincide. $\square$

Corollary. *If the integrable system is topologically stable, then all fibers of the Seifert fibration can be canonically oriented in such a way that this orientation on the critical circles $S_1, \ldots, S_k$ coincides with the orientation given by the Hamiltonian flow $v = \text{sgrad} \, H$.*

How do the integral trajectories of sgrad H behave on the singular leaf L?

Consider L and remove all the critical circles $S_1, \ldots, S_k$ (i.e., all the critical periodic solutions) from it. The leaf L will be divided into a disjoint union of several annuli. Consider the flow v on one of them.

Proposition 3.7 (see [53]).   *Only the following three cases are possible*
(up to a diffeomorphism).

a) *All the integral curves of v are closed on the annulus (Fig. 3.16(a)). This*
case is called resonant.

b) *All the integral curves are open, and each of them winds (as a spiral)*
from one boundary of the annulus onto the other, while preserving the orientation
(Fig. 3.16(b)).

c) *All the integral curves are open, and each of them winds from one boundary*
of the annulus onto the other, while reversing the orientation (Fig. 3.16(c)).

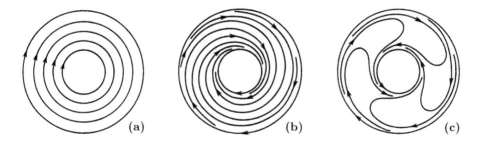

(a)   (b)   (c)

Figure 3.16

Proof.   Consider the vector field $v = \text{sgrad} \, H$ and the periodic integral F
on the given annulus. Then, in the interior of this annulus, we have $u = \text{sgrad} \, F =$
$\lambda \, \text{sgrad} \, H + \mu \, \text{sgrad} \, f$, where $\lambda, \mu \in \mathbb{R}$. Since the trajectories of u are all closed
with period 2π, we can introduce natural coordinates (t, φ) on the annulus,
where $t \in [0, 1]$, $\varphi \in \mathbb{R} \, \text{mod} \, 2\pi$, such that $u = \partial/\partial\varphi$. Two cases are possible:
$\mu = 0$ and $\mu \neq 0$. In the first case, the trajectories of $v = \text{sgrad} \, H$ are all closed
and we get the situation shown in Fig. 3.16(a). In the second case, we have

$$v = \text{sgrad} \, H = a(t)\frac{\partial}{\partial t} + b(t)\frac{\partial}{\partial\varphi} \,,$$

where $a(t)$ and $b(t)$ are some smooth functions on the segment $[0, 1]$. These
functions do not depend on φ, because v and $u = \partial/\partial\varphi$ commute. Note that,
in this case, $a(t)$ nowhere vanishes on $(0, 1)$, since $\text{sgrad} \, H$ and $\text{sgrad} \, F = \partial/\partial\varphi$
are linearly independent. The trajectories of $v = \text{sgrad} \, H$ can now be presented
by the explicit formula:

$$\varphi(t) = \int_{t_0}^{t} \frac{b(t)}{a(t)} \, dt + \text{const} \,.$$

Let us analyze it. The function $a(t)$ vanishes at the endpoints of the segment $[0, 1]$, i.e., on the boundary circles of the annulus, since these circles
are trajectories of $\text{sgrad} \, H$. In other words, $\text{sgrad} \, H$ is proportional to $\partial/\partial\varphi$
on the boundary.   The same argument shows that $b(t)$ takes finite non-zero
values at the endpoints of $[0, 1]$. Thus, the function $\varphi(t)$ is defined on the whole

interval $(0, 1)$ and tends to infinity as $t \to 0$ or $t \to 1$. If the "signs of these infinities" coincide, then we have the case (b) (Fig. 3.16(b)); if they are opposite, then we have the case (c) (Fig. 3.16(c)). $\square$

Proposition 3.8 (see [53]). *If the system v is topologically stable, then L has no annuli of type* (c).

Proof. The boundary circles of the annulus are evidently closed trajectories of $v = \mathrm{sgrad}\, H$. Clearly, in the case (c), the orientations of the boundary circles given by the field v are opposite (Fig. 3.16(c)). However, according to Proposition 3.8, the orientations of all critical circles lying on L must coincide in the case of topologically stable systems. This contradiction proves our statement. $\square$

COMMENT. Let L be a singular leaf of an integrable topologically stable system. We assert that all the annuli lying on L have simultaneously either type (a) or type (b).

Indeed, it suffices to use the relation

$$ u = \mathrm{sgrad}\, F = \lambda\, \mathrm{sgrad}\, H + \mu\, \mathrm{sgrad}\, f , $$

which holds simultaneously for all the annuli. If $\mu = 0$, then the trajectories of sgrad H coincide with those of sgrad F and, therefore, are closed on all the annuli of L. On the contrary, if $\mu \neq 0$, then the trajectories of sgrad H are not closed simultaneously on all the annuli.

3.4. EXAMPLE OF A TOPOLOGICALLY UNSTABLE INTEGRABLE SYSTEM

Here we give an example of an *unstable* integrable Hamiltonian system whose singular leaf has an annulus of type (c) (see Fig. 3.16(c)).

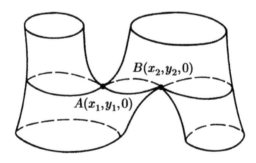

Figure 3.17

Consider the surface P in $\mathbb{R}^3\,(x, y, z)$ illustrated in Fig. 3.17. The height function $f = z$ on P has the only critical value $z = 0$ and two critical points A and B lying on this level. Let $H(x, y, z)$ be a smooth function for which P is a regular level surface, i.e., $P = \{H = 0\}$ and $dH|_P \neq 0$.

As a symplectic manifold, we now consider the direct product $\mathbb{R}^3 \times S^1$ with the symplectic form $\omega = dx \wedge dy + dz \wedge d\varphi$. It is easy to see that the functions H and f commute and, consequently, the Hamiltonian system $v = \operatorname{sgrad} H$ is integrable.

Consider the singular leaf L of the Liouville foliation given by two equations $H = 0$ and $f = 0$. The leaf L contains two critical circles $\{A\} \times S^1$ and $\{B\} \times S^1$. Let us compare the orientations on the circles given by the flow $v = \operatorname{sgrad} H$. To this end, note that the trajectories of the flow $\operatorname{sgrad} f$ are closed and define the structure of an oriented Seifert fibration in the neighborhood $U(L)$ (in other words, f is a periodic integral of the system). That is why, to compare the orientations on the circles, one needs to compare the directions of the flows $\operatorname{sgrad} f$ and $\operatorname{sgrad} H$. Suppose, for definiteness, that the gradient of H at the point $A \in P \subset \mathbb{R}^3$ is directed upward, i.e., has the same direction as the gradient of $f = z$. Then, as is seen from Fig. 3.17, at the point B, the gradient of H will be directed downward so that the directions of $\operatorname{grad} H(B)$ and $\operatorname{grad} f(B)$ will be opposite. Thus, on the circle $\{A\} \times S^1$, the flows $\operatorname{sgrad} H$ and $\operatorname{sgrad} f$ have the same direction, but their directions are opposite on $\{B\} \times S^1$. Consequently, the orientations given on the critical circles $\{A\} \times S^1$ and $\{B\} \times S^1$ by the flow $v = \operatorname{sgrad} H$ are different.

Besides two critical circles, the singular leaf L contains four two-dimensional orbits, each of which is homeomorphic to the annulus. Let us look at the behavior of the trajectories of v on these annuli. Among these four annuli, there are two adjacent to both critical circles $\{A\} \times S^1$ and $\{B\} \times S^1$. Since the flow v runs on these circles in opposite directions, the behavior of trajectories on these annuli is as shown in Fig. 3.16(c). On the two remaining annuli, the behavior of trajectories corresponds to Fig. 3.16(b).

Finally, observe that the constructed Hamiltonian system is not topologically stable on the isoenergy level $\{H = 0\}$. Indeed, the singular points of the height function $f = z$ on the surface $\{H = \varepsilon\} \subset \mathbb{R}^3$ turn out to be on different levels for $\varepsilon \neq 0$. From the point of view of the Hamiltonian system, this means that, under small variation of the energy level, the critical circles pass to different singular leaves of the Liouville foliation. As a result, the singular leaf L splits into two simpler leaves and the structure of the foliations changes.

3.5. 2-ATOMS AND 3-ATOMS

Consider a topologically stable integrable system with a Bott integral f on an isoenergy 3-surface Q and take some singular leaf L of the corresponding Liouville foliation on Q.

Consider a neighborhood of this leaf, i.e., a three-dimensional manifold $U(L)$ with the Liouville foliation structure and fixed orientation. By analogy with the two-dimensional case, as the neighborhood $U(L)$, we take the connected component of the set $f^{-1}(c - \varepsilon, c + \varepsilon)$ that contains the singular leaf L (here

$c = f(L)$ is the critical value of f). Such an object is naturally called a 3-*atom*. However, from the formal viewpoint, we have to be more careful. We shall assume two such 3-manifolds $U(L)$ and $U'(L')$ with the structure of the Liouville foliation to be fiberwise equivalent if

1) there exists a diffeomorphism between them that maps the leaves of the first Liouville foliation into those of the second one,

2) this diffeomorphism preserves both the orientation on 3-manifolds and the orientation on the critical circles defined by the Hamiltonian flows.

Definition 3.4. The equivalence class of the three-dimensional manifold $U(L)$ is called a 3-*atom*. The number of critical circles in the 3-atom is called its *atomic weight* (or *complexity*).

Note that a 3-atom (i.e., its representative $U(L)$) is always oriented and the orientation is assumed to be fixed. By changing the orientation, we may obtain, generally speaking, a different atom.

There appears a natural problem to classify all 3-atoms. It turns out that this classification can be obtained in terms of 2-atoms. To do this, we need first some additional construction.

Consider a Morse function f on an oriented surface; let c be its critical value. Recall that by an atom we mean a neighborhood P^2 of the critical level $K = f^{-1}(c)$ (given by the inequality $c - \varepsilon \leq f \leq c + \varepsilon$ for sufficiently small ε) foliated into level lines of the function f and considered up to fiber equivalence.

It is convenient to define the atom to be the pair (P^2, K) so that two atoms (P^2, K) and (P'^2, K') are considered as identical if there exists an orientation preserving homeomorphism which sends P'^2 onto P^2 and K' onto K.

We emphasize that from now on we shall consider only oriented atoms, i.e., such that the surface P^2 is orientable and the orientation on it is fixed.

Side by side with the previous atoms, we consider one more simple atom, which is obtained as follows. Take an annulus P and assume some of its axial circles to be the graph K (Fig. 3.18). It is clear that this atom can be considered as a neighborhood of a non-singular level of f.

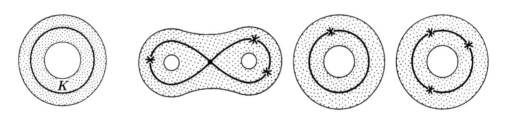

Figure 3.18   **Figure 3.19**

We want now to extend the store of atoms by adding new objects, namely, the so-called atoms with stars. Take an arbitrary atom (P^2, K) and consider its graph $K = \{f = c\}$. Fix some interior points on edges of the graph K, declare them to be new vertices of K and denote by stars (see examples in Fig. 3.19).

Definition 3.5. An atom (P^2, K) which contains at least one star-vertex is called an *atom with stars*. If there are no such vertices, then we shall speak of an *atom without stars*.

Definition 3.6. Now a 2-*atom* is defined to be an oriented atom (P^2, K) with or without stars. Note that the previous atoms (see Definition 2.6) are identified with 2-atoms without stars. In what follows, to simplify the terminology, when speaking of atoms, we shall mean that atoms can be both with and without stars.

Consider a 3-atom $U(L)$ with the structure of a Seifert fibration on it. Let

$$\pi : U(L) \to P^2$$

denote its projection onto a two-dimensional base P^2 with the embedded graph $K = \pi(L)$. Let us mark those points on the base P^2, into which the singular fibers of the Seifert fibration (i.e., the fibers of type $(2,1)$) are projected. Recall that the base P^2 has a canonical orientation. The point is that an orientation is already fixed on $U(L)$, as well as on the fibers of the Seifert fibration. It is clear that, as a result, we obtain some oriented 2-atom (P^2, K).

Theorem 3.4 [65].

a) *Under the projection* $\pi : U(L) \to P^2$, *the 3-atom* $U(L)$ *turns into the 2-atom* (P^2, K), *and moreover, the singular fibers of the Seifert fibration on the 3-atom are in one-to-one correspondence with the star-vertices of the 2-atom.*

b) *This correspondence between 3-atoms and 2-atoms is a bijection.*

Proof. The first statement is evident. To prove the second one, we shall construct the inverse mapping which assigns a certain 3-atom to every 2-atom. Take a 2-atom (P^2, K) and construct a Morse function f on P^2 such that its single critical level coincides with K. Such a function is defined uniquely up to fiber equivalence. Then P^2 is naturally foliated into level lines of f. Theorem 3.1 implies that it is possible to reconstruct (uniquely up to fiber equivalence) the 3-manifold $U(L)$ with the structure of a Seifert fibration over the base P^2 (with fixed star-vertices that correspond to singular fibers). To obtain a 3-atom, we need to define the structure of a Liouville foliation on $U(L)$. Let us use the function f which is already defined on the base P^2. Taking the pull-back of f under the projection $\pi : U(L) \to P^2$, we obtain the function $\tilde{f} = f \circ \pi$ on $U(L)$, whose regular level surfaces are diffeomorphic to the torus. Clearly, $\tilde{f}$ is a Bott function on $U(L)$. Its critical circles are exactly the preimages of the vertices of K including the star-vertices. This reconstruction process is unambiguous, since changing the function f on the base P^2 leads to a Liouville foliation on $U(L)$ which is fiber equivalent to the initial one.

Having constructed the foliation of $U(L)$ into 2-tori, we now have to represent it as the Liouville foliation related to some integrable Hamiltonian system on a suitable symplectic manifold. To this end, we need to introduce the symplectic structure on the direct product $V(L) = U(L) \times I$ for which our foliation on $U(L)$ is Lagrangian. It can actually be done, as we shall see below when proving the general realization theorem (Theorem 4.2).

Thus, we have constructed the correspondence $(P^2, K) \to (U(L), L)$ between 2-atoms and 3-atoms which, as is easy to see, is inverse to the correspondence established by the projection π. This completes the proof. $\square$

We now summarize the results obtained.

1. *3-atom A*. Topologically, this 3-atom is presented as a solid torus foliated into concentric tori, shrinking into the axis of the solid torus. In other words, the 3-atom A is the direct product of a circle and a disc foliated into concentric circles (see Fig. 3.20). From the viewpoint of the corresponding dynamical system, A is a neighborhood of a stable periodic orbit.

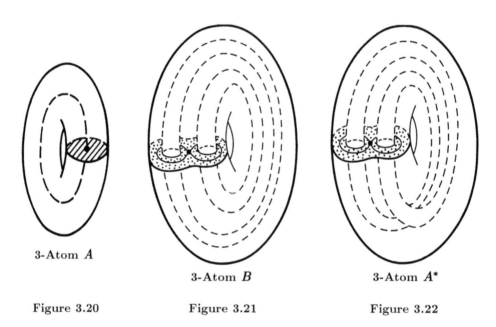

3-Atom A

3-Atom B 3-Atom A^*

Figure 3.20   **Figure 3.21**   **Figure 3.22**

2. *Saddle 3-atoms without stars*. Consider an arbitrary 2-atom without stars, i.e., a two-dimensional oriented compact surface P with a Morse function $f: P \to \mathbb{R}$ having just one critical value. The corresponding 3-atom is the direct product $U = P \times S^1$. The Liouville foliation on it is determined by the function f extended onto Q in the natural way:

$$f(x, \varphi) = f(x), \qquad x \in P, \quad \varphi \in S^1.$$

An example is shown in Fig. 3.21; this is the simple 3-atom B.

3. *Saddle 3-atoms with stars*. As in the previous case, first we consider a 2-surface $\widehat{P}$ with a Morse function $\widehat{f}$ on it. Assume that, on the surface, there is an involution, i.e., a smooth mapping $\tau: \widehat{P} \to \widehat{P}$ with the following properties:

1) $\tau^2 = \mathrm{id}$;
2) τ preserves the function $\widehat{f}$, i.e., $\widehat{f}(\tau(x)) = \widehat{f}(x)$ for any $x \in \widehat{P}$;
3) τ preserves the orientation;
4) the fixed points of τ are some of the critical points of $\widehat{f}$.

To construct a 3-atom, consider the cylinder $P \times [0, 2\pi]$ and glue its feet by the involution τ identifying the points $(x, 2\pi)$ and $(\tau(x), 0)$. As a result, we obtain an orientable 3-manifold U with boundary. The function $\widehat{f}$ is naturally extended to U, since $\widehat{f}(\tau(x)) = \widehat{f}(x)$ and its level surfaces determine the structure of the Liouville foliation on U with only one critical leaf. Note that, from the topological viewpoint, the manifold U is a fiber bundle over a circle with fibers homeomorphic to $\widehat{P}$.

An example is presented in Fig. 3.22; this is the simple 3-atom A^*.

Note that the corresponding 2-atom (P, K) (where $K = \{f = c\}$) is obtained from the pair $(\widehat{P}, \widehat{K})$ by factorization with respect to the involution τ. Here $\widehat{K} = \{\widehat{f} = c\}$.

Definition 3.7. The pair $(\widehat{P}, \widehat{K})$ is called the *double* of the 2-atom (P, K) with stars.

It is clear that the double $(\widehat{P}, \widehat{K})$ is a two-sheeted branching covering over the 2-atom (P, K). Moreover, its branching points are exactly the star-vertices of the atom (P, K).

REMARK. It should be taken into account that the same 2-atom can have several different doubles (i.e., non-homeomorphic to each other). Therefore, different doubles $(\widehat{P}_1, \widehat{K}_1)$ and $(\widehat{P}_2, \widehat{K}_2)$ can generate the same 3-atom with stars. Let us give one of the simplest examples. Consider the 2-atoms C_1 and C_2 shown in Fig. 3.23 (see Table 2.1). On each of them, there is a natural involution, which is

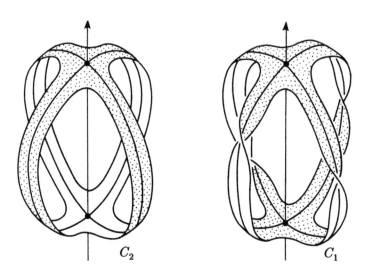

Figure 3.23

the symmetry with respect to the axis passing through the vertices of the atoms. Using the above construction for each of these atoms, we obtain two isomorphic 3-atoms of type A^{**} (see Table 3.1 below). In particular, in this example, two atoms C_1 and C_2 appear to be two different doubles of the same 2-atom A^{**}.

3.6. CLASSIFICATION OF 3-ATOMS

By virtue of Theorem 3.4, the classification of 3-atoms is reduced to the classification of 2-atoms (with and without stars). The classification of 2-atoms without stars has been obtained above (see Table 2.1 in Chapter 2). To obtain an analogous classification table for 2-atoms with stars, we should proceed as follows. We put one or several star-vertices on edges of the graphs K of atoms without stars and add one more series of 2-atoms represented as an annulus with the axial circle on which we put an arbitrary number of star-vertices (Fig. 3.19). The beginning of the list is shown in Table 3.1 (see [65], [123]). Recall that 2-atoms are presented here as immersions of their skeletons into the 2-sphere. Mirror symmetric atoms are considered as different ones. Note that the first examples of non-mirror atoms occur in complexity 3. These are two atoms D_{22}^* and D_{23}^*. By the way, they turn into each other under the mirror reflection.

Table 3.1. 3-Atoms of complexity $1,2,3$

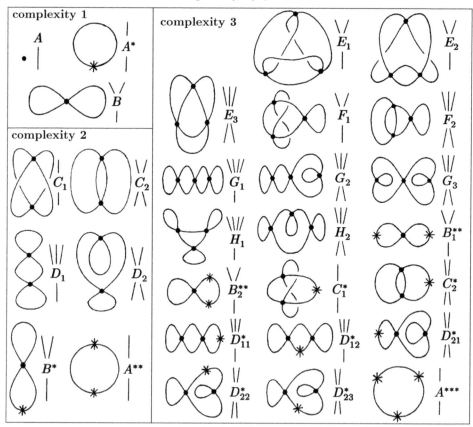

REMARK. In specific integrable systems from classical mechanics and mathematical physics, the most prevalent atoms are A, B, A^*, C_2, D_1 (see examples in Chapters 12–16).

3.7. 3-ATOMS AS BIFURCATIONS OF LIOUVILLE TORI

Consider an integrable Hamiltonian system $v = \operatorname{sgrad} H$ on M^4 and the momentum mapping $\mathcal{F}: M^4 \to \mathbb{R}^2$. Let Σ be the bifurcation diagram, and let $y \in R^2$ be an arbitrary regular point in the image of the momentum mapping. If the inverse image of y is compact, then it consists of one or several Liouville tori. By moving the point y on the plane, we make these tori to move somehow inside M^4. As long as y remains regular, the Liouville tori are transformed by means of some diffeomorphism. But at the instant when y meets the bifurcation diagram Σ and intersects it, the Liouville tori undergo, generally speaking, a non-trivial bifurcation. It is a natural question how to describe typical bifurcations of Liouville tori. The typical situation is that the bifurcation diagram consists of piecewise-smooth curves and the moving point y intersects one of them transversely at an interior point y^* (Fig. 3.24). Suppose the inverse image of the path γ from y to y' is compact in M^4. Then the inverse image of every point of γ (except for y^*) consists of a certain number of Liouville tori.

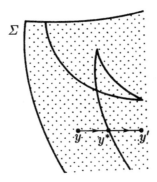

Figure 3.24

Assume that a curve of Σ that contains y^* corresponds to non-degenerate only critical points at which the rank of the momentum mapping is equal to 1. Let such bifurcations of Liouville tori be *non-degenerate*. They are stable in the sense that, under a small perturbation of the path $\gamma = yy'$, the type of the bifurcation does not change.

All such bifurcations of Liouville tori are described by means of 3-atoms.

Namely, there exists a one-to-one correspondence between the set of 3-atoms and that of non-degenerate bifurcations of Liouville tori.

Let us comment on this statement. Consider a connected component of the inverse image of the path yy' in M^4. It is a smooth three-dimensional manifold with boundary, which can be thought as the 3-atom $U(L)$. As for the function f on it, we have to take the parameter t on the path yy' (which defines the motion of a point along the path). The singular leaf L is the inverse image of y^* under the momentum mapping. Then the non-degeneracy of the bifurcation of Liouville tori is equivalent to the fact that f is a Bott function on $U(L)$.

3.8. THE MOLECULE OF AN INTEGRABLE SYSTEM

Let Q^3 be a compact isoenergy surface of an integrable system $v = \operatorname{sgrad} H$, and let f be an additional integral independent of H.

In what follows, we shall consider integrable systems that satisfy the following natural conditions:

1) the isoenergy surface Q is compact and regular;
2) the system $v = \operatorname{sgrad} H$ is non-resonant on Q;
3) the system v possesses a Bott integral f on Q;
4) the system v is topologically stable on Q.

In addition to these conditions, we assume for a while that the integral f has no critical Klein bottles and consider critical tori (if they exist) as regular leaves. The construction that includes critical Klein bottles will be described in the next chapter.

Now the notion of a molecule can be introduced in two different ways.

The first way. Consider the Liouville foliation on Q. First, we construct an analog of the Reeb graph. To this end, we take the base of the Liouville foliation, i.e., the quotient space Q/ρ, where ρ is the natural equivalence relation which identifies the points that belong to the same leaf of the Liouville foliation. We obtain some graph. Obviously, its vertices correspond to the singular leaves of the Liouville foliation, and its edges represent one-parameter families of Liouville tori (without bifurcations).

It is easily seen that each vertex of the graph corresponds to a certain 3-atom, i.e., to a certain bifurcation of Liouville tori. Therefore, we can endow each vertex with the symbol corresponding to this 3-atom. It is convenient to use the symbolic notation from Table 3.1 (see above). We have found that there exists a natural one-to-one correspondence between 3-atoms and 2-atoms. That is why, instead of 3-atoms, we assign the corresponding 2-atoms to the vertices of the graph. Recall that a 3-atom is just the type of the Liouville foliation on the oriented 3-manifold that is a neighborhood of a singular leaf. As a result, we obtain a graph whose vertices are 2-atoms. Every such 2-atom is considered to be a two-dimensional surface with boundary whose boundary circles are in a natural one-to-one correspondence with the edges of the graph adjacent to the vertex. We suppose that such a correspondence between the edges and boundary circles is fixed for every vertex of the graph.

Definition 3.8. The graph W obtained is called the *molecule of the integrable system* on the given isoenergy 3-surface Q.

From the formal viewpoint, the notion of the molecule just introduced coincides with that from Chapter 2. The only difference is that now its atoms can have star-vertices. The notion of coincidence for two molecules is also similar to that in Definition 2.22.

The molecule W of an integrable system represents the structure of the Liouville foliation on a given isoenergy 3-surface Q. The molecule W is obviously an invariant of a system from the viewpoint of the Liouville equivalence. In other words, the molecules of two Liouville equivalent integrable systems coincide.

Note that the molecule W does not depend on the choice of an integral f.

The second way (to define the molecule). Consider an integral f on Q and all singular leaves L_i of the Liouville foliation, i.e., connected components of singular levels of f. By cutting Q along regular leaves, i.e., along Liouville tori, one can divide Q into three-dimensional pieces $U(L_i)$ each of which contains exactly one singular leaf L_i (Fig. 3.25).

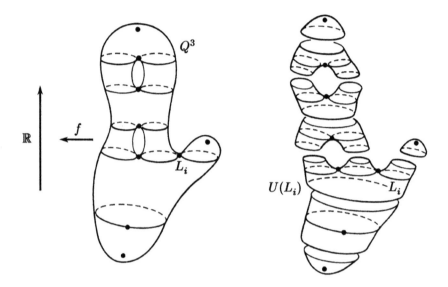

Figure 3.25

It is clear that every 3-manifold $U(L_i)$ is a regular neighborhood of the singular leaf L_i. Therefore, every $U(L_i)$ has the structure of a certain 3-atom. We construct a graph whose vertices are these 3-atoms and whose edges correspond to those Liouville tori along which we cut Q. Let us notice that each 3-atom has ends; and we just connect those of them which correspond to the pairs of glued Liouville tori.

Thus, the molecule W describes the decomposition of Q into a union of Seifert components (i.e., 3-atoms). In other words, having known the molecule, we know from which Seifert manifolds the given 3-manifold Q is glued, and in what order the boundary tori of these Seifert pieces should be glued. Although it is impossible in general to reconstruct the topology of Q by means of the molecule W, it carries the most essential part of information about the Liouville foliation on Q. This means that the molecule W describes the type of the Liouville foliation up to the rough Liouville equivalence.

Recall that two systems v on Q and v' on Q' are called roughly Liouville equivalent if one Liouville foliation can be obtained from the other by a sequence of twisting operations (see Chapter 1). Recall that the twisting operation means the cut of Q along a Liouville torus and gluing back by using another diffeomorphism of boundary tori. In addition, we require that this diffeomorphism preserves the orientation.

Theorem 3.5 (see [65], [123], [135]). *Let (v, Q) and (v', Q') be two integrable system, and let W and W' be the corresponding molecules. Then the systems v and v' are roughly Liouville equivalent (with preserving orientation) if and only if the molecules W and W' coincide.*

Proof. This fact follows immediately from the definition of W. Indeed, the coincidence of the molecules means that the Liouville foliations on Q and Q' are glued from the same components. The only difference is that the boundary tori of these components may be glued by means of different diffeomorphisms. But such a difference can be avoided by suitable twisting operations.   □

3.9. COMPLEXITY OF INTEGRABLE SYSTEMS

Let, as before, v be an integrable Hamiltonian system on an isoenergy surface Q, and let $f: Q \to \mathbb{R}$ be its Bott integral. By m we denote the total number of critical circles of f. Then we remove all singular leaves L_i from Q. As a result, Q splits into a disjoint union of one-parameter Liouville tori. Let n denote the number of such families.

Definition 3.9. The pair of integers (m, n) is called the *complexity* of the integrable system v on Q.

The same complexity can be calculated starting from the molecule W corresponding to (v, Q). It is clear that m is the molecular weight of W, i.e., the sum of atomic weights of its atoms. And n is simply the number of the edges of W.

Clearly, the complexity (m, n) is invariant in the sense of rough Liouville equivalence.

Theorem 3.6. *The number of different molecules of a fixed complexity is finite.*

Proof. This statement follows immediately from the fact that the number of atoms with fixed atomic weight is finite. Therefore, if the complexity of a system is fixed, then there is only a finite number of atoms from which its molecule can be glued. But, for a finite number of atoms, there is only a finite number of possibilities for gluing.   □

Theorem 3.6 allows us to obtain the complete list of all possible molecules, i.e., to enumerate algorithmically all the molecules according to their complexity. To this end, it suffices to take the complete list of atoms (obtained above) and to start to link their ends so that no free edges remain. Of course, here we use the realization theorem (not yet proved) which states that any abstract molecule composed of arbitrary atoms is actually admissible, i.e., can be realized as the molecule of a certain integrable system. We shall prove this theorem later, and here we only refer to it without discussing the question of realizability. In other words, we may so far consider the molecules as abstract objects composed from atoms. From this point of view, we can list them, compare one with another, etc.

Let $\lambda(m, n)$ denote the number of all abstract molecules of complexity (m, n). By virtue of Theorem 3.6, this number is always finite. In Table 3.2 we show the values of the function $\lambda(m, n)$ for $m \leq 4$. This result was obtained by S. V. Matveev by computer analysis. As is seen from this experimental table, the function $\lambda(m,n)$ vanishes for sufficiently large n (at least if $m \leq 4$). This reflects the following general fact.

Theorem 3.7. *Let* $\Lambda(m) = \max\limits_{\lambda(m,n)\neq0} n$. *Then*

$$\Lambda(m) = [3m/2].$$

Proof. Suppose that a molecule of complexity (m,n) consists of k atoms (P_i, K_i) with atomic weight m_i and valency n_i. Recall that the valency of an atom (P_i, K_i) is the number of its ends, i.e., boundary circles of P_i. Let w_i and v_i denote the number of vertices of the graph K_i of degree 4 and 2 respectively. Recall that the vertices of degree 2 are star-vertices. Let χ_i be the Euler characteristic of the closed surface $\tilde{P}_i$ obtained from the initial surface P_i by gluing discs to each of its boundary circles. It is easy to see that

$$n_i = \chi_i + w_i.$$

Summing these equalities over i and taking into account that

$$\chi_i \leq 2, \qquad m_i = v_i + w_i, \qquad \sum_{i=1}^{k} m_i = m, \qquad m_i \geq 1,$$

we obtain

$$n = \frac{1}{2}\sum_{i=1}^{k} n_i \leq \sum_{i=1}^{k}(2 + m_i) = k + \frac{m}{2} \leq \frac{3m}{2}.$$

Thus, $\Lambda(m) \leq [3m/2]$.

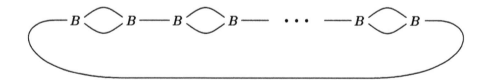

Figure 3.26

The existence of a molecule W of complexity $(m, [3m/2])$ for even m is proved by its explicit construction in Fig. 3.26. For odd m, we should only insert the atom A^* into one of the edges of this molecule. This completes the proof. □

In Table 3.3 we collect all the molecules of low complexity. It is seen from it that the ends of atoms actually differ. Different connections between the same atoms can lead to different molecules.

Table 3.2. The number of molecules of complexity (m,n) for $m \leq 4$

m \ n	1	2	3	4	5	6	≥ 6
1	1	0	0	0	0	0	0
2	3	10	5	0	0	0	0
3	2	24	24	11	0	0	0
4	8	128	561	530	247	54	0

Table 3.3. Molecules of complexity (m,n) for $m \leq 2$

Chapter 4

Liouville Equivalence of Integrable Systems with Two Degrees of Freedom

Now we make more precise the definition of the Liouville equivalence for integrable Hamiltonian systems. From now on, we shall assume that two Liouville foliations are Liouville equivalent if and only if there exists a diffeomorphism that sends the leaves of the first foliation to those of the second one and satisfies two conditions related to the orientation. Namely, it preserves the orientation on 3-manifolds Q and Q', and moreover, it also preserves the orientation on the critical circles given by the Hamiltonian flows.

4.1. ADMISSIBLE COORDINATE SYSTEMS ON THE BOUNDARY OF A 3-ATOM

The molecule W contains a lot of essential information on the structure of the Liouville foliation on Q^3. However, this information is not quite complete. Indeed, the molecule of the form $A-A$, for example, informs us that the manifold Q^3 is glued from two solid tori foliated into concentric tori in a natural way. However, it does not tell us how this gluing is made, and what three-dimensional manifold is obtained as a result. Therefore, we have to add some additional information to the molecule W, namely, the rules that clarify how to glue the isoenergy surface Q^3 from individual 3-atoms.

To this end, cut every edge of the molecule in the middle. The molecule will be divided into individual atoms. From the point of view of the manifold Q^3, this operation means that we cut it along some Liouville tori into 3-atoms. Imagine that we want to make the backward gluing. The molecule W tells us which pairs of boundary atoms we have to glue together. To realize how exactly they should be glued, for every edge of W, we have to define the gluing matrix C, which determines the isomorphism between the fundamental groups of the two glued tori. To write down this matrix, we have to fix some coordinate systems on the tori. As usual, by a coordinate system on the torus, we mean a pair of independent oriented cycles (λ, μ) that are generators of the fundamental group $\pi_1(T^2) = \mathbb{Z} \oplus \mathbb{Z}$ (or, what is the same in this case, of the one-dimensional homology group). Geometrically, this simply means that the cycles λ and μ are both non-trivial and are intersected transversely at a single point.

Consider now an individual 3-atom and introduce a special coordinate system on its boundary tori, which will be called *admissible*.

Case 1. Let $U(L)$ be a 3-atom of type A, i.e., a solid torus. Then, as the first basis circle λ, we take the meridian of the solid torus, i.e., the contractible cycle. As the second cycle μ, we take an arbitrary cycle that complements λ up to a basis. Note that μ can be considered as a fiber of the Seifert fibration (Fig. 4.1). (Recall that the Seifert fibration structure on the solid torus is not uniquely defined.) The fibers of the Seifert fibration have the natural orientation given by the Hamiltonian vector field. More precisely, only one of these fibers is a trajectory of the Hamiltonian vector field, namely, the critical circle of the additional integral f (i.e., the axis of the solid torus). But its orientation allows us to define uniquely the orientation on the cycle μ.

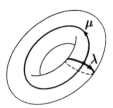

Figure 4.1

Moreover, we have an orientation on the whole 3-atom and, consequently, on its boundary torus. Therefore, we can uniquely define the orientation of the first basis cycle λ by requiring the pair (λ, μ) to be positively oriented on the boundary torus. It is easy to see that these conditions determine λ uniquely, whereas μ is defined up to the following transformations: $\mu' = \mu + k\lambda$, where $k \in \mathbb{Z}$.

Lemma 4.1. *Let (λ, μ) be an admissible coordinate system on the boundary of the atom A (i.e., of the solid torus). In order for another coordinate system (λ', μ') to be admissible it is necessary and sufficient that*

$$\lambda' = \lambda,$$
$$\mu' = \mu + k\lambda, \quad \text{where } k \text{ is an integer number}.$$

Proof. The first basis cycle λ is evidently uniquely defined (up to isotopy), since λ is the disappearing cycle of the solid torus, i.e., shrinks into a point (see Fig. 4.1). The second relation follows from the definition of μ. $\square$

Although admissible coordinate systems are not uniquely defined, all of them are absolutely equivalent because of the following lemma.

Lemma 4.2.

a) *Any two admissible coordinate systems (λ, μ) and (λ', μ') on the boundary of the atom A can be transferred one to another by means of a suitable automorphism of the atom A.*

b) *Conversely, any automorphism of the atom A transfers an admissible coordinate system (λ, μ) to another admissible coordinate system (λ', μ').*

Proof. First we explain that, speaking about an automorphism of the atom A, we mean a diffeomorphism ξ of the solid torus onto itself which preserves

1) the structure of the Liouville foliation,
2) the orientation of the solid torus,
3) the orientation of the axis of the solid torus.

It is easy to see that such an automorphism can be described in the following way. First we cut the solid torus along a meridional disc (Fig. 4.2) and then

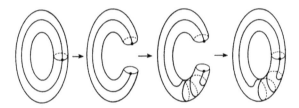

Figure 4.2

identify the two obtained boundary discs again after rotating one of them through the angle $2\pi k$. Such a diffeomorphism is called a *twisting operation* along a given disc (Fig. 4.2). After this observation, our assertion immediately follows from the description of ξ.   □

Case 2. Let $U(L)$ be a saddle 3-atom that has the structure of a trivial S^1-fibration over a surface (i.e., 2-atom) P. Then, as the first basis cycle λ_i on each of boundary tori T_i, we take a fiber of this fibration. Additional cycles μ_i are chosen in the following way. Consider an arbitrary section $P \subset U(L)$. On each boundary tori T_i, it cuts out some cycle μ_i, which we just take as the second basis cycle on T_i (Fig. 4.3). Note that, on every single boundary torus T_i, the cycle μ_i

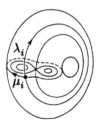

Figure 4.3

can be chosen in an arbitrary way. However, all together they should be connected by the condition of the existence of a global section $P \subset U(L)$ such that $\mu_i = P \cap T_i$. The orientation on the basis cycles is uniquely chosen in the same way as above.

Two different collections of admissible coordinate systems $\{(\lambda_i, \mu_i)\}$ and $\{(\lambda'_i, \mu'_i)\}$ will be connected by the following relations:

$$\begin{cases} \lambda'_i = \lambda_i, \\ \mu'_i = \mu_i + k_i \lambda_i, \end{cases}$$

where $\sum k_i = 0$ (see Lemma 4.3 below). The ambiguity of admissible coordinate systems in this case is explained by the fact that the section $P \subset U(L)$ can be defined in several essentially different ways.

Case 3.   Finally, consider the last case, where the 3-atom $U(L)$ contains saddle critical circles with non-orientable separatrix diagrams and has, therefore, a non-trivial Seifert fibration structure. In this case, the first basis cycle λ_i is defined to be a fiber of the Seifert fibration as before. After that, we would like to proceed similarly, but, unfortunately, this fibration does not have any global section (such that every fiber intersects it only once). However, we can construct such a section after removing small neighborhoods of singular fibers. We only need to fix this section near each critical fiber in some natural way. It turns out that it is in fact possible. Consider a tubular neighborhood of a singular fiber that is homeomorphic to a solid torus. On its boundary torus, there are two uniquely defined cycles: one of them is the fiber λ of the Seifert fibration, and the other is the meridian $\varkappa$ of this solid torus. Let $\varkappa$ be oriented in such a way that these cycles together form a positively oriented pair $(\lambda, \varkappa)$, which is not, however, a basis, since

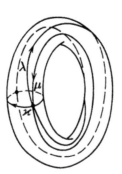

Figure 4.4

the cycles have two points of intersections. Now, having the pair of oriented cycles, we can define one more cycle μ that complements λ up to a basis by the following relation (Fig. 4.4):

$$\lambda = \varkappa - 2\mu.$$

Return to the 3-atom $U(L)$ as a whole.   Remove tubular neighborhoods of the singular fibers from it.   As a result, we obtain some new manifold with the structure of a trivial S^1-fibration whose boundary contains several new tori. On each of these tori, we have a uniquely defined cycle μ. Consider the sections of this trivial fibration that pass through the cycles μ. By this condition, we fix the section P near each singular fiber. Such sections are called *admissible*. Now, as in the case 2, the basis cycles μ_i on the boundary tori T_i are defined as $\mu_i = P \cap T_i$.

In what follows, we shall need another description of admissible coordinate systems on boundary tori of 3-atoms in the case 3. The idea is the following. It turns out that the cycles of an admissible coordinate system can be naturally constructed by using the double $\widehat{P}$ of the base P of the Seifert fibration corresponding to a given atom (see Definition 3.7). We shall use the fact that the Seifert fibration in the third case possesses a "doubled" section, i.e., it is possible to embed $\widehat{P}$ into $U(L)$ in such a way that each regular fiber of the Seifert fibration intersects $\widehat{P}$ exactly at two points, while singular fibers intersect it only once. Such an embedding defines a natural involution $\tau: \widehat{P} \to \widehat{P}$ such that the base P of the Seifert fibration is the quotient space $P = \widehat{P}/\tau$ (see Fig. 4.5(a,b)). In real examples, such a section can often be constructed explicitly.

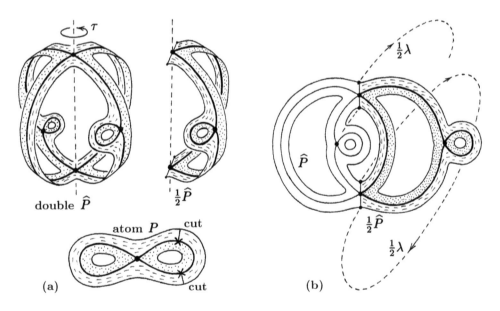

Figure 4.5

Consider an embedded double $\widehat{P} \subset U(L)$ and its boundary $\partial \widehat{P} = \widehat{P} \cap \partial U(L)$. Let $\widehat{\mu}_i = \widehat{P} \cap T_i$ be the part of the boundary lying on the torus $T_i \subset \partial U(L)$.

Two cases are possible. The first possibility is that $\widehat{\mu}_i$ represents a union of two separate cycles each of which intersects with the fiber λ_i of the Seifert fibration at one point and, consequently, is a section of the Seifert fibration on the boundary torus T_i. In the second case, $\widehat{\mu}_i$ is a connected cycle which has intersection index 2 with the fiber λ.

Let us try to construct the desired cycles μ_i of an admissible coordinate system from the cycles $\widehat{\mu}_i$. We can proceed, for example, as follows. In the first case, as the cycle μ_i we may take one of two connected components of $\widehat{\mu}_i$. In the second case, we set $\mu_i = \dfrac{1}{2}(\widehat{\mu}_i + \lambda_i)$. Locally, the constructed cycles μ_i will satisfy all required properties, i.e., will be sections of the Seifert fibration on each boundary torus T_i. However, as a whole, this construction may differ from the original method for constructing an admissible coordinate system. In order for the definitions of cycles μ_i to be equivalent, one of these cycles should be corrected by adding to it a cycle of the form $k\lambda$, $k \in \mathbb{Z}$. Here k must be chosen so that the following relation holds:

$$\sum_i \mu_i = \frac{1}{2}\left(\sum_i \widehat{\mu}_i + s\lambda\right) = \frac{\partial\widehat{P} + s\lambda}{2},$$

where s denotes the number of critical circles in $U(L)$ with non-orientable separatrix diagrams.

COMMENT. This relation has a natural homological meaning. To explain it, we consider the following construction. Remove small neighborhoods of singular fibers (i.e., critical circles with non-orientable separatrix diagrams) and consider two embedded surfaces $\dot{P}$ and $\widehat{P}$ in $\dot{U}(L)$, where $\dot{P}$ is a real section of the Seifert fibration which was used in the first definition of μ_i, and $\widehat{P}$ is the embedded double. Recall that $\mu_i = \dot{P} \cap T_i$ and $\widehat{\mu}_i = \widehat{P} \cap T_i$. The behavior of the surfaces $\dot{P}$ and $\widehat{P}$ near a singular fiber is described by the relation $\lambda = \varkappa - 2\mu$ (see above), where μ is the cycle defined by $\mu = \dot{P} \cap T$, and analogously, $\varkappa = \widehat{P} \cap T$ (here T denotes the torus around the singular fiber, i.e., the boundary of a tubular neighborhood of this fiber). Taking into account this equality, it is easy to see that the relation under consideration is equivalent to the following:

$$\partial\dot{P} = \frac{1}{2}\partial\widehat{P}.$$

From the topological point of view this relation means, in particular, that the intersection index of $\partial\dot{P}$ and $\partial\widehat{P}$ on the boundary $\partial\dot{U}(L)$ is zero. This condition must be, of course, satisfied (see Fig. 4.6 and the comment to it).

We now turn to the following question. How admissible coordinate systems are connected with each other? The following lemma gives an answer for the second and third cases.

Lemma 4.3. Let (λ_i, μ_i) be an admissible coordinate system on the boundary tori T_i of a saddle 3-atom. In order for another coordinate system (λ_i', μ_i') to be admissible, it is necessary and sufficient that for any i the following relations hold:

$$\lambda_i' = \lambda_i, \qquad \mu_i' = \mu_i + k_i\lambda_i, \qquad \text{where} \quad \sum_i k_i = 0. \tag{1}$$

Proof. Let us be given two admissible coordinate systems. First we prove that they satisfy the above relations. The first equality is evident, since λ_i is the uniquely defined fiber of the Seifert fibration on $U(L)$. The relation $\mu_i' = \mu_i + k_i\lambda_i$ is also evident, and we only need to prove the equality $\sum_i k_i = 0$.

Consider two admissible sections P and P' of the given 3-atom. On the boundary of the 3-atom, these sections cut out two collections of cycles $\{\mu_i\}$ and $\{\mu'_i\}$, which complement the cycles $\{\lambda_i\}$ and $\{\lambda'_i\}$ respectively up to an admissible coordinate system on the boundary tori. The integer numbers k_i can now be interpreted as the intersection indexes of the cycles μ_i and μ'_i, and their sum $\sum\limits_i k_i$ as the total intersection index of the boundaries of two sections P and P'. Recall that, in 3-atoms related to the case 3 (atoms with stars in our terms), we removed tubular neighborhoods of singular fibers. On the boundary tori of such neighborhoods, the sections P and P' coincide up to isotopy. This means that such tori give no contribution to the total intersection index, and we can forget about them.

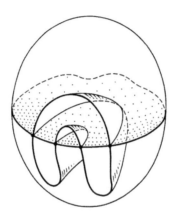

Figure 4.6

Now recall the following general fact from 3-topology. Let us be given two oriented surfaces P and P' lying inside an oriented 3-manifold U with boundary ∂U in such a way that ∂P and $\partial P'$ are embedded into ∂U as two smooth curves. Then the intersection index of these curves ∂P and $\partial P'$ is equal to zero (see Fig. 4.6).

Using this assertion in our case, we immediately obtain that $\sum\limits_i k_i = 0$, as required.

We now prove the converse statement. Let us be given an admissible coordinate system (λ_i, μ_i) and another system of cycles (λ'_i, μ'_i) defined by (1). We must prove that these cycles form an admissible coordinate system. It suffices to verify that there exists an admissible section P' such that $\{\mu'_i\} = P' \cap \partial U(L)$. We shall construct the desired section P' from the initial section P by means of several sequential steps using twisting operations.

Take two different boundary tori T_i and T_j and connect them inside the 3-atom by a simple curve a lying entirely on the section P (Fig. 4.7(a)). Let

$$\pi : U(L) \to P$$

be the projection of the Seifert fibration. Consider the preimage $\pi^{-1}(a)$ of the curve a. This is an annulus in $U(L)$ intersecting the section P along the curve a. Let us cut $U(L)$ along the annulus $\pi^{-1}(a)$ and twist one of the coasts by the angle 2π. After this we glue this cut back as shown in Fig. 4.7(b). As a result, we shall obtain a fiber diffeomorphism of $U(L)$ onto itself which maps the admissible section P to another admissible section P' (see Fig. 4.7(c)).

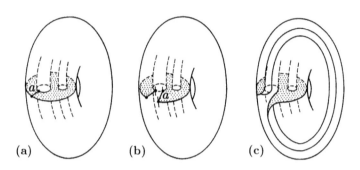

Figure 4.7

For these two sections the number k_i is equal to 1, whereas k_j is equal to -1, the others k_s are all zeros. It is clear that, by using such operations, we can realize any collection of numbers $\{k_i\}$ provided $\sum k_i = 0$. This completes the proof. $\quad\square$

Let us emphasize that all admissible coordinate systems on the boundary tori are absolutely equivalent; and conversely, all the other (i.e., not admissible) coordinate systems are not equivalent to them. Speaking of the equivalence of coordinate systems, we mean the existence of a diffeomorphism of the 3-atom onto itself that maps one coordinate system into the other and, in addition, preserves the Liouville foliation structure, the orientation of the atom itself, and the orientation of its critical circles. Such a diffeomorphism can be naturally regarded as an *automorphism* of the 3-atom.

Lemma 4.4.

a) *Any two admissible coordinate systems* $\{(\lambda_i, \mu_i)\}$ *and* $\{(\lambda'_i, \mu'_i)\}$, *where* $i = 1, 2, \ldots$, *on the boundary of a saddle atom* $U(L)$ *can be transferred one to another by means of a suitable automorphism of the atom* $U(L)$.

b) *Conversely, any automorphism of the atom* $U(L)$ *maps an admissible coordinate system* $\{(\lambda_i, \mu_i)\}$ *to another admissible coordinate system* $\{(\lambda'_i, \mu'_i)\}$.

Proof. This assertion easily follows from the fact that any automorphism of a saddle atom is generated by twisting operations described above. In other words, an automorphism is uniquely defined (up to isotopy) by the image of a certain admissible section P. $\quad\square$

4.2. GLUING MATRICES AND SUPERFLUOUS FRAMES

Thus, we have defined admissible coordinate systems on the boundary tori of every atom. Consider now an arbitrary edge e_i of the molecule W and fix a certain orientation on it (for example, according to the increase of f).

We cut this edge along a Liouville torus and define admissible coordinate systems on the two boundary tori obtained, which are denoted now by (λ_i^-, μ_i^-) and (λ_i^+, μ_i^+) (the sign "$-$" corresponds to the beginning of the edge, and "$+$" corresponds to its end). Considering these pairs of cycles as bases in the one-dimensional homology group, we obtain a natural gluing matrix

$$C_i = \begin{pmatrix} \alpha_i & \beta_i \\ \gamma_i & \delta_i \end{pmatrix}$$

such that

$$\begin{pmatrix} \lambda_i^+ \\ \mu_i^+ \end{pmatrix} = \begin{pmatrix} \alpha_i & \beta_i \\ \gamma_i & \delta_i \end{pmatrix} \begin{pmatrix} \lambda_i^- \\ \mu_i^- \end{pmatrix}.$$

It is clear that C_i is an integer-valued matrix whose determinant is equal to -1. In all other respects, this matrix can be absolutely arbitrary.

It is easy to see that, assigning all these matrices, we uniquely define the topology of the whole Liouville foliation. However, these matrices are not uniquely defined themselves, since we can change admissible coordinate systems. That is why we introduce the following important notion.

Definition 4.1. The set of all gluing matrices $\{C_i\}$ is called the *superfluous frame* of the molecule W.

Consider a transformation of admissible coordinate systems. It is easy to see that all such transformations form a group $\mathbb{G}$, which naturally acts on the set of superfluous frames of the given molecule.

Definition 4.2. Two superfluous frames $\{C_i\}$ and $\{C_i'\}$ of the molecule W are called *equivalent* if one of them can be obtained from the other by changing admissible coordinate systems on the atoms of the molecule.

Proposition 4.1. *Two integrable system v on Q and v' on Q' are Liouville equivalent if and only if the following two conditions hold:*

1) their molecules W and W' coincide, i.e., the systems are roughly Liouville equivalent;

2) the corresponding superfluous frames of the molecule $W = W'$ are equivalent.

COMMENT. This proposition can also be reformulated as follows. There exists a natural one-to-one correspondence between the Liouville equivalence classes of integrable systems and the equivalence classes of superfluous frames of molecules.

Proof. In one direction, this statement is evident: if two systems are Liouville equivalent, then their molecules coincide, and their superfluous frames are equivalent. Let us prove the converse.

Suppose two systems have equivalent superfluous frames on the same molecules. Then, by choosing suitable admissible coordinate transformations, one can achieve the coincidence of all gluing matrices for these two systems. The coincidence of the molecules themselves guarantees that the two Liouville foliations are glued from the same pieces, i.e., from the same 3-atoms. The coincidence of the gluing matrices means that these pieces are glued in the same way. This evidently yields the same result: we obtain the same foliation on the same three-dimensional manifold. □

Thus, the Liouville classification of integrable systems on isoenergy 3-surfaces is reduced to the description of the invariants of the action of $\mathbb{G}$ on the set of superfluous frames, where $\mathbb{G}$ is the group of admissible coordinate system transformations. This problem is in fact algebraic. We now describe a complete system of invariants for this action.

4.3. INVARIANTS (NUMERICAL MARKS) r, ε, AND n

Without giving the complete proof yet, we now formulate the final result, namely, we describe the complete list of invariants of the action of $\mathbb{G}$. The invariants in question are some numerical marks, which are explicitly calculated from the matrices C_i and have the property that, using these marks, we can uniquely reconstruct all the gluing matrices up to an admissible coordinate system transformation (i.e., up to equivalence).

4.3.1. Marks r_i and ε_i

To the matrix C_i, we assign two following numerical marks.

Definition 4.3. The mark r_i on the edge e_i of the molecule W is

$$
r_i = \begin{cases} \dfrac{\alpha_i}{\beta_i} \bmod 1 \in \mathbb{Q}/\mathbb{Z} & \text{if } \beta_i \neq 0, \\ \text{symbol } \infty & \text{if } \beta_i = 0. \end{cases}
$$

Definition 4.4. The mark ε_i on the edge e_i of the molecule W is

$$
\varepsilon_i = \begin{cases} \operatorname{sign} \beta_i & \text{if } \beta_i \neq 0, \\ \operatorname{sign} \alpha_i & \text{if } \beta_i = 0. \end{cases}
$$

Lemma 4.5. *The marks r_i and ε_i do not change under admissible coordinate system transformations, i.e., are invariants of the action of $\mathbb{G}$ on the set of superfluous frames.*

Proof. It is easy to see that, under admissible coordinate changes, each matrix C_i transforms according to the following rule:

$$
C_i = \begin{pmatrix} \alpha_i & \beta_i \\ \gamma_i & \delta_i \end{pmatrix} \quad \rightarrow \quad C_i' = \begin{pmatrix} 1 & 0 \\ -k_i^+ & 1 \end{pmatrix} \begin{pmatrix} \alpha_i & \beta_i \\ \gamma_i & \delta_i \end{pmatrix} \begin{pmatrix} 1 & 0 \\ k_i^- & 1 \end{pmatrix},
$$

where k_i^+ and k_i^- are some integers. Hence

$$
C_i' = \begin{pmatrix} \alpha_i + k_i^- \beta_i & \beta^i \\ \gamma_i + k_i^- \delta_i - k_i^+ \alpha_i - k_i^+ k_i^- \beta_i & \delta_i - k_i^+ \beta^i \end{pmatrix}.
$$

This explicit formula immediately implies that, under admissible coordinate transformations, the marks r_i and ε_i are not actually changed, as required. $\square$

4.3.2. Marks n_k and Families in a Molecule

The mark n can be introduced in several slightly different ways. Such a situation is well known in the theory of invariants. An invariant can be chosen in different ways, and a concrete choice is usually defined by the specific character of the problem under consideration. As the basic definition of n, we choose that one which is more convenient for the general theory. Later on, we shall give another definition of this "invariant" and indicate the explicit formula which connects these two marks.

First, we need some preliminary construction. An edge of the molecule with mark r_i equal to ∞ is said to be an infinite edge. The other edges are called finite. Let us cut the molecule along all the finite edges. As a result, the molecule splits into several connected pieces.

Definition 4.5. Those pieces which do not contain atoms of type A are said to be *families*. For example, if all the edges of a molecule are finite, then each of its saddle atoms is a family by definition.

Consider a single family $U = U_k$. All its edges can be divided into three classes: incoming, outgoing, and interior (Fig. 4.8).

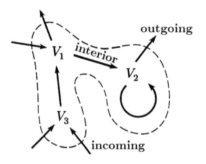

Figure 4.8

To each of these edges e_i, we assign an integer number Θ_i by the following rule:

$$\Theta_i = \begin{cases} [\alpha_i/\beta_i] & \text{if } e_i \text{ is an outgoing edge,} \\ [-\delta_i/\beta_i] & \text{if } e_i \text{ is an incoming edge,} \\ -\gamma_i/\alpha_i & \text{if } e_i \text{ is an interior edge.} \end{cases}$$

Definition 4.6. For every family U_k, we define an integer number n_k by setting

$$n_k = \sum \Theta_i \,,$$

where the sum is taken over all edges of the given family, and k is the number of the family. Note that n_k is an integer number, since, for any interior edge, we always have $\alpha_i = \pm 1$.

Lemma 4.6.   *The number n_k does not change under admissible coordinate system transformations.*

Proof. Roughly speaking, the point is that the numbers Θ_i are chosen in such a way that each of them transforms according to the following simple rule:

$$\Theta_i \to \Theta_i' = \Theta_i + k_i \,.$$

But the sum of k_i's over all the edges both outgoing from and incoming to a certain atom is zero (see above the formulas for admissible coordinate transformations on an atom). This implies that the number n_k is invariant under admissible coordinate transformations.

Let us now verify the assertion of the lemma by straightforward calculation. Using the explicit form for the transformation of the gluing matrix C_i, we obtain

$$\Theta_i' = \begin{cases} \Theta_i + k_i^- & \text{if } e_i \text{ is an outgoing edge,} \\ \Theta_i + k_i^+ & \text{if } e_i \text{ is an incoming edge,} \\ \Theta_i + k_i^- + k_i^+ & \text{if } e_i \text{ is an interior edge.} \end{cases}$$

Taking the sum over all the edges of a given family, we see that the mark n changes by adding the sum of all numbers k_s, where s indexes the boundary tori of all atoms occurring in a given family. Summing all Θ_i', we sum the numbers k_s, which can be arranged into groups each of which corresponds to a certain atom of the family. But, for each atom, the sum of k_s is zero (see Lemma 4.3). Thus, $\sum_i \Theta_i' = \sum_i \Theta_i$, as required.   □

The mark n admits an interesting topological interpretation. In some sense, n describes the obstruction to extending a section inside the family from its boundary. The point is that each family has the natural structure of a Seifert fibration. Indeed, the definition of a family means exactly that we include two neighboring saddle atoms into a family if and only if the Seifert fibrations (which are already defined on each single atom) coincide on the common boundary torus of these two atoms and, consequently, can be extended onto the whole family.

On each boundary torus of the family there is an "exterior" cycle μ that comes from the neighboring exterior (with respect to the given family) atom. Each such cycle μ intersects transversally the fibers of the Seifert fibration and, consequently, defines a "multivalued" section of this fibration on the boundary of the family. Roughly speaking, if we try to extend this "section" inside the family, we shall discover a certain obstruction which is just defined by the integer number n. We shall not discuss this representation of n in detail, since we do not need it in what follows. We only note that, in the particular case when the whole molecule W is a single family whose atoms contain no star-vertices, the global Seifert fibration has the naturally defined Euler class which is given exactly by the integer number n.

4.4. THE MARKED MOLECULE IS A COMPLETE INVARIANT OF LIOUVILLE EQUIVALENCE

Definition 4.7. The molecule W endowed with the marks r_i, ε_i, and n_k is called a *marked molecule*. We denote it by

$$W^* = (W, r_i, \varepsilon_i, n_k).$$

Proposition 4.2. *The collection of marks r, ε, n is a complete set of invariants of the action of $\mathbb{G}$ on superfluous frames (i.e., on the gluing matrices C_i). In other words, two superfluous frames of the molecule W are equivalent if and only if the corresponding collections of marks r, ε, n coincide.*

Proof. Let $\{C_i\}$ and $\{C_i'\}$ be two superfluous frames with identical collections of invariants r, ε, n. Note, first of all, that this collection uniquely defines the decompositions of the molecule W into families, since the marks r define finite and infinite edges of W. Consider a single family Y. Using the superfluous frames $\{C_i\}$ and $\{C_i'\}$, we can define integer numbers Θ_i and Θ_i' on each edge of this family (see the explicit formula before Definition 4.6).

Consider the difference $\Theta_i' - \Theta_i$ on each edge e_i. Since the mark n is the same for the frames under consideration, it follows that $\sum(\Theta_i' - \Theta_i) = 0$. Consider the boundary tori of all atoms from the given family and associate an integer number with each of these tori by the following rule. If a torus corresponds to an exterior edge e_i (incoming or outgoing), then we assign the number $\Theta_i' - \Theta_i$ to it. If e_i is an interior edge, then, on the two boundary tori T_i^- and T_i^+ (corresponding to the beginning and end of the edge), we put numbers k_i^- and k_i^+ in such a way that $k_i^- + k_i^+ = \Theta_i' - \Theta_i$. It can be done in different ways. However, using the condition $\sum(\Theta_i' - \Theta_i) = 0$, we can always obtain the situation when, for each atom from the family, the sum of integer numbers associated with its boundary tori is zero.

Thus, now, on the boundary tori of each atom, we have a collection of integer numbers with the sum equal to zero. This allows us to make an admissible coordinate transformation by the formula from Lemma 4.3. As a result (see the above formulas for transformation of numbers Θ_i), we achieve the equality $\Theta_i' = \Theta_i$ on each edge of the family under consideration.

Let us note another important fact. Coordinate changes on the atoms of a given family have no influence on the numbers Θ_i for other families. Therefore, we may carry out the above procedure for each family separately.

Moreover, the same procedure can be carried out on other pieces of the molecule W obtained by cutting W along all the finite edges. For such pieces, the numbers Θ_i can be defined just by the same rule as for families. In the case of a family, we used the equality $\sum(\Theta_i' - \Theta_i) = 0$, which does not hold any more. However, such an equality can be made in an artificial way. Indeed, the piece of W under consideration, being different from a family, necessarily contains an atom A. Using the change $\lambda' = \lambda$, $\mu' = \mu + k\lambda$ on its boundary torus, we see that the number Θ on the edge adjacent to this atom transforms into $\Theta + k$, while the remaining numbers Θ_i do not change. Clearly, by choosing k in an appropriate way, we can achieve the equality $\sum(\Theta_i' - \Theta_i) = 0$. After this, we just repeat the same argument as before (i.e., in the case of a family).

We assert that, after the above coordinate transformations, all the gluing matrices will coincide. Indeed, we have already obtained the equalities

$$[\alpha_i/\beta_i] = [\alpha_i'/\beta_i'] \quad \text{and} \quad [-\delta_i/\beta_i] = [-\delta_i'/\beta_i']$$

on the finite edges of the molecule. Besides, by our assumption, $(\alpha_i/\beta_i) \bmod 1 = (\alpha_i'/\beta_i') \bmod 1$ and $\mathrm{sign}\,\beta_i = \mathrm{sign}\,\beta_i'$. This evidently implies that $\alpha_i = \alpha_i'$ and $\beta_i = \beta_i'$. Thus, the first rows of the matrices C_i and C_i' coincide. Since $\det C_i' = \det C_i = -1$, their second rows can differ only by adding the first row with some multiplicity. Therefore, the condition $-[\delta_i/\beta_i] = [-\delta_i'/\beta_i']$ implies that the rows actually coincide.

Finally, consider an infinite edge. In view of the coincidence of the invariants r and ε on this edge, we see that all entries of the matrices C_i and C_i' coincide except, perhaps, for γ_i and γ_i'. However, we have the additional condition that $\Theta_i = -\gamma_i/\alpha_i = \Theta_i' = -\gamma_i'/\alpha_i'$, which guarantees the equality $\gamma_i = \gamma_i'$.

Thus, after the coordinate transformations we have made, the superfluous frames coincide. This completes the proof. □

This statement implies the main theorem of this chapter.

Theorem 4.1 [65], [123], [135]. *Two integrable systems (v, Q) and (v', Q') are Liouville equivalent if and only if their marked molecule W^* and $W^{*\prime}$ coincide.*

Proof. The statement of the theorem is a direct consequence of Propositions 4.1 and 4.2. □

4.5. THE INFLUENCE OF THE ORIENTATION

While constructing the marked molecule W^*, we use the orientation of Q^3 as well as the orientation of the critical circles of an integral f and that of the edges of a molecule. If we change some of these orientations, the marked molecule W^* is also changed in general. In this section, we describe the formal rules that show what happens to the marked molecule under the change of orientation.

4.5.1. Change of Orientation of an Edge of a Molecule

Under the change of orientation of an edge of a molecule we obtain the following.

a) The gluing matrix C is replaced by its inverse C^{-1}.

b) In the case of an infinite edge, the marks $r = \infty$ and ε do not change (in particular, the edge remains infinite). In the case of a finite edge, the mark $r = (\alpha/\beta) \bmod 1$ is replaced by $r^* = (\delta/\beta) \bmod 1$, where δ is uniquely defined by the condition $(\alpha\delta - 1) \bmod \beta = 0$ (here we assume, of course, that α and β are relatively prime). The mark ε does not change. Since the gluing matrix changes only on a single edge of the molecule, it does not affect the marks r and ε on the other edges.

c) The invariants n on families of the molecule do not change.

Thus, any change of the orientation on an edge of a molecule causes a change of only one mark r (assigned to this edge). In order to avoid additional difficulties, we assume that two marked molecules obtained from each other by changing the orientation on some edges are equivalent (or just coincide).

4.5.2. Change of Orientation on a 3-Manifold Q

A change in the orientation of Q transforms the molecule W in the following way.

a) Each atom of W is replaced by the mirror symmetric atom. In other words, we have to consider the same atoms but with the opposite orientation.

b) The admissible coordinate systems also change. Namely, for saddle atoms the sign of the second basis cycle μ changes, while for atoms of type A the sign of the first basis cycle λ changes. As a result, on the edges between two saddle atoms and on the edges between two atoms A, the gluing matrix $C = \begin{pmatrix} \alpha & \beta \\ \gamma & \delta \end{pmatrix}$ becomes $C' = \begin{pmatrix} \alpha & -\beta \\ -\gamma & \delta \end{pmatrix}$. On the edges between saddle atoms and between those of type A, the matrix C becomes $C' = \begin{pmatrix} -\alpha & \beta \\ \gamma & -\delta \end{pmatrix}$.

Let us distinguish the following cases.

1) Let an edge connect two atoms of the same type, i.e., either A with A or a saddle atom with a saddle one. Then, in the case of a finite edge (i.e., with $\beta \neq 0$), the marks r and ε change their signs; in the case of an infinite edge (i.e., with $\beta = 0$), the marks r and ε do not change.

2) Let an edge connect atoms of different types, i.e., an atom A with a saddle. Then, in the case of a finite edge, the mark r changes its sign, and ε does not change; in the case of an infinite edge, on the contrary, the mark r does not change, but ε changes sign.

By the way, it follows from this that the decomposition of a molecule into families remains the same.

c) The mark n assigned to a family of a molecule undergoes the following transformation. At first, assume that the family has no atoms with stars. Then, under a change in the orientation of Q, the numbers α_i/β_i, δ_i/β_i, and γ_i/α_i (see the definition of the n-mark) change their signs. Taking into account a simple formula

$$[-x] = \begin{cases} -[x] & \text{if } x \text{ is integer,} \\ -[x] - 1 & \text{if } x \text{ is not integer,} \end{cases}$$

we obtain the transformation formula for n (under a change in the orientation on Q^3):

$$n' = -n - l,$$

where l denotes the number of those exterior edges of the family for which $r \neq 0$.

Recall now that saddle atoms can have star-vertices. In this case, the admissible coordinate systems $\{(\lambda_i, \mu_i)\}$ defined on the boundary tori of the atom are transformed in the following way. Choose some boundary torus; let i_0 be its number. Then the transformation formulas are as follows:

$$\lambda'_i = \lambda_i , \qquad \begin{aligned} \mu'_i &= -\mu_i \quad \text{for } i \neq i_0 , \\ \mu'_{i_0} &= -\mu_{i_0} - s\lambda_{i_0} , \end{aligned}$$

where s is the number of star-vertices in the given atom, i.e., the number of critical circles with a non-orientable separatrix diagrams (if there are no star-vertices, then $s = 0$). This is easily implied by the definition of admissible coordinates on atoms with stars (see above). Thus, the gluing matrices C_i are changed, and as a result, the mark n becomes

$$n' = -n - l - \sum s ,$$

where the sum is taken over all the atoms from the given family.

4.5.3. *Change of Orientation of a Hamiltonian Vector Field*

Under such a change, the flow $v = \operatorname{sgrad} H$ is replaced by $-v = \operatorname{sgrad}(-H)$. Since the orientation of Q is assumed to be the same, all the atoms of the molecule W should be changed by the mirror symmetric atoms. The numerical marks of the molecule are not changed. The point is that every admissible coordinate system (λ, μ) is replaced by $(-\lambda, -\mu)$. As a result, the gluing matrices C remain the same.

4.6. REALIZATION THEOREM

Here we discuss the question whether an abstract molecule W^* can be realized as the molecule of an integrable system. The point is that the marked molecule W^* can be defined in an abstract way as a graph whose vertices are atoms, and the edges and families are endowed with some marks r (rational numbers from 0 to 1, or ∞), marks $\varepsilon = \pm 1$, and n (integer numbers). The question is if there exists an integrable system $v = \operatorname{sgrad} H$ on an appropriate isoenergy 3-manifold Q in a symplectic 4-manifold M such that its marked molecule coincides with W^*. The answer is positive.

Theorem 4.2 [65] (Realization theorem). *Any abstract marked molecule W^* is realized as a marked molecule of some integrable Hamiltonian system.*

Proof.

Step 1. Consider a marked molecule W^* and reconstruct the corresponding 3-manifold Q^3. To this end, take all the 3-atoms that are contained in W^*. Using the collection of marks of W^*, one can reconstruct the superfluous frame of the molecule, i.e., the set of gluing matrices. This can be done uniquely up to the natural equivalence relation introduced above (see Proposition 4.2). Using these gluing matrices, we now glue the 3-manifold Q^3 from the 3-atoms. It is easy to see that, as a result, the natural structure of a Liouville foliation appears on Q^3. Indeed, each 3-atom has such a structure by definition and, moreover, each of its boundary tori is a leaf of the foliation. Therefore, gluing the atoms among themselves, we naturally extend this foliation from single atoms onto the whole of Q^3.

Now consider the 4-manifold $M^4 = Q^3 \times I$, where I is an interval, and extend the foliation from Q^3 onto M^4 in the natural way. Fix a certain orientation on $Q^3 \times I$. To complete the proof, we only need to introduce a symplectic structure on M^4 in such a way that the foliation obtained is Lagrangian (i.e., its leaves are all Lagrangian submanifolds).

Step 2. First, we define a symplectic structure on each "4-atom" $U \times I$, where U is a 3-atom from W^*. We shall assume that these 3-atoms are contained in Q^3 as subsets. As we know, from the topological point of view, the manifold U can be of two types: it is either the direct product $P^2 \times S^1$ or the skew product $P^2 \widetilde{\times} S^1$. In the first case, the structure of the Liouville foliation on $P^2 \times S^1$ is in fact given by a Morse function on the surface P^2. Namely, if x and y are coordinates on P^2, and φ is a periodic coordinate on the circle S^1, then the integral $f(x, y, \varphi)$ does not really depend on φ (and, consequently, can be considered as a Morse function on P). Let t be a coordinate on the interval I. Then the symplectic structure on the "4-atom" $U \times I$ can be defined by the following explicit formula:

$$\Omega = \omega(x, y) + d\varphi \wedge dt,$$

where $\omega(x, y)$ is a symplectic structure on P^2 chosen in such a way that the orientation on $U \times I$ defined by the form Ω coincides with the above fixed orientation on M^4.

In the case of a skew product $P^2 \widetilde{\times} S^1$, the 3-atom U is the result of gluing the two bases of the cylinder $P^2 \times [0, 2\pi]$ by an involution $\tau \colon P^2 \to P^2$ which preserves the function $f(x, y)$ on P^2. In this case, the symplectic structure on $U \times I$ is given by the same formula

$$\Omega = \omega(x, y) + d\varphi \wedge dt,$$

where φ is a coordinate on the segment $[0, 2\pi]$, and t is a coordinate on the interval I. The only difference is that the symplectic structure $\omega(x, y)$ on P^2 must be invariant under the involution τ. Such a form ω always exists. It suffices to take the form $\omega = \alpha + \tau^* \alpha$, where α is any non-degenerate 2-form on P^2. Since ω is τ-invariant, Ω is well-defined on the whole of $U \times I$.

As a result, we have constructed a symplectic structure in a neighborhood of each singular leaf of the Liouville foliation. It remains to extend it onto that part of the 4-manifold M^4 which corresponds to the edges of the molecule W, i.e., onto the families of tori of the form $T^2 \times E \times I$. Here the intervals E correspond to the edges of W, i.e., one-parameter families of Liouville tori. Recall that these families are obtained from Q^3 by removing (the neighborhoods of) the singular leaves of the Liouville foliation.

Step 3. We now need to sew together the symplectic structures constructed on individual 4-atoms $U \times I$. It is clear that we may assume the symplectic structure to be already given on that part of the direct product $T^2 \times E \times I$ which corresponds to neighborhoods of the endpoints of E (Fig. 4.9).

Lemma 4.7 (Sewing lemma).

Consider a two-parameter family $X = T^2 \times [a,b] \times [-\varepsilon, \varepsilon]$ *of 2-tori; and suppose we are given two symplectic structures* Ω_1 *and* Ω_2 *on the subsets* $X_a = T^2 \times [a, a+\delta] \times [-\varepsilon, \varepsilon]$ *and* $X_b = T^2 \times [b-\delta, b] \times [-\varepsilon, \varepsilon]$ *respectively such that the foliation into 2-tori is Lagrangian with respect to these symplectic forms. Suppose the orientations on* $T^2 \times [a,b] \times [-\varepsilon, \varepsilon]$ *canonically defined by the forms* Ω_1 *and* Ω_2 *are the same. Then there exists a symplectic structure* Ω *on the whole family* $T^2 \times [a,b] \times [-\varepsilon, \varepsilon]$ *such that* $\Omega|_{X_a} = \Omega_1$, $\Omega|_{X_b} = \Omega_2$, *and the foliation into 2-tori on* X *is Lagrangian.*

Proof. Let us fix a basis on an individual torus T^2 from the family under consideration and extend this basis (by continuity) to all tori. According to the Liouville theorem, on the subsets X_a and X_b, there exist action-angle variables $s_1, s_2, \varphi_1, \varphi_2$ related to the fixed basis on the tori. Since we define the same basis on all tori, we can smoothly extend the angle coordinates φ_1 and φ_2 to each torus from the family X. We now wish to extend the action functions s_1 and s_2 to the whole family X. Since s_1 and s_2 must be constant on 2-tori, it suffices in fact to define them as functions on the two-dimensional space of parameters, i.e., on the rectangle $[a,b] \times [-\varepsilon, \varepsilon]$. Consequently, the problem is reduced to the following. One needs to extend the mapping given on two narrow strips $[a, a+\delta] \times [-\varepsilon, \varepsilon]$ and $[b-\delta, b] \times [-\varepsilon, \varepsilon]$ up to a mapping given on the whole rectangle $[a,b] \times [-\varepsilon, \varepsilon]$ (see Fig. 4.9). The desired mapping must have no singularities and be an immersion of the rectangle into the (s_1, s_2)-plane (Fig. 4.10). It can be done, since the signs of the Jacobians of the two mappings given on the strips coincide. This condition, as is easy to see, is exactly equivalent to the fact that the symplectic forms Ω_1 and Ω_2 define the same orientation on X.

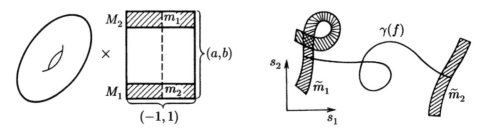

Figure 4.9   Figure 4.10

Thus, we obtain an immersion of the rectangle $[a,b] \times [-\varepsilon, \varepsilon]$ into the plane. As a result, we can consider s_1 and s_2 at each point as regular coordinates and define the symplectic structure Ω on X by the natural formula:

$$\Omega = ds_1 \wedge d\varphi_1 + ds_2 \wedge d\varphi_2 .$$

This formula evidently satisfies all required conditions.   □

Step 4. Thus, using the sewing lemma, we can construct a symplectic structure in the whole manifold $M^4 = Q^3 \times I$ in such a way that the foliation on M^4 becomes Lagrangian. Now to construct the desired Hamiltonian system, it suffices to take the parameter t on the interval I as the Hamiltonian H. The marked molecule corresponding to the isoenergy surface coincides with W^* by construction, as required.

REMARK.   The constructed system is in fact resonant in neighborhoods of singular fibers.   To avoid this defect, we just need to perturb the system, for instance, in the following way:

$$H \to H + \varepsilon f,$$

where f is an arbitrary function which is constant on the leaves of the folia-tion.   It is clear that, by arbitrarily small perturbation of this kind, we can make the system non-resonant without changing the topology of the isoenergy surface Q^3 and the corresponding Liouville foliation on it (that is, without changing the molecule W^*).

This remark completes the proof of the realization theorem.   □

Thus, Theorems 4.1 and 4.2 give a complete Liouville classification of all integrable Hamiltonian system (of the above type).

Corollary.

1) *There exists a one-to-one correspondence between the Liouville equivalence classes of integrable systems and marked molecules.   In particular, the set of Liouville equivalence classes of integrable systems is discrete (countable) and has no continuous parameters.*

2) *There exists an enumeration algorithm for marked molecules (i.e., classes of integrable systems).*

3) *There exists an algorithm for comparison of marked molecules, i.e., the al-gorithm that gives an answer to the question whether two integrable systems corresponding to given molecules are Liouville equivalent or not.*

4.7. SIMPLE EXAMPLES OF MOLECULES

Proposition 4.3.

1) *The molecule A——A determines a 3-manifold that is glued from two solid tori (in particular, the corresponding Heegard diagram has genus 1). Each of these solid tori is foliated into 2-tori in a standard way; and this gives the foliation of the whole 3-manifold into Liouville tori with exactly two singular leaves (namely, the axes of the solid tori).*

2) *The molecule A——A with the mark $r = 0$ corresponds to the three-dimensional sphere S^3.*

3) *The molecule A——A with the mark $r = 1/2$ corresponds to the three-dimensional projective space $\mathbb{RP}^3$.*

4) *The molecule A——A with the mark $r = \infty$ corresponds to the direct product $S^1 \times S^2$.*

5) *The molecule A——A with the mark $r = q/p$, where $q < p$ and $p \geq 3$, corresponds to the lens space $L_{p,q}$.*

Proof.

1. Since the 3-atom A is topologically presented as a solid torus, the molecule A—A evidently defines a 3-manifold which is the result of gluing two solid tori along their boundary tori. Thus, the first assertion simply follows from the definition of the molecule.

In view of Theorem 4.1 that establishes a one-to-one correspondence between marked molecules and Liouville foliations, to prove the remaining assertions, we shall construct model Liouville foliations on S^3, $\mathbb{RP}^3$, $S^1 \times S^2$, and $L_{p,q}$ with the desired marked molecules.

2. Define admissible bases λ^-, μ^- and λ^+, μ^+ on the boundary of two solid tori. The equality $r = 0$ means that the meridian of the first solid torus is identified with some parallel of the second torus, and vice versa. Therefore, by an admissible coordinate change, the gluing matrix can be reduced to the form

$$\begin{pmatrix} 0 & \pm 1 \\ \pm 1 & 0 \end{pmatrix}.$$

As a result, we obtain a 3-sphere. Indeed, consider the 3-sphere S^3 to be embedded into two-dimensional complex space $\mathbb{C}^2(z, w)$ and given by the equation $|z|^2 + |w|^2 = 1$. Let $f = |z|^2 - |w|^2$ be a smooth function on S^3. Its zero level surface $\{f = 0\}$ is a two-dimensional torus $T^2 = \{|z| = \sqrt{2}/2, |w| = \sqrt{2}/2\}$. This torus divides the sphere into two solid tori $A_- = \{|z| \leq \sqrt{2}/2, |w| = 1 - |z|^2\}$ and $A_+ = \{|z| = 1 - |w|^2, |w| \leq \sqrt{2}/2\}$.

Each of them is foliated into concentric tori presented as level surfaces $\{f = \text{const} \neq \pm 1\}$. The level $\{f = +1\}$ is the axis of one of the solid tori, on which f has a maximum. The level $\{f = -1\}$ is the axis of the other solid torus, which is a critical (namely, minimal) circle of f. On each torus of the foliation, there is one disappearing cycle (the first basis cycle). As the torus tends to the axis circle, the other basis cycle becomes this circle.

The disappearing cycle on each torus from A_+ is given by the equation $z = \text{const}$. It shrinks into a point as f tends to the maximum $f_{\max} = 1$. On the contrary, as f tends to the minimum $f_{\min} = -1$, this cycle approaches the axis of the solid torus A_-. On the other hand, the disappearing cycle on the tori of the second solid torus A_- is given by $w = \text{const}$. It shrinks into a point as $f \to -1$ and becomes the axis of A_+ as $f \to +1$.

Therefore, the gluing matrix takes the form

$$\begin{pmatrix} 0 & 1 \\ 1 & 0 \end{pmatrix}.$$

It is easy to see that the sign in front of 1 in this matrix can be changed (if necessary) by changing the orientation on basis cycles. This, of course, has no influence on the topology of the foliation.

3. Now consider the case A—A with $r = 1/2$. Here it is convenient to realize the projective space $\mathbb{RP}^3$ as the unit tangent vector bundle over the standard 2-sphere $S^2 \subset \mathbb{R}^3$. Consider the standard height function on the sphere and lift it up

to $\mathbb{R}P^3$ in the natural way, assuming it to be constant on fibers of the bundle $\mathbb{R}P^3 \to S^2$. We obtain a smooth Bott function f, whose critical submanifolds are just the S^1-fibers over the North and South poles (that correspond to the maximum and minimum of f respectively). The other level surfaces of f are two-dimensional tori, since they are simply direct products of parallels of the sphere by an S^1-fiber. Consider the foliation of $\mathbb{R}P^3$ into level surfaces of f. Evidently, the molecule related to this foliation has the desired form $A \!-\! A$. It remains to calculate the value of the r-mark. Let us formulate the following useful statement.

Lemma 4.8. *Consider an arbitrary edge e of a molecule W; let (λ^+, μ^+) and (λ^-, μ^-) be two admissible coordinate systems related to the atoms connected by this edge. Suppose that all these cycles lie on the same Liouville torus (in the middle of the edge).*

a) If the cycles λ^+ and λ^- are not intersected (i.e., are isotopic on the torus), then $r = \infty$.

b) If the cycles λ^+ and λ^- are intersected at exactly one point, then $r = 0$.

c) If the cycles λ^+ and λ^- have the index of intersection 2, then $r = 1/2$.

In these three cases, the mark r does not depend on the choice of the orientation on Q^3, on the edges of the molecule, and on the critical circles.

The proof of the lemma immediately follows from the definition of r (Definition 4.3). □

Let us apply this lemma to compute the r-mark in the case of $\mathbb{R}P^3$. Represent $\mathbb{R}P^3$ as the result of gluing two solid tori A_+ and A_-, which are direct products of two hemi-spheres S^2_+ and S^2_- by an S^1-fiber of the fiber bundle $\mathbb{R}P^3 \to S^2$. In other words, by cutting the base S^2 into halves, we obtain the trivial fibration over the hemi-spheres $A_+ \to S^2_+$ and $A_- \to S^2_-$. The common boundary torus of A_+ and A_- is the direct product of the equator by S^1-fiber. On this torus, we need to draw the cycles λ^+ and λ^-. Clearly, for this we need to define two smooth unit vector fields on the equator. Recall that the cycles λ^+ and λ^- must shrink into a point (each inside its own solid torus). Consequently, the desired unit vector fields must be extendable from the equator onto the corresponding hemi-sphere (without singularities). Such two fields are shown in Fig. 4.11. They are translated onto a plane by means of stereographic projection. The first field can be evidently extended inside the shaded disc that is the image of the lower hemi-sphere. The second field is defined on the complement to the first disc and can be evidently extended onto the upper hemi-sphere. The same fields are illustrated in Fig. 4.11 on the corresponding hemi-spheres. Comparing these two fields on the equator, we see that they coincide exactly at two points (Fig. 4.11). This means that the intersection index of the corresponding cycles λ^+ and λ^- is equal to 2. (If this index would have been equal to zero, we could have deformed the first field to the second one on the equator, but this is impossible, since there is no continuous vector field without zeros on the sphere.) Note by the way that the occurring number 2 is nothing else than the Euler number of the tangent fiber bundle over S^2. Thus, since the intersection index of λ^+ and λ^- is equal to 2, it follows that $r = 1/2$, as was to be proved. As a result, we have constructed the Liouville foliation on $\mathbb{R}P^3$ with the desired molecule.

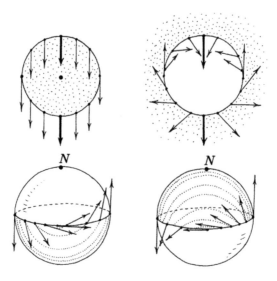

Figure 4.11

4. Consider the direct product $S^1 \times S^2$ and the Liouville foliation on it generated by the usual height function on the sphere S^2 embedded into $\mathbb{R}^3$ in the standard way. Its regular level lines on S^2 are parallels, while the North and South poles are critical points (maximum and minimum). Multiplying the level lines of f by the S^1-fiber, we obtain leaves of the Liouville foliation. The disappearing cycle in the first solid torus and that in the second one are in fact the same. Namely, this is the equator of the sphere multiplied by a fixed point of the S^1-fiber. Therefore, in this case $r = \infty$.

5. Let us define the Liouville foliation with the desired molecule W^* on the lens space $L_{p,q}$. The lens space $L_{p,q}$ can be obtained as the quotient space of the sphere $S^3 = \{|z|^2 + |w|^2 = 1\}$ with respect to the action of $\mathbb{Z}_p$ whose generator ξ acts as follows: $\xi: (z, w) \to (ze^{-2\pi iq/p}, we^{2\pi ip})$. Recall that the sphere S^3 is presented as the union of two solid tori whose common boundary torus is given as the zero level of the function $f(z, w) = |z|^2 - |w|^2$. Then the solid tori A_+ and A_- are given by the relations

$$A_- = \{|z| \le \sqrt{2}/2, \ |w| = 1 - |z|^2\} \quad \text{and} \quad A_+ = \{|z| = 1 - |w|^2, \ |w| \le \sqrt{2}/2\}.$$

Evidently, these solid tori are invariant under the action of $\mathbb{Z}_p$. Moreover, after taking the quotient of the sphere, they are again transformed into solid tori into which the lens space $L_{p,q}$ splits. The function $f(z, w)$ defined initially on the sphere generates a smooth function on $L_{p,q}$. The levels of this function define some foliation on the lens space. As we shall show now, this is the desired foliation, i.e., the corresponding molecule A—A has the mark r equal to q/p. To this end, we examine the action of $\mathbb{Z}_p$ on the boundary torus $T^2 = \{|z| = \sqrt{2}/2, \ |w| = \sqrt{2}/2\}$. We need to take the quotient torus $\tilde{T}^2 = T^2/\mathbb{Z}_p$ and consider two admissible coordinate systems on it (λ^+, μ^+) and (λ^-, μ^-). Clearly, as disappearing cycles

λ^+ and λ^-, we can take the images of the meridians of the solid tori A_+ and A_-, i.e., the images of the cycles $\{z = \text{const}, w = w_0 e^{i\varphi}\}$ and $\{w = \text{const}, z = z_0 e^{i\varphi}\}$ respectively.

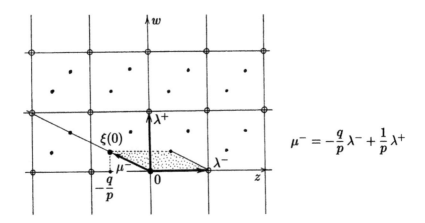

$$\mu^- = -\frac{q}{p}\lambda^- + \frac{1}{p}\lambda^+$$

Figure 4.12

Let us represent the torus $T^2 = \{|z| = \sqrt{2}/2, |w| = \sqrt{2}/2\}$ as the quotient space of the Euclidean plane $\mathbb{R}^2$ by the lattice $\mathbb{Z} \oplus \mathbb{Z}$ (see Fig. 4.12). In the same figure we represent the action of $\mathbb{Z}_p$ by drawing its fundamental domain. This domain is shown as the shaded parallelogram in Fig. 4.12 (we illustrate the case when $p = 3$, $q = 2$). As a result, there appear two lattices on $\mathbb{R}^2$: the initial one that generates the torus T^2, and the new one, more shallow, generating the quotient torus $\widetilde{T}^2 = T^2/\mathbb{Z}_p$. We can imagine the bases (λ^+, μ^+) and (λ^-, μ^-) by means of the second lattice. The two fat orthogonal vectors shown in Fig. 4.12 present the basis of the initial lattice, namely,

$$\{z = \text{const}, w = w_0 e^{i\varphi}\} \quad \text{and} \quad \{w = \text{const}, z = z_0 e^{i\varphi}\}.$$

These two cycles are projected from T^2 onto the quotient torus $\widetilde{T}^2$ without self-intersections. Therefore, we can take their projections as the cycles λ^+ and λ^- on $\widetilde{T}^2$.

Let us describe the fundamental domain for the torus $\widetilde{T}^2$. The action of $\mathbb{Z}_p$ on the plane is as follows: the generator ξ shifts the plane by the vector $-\frac{q}{p}\lambda^- + \frac{1}{p}\lambda^+$. Therefore, as the fundamental domain, we can take the parallelogram spanned on the vectors

$$\lambda^- \quad \text{and} \quad \mu^- = -\frac{q}{p}\lambda^- + \frac{1}{p}\lambda^+,$$

which evidently form a basis on the torus $\widetilde{T}^2$. Hence $\lambda^+ = q\lambda^- + p\mu^-$. Consequently, $r = q/p$, as required.

This completes the proof of Proposition 4.3. $\square$

Let us give some more examples of the molecules that describe important and interesting Liouville foliations.

Proposition 4.4. *Consider a molecule W shown in Fig. 4.13. All of its edges between saddle atoms have the r-mark equal to infinity, and those adjacent to atoms A have the r-mark equal to 0. Assume that the molecule has no atoms with stars. Then the corresponding 3-manifold Q is a locally trivial S^1-fibration over a closed two-dimensional surface P^2 (orientable or non-orientable):*

$$\pi: Q \xrightarrow{S^1} P^2 .$$

The Liouville foliation is generated by some Morse function f on the base (to obtain two-dimensional leaves in Q, we only need to lift this function to Q).

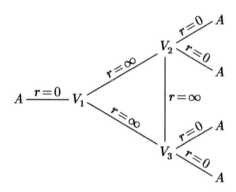

Figure 4.13

COMMENT. If the molecule W contains atoms with stars, then the manifold Q^3 is a global Seifert fibration whose singular fibers correspond exactly to the critical circles with non-orientable separatrix diagrams. If we replace, in addition, the zero r-marks by arbitrary rational numbers p/q on the edges of the molecule W adjacent to atoms A, then the manifold Q still remains a global Seifert fibration. But, in this case, some new singular fibers appear that correspond to the axial circles of those atoms A which are incident to the edges with $r = p/q$.

Proof. On each individual 3-atom without stars we already have the structure of a trivial S^1-fibration (Theorem 3.3). Gluing the boundary tori of neighboring 3-atoms, we see that the fibers of these fibrations are compatible on each boundary torus. The point is that the r-mark is equal to infinity when both atoms are saddle, and is equal to zero when one of these atoms has type A. This exactly means that the fibers coming to the boundary torus from the neighboring atoms coincide (more precisely, are isotopic) and, consequently, one can sew together the S^1-fibrations given on neighboring atoms. As a result, we obtain the structure of a locally trivial S^1-fibration defined globally on Q. Note that orientability or non-orientability of the base of the constructed fibration depends on the ε-marks on edges of W. □

COMMENT. In the above molecule (see Proposition 4.3), the saddle atoms form a family. The point is that all the edges between them carry an infinite r-mark. As was shown above, in this case, one more invariant appears, namely, the integer-valued mark n, which is assigned to this family. Its formal definition was given above. On the other hand, according to Proposition 4.3, the corresponding 3-manifold Q is an S^1-fibration over P^2. Suppose that its base is orientable. Then, for this fibration, we can define the well-known invariant called the Euler number. It turns out that this number and our mark n coincide. This fact is quite natural, since both invariants have the same topological nature: they can be regarded as obstructions to the extendability of a certain section.

Proposition 4.5. *Consider the molecule W shown in Fig. 4.14, which consists of one saddle atom V without stars and several atoms A connected with V by edges. Let all r-marks on these edges be equal to infinity. Then the 3-manifold Q corresponding to this molecule is homeomorphic to the connected sum of $k+1$ copies of $S^1 \times S^2$, where k is the complexity of atom V (i.e., the number of its vertices).*

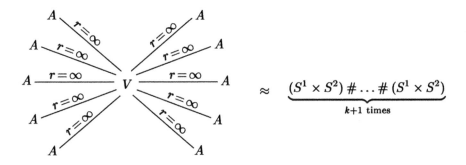

Figure 4.14

Proof. By our assumption, the atom V contains no star-vertices. Therefore, the corresponding 3-manifold is a direct product $V = P^2 \times S^1$. The boundary of this manifold consists of several tori. Since all the edges outgoing from V end with atoms A, each of these boundary tori must be glued by a solid torus. The r-marks are all equal to infinity. This means that each solid torus is glued in the following way. Consider an arbitrary boundary torus $T^2 \subset \partial V$. As we know, there is the canonical structure of a trivial S^1-fibration on it. As a result of gluing, the fiber of this fibration is identified with the disappearing cycle of the solid torus. In other words, each fiber lying on the boundary ∂V shrinks into a point.

Therefore, the 3-manifold Q can be presented as follows. First, we multiply P_2 by the circle S^1, and then we contract each circle $S^1 \times \{x\}$ lying on the boundary ∂V into a point $x \in \partial P_2$. Since P_2 can be considered as a narrow strip, we can cut P_2 on each edge (Fig. 4.15). Each cross-cut on a strip generates a cross-cut in Q along a 2-sphere. Reconstructing the cross-cut is equivalent to taking the connected sum with the manifold $S^1 \times S^2$. Consequently, Q can be cut along several 2-spheres in such a way that, as a result, we obtain the three-dimensional sphere from which $2(k + 1)$ three-dimensional non-intersecting balls are removed.

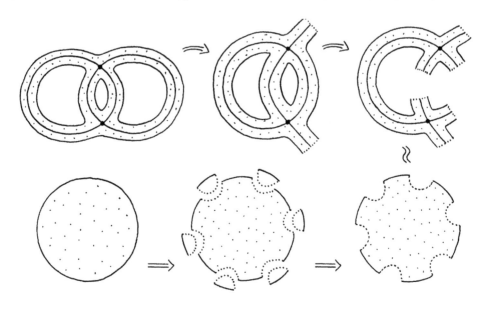

Figure 4.15

Indeed, after cutting P^2 at each edge, we obtain the 2-disc, shown in Fig. 4.15, from which $k + 1$ half-discs are removed. By multiplying the 2-disc with the circle S^1 and contracting each circle over its boundary into a point, we obtain the 3-sphere. By removing the half-discs from the 2-disc, we remove three-dimensional balls from the 3-sphere. Therefore, the desired 3-manifold Q is obtained from the 3-sphere with $2(k+1)$ balls removed by identifying the occurring boundary 2-spheres in pairs. Each such identification is evidently equivalent to taking the connected sum with the 3-manifold $S^1 \times S^2$.

This completes the proof. □

4.8. HAMILTONIAN SYSTEMS WITH CRITICAL KLEIN BOTTLES

Basically, in applications, there appear integrable systems that have neither critical tori nor Klein bottles on isoenergy 3-surfaces. Nevertheless, the critical tori and Klein bottle occur, for example, in the theory of integrable geodesic flows. We do not pay attention to critical tori, since they do not change the topology of the Liouville foliation. The critical Klein bottles, on the contrary, deserve consideration, since the topology of the foliation is really changed near them. This work has been done by P. Topalov in [343].

Let K be a critical Klein bottle. A neighborhood $U(K)$ of the Klein bottle K in the 3-manifold Q can be considered as a 3-atom of a special type. This atom will be denoted by the letter K with one outgoing edge. Its topological structure can be described in the following way. Since K is a smooth submanifold in Q^3, its tubular neighborhood $U(K)$ can be considered as a normal bundle, namely, the bundle over K with the fiber D^1 (one-dimensional disc). The boundary $\partial U(K) = K^2 \widetilde{\times} \partial D^1$ is a Liouville torus T^2, and the natural projection $\pi \colon T^2 = \partial U(K) \to K^2$ is a two-sheeted covering. The uniqueness of such an atom follows from the fact that there exists only one two-sheeted covering of the Klein bottle by the torus.

Let us choose an admissible coordinate system on the boundary torus T^2. To this end, we first prove the following statement.

Proposition 4.6. *All integral trajectories of an integrable Hamiltonian flow on a critical Klein bottle are closed.*

Proof. Let K be a critical Klein bottle of an integrable Hamiltonian system v. Since K lies inside the 3-atom $U(K)$, we can consider a two-sheeted covering over $U(K)$ which unfolds the Klein bottle K into a torus $\widetilde{T}^2$. Here $U(K)$ is covered by an orientable 3-manifold $\widetilde{U}(\widetilde{T}^2)$. Take the torus $\widetilde{T}^2$ and consider the involution ξ on it corresponding to the given covering $\widetilde{T}^2 \to K$. On the plane $\mathbb{R}^2$ that covers the torus $\widetilde{T}^2$, we introduce standard coordinates (x, y) connected with the lattice of the torus. Without loss of generality, we may assume that in terms of these coordinates the involution ξ is given by the formula

$$(x, y) \to (x + 1/2, -y).$$

The integral trajectories of v can be lifted from the Klein bottle to the torus $\widetilde{T}^2$. As a result, they transform into integral trajectories of the covering integrable system $\widetilde{v}$. The torus $\widetilde{T}^2$ can be regarded as a regular Liouville torus in $\widetilde{U}(\widetilde{T}^2)$. Therefore, the integral trajectories of the covering system must define a rectilinear winding on $\widetilde{T}^2$. We assert that this winding cannot be irrational, i.e., the integral trajectories are all closed on $\widetilde{T}^2$. Moreover, the integral trajectories are defined uniquely up to isotopy. Indeed, the covering vector field $\widetilde{v}$ on the torus $\widetilde{T}^2$ must be invariant under the involution ξ. Consider an integral trajectory $\gamma(t) = (x(t), y(t))$ of $\widetilde{v}$ on the covering plane $\mathbb{R}^2$. For a Hamiltonian vector field on a Liouville torus, one can define a pair of numbers (ω_1, ω_2) (called *frequencies*) by the formulas

$$\omega_1 = \lim_{t \to \infty} \frac{x(t)}{t}, \qquad \omega_2 = \lim_{t \to \infty} \frac{y(t)}{t}.$$

They do not depend on the choice of a specific trajectory on a given torus. Since, in our case, $\widetilde{v}$ is ξ-invariant, the trajectory $\gamma(t)$ is mapped under the action of ξ into a certain trajectory of the same vector field $\widetilde{v}$. This new trajectory is

$$\xi\gamma(t) = (x(t) + 1/2, -y(t)).$$

Hence

$$\omega_2 = \lim_{t \to \infty} \frac{y(t)}{t} = -\lim_{t \to \infty} \frac{y(t)}{t} = -\omega_2,$$

that is, $\omega_2 = 0$. Since $\widetilde{v}$ is a Hamiltonian vector field, it follows from $\omega_2 = 0$ that the trajectories of $\widetilde{v}$ are all closed on $\widetilde{T}^2$ and isotopic to the first basis cycle. It remains to observe that, if the trajectories of the covering system $\widetilde{v}$ are closed on $\widetilde{T}^2$, then the same is evidently true for the trajectories of the initial system v on the Klein bottle K, as was to be proved.   □

REMARK. It is an interesting fact that any smooth vector field on the Klein bottle without equilibrium points always has at least two closed trajectories.

Choose one of these closed trajectories and lift it from the Klein bottle K^2 to the boundary torus $T^2 = \partial U(K)$. As a result, we obtain a cycle on T^2, which we denote by λ and take as the first basis cycle of an admissible coordinate system.

Note that λ is a fiber of a Seifert fibration defined on $U(K)$. The base of this fibration is a two-dimensional disc with two singular points. Each of them corresponds to a singular fiber of the Seifert fibration of type $(2,1)$. Besides that, on $U(K)$ there is another structure of a Seifert fibration. The base of this Seifert fibration is the Möbius strip, and its fiber is a circle. Moreover, this second Seifert fibration is a usual locally trivial S^1-fiber bundle (i.e., without singular fibers).

It turns out that the structures of these two Seifert fibrations on $U(K)$ are uniquely defined (up to a fiber isotopy).

As the second basis cycle μ on the boundary torus $T^2 = \partial U(K)$, we take now the fiber of the second Seifert fibration $U(K) \to$ (Möbius strip). Thus, we have constructed an admissible coordinate system (λ, μ) on the boundary torus. Note that the orientation of λ is already given by the Hamiltonian flow v. As above, the orientation μ is chosen so that the pair of cycles (λ, μ) is positively oriented on the torus $T^2 = \partial U(K)$ (we assume here that T^2 has a canonical orientation induced by that of $U(K)$).

By analogy with the case of usual 3-atoms, we define numerical marks r and ε on the edge incident to the atom K. As before, using the constructed admissible coordinate system, first we take the gluing matrix $C = \begin{pmatrix} \alpha & \beta \\ \gamma & \delta \end{pmatrix}$ on the given edge.

After this, the rational mark r on the edge e of the molecule W incident to the atom K (related to the Klein bottle) is defined to be

$$r = \begin{cases} \dfrac{\alpha_i}{\beta_i} \bmod 1 \in \mathbb{Q}/\mathbb{Z} & \text{if } \beta_i \neq 0, \\ \text{symbol } \infty & \text{if } \beta_i = 0. \end{cases}$$

The integer mark ε on this edge is defined to be

$$\varepsilon = \begin{cases} \operatorname{sign} \beta_i & \text{if } \beta_i \neq 0, \\ \operatorname{sign} \alpha_i & \text{if } \beta_i = 0. \end{cases}$$

It remains to define the mark n. It is defined here in the same way as above. In particular, if the edge is infinite (i.e., $r = \infty$), then the atom K is included into the family adjacent to this edge. If the r-mark on the edge is finite, then the atom K is a family itself. After having fixed an admissible coordinate system, all definitions of numerical marks will be the same as before.

However, we have to distinguish the case where the molecule W has the form $K - K$. In this case, two admissible coordinate systems related to the atoms K

are uniquely defined. Therefore, as a mark that should be put on the single edge of the molecule, we can consider the gluing matrix itself. There are no other marks here.

Thus, endowing the molecule W with the collection of the numerical marks listed above, we obtain the marked molecule W^* for which the above Theorems 4.1 and 4.2 remain valid.

Thus, the Hamiltonian systems with critical Klein bottles are naturally included into the general classification theory without any essential distinctions.

However, one should take into account what we mean here by the Liouville equivalence of systems with critical Klein bottles. Two such systems v and v' are considered to be Liouville equivalent if there exists a diffeomorphism between the isoenergy 3-manifolds Q and Q' that preserves the Liouville foliation structure and, moreover, maps the closed oriented trajectories of v (which lie on the critical Klein bottle $K \subset Q$) to those of v' (which lie on the corresponding critical Klein bottle $K' \subset Q'$).

4.9. TOPOLOGICAL OBSTRUCTIONS TO INTEGRABILITY OF HAMILTONIAN SYSTEMS WITH TWO DEGREES OF FREEDOM

We have described the topology of isoenergy 3-surfaces of integrable Hamiltonian systems. Roughly speaking, all such 3-manifolds are obtained by gluing 3-atoms along their boundary tori. Which manifolds can be obtained in this way? Or, in other words, what are topological obstructions to the integrability of integrable Hamiltonian systems with two degrees of freedom (in terms of the topology of isoenergy surfaces)? Now we are able to answer this question. It turns out that not every three-dimensional manifold can be an isoenergy surface of an integrable system. Such manifolds form a "thin" subset in the set of 3-manifolds, and their topology can be described. It turns out that the class of such manifolds coincides with the class of graph-manifolds well-known in 3-topology, introduced by F. Waldhausen. Thus, if we are given a Hamiltonian system one of whose isoenergy surfaces is not a graph-manifold, then this system is certainly not integrable in the class of Bott integrals (at least, in a neighborhood of this isoenergy surface).

4.9.1. The Class (M)

Let (M) denote the class of all connected orientable closed 3-manifolds. It turns out that there are no obstructions for a manifold from the class (M) to be an isoenergy surface of some Hamiltonian (but not necessarily integrable) system.

Proposition 4.7 (S. V. Matveev, A. T. Fomenko). *Let X^3 be an arbitrary manifold from the class (M); then the direct product $M^4 = X^3 \times D^1$ (where D^1 is a segment) is a symplectic manifold.*

This assertion immediately implies the following fact.

Corollary.   *The manifold $X^3 \in (M)$ is an isoenergy surface of the natural Hamiltonian system on $M^4 = X^3 \times D^1$ given by the Hamiltonian $H = t$, where t is a coordinate on the segment D^1.*

Proof (of Proposition 4.7).   The existence of a symplectic structure on $M^4 = X^3 \times D^1$ follows from the well-known topological theorem which asserts that, for any orientable closed 3-manifold X^3, there is an immersion $i: X^3 \to \mathbb{R}^4$. Taking a tubular neighborhood U of the immersed manifold $i(X^3)$ (Fig. 4.16), we obtain

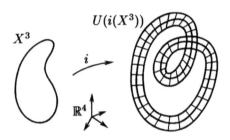

Figure 4.16

an immersion of a certain 4-manifold. It is easily seen that this 4-manifold is our direct product $M^4 = X^3 \times D^1$, because the normal fiber bundle of an immersed oriented manifold of codimension 1 in $\mathbb{R}^4$ is always trivial. Then, since on $\mathbb{R}^4$ there is the canonical symplectic structure $\omega = \sum dp_i \wedge dq_i$, we can take its pull-back $i^*\omega$ as a symplectic structure on M^4, as required.   □

4.9.2. The Class (H)

Definition 4.8. By (H) we denote the class of all orientable closed 3-manifolds that are isoenergy surfaces of integrable (by means of Bott integrals) Hamiltonian systems with two degrees of freedom.

The class (H) forms some subset in (M). A natural question is whether (H) coincides with (M) or not? As we already explained, this question is interesting, since the negative answer means the existence of topological obstructions to integrability. As we shall see soon, the class (H) is indeed much smaller than the class (M).

4.9.3. The Class (Q)

Consider two quite simple three-dimensional manifolds with boundary A^3 and B^3. They are described as follows.

The manifold A^3 is diffeomorphic to the direct product of the 2-disc by the circle, i.e., $A^3 = D^2 \times S^1$ (Fig. 4.17). The boundary of A^3 is diffeomorphic to the torus T^2. In other words, A^3 is the solid torus.

The manifold B^3 is diffeomorphic to the direct product of the disc with two holes N^2 by the circle, i.e., $B^3 = N^2 \times S^1$. Its boundary consists of three tori (Fig. 4.17).

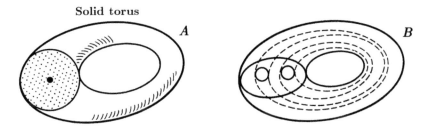

Figure 4.17

Definition 4.9. By (Q) we denote the class of all orientable closed 3-manifolds that can be represented in the form

$$Q^3 = aA^3 + bB^3 \,,$$

where $a \geq 0$ and $b \geq 0$ are integers, and the sign $+$ denotes gluing of manifolds by some diffeomorphisms of boundary tori. In other words, Q^3 is obtained by gluing a copies of the manifold A^3 with b copies of the manifold B^3 by some pairwise identifications of their boundary tori (so that finally we obtain a manifold without boundary).

Clearly, the numbers a and b cannot be arbitrary; they must satisfy the simple relation:

$$a + 3b \text{ is an even number}.$$

The point is that the total number of boundary tori must be even (in order for the resulting manifold to be closed).

4.9.4. The Class (W) of Graph-Manifolds

Definition 4.10. By (W) we denote the class of orientable closed 3-manifolds W satisfying the following property: W contains a finite set of non-intersecting tori such that, after removing these tori, W splits into a disjoint union of connected 3-manifolds each of which is a Seifert fibration over some two-dimensional surface (may be non-orientable).

This class of 3-manifolds was introduced by F. Waldhausen [357]. They were called *Graphenmannigfaltigkeiten* (graph-manifolds) and appeared in the papers by F. Waldhausen from deep problems in 3-topology without any connection with Hamiltonian mechanics and symplectic geometry. F. Waldhausen classified all such manifolds and, as we shall see below, this classification turns out to be closely connected with the classification of integrable Hamiltonian systems.

4.9.5. *The Class* (H') *of Manifolds Related to Hamiltonians with Tame Integrals*

Although the first integral is a Bott function on almost all energy levels $Q^3 = \{H = \text{const}\}$ in most physical systems, for some special values of energy (filling a set of measure zero) the integral f loses this property. Some singularities may appear which are more complicated than those satisfying the Bott property. It is natural to ask what happens to the class (H) of isoenergy surfaces of integrable systems if we expand the class of first integrals by admitting not only Bott functions? Of course, although now we are interested in integrable systems with non-Bott integrals, we shall assume these integrals not to be too pathological. Referring to some experience in the analysis of real physical systems, we shall consider the class of Hamiltonian systems admitting the so-called tame integrals.

Definition 4.11. A smooth integral f is said to be *tame* (on a given isoenergy 3-manifold Q^3) if for each critical value c of the function f the corresponding level surface $f^{-1}(c)$ is tame. This means that there exists a homeomorphism of Q^3 onto itself which maps the set $f^{-1}(c)$ into a polyhedron.

COMMENT. By a polyhedron, we mean a simplicial subcomplex in Q^3 each of whose simplex is smoothly embedded into Q^3.

Thus, although a tame integral is not necessarily a Bott function any more, it is not too awful yet: all of its level surfaces are in fact polyhedrons in Q.

Definition 4.12. By (H') we denote the class of oriented closed 3-manifolds that are isoenergy surfaces of Hamiltonian systems integrable by means of tame integrals.

It is clear that any Bott integral is tame (the converse is not true). Therefore, we have the trivial inclusion: the class (H) is contained in the class (H'). Thus, by expanding the class of integrable systems, we may *a priori* expand the class of isoenergy manifolds. Does it really happen?

4.9.6. *The Coincidence of the Four Classes of 3-Manifolds*

Thus, we have introduced the following four classes of 3-manifolds:

$$(H), \quad (Q), \quad (W), \quad (H').$$

In what follows, we also need the notion of a connected sum of manifolds and the notion of an irreducible manifold.

Let M and N be two smooth manifolds of the same dimension n. By removing an open ball D^n from each of them, we obtain two manifolds $M \setminus D$ and $N \setminus D$ with the boundary homeomorphic to the sphere S^{n-1}. Let us construct a new manifold by gluing $M \setminus D$ and $N \setminus D$ by some diffeomorphism of their boundary spheres. It is easy to show that the manifold obtained is smooth (i.e., can be endowed with a natural smooth structure).

Definition 4.13. The n-manifold obtained is usually denoted by $M \# N$ and is called the *connected sum* of the manifolds M and N. A manifold is called *prime* if it cannot be presented as the connected sum of two other manifolds each of which is different from the sphere. A three-dimensional manifold is called *irreducible* if each two-dimensional sphere embedded into it bounds a three-dimensional ball.

In what follows, we restrict ourselves with orientable 3-manifolds only.

Theorem 4.3 (A. V. Brailov, S. V. Matveev, A. T. Fomenko, H. Zieschang).
a) *The four above described classes of 3-manifolds coincide, i.e.,*

$$(H) = (Q) = (W) = (H') .$$

b) *The class* (H) *is strictly less than* (*i.e., does not exhaust*) *the class* (M) *of all 3-manifolds.*
c) *If* Q' *and* Q'' *are two arbitrary manifolds from the class* (H), *then their connected sum* $Q = Q' \# Q''$ *also belongs to the class* (H).
d) *If* $Q \in (H)$ *is reducible, i.e, is presented as the connected sum of some manifolds* Q' *and* Q'' *different from the 3-sphere, then both of the manifolds* Q' *and* Q'' *belong to* (H).

From Theorem 4.3, we immediately obtain, in particular, the following corollary.

Proposition 4.8. *Not every orientable closed 3-manifold can be an isoenergy surface of a Hamiltonian system integrable by means of a Bott* (*or just tame*) *integral.*

One can give an example of 3-manifolds which do not belong to the class (H). Recall that a 3-manifold is called *hyperbolic* if it can be endowed with a complete Riemannian metric with constant negative sectional curvature.

It turns out that the class (H) contains no hyperbolic manifolds [235]. Therefore, any Hamiltonian system which has a hyperbolic manifold as one of its isoenergy surfaces is non-integrable (on this isoenergy surface) in the class of Bott integrals (and, moreover, even in the class of tame integrals).

Proof (of Theorem 4.3).
The coincidence of (H) *and* (Q).
We first prove that $(H) \subset (Q)$. By definition, a 3-manifold Q from the class (H) is a closed isoenergy surface of a certain integrable system. As was already proved, this manifold is presented as the result of gluing some 3-atoms. Thus, it suffices to verify that each atom can be obtained by gluing some number of solid tori $A^3 = D^2 \times S^1$ and 3-manifolds $B^3 = N^2 \times S^1$, where N^2 denotes the 2-disc with two holes (Fig. 4.17). If a 3-atom contains no star-vertices, then topologically it is the direct product of a 2-atom P by the circle S^1. Clearly, every 2-atom P, being an oriented 2-surface with boundary, can be obtained by gluing some copies of the surface N, i.e., $P = N + N + \ldots + N$. Multiplying this decomposition by the circle, we obtain the proof in the case of atoms without stars. If a 3-atom V contains star-vertices, then, on the base P of the corresponding Seifert fibration $V \xrightarrow{S^1} P$, there are singular points indicating the singular fibers of the Seifert fibrations of type $(2, 1)$. By removing small discs around these points on the base P, we remove the solid tori from Q, i.e., manifolds of type A^3 that are projected

onto these discs.   As a result, the initial atom V is presented in the form
$V = V' + A^3 + \ldots + A^3$, where the 3-manifold V' has the structure of the direct
product $P' \times S^1$. Taking into account our argument in the previous case, we obtain
the desired assertion. Thus, we have proved the inclusion $(H) \subset (Q)$.

Let us prove the converse inclusion: $(H) \supset (Q)$. Since any 3-manifold from
the class (Q) is glued from solid tori $A^3 = D^2 \times S^1$ and manifolds of type
$B^3 = N^2 \times S^1$, the desired inclusion immediately follows from the realization
theorem (Theorem 4.2). Here we use the simple observation that the manifolds
A^3 and B^3 are just topological realizations of the 3-atoms A and B respectively.

The coincidence of (Q) and (W).

Let us prove that $(W) \subset (Q)$. To this end, it suffices to verify that every Seifert
fibration U^3 can be obtained by gluing some number of copies of A^3 and B^3.
Surrounding the singular fibers of the Seifert fibration by solid tori and removing
them from U^3, we obtain a 3-manifold U' which is a locally trivial S^1-fiber bundle
over a 2-surface P' with boundary (if its boundary is empty, then we just cut
a fibered solid torus from U' in order for the boundary to occur). If P' is orientable,
then the fiber bundle $U' \xrightarrow{S^1} P'$ is trivial and, consequently (see above), can be
obtained by gluing some copies of A^3 and B^3. If P' is not orientable, then we
first remove all Möbius strips from P' to obtain an orientable base. After this,
we proceed with this base in the same way as before. With the Möbius strips
we proceed in the following way. There are only two S^1-fibrations over the Möbius
strips μ: these are the direct product $\mu \times S^1$ and $\mu \, \tilde{\times} \, S^1$, where the "tilde" means
the skew product which will be described below. The case of the direct product
must be excluded because of the simple reason that $\mu \times S^1$ is a non-orientable
3-manifold, which we do not consider.

Lemma 4.9. *The skew product $\mu \, \tilde{\times} \, S^1$ can be presented as a Seifert fibration
over the 2-disc with two singular fibers of type $(2,1)$.*

Proof.   Consider the fat cylinder $S^1 \times [-1,1] \times D^1$ and identify its bases
$S^1 \times [-1,1] \times \{0\}$ and $S^1 \times [-1,1] \times \{1\}$ by the diffeomorphism τ that is
the superposition of the symmetry relative to the circle and the symmetry relative
to its diameter (Fig. 4.18). The symmetry τ is an involution with two fixed points.

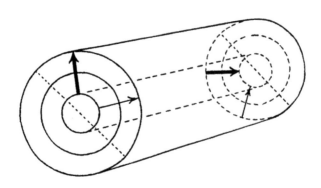

Figure 4.18

We now show that, on the 3-manifold X obtained, it is possible to introduce two different structures of a Seifert fibration. The first case is as follows: the fat cylinder is decomposed into circles of the form $S^1 \times \{*\} \times \{*\}$. This decomposition induces on X the structure of a Seifert fibration without singular fibers and with the Möbius strip as its base. In other words, $X = \mu \widetilde{\times} S^1$. On the other hand, the fat cylinder can be decomposed into segments of the form $\{*\} \times \{*\} \times D^1$, which turn into circles after gluing the bases of the cylinder. Such a decomposition induces another structure of a Seifert fibration on X over the disc D^2 with two singular fibers of type $(2, 1)$ that correspond to the fixed points of the involution τ.   □

It follows from this that the manifold $\mu \widetilde{\times} S^1$ can be glued from one copy of B^3 and two solid tori A^3; and, consequently, it gets into the class (Q). Thus, we have proved that $(W) \subset (Q)$.

The converse inclusion $(Q) \subset (W)$ evidently follows from the definitions of these classes.

Thus, we have $(H) = (Q) = (W)$.

The coincidence of the classes (H) and (H') was proved in the paper by S. V. Matveev and A. T. Fomenko [236]. This proof is also presented in our book [62].

The class (H) is strictly less than the class (M).

We have already shown that $(H) = (W)$. At the same time, according to the theory of graph-manifolds due to F. Waldhausen [357], the class (W) does not exhaust the class (M) of all orientable closed 3-manifolds. In particular, as remarked above, some interesting classes of 3-manifolds (for instance, the class of hyperbolic manifolds) do not intersect with the class (H).   □

Chapter 5

Orbital Classification
of Integrable Systems with
Two Degrees of Freedom

5.1. ROTATION FUNCTION AND ROTATION VECTOR

As above, let $v = \operatorname{sgrad} H$ be an integrable Hamiltonian system restricted to the compact isoenergy surface Q^3, and let W^* be its marked molecule.

Consider an arbitrary edge e of the molecule W^*. Recall that it represents a one-parameter family of tori. Suppose that, on some Liouville torus from this family, we have chosen and fixed an arbitrary basis in its fundamental group, i.e., a pair of cycles (λ, μ). According to the Liouville theorem, the trajectories of the Hamiltonian system on the torus are windings (rational or irrational). This means that there exists a coordinate system

$$(\varphi_1 \bmod 2\pi, \varphi_2 \bmod 2\pi)$$

on the torus in which v is straightened and takes the form

$$v = a\,\frac{\partial}{\partial\varphi_1} + b\,\frac{\partial}{\partial\varphi_2}.$$

Moreover, the coordinate lines of this coordinate system $\{\varphi_1 = \text{const}\}$ and $\{\varphi_2 = \text{const}\}$ are homologous to the basis cycles λ and μ, respectively.

Recall that the rotation number of the Hamiltonian system on the torus with respect to the basis (λ, μ) is defined to be the ratio $\rho = a/b$. If $b = 0$, then we set $\rho = \infty$ by definition.

It is easy to see that the rotation number is a complete orbital invariant of an integrable system on a single Liouville torus (see, for example, [9], [16]).

We can assume that the basis (λ, μ) is smoothly extended to all the other Liouville tori of the given edge of the molecule. This extension is uniquely defined up to an isotopy, which does not influence later arguments. We fix such a basis on each Liouville torus of the given edge.

Assume that the family of tori is parameterized by a parameter t that varies from 0 to 1, where t increases in the direction of the arrow assigned previously to the edge e. Denote by $T^2(t)$ the Liouville torus corresponding to the value of t. As before, we denote the basis cycles on this torus by λ and μ (not by $\lambda(t)$ and $\mu(t)$), since they are uniquely defined (up to isotopy) on all the tori of the family.

When the torus moves along the edge, the rotation number changes, and as a result, we obtain a function $\rho(t)$ which is defined on the interval $(0,1)$, where $\rho(t)$ is the value of the rotation number on the torus $T^2(t)$ with respect to the basis (λ, μ).

Definition 5.1. The function $\rho(t)$ is called the *rotation function* of the given integrable system.

Lemma 5.1. *The rotation function $\rho(t)$ is well-defined almost everywhere on the interval $(0,1)$ (i.e., except for the points at which ρ goes to infinity) and is smooth in a neighborhood of each finite value of it.*

Proof. This assertion is evident but we shall comment on it by recalling one method for computing the rotation function.

Consider a four-dimensional neighborhood $\mathcal{U}$ of the given one-parameter family of tori in the symplectic manifold (M^4, ω). Without loss of generality, we may assume that this neighborhood is a two-parameter family of Liouville tori of the form $\mathcal{U} = T^2 \times D^2$. Since the Liouville tori are Lagrangian submanifolds, i.e., $\omega|_{T^2} = 0$, it follows that ω is exact in $\mathcal{U}$. Therefore, there exists a 1-form $\varkappa$ such that $\omega = d\varkappa$.

Consider the standard action variables s_1 and s_2 defined for all points $p \in \mathcal{U}$ by the formulas:

$$s_1(p) = \frac{1}{2\pi} \int_\lambda \varkappa, \qquad s_2(p) = \frac{1}{2\pi} \int_\mu \varkappa,$$

where the integral is taken along the cycles λ and μ lying on the torus that contains the point p. In particular, the functions s_1, s_2 are constant on the tori and can be regarded as parameters of the two-parameter family of tori $\mathcal{U}$.

According to the Liouville theorem, the action variables are independent, $H = H(s_1, s_2)$, and the Hamiltonian vector field v can be presented in the form

$$v = \operatorname{sgrad} H = a \operatorname{sgrad} s_1 + b \operatorname{sgrad} s_2,$$

where $a = \dfrac{\partial H}{\partial s_1}$ and $b = \dfrac{\partial H}{\partial s_2}$ are smooth functions of s_1 and s_2, which are constant on each Liouville torus. For the family $\{T^2(t)\}$, in particular, a and b are smooth functions of the parameter t.

It is easy to see that now the rotation function can be written as $\rho(t) = \dfrac{a(t)}{b(t)}$.
Evidently, $\rho(t)$ is a smooth function on the interval $(0,1)$ everywhere except for those points where $b(t) = 0$, i.e., $\rho(t) = \infty$. □

Consider the rotation function $\rho(t)$ on the edge, i.e., on the interval $(0,1)$. In what follows, we shall consider the class of integrable system whose rotation functions are "good". More precisely, this means the following. We assume that the rotation functions on all the edges of the molecule W satisfy the following conditions.

1) All the critical points of the function $\rho(t)$ are isolated, and there is a finite number of them.

2) The function $\rho(t)$ is smooth, except for a finite number of points at which it is infinite. These points will be called *poles* (ρ can have no poles in general).

3) In a neighborhood of each pole, the function $1/\rho$ is also smooth.

REMARK.   It follows from these properties (1)–(3) that $\rho(t)$ has a limit as t tends to endpoints of the interval $(0,1)$. This limit can, certainly, be infinite. The function ρ is monotone in a neighborhood of the endpoints.

In particular, we note that a function ρ satisfying (1)–(3) cannot be constant on any interval.

REMARK.   The above conditions (1)–(3) do not depend on the choice of the basis (λ, μ) inside the given family of tori. This follows immediately from Proposition 1.7.

Proposition 5.1. *Any such function ρ is realized as the rotation function of some integrable Hamiltonian system.*

This statement is almost evident and follows formally from the general realization theorem (Theorem 8.1) which we shall prove below.

The class of "good" functions ρ is quite natural. One of the reasons is the following. We can consider the function $\arctan \rho(t)$, which maps the unit interval into the circle. Then conditions (1)–(3) listed above simply mean that this mapping into the circle has a limit at each endpoint of the interval, and the set of its critical points is finite.

Definition 5.2. Two rotation functions ρ_1 and ρ_2 on the interval $(0,1)$ are said to be *continuously* (*smoothly*) *conjugate* if there exists an orientation preserving homeomorphism (diffeomorphism) $\tau: (0,1) \to (0,1)$ such that $\rho_2(t) = \rho_1(\tau(t))$.

In other words, the functions are conjugate if they are mapped into each other under some monotone change of the parameter t (continuous or smoothly depending on the conjugacy type).

Consider a rotation function $\rho(t)$, all its poles, and local minima and maxima. Construct a vector (finite sequence) consisting of real number and the symbols "plus infinity" and "minus infinity".   The first element of this sequence is, by definition, the limit of ρ at zero (infinite or finite). Then, varying t from 0 to 1, we successively write out the values of ρ at all its poles, local minima, and maxima. Each pole here is depicted by two symbols: we indicate the left and right limits of the function at the pole. Finally, the last element of the sequence will be the limit of ρ at the point 1. As a result, we obtain an ordered set of numbers and symbols $\pm\infty$, which is denoted by R (see Fig. 5.1).

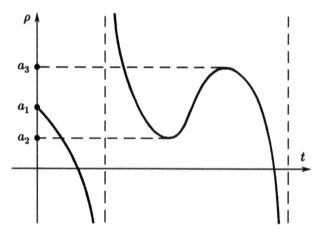

$$R = (a_1, -\infty, +\infty, a_2, a_3, -\infty)$$

Figure 5.1

Definition 5.3. The set R is called the *rotation vector* or *R-vector* of the integrable system on a given one-parameter family of tori (or on a given edge of the molecule W^*) relative to the given basis (λ, μ).

Proposition 5.2. *Rotation functions on the interval $(0,1)$ are conjugate if and only if the corresponding rotation vectors are the same.*

Proof. Consider two rotation functions on the same interval $(0,1)$. Suppose that the two corresponding rotation vectors coincide. For each of the functions we write down the sequence of those values of t where f has poles, local minima, and local maxima. For the function ρ_1 we obtain a sequence $(x_1, \ldots, x_N)$, and for ρ_2 we obtain $(y_1, \ldots, y_N)$. These sequences have the same length, because the rotation vectors coincide. In particular, $\rho_1(x_i) = \rho_2(y_i)$ for all i. On each interval $[x_i, x_{i+1}]$ and $[y_i, y_{i+1}]$ the functions $\rho_1(t)$ and $\rho_2(t)$ are simultaneously strictly increasing or strictly decreasing. We now construct a continuous monotone change of parameter t that makes these functions coincide. It suffices to construct this change for each of the indicated segments separately. The corresponding change is defined by the following simple formula:

$$\tau(t) = \rho_2^{-1}\rho_1(t) \qquad \text{for} \quad t \in [x_i, x_{i+1}].$$

These changes $\tau \colon [x_i, x_{i+1}] \to [y_i, y_{i+1}]$ can be sewn then into a global continuous change $\tau \colon (0,1) \to (0,1)$ because of the condition $\rho_1(x_i) = \rho_2(y_i)$.

The proof of the converse is obvious. □

COMMENT. Thus, the R-vector classifies the rotation functions satisfying conditions (1)–(3) up to a continuous conjugacy. In the smooth case, one should be more careful and look after the character of ρ at its critical points. However, if we require in advance that all its critical points are non-degenerate (more precisely, one should require the same condition for the function $\operatorname{arc cotan} \rho \colon (0,1) \to S^1$), then the same R-vector will classify such functions up to a smooth conjugacy.

By means of rotation functions and rotation vectors, we can now give the orbital classification of systems "on an edge of the molecule".

Proposition 5.3. *Let v and v' be two integrable systems given on symplectic 4-manifolds M and M'. Consider two one-parameter families E and E' of Liouville tori in M and M'. Then the systems (v, E) and (v', E') are topologically (smoothly) orbitally equivalent if and only if, for each of these families, there exist bases (λ, μ) and (λ', μ') relative to which the rotation functions ρ and ρ' are continuously (smoothly) conjugate.*

Proof. Suppose the rotation functions $\rho(t)$ and $\rho'(t')$ corresponding to the bases (λ, μ) and (λ', μ') are conjugate.

For each family of tori E and E' we construct the angle variables (φ_1, φ_2) and (φ_1', φ_2') related to the chosen bases. Each point of the one-parameter family of tori is defined by the coordinates $(t, \varphi_1, \varphi_2)$ (resp. $(t', \varphi_1', \varphi_2')$). The desired continuous mapping $\xi \colon E \to E'$ can now be defined by the following formula:

$$\xi(t, \varphi_1, \varphi_2) = (\tau(t), \varphi_1, \varphi_2),$$

i.e., $t' = \tau(t)$, $\varphi_1' = \varphi_1$, $\varphi_2' = \varphi_2$. Here τ denotes the mapping that conjugates the rotation functions, i.e., $\rho'(\tau(t)) = \rho(t)$. It is easy to see that ξ is continuous and maps trajectories to trajectories, as required.

The proof of the converse is obvious.   □

Corollary.   *Under the assumptions of Proposition 5.3, two systems are topologically orbitally equivalent if and only if the corresponding rotation vectors R and R' (related to appropriate bases) are the same.*

5.2. REDUCTION OF THE THREE-DIMENSIONAL ORBITAL CLASSIFICATION TO THE TWO-DIMENSIONAL CLASSIFICATION UP TO CONJUGACY

5.2.1. *Transversal Sections*

We study now the behavior of trajectories of an integrable Hamiltonian system in a neighborhood of a singular leaf, i.e., on a 3-atom in our notation. Let $L = L_c = f^{-1}(c)$ be a singular leaf of the Liouville foliation, where, as before, f is a Bott integral of the system, and $c \in \mathbb{R}$ is its critical value. According to the above description of the structure of the 3-atom $U(L)$, the leaf L can be considered as a Seifert type fibration over the graph K, which is embedded into the two-dimensional surface P that is the base of the Seifert fibration $\pi \colon U(L) \to P$ (see Chapter 3). We distinguish the following two cases:

a) the atom $U(L)$ does not have star-vertices (corresponding to the saddle critical circles with non-orientable separatrix diagrams);

b) the atom $U(L)$ has at least one star-vertex.

As was seen above, in case (a), the base P^2 can be realized as a section of π. In case (b), such a section does not exist. However, instead of the base P, we can consider a doubled surface $\widehat{P}$ with involution τ such that $P = \widehat{P}/\tau$, and fixed points of the involution are exactly the star-vertices. On the surface $\widehat{P}$, there appears a natural graph $\widehat{K}$, and the pair $(\widehat{P}, \widehat{K})$ can be considered as a 2-atom without stars. All the star-vertices of K are turned into vertices of degree 4 after duplication. The simplest example is shown in Fig. 5.2.

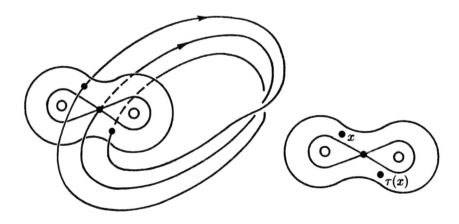

Figure 5.2

The surface $\widehat{P}$ can be now embedded into the 3-atom $U(L)$ in such a way that each fiber of the Seifert fibration transversely intersects $\widehat{P}$. Regular fibers do this twice, while singular fibers only once. We still call this embedded surface $\widehat{P} \subset U(L)$ a *section* (although literally it is not a section of the Seifert fibration). Figure 5.2 shows what this section looks like in the case of the simplest atom A^*. Here the leaf L is obtained as the surface which is swept out by the figure eight curve as its center moves along the circle, and the figure eight curve itself is turned finally by the angle π. The manifold $U(L)$ is obtained by a solid torus by removing another (narrower) solid torus that turns twice along the axis of the first solid torus.

Proposition 5.4. *For topologically stable integrable systems, the two-surfaces P and $\widehat{P}$ can always be chosen so that they are transversal to integral curves of v in a neighborhood of the singular leaf L.*

Proof. We begin with studying the properties of integral curves on the singular leaf L. We remove from L all the critical circles of the integral f, i.e., all critical periodic solutions. The leaf L splits into a disjoint union of a number of annuli each of which is foliated into integral curves of the field v. The behaviour of these curves can be of the three types (a), (b), and (c) described in Chapter 3, Proposition 3.7 and illustrated in Fig. 3.16.

In Chapter 3 we also showed that if an integrable system is topologically stable, then the singular leaf L has no annuli of type (c).

Now consider an embedding of the surface P or of the double $\widehat{P}$ into the 3-atom $U(L)$ which is transversal to the fibers of the Seifert fibration. Consider the intersection of this surface with the singular leaf. This will be a certain graph $K = P \cap L$ (resp. $\widehat{K} = \widehat{P} \cap L$). Since there are no annuli of the third type (c), it is possible to deform this embedded graph inside l so that it becomes transversal to the flow v on L. Indeed, thinking of the vertices of the graph as fixed, we reduce the problem to the analogous problem for each annulus separately. On a single annulus, such a deformation can be found provided this annulus has type (a) or (b). (On the contrary, it cannot be done for annuli of type (c).) As a result, we obtain a transversal embedding of the graph K (or $\widehat{K}$) into the singular leaf L.

Now this embedding of the K can be extended to the nearby Liouville tori up to an embedding of the whole surface P which can be viewed as a regular neighborhood of the graph. The transversality condition, being generic, remains satisfied for the whole of P (at least if P is sufficiently narrow). In the case of the double $\widehat{P}$, the argument is just the same. $\square$

Definition 5.4. The two-dimensional surface in the atom $U(L)$ constructed in Proposition 5.4 is called a *transversal section* of the atom $U(L)$. We denote this surface by P_{tr}.

Sometimes such surfaces satisfying the transversality condition are called *Poincaré sections*.

5.2.2. Poincaré Flow and Poincaré Hamiltonian

We now define the Poincaré map σ on the transversal section P_{tr}. Let x be an arbitrary point of P_{tr}. We emit an integral curve of the vector field v from it. At some instant of time, it will first hit the section P_{tr} and pierce it at some point y. Denote the mapping $x \to y$ by $\bar{\sigma}$ and define the mapping σ as follows:

$$\sigma = \begin{cases} \bar{\sigma} & \text{in the cases of an atom } A \text{ or of a saddle atom without stars;} \\ (\bar{\sigma})^2 & \text{in the case of a saddle atom with stars.} \end{cases}$$

Definition 5.5. The mapping $\sigma: P_{\mathrm{tr}} \to P_{\mathrm{tr}}$ is called the *Poincaré map* of the atom $U(L)$.

Note that the points of intersection of P_{tr} with critical circles of the integral f (which represent periodic trajectories of v) are fixed points of the Poincaré map. We denote these points by $S_1, \ldots, S_k$.

Since the section P_{tr} is realized in Q, we can restrict the symplectic structure ω from Q to P_{tr}. We obtain a non-degenerate closed 2-form (symplectic structure) on the two-surface P_{tr}. Denote this 2-form by ω as before. The non-degeneracy of ω on P_{tr} follows from the transversality of P_{tr} to all the integral curves of v, since the kernel of the form $\omega|_{Q^3}$ at each point of Q is generated by the vector v. The following statement is well known.

Lemma 5.2. *The Poincaré map σ preserves the symplectic form ω restricted to the transversal section P_{tr}.*

It turns out that the Poincaré map allows us to define a natural Hamiltonian system (with one degree of freedom) on P_{tr}.

Proposition 5.5. *On the smooth transversal section P_{tr}, there exists a Hamiltonian (with respect to ω) vector field $w = \operatorname{sgrad} F$ with Hamiltonian $F \colon P_{\mathrm{tr}} \to \mathbb{R}$ possessing the following properties.*

a) *The Poincaré map $\sigma \colon P_{\mathrm{tr}} \to P_{\mathrm{tr}}$ is a translation along the integral curves of the vector field $w = \operatorname{sgrad} F$ by time $t = 1$.*

b) *The original Bott integral f of the system v is also an integral of the Hamiltonian field w.*

c_1) *In the case of a saddle atom $U(L)$, the field w with properties (a) and (b) is uniquely determined. If the differential of the Poincaré map is not the identity mapping at the vertices of the graph K (i.e., at fixed points of the Poincaré map), then the Poincaré Hamiltonian F is a Morse function on the transversal section.*

c_2) *In the case of an atom A, the field w is defined uniquely up to addition of the field $2\pi k \dfrac{\partial}{\partial \varphi}$ to it, where φ is the angle variable on the two-dimensional section P_{tr}, which is a disc, and k is an arbitrary integer. The Poincaré Hamiltonian F is defined here uniquely up to $2\pi k s$, where s is the action variable on the disc P_{tr}.*

COMMENT. In the case of an atom A, the section P_{tr} is a disc foliated into circles, the level curves of the integral f. For such a foliation, one can define the standard action-angle coordinate system (s, φ). These are the functions that appear in item (c_2) of the proposition.

Proof. The proof of this assertion follows from the following general fact (see Proposition 1.6 in the book by J. Moser [248], [250]): a symplectomorphism σ on a two-dimensional manifold can be presented in the form σ^1, where σ^t is a Hamiltonian flow, if and only if σ possesses a non-trivial first integral. $\square$

It turns out that, as P. Topalov [341] remarked, it is possible to write down a simple explicit formula for the Hamiltonian F of the vector field w.

Proposition 5.6. *The Hamiltonian F coincides with the function $-2\pi s_1$ restricted to P_{tr}, where s_1 is the action variable related to the cycle that is the fiber of the Seifert fibration on $U(L)$. If this cycle is denoted by ν, then*

$$F = -\oint_{\nu} \varkappa ,$$

where $\varkappa$ is a differential 1-form in a neighborhood of the singular leaf L that satisfies the condition $d\varkappa = \omega$.

REMARK. This function F appeared above in Chapter 3 and was called the periodic integral.

Proof. Consider an arbitrary point x on the transversal section and show that $\sigma(x)$ coincides with the translation of this point by time $t = 1$ along the vector field $\operatorname{sgrad} F$. The operator sgrad is considered here in the sense of the symplectic structure ω restricted to P_{tr}. Clearly, it suffices to verify this condition only for points lying on Liouville tori. Moreover, if we consider two isotopic sections, then the verification can be carried out for any of them, since the symplectic structure and Hamiltonian F are preserved under the translations along $v = \operatorname{sgrad} H$. That is why we can choose a section in the most convenient way.

Let ν be a fiber of the Seifert fibration, and let μ be the cycle on a Liouville torus T^2 defined by $\mu = P_{\mathrm{tr}} \cap T^2$. Consider the action-angle variables $(s_1, s_2, \varphi_1, \varphi_2)$ related to these cycles. In particular, $s_1 = -\dfrac{F}{2\pi}$.

As a new transversal section P_{tr}, we now choose a two-dimensional surface given (in a neighborhood of some fixed Liouville torus T^2) by the equations $H = \mathrm{const}$ and $\varphi_1 = 0$. Recall that the Hamiltonian is a function of the action variables; moreover, in our case, $\dfrac{\partial H}{\partial s_1} \neq 0$. (Otherwise the integral curves of the field $v = \operatorname{sgrad} H$ would have been closed on T^2 and homologous to the cycle μ, but this is impossible in view of the transversality condition.) Hence, as local coordinates on this section we can choose s_2 and φ_2. Since H is fixed, the action variable s_1 can be considered on the section as a function $S(s_2)$ of s_2.

It is easy to see that the symplectic structure on P_{tr} has the form $ds_2 \wedge d\varphi_2$, and $F = -2\pi s_1 = -2\pi S(s_2)$. Then the vector field $\operatorname{sgrad} F$ becomes

$$\operatorname{sgrad} F = -2\pi \left(\frac{\partial S}{\partial s_2} \right) \frac{\partial}{\partial \varphi_2}\,,$$

and the translation along this field by time $t = 1$ takes the form

$$(s_2, \varphi_2) \to \left(s_2, \varphi_2 - 2\pi \frac{\partial S}{\partial s_2} \right).$$

Now look at what happens to a point under the Poincaré map. The vector field v in terms of the action-angle variables has the form

$$v = \frac{\partial H}{\partial s_1} \frac{\partial}{\partial \varphi_1} + \frac{\partial H}{\partial s_2} \frac{\partial}{\partial \varphi_2}\,.$$

Since in angle variables the Hamiltonian flow straightens, the Poincaré map sends the point $x = (0, \varphi_2) \in T^2$ to the point $x + \alpha v$, where α is chosen so that the first coordinate gets increment of 2π in order for the point to occur on the same section. Clearly, the second coordinate changes by adding the quantity

$$2\pi \frac{\partial H / \partial s_1}{\partial H / \partial s_2}\,.$$

In other words, the Poincaré map takes the form

$$(s_2, \varphi_2) \to \left(s_2, \varphi_2 + 2\pi \frac{\partial H / \partial s_1}{\partial H / \partial s_2} \right).$$

Taking into account that $H(s_1, s_2) = H(S(s_2), s_2) = \mathrm{const}$ on the section P_{tr}, we see that the magnitudes of the two translations $-2\pi \dfrac{\partial S}{\partial s_2}$ and $2\pi \dfrac{\partial H / \partial s_1}{\partial H / \partial s_2}$ coincide. This leads us to the desired result. $\square$

As we see, the Poincaré vector field w is defined uniquely on saddle atoms, unlike atoms A, where w is defined up to addition of an arbitrary multiple of the field $2\pi\dfrac{\partial}{\partial\varphi}$. On the other hand, in the saddle case, the section P_{tr} itself is not uniquely defined, while it is uniquely defined in the case of atoms A (since the meridian of the solid torus is always uniquely defined).

Definition 5.6. The one-parameter group of diffeomorphisms $\sigma^t\colon P_{\mathrm{tr}} \to P_{\mathrm{tr}}$ corresponding to the Hamiltonian vector field w is said to be the *Poincaré flow* on the transversal section P_{tr}.

It is clear that $\sigma^1 = \sigma$ ($=$ Poincaré map), and σ^t preserves the symplectic form ω on P_{tr}.

Consider an arbitrary saddle atom that contains at least one star-vertex. Take an arbitrary smooth transversal section P_{tr} and construct a natural involution $\chi = \bar{\sigma}\sigma^{-1/2}$ on it, where $\bar{\sigma}$ has already been defined above, $\sigma^{-1/2}$ is the diffeomorphism σ^t for $t = -1/2$.

Let us verify that $\chi\colon P_{\mathrm{tr}} \to P_{\mathrm{tr}}$ is indeed an involution. Consider the flow $g^t = (\bar{\sigma})^{-1}\sigma^t\bar{\sigma}$. Clearly, g^t preserves the symplectic structure on the section P_{tr}, i.e., is Hamiltonian, and moreover, for $t = 1$ we have

$$g^1 = (\bar{\sigma})^{-1}\sigma^1\bar{\sigma}$$
$$(\text{since } \sigma^1 = \sigma = (\bar{\sigma})^2)$$
$$= (\bar{\sigma})^{-1}(\bar{\sigma})^2\bar{\sigma} = (\bar{\sigma})^2 = \sigma.$$

Thus, $g^1 = \sigma$. However, according to Proposition 5.5, such a Hamiltonian flow g^t is uniquely defined for a saddle atom and coincides with the Poincaré flow σ^t. Therefore, $g^t = \sigma^t$, i.e., $\sigma^t = (\bar{\sigma})^{-1}\sigma^t\bar{\sigma}$, i.e., σ^t and $\bar{\sigma}$ commute for all t. Hence, $\chi^2 = \bar{\sigma}\sigma^{-1/2}\bar{\sigma}\sigma^{-1/2} = (\bar{\sigma})^2\sigma^{-1} = \mathrm{id}$, i.e., χ is an involution. Besides, χ commutes with σ^t, i.e., preserves the Poincaré flow σ^t on the section $P_{\mathrm{tr}} = \widehat{P}$.

Note that the involution χ is uniquely determined by the vector field v itself without using the symplectic structure and action variables.

5.2.3. Reduction Theorem

Recall that two dynamical systems (g^t, X) and (g'^t, X') are called topologically (smoothly) conjugate if there exists a homeomorphism (resp. diffeomorphism) $\xi\colon X \to X'$ that maps the first system to the second one, i.e.,

$$g'^t = \xi g^t \xi^{-1}.$$

In the case where the manifolds X and X' are oriented, we shall, in addition, assume that ξ preserves the orientation.

The following reduction theorem shows us that the orbital classification of integrable Hamiltonian systems on 3-atoms is reduced to the classification of the corresponding Poincaré flows on the transversal sections up to a conjugacy.

Theorem 5.1 [53].

a) *Let two integrable systems be topologically (smoothly) orbitally equivalent. Consider the atoms $U(L)$ and $U'(L')$ corresponding to each other under this equivalence; let $P_{\mathrm{tr}} \subset U(L)$ be an arbitrary smooth transversal 2-section. Then there exists a smooth transversal 2-section $P'_{\mathrm{tr}} \subset U'(L')$ such that the Poincaré flows on P_{tr} and on P'_{tr} are topologically (smoothly) conjugate. Moreover, in the case of a saddle atom with stars, the conjugating homeomorphism (diffeomorphism) also conjugates the involutions χ and χ'.*

b) *Conversely, let a system v on a 3-atom $U(L)$ be given, and let a system v' be defined on a 3-atom $U'(L')$. Suppose that, inside each of these atoms, there exist transversal 2-sections P_{tr} and P'_{tr} such that the Poincaré flows on these sections are topologically (smoothly) conjugate, and in addition, in the case of atoms with stars, the conjugating homeomorphism (diffeomorphism) conjugates the involutions χ and χ'. Then the systems v and v' are topologically (smoothly) orbitally equivalent on the given atoms.*

Proof.

a) First, consider the continuous case. Let $\overline{P}_{\mathrm{tr}}$ denote the image of the section P_{tr} under the orbital isomorphism $U(L) \to U'(L')$. Generally speaking, $\overline{P}_{\mathrm{tr}}$ is not a smooth surface in $U'(L')$. But we need a smooth section. Therefore, instead of $\overline{P}_{\mathrm{tr}}$ we take any smooth section P'_{tr} isotopic to it. Since v and v' are orbitally equivalent, it follows that the Poincaré maps σ and σ' are conjugate on the sections P_{tr} and P'_{tr}. Here we use the fact that under an isotopy of the section, the conjugacy class of the Poincaré map does not change.

We now have to show that the conjugacy of the Poincaré maps implies that of the corresponding Poincaré flows. Recall that we consider non-resonant systems only. We now show that, under this assumption, the condition $\sigma' = \xi^{-1}\sigma\xi$ automatically implies that $\sigma'^t = \xi\sigma^t\xi^{-1}$. Indeed, in terms of the Poincaré map, the condition that v is not resonant means that for almost any point $x \in P_{\mathrm{tr}}$ the closure of its orbit under σ is a whole level line of the additional integral f homeomorphic to a circle. In view of the fact that σ^t is a Hamiltonian flow, the restriction of σ to this circle is conjugate to the rotation through a certain angle $2\pi\alpha$, where α is some irrational number. How do we find the point $\sigma^t(x)$ if we know the images of x under σ^n only, where n is integer? The answer is as follows. Since α is irrational, there exists a sequence of integer numbers n_k for which $(\alpha n_k - t) \bmod 1$ tends to 0. Therefore, $\sigma^t(x)$ can be characterized as the following limit:

$$\sigma^t(x) = \lim_{k \to \infty} \sigma^{n_k}(x).$$

On the transversal section P'_{tr}, we have the same situation for the point $y = \xi(x)$. The numbers α and α' must coincide in view of the conjugacy of the Poincaré maps σ and σ'. Therefore,

$$\sigma'^t(\xi(x)) = \lim_{k \to \infty} \sigma'^{n_k}(\xi(x)).$$

Taking the limit in the equality $\sigma'^{n_k}(\xi(x)) = \xi\sigma^{n_k}(x)$, we obtain

$$\sigma'^t(\xi(x)) = \xi(\sigma^t(x)).$$

Since irrational points are everywhere dense, the continuity argument shows that this relation will hold identically, as required.

Note that, although we used the condition that v is non-resonant, the statement of the theorem remains true in the general case.

In the case of atoms with stars, the statement about the compatibility of the conjugating homeomorphism ξ with the involutions χ and χ' follows from the fact that these involutions are uniquely determined by the trajectories of the given systems.

b) We now prove the converse. Suppose we are given a homeomorphism ζ between the sections P_{tr} and P'_{tr} which conjugates the Poincaré flows (and maps χ into χ' in the case of atoms with stars). Consider an arbitrary point x in the atom $U(L)$. Let γ be an integral curve of the vector field $\mathrm{sgrad}\,H$ passing through it. Moving along it in the reverse direction from the point x, at some moment of time t we first hit the section P_{tr} at some point y. Consider the corresponding point $\zeta(y) \in P'_{\mathrm{tr}}$ and then move along the trajectory γ' of the vector field v' in time $t' = tc'/c$. Here c denotes the first return time for the point y. In other words, c is the length of the piece of trajectory between y and $\bar{\sigma}(y)$. The number c' is defined in the same way. As a result, we obtain a certain point on the trajectory γ' which we denote by $\xi(x)$.

The mapping thus constructed $\xi\colon U(L) \to U'(L')$ is continuous and preserves trajectories. Indeed, the continuity must be verified only on the transversal section P_{tr} itself, but this is guaranteed by the condition that ζ conjugates the Poincaré maps. In the case of a saddle atom with stars, we actually need the mapping $\zeta\colon P_{\mathrm{tr}} \to P'_{\mathrm{tr}}$ to conjugate $\bar{\sigma}$ and $\bar{\sigma}'$. But this immediately follows from the fact that $\bar{\sigma} = \chi\sigma^{1/2}$, and both the mappings in the right hand side are preserved under the action of ζ.

Thus, we have constructed the homeomorphism $\xi\colon U(L) \to U'(L')$, which is obviously the desired equivalence.

In the smooth case the proof is practically repeated word by word. We only need to guarantee, in addition, the time along trajectories to be changed smoothly. This completes the proof. $\square$

Thus, the topological (smooth) orbital classification of integrable Hamiltonian systems on three-dimensional atoms can be reduced to the classification of Hamiltonian systems on two-dimensional surfaces up to topological (smooth) conjugacy. The latter problem is not trivial. However, for systems with simple bifurcations (atoms), the information on the rotation functions turns out to be sufficient for the orbital classification. But, in the general case, we really need the description of conjugacy invariants for Hamiltonian systems on two-dimensional atoms. See [46], [47], and [53]. We shall not discuss this subject here, but speak only about the general strategy.

5.3. GENERAL CONCEPT OF CONSTRUCTING ORBITAL INVARIANTS OF INTEGRABLE HAMILTONIAN SYSTEMS

Thus, in the previous sections, we have discussed the orbital structure of an integrable Hamiltonian system on natural pieces, of which the isoenergy manifold Q^3 consists, namely, of the edges and atoms of the molecule. Now, after we have made sure of the principal possibility of describing this structure on separated pieces of the surface Q^3, we can roughly imagine how to draw the orbital portrait of an integrable Hamiltonian system as a whole, and how it will finally look. The process of constructing the orbital portrait of a system can be divided into a number of natural steps.

Step 1. *The molecule.* First, we have to solve a more rough problem and describe the structure of the Liouville foliation on Q^3. In other words, we have to find the so-called marked molecule W^* of the system. As a result, we obtain, in particular, the decomposition of Q^3 into natural components: the edges and narrow atoms. Recall that the edges are just one-parameter families of Liouville tori (without singularities) into which the isoenergy surface is decomposed after removing all singular leaves. On the contrary, the atoms are regular neighborhoods of these singular leaves (sufficiently narrow in order for a transversal section to exist).

Step 2. *Edge invariants.* After having described the structure of the Liouville foliation, we have to pass to the description of trajectories on the tori and singular leaves (more precisely, on the edges and atoms). Therefore, the next step is the description of the orbital structure on every edge. As was shown above, to do this, we need to calculate the rotation function on every edge of the molecule and consider its conjugacy class (with respect to a smooth or continuous change of parameter depending on what kind of classification we are interested in). If the rotation function is "good" enough (see above), then its conjugacy class can be completely described by means of the rotation vector introduced above. However, we note that, at this step, there is some ambiguity in the choice of a basis on the Liouville tori. That is why we shall need to avoid it afterwards in order to recognize which rotation function should be chosen for conjugacy testing. Nevertheless, we can suppose that the edge invariants have been described in essence.

Step 3. *Atomic invariants.* According to the reduction theorem, instead of considering the Hamiltonian system on a 3-atom $U(L)$, we can take a transversal 2-section P^2_{tr} in $U(L)$ and describe the invariants of the corresponding reduced system with one degree of freedom (i.e., the Poincaré flow). However, for reduction to one degree of freedom, we have to pay by passing from the orbital classification to the classification up to a conjugacy. Thus, orbital atomic invariants coincide with the conjugacy invariants of the reduced Hamiltonian system with one degree of freedom.

Step 4. *The framed molecule.* At the previous steps, for each edge and for each atom, we have described separately the corresponding orbital invariants. Is this information sufficient to describe completely the orbital structure of the system on the isoenergy surface as a whole? Both yes and no. Yes, because no other essential invariants exist. And no, because the calculated invariants are not well-defined. For example, the rotation function on an edge depends on the choice of basis cycles on Liouville tori. The analogous ambiguity takes place also for atomic invariants. The point is that the reduced system (the Poincaré flow) substantially depends on the choice of a transversal section $P_{tr} \subset U(L)$ (more precisely, of its homotopy type). Therefore, we have to assume transversal sections to be still fixed. By the way, this makes it possible for us to fix basis cycles on the edges adjacent to the atom $U(L)$ and to compute the rotation function and rotation vector with respect to special bases connected with the fixed transversal sections. Moreover, fixing transversal sections, we arrive at the appearance of a gluing matrix on each edge of the molecule; this in fact shows the mutual location of neighboring sections.

In our opinion, this approach has a natural analogy with many standard constructions in mathematics. For example, if we want to determine some object on a smooth manifold (for instance, a vector field), we can choose a certain atlas of charts and write down this vector field in the corresponding local coordinates. We also need to indicate the transition functions between these charts. As a result, the pair (manifold, vector field) will be completely determined. This procedure, however, is ambiguous, since it depends on the choice of atlas. Our situation is similar. The atlas of charts is a collection of transversal sections. The transition functions are the corresponding gluing matrices. And we try to study some object by writing it in a fixed atlas.

Thus, at this step, we suppose the collection of transversal sections to be fixed. This allows us to compute all the invariant (both atomic and edge ones) uniquely. We collect all of them together with the gluing matrices and add to the molecule W as the so-called t-frame. As a result, we obtain the molecule endowed with some additional information about the trajectories.

Step 5. *Group $G\mathbb{P}$ and its action.* If one uses another collection of transversal sections in the previous steps, of course, one obtains another t-frame of the molecule. It is a natural question: how are two frames corresponding to the same system related to each other if they have been computed with respect to different collections of transversal sections? It turns out that this relationship can be explicitly described, and as a result, we obtain the action of the discrete group $G\mathbb{P}$ of substitutions of transversal sections on the set of t-frames of the given molecule.

It would be worth indicating again the same analogy as above: to describe some object (for example, a vector field) on a smooth manifold, it is useful to know how its coordinate representation is changed under the transformation of an atlas.

Step 6. *Invariants of the group $G\mathbb{P}$, t-molecule and st-molecule.* This is the last step. We are interested in the orbital invariants of the system by themselves, i.e., without any connection with the choice of a collection of transversal sections. Therefore, instead of framed molecules, which are not well-defined, we have to consider the corresponding invariants of the group of $G\mathbb{P}$. Endowing the molecule W with a complete set of such invariants, we obtain a final orbital portrait of the system, which contains all necessary information.

This portrait is called the *t-molecule* in the topological case and *st-molecule* in the smooth one. It should be pointed out that in general, an explicit description of a complete set of invariants (i.e., a set that distinguishes any two orbits) can be a rather non-trivial problem, which can be treated in different ways and even can have no reasonable final solution (for example, if the orbit space is not a Hausdorff one). That is why, from the formal viewpoint, we can define the *t*-molecule (*st*-molecule) just as an element of the corresponding orbit space. On the other hand, the structure of molecules that occur in real problems is not very complicated, and for them, it is possible to obtain a final answer in the form of a molecule endowed with a finite number of numerical parameters.

Thus, we have briefly described the general scheme for constructing a complete set of orbital invariants for an integrable Hamiltonian system. This program will be carried out in the next chapters.

Chapter 6

Classification of Hamiltonian Flows on Two-Dimensional Surfaces up to Topological Conjugacy

6.1. INVARIANTS OF A HAMILTONIAN SYSTEM ON A 2-ATOM

In this section we produce a complete set of invariants giving the classification of Hamiltonian systems with one degree of freedom in a neighborhood of a singular level of the Hamiltonian up to topological conjugacy. When we speak of topological conjugacy of such systems, we mean the existence of a homeomorphism that not only conjugates flows, but also preserves orientation.

Consider a Hamiltonian system $w = \operatorname{sgrad} F$ with one degree of freedom on a two-dimensional symplectic manifold (X, ω). Let $\sigma^t : X \to X$ denote the corresponding Hamiltonian flow (i.e., the one-parameter group of diffeomorphisms generated by the Hamiltonian vector field $\operatorname{sgrad} F$). Here we assume that the Hamiltonian $F : X \to \mathbb{R}$ is a Morse function, that is, all of its critical points are non-degenerate. Let c be a critical value of F, and let $K = F^{-1}(c)$ be the corresponding singular level of the Hamiltonian which we assume, without loss of generality, to be connected. Consider a sufficiently small regular neighborhood P of the singular level K. As such a neighborhood, it is convenient to take the set $P = F^{-1}[c - \varepsilon, c + \varepsilon]$, where ε is sufficiently small in order to avoid any additional singular points from the neighborhood P (except for those which belong to the singular fiber K). Our goal is to describe the complete set of invariants of the Hamiltonian system in the neighborhood of K, or, in our terminology (see Chapter 2), on the atom (P, K). The most interesting case for us is when (P, K) is a saddle atom, and we shall assume this below. In this case, K is a graph all of whose vertices have degree 4 and coincide with the singular points of the Hamiltonian. In addition, without loss of generality, we shall assume that the critical value c is equal to zero, i.e., $F(K) = 0$.

6.1.1. Λ-Invariant

Consider all the critical points $S_1, \ldots, S_n$ of the function F on P, i.e., the vertices of the graph K. At each critical point S_i, we can consider the linearization of the Hamiltonian vector field $w = \operatorname{sgrad} F$ and the eigenvalues λ_i and μ_i of the linearized system. Since our vector field is Hamiltonian, it follows that $\lambda_i = -\mu_i$, and moreover, by virtue of the non-degeneracy of the singular point, we have $\lambda_i > 0$. It is well known that λ_i is a smooth invariant of w at the singular point S_i. However, generally speaking, it is not preserved under homeomorphisms. In other words, each of numbers λ_i, considered separately, is not an invariant of the flow in the sense of topological conjugacy. Nevertheless, by taking all these numbers together, we can produce a topological invariant from them. Instead of eigenvalues of the linearized vector field, it will be more convenient to consider their inverses $\Lambda_i = \lambda_i^{-1}$.

Definition 6.1. The set of real numbers $\{\Lambda_1 : \Lambda_2 : \ldots : \Lambda_n\}$ considered up to a common non-zero positive scalar factor (i.e., up to proportionality) is called the Λ-*invariant* of the given Hamiltonian system $w = \operatorname{sgrad} F$ on the atom (P, K).

COMMENT. If (x^1, x^2) is a local coordinate system in a neighborhood of a singular point S_i, then the number Λ_i can be computed by the following explicit formula

$$\Lambda_i = \left(-\det \left(\sum_i \omega^{ij} \frac{\partial^2 F}{\partial x^i \partial x^k}(S_i) \right) \right)^{-1/2},$$

where $\Omega = (\omega_{kl}(S_i))$ is the matrix of the symplectic form at the point S_i, and (ω^{ij}) is its inverse.

Proposition 6.1. *The Λ-invariants of topologically conjugate Hamiltonian systems (given on two copies of the same atom (P, K)) coincide.*

Proof. We start with a technical, but important lemma, which will be used more than once in what follows.

Removing the singular fiber K from P, we turn P into a disjoint union of annuli $C_1, \ldots, C_l$, each of which is naturally foliated into closed integral curves of the Hamiltonian field w. On each of these annuli we can introduce canonical "action-angle" variables s and φ. We are now interested in the angle variable φ. This variable is a smooth function on an annulus, and, thus, we can consider its level curves. The level lines of φ are not uniquely defined, since the angle on each non-singular circle (which is a level line of the action variable s) is defined only up to translation. Therefore, if we wish to draw the level lines of φ, we need to choose and fix a "reference point" on each circle. This can be done by setting $\varphi|_N = 0$, where N is a certain smooth segment joining a pair of points on the outer and inner boundary of the annulus and transversally intersecting all the circles, i.e., integral curves of w (Fig. 6.1). After this, the function φ will be uniquely defined. Figure 6.1 shows a qualitative picture of the behavior of the level lines of φ. More precisely, the following statement holds.

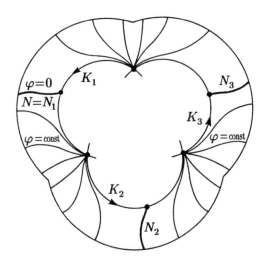

Figure 6.1   Figure 6.2

Lemma 6.1. *Let $C = C_m$ be an arbitrary annulus of the atom (P, K). Let $K_1, \ldots, K_p$ be the edges of the graph K adjoining (incident) to the given annulus C. Let S_{m_i} be the vertex of K which is the endpoint of the edge K_i $(i = 1, \ldots, p)$.*

a) On each edge K_i there exists a unique interior point x_i that is a limit point of some smooth level line $N_i = \{\varphi = \alpha_i\}$ of the angle variable φ on C (Fig. 6.1). Here the initial segment N coincides with N_1.

b) The segments N_i divide the annulus C into a sum of "rectangles" Z_i on each of which the level lines of φ behave qualitatively as shown in Fig. 6.2 (see also Fig. 6.1). In other words, all the remaining level lines $\{\varphi = \text{const}\}$ (except for the segments N_i) are pierced at vertices S_{m_i}.

c) The following formulas hold:

$$N_1 = \{\varphi = 0\},$$

and, for $i = 1, \ldots, p$,

$$N_{i+1} = \left\{\varphi = 2\pi \left(\sum_{j=1}^{i} \Lambda_{m_j}\right) \Big/ \left(\sum_{j=1}^{p} \Lambda_{m_j}\right)\right\}.$$

In other words the increment of the angle φ inside the domain Z_i is proportional to the number Λ_{m_i} corresponding to the vertex S_{m_i}.

We shall call the segments N_i constructed in this lemma *separation segments*.

Proof. First, we shall prove the following useful assertion which describes the velocity of a flow in a neighborhood of a saddle singularity.

On the Euclidean plane (u, v), we consider the function $F = uv$, an arbitrary symplectic structure $\omega = \omega(u, v)\, du \wedge dv$, and the corresponding Hamiltonian vector field $w = \operatorname{sgrad} F$. Consider the domain G shown in Fig. 6.3. It is bounded by the positive half-axes of the coordinates u and v, the hyperbola $F = uv = \varepsilon_0$, and the two segments $\gamma_1 = \{u = 1\}$ and $\gamma_2 = \{v = 1\}$ that intersect the level lines $\{F = \text{const}\}$ transversally.

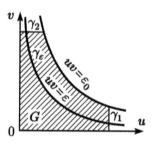

Figure 6.3

Consider the function $\Pi(\varepsilon)$ which is the time of motion along the piece γ_ε of the level line $\{F = \varepsilon\}$ lying between the arcs γ_1 and γ_2 (Fig. 6.3).

Lemma 6.2. *For any $n \in \mathbb{N}$, we have the equality*

$$\Pi(\varepsilon) = -P_n(\varepsilon)\ln\varepsilon + c(\varepsilon),$$

where P_n is a polynomial of degree n, and $c(\varepsilon)$ is a function of class C^n on the interval $[0, \varepsilon_0]$. Moreover, the coefficients a_i of the polynomial

$$P_n(\varepsilon) = a_0 + a_1\varepsilon + a_2\varepsilon^2 + \ldots + a_n\varepsilon^n$$

coincide with the coefficients a_{ii} in the Taylor expansion

$$\omega(u, v) \simeq \sum_{i=0}^{\infty} a_{ij} u^i v^j.$$

In particular, $a_0 = \omega(0, 0)$.

Proof. We assert that the function $\Pi(\varepsilon)$ can be calculated by the formula

$$\Pi(\varepsilon) = \int_{\gamma_\varepsilon} \omega(u, v)\, \frac{u\, du - v\, dv}{u^2 + v^2}.$$

Indeed, let us parameterize γ_ε as a trajectory of the vector field w. Then $\gamma_\varepsilon = (u(t), v(t))$, where $t \in [0, \Pi(\varepsilon)]$ and

$$\left(\frac{du}{dt}, \frac{dv}{dt}\right) = w = \omega^{-1}(dF) = \left(\frac{u}{\omega(u, v)}, -\frac{v}{\omega(u, v)}\right).$$

Substituting into the integral, we obtain

$$\int_{\gamma_\varepsilon} \omega(u, v)\, \frac{u\, du - v\, dv}{u^2 + v^2} = \int_0^{\Pi(\varepsilon)} dt = \Pi(\varepsilon).$$

We now parameterize the same curve in another way:

$$\gamma_\varepsilon = (\varepsilon e^\tau, e^{-\tau}), \qquad \text{where} \quad \tau \in [0, -\ln \varepsilon].$$

Integrating, we obtain

$$\Pi(\varepsilon) = \int_0^{-\ln \varepsilon} \omega(\varepsilon e^\tau, e^{-\tau})\, d\tau.$$

Since ω is a smooth function, we have the following representation:

$$\omega(u, v) = a_{00} + u g_0(u) + v h_0(v) + u v l_0(u, v),$$

where g_0, h_0, and l_0 are smooth functions. Applying such a representation for the function l_0 and iterating this procedure several times, we obtain

$$\omega(\varepsilon e^\tau, e^{-\tau}) = \sum_{k=0}^{n} a_{kk}\varepsilon^k + \varepsilon e^\tau g_n(\varepsilon e^\tau) + e^{-\tau} h_n(e^{-\tau}) + \varepsilon^{n+1} l_n(\varepsilon e^\tau, e^{-\tau}),$$

where g_n, h_n, and l_n are smooth functions. Integrating this expression on τ, we get

$$\Pi(\varepsilon) = -\left(\sum_{k=0}^{n} a_{kk}\varepsilon^k\right) \ln \varepsilon + c(\varepsilon),$$

where $c(\varepsilon)$ is a function of class C^n on the interval $[0, \varepsilon_0]$ as required. Lemma 6.2 is proved. □

Corollary. *The function $\Pi(\varepsilon)$ admits the representation*

$$\Pi(\varepsilon) = -A(\varepsilon) \ln \varepsilon + B(\varepsilon),$$

where $A(\varepsilon)$ and $B(\varepsilon)$ are C^∞-smooth functions on the interval $[0, \varepsilon_0]$.

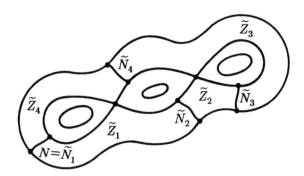

Figure 6.4

We now return to the proof of Lemma 6.1. On an annulus C, we consider smooth segments $\tilde{N}_i$ dividing the annulus into rectangles $\tilde{Z}_i$ as shown in Fig. 6.4. Exactly one segment $\tilde{N}_i$ corresponds to each edge K_i. Recall that the annulus C is foliated into closed integral curves of the flow σ^t, and each such curve is uniquely defined by the value of F on it (we shall denote it by γ_F). Let $\Pi_i(F)$ be the passage time of a point inside the rectangle $\tilde{Z}_i$ from its left side $\tilde{N}_i$ to the right side $\tilde{N}_{i+1}$ under the action of the flow σ^t along the integral curve γ_F.

Note that instead of segments γ_1 and γ_2 in Lemma 6.2 we can consider any other smooth curves transversal to the flow. The formula for $\Pi(\varepsilon)$ remains the same; the only thing we must do is to add some smooth function to $c(\varepsilon)$. Therefore, using the Morse lemma, we can apply Lemma 6.2 to the rectangle $\tilde{Z}_i$. As a result (for $n = 0$), we obtain the following asymptotic representation for $\Pi_i(F)$:

$$\Pi_i(F) = -\Lambda_{m_i} \ln F + c_i(F),$$

where $c_i(F)$ is a continuous function on the whole interval $[0, F_0]$ (including zero).

Let $\Pi(F)$ denote the full period of the trajectory γ_F. The function $\Pi(F)$ will appear below many times, and we shall call it the *period function* (related to the given annulus C). For each i, we consider the function

$$\Theta_i(F) = d_i \Pi(F) - \sum_{j=1}^{i} \Pi_j(F),$$

where

$$d_i = 2\pi \cdot \frac{\sum\limits_{j=1}^{i} \Lambda_{m_j}}{\sum\limits_{j=1}^{p} \Lambda_{m_j}}.$$

We claim that this function is continuous on the whole segment $[0, F_0]$ and smooth everywhere except, perhaps, at zero. Indeed, all the "pure logarithms" included in the expressions for periods are canceled, and, as a result, we get the expression

$$\Theta_i(F) = d_i \left(\sum_{j=1}^{m} c_j(F) \right) - \sum_{j=1}^{i} c_j(F),$$

which is obviously a continuous function of F on the whole segment $[0, F_0]$ (including zero).

Note that, if the functions Θ_i were identically equal to zero, we would obtain that $\sum_{j=1}^{i} \Pi_j(F) = d_i \Pi(F)$. This would mean that the increment of φ in each rectangle $\tilde{Z}_i$ would be equal to $2\pi \cdot \dfrac{\Lambda_{m_i}}{\sum\limits_{j=1}^{p} \Lambda_{m_j}}$. In other words, the segments $\tilde{N}_i$ would be the level lines of φ and would coincide with the desired segments N_i. Here we use the fact that, for each curve γ_F, its natural parameter t and the angle variable φ are connected by the simple relation $d\varphi = \dfrac{2\pi}{\Pi(F)}\, dt$, that is, are simply proportional with a constant (on the curve) coefficient.

However, in general, Θ_i are different from zero. But they are continuous functions of F, so that it suffices to consider new segments N_i obtained from the original segments $\tilde{N}_i$ by some translations. Namely, it is necessary to translate each point of the segment $\tilde{N}_i$ by Θ_i. More precisely, each point on $\tilde{N}_i$ is defined by some value of F, and one needs to move it along the trajectory of the flow σ^t by $\Theta_i(F)$. It is clear that, for the new segments N_i constructed in this way, the new functions Θ_i will be identically equal to zero; and then the above remark becomes valid. The new segments N_i are smooth on the open annulus, and each of them has a limit point on the inner boundary of the annulus which we take as x_i.

Thus, we have constructed segments N_i so that the formula of part (c) holds. It remains to prove the uniqueness of the point x_i (see part (a) of the lemma) and the fact that all the remaining trajectories behave as indicated in part (b). In other words, it suffices to prove that, in the rectangle Z_i, all the remaining level lines of φ hit the vertex S_{m_i}. But this fact easily follows from the already used relation

$$ d\varphi = \frac{2\pi}{\Pi(F)} \, dt \, . $$

Indeed, if we move slightly away from the segment $N_i = \{\varphi = \alpha_i\}$ into the rectangle Z_i for a time $\Delta\varphi$ (i.e., we consider the increment $\varphi \to \varphi + \Delta\varphi$), then near the graph K we go away from the segment N_i for an arbitrary long time in the sense of the flow σ^t, since $\Pi(F) \to \infty$ as $F \to 0$. Therefore, the only possible limit point of the translated segment $\{\varphi = d_i + \Delta\varphi\}$ is the vertex S_{m_i}. This completes the proof of Lemma 6.1. $\square$

COMMENT. The initial segment $N = N_1$ has been assumed to be *smooth*; however, it is clear that one could take any *continuous* segment on the annulus C that joins a pair of points on the opposite boundaries of the annulus and intersects each integral curve of the flow σ^t once. The only difference from Lemma 6.1 will then consists in the fact that all the remaining segments N_i (constructed from N_1) will also be continuous arcs joining pairs of points on the opposite boundaries of the annulus and intersecting each integral curve of the flow once.

We return to the proof of Proposition 6.1. Suppose we are given topologically conjugate Hamiltonian systems w and w' on two copies (P, K) and (P', K') of the same atom V. Let $\xi \colon P \to P'$ be the conjugating homeomorphism. Denote by Λ_i and Λ'_i the values of the Λ-invariant of the first and second systems respectively. Here by the same indices we enumerate singular points S_i of the first system and their images $S'_i = \xi(S_i)$.

The graph K divides P into the union of annuli. Let C be any one of them, and let C' be the annulus corresponding to it under the homeomorphism ξ. We consider on C the system of separation segments N_i constructed via Lemma 6.1 and corresponding to some angle variable φ.

Consider the images $N'_i = \xi(N_i)$ of the segments N_i under ξ. They divide the annulus C' into a union of rectangles Z'_i. These segments will be level lines of the angle φ' provided that $\xi(N_1)$ is taken to be the initial segment (corresponding to $\varphi' = 0$). This follows immediately from the topological conjugacy of w and w'. Moreover, the function φ' takes the same values on the segments N_i' as the function φ on N_i. Thus, N_i' are separation segments for the annulus C' and possess all the properties listed in Lemma 6.1. In particular, the formula of part (c) is true for them.

Since $\varphi_i|_{N_i} = \varphi_i'|_{N_i'}$, it follows that

$$\frac{\sum_{j=1}^{i} \Lambda_{m_j}}{\sum_{j=1}^{p} \Lambda_{m_j}} = \frac{\sum_{j=1}^{i} \Lambda_{m_j}'}{\sum_{j=1}^{p} \Lambda_{m_j}'}.$$

This relation implies that the sets of numbers $\{\Lambda_{m_i}\}$ and $\{\Lambda_{m_i}'\}$ (for each annulus C and its image C') coincide up to proportionality. Carrying out this argument for all the annuli, we obtain the assertion of Proposition 6.1. □

6.1.2. Δ-Invariant and Z-Invariant

We shall again use Lemma 6.1 to construct two new invariants of the Hamiltonian system w in the neighborhood P of the singular fiber K. We assume that the orientation on P is defined by the symplectic structure ω. The graph $K = F^{-1}(0)$ divides P into annuli $C_1, \ldots, C_l$.

Definition 6.2. An annulus $C = C_m$ is said to be *positive* if the function F is greater than zero on C; otherwise it is called *negative*.

COMMENT. This definition is equivalent to the following. An annulus is positive if the flow σ^t has positive direction on the outer boundary of the annulus. Here a boundary of an annulus is said to be inner if it adjoins the graph K, and outer otherwise. We assume that the orientation on the outer boundary is induced by the orientation of the atom via an outer normal. Therefore, the positivity or negativity of an annulus is preserved under topological conjugacy of flows.

Consider all the edges K_i of $K = F^{-1}(0)$. Exactly two annuli of the atom V adjoin each edge K_i: one positive and one negative. Following Proposition 8.1, on each of these annuli, we define action-angle variables s and φ and the separation segments (i.e., the level lines of φ that hit edges K_i at a certain interior point x_i). As a result, for each K_i, we obtain a pair of points, which we denote by x_i^+ and x_i^- (for positive and negative annuli, respectively).

In what follows, we shall call the points x_i^+ and x_i^- *positive* and *negative* separation points on the edge K_i of the graph K.

Since the Hamiltonian F was assumed to be a Morse function, our field $w = \operatorname{sgrad} F$ is non-zero at all interior points of the edges of K. Therefore, the flow σ^t is not the identity on any edge of the graph K. Let t_i denote the time that is necessary for a point to pass from x_i^- to x_i^+ under the action of the flow σ^t (Fig. 6.5(a)). In other words, t_i is uniquely determined by the relation $x_i^+ = \sigma^{t_i}(x_i^-)$.

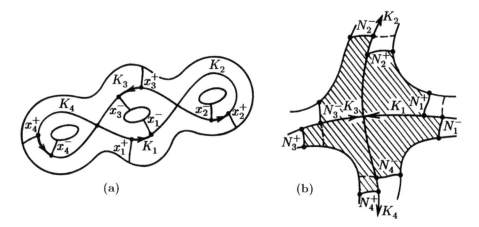

Figure 6.5

We now consider a formal linear combination $l = \sum t_i K_i$ as a one-dimensional chain l (in the sense of real homology) on the graph K. Here the edges K_i are regarded as one-dimensional basic cells. The orientation on each of them is determined by the flow σ^t.

It is clear that this chain is not an invariant of our Hamiltonian system, since the separation segments are not uniquely defined. However this ambiguity is easily controlled. Indeed, on each annulus we can move these segments by the same value. The corresponding separation points will simultaneously move, but the size of translation (in the sense of the flow σ^t) for points lying on the same annulus will be the same.

What does this mean in terms of the chain l? To answer this question, let us consider the closed surface $\widetilde{P}$ obtained from P by gluing 2-discs to all of its boundary circles. The graph K evidently determines a cell decomposition of $\widetilde{P}$; and, therefore, we can define the groups $C_k(\widetilde{P})$, $Z_k(\widetilde{P})$, and $B_k(\widetilde{P})$ of real chains, cycles, and boundaries respectively, which are related to this decomposition, by considering formal linear combinations of k-cells ($k = 0, 1, 2$).

Using these homological terms, it is easy to see that the ambiguity in the choice of separation segments affects the 1-chain $l \in C_1(\widetilde{P})$ in the following way: this chain is defined modulo the subspace $B_1(\widetilde{P})$ of 1-boundaries, i.e., its class $[l]$ in the quotient space $C_1(\widetilde{P})/B_1(\widetilde{P})$ is well-defined.

As we saw in the proof of Proposition 6.1, separation segments are mapped to separation segments under a conjugating homeomorphism ξ. Therefore, under such a homeomorphism ξ, the separation points x_i^+ and x_i^- are mapped to some separation points $x_i'^+ = \xi(x_i^+)$ and $x_i'^- = \xi(x_i^-)$. Besides, since ξ conjugates the flows σ^t and σ'^t, the condition $x_i^+ = \sigma^{t_i}(x_i^-)$ implies that $x_i'^+ = \sigma'^{t_i}(x_i'^-)$. In other words, ξ preserves the coefficients of the chain l.

This argument shows that the class $[l] \in C_1(\widetilde{P})/B_1(\widetilde{P})$ is a well-defined invariant of an integrable Hamiltonian system on an atom (in the sense of topological conjugacy).

It will be more convenient for us to divide $[l]$ into two simpler invariants.

To this end, we use the following formal isomorphism:

$$C_1(\widetilde{P})/B_1(\widetilde{P}) \simeq C_1(\widetilde{P})/Z_1(\widetilde{P}) + Z_1(\widetilde{P})/B_1(\widetilde{P}) \simeq B_0(\widetilde{P}) + H_1(\widetilde{P}),$$

where $B_0(\widetilde{P})$ is the group of zero-dimensional boundaries, and $H_1(\widetilde{P})$ is one-dimensional real homology group of the closed surface $\widetilde{P}$. The indicated isomorphism is not natural, but it can be defined explicitly by introducing an inner product in the space of one-chains $C_1(\widetilde{P})$. Let, for definiteness, the elementary chains of the form $1 \times K_i$ form an orthonormal basis in $C_1(\widetilde{P})$ (here K_i, as before, denote edges of the graph K, i.e., 1-cells).

From the 1-chain l, we now construct two new objects. Let us project l orthogonally to the space $Z_1(\widetilde{P})$ of 1-cycles and consider the homology class of the cycle $z = \pi(l)$ obtained, where $\pi \colon C_1(\widetilde{P}) \to Z_1(\widetilde{P})$ is the orthogonal projection.

Definition 6.3. We denote the homology class $[z] \in H_1(\widetilde{P}) = Z_1(\widetilde{P})/B_1(\widetilde{P})$ by Z and call it the Z-*invariant* of the Hamiltonian system w (on the atom (P, K)).

Further, consider the boundary $\partial(l) \in B_0(\widetilde{P})$ of the 1-chain l. Here $\partial \colon C_1(\widetilde{P}) \to B_0(\widetilde{P})$ is the standard boundary operator.

Definition 6.4. We denote the boundary $\partial(l) \in B_0(\widetilde{P})$ of the chain l by Δ and call it the Δ-*invariant* of the Hamiltonian system w (on the atom (P, K)).

It is easy to see that Z and Δ do not change when l is changed by adding an arbitrary 1-boundary. Thus, to each homology class $[l] \in C_1(\widetilde{P})/B_1(\widetilde{P})$ we assign the pair (Δ, Z). This correspondence defines the above isomorphism $C_1(\widetilde{P})/B_1(\widetilde{P}) \simeq B_0(\widetilde{P}) + H_1(\widetilde{P})$, and, therefore, the pair (Δ, Z) contains just the same information about the system as the initial class $[l] \in C_1(\widetilde{P})/B_1(\widetilde{P})$. Moreover, since $[l]$ is an invariant of the Hamiltonian system, so are Δ and Z. In other words, we have the following assertion.

Proposition 6.2. *The Δ-invariants and Z-invariants of topologically conjugate systems (given on two copies of the same atom) coincide.*

We now give another interpretation of coefficients of $\Delta = \sum \Delta_i S_i$ that is extremely useful for what follows. (Here S_i are the vertices of K, i.e., zero-dimensional cells.) Note that the zero-dimensional cycle Δ can be understood as a set of real numbers on the vertices of K whose total sum is zero. It turns out that the numbers Δ_i can be defined by explicit formulas, expressing Δ_i in terms of the Λ-invariant and the full periods of the flow σ^t on the annuli of the atom.

Consider an arbitrary vertex $S = S_j$ of the graph K and the four edges of the graph incident to it: K_1, K_2, K_3, K_4. On each K_i, two separation points x_i^+ and x_i^- are distinguished. The corresponding separation segments N_i^+ and N_i^- pierce the edges at these points (Fig. 6.5(b)). Consider the region $U = U(S_j)$ bounded by them and shown in Fig. 6.5(b). It consists of four sectors bounded by the edges of K, the separation segments, and the level lines $F = \pm\varepsilon_0$ of the Hamiltonian F. To each of these sectors we can apply the assertion of Lemma 6.2 for $n = 0$. As a result, in each of these sectors, there is a continuous function $c_i(F)$ included in the formula for the function $\Pi_i(F)$ that gives the passage time for a point moving inside the sector from one separation segment to the other. Consider the values of these four functions at zero, i.e., the four numbers $c_i(0) = c_i$.

Four annuli adjoin the vertex S. We denote these annuli by C_I, C_{II}, C_{III}, C_{IV} (in general, some of them may coincide). Let $\Pi_I(F)$, $\Pi_{II}(F)$, $\Pi_{III}(F)$, $\Pi_{IV}(F)$ be the corresponding period functions. Then we have

$$\Pi_I(F) = -\Lambda_I \ln|F| + c_I(F),$$

where $c_I(F)$ is a function continuous at zero, and Λ_I is the sum of Λ_i over all vertices of K that belong to the boundary of the annulus C_I (including multiplicities). The function $c_I(F)$ will be sometimes called the *finite part* of the period function $\Pi_I(F)$. Set $c_I = c_I(0)$. The numbers c_{II}, c_{III}, c_{IV} can be analogously defined as the finite parts of the period functions Π_{II}, Π_{III}, Π_{IV} for $F = 0$.

Proposition 6.3. *At each vertex $S = S_j$ of the graph K we have the following equalities*:
   a) $\Delta_j = c_1 + c_3 - c_2 - c_4$;
   b) $c_1 = (\Lambda_j/\Lambda_I)c_I$,
      $c_2 = (\Lambda_j/\Lambda_{II})c_{II}$,
      $c_3 = (\Lambda_j/\Lambda_{III})c_{III}$,
      $c_4 = (\Lambda_j/\Lambda_{IV})c_{IV}$.

Proof. We start with part (a). Consider the region U in Fig. 6.5(b) and manufacture from it a usual cross $\widetilde{U}$ by extending the separation segments N_i^- inside the positive annuli (dotted lines in Fig. 6.5(b)). For the cross $\widetilde{U}$, we consider the corresponding new quantities $\tilde{c}_1, \tilde{c}_2, \tilde{c}_3, \tilde{c}_4$ defined in the same way as c_1, c_2, c_3, c_4. Let t_1, t_2, t_3, t_4 denote the coefficients of the chain l associated with the edges K_1, K_2, K_3, K_4. Recall that they are determined from the relations

$$\sigma^{t_i}(x_i^-) = x_i^+ \qquad \text{(for } i = 1, 2, 3, 4).$$

It easily follows from the definition of the regions U and $\widetilde{U}$ that

$$\tilde{c}_1 - c_1 = t_2 - t_1, \qquad \tilde{c}_2 - c_2 = 0,$$
$$\tilde{c}_3 - c_3 = t_4 - t_3, \qquad \tilde{c}_4 - c_4 = 0.$$

We add these four equalities (changing signs in the second and fourth) so as to obtain the following expression:

$$(\tilde{c}_1 + \tilde{c}_3 - \tilde{c}_2 - \tilde{c}_4) - (c_1 + c_3 - c_2 - c_4) = t_2 + t_4 - t_1 - t_3.$$

Lemma 6.3. *Under the above assumptions, we have*

$$\tilde{c}_1 + \tilde{c}_3 - \tilde{c}_2 - \tilde{c}_4 = 0.$$

Proof. By our construction, the cross $\widetilde{U}$ is bounded by the level lines of the Hamiltonian $F = \pm\varepsilon_0$ and the four smooth curves that are extensions of the separation segments N_i^- inside positive annuli. It is easy to see that the quantity $\tilde{c}_1 + \tilde{c}_3 - \tilde{c}_2 - \tilde{c}_4$ remains constant under changing these curves (it is only required that they are smooth and transversal to the flow). Besides, in our calculations we may use the Morse–Darboux lemma (see Chapter 8 below), which states that, in a neighborhood of a saddle singular point $S = S_j$, there exist local coordinates (u, v) such that $F = uv$ and $\omega = \omega(uv)\, du \wedge dv$.

Thus, without loss of generality, we may assume that, in terms of coordinates (u, v), the cross $\tilde{U}$ is given by the relations $|uv| < \varepsilon_0$, $|u| < 1$, $|v| < 1$, i.e., is standard. For such a cross, the calculations can be done explicitly (see the proof of Lemma 6.2 above). Having done this, we see that, for the standard cross, all $\tilde{c}_i$'s are just equal to zero. This completes the proof of Lemma 6.3.   □

Now, to complete the proof of part (a), it suffices to observe that the alternated sum $t_1 - t_2 + t_3 - t_4$ coincides, by definition, with the coefficient Δ_j of the zero-dimensional chain ∂l.

Let us now prove the relation $c_1 = (\Lambda_j / \Lambda_I) c_I$ from part (b). To this end, we consider the rectangle bounded on the annulus C_I by the separation segments N_1^+ and N_2^+. Let $\Pi_1(F)$ denote the time along the piece of trajectory γ_F with a fixed value of the Hamiltonian F between N_1^+ and N_2^+ (that is, the increment of time inside the rectangle). As we already see, the increment of φ on any piece of the trajectory and that of the time t (in the sense of the flow w) are connected by the relation $\dfrac{\Delta\varphi}{\pi} = \dfrac{\Delta t}{\Pi_I(F)}$. But, according to Lemma 6.2, we have $\Delta\varphi = 2\pi(\Lambda_j / \Lambda_I)$. Therefore, $\Pi_1(F) = (\Lambda_j / \Lambda_I)\Pi_I(F)$. Taking the "finite parts" c_1 and c_I of the functions $\Pi_1(F)$ and $\Pi_I(F)$ in this relation, we obtain the desired equality $c_1 = (\Lambda_j / \Lambda_I) c_I$. This completes the proof.   □

6.2. CLASSIFICATION OF HAMILTONIAN FLOWS WITH ONE DEGREE OF FREEDOM UP TO TOPOLOGICAL CONJUGACY ON ATOMS

Thus, to each Hamiltonian system on a given atom (i.e., in a regular neighborhood of a singular level of the Hamiltonian) we have assigned a triple of invariants (Λ, Δ, Z). It turns out that invariants form a complete set, i.e., are sufficient for the classification of systems up to topological conjugacy.

Theorem 6.1. *Suppose we are given two smooth Hamiltonian systems w and w' with Morse Hamiltonians F and F' on two-dimensional compact oriented surfaces X and X'. Let $K = F^{-1}(0)$ and $K' = F'^{-1}(0)$ be connected singular levels of the Hamiltonians which are homeomorphic together with some regular neighborhoods (i.e., correspond to the same atom). Then the two following conditions are equivalent:*

1) for some neighborhoods $P = U(K)$ and $P' = U'(K')$ of these singular levels there exists a homeomorphism $\xi: P \to P'$ conjugating the Hamiltonian systems w and w' and preserving orientation;

2) the corresponding triples of invariants (Λ, Δ, Z) and (Λ', Δ', Z') coincide.

Four annuli adjoin the vertex S. We denote these annuli by C_I, C_{II}, C_{III}, C_{IV} (in general, some of them may coincide). Let $\Pi_I(F)$, $\Pi_{II}(F)$, $\Pi_{III}(F)$, $\Pi_{IV}(F)$ be the corresponding period functions. Then we have

$$\Pi_I(F) = -\Lambda_I \ln|F| + c_I(F),$$

where $c_I(F)$ is a function continuous at zero, and Λ_I is the sum of Λ_i over all vertices of K that belong to the boundary of the annulus C_I (including multiplicities). The function $c_I(F)$ will be sometimes called the *finite part* of the period function $\Pi_I(F)$. Set $c_I = c_I(0)$. The numbers c_{II}, c_{III}, c_{IV} can be analogously defined as the finite parts of the period functions Π_{II}, Π_{III}, Π_{IV} for $F = 0$.

Proposition 6.3. *At each vertex $S = S_j$ of the graph K we have the following equalities:*

a) $\Delta_j = c_1 + c_3 - c_2 - c_4$;

b) $c_1 = (\Lambda_j/\Lambda_I)c_I$,
$\quad c_2 = (\Lambda_j/\Lambda_{II})c_{II}$,
$\quad c_3 = (\Lambda_j/\Lambda_{III})c_{III}$,
$\quad c_4 = (\Lambda_j/\Lambda_{IV})c_{IV}$.

Proof. We start with part (a). Consider the region U in Fig. 6.5(b) and manufacture from it a usual cross $\widetilde{U}$ by extending the separation segments N_i^- inside the positive annuli (dotted lines in Fig. 6.5(b)). For the cross $\widetilde{U}$, we consider the corresponding new quantities $\widetilde{c}_1, \widetilde{c}_2, \widetilde{c}_3, \widetilde{c}_4$ defined in the same way as c_1, c_2, c_3, c_4. Let t_1, t_2, t_3, t_4 denote the coefficients of the chain l associated with the edges K_1, K_2, K_3, K_4. Recall that they are determined from the relations

$$\sigma^{t_i}(x_i^-) = x_i^+ \qquad \text{(for } i = 1, 2, 3, 4).$$

It easily follows from the definition of the regions U and $\widetilde{U}$ that

$$\widetilde{c}_1 - c_1 = t_2 - t_1, \qquad \widetilde{c}_2 - c_2 = 0,$$
$$\widetilde{c}_3 - c_3 = t_4 - t_3, \qquad \widetilde{c}_4 - c_4 = 0.$$

We add these four equalities (changing signs in the second and fourth) so as to obtain the following expression:

$$(\widetilde{c}_1 + \widetilde{c}_3 - \widetilde{c}_2 - \widetilde{c}_4) - (c_1 + c_3 - c_2 - c_4) = t_2 + t_4 - t_1 - t_3.$$

Lemma 6.3. *Under the above assumptions, we have*

$$\widetilde{c}_1 + \widetilde{c}_3 - \widetilde{c}_2 - \widetilde{c}_4 = 0.$$

Proof. By our construction, the cross $\widetilde{U}$ is bounded by the level lines of the Hamiltonian $F = \pm\varepsilon_0$ and the four smooth curves that are extensions of the separation segments N_i^- inside positive annuli. It is easy to see that the quantity $\widetilde{c}_1 + \widetilde{c}_3 - \widetilde{c}_2 - \widetilde{c}_4$ remains constant under changing these curves (it is only required that they are smooth and transversal to the flow). Besides, in our calculations we may use the Morse–Darboux lemma (see Chapter 8 below), which states that, in a neighborhood of a saddle singular point $S = S_j$, there exist local coordinates (u, v) such that $F = uv$ and $\omega = \omega(uv)\, du \wedge dv$.

Thus, without loss of generality, we may assume that, in terms of coordinates (u, v), the cross $\widetilde{U}$ is given by the relations $|uv| < \varepsilon_0$, $|u| < 1$, $|v| < 1$, i.e., is standard. For such a cross, the calculations can be done explicitly (see the proof of Lemma 6.2 above). Having done this, we see that, for the standard cross, all $\widetilde{c}_i$'s are just equal to zero. This completes the proof of Lemma 6.3.   □

Now, to complete the proof of part (a), it suffices to observe that the alternated sum $t_1 - t_2 + t_3 - t_4$ coincides, by definition, with the coefficient Δ_j of the zero-dimensional chain ∂l.

Let us now prove the relation $c_1 = (\Lambda_j/\Lambda_I)c_I$ from part (b). To this end, we consider the rectangle bounded on the annulus C_I by the separation segments N_1^+ and N_2^+. Let $\Pi_1(F)$ denote the time along the piece of trajectory γ_F with a fixed value of the Hamiltonian F between N_1^+ and N_2^+ (that is, the increment of time inside the rectangle). As we already see, the increment of φ on any piece of the trajectory and that of the time t (in the sense of the flow w) are connected by the relation $\dfrac{\Delta\varphi}{\pi} = \dfrac{\Delta t}{\Pi_I(F)}$. But, according to Lemma 6.2, we have $\Delta\varphi = 2\pi(\Lambda_j/\Lambda_I)$. Therefore, $\Pi_1(F) = (\Lambda_j/\Lambda_I)\Pi_I(F)$. Taking the "finite parts" c_1 and c_I of the functions $\Pi_1(F)$ and $\Pi_I(F)$ in this relation, we obtain the desired equality $c_1 = (\Lambda_j/\Lambda_I)c_I$. This completes the proof.   □

6.2. CLASSIFICATION OF HAMILTONIAN FLOWS WITH ONE DEGREE OF FREEDOM UP TO TOPOLOGICAL CONJUGACY ON ATOMS

Thus, to each Hamiltonian system on a given atom (i.e., in a regular neighborhood of a singular level of the Hamiltonian) we have assigned a triple of invariants (Λ, Δ, Z). It turns out that invariants form a complete set, i.e., are sufficient for the classification of systems up to topological conjugacy.

Theorem 6.1. *Suppose we are given two smooth Hamiltonian systems w and w' with Morse Hamiltonians F and F' on two-dimensional compact oriented surfaces X and X'. Let $K = F^{-1}(0)$ and $K' = F'^{-1}(0)$ be connected singular levels of the Hamiltonians which are homeomorphic together with some regular neighborhoods (i.e., correspond to the same atom). Then the two following conditions are equivalent:*

1) for some neighborhoods $P = U(K)$ and $P' = U'(K')$ of these singular levels there exists a homeomorphism $\xi\colon P \to P'$ conjugating the Hamiltonian systems w and w' and preserving orientation;

2) the corresponding triples of invariants (Λ, Δ, Z) and (Λ', Δ', Z') coincide.

COMMENT. More precisely, the coincidence of these invariants means that the pair (P, K) can be homeomorphically mapped onto (P', K') so that
1) the orientation is preserved,
2) the sign of the Hamiltonian is preserved,
3) the triple (Λ, Δ, Z) transfers to the triple (Λ', Δ', Z').
The last condition is in fact combinatorial. This must be taken into account when the atoms corresponding to the singularities in question admit non-trivial symmetries.

Proof (of Theorem 6.1).

a) Suppose the systems w and w' are topologically conjugate in some neighborhoods of the singular levels K and K'. Then the coincidence of the triples of invariants follows from Propositions 6.1, 6.2.

b) Now suppose the triples of invariants (Λ, Δ, Z) and (Λ', Δ', Z') coincide. As neighborhoods P and P' we take again the subsets of the form $P = F^{-1}[-\varepsilon, \varepsilon]$ and $P' = F'^{-1}[-\varepsilon', \varepsilon']$ so that the pairs (P, K) and (P', K') have the structure of two homeomorphic atoms.

REMARK. In some sense, the quantity ε' defines the width of an atom. This width in fact depends on ε and, moreover, depends on an annulus of the atom P'. This dependence follows from the fact that we have to make the period functions equal. The choice of ε' for each annulus will be commented on below (see Step 3).

The construction of a conjugating homeomorphism $\xi \colon P \to P'$ is divided into several steps.

Step 1. Let us choose separation segments on each annulus of the atom (P, K) just in the same way as we did it while constructing Δ and Z. On each edge K_i we take the corresponding separation points x_i^+ and x_i^- in terms of which we construct the chain l (see above). Then we do the same for the atom (P', K'). We have already some homeomorphism $\xi_0 \colon (P, K) \to (P', K')$ transferring the triple of invariants (Λ, Δ, Z) to the triple (Λ', Δ', Z'). It is clear that the separation segments for the second atom P' can be chosen so that ξ_0 transfers l into l'.

Step 2. Then we construct a new homeomorphism ξ_C from each annulus C of the atom P onto the corresponding annulus $C' = \xi_0(C) \subset P'$. Suppose, for definiteness, that the annuli C and C' are negative. Take an arbitrary initial separation segment N_C on the annulus C one of whose endpoints is the separation point x_i^- and the corresponding separation segment $N_{C'}$ on the annulus C' (this segment has been already constructed).

Let us introduce natural coordinate systems on the annuli C and C'. As the first coordinate on C we take the function F (increasing along the segment N_C) such that $F \in [-\varepsilon, 0)$. As the second coordinate t we take the time defined by the flow σ^t on each level line of F (that is an integral curve γ_F). Here we measure time from an initial point on N_C. As a result, we obtain smooth coordinates (F, t) on the open annulus C. We do the same thing to C' and obtain smooth coordinates (F', t') on the open annulus C'. It is clear that the coordinate t (resp. t') is defined modulo the period $\Pi(F)$ (resp. $\Pi'(F')$).

Step 3. It follows from Lemma 6.2 that both period functions $\Pi(F)$ and $\Pi'(F')$ tend monotonically to infinity as $F \to 0$ and $F' \to 0$. Hence the period functions $\Pi(F)$ and $\Pi'(F')$ are conjugate in a neighborhood of zero, i.e., there exists a continuous monotone change of variables $F' = \tau(F)$ such that $\Pi(F) = \Pi'(\tau(F))$. Without loss of generality, we shall assume that this change transfers the interval $[-\varepsilon, 0]$ exactly onto the interval $[-\varepsilon', 0]$.

We now consider the homeomorphism $\xi_C : C \to C'$ defined in coordinates by the following explicit formula:

$$\xi_C(F, t) = (\tau(F), t)$$

or, equivalently,

$$F' = \tau(F), \qquad t' = t.$$

This homeomorphism $\xi_C : C \to C'$ is well defined, because we have first equated the periods of the flows on the corresponding integral curves. Furthermore, the homeomorphism ξ_C conjugates the flows σ^t and σ'^t, since we have set $t' = t$.

Step 4.

Lemma 6.4.

a) *The homeomorphism ξ_C constructed above can be continuously extended to the boundary of the annulus C.*

b) *The homeomorphisms of the type of ξ_C constructed for neighboring annuli are sewn continuously into a single homeomorphism of the atom (P, K) onto the atom (P', K').*

Proof. Consider the standard action-angle variables on the annuli C and C', where we assume that the angle is measured from the initial segments N_C and $N_{C'}$. Then, under the mapping ξ_C, the level lines of the angle variable for the system w are mapped to the level lines of the angle variable for w' with the same values of the angle. We now use Lemma 6.1 and the equality of the Λ-invariants. By definition, the separation segments on the annuli C and C' are level lines of the angle variables φ and φ'. By Lemma 6.1, the value of the angle on them is determined uniquely by the Λ-invariant. Therefore (since the Λ-invariants coincide), ξ_C maps separation segments to separation segments.

Thus, the mapping ξ_C can be well defined at those points of the graph K that are endpoints of the separation segments on C and C' (i.e., at the separation points).

Take an arbitrary edge K_i of K which belongs to the boundary of C. This edge is an integral curve of w (a separatrix). Since the mapping ξ_C has already been defined at one point of this edge, we can extend ξ_C to all the remaining points of K_i up to a homeomorphism from K_i onto K_i' using the uniquely determined parametrization of the points on these edges regarded as integral curves of our Hamiltonian flows.

Doing the same thing for all the edges belonging to the boundary of C, we extend the mapping ξ_C to the inner boundary of C. The continuity of the mapping obtained is obvious. This is where we use the coincidence of Λ-invariants of the flows under consideration. This completes the proof of part (a) of Lemma 6.4.

We now prove part (b). Take two neighboring annuli C and D (positive and negative) adjoining a critical level of the Hamiltonian F (Fig. 6.6). Conjugating homeomorphisms ξ_C and ξ_D have already been constructed on them (including their inner boundaries). Consider an edge K_i which simultaneously adjoins the annuli C and D. We must prove that ξ_C and ξ_D coincide on this edge. We have two separation points x_i^+ and x_i^- on K_i; an analogous pair of points x'^+_i and x'^-_i is defined on the image K'_i. By our construction, $\xi_C(x_i^-) = x'^-_i$ and $\xi_D(x_i^+) = x'^+_i$.

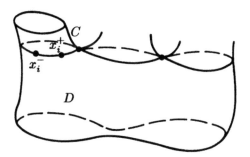

Figure 6.6

Now it is easy to see that the assertion to be proved follows from the coincidence of l and l'. Indeed, $x_i^+ = \sigma^{t_i}(x_i^-)$ and $x'^+_i = \sigma'^{t_i}(x'^-_i)$, where t_i is the coefficient of the chains l and l' corresponding to the edges K_i and K'_i. But then, since ξ_C is a conjugation, we obtain that

$$\xi_C(x_i^+) = \xi_C(\sigma^{t_i}(x_i^-)) = \sigma'^{t_i}(\xi_C(x_i^-)) = \sigma'^{t_i}(x'^-_i) = x'^+ = \xi_D(x_i^+).$$

Analogously, $\xi_C(x_i^-) = \xi_D(x_i^-) = x'^-_i$. It is clear that, coinciding at least at one point of the edge K_i, the homeomorphisms ξ_C and ξ_D will coincide on the whole edge K_i. This completes the proof of Lemma 6.4. □

The proof of this lemma in fact completes the proof of Theorem 6.1. Indeed, by Lemma 6.4, the constructed conjugating homeomorphisms $\xi_C \colon C \to C'$ are well sewn into a single conjugating homeomorphism $\xi \colon P \to P'$. □

COMMENT. It is easily seen from the proof that the constructed homeomorphism ξ, which conjugates the flows, is in fact smooth everywhere except, perhaps, for the points of the graph K.

COMMENT. In the case of the simplest singular fiber (like a figure eight curve), which contains the only singular point of the Hamiltonian (i.e., in the case of the atom B in our terms), all the constructed invariants are trivial. Thus, in this case, any two Hamiltonian systems are topologically conjugate in some neighborhoods of their singular leaves.

Thus, we have described a complete set of atomic invariants for Hamiltonian systems with one degree of freedom. In view of the reduction theorem (Theorem 5.1), we can now give a (topological) orbital classification of integrable systems on 3-atoms. Recall that for this we must consider, instead of the original Hamiltonian flow v on a 3-atom, the corresponding Poincaré flow on a transversal section. Note, however, that the transversal section is not uniquely chosen (even up to isotopy). That is why the Λ-, Δ-, and Z-invariants of the Poincaré flow will in general depend on a choice of a transversal section. Can one describe this dependence explicitly? A positive answer is obtained below. It turns out that it can be formulated in terms of a quite natural operation on the set of Hamiltonian systems defined on a fixed 2-atom (see Section 6.4).

In conclusion we note that these results allow us to classify Hamiltonian systems with one degree of freedom up to topological conjugacy not only on individual atoms, but also on closed two-dimensional surfaces. For this we must add to the invariants discussed above another one, which is an analog of the R-invariant on an edge of a molecule. Here, instead of the R-vectors, we must take the Π-vectors constructed on the basis of the period functions $\Pi(t)$ defined on the cylinders joining different atoms. As a result, we obtain the Reeb graph Y of the Hamiltonian endowed with some additional information. The coincidence of such graphs is a necessary and sufficient condition for the topological conjugacy of systems with one degree of freedom on closed surfaces (see [62] for details).

6.3. CLASSIFICATION OF HAMILTONIAN FLOWS ON 2-ATOMS WITH INVOLUTION UP TO TOPOLOGICAL CONJUGACY

In the reduction theorem we distinguished two natural cases. The first one is the case of a 3-atom without critical circles with non-orientable separatrix diagram. The corresponding 2-atom in this case has no star-vertices. In the second case, such critical circles exist. Theorem 6.1 gives a complete description of atomic invariants for 3-atoms in the first case. We now turn to a description of invariants for 3-atoms $U(L)$ that contain critical circles with non-orientable separatrix diagram. In this case, as we showed in the preceding chapter, the reduction theorem also holds, but now we must consider flows on 2-atoms P_{tr} with an involution χ. Recall that P_{tr} is a transversal section in the 3-atom $U(L)$. The quotient space of P_{tr} with respect to the involution χ is exactly the 2-atom P with stars corresponding to the 3-atom $U(L)$. In this case, P_{tr} is a double of the atom P.

According to the reduction theorem, we need to solve the following problem. Suppose we are given an involution χ on a surface P_{tr} and a Hamiltonian flow σ^t which corresponds to the Hamiltonian vector field $w = \operatorname{sgrad} F$ and is invariant with respect to χ. We need to classify such flows σ^t up to topological conjugations compatible with the involution χ. In other words, we consider two triples $(P_{\text{tr}}, \sigma^t, \chi)$ and $(P'_{\text{tr}}, \sigma'^t, \chi')$ to be equivalent if there exists a homeomorphism $\xi \colon P_{\text{tr}} \to P'_{\text{tr}}$ such that $\chi' = \xi^{-1}\chi\xi$ and $\sigma'^t = \xi^{-1}\sigma^t\xi$. For the classification we need appropriate invariants. It is natural to try to produce them from the above described invariants (Λ, Δ, Z) taking into account the involution χ.

Let $(\Lambda_{\text{tr}}, \Delta_{\text{tr}}, Z_{\text{tr}})$ be invariants of the flow σ^t on the double P_{tr}. Since σ^t is invariant under the involution χ, these invariants withstand the action of this involution. That is why formally we may suppose that they take their values on the quotient space $P = P_{\text{tr}}/\chi$, and we denote them by (Λ, Δ, Z).

In other words, we consider the invariant Λ as a set of numbers on the vertices of K. The invariant Δ is an element of the zero-dimensional boundary group $B_0(\widetilde{P})$. The invariant Z is an element of the homology group $H_1(\widetilde{P})$, where $\widetilde{P}$ is a closed surface obtained from the atom P by gluing 2-discs along all boundary circles.

Thus, the orbital invariants of a Hamiltonian system on a 3-atom $U(L)$ with critical circles with non-orientable separatrix diagrams take their values on the corresponding 2-atom P with stars, but not on its double P_{tr}. And this is very nice, since the atom P is well defined, but the double P_{tr} is not. Let us describe this construction in greater detail.

Take an arbitrary transversal section P_{tr} for the given 3-atom $U(L)$. The involution χ acts on P_{tr} as was described in the previous chapter. Consider the projection

$$(P_{\text{tr}}, K_{\text{tr}}) \to (P, K) = (P_{\text{tr}}, K_{\text{tr}})/\chi.$$

On the vertices of the graph K_{tr} we already have the numbers Λ_i that form the invariant $\Lambda_{\text{tr}} = \{\Lambda_1 : \Lambda_2 : \ldots : \Lambda_m\}$ of the Poincaré flow on P_{tr}. We now take an arbitrary vertex of the graph K and assign to it the number Λ_i associated with its preimage in the graph K_{tr}. A vertex of K has either one preimage (then this is a star-vertex) or two. In the case of two preimages, they correspond to the same number Λ_i, since the flow σ^t is χ-invariant. Thus, we have defined a certain set of numbers $\{\Lambda_i\}$ on the vertices of the graph K. We denote it by Λ.

We now define the invariants Δ and Z. To this end, we consider all the annuli of the double P_{tr}. They can be divided into two classes. The first class includes the annuli that are mapped onto themselves under the action of the involution χ. The second one consists of pairs of the annuli transferring to each other under the action of χ.

On the annuli of the first type we choose separation segments in an arbitrary way. Note that such a set of separation segments will automatically be invariant with respect to χ.

On the annuli of the second type we proceed as follows. Take a pair of annuli which are interchanged under the action of χ. On one of them, we choose separation segments arbitrarily. On the other, as separation segments we take their images under χ.

Now, just in the same way as before, we construct the one-dimensional chain l_{tr}. It is easy to see that this chain is χ-invariant. The point is that the set of separation segments itself has been made χ-invariant from the beginning. Now consider an arbitrary edge of the graph K. It has exactly two preimages in the graph K_{tr}. The coefficients of the chain l_{tr} corresponding to these two edges are the same. We assign their common value to the edge of K. As a result, we obtain some 1-chain, which we denote by l. In other words, we just identify the set of χ-invariant 1-chains of the graph K_{tr} with the set of 1-chains of K.

Now, in the same way as before, starting from l, we produce the invariants Δ and Z for the χ-invariant Hamiltonian flow σ^t.

Thus, to each χ-invariant flow σ^t on the double P_{tr}, we have assigned the triple of invariants (Λ, Δ, Z), which take their values on the 2-atom $P = P_{\mathrm{tr}}/\chi$. Note that on the atom (P, K) there is no Hamiltonian flow any more. The symplectic structure cannot be descended from P_{tr} to P, since the projection $P_{\mathrm{tr}} \to P$ is not a local diffeomorphism at star-vertices. However, from a formal viewpoint, the desired invariants (Λ, Δ, Z) appear finally on the atom (P, K), but not on its double $(P_{\mathrm{tr}}, K_{\mathrm{tr}})$.

This triple (Λ, Δ, Z) gives a complete set of invariants for a Hamiltonian system on a 2-atom with involution (or, equivalently, a complete set of orbital invariants on a 3-atom which has critical circles with non-orientable separatrix diagrams). We emphasize that, speaking of 2-atoms with involution, we mean an involution of a rather special type: it is symplectic, it preserves the Hamiltonian flow, and its fixed points are some of vertices of the graph K_{tr}. Besides (recall again) we consider Hamiltonian systems with Morse Hamiltonians only.

Theorem 6.2. *Suppose we are given two smooth Hamiltonian flows on atoms with involutions $(\sigma^t, P_{\mathrm{tr}}, \chi)$ and $(\sigma'^t, P'_{\mathrm{tr}}, \chi')$. These flows are topologically conjugate by means of a homeomorphism compatible with the involutions χ and χ' if and only if the corresponding invariants (Λ, Δ, Z) and (Λ', Δ', Z') coincide.*

Proof. Theorem 6.2 can be proved by analogy with Theorem 6.1. Indeed, using the invariants (Λ, Δ, Z) and (Λ', Δ', Z'), we can uniquely reconstruct the usual invariants $(\Lambda_{\mathrm{tr}}, \Delta_{\mathrm{tr}}, Z_{\mathrm{tr}})$ and $(\Lambda'_{\mathrm{tr}}, \Delta'_{\mathrm{tr}}, Z'_{\mathrm{tr}})$ of these systems. By Theorem 6.1, the coincidence of these invariants implies that the flows σ^t and σ'^t are topologically conjugate without taking into account the involutions χ and χ'. But this defect can easily be avoided by taking into account the symmetry of the flows with respect to the involutions. $\square$

6.4. THE PASTING-CUTTING OPERATION

Consider an arbitrary saddle atom (P^2, K) with a Hamiltonian system $w = \mathrm{sgrad}\, F$. As before, without loss of generality, we assume that $K = F^{-1}(0)$ and $P^2 = F^{-1}[-\varepsilon, \varepsilon]$.

We now introduce an important operation allowing us to rearrange the original system on the atom. This operation will vary the conjugacy class of the system, and our goal is to understand this variation.

Take an arbitrary edge K_i of the graph K and cut the surface P along a smooth segment that is transversal to the edge K_i and to integral curves of the vector field w (an example is shown in Fig. 6.7). Consider the rectangle $M_i = [0, m_i] \times [-\varepsilon, \varepsilon]$, where m_i is a certain positive real number, and 2ε is the "width" of the atom (in other words, $-\varepsilon$ and ε are the limits of variation of the Hamiltonian F inside the atom). Introduce natural coordinates (u, f) on M_i, where $u \in [0, m_i]$ and $f \in [-\varepsilon, \varepsilon]$, and consider the vector field $\partial/\partial u$. One can assume this vector field to be Hamiltonian with respect to the form $du \wedge df$ with the Hamiltonian f. The integral curves of this field foliate the rectangle in the horizontal direction, and the passage time is the same for each integral curve and is equal to m_i.

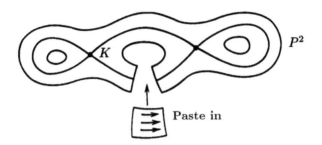

Figure 6.7

We now paste the rectangle M_i into the cut surface P as shown in Fig. 6.7: the vertical sides of the rectangle are pasted to the two sides of the cut surface so that the lines $\{f = f_0\}$ become parts of the level lines $\{F = f_0\}$. In other words, we increase the passage time along the edge K_i (as well as along all neighboring trajectories) by m_i forcing the flow to pass the additional piece M_i. Using Darboux's theorem, we can sew the smooth structures on the rectangle and on the cut surface so that we obtain a new smooth Hamiltonian system $\widetilde{w}$ on the same atom.

The described operation Φ is called *pasting a new piece* into the original flow σ^t on the edge K_i of the graph K.

We now consider the inverse operation. As above, we choose a transversal segment and consider the translation of this segment along the Hamiltonian flow by time m_i. As a result, we obtain another transversal segment. Then we cut from the surface the rectangle contained between these segments and glue the two sides of the cut together in the natural way (i.e., the original transversal segment and its image under translation by time m_i).

This inverse operation Φ^{-1} is called *cutting out a piece* of the original flow on the edge K_i of the graph K.

Now consider the general operation which is the composition of the operations of cutting out of a flow and pasting into a flow. With each edge K_i of the graph K we associate an arbitrary (positive or negative) real number m_i. This set of numbers can be treated as a real one-dimensional cochain m_i on the graph K. If m_i is positive, then we apply the pasting operation Φ on the corresponding edge; if m_i is negative, then we apply the operation Φ^{-1} of cutting out a piece of the flow on K_i. Let Φ_m denote the resulting operation (which is the compositions of the above elementary operations). It is clear that the result does not depend on the order of elementary operations.

Definition 6.5. The operation Φ_m is called *pasting-cutting of a piece of a flow* corresponding to a given 1-cochain m.

The operation Φ_m has the following two properties:
1) $\Phi_{m_1} \circ \Phi_{m_2} = \Phi_{m_2} \circ \Phi_{m_1} = \Phi_{m_1 + m_2}$,
2) $\Phi_{-m} = \Phi_m^{-1}$.
This immediately implies that we obtain an action of the group of one-dimensional cochains of the graph K on the space of all Hamiltonian systems on the given atom with Morse Hamiltonians.

Our aim is to understand what happens to the system as a result of applying the operation Φ_m. To this end, we must in fact compute the action of this operation on the invariants Λ, Δ and Z of the given system. We denote this induced action by Φ_m^*. It is easy to see that the action Φ^* is indeed well defined, because, under the action of Φ_m, conjugate systems are mapped to conjugate systems.

First of all we observe that Φ_m does not change the Λ-invariant. Indeed, all changes happen far from the singular points of the Hamiltonian.

The changes of Δ- and Z-invariants under the action of Φ_m are non-trivial. It is not difficult to show (see [53]) that this action admits the following representation:

$$\Phi_m^*(\Delta) = \Delta + \phi_1(m),$$
$$\Phi_m^*(Z) = Z + \phi_2(m),$$

where $\phi_1: C^1(\widetilde{P}) \to B_0(\widetilde{P})$ and $\phi_2: C^1(\widetilde{P}) \to H_1(\widetilde{P})$ are some linear operators (depending, in general, on the value of the Λ-invariant).

Notice that we do not know yet what values the invariants Δ and Z can take on the given atom $V = (P, K)$. Consider all possible Hamiltonian systems on this atom with the same value of the Λ-invariant.

Let $\Delta(V)$ and $Z(V)$ denote the subsets in $B_0(\widetilde{P})$ and in $H_1(\widetilde{P})$ respectively that consists of all possible values of the Δ- and Z-invariants for such systems.

Definition 6.6. $\Delta(V)$ (resp. $Z(V)$) is called the set of *admissible* Δ-invariants (resp. Z-invariants).

Let us emphasize that these sets depend on the choice of the Λ-invariant (fixed in advance).

From the formal viewpoint, the action Φ_m^* is defined only on the subsets $\Delta(V)$ and $Z(V)$, but, using the explicit formulas for this action, we can extend it to the whole spaces $B_0(\widetilde{P})$ and $H_1(\widetilde{P})$. After this, the subsets $\Delta(V)$ and $Z(V)$ remain, of course, invariant. Let us analyze their structure.

Proposition 6.4. *The action Φ^* is transitive on the sets of admissible invariants $\Delta(V)$ and $Z(V)$. These sets coincide with the images of the operators ϕ_1 and ϕ_2 respectively and are, in particular, vector subspaces.*

Proof. To prove this assertion, it suffices to verify that, by using an appropriate operation Φ_m, we can transform any system into a system with zero invariants Δ and Z. Let us do it.

Thus, we are given a Hamiltonian system w on the atom $V = (P, K)$. Consider one of the vertices S_j of the graph K. On each of the four edges that are incident to S_j we choose a point and draw transversal segments N_1, N_2, N_3, N_4 through them. As a result, we obtain a cross centered at S_j (Fig. 6.8). Recall that, on each of the four rectangles Z_i (into which this cross is divided by the graph K), we have a function (see above)

$$\Pi_i(F) = -\Lambda_j \ln|F| + c_i(F),$$

where $c_i(F)$ is a continuous function having some finite limit $c_i = c_i(0)$ at zero. It is easy to see that, moving the segments N_i, we can achieve the equalities $c_i = 0$ ($i = 1, 2, 3, 4$). Obviously, the quantities c_i depend only on the points of intersection of the segments N_i with the edges K_i.

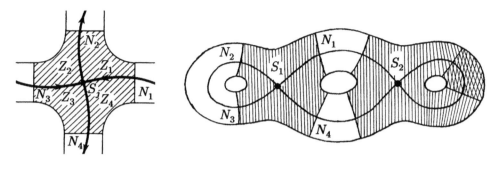

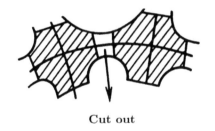

Figure 6.8   Figure 6.9

By continuing this procedure for all the vertices of the graph K, we obtain a set of crosses surrounding the vertices of K. As a result, we get the picture shown in Fig. 6.9. Note that some of the constructed crosses may intersect overlapping each other.

For each edge of the graph K, we now consider the region bounded by the pair of boundary segments of two neighboring crosses. Then we change these boundary segments (without changing the intersection points with K) so that this region becomes a rectangle in the sense that the passage time of the flow through this region is constant. This modification does not affect the main property of the crosses we need: all c_i's are equal to zero.

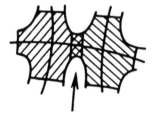

Cut out

Paste in a duplicate

Figure 6.10

It remains to notice that these rectangles constructed above are exactly the ones in the definition of the operation Φ_m. Let us carry out this procedure: we cut out those regions which are not covered by crosses and, on the contrary, paste their duplicates on those edges where these rectangles are the intersections of neighboring crosses (see Fig. 6.10). It is easy to see that, as a result, we obtain a system that is glued directly from the indicated crosses (with no overlaps or gaps).

It is easy to see that this system has zero Δ- and Z-invariants.   □

Among all the systems on V with zero Δ- and Z-invariants we distinguish one special system called a 0-*model*. As we have just seen, a Hamiltonian system on the atom V can be "glued" from systems given on individual crosses that surround the vertices of the graph K. Consider the simplest systems on such crosses.

Namely, represent each cross as the standard domain in the plane $\mathbb{R}^2(u, v)$ given by

$$|u| \leq 1, \quad |v| \leq 1, \quad |F| \leq \varepsilon_0, \qquad \text{where } F = uv.$$

Consider the differential form $\Lambda_j\, du \wedge dv$ on this cross and the Hamiltonian vector field $w = \operatorname{sgrad} F$. Note that, for such a system, the function $\Pi_i(F)$ giving the passage time inside the cross (in the i-th quadrant) will be very simple:

$$\Pi_i(F) = -\Lambda_j \ln |F|.$$

Here Λ_j is the value of the Λ-invariant on the corresponding vertex S_j of K. We now glue the surface P from these standard crosses along the boundary segments each of which is parametrized by the function F. Of course, we identify points with the same values of F so that, as a result, the functions F's given on individual crosses are sewn into a single smooth function F on the atom P. Besides (using Darboux's theorem) we can smoothly sew the symplectic structures and, consequently, the Hamiltonian flows. As a result, we obtain a smooth Hamiltonian system $w = \operatorname{sgrad} F$ given on the whole atom $V = (P, K)$. We shall call it a 0-*model* corresponding to the given atom V. Note that, in this case, the segments along which we glue crosses are exactly separation segments for w.

We now study the properties of the representation Φ^* in more detail. Note that our interpretation of the set of numbers $m = \{m_i\}$ as a one-dimensional cochain still remains mysterious. However, now we shall see that this is quite natural.

Below we consider several natural objects:

$K(\widetilde{P})$ is the cell decomposition (cell complex) of the surface $\widetilde{P}$ generated by the graph K;

$K^*(\widetilde{P})$ is the dual cell decomposition of $\widetilde{P}$.

As above, C_i, B_i, and Z_i are respectively the spaces of i-dimensional chains, boundaries, and cycles corresponding to the complex $K(\widetilde{P})$. We denote the analogous spaces for the dual complex $K^*(\widetilde{P})$ by C_i^*, B_i^*, Z_i^*. Finally, C^i, B^i, and Z^i are respectively the spaces of i-dimensional cochains, coboundaries, and cocycles for the cochain complex corresponding to the cell complex $K(\widetilde{P})$. We note the standard natural isomorphisms $C_i^* \simeq C^{2-i}$, $B_i^* \simeq B^{2-i}$, $Z_i^* \simeq Z^{2-i}$. The boundary and coboundary operators will be denoted by ∂ and δ.

Lemma 6.5. *If a 1-cochain m is a coboundary, then the operation Φ_m does not change the topological conjugacy class of a system (i.e., transforms each system into a conjugate one). In terms of invariants this means that the following inclusions holds: $B^1 \subset \ker \phi_1$, $B^1 \subset \ker \phi_2$.*

Proof. We prove this assertion for basis coboundaries. We interpret 1-cochains as linear combinations of the form $\sum m_i K_i^*$, where K_i^* are edges of the dual graph. Consider an arbitrary vertex S of the graph K. Let K_{i_1}, K_{i_3} be the edges entering it, and K_{i_2}, K_{i_4} the edges exiting from it. Then the basis 1-coboundary (related to the vertex S) can be written as

$$m = K_{i_1}^* + K_{i_3}^* - K_{i_2}^* - K_{i_4}^*.$$

We now apply the operation Φ_m to a 0-model. Clearly, this operation can be carried out on the cross corresponding to the vertex S. Namely, we must cut off two rectangles from two opposite sides of the cross, and then paste just the same rectangles to the two other opposite sides of the cross (see Fig. 6.11).

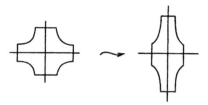

Figure 6.11

As a result, the symmetric cross becomes flattened in the horizontal direction and stretched in the vertical one. But the difference between these "symmetric" and "asymmetric" crosses can be observed only from the "Euclidean" point of view. From the "Hamiltonian" (or symplectic) point of view these crosses are absolutely identical, since they are obtained from each other by a symplectic transformation (by a translation by time ± 1 along the flow w). Thus, without changing the cross, we do not change the initial system, i.e., $\phi_1(m) = 0$ and $\phi_2(m) = 0$. ☐

Corollary. *The linear operator*

$$\phi_2': H^1(\widetilde{P}) \to H_1(\widetilde{P})$$

satisfying the relation $\phi_2'[m] = [\phi_2(m)]$ *for any* 1*-cocycle* $m \in Z^1$ *is well defined.*

Lemma 6.6. *If a* 1*-cochain* m *is a cocycle, then the operation* Φ_m *does not change the* Δ*-invariant of a system. In other words,* $Z^1 \subset \ker \phi_1$.

Proof. Let us apply the operation Φ_m to a 0-model for an arbitrary cocycle $m = \sum m_i K_i^*$. This means that, for each annulus of the atom $V = (P, K)$ (which is glued up by a disc to obtain the closed surface $\widetilde{P}$), the sum of numbers m_i over all the edges of K adjoining this annulus is zero. From the point of view of the operation Φ_m, this is equivalent to the fact that the sum of lengths of all rectangles pasted on the given annulus is also zero. This means, in particular, that the period of each trajectory of w does not change. Thus, all the period functions are preserved. But we know (see Proposition 6.3) that these functions allow us to compute the value of the Δ-invariant by an explicit formula. Therefore, the Δ-invariant does not change, and, consequently, $m \in \ker \phi_1$, as required. ☐

Since the quotient space C^1/Z^1 is canonically isomorphic to the space of 2-coboundaries B^2, Lemma 6.6 implies the following statement.

Corollary. *The linear operator*

$$\phi_1': B^2 = B_0^* \to B_0$$

satisfying the relation $\phi_1 = \phi_1' \circ \delta$ *is well defined.*

Clearly, the set $\Delta(V)$ of admissible Δ-invariants coincides with the image of ϕ_1'. This operator transforms the set of numbers $b = \{b_k\}$ associated with the annuli of the atom $V = (P, K)$ into the set of numbers $\Delta = \{\Delta_j\}$ associated to the vertices of K.

The next assertion gives an explicit formula for the operator ϕ_1'. Consider an arbitrary vertex S_j of the graph K and the four annuli $C_I, C_{II}, C_{III}, C_{IV}$ adjacent to it (see an analogous construction and notation in Proposition 6.3). Take the four real numbers $b_I, b_{II}, b_{III}, b_{IV}$ that are the coefficients of the zero-dimensional cochain $b \in B_0^*$ associated with these annuli.

Lemma 6.7. *The coefficient Δ_j of the zero-dimensional boundary chain $\Delta = \phi_1'(b) = \phi_1(m)$ that corresponds to the vertex S_j can be calculated by the following formula:*

$$\Delta_j = \Lambda_j \left(\frac{b_I}{\Lambda_I} + \frac{b_{II}}{\Lambda_{II}} + \frac{b_{III}}{\Lambda_{III}} + \frac{b_{IV}}{\Lambda_{IV}} \right).$$

Proof. In fact, this formula is already known (see Proposition 6.3). To show this, consider an arbitrary 1-cochain m such that $\delta m = b$ and apply the pasting-cutting operation Φ_m to a 0-model. It is clear that, for the initial 0-model, the finite parts of all period functions were equal to zero. Now, after applying the pasting-cutting operation, the finite parts of the period functions on each annulus are changed by the total length of the rectangles pasted on the given annulus. But the corresponding sum is equal up to a sign to the coefficient of the coboundary δm on the annulus under consideration. More precisely, the coefficients of the coboundary and the finite parts of the periods are the same on positive annuli and differ by sign on the negative annuli.

In other words, if a certain Hamiltonian system is obtained from a 0-model by the operation Φ_m, then the finite parts of the period functions of this system coincide up to sign with the coefficients of the cochain $b = \delta m$.

After this remark, the formula to be proved follows directly from Proposition 6.3. $\square$

REMARK. The total sum of finite parts b_i of all period functions (taken with sign $+$ or $-$ depending on the sign of an annulus) is equal to zero. It follows from the above interpretation of the set $\{b_i\}$ as the coboundary of m. The same assertion can be obtained from Lemma 6.3.

6.5. DESCRIPTION OF THE SETS OF ADMISSIBLE Δ-INVARIANTS AND Z-INVARIANTS

We have already introduced the sets $\Delta(V)$ and $Z(V)$ of admissible values of the Δ- and Z-invariants for a given atom $V = (P, K)$. Our aim is to describe these sets explicitly.

It turns out that the structure and dimension of the space $\Delta(V)$ depend essentially on the topology of the atom V. The relationship between them can be analyzed in the following way. First of all, we note that the space $\Delta(V)$ coincides with the image of the operator $\phi_1': B_0^* \to B_0$ (see the previous section), which determines explicitly the action Φ^* on Δ-invariants. The matrix of this operator can be written explicitly and analyzed from the point of view of its rank, kernel, and image. This scheme is realized in detail in [53], [62]; and here we only give the final result.

First we introduce a new interesting object, namely, the set of atomic circles corresponding to a given atom $V = (P, K)$. Consider the graph K and one of its edges. Then we start to move toward one of two vertices of K that are the endpoints of the edge. Upon reaching the vertex, we can uniquely exit from it along the opposite edge of the cross (i.e., not detouring from the path). Thus, we move over the graph K (generally, with self-intersections) until we return to the initial edge. It is clear that as a result we have described a circle immersed in the surface P. Then we take one of the remaining edges (if there are any) and repeat the process. Thus, we represent the graph K as a union of a number of circles $\gamma_1, \ldots, \gamma_q$ immersed to P.

Definition 6.7. The circles $\gamma_1, \ldots, \gamma_q$ constructed in the manner described above will be called *atomic circles.*

The atomic circles are a very natural object: each graph K can be represented as the result of "superimposing" several immersed circles (Fig. 6.12).

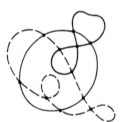

Figure 6.12

Each atomic circle γ_i realizes some 1-cycle $[\gamma_i]$ in the one-dimensional real homology group $H_1(\tilde{P})$. Let $\gamma H_1(\tilde{P})$ denote the subgroup in $H_1(\tilde{P})$ generated by all the cycles $[\gamma_1], \ldots, [\gamma_q]$.

Proposition 6.5. *The dimension of the space $\Delta(V)$ of admissible values of the Δ-invariant on a given atom $V = (P, K)$ is equal to*

$$\dim \Delta(V) = n - 2g - q + \dim \gamma H_1(\tilde{P}),$$

where n is the number of vertices of the graph K (i.e, the number of singular points of the Hamiltonian on the singular level), g is the genus of the closed surface $\tilde{P}$, and q is the number of atomic circles.

Proof. See [53], [62]. □

The space $B_0(\tilde{P})$ of zero-dimensional chains can be decomposed into the direct sum of two subspaces: $\Delta(V) \oplus \Delta^*(V)$, where $\Delta^*(V)$ is the orthogonal complement of $\Delta(V)$ in $B_0(\widetilde{P})$. Here we use a natural inner product in $C_0(\widetilde{P})$ such that the basis consisting of elementary chains of the form $1 \times S_j$ (where S_j is a vertex of K) is orthonormal.

We now describe explicitly a basis in $\Delta^*(V)$. In other words, we indicate linear relations which distinguish admissible Δ-invariants. Actually, these relations depend on the value of the Λ-invariant. So, for simplicity, we shall assume that $\Lambda = (1 : 1 : \dots : 1)$.

It follows from Proposition 6.5 that the dimension of $\Delta^*(V)$ is equal to $(q - 1) + (2g - \dim \gamma H^1(\widetilde{P}))$. First we shall indicate the part of the basis that consists of $(q - 1)$ elements and corresponds to the atomic circles. We take an arbitrary atomic circle γ_i, and for it we construct the corresponding basis element $b(\gamma_i) \in \Delta^*(V)$. Moving along this circle, we alternately place the numbers $+1$ or -1 on the vertices of K that we encounter. If we pass through some vertex two times, then it turns out that we place numbers of different signs on it (i.e., $+1$ and -1). In this case, we take the sum of these numbers, i.e., zero. To demonstrate this effect, we shall place numbers in a more definite way. Imagine that we move not just along the atomic circle, but slightly from the right of it. This sentence has a natural meaning, because our atom is oriented. Then, we successively pass from positive annuli to negative annuli and vice versa. We could assume for definiteness that we place $+1$ in passing from a positive to a negative annulus, and -1 otherwise. It is easy to see that, after this agreement, the rule of placing signs becomes well-defined and does not depend on the direction of moving along the circle.

Proposition 6.6 [53]. *The set of 0-chains $b(\gamma_1), \dots, b(\gamma_{q-1})$ gives a set of $q-1$ linearly independent elements in the space $\Delta^*(V)$.*

REMARK. If we take all the chains $b(\gamma_1), \dots, b(\gamma_q)$, then they will satisfy the linear relation $\sum_{i=1}^{q} b(\gamma_i) = 0$. But, if we remove at least one element from this set, then the remaining 0-chains become linearly independent.

We now proceed to an explicit description of the second part of the basis in the space $\Delta^*(V)$. Consider the subgroup $\gamma H_1(\widetilde{P})$ generated by the homology classes $[\gamma_i]$ of the atomic circles, $1 \leq i \leq q$. Let $\alpha_1, \dots, \alpha_p$ denote 1-cocycles from $H^1(\widetilde{P})$ which are orthogonal to the subspace $\gamma H_1(\widetilde{P})$ (i.e., orthogonal to each atomic circle). There are p of them, where $p = 2g - \dim \gamma H_1(\widetilde{P})$. For each of them, we now construct a certain element $b(\alpha_i) \in \Delta^*(V)$. We may assume that the 1-cocycle $\alpha = \alpha_i$ is realized as a circle immersed smoothly into the surface $\widetilde{P}$. The condition that α is orthogonal to all atomic circles means that its intersection index with each of them is zero.

We divide the construction of $b(\alpha)$ into several steps.

Step 1. Consider an arbitrary atomic circle γ_i and its intersection points with the cocycle α. Assuming the atomic circle γ_i and the cocycle α to be oriented somehow, we can assign $+$ or $-$ to each of their intersection points (following the same rule as that usually used to define the intersection index). Since the intersection index of α and γ_i is equal to zero, the number of intersection point is even, and they can be divided into pairs of points with different signs. Consider one of such pairs x and x'.

Step 2. Then we move from the point x along the atomic circle γ_i (we may choose either of two possible directions) and place alternately numbers $+1$ and -1 on all the vertices that we encounter until we come to the point x'. The first number is placed according to the following rule. The point x belongs to some edge of the graph K. This edge has its own canonical orientation (given by the flow). If the intersection index for this edge and the cocycle α is positive at the point x, then we place $+1$ on the first vertex of K that we meet, otherwise we place -1. Then the signs for the remaining points alternate. If we pass through a certain vertex twice, then we take the sum of those numbers which we must place on it. Thus, we obtain some set of numbers $b(x, x')$ on one of two halves of the atomic circle γ_i into which it is divided by the points x and x'.

Step 3. We now carry out this procedure for the remaining pairs of points of intersection of α and γ_i. And then we repeat the same for each atomic circle. If we pass through some vertex of the graph several times, then all the numbers placed on it are added together. In other words, we sum all the sets $b(x, x')$ (considered as 0-chains).

As a result, we obtain some set of numbers placed on the vertices of the graph K, i.e., a 0-chain, which we denote by $b(\alpha)$.

Proposition 6.7 [53]. *The 0-chains $b(\alpha_1), b(\alpha_2), \ldots, b(\alpha_p)$ constructed above give the "second half" of the desired basis in $\varDelta^*(V)$. More precisely, the set of 0-chains $b(\gamma_1), \ldots, b(\gamma_{q-1}), b(\alpha_1), \ldots, b(\alpha_p)$ forms a basis of $\varDelta^*(V)$.*

Thus, we can define the space $\varDelta(V)$ of admissible $\varDelta$-invariants as the orthogonal complement to the subspace $\varDelta^*(V)$, whose basis we have just described. There is, of course, another way to describe $\varDelta(V)$. We can just take the image of the operator $\phi_1': B_0^* \to B_0$, which is given by an explicit formula (see the previous section).

Finally we describe the space $Z(V)$ of admissible Z-invariants. As we see from the corollary of Lemma 6.5 (see the previous section), the space $Z(V)$ can be interpreted as the image of the operator $\phi_2': H^1(\widetilde{P}) \to H_1(\widetilde{P})$. It turns out that this operator has a very natural topological meaning, which, in particular, immediately gives the desired description of $Z(V)$.

Proposition 6.8 [53]. *The operator $\phi_2': H^1(\widetilde{P}) \to H_1(\widetilde{P})$ is the Poincaré duality isomorphism. In particular, the space $Z(V)$ coincides with the whole homology group $H_1(\widetilde{P})$. In other words, any abstract Z-invariant is admissible.*

The following statement summarizes the above results.

Proposition 6.9. *Suppose we are given arbitrary admissible values of the invariants $\varLambda$, $\varDelta$, and Z on a saddle atom $V = (P, K)$. Then, on this atom, there exists a Hamiltonian system (obtained by pasting-cutting operation from the 0-model) with the given values of the invariants $\varLambda$, $\varDelta$, and Z.*

In conclusion we discuss the following natural question. Suppose we are given some saddle atom $V = (P, K)$ and we are interested in the space of all Hamiltonian systems on it considered up to topological conjugacy. In other words, we consider topologically conjugate systems to be identical. What is the dimension of this space? What is the relationship between this dimension and the topology of a given atom?

To answer this question, we only need to compute the total dimension of the spaces of admissible invariants $\varLambda$, $\varDelta$, and Z, which can be naturally considered as parameters in the space of Hamiltonian systems on an atom $V = (P, K)$. As we already noticed, the Λ-invariant can be arbitrary, and, consequently, the dimension of the "space of Λ-invariants" is $n - 1$, where n is the number of vertices of the graph K (i.e., the complexity of the atom V). The dimension of the space $\Delta(V)$ is given by Proposition 6.5, and the dimension of $Z(V)$ coincides with the first Betti number of the surface $\widetilde{P}$, i.e., is equal to $2g$, where g is the genus of V. Summing these three numbers, we obtain the following result.

Proposition 6.10. *The dimension of the space of Hamiltonian systems on an atom $V = (P, K)$ is equal to*

$$2n - q - 1 + \dim \gamma H_1(\widetilde{P}),$$

where n is the complexity of V, q is the number of atomic circles, and $\gamma H_1(\widetilde{P})$ is the subspace in $H_1(\widetilde{P})$ generated by the atomic circles.

In particular, for the simplest atoms B, C_1, C_2, D_1, D_2, this dimension is equal to 0, 3, 1, 2, 2, respectively.

Chapter 7

Smooth Conjugacy of Hamiltonian Flows on Two-Dimensional Surfaces

7.1. CONSTRUCTING SMOOTH INVARIANTS ON 2-ATOMS

In this chapter we study the question on the classification of Hamiltonian vector fields with one degree of freedom in the smooth case. Of course, all the topological invariants constructed in the preceding chapter remain smooth invariants, but this is not enough for the smooth classification. However, although the set of smooth invariants is much bigger than that of topological invariants, the construction of smooth invariants is simpler and more natural.

In our book, we confine ourselves to semi-local (i.e., "atomic") invariants and do not discuss the global classification of Hamiltonian flows on two-dimensional closed surfaces.

Let, as before, (P, K) be an atom with a smooth Hamiltonian vector field $w = \operatorname{sgrad} f$ on it. Without loss of generality, we shall assume in this section that $K = f^{-1}(0)$.

REMARK. In this chapter, the Hamiltonian of the system shall be denoted by f instead of F. Recall that we used the notation F for a Poincaré Hamiltonian on a transversal section, and f denoted an integral of the initial Hamiltonian system on M^4. Generally speaking, the functions f and F are, of course, different. But on the other hand, by changing the symplectic structure on the transversal section, we can consider the additional integral f as the Hamiltonian of the Poincaré flow. We can make such a change, since we are interested in the Hamiltonian flow w itself, but not in its representation in the form $w = \operatorname{sgrad} f$ by means of a symplectic structure and a Hamiltonian. Besides, in this chapter, we shall discuss Hamiltonian flows on two-dimensional surfaces as an independent object, without any connection with the reduction theorem (Theorem 5.1), which clarifies an important role of such flows for the orbital classification of integrable Hamiltonian systems with two degrees of freedom.

First of all, it is natural to study the question on the local classification of Hamiltonian vector fields in a neighborhood of a singular point, i.e., near a vertex of an atom.

If an atom has type A (that is, the Hamiltonian has a local minimum or a local maximum at the singular point), then the study of this case does not lead to any difficulties. Trajectories of this vector field are circles around the singular point, and the field is completely characterized by the period function for these trajectories. That is why we shall consider the case of a saddle singular point.

Since the Hamiltonian vector field is completely determined by the symplectic structure and the Hamiltonian, the problem under consideration can be solved with the help of the following lemma, proved in [86], on the canonical form of the Hamiltonian.

Lemma 7.1. *Let S be a non-degenerate singular point of the Hamiltonian f. Then there is a regular coordinate system (x, y) such that*

a) $\omega = dx \wedge dy$,

b) $f = f(z)$, *where* $z = xy$.

This assertion may be considered as a natural generalization of two classical results: Morse's lemma and Darboux's theorem. It will be convenient for us to restate it in the following equivalent form.

Lemma 7.2. *Let S be a non-degenerate saddle singular point of the Hamiltonian f. Then there is a regular coordinate system (x, y) such that*

a) $f = xy$,

b) $\omega = \omega(z)dx \wedge dy$, *where* $z = xy$.

As well as this lemma, we also need to know the answer to the following question: how to find the function $\omega(z)$ without finding the canonical coordinate system explicitly from Lemma 7.2. To answer this question we introduce the Λ^*-invariant of a singular point, which has the same nature as the Λ-invariant defined in Chapter 6.

Choose an arbitrary coordinate system (u, v) in a neighborhood of the singular point S in terms of which the Hamiltonian has the form $f = uv$; and let $\omega = \omega(u, v) \, du \wedge dv$. Expand the function $\omega(u, v)$ as a Taylor series about the point S:

$$\omega(u, v) \simeq \sum_{i,j=0}^{\infty} a_{ij} u^i v^j \,.$$

Then we put

$$\Lambda^*(S) = \Lambda_f^*(S) = \sum_{k=0}^{\infty} \lambda_k z^k \,,$$

where $\lambda_k = a_{kk}$.

Definition 7.1. The series $\Lambda^*(S)$ is called the Λ^*-*invariant* of the saddle singular point S.

Let us show that the Λ^*-invariant does not depend on the choice of the coordinate system (u, v). To this end, we give another coordinate-free definition of this invariant via some natural characteristics of the vector field w. Consider the domain (cross) $U(S)$ surrounding the singular point S as indicated in Fig. 7.1. The curves N_1, N_2, N_3, N_4 are smooth and intersect the trajectories of the field w transversally. Clearly, the qualitative shape of this domain is not changed under any smooth diffeomorphisms.

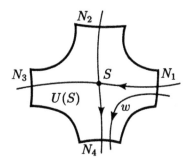

Figure 7.1

Let γ_f denote the piece of the trajectory of the Hamiltonian field w that lies in $U(S)$, and on which the value of f is fixed.

As before, we consider the function $\pi(f)$ which assigns the time taken to go along the piece γ_f under the action of the Hamiltonian flow (Fig. 7.1). In Lemma 6.2, we obtained the following representation for $\pi(f)$:

$$\pi(f) = -\left(\sum_{k=0}^{\infty} a_{kk} f^k\right) \ln f + c_n(f),$$

where $c_n(f)$ is a C^n-smooth function on the segment $[0, f_0]$. It is easy to see that the polynomial $\sum_{k=0}^{\infty} a_{kk} f^k$ in this representation is well defined. Indeed, what happens to the function $\pi(f)$ if we change the domain $U(S)$? It is clear that N_i are replaced by some other transversal segments N_i', but this influences only the function $c_n(f)$, namely, the result is to add some C^∞-smooth function to $c_n(f)$ that characterizes the distance between the new and old segments.

Thus, the coefficients of the polynomial $\sum_{k=0}^{\infty} a_{kk} f^k$ can be defined in an invariant way as the coefficients in the asymptotics of the function $\pi(f)$ as $f \to 0$. On the other hand, the set of these coefficients gives the Λ^*-invariant of the singular point S.

COMMENT. As was indicated in the Corollary of Lemma 6.2, the function $\pi(f)$ admits the following representation:

$$\pi(f) = -\lambda(f) \ln|f| + c(f), \qquad (*)$$

where $\lambda(f)$ and $c(f)$ are C^∞-smooth functions on $[0, f_0]$. It is easily seen that the Λ^*-invariant of the Hamiltonian vector field at the singular point S is exactly the Taylor series of the function $\lambda(f)$.

Thus, in particular, we have shown that the canonical form of the symplectic structure (see Lemma 7.2) is defined almost uniquely. More precisely, the Taylor expansion of $\omega(z)$ at zero will be defined uniquely. It is not difficult to show that, on the other hand, any function $\tilde{\omega}(z)$ with the right Taylor expansion (that is, coinciding with the Λ^*-invariant) may be realized by choosing a suitable coordinate system. Notice also that in all our arguments the Hamiltonian f is assumed to be fixed. Therefore, as a result, we end up with the following assertion.

Proposition 7.1. *Let S_i be a non-degenerate singular saddle point of the Hamiltonian f_i, and let ω_i be the corresponding symplectic structure $(i = 1, 2)$. If the Λ^*-invariants of the points S_1 and S_2 coincide, then there is a local diffeomorphism $\xi: U_1(S_1) \to U_2(S_2)$ such that $\xi^*(f_2) = f_1$ and $\xi^*(\omega_2) = \omega_1$. Conversely, if there is such a diffeomorphism, then the Λ^*-invariants of the singular points S_1 and S_2 coincide.*

In fact, we are interested not in the symplectic structure but in the Hamiltonian flow induced by it. However, since this flow is well defined by f and ω, we can reformulate Proposition 7.1 in the following way.

Corollary. *Let S_i be a non-degenerate singular point of the Hamiltonian f_i, and let $w_i = \operatorname{sgrad} f_i$ be the corresponding Hamiltonian vector field $(i = 1, 2)$. If the Λ^*-invariants of S_1 and S_2 coincide, then there is a local diffeomorphism $\xi: U_1(S_1) \to U_2(S_2)$ such that $f_2 \circ \xi = f_1$ and $d\xi(w_1) = w_2$. Conversely, if there is such a diffeomorphism, then the Λ^*-invariants of the singular points S_1 and S_2 coincide. In other words, the Hamiltonian vector fields w_1 and w_2 are locally C^∞-conjugate (with the additional condition that the Hamiltonian is invariant) if and only if their Λ^*-invariants coincide.*

The condition that the Hamiltonian is ξ-invariant (i.e., $f_2 \circ \xi = f_1$) is, however, superfluous and we can avoid it in a natural way. To this end, we should control the dependence of Λ^* on the choice of the Hamiltonian.

Let g be some other Hamiltonian of a given vector field $w = \operatorname{sgrad} f$ in a neighborhood of S (of course, the symplectic structure is then also different). Let $g(S) = 0$. Then the Hamiltonians are expressed one from the other and we can expand f at the point S into a power series with respect to g:

$$f \simeq \sum_{k=0}^{\infty} b_k g^k, \qquad b_1 \neq 0.$$

It is easy to see that, as a result, we get a new series $\Lambda_g^*(S) = \sum_{k=0}^{\infty} \tilde{\lambda}_k \tilde{z}^k$ obtained from $\Lambda_f^*(S) = \sum_{k=0}^{\infty} \lambda_k z^k$ by the formal substitution $z = \sum_{k=1}^{\infty} b_k \tilde{z}^k$.

Thus, the invariant Λ^* is defined, generally speaking, modulo formal transformations, and we can formulate the final result as follows.

Proposition 7.2. *Two Hamiltonian vector fields are locally C^∞-conjugate in neighborhoods of singular saddle points if and only if their Λ^*-invariants are formally conjugate.*

It is not difficult to describe all conjugate classes of Hamiltonian vector fields at a singular saddle point (for one degree of freedom). The number of such classes is that of formally conjugate power series in the above sense. For power series (in one variable) we can explicitly indicate a canonical representative in each class.

Lemma 7.3. *A power series in one variable is formally conjugate to one of the following polynomials:*

$$\lambda,$$
$$\lambda + z,$$
$$\lambda + z^2, \quad \lambda - z^2,$$
$$\lambda + z^3,$$
$$\lambda + z^4, \quad \lambda - z^4,$$
$$\cdots$$
$$\lambda + z^{2k-1},$$
$$\lambda + z^{2k}, \quad \lambda - z^{2k},$$
$$\cdots$$

where λ is some real non-zero number. None of the listed polynomials are conjugate.

Proof. Consider an arbitrary power series of the form

$$\lambda + a_n w^n + a_{n+1} w^{n+1} + \ldots .$$

Let n be even, and let $a_n > 0$. Then this series is formally conjugate to the polynomial $\lambda + z^n$. The formula of the corresponding change is as follows:

$$z = w(a_n + a_{n+1}w + \ldots)^{1/n}.$$

Here, of course, we mean the formal Taylor expansion of the radical into power series in w. If $a_n < 0$, then we set $z = -w(-a_n - a_{n+1}w - \ldots)^{1/n}$.

If n is odd, then the formula is analogous. $\quad\square$

COMMENT. In the general case (not necessarily Hamiltonian), the smooth classification of vector fields in a neighborhood of a saddle singular point can be obtained by using the Chen theorem, which reduces the smooth classification of vector fields to the smooth classification of power series, and the Poincaré–Dulac theorem, which determines the canonical form of the corresponding power series (see [16]). Our last assertion can also be obtained with the help of these classical results. However, in what follows, we shall need the condition that the Hamiltonian is preserved; that is why we proceeded in some other way.

In what follows, we shall also need another statement related to the behaviour of the system about a singular saddle point. Consider again the cross $U(S)$ surrounding the singular point (Fig. 7.1). It is easily seen that by changing the boundary of the cross, i.e., the segments N_1, N_2, N_3, N_4, we may reduce the four functions $\pi_i(f)$ $(i = 1, 2, 3, 4)$ to the form

$$\pi_i(f) = -\lambda(f) \ln |f|,$$

i.e., make all the functions $c(f)$ from the representation (*) to be equal zero identically. A cross $U(S)$ satisfying this condition shall be called *canonical*. We emphasize that this notion depends on the choice of a Hamiltonian.

Now consider the Hamiltonian vector field w on the whole atom, i.e., in some neighborhood P of the critical fiber $K = f^{-1}(0)$ of the Hamiltonian f. If we remove the graph K from the surface P, then P splits into a disjoint union of annuli. Each annulus C_n can be regarded as a one-parameter family of closed trajectories whose parameter is the function f itself. Since each trajectory is closed, it has a certain period. Thus, for every annulus C_n, we can define the natural period function $\Pi_n(f)$ that indicates the period of the trajectory with a given value f on it.

It is clear that the period functions are invariants of the vector field w (up to conjugacy). In the continuous case, by the way, they are invariants too, but there we had to consider conjugacy by homeomorphisms; so this invariant turned out to be trivial. Indeed, all period functions are monotonically increasing up to infinity as f tends to a critical value. Therefore, from the topological viewpoint, any two of them are conjugate. In the smooth case, the situation is different.

First consider the period function on a single annulus. It has, as we already saw, the representation

$$\Pi_n(f) = -A(f) \ln |f| + B(f),$$

where A and B are some smooth functions on $[0, \varepsilon_0]$. These functions are not uniquely determined themselves, but their Taylor expansions at zero are. Moreover, the Taylor expansion of $A(f)$ coincides with the sum of Λ^*-invariants of the vertices of K (i.e., singular points of w) lying on the boundary of the given annulus.

Consider two functions of this type:

$$\Pi\,(f) = -A\,(f)\ln|f| + B\,(f),$$
$$\Pi'(f') = -A'(f')\ln|f'| + B'(f').$$

What is it possible to say about their smooth conjugacy? In other words, when does there exist a smooth change $f' = \tau(f)$ (on the whole segment $[0, \varepsilon_0]$ including zero) such that
$$\Pi(f) = \Pi'(\tau(f))\,?$$

It turns out that the smoothness condition imposes very serious restrictions on the pair of these functions. The question is reduced to the classification of pairs $(\widetilde{A}(f), \widetilde{B}(f))$ of the Taylor expansions of A and B at zero. It is not difficult to write down formal transformations of these series under a formal change $f = \tau(f')$. We shall not do this, but shall indicate some (also formal) method for choosing a certain canonical representative in each equivalence class.

Lemma 7.4. *For any function of the form* $\Pi(f) = -A(f)\ln|f| + B(f)$, *there exists a smooth change* $f = \tau(f')$ *on the segment* $[0, \varepsilon_0]$ *which reduces this function to the form* $\Pi(\tau(f')) = -A'(f')\ln|f'|$ *(that is, totally eliminates the finite part of this representation). Two functions*

$$-A'(f')\ln|f'| \quad and \quad -A''(f'')\ln|f''|$$

are smoothly conjugate on a sufficiently small segment $[0, \varepsilon_0]$ *if and only if the Taylor expansions of* A' *and* A'' *coincide.*

Thus, the smooth conjugacy class of a period function (in a neighborhood of a singular fiber) can be parameterized by a certain power series. In particular, the number of such classes is infinite. More precisely, we have a countable number of real-valued parameters as invariants of smooth conjugacy of period functions.

Note that our case is even more complicated. We have not one but several period functions for the same atom, which correspond to different annuli. Speaking of the conjugacy of two collections of period functions, we must remember that the conjugating change $f' = \tau(f)$ must be the same for all period functions. As a result, the number of invariants is increasing. From the formal point of view, the conjugacy problem for two collections of period functions can obviously be solved by using Lemma 7.4.

In order not to discuss the formal part of the problem, below we shall assume the period functions to be fixed. Do there exist any other invariants except period functions and Λ^*-invariants?

It turns out that, if an atom is planar, then the above invariants are sufficient for classification. If not, then another Z^*-invariant appears, which is a smooth analog of the topological Z-invariant. Let us describe this construction. The idea completely corresponds to the proof of the fact that any system can be constructed from a system with zero invariants Δ and Z by means of the pasting-cutting operation (see Proposition 6.4).

Thus, consider an arbitrary smooth system $w = \operatorname{sgrad} f$ on an atom $V = (P, K)$. Consider all vertices of the graph K and surround them by canonical crosses. Now cut the atom along the boundary segments of these crosses. As a result, the atom splits into crosses and "rectangles" (see Fig. 7.2). We use the term "rectangle" only by convention. More precisely, we should rather speak of a part of the atom bounded by two segments N_i^+ and N_i^- transversally intersecting an edge of the graph K.

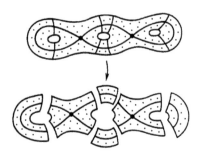
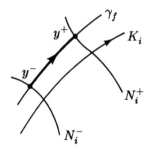

Figure 7.2   Figure 7.3

These segments are parameterized by the value of the Hamiltonian f at their points. Consider the function $m_i(f)$ that measures the distance between these segments. More precisely, if γ_f is an integral curve of the vector field w with the given value of the Hamiltonian f on it, and $y^+ = N_i^+ \cap \gamma_f$, $y^- = N_i^- \cap \gamma_f$ (see Fig. 7.3), then $m_i(f)$ is the distance between these points in the sense of the Hamiltonian flow σ^t, i.e.,

$$y^+ = \sigma^{m_i(f)}(y^-).$$

Thus, on each edge of K, a certain smooth function $m_i(f)$ occurs. We now consider the formal cochain $m^* = \Sigma \tilde{m}_i K_i^*$, where K_i^* denote the edges of the conjugate graph, which we interpret as a basis for the space of one-dimensional cochains, and where $\tilde{m}_i$ is the formal Taylor series of the function $m_i(f)$ at zero. This cochain is just a smooth analog of the cochain m which has occurred in the definition of the pasting-gluing operation. In the smooth case, there is a natural analog of this operation. However, here it is natural to consider the sides of the "rectangles" to be curvilinear so that the "rectangle" has varying width measured just by the function m_i.

Clearly, the cochain m^* is not uniquely determined, since it depends on the choice of canonical crosses surrounding the vertices of K. However, we assert that (if the Hamiltonian is considered to be fixed) the class of this cochain $[m^*]$ modulo the space of coboundaries $B^1(V, \mathbb{R}[f])$ (with coefficients in the ring of formal power series $\mathbb{R}[f]$).

Let us prove this. To this end, it suffices to verify that the class $[m^*]$ does not depend on the choice of canonical crosses.

The fact that, under change of m^* by adding a coboundary, the system does not change has actually been proved in Lemma 6.5. The only difference is that, in the topological case, the coefficients were real numbers, whereas now we replace them by formal power series.

Let us look more carefully what arbitrariness in the choice of a canonical cross $U(S)$ surrounding the vertex $S \in K$ is. In other words, which transformations can such a cross undergo while preserving the property to be canonical.

First of all, we can shift the cross along the vector field w, i.e., consider a transformation of the form σ^{t_0}. Since the flow σ^t has an integral f, we may generalize this transformation in the following way. Let $g(f)$ be an arbitrary smooth function. Then we move each point x along its trajectory by an amount $g(f(x))$. In other words, the transformation is given by the following formula:

$$A_g(x) = \sigma^{g(f(x))}(x).$$

This transformation obviously preserves all properties of a canonical cross. What happens to the cochain m^* under this transformation? It is easy to see that, as a result, one adds to the cochain m^* the coboundary of the form $\widetilde{g}\delta(S^*)$, where $\widetilde{g}$ is the Taylor series of $g(f)$ at zero, and S^* is the elementary cochain concentrated at the vertex S^*. The transformation $A_g(x)$ is illustrated in Fig. 7.4.

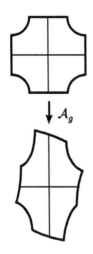

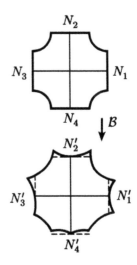

| Figure 7.4 | Figure 7.5 |

Another transformation B of a canonical cross is that the boundary segments N_1, N_2, N_3, N_4 are replaced by some new segments N_1', N_2', N_3', N_4' which differ from the initial ones by functions with smallness of infinite order at zero (see Fig. 7.5). In other words, N_i and N_i' have tangency of infinite order at the point of the graph K. It is clear that the functions $m_i(f)$ on the edges of K, in general, change, but their Taylor expansions at zero remain the same; and, therefore, the cochain m^* does not change at all.

It is easy to see that any transformation of a canonical cross is reduced to a superposition of the transformations A and B just described.

Thus, the class of the cochain m^* modulo the subspace of coboundaries is well defined and is, consequently, an invariant of the flow (in the sense of smooth conjugacy).

Now, as it was done in the topological case for 1-chain l, we can produce two invariants from the cochain m^*. First of all, we can take its coboundary $\Delta^* = \delta m^*$. This is actually a smooth invariant. But such an invariant is not new. Indeed, summing the coefficients of the 1-cochain m^* over each annulus C_n (this is just equivalent to taking coboundary), we obtain the infinite parts of the period functions

$$\Pi_n(f) = -A_n(f) \ln |f| + B_n(f),$$

more precisely, the Taylor expansions of the functions $B_n(f)$ at zero. Indeed, all terms that contain "logarithms" are placed into canonical crosses, and all other terms are distributed among the coefficients of m^*. Thus, the Δ^*-invariant is known from the period functions.

Besides, we can take the orthogonal projection of the cochain m^* to the space of 1-cocycles $Z^1(\widetilde{V}, \mathbb{R}[f])$ and consider the corresponding cohomology class $Z^* \in H^1(\widetilde{V}, \mathbb{R}[f])$.

Definition 7.2. The cohomology class Z^* is said to be the Z^*-*invariant* of the integrable system w on the atom V.

As we see, in the smooth case all types of invariants are, in essence, preserved: the Λ-invariant transforms into the Λ^*-invariant; the Δ-invariant (0-boundary) transforms into Δ^* (presented as a 2-coboundary or, which is the same, a 0-boundary of the conjugate graph); the 1-homology class Z transforms into the 1-cohomology class Z^*. In the smooth case, the elements (coefficients) of these invariants are formal power series, but not real numbers as in the topological case. This is very natural, since, speaking informally, we must sew the derivatives of all orders. Moreover, if we want to obtain a C^k-classification, it suffices just to cut these series on a certain step.

7.2. THEOREM OF CLASSIFICATION OF HAMILTONIAN FLOWS ON ATOMS UP TO SMOOTH CONJUGACY

Thus, all smooth invariants have been described and we can formulate the theorem of smooth classification of Hamiltonian flows on an atom.

For simplicity, we first require that the conjugating diffeomorphism preserves the Hamiltonian of a system. Then we can simply assume that we are given two Hamiltonian systems on the same surface P and with the same Hamiltonian f, but the related symplectic structures are different.

Theorem 7.1. *Let two Hamiltonian flows σ_1^t and σ_2^t be given on the atom P (with the same Hamiltonian but with different symplectic structures). Suppose that the corresponding period functions, Λ^*-invariants, and Z^*-invariants coincide. Then the flows σ_1^t and σ_2^t are smoothly conjugate, i.e., there exists a diffeomorphism $\xi : P \to P$ such that $\xi \circ \sigma_1^t = \sigma_2^t \circ \xi$. Moreover, the diffeomorphism may be chosen so that the Hamiltonian is preserved.*

Proof. First, we observe that the coincidence of Λ^*-invariants guarantees the existence of a conjugating diffeomorphism ξ in a neighborhood of each singular point of the Hamiltonian f (moreover, the diffeomorphism may be chosen so that f is preserved). Our problem is to sew these local diffeomorphisms into a single diffeomorphism defined on the whole surface P. This can in fact be done, since the period functions coincide. Let us prove this.

Choose a canonical cross $U_1(S_j)$ (in the sense of the flow σ_1^t) for each vertex S_j of the graph K and consider a local conjugating diffeomorphism ξ_j in a neighborhood of this vertex. Let $U_2(S_j) = \xi(U_1(S_j))$ be the image of the cross $U_1(S_j)$. Clearly, $U_2(S_j)$ is a canonical cross for the flow σ_2^t. As a result, we obtain two decompositions of the surface P into canonical crosses and "rectangles" which correspond to the Hamiltonian flows under consideration. For each "rectangle", we define the function $m_i(f)$ which measures its width (see the construction of the cochain m^* above). In our case, we denote these functions by $m_{1i}(f)$ and $m_{2i}(f)$, where $m_{ki}(f)$ is the width of the "rectangle" related to the i-th edge of the graph K and to the k-th system ($k = 1, 2$). These decompositions define, in particular, the cochains m_1^* and m_2^*.

When is it possible to extend the local diffeomorphisms ξ_j up to a global conjugating diffeomorphism defined on the whole atom? Evidently, this can be done if and only if $m_{1i}(f) \equiv m_{2i}(f)$ for all i. It turns out that this condition can be obtained by changing local diffeomorphisms ξ_j. Indeed, instead of ξ_j, we can consider conjugating diffeomorphisms of the form $A_g \circ \xi_j$, where, as before, $A_g(x) = \sigma_2^{g(f(x))}(x)$, and g is some smooth function. As a result, we can change the cochain m_2^* by adding an arbitrary coboundary and achieve, as a result, the equality $m_1^* = m_2^*$. Here we use the fact that m_1^* and m_2^* coincide modulo the subspace of 1-coboundaries, since the period functions and Z^*-invariants coincide by our assumption.

We now need to make the functions $m_{1i}(f)$ and $m_{2i}(f)$ equal identically. This procedure can be carried out for each annulus of the atom P separately. To this end, it suffices to use $\mathcal{B}$-transformations of canonical crosses.

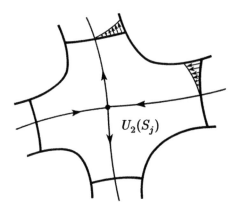

Figure 7.6

Consider a cross $U_2(S_j)$ (Fig. 7.6). The separatrices of the flow σ_2^t entering and leaving the vertex S_j divide this cross into four pieces. Take a C^∞-smooth function which is identically zero on three of these parts, but on the fourth piece it has the form $h(f)$, where h has zero of infinite order for $f = 0$ (for example, $h = m_{1i}(f) - m_{2i}(f)$). Thus, $h(x)$ is a smooth function on $U_2(S_j)$ which is an integral of the flow σ_2^t. Now consider the transformation of the cross $U_2(S_j)$ given by the formula

$$\mathcal{B}_h(x) = \sigma_2^{h(x)}(x).$$

This transformation is identical on three of four pieces of the cross, but the fourth part undergoes a shift along the flow which is "almost identical" near the graph K. By means of such a transformation, we can correct the functions $m_{2i}(f)$ and achieve the equalities $m_{1i}(f) = m_{2i}(f)$. It is easy to see that, using the coincidence of the period functions, we can do it for all i simultaneously. As a result, we have sewn the local diffeomorphisms into a global conjugating diffeomorphism, as required. $\square$

COMMENT.   The proof of Theorem 7.1 can be obtained in a different way. The idea is to compare not the Hamiltonian flows σ_1 and σ_2 themselves but the corresponding symplectic structures ω_1 and ω_2 on the surface P (if the Hamiltonian is fixed, then these problems are equivalent). Then a diffeomorphism ξ sending ω_1 into ω_2 can be constructed by using a method by J. Moser [247]. However, here we have to take into account that the Hamiltonian must be preserved. This idea was realized by B. S. Kruglikov [204], who also obtained some generalizations of Theorem 7.1 (in particular, to the C^k-smoothness case). In paper [103], J.-P. Dufour, P. Molino, and A. Toulet studied a question on the classification of triples $(P^2, \omega, \mathcal{F})$, where ω is a symplectic structure on a surface P^2, and $\mathcal{F}$ is a one-dimensional foliation with singularities generated by some Morse function on the surface P^2.

How can we avoid the condition that the Hamiltonian is preserved? To answer this question, it suffices to look at what happens to the invariants of a fixed Hamiltonian system if we change its Hamiltonian (changing the symplectic structure at the same time so that the system itself remains the same). In principle, we can explicitly formulate some formal rule of changing the invariants. For example, for the Λ^*-invariant, it will be formal conjugation of power series. The other invariants can also be presented in the form of power series, and for them the analogous (but more complicated) rule "of formal conjugation" can also be indicated. As a result, the final formulation of the classification theorem will be as follows: two Hamiltonian systems w_1 and w_2 on an atom P are smoothly conjugate if and only if the corresponding sets of their invariants $(\Lambda_1^*, \Delta_1^*, Z_1^*)$ and $(\Lambda_2^*, \Delta_2^*, Z_2^*)$ are formally conjugate.

However, the rule "of formal conjugation" will be rather awkward; that is why we shall proceed in another way, and, first of all, once more recall a method which allows us to test two given systems from the point of view of their conjugacy.

Thus, suppose we are given two Hamiltonian systems on the same atom $V = (P, K)$. Consider the period functions of these systems on the annuli of the atom. Take one of these annuli and compare the period functions $\Pi_1(f_1)$ and $\Pi_2(f_2)$ on it. They, of course, need not coincide, since their arguments (i.e., the Hamiltonians f_1 and f_2 of the systems) are different and, actually, in no way connected. According to Lemma 7.4, we can change the Hamiltonians f_1 and f_2 so that the period functions become

$$\Pi_i(f_i') = -A_i(f_i') \ln |f_i'| \, .$$

After such a change of the Hamiltonians, the systems are conjugate if and only if the corresponding sets of invariants $(\Lambda_1^*, \Delta_1^*, Z_1^*)$ and $(\Lambda_2^*, \Delta_2^*, Z_2^*)$ coincide. Thus, if we consider one of the annuli to be fixed, we can define invariants of a Hamiltonian flow which do not depend on the choice of a Hamiltonian.

REMARK. From the formal viewpoint, this procedure can be interpreted in the following way. To every Hamiltonian system with a fixed Hamiltonian, we can assign a set of its invariants. By changing a Hamiltonian, we change these invariants. As a result, on the set of all invariants we can introduce the action of the group of changes of Hamiltonians. A real invariant of a Hamiltonian flow, which does not depend on the choice of a Hamiltonian, is an orbit of this action. The above procedure means exactly that we indicate a certain representative for each orbit. This representative is distinguished by the condition that the value of the Δ^*-invariant on the chosen annulus is zero.

We can now, in particular, say how many parameters parameterize the set of classes of smoothly conjugate Hamiltonian systems on a given atom $V = (P^2, K)$. Recall that, in the topological case, this set is finite-dimensional. In the smooth case, the set of conjugacy classes is parameterized by $2k$ formal power series, where k is the number of vertices of the atom.

Finally, it is useful to study the invariants in the case of simplest atoms. In the case of the atom B (Fig. 7.7) there is one vertex and, consequently,

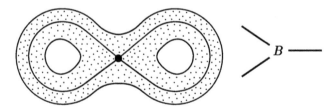

Figure 7.7

the invariants are two power series. The most natural invariants are the period functions on the annuli of the atom. The classification theorem for this atom is very natural: two systems on the atom B are smoothly conjugate if and only if the corresponding period functions are conjugate. We should emphasize one very essential circumstance. In this context, the conjugacy (of period functions) means not only that the conjugating diffeomorphisms are smooth up to zero (zero is included), but also that these diffeomorphisms are smoothly sewn at zero. In other words, the three diffeomorphisms must have the same Taylor expansion at zero. Also note that the period functions on different annuli cannot be independent of each other. A necessary condition is as follows. If we consider the sum of period functions over all negative annuli and the analogous sum over all positive annuli

$$\Pi^-(f) = -A^-(f)\ln|f| + B^-(f), \quad f \in [-\varepsilon_0, 0),$$
$$\Pi^+(f) = -A^+(f)\ln|f| + B^+(f), \quad f \in (0, \varepsilon_0],$$

then the Taylor expansions of the functions A^- and B^- must coincide with those of A^+ and B^+ respectively.

Finally, note that there exists a Hamiltonian for which the period function on the outer annulus takes the form

$$\Pi_1(f) = -A(f)\ln|f|.$$

Then, on the two remaining annuli, the period functions become

$$\Pi_2(f) = -\frac{1}{2}A(f)\ln|f| + B(f),$$
$$\Pi_3(f) = -\frac{1}{2}A(f)\ln|f| - B'(f),$$

where $B(f)$ and $B'(f)$ do not necessarily coincide, but have the same Taylor expansions at zero. Then the well-defined invariants are the Taylor expansions of $A(f)$ and $B(f)$ at zero.

We also consider another important case of the atom C_2 (Fig. 7.8), which often occurs in applications. This atom admits a natural involution, namely, the central symmetry in $\mathbb{R}^3$ (see Fig. 3.23(a)). Assume that this involution changes the sign

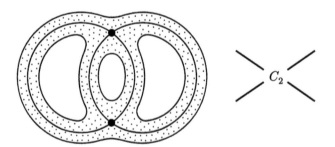

Figure 7.8

of the Hamiltonian vector field v given on this atom, that is, sends v into $-v$. This condition is often fulfilled in concrete problems (for example, as we shall see below, in the case of integrable geodesic flows on the 2-sphere). Due to such a symmetry, the Λ^*-invariants of the vertices coincide as well as the period functions on the annuli of the same sign. Let us compute the number of essential invariants in this case. We choose again the Hamiltonian in such a way that the period function on one of the two positive annuli takes the form

$$\Pi_1(f) = -A(f)\ln|f|.$$

Then on the other positive annulus (in view of the symmetry) the period function will be the same

$$\Pi_2(f) = -A(f) \ln |f| \, .$$

Analogously, on the remaining two negative annuli the period functions will coincide:

$$\Pi_3(f) = -A(f) \ln |f| + B(f) \, ,$$
$$\Pi_4(f) = -A(f) \ln |f| + B(f) \, .$$

The coefficients before the logarithm coincide, since each annulus passes by the same singular points. Moreover, according to the condition formulated above, the Taylor expansion of the function $B(f) + B(f) = 2B(f)$ must be identically zero (since $B(f)$ is the finite part of the period functions Π_3 and Π_4). This means that $B(f)$ has zero of infinite order and does not give any invariant. Thus, if the symmetry condition is fulfilled, then the only smooth invariant is the Taylor series of the function $A(f)$ at zero.

Finally, the last remark is that, in the smooth case, there are no non-trivial restrictions to the invariants (similar to those which were discussed in Section 6.3). The Λ^*-invariant can be absolutely arbitrary with the only condition that its first term is positive. There are no restrictions to the Z^*-invariant at all. And the period functions $\Pi_n(f)$ must satisfy two obvious properties:

1) the coefficient before the logarithm must be equal to the sum of Λ^*-invariants of those vertices of the graph K which belong to the boundary of the corresponding annulus;

2) the sum of Taylor expansions of their finite parts over all positive annuli is equal to the analogous sum over all negative annuli.

Chapter 8

Orbital Classification of Integrable Hamiltonian Systems with Two Degrees of Freedom. The Second Step

Here we present a general scheme for constructing orbital invariants of integrable Hamiltonian systems with two degrees of freedom.

Thus, let v be an integrable Hamiltonian system on a three-dimensional isoenergy surface $Q = Q^3$, f an additional Bott integral of the system, and $Q = \sum Q_c$ the canonical decomposition into components each of which contains exactly one singular leaf of the Liouville foliation (atomic decomposition). Recall that for Q_c we also used the notation $U(L)$, where $L \subset f^{-1}(c)$ is a singular leaf corresponding to a critical value c of the integral f on Q.

One of the difficulties that appears in constructing invariants of integrable Hamiltonian systems on isoenergy 3-surfaces is that many objects, which naturally occur in the framework of this theory, depend on the choice of basis cycles on Liouville tori. First of all, we mean the gluing matrices and rotation functions. That is why we divide the solution of the problem of invariants into two parts. First, assuming that bases on Liouville tori are fixed, we define all necessary invariants and show that if these invariants coincide, then the systems are equivalent. After this, we analyze what happens under change of bases on Liouville tori and make invariants well-defined, i.e., independent of the choice of bases.

We emphasize that the construction presented below can be applied for arbitrary atoms. However, in the smooth case, we shall assume for simplicity that all atoms are planar and, in addition, have no critical circles with non-orientable separatrix diagrams.

8.1. SUPERFLUOUS t-FRAME OF A MOLECULE (TOPOLOGICAL CASE). THE MAIN LEMMA ON t-FRAMES

We begin with choosing and fixing a certain transversal section P_{tr} for each atom $U(L)$ as well as admissible coordinate systems on the boundary tori of $U(L)$. Here transversal sections and admissible coordinate systems are assumed to be compatible. To clarify what this mean, we have to consider three types of atoms separately.

Case 1: Atom A.

Case 2: A saddle 3-atom all of whose critical circles have orientable separatrix diagrams (an atom without stars).

Case 3: A saddle 3-atom that has critical circles with non-orientable separatrix diagrams (an atom with stars).

We begin with the case of the atom A. In this case, the transversal section P_{tr} is defined uniquely up to isotopy and is homeomorphic to a 2-disc. Its boundary is the contractible cycle λ on the boundary torus. This is the first cycle of an admissible coordinate system. The second basis cycle μ can be chosen arbitrarily provided the pair of cycles (λ, μ) form a basis on the boundary torus.

Let us consider the second case, i.e., a saddle 3-atom Q_c without stars. Here we choose admissible coordinate systems (λ_j, μ_j) on the boundary of Q_c in the following way. Recall that the index j enumerates the boundary tori of Q_c. As the first cycle λ_j, we take a fiber of the trivial Seifert fibration on Q_c. The second basis cycle μ_j is taken to be the intersection of the boundary torus T_j and the transversal section P_{tr}. Note that, if the atom is planar (i.e., P_{tr} can be embedded into a plane), then P_{tr} is defined uniquely up to isotopy by the collection of cycles $\{\mu_j\}$, i.e., by its boundary. However, in general this is not true, i.e., it is impossible, generally speaking, to reconstruct a transversal section in a unique way from a given admissible coordinate system.

Now consider the last case, i.e., a saddle 3-atom with stars. In this case, the basis cycles $\{\mu_j\}$ of an admissible coordinate system do not form the boundary of a transversal section $P_{\mathrm{tr}} \subset Q_c$. Roughly speaking, these cycles $\{\mu_j\}$ give only a half of the boundary of the section P_{tr}. Moreover, in the case of atoms with stars, there are very many different transversal sections. These sections can have different topology and be non-homeomorphic. That is why first we have to choose and fix the topological type of transversal sections in such atoms.

Suppose we are given a 3-atom Q_c having critical circles with non-orientable separatrix diagrams. As shown in Chapter 3, the 3-atom Q_c corresponds to some 2-atom with stars $V = (P, K)$ that is the base of the Seifert fibration on Q_c. As a transversal section P_{tr}, we must take a certain double of this atom, i.e., a two-dimensional surface $\widehat{P}$ with an involution χ such that $P = \widehat{P}/\chi$ (see Section 3.5 for details). We wish to choose a canonical type of these double. To this end, consider all star-vertices of the atom (P, K). Then we connect each of them by a segment with the positive boundary circle of P that passes by this

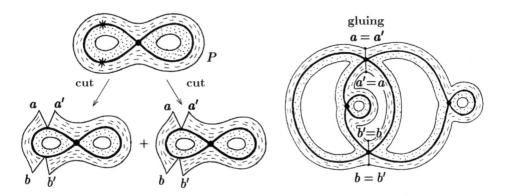

Figure 8.1

vertex, and cut P along this segment. In other words, we make cuts of the 2-atom in the same direction: from the star-vertices to the positive boundary of P. After this, we take two copies of the surface obtained and glue a double $\widehat{P}$ from them by identifying the equivalent cuts, i.e., corresponding to the same star-vertex (see Fig. 8.1). On this double, there is a natural involution χ whose fixed points are exactly star-vertices.

Then we construct an embedding

$$\alpha \colon \widehat{P} \to Q_c$$

of the double $\widehat{P}$ into the 3-atom Q_c to obtain a transversal section $P_{\mathrm{tr}} = \alpha(\widehat{P})$. This embedding must satisfy the following property: the diagram

is commutative.

The existence of such a section has been in fact proved in Chapters 3 and 5.

From now on, in the case of atoms with stars, we shall consider only such transversal sections $\alpha(\widehat{P}) = P_{\mathrm{tr}}$.

We now explain how, using the section P_{tr}, we can construct admissible coordinate systems on the boundary tori of the 3-atom Q_c. Consider an arbitrary boundary torus $T_j \subset \partial Q_c$. As the first basis cycle we take, by definition, the fiber of the Seifert fibration. Then, as we did it in Chapter 4, we set $\widehat{\mu}_j = P_{\mathrm{tr}} \cap T_j$. All the tori T_j can naturally be divided into positive and negative ones (depending on the sign of the rotation function; see Lemma 8.5 below). It is easy to see that, in the case of a negative torus, $\widehat{\mu}_j$ represents a disjoint union of two homologous cycles each of which is a section of the Seifert fibration. As the second basis cycle μ on T_j, we take one of them. If the torus is positive, then two cases are possible: either $\widehat{\mu}_j$ is a disjoint union of two cycles, or $\widehat{\mu}_j$ is a single cycle which has intersection index 2 with fibers of the Seifert fibration. It is not difficult to understand when each of these cases happens. For this, we consider the projection $\pi \colon P_{\mathrm{tr}} \to P$ and the image $\pi(\widehat{\mu}_j)$. This image is evidently one of the boundary circles of P. Let s_j be the number of star-vertices by which this boundary circle passes (or, equivalently, the number of cuts corresponding to this circle that we made to construct the double $\widehat{P}$). If s_j is even, then $\widehat{\mu}_j$ is a pair of disjoint cycles and, on the contrary, if s_j is odd, then $\widehat{\mu}_j$ is a single cycle.

In both cases, we define the second basis cycle μ_j from the relation

$$\mu_j = \frac{1}{2}(\widehat{\mu}_j + s_j \lambda)\,.$$

This relation makes sense, since the expression in brackets is a double cycle. Thus, we have indicated explicit formulas which connect the boundary of the transversal section $\partial P_{\mathrm{tr}} = \{\widehat{\mu}_j\}$ with the basis cycles of the admissible coordinate system.

Now for each saddle 3-atom Q_c (with or without stars) we fix a transversal section P_{tr}. For the atoms of type A, we fix the structure of a trivial S^1-fibration by choosing a fiber μ on its boundary torus. Recall that, unlike saddle atoms, in this case, the ambiguity is in the choice of a trivial S^1-fibration on A, whereas the transversal section is uniquely defined. The set of fixed sections for the saddle atoms and fibers for the atoms A will be denoted by $\mathbb{P}$.

COMMENT. In what follows, speaking of the set $\mathbb{P}$, we shall use the term "set of section", although, for the atoms of type A, we choose not a section, but a fiber. This should not be confusing, since the "theory of transversal sections" actually plays the main role in our construction.

Thus, suppose we are given a concrete set of sections $\mathbb{P}$. Then we can compute many natural objects. Namely, we can compute all the gluing matrices, all the rotation vectors, all the Λ-, Δ-, and Z-invariants of Poincaré flows for each transversal section $P_{\mathrm{tr}} \subset Q_c$. To do this, we introduce the following notation:

e_j is an edge of the molecule W;

(λ_j^-, μ_j^-) and (λ_j^+, μ_j^+) are admissible coordinate systems at the beginning and end of the edge e_j respectively, which, of course, depend on $\mathbb{P}$;

$C_j(\mathbb{P})$ is the corresponding gluing matrix on the edge e_j;

$R_j^-(\mathbb{P})$ and $R_j^+(\mathbb{P})$ are the rotation vectors of the Hamiltonian system v on e_j in these coordinate systems;

$\Lambda_c(\mathbb{P})$, $\Delta_c(\mathbb{P})$ and $Z_c(\mathbb{P})$ are Λ-, Δ-, and Z-invariants of the Poincaré flows for each atom Q_c (for a given choice of the set of sections $\mathbb{P}$).

COMMENT. Recall that all saddle critical circles of the integral f on Q^3 are assumed to be hyperbolic. This guarantees that the Poincaré Hamiltonian on a transversal section is a Morse function (see Proposition 5.5). Therefore, the Λ-, Δ-, and Z-invariants are well-defined for each saddle atom.

Definition 8.1. The set of objects

$$\mathbb{T} = \{C_j(\mathbb{P}), R_j^-(\mathbb{P}), R_j^+(\mathbb{P}), \Lambda_c(\mathbb{P}), \Delta_c(\mathbb{P}), Z_c(\mathbb{P})\}$$

is called a *superfluous t-frame* of the molecule W.

In other words, by considering a superfluous t-frame, we collect together all information about atomic and edge invariants. The next statement shows that this information is sufficient for the orbital classification.

Lemma 8.1 (Main lemma). *Let v_1 and v_2 be two integrable Hamiltonian systems on isoenergy 3-surfaces Q_1 and Q_2 respectively. Suppose their molecules are equal. These systems are orbitally topologically equivalent if and only if there exist sets of sections $\mathbb{P}_1$ and $\mathbb{P}_2$ (for v_1 and v_2, respectively) such that the corresponding superfluous t-frames coincide.*

COMMENT. The lemma shows that the set of orbital invariants included into a t-frame is complete. Therefore, we need not look for any other invariants (for orbital classification). At the same time, it should be noted that the deficiency of the discovered invariants is that they are not uniquely defined but depend on the choice of transversal sections. However this ambiguity can be avoided by some formal procedure. Roughly speaking, one needs to factorize superfluous t-frames with respect to the action of the group of transformations of transversal sections. This will be done in the next section.

Proof. In one direction, this claim is evident. Indeed, if the systems are equivalent, then, for any set $\mathbb{P}_1$, we may consider its image under the orbital homeomorphism ξ as a set of sections $\mathbb{P}_2$ for the second system. After this, all invariants included in the t-frames will obviously coincide.

Let us prove the converse. We need to show that the two systems v_1 and v_2 with the same superfluous t-frames are orbitally equivalent. Consider the given sets of sections $\mathbb{P}_1$ and $\mathbb{P}_2$. The coincidence of the Λ-, Δ-, and Z-invariants for v_1 and v_2 implies that the corresponding Poincaré flows w_1 and w_2 are topologically conjugate on the given sections (see Theorems 6.1 and 6.2). Hence the systems v_1 and v_2 are orbitally equivalent on the corresponding saddle 3-atoms (see Theorem 5.1).

In the case of atoms A, the situation is analogous and even simpler, since there are no atomic invariants. It is sufficient to know the behavior of the rotation function near the atom A. Namely, the following lemma holds.

Lemma 8.2. *Suppose we are given two integrable systems v_1 and v_2 on the 3-atom A, i.e., in a neighborhood of a stable periodic trajectory. Suppose the rotation functions ρ_1 and ρ_2 of these systems are conjugate (continuously or smoothly) in this neighborhood. Then v_1 and v_2 are orbitally equivalent (topologically or smoothly, respectively).*

Proof. This assertion is a corollary of the reduction theorem (Theorem 5.1). As we shall see below, the rotation function of the system v_i on the 3-atom A coincides with the period function Π of the reduced system on the 2-atom A, i.e., on a two-dimensional disc. As a result, the question of whether the given systems are orbitally equivalent is reduced to the proof of the fact that the reduced systems w_1 and w_2 are conjugate on a two-dimensional disc provided the period functions are conjugate. But, under this assumption, we can write an explicit formula which gives the desired conjugating homeomorphism (or diffeomorphism, respectively). It is easy to see (see, for example, [110]) that, for each reduced system w_i (where $i = 1, 2$), there exist local canonical coordinates p_i, q_i on the disc such that the Hamiltonian F_i of the system w_i takes the form $F_i = F_i(p_i^2 + q_i^2)$.

Moreover, the period functions $\Pi_i(s_i)$ become

$$\Pi_i(s_i) = 2\pi \left(\frac{\partial F_i}{\partial s_i}\right)^{-1},$$

where $s_i = p_i^2 + q_i^2$. If the period functions are conjugate by means of a change $s_2 = s_2(s_1)$, then the homeomorphism (resp. diffeomorphism) which conjugates the systems w_1 and w_2 can be written in the simple form: $\varphi_2 = \varphi_1$, $s_2 = s_2(s_1)$, where φ_i are polar angles corresponding to the Cartesian coordinates p_i, q_i. Therefore, the Poincaré flows are conjugate and, consequently, the initial systems v_1 and v_2 are orbitally equivalent on the 3-atom A. □

Thus, the systems v_1 and v_2 are orbitally equivalent near singular fibers (i.e., on 3-atoms). Besides, by our assumption, the rotation vectors of these systems coincide on the corresponding edges of the molecules W_1 and W_2. Then it follows from Propositions 5.2 and 5.3 that the systems v_1 and v_2 are orbitally equivalent on each edge. It remains to sew the existing orbital isomorphisms on atoms and edges into a single orbital isomorphism $Q_1 \to Q_2$.

Thus, consider an arbitrary edge e adjoining a certain atom V. Near this atom, on a one-parameter family of Liouville tori $T^2 \times [a, b]$, we have two different orbital isomorphisms:

$$\xi, \eta : T^2 \times [a, b] \to T^2 \times [a', b'].$$

To sew these isomorphisms, we must construct a new orbital isomorphism $\zeta : T^2 \times [a, b] \to T^2 \times [a', b']$ which coincides with ξ in a neighborhood of one boundary torus, i.e., on the set $T^2 \times [a, a + \varepsilon]$, and coincides with η in a neighborhood of the other boundary torus, i.e., on the set $T^2 \times [b - \varepsilon, b]$. Note that, in our situation, ξ and η are both fiber homeomorphisms, i.e., the image of a Liouville torus from the family $T^2 \times [a, b]$ is a certain Liouville torus from $T^2 \times [a', b']$; moreover, this torus is the same for both isomorphisms. The point is that, under orbital isomorphisms, the rotation number must be preserved. But it changes monotonically near a saddle atom; therefore, the image of each Liouville torus is uniquely defined. Besides, ξ and η are homotopically equivalent, since the homotopy type of these mappings is determined by the images of the basis cycles, which define admissible coordinate systems. In our case, the images of the basis cycles are fixed, since we have fixed the sets of sections. The desired sewing is possible due to the following sewing lemma, which holds both in the smooth and topological cases.

Lemma 8.3. *Suppose we are given two orbital isomorphisms between two integrable Hamiltonian systems v_1 and v_2 restricted to one-parameter families of Liouville tori:*

$$\xi, \eta : T^2 \times [a, b] \to T^2 \times [a', b'].$$

Let $\xi(T^2 \times \{c\}) = \eta(T^2 \times \{c\})$, and let ξ and η be homotopic. Then there exists a sewing orbital isomorphism ζ such that

$$\zeta = \begin{cases} \xi & \text{on the set } T^2 \times [a, a + \varepsilon], \\ \eta & \text{on the set } T^2 \times [b - \varepsilon, b]. \end{cases}$$

Proof. Without loss of generality, we may assume that the rotation function does not change very much on our family of tori. Otherwise we can divide $[a, b]$ into small segments and prove the lemma separately for each of them. This condition is needed for the existence of a transversal section. In our case, however, it will be fulfilled automatically, since near an atom such a section always exists.

Thus, consider an arbitrary transversal section $P = S^1 \times [a, b] \subset T^2 \times [a, b]$ for the first vector field v_1 and construct the transversal section P' in the family $T^2 \times [a', b']$ so that P' coincides with the image $\xi(P)$ near the boundary torus $T^2 \times \{a\}$, and it coincides with the image $\eta(P)$ near the other boundary torus $T^2 \times \{b\}$. It is easy to see that such a transversal section P' exists.

Consider the Poincaré flows on the sections P and P'. They are conjugate by the reduction theorem. The isomorphism $\varkappa$ conjugating these flows is not uniquely defined. The ambiguity can be explained as follows. Let N be a curve which joins a pair of points on the two components of the boundary of the annulus P and is transversal to the integral curves of the Poincaré flow (Fig. 8.2). Then, as the image of N under the mapping $\varkappa$, we can take any analogous transversal curve N' on the annulus P'. It is easy to see that, if the image $\varkappa(N)$ is fixed, then $\varkappa$ can be uniquely reconstructed.

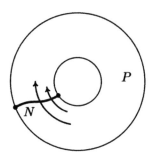

Figure 8.2

In our case, as a transversal curve $N' \subset P'$, we take such a curve which coincides with $\xi(N)$ near one boundary component of P' and coincides with $\eta(N)$ near the other boundary component.

Inside the annulus, the curve N' can be chosen arbitrarily provided it is smooth and transversal to the flow. After this, we can uniquely reconstruct the isomorphism $\varkappa: P \to P'$ which sends the Poincaré flow σ^t on P into the Poincaré flow σ'^t on P'. It is clear that, by construction, $\varkappa$ coincides with the restriction $\xi|_P$ near the first boundary circle and coincides with $\eta|_P$ near the other boundary circle.

We now need to extend the mapping $\varkappa: P \to P'$ to an orbital isomorphism ζ between the families of Liouville tori $T^2 \times [a, b]$ and $T^2 \times [a', b']$. Such an extension is always possible, and we now give a formal construction.

We make a smooth change of time on the trajectories of the flows v and v' so that the passage time from each point $x \in P$ (resp. $x' \in P'$) up to the point $\sigma(x) \in P$ (resp. $\sigma'(x') \in P'$) is equal to 1. Then we get natural coordinate systems (x, t) and (x', t') on the families of tori $T^2 \times [a, b]$ and $T^2 \times [a', b']$, where x and x'

are points on the sections P and P' respectively, and t and t' are new times along the flows measured from these sections (in other words, we assume that $t|_P = 0$ and $t'|_{P'} = 0$). Here the points $(x, t + 1)$ and $(\sigma(x), t)$ are identified. (Similarly, $(x', t' + 1) = (\sigma'(x'), t')$.)

The isomorphism ζ in terms of these coordinates can now be written as follows:

$$\zeta(x, t) = (x', t') = (\varkappa(x), t'(x, t)),$$

where
1) t' is monotone (as a function of t),
2) $t'(x, t + 1) = t'(\sigma(x), t) + 1$,
3) $t'(x, 0) = 0$, $t'(x, 1) = 1$.
A mapping satisfying these conditions obviously exists, since the Poincaré flows σ^t and σ'^t are conjugate.

Let us choose such a mapping $\tilde{\zeta}$ (we denote the corresponding function $t'(x, t)$ by $\tilde{t}'(x, t)$) and "deform" it so that near the boundary tori it coincides with ξ and η respectively. It is clear that the formulas for the isomorphisms ξ and η, in the coordinates (x, t), have the same form as the one given above for ζ. The only difference is in the choice of the function $t'(x, t)$. We denote these functions for ξ and η by t'_ξ and t'_η respectively. We now define a new function t' by the formula

$$t'(x, t) = \begin{cases} (1 - c(f))t'_\xi(x, t) + c(f)\tilde{t}'(x, t) & \text{if } f \in [a, a + 2\varepsilon], \\ \tilde{t}'(x, t) & \text{if } f \in [a + 2\varepsilon, b - 2\varepsilon], \\ (1 - c(f))t'_\eta(x, t) + c(f)\tilde{t}'(x, t) & \text{if } f \in [b - 2\varepsilon, b]. \end{cases}$$

Here $c(f)$ is a smooth function with the graph shown in Fig. 8.3.

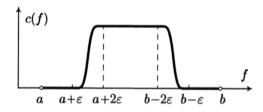

Figure 8.3

It is easily seen that $t'(x, t)$ satisfies all the required conditions. The sewing lemma is proved. □

REMARK. It is not difficult to see that the proof of the sewing lemma works also in the smooth case. Moreover, the proof is actually constructed for the case when all the objects are smooth. This is very natural since all non-smooth effects are located on singular leaves, but we make sewing at a certain distance from them.

Thus, by sewing the existing orbital isomorphisms given on individual atoms and edges, we obtain a global orbital isomorphism, as required.

Thus, the main lemma is proved in the topological case. □

8.2. THE GROUP OF TRANSFORMATIONS OF TRANSVERSAL SECTIONS. PASTING-CUTTING OPERATION

From Lemma 8.1 we immediately get a method for constructing orbital invariants. We only need to select combinations of elements of superfluous t-frames that do not depend on the choice of the set of sections $\mathbb{P}$.

From the formal point of view, we can consider the following construction. Let $\mathbb{T}$ be the superfluous t-frame corresponding to a certain set of transversal sections $\mathbb{P}$. If we change the set of sections, then the superfluous t-frame also changes according to some definite rule. This actually means that, on the set $\{\mathbb{T}\}$ of all possible superfluous t-frames, there is an action of the group of transformations of transversal sections, which we denote by $G\mathbb{P}$. It turns out that, as $G\mathbb{P}$, we should take the direct sum of the integer 1-cohomology groups $H^1(P_c, \mathbb{Z})$ for all saddle atoms of the given molecule and, in addition, s copies of $\mathbb{Z}$ (each copy corresponds to some atom A).

Thus, $G\mathbb{P} = \left(\bigoplus_c H^1(P_c, \mathbb{Z}) \right) \oplus \mathbb{Z}^s$.

This assertion is actually a general fact from 3-topology. We now comment on it. Consider two different transversal sections P_{tr} and P'_{tr} for some saddle 3-atom Q_c. What is the difference between them? To answer this question we have to consider two cases separately: atoms with stars and atoms without stars.

We begin with atoms without stars. In this case, a 3-atom has the type of direct product $(Q_c, f^{-1}(c)) = (P_c, K_c) \times S^1$. Consider two transversal sections

$$P_{\mathrm{tr}} = j(P_c) \quad \text{and} \quad P'_{\mathrm{tr}} = j'(P_c), \qquad \text{where} \quad j, j' : P_c \to Q_c.$$

Since both the sections are defined up to isotopy, we may assume without loss of generality that the sections P_{tr} and P'_{tr} intersect the critical circles of the atom at the same points. This means that the images of each vertex of the graph K_c under the mappings j and j' coincide. Consider an arbitrary edge K_i of the graph K_c and its images $j(K_i)$ and $j'(K_i)$. Clearly, $j(K_i) = P_{\mathrm{tr}} \cap L_i$ and $j'(K_i) = P'_{\mathrm{tr}} \cap L_i$, where L_i is one of the annuli that form the critical level $f^{-1}(c)$ of the integral f. Thus, we have two edges $j(K_i)$ and $j'(K_i)$ on the annulus L_i with coinciding endpoints lying on the opposite boundary circles of the annulus.

Consider the oriented cycle λ on the annulus L_i that is the fiber of the Seifert fibration. The edges $j(K_i)$ and $j'(K_i)$ also have natural orientations given by the Poincaré flows w and w' (induced by the same Hamiltonian vector field v on the annulus L_i). Hence we can uniquely write down the following decomposition: $j'(K_i) = j(K_i) + m_i \lambda$, where m_i is an integer. (Indeed, the difference $j'(K_i) - j(K_i)$ is obviously a 1-cycle on the annulus L_i; therefore, in the homological sense, this cycle is a multiple of λ with some integer coefficient m_i.)

Thus, as a result of comparing the transversal sections P_{tr} and P'_{tr}, we assign an integer m_i to each edge K_i. We interpret this set of integer numbers as a 1-cochain m, which we call a *difference cochain*. It is easy to observe that the cochain m is defined up to an integer coboundary. Indeed, consider the image of a vertex S_j of the graph K_c under the mapping j'. Since we are interested in transversal sections up to isotopy only, we can move this vertex along the fiber λ until it returns to the initial position (the other vertices remain fixed). On the one hand, such operation transforms the section into an isotopic one. But on the other hand, it is easily seen that the difference cochain m is changed by the elementary coboundary corresponding to the vertex S_j (regarded as a 0-cochain). Thus, a well-defined difference of two sections is an element of the integer cohomology group $C^1(K_c, \mathbb{Z})/B^1(K_c, \mathbb{Z}) = H^1(K_c, \mathbb{Z})$. Taking into account that the surface P_c can be contracted to the graph K_c, we can interpret the difference m as an element of the group $H^1(P_c, \mathbb{Z})$.

Conversely, let $P_{\mathrm{tr}} = j(P_c)$ be a transversal section, and let $m \in H^1(P_c, \mathbb{Z})$. Then we can uniquely (up to isotopy) reconstruct a new section $P'_{\mathrm{tr}} = j'(P_c)$ such that the difference cochain between P_{tr} and P'_{tr} will be equal to m.

Now consider the case of atoms with stars. The general scheme remains the same. As transversal sections, we must consider embeddings of the double $\widehat{P}_c$ corresponding to a 2-atom (P_c, K_c) with stars. Thus, suppose we are given two embeddings

$$j: \widehat{P}_c \to Q_c, \qquad j': \widehat{P}_c \to Q_c.$$

Recall that we consider only those embeddings for which the diagram

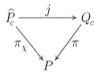

is commutative.

It is easy to see that, by means of a suitable isotopy, we can achieve the situation when these embeddings coincide on small neighborhoods of the vertices of the double $\widehat{P}_c$. Let $\widehat{K}_c \subset \widehat{P}_c$ denote the graph that is the double of K_c. Then the same argument as above gives the relation

$$j'(\widehat{K}_i) = j(\widehat{K}_i) + m_i \lambda,$$

where $\{m_i\}$ are integers, and $\widehat{K}_i$ are edges of $\widehat{K}_c$.

In this case, each annulus L_i of the singular leaf L intersects a transversal section not by one segment (as it was in the case of atoms without stars), but by two segments. These segments are mapped into each other under the involution τ defined on the double. Let us denote them by $\widehat{K}_i$ and $\tau(\widehat{K}_i)$. Then the relation analogous to that obtained for the segment $\widehat{K}_i$ will hold also for $\tau(\widehat{K}_i)$. Namely,

$$j'(\tau(\widehat{K}_i)) = j(\tau(\widehat{K}_i)) + m_i \lambda,$$

where m_i is the same as the one in the formula for the edge $\widehat{K}_i$. The obtained set of integers $\{m_i\}$ can be interpreted as a τ-symmetric 1-cochain, i.e., an element of the group $C^1(\widehat{K}_c, \mathbb{Z})$. In other words, this cochain takes the same values on the edges which are mapped into each other under the involution τ. Therefore, the set $\{m_i\}$ can be interpreted as a 1-cochain on the graph $K_c = \widehat{K}_c/\tau$. Thus, actually, the 1-cochain $\{m_i\}$ lies in the group $C^1(K_c, \mathbb{Z})$. As in the previous case, it is easily seen that this cochain is defined modulo the space of coboundaries, and we come to the same result: the difference of two sections is an element of the one-dimensional cohomology group $H^1(P_c, \mathbb{Z})$. Conversely, if we are given an arbitrary element $m \in H^1(P_c, \mathbb{Z})$, then it is always possible to construct a new transversal section $P'_{\text{tr}} = j'(\widehat{P}_c)$ in the 3-atom Q_c which differs from the section $P_{\text{tr}} = j(\widehat{P})$ by the cocycle m.

REMARK. From the formal point of view, the group of transformations of transversal sections has the same structure both for atoms with stars and for atoms without stars. This is the one-dimensional cohomology group $H^1(P_c, \mathbb{Z})$. Formally, the invariants (Λ, Δ, Z) of the reduced system also have the same nature both for atoms with stars and for atoms without stars. These two observations by A. B. Skopenkov [80] allow us to include atoms with stars into the general orbital classification theory.

In the case of an atom of type A, an analog of a difference cochain is introduced in the following way. Recall that topologically an atom A is represented by a solid torus, and the ambiguity in this case consists in the choice of a fiber μ of a trivial Seifert fibration on the boundary torus. Clearly, two such fibers are connected by the relation $\mu' = \mu + m\lambda$, where λ is the meridian of the solid torus, and m is some integer number. This number is just an analog of a difference cocycle; for convenience, we shall use this term both for saddle atoms and for atoms A.

Thus, if we have a set of transversal sections $\mathbb{P} = \{P_{\text{tr}}\}$ and a set of cocycles $\mathbb{M} = \{m_c\} \in G\mathbb{P}$, then we can construct in a natural way a new set of sections $\mathbb{P}'$, which is the result of the action of $\mathbb{M}$ on $\mathbb{P}$.

Thus, we can define the action of the group $G\mathbb{P}$ (the group of transformations of transversal sections) on the set of all transversal sections $\{\mathbb{M}\}$. The structure of the group $G\mathbb{P}$ is very simple: this is a free Abelian group $\mathbb{Z}^{n+k}$, where n is the number of closed hyperbolic trajectories of the system with oriented separatrix diagram, and k is the total number of all singular leaves (i.e., atoms). Here we use the fact that the dimension of the cohomology group $H^1(P_c, \mathbb{Z})$ for each individual atom can be calculated by the following simple formula: $\dim H^1(P_c, \mathbb{Z}) = s_c + 1$, where s_c is the number of vertices of the graph K_c of degree 4 (i.e., the number of hyperbolic trajectories with oriented separatrix diagrams that belong to this atom).

The next assertion shows what happens to the Poincaré flow under a transformation of the transversal section.

Proposition 8.1.

a) Let Q_c be a 3-atom without stars; let w and w' be the Poincaré flows generated by the system v on sections $P_{\text{tr}} = j(P_c) \subset Q_c$ and $P'_{\text{tr}} = j'(P_c) \subset Q_c$. Let $m \in H^1(P_c, \mathbb{Z})$ be a difference cocycle for P_{tr} and P'_{tr}. Then the flow w' on the 2-atom P_c is obtained from w by the pasting-cutting operation Φ_{-m}.

b) Let Q_c be a 3-atom with stars; let w and w' be the Poincaré flows generated by the system v on sections $P_{\text{tr}} = j(\widehat{P}_c)$ and $P'_{\text{tr}} = j'(\widehat{P}_c)$, where $\widehat{P}_c$ is the canonical double of the 2-atom P_c. Let $m \in H^1(P_c, \mathbb{Z})$ be the difference cocycle between these sections. Then the system w' on the double $\widehat{P}_c$ is obtained from w by the pasting-cutting operation $\Phi_{-\widehat{m}}$, where the 1-cocycle $\widehat{m}$ denotes the lift of m from $P_c = \widehat{P}_c/\tau$ to its double $\widehat{P}_c$. In other words, $\widehat{m}$ is the τ-symmetric 1-cocycle which takes the same values on those edges of the graph $\widehat{K}_c$ that are mapped into each other under the involution τ. More precisely, if m takes a value m_i on an edge $K_i \in K_c$, then $\widehat{m}$ takes the same value m_i both on $\widehat{K}_i$ and on $\tau(\widehat{K}_i)$.

Proof. We begin with the case of atoms without stars. It suffices to prove Proposition 8.1 for an elementary 1-cocycle m, namely for the cocycle that is equal to zero on all the edges of K_c except for one edge K_i on which it is equal to 1.

Consider an annulus L_i (Fig. 8.4) with two transversal smooth curves $j(K_i)$ and $j'(K_i)$ (bold lines in Fig. 8.4). Here $j'(K_i)$ is obtained from $j(K_i)$ by a winding with the coefficient $m_i = 1$ along the axis of the annulus. By a thin line we show a trajectory of the vector field $v = \operatorname{sgrad} H$ whose behavior, according to our assumptions, corresponds to Fig. 3.16(b). Recall that the edges $j(K_i)$ and $j'(K_i)$ are the intersections of the transversal sections P_{tr} and P'_{tr} with the annulus L_i. We have chosen them in a special way to make the proof more visual. We are allowed to do this, since a section can undergo an isotopy without changing any of its properties.

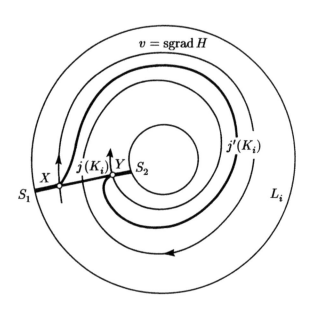

Figure 8.4

On the segments $S_1 X$ and $Y S_2$, the edges $j(K_i)$ and $j'(K_i)$ coincide; they differ only on the interval XY. Therefore, on the segments $S_1 X$ and $Y S_2$, the Poincaré flows σ^t and σ'^t coincide. What happens between points X and Y? It is easy to see by our construction that $Y = \sigma^2(X)$ and, at the same time, $Y = \sigma'^1(X)$. The situation on all neighboring tori will be similar. Thus, the flow σ^t requires more time to pass the interval from X to Y than the flow σ'^t. Moreover, the difference is exactly 1. But this just means that the flow σ'^t is obtained from σ^t by cutting out a piece of length 1, as was to be proved.

Part (b) is proved similarly. We only need to repeat the above arguments for each pair of corresponding edges $\widehat{K}_i$ and $\tau(\widehat{K}_i)$. $\square$

8.3. THE ACTION OF $G\mathbb{P}$ ON THE SET OF SUPERFLUOUS t-FRAMES

It is remarkable that the action of $G\mathbb{P}$ on the set of sections $\{\mathbb{P}\}$ induces a natural action on the set $\{\mathbb{T}\}$ of (admissible) superfluous t-frames. Suppose we are given a Hamiltonian system v and a set of sections $\mathbb{P} = \{P_{tr}\}$. Then the superfluous t-frame $\mathbb{T}$ corresponding to this set is well defined. We act on $\mathbb{P}$ by an element $\mathbb{M}$ of the group of transformations $G\mathbb{P}$. After this, we obtain a new set of sections $\mathbb{P}'$ with a new superfluous t-frame $\mathbb{T}'$. By definition, we shall assume that $\mathbb{T}'$ is the result of the action of $\mathbb{M}$ on $\mathbb{T}$.

We stress that non-triviality of this construction consists of the fact that $\mathbb{T}'$ depends only on $\mathbb{T}$ and $\mathbb{M}$ and does not depend on the concrete choice of v and $\mathbb{P}$.

We shall now prove this theorem, resting on the theory already developed. We shall see this via explicit formulas that describe this action.

Besides the difference cochain m_c, we shall consider a difference 2-cochain $k_c = \delta m_c$ which is the coboundary of m_c in the following sense. Notice that m_c is a cocycle from the point of view of the graph K_c (or the surface P_c with boundary). However, from the point of view of the closed surface $\tilde{P}_c$ obtained from P_c by gluing discs to all of its boundary circles, the cochain m_c will have a non-trivial coboundary δm_c, which we denote by k_c.

The cochain k_c assigns a certain integer number k_n to each boundary torus of the atom Q_c. In fact, this number has occurred many times in the formulas for the change of admissible coordinate systems on the boundary tori of the atom (see Chapter 4):

$$\lambda_n = \lambda_n', \qquad \mu_n = \mu_n' + k_n \lambda_n'.$$

Let us note that, in terms of the sections $\mathbb{P}$ and $\mathbb{P}'$, the difference 2-cochain k_c has a very natural meaning: in the case of atoms without stars, k_c shows how the boundaries of the sections P_{tr} and P'_{tr} differ (while the difference 1-cochain m_c describes the difference between the sections themselves, i.e., contains more precise information). Let $\mathbb{K}$ denote the set $\{k_c\}$. In the case of atoms with stars, the interpretation of the 2-cochain k_c actually remains the same. But since we have to consider the doubles, the boundaries of the transversal sections $P_{tr} = j(\widehat{P}_c)$ and $P'_{tr} = j'(\widehat{P}_c)$ differ by $2k_c$.

The above arguments were related to saddle atoms. In the case of an atom of type A, as an analog of 2-cochain k_c we shall consider one integer number k assigned to the boundary torus. As such a number, we take the number m itself defined by the relation $\mu' = \mu + m\lambda$. In other words, in this case, the same number m plays both the role of the difference 1-cochain m_c and that of the difference 2-cochain $k_c = \delta m_c$.

Thus, on each boundary torus of each atom Q_c there is an integer k_n. Therefore, we may assume that these numbers are located in the beginning and at the end of each edge e_j of the molecule W. That is why the indexing of these numbers can be produced in two ways:

1) k_j^- and k_j^+, where j enumerates the edges of the molecule W (the minus corresponds to the beginning and the plus corresponds to the end of the edge e_j);

2) k_n, where n enumerates the boundary tori of an atom Q_c (in this case, the k_n's can be considered as the coefficients of the difference 2-cochain k_c).

An analogous convention will be used for indexing the rotation vectors R^-, R^+ and admissible coordinate systems (λ, μ) on Liouville tori. In this approach, many formulas are simplified.

Proposition 8.2. *Let $\mathbb{P}$ be an arbitrary set of transversal sections for some integrable system v on an isoenergy surface Q^3. Let $\mathbb{M}$ be an arbitrary element of the group $G\mathbb{P}$, and let $\mathbb{K} = \delta\mathbb{M}$. In other words, $\mathbb{M}$ and $\mathbb{K}$ are the sets of difference 1- and 2-cochains respectively. Let $\mathbb{P}'$ be the set of sections obtained from $\mathbb{P}$ by the transformation $\mathbb{M}$, and let $\mathbb{T}$ and $\mathbb{T}'$ be the superfluous t-frames corresponding to the sets $\mathbb{P}$ and $\mathbb{P}'$. Then the elements of these t-frames are connected with each other as follows:*

1) $C'_j = \begin{pmatrix} \alpha'_j & \beta'_j \\ \gamma'_j & \delta'_j \end{pmatrix} = \begin{pmatrix} 1 & 0 \\ -k^+_j & 1 \end{pmatrix} \begin{pmatrix} \alpha_j & \beta_j \\ \gamma_j & \delta_j \end{pmatrix} \begin{pmatrix} 1 & 0 \\ k^-_j & 1 \end{pmatrix} = (A^+_j)^{-1} C_j A^-_j$, *where* $A^\pm_j = \begin{pmatrix} 1 & 0 \\ k^\pm_j & 1 \end{pmatrix}$,

2) $(R^-_j)' = R^-_j + k^+_j$ *and* $(R^+_j)' = R^+_j + k^-_j$,

3) $\Lambda'_c = \Lambda_c$,

4) $\Delta'_c = \Delta_c + \phi'_1(k_c)$ *or, equivalently,* $\Delta'_c = \Delta_c + \phi_1(m_c)$,

5) $Z'_c = Z_c + \phi_2(m_c)$.

Proof. Formula (1) is proved in Chapter 4. Formula (2) follows from Proposition 1.14. Formulas (3), (4), (5) follow from the properties of the pasting-cutting operation (see Section 6.4) and Proposition 8.1, which interprets the action of the element m_c as a pasting-cutting. ☐

Corollary. *The action of the group of transformations of transversal sections on the set of superfluous t-frames is well defined. In particular, this action does not depend on the choice of a concrete Hamiltonian system and a set of transversal sections realizing the given t-frame.*

8.4. THREE GENERAL PRINCIPLES FOR CONSTRUCTING INVARIANTS

We can now state some general principles of the construction of orbital invariants.

8.4.1. First General Principle

Let g be a function on the set of superfluous t-frames that is invariant under the above described action of the group $G\mathbb{P}$ (g takes values in some reasonable set):

$$g: \{\mathbb{T}\} \to X, \quad \text{and} \quad g(\mathbb{T}) = g(\mathbb{T}') \quad \text{if} \quad \mathbb{T}' = \mathbb{M}(\mathbb{T}),$$

where X denotes the set of possible values of the function g (for example, real numbers, projective space, chains, cochains, etc.).

Then g (now as a function on the set of integrable Hamiltonian systems) is a topological orbital invariant of integrable Hamiltonian systems on isoenergy surfaces.

8.4.2. Second General Principle

Let $g_1, \ldots, g_p$ be a set of topological orbital invariants (see the first general principle), and assume that this set is complete, i.e., allows us to distinguish the orbits of the action of the group $G\mathbb{P}$ on the set of superfluous t-frames. Then the object $(W, g_1, \ldots, g_p)$ (which can be called a t-molecule, interpreting $g_1, \ldots, g_p$ as some new marks attached to the molecule W) is a complete topological orbital invariant of integrable Hamiltonian systems on isoenergy surfaces. In other words, two integrable Hamiltonian systems are topologically orbitally equivalent if and only if the corresponding t-molecules coincide.

Thus, our problem reduces to a rather formal search for invariants of the action of $G\mathbb{P}$ on the set $\{\mathbb{T}\}$.

8.4.3. Third General Principle

For a proper approach to the classification problem, we consider the set of all superfluous t-frames and the action of the group of transformations of transversal sections. As a result, we obtain an orbit space (perhaps not very nice if the Hamiltonian system is sufficiently complicated). After this, we must consider this orbit space and define a set of "functions" separating the orbits. Our third principle is the following.

A complete set of invariants can be chosen in a non-unique way. Each choice of invariants is determined by specific properties of the molecule of a given system. This means that t-molecules can be determined in different ways. However, any choice of a specific form of a t-molecule must take into account its underlying "compulsory part", namely, the marked molecule W^*, which can be "extended" by new invariant parameters by different methods.

For example, working only with simple molecules, we can add to W^* only the b-invariant introduced below and the rotation vectors; if complex molecules are considered, it will be necessary to add the more delicate $\widetilde{\Delta Z}[\widetilde{\Theta}]$-invariant.

8.5. ADMISSIBLE SUPERFLUOUS t-FRAMES AND A REALIZATION THEOREM

8.5.1. Realization of a Frame on an Atom

Let P_c^2 be a saddle 2-atom with or without stars, and let Q_c^3 be the corresponding 3-atom.

Lemma 8.4.

a) *Suppose we are given an arbitrary Hamiltonian system* $w = \operatorname{sgrad} F$ *on the 2-atom* P_c *(without stars) with a Morse Hamiltonian* F. *Then this system can be realized as the Poincaré flow for some integrable Hamiltonian system* $v =$

sgrad H *with two degrees of freedom on the symplectic manifold M^4 diffeomorphic to the direct product $Q_c^3 \times (-1, 1)$.*

b) *Suppose we are given a Hamiltonian system $w = \operatorname{sgrad} F$ on the double $\widehat{P}_c$ of the 2-atom P_c (with stars) with a Morse Hamiltonian H which is invariant under the involution $\chi \colon \widehat{P}_c \to \widehat{P}_c$. Then this system can be realized as the Poincaré flow of some integrable Hamiltonian system $v = \operatorname{sgrad} H$ with two degrees of freedom on the symplectic manifold M^4 diffeomorphic to the direct product $Q_c^3 \times (-1, 1)$. Here Q_c^3 is the 3-atom corresponding to the 2-atom $P_c = \widehat{P}_c / \chi$.*

Proof. a) Let ω be the symplectic structure on P_c that corresponds to the Hamiltonian F and the field w. Consider the 4-manifold $\widetilde{M} = P_c \times [0, 2\pi] \times (-1, 1)$ with the symplectic structure $\widetilde{\Omega} = \omega + dH \wedge d\varphi$, where H and φ are the natural coordinates on $(-1, 1)$ and $[0, 2\pi]$ respectively. This is, clearly, a symplectic structure, and F is an integral of the Hamiltonian vector field $\operatorname{sgrad} H = \partial/\partial\varphi$.

We now identify the bases of the cylinder $P_c \times (-1, 1) \times \{0\}$ and $P_c \times (-1, 1) \times \{2\pi\}$ via the diffeomorphism $g(p, H, 2\pi) = (\sigma(p), H, 0)$, where $\sigma = \sigma^1$ is the shift by unit time along the vector field w. Here (p, H) denotes a point of $P_c \times (-1, 1)$.

As a result, we obtain a manifold $M^4 = P_c \times S^1 \times (-1, 1)$; and the symplectic structure $\widetilde{\Omega}$ turns (upon sewing) into a good symplectic structure on M (since the symplectic structure ω is preserved under the mapping σ).

It is clear that the mapping $\sigma \colon P_c \to P_c$ will be the Poincaré map of the Hamiltonian flow $v = \operatorname{sgrad}(H)$ on each isoenergy surface, as required.

b) In the case of atoms with stars, we proceed similarly. Namely, we consider the cylinder $\widetilde{M} = \widehat{P}_c \times [0, \pi] \times (-1, 1)$ and identify its bases $\widehat{P}_c \times (-1, 1) \times \{0\}$ and $\widehat{P}_c \times (-1, 1) \times \{\pi\}$ via the diffeomorphism of the form $g(p, H, \pi) = (\chi\sigma^{1/2}(p), H, 0)$. As in the previous case, we obtain a symplectic manifold with the Hamiltonian flow $\operatorname{sgrad} H$, which is transversal to the section $\widehat{P}_c$. Here, the standard Poincaré map (of multiplicity 1) on this section will be of the form $\bar{\sigma} = \chi\sigma^{1/2}$. But, according to our terminology, as the Poincaré map on atoms with stars consider the repeated mapping, which in this case takes the following form:

$$\sigma = (\bar{\sigma})^2 = \chi\sigma^{1/2}\chi\sigma^{1/2} = \chi^2\sigma^1 = \sigma^1 \,,$$

as required. This completes the proof. $\square$

This lemma shows, in particular, that we can realize any admissible triple (Λ, Δ, Z) (see Section 6.3) as the triple of atomic invariants for an integrable system with two degrees of freedom, i.e., as an element of some superfluous t-frame.

Let us note another important link between a system on a 3-atom and its reduction on the 2-atom. In Chapter 6, we considered many times the period function Π which assigned the period to each closed trajectory μ of a system given on 2-atom. In our case, each closed trajectory of the reduced system on P corresponds to a certain Liouville torus, on which we can define the rotation number ρ. It is easily seen that there is a very natural relationship between the numbers ρ and Π. To formulate the answer in a convenient form, we shall take the period function with sign. Namely, we take "plus" if the trajectory is located on a positive annulus, and "minus" if the trajectory is located on a negative annulus.

Note that the trajectory μ, being the intersection of the torus with the transversal section P, can be considered as one of the basis cycles of an admissible coordinate system that is additional to the first cycle λ, which is a fiber of the Seifert fibration. However, two natural orientations on μ (as a trajectory of the reduced system and as a basic cycle of the admissible coordinate system) may differ. This exactly means that the period of μ should be taken with the minus sign. Note that, in the notation we use now, an admissible coordinate system on an atom A is the pair $(-\mu, \lambda)$, but not (λ, μ).

Thus, suppose we have the orientation on μ considered as the second basis cycle of the admissible coordinate system. Then one can easily verify the following assertion which is sometimes taken as a definition of the rotation number.

Lemma 8.5. *Let Q_c be an arbitrary atom, and let Π be the period of a closed trajectory μ of the Poincaré vector field w on a transversal section $P_{\mathrm{tr}} \subset Q_c$. Let T be the Liouville torus in Q_c corresponding to this trajectory, and let ρ be the rotation number of v on the torus T relative to the coordinate system consisting of the following two cycles: the first is a fiber of the Seifert fibration on Q_c, and the second is the intersection of P_{tr} with the given torus, i.e., μ. Then $\rho = \Pi$.*

Note that this lemma allows us to divide all the annuli of an atom into positive and negative ones by a very natural way: the sign of an annulus is determined by the sign of the rotation function of the family of Liouville tori corresponding to the annulus.

Corollary. *Let Q_c be a saddle atom (with or without stars). Then the rotation function ρ written in any admissible coordinate system tends to infinity as the torus approaches the singular leaf $L = f^{-1}(c)$. In the case of an atom A, the limit of the rotation function ρ, as the Liouville torus shrinks into a stable closed trajectory, can be an arbitrary real number and cannot be equal to infinity.*

We also indicate an important link between the multipliers of the Poincaré map and the rotation function. Consider a 3-atom A represented as a neighborhood of a stable periodic trajectory γ foliated into Liouville tori. Let ν be the multiplier of γ. Recall that ν is, by definition, an eigenvalue of the linearized Poincaré map on a transversal section P_{tr}. Here there are two eigenvalues ν and ν^{-1}. Let ρ_0 be the limit of the rotation function ρ as the torus tends to γ. We assume that the rotation function is calculated in an admissible coordinate system related to the atom A: the first basis cycle lies on the transversal section P_{tr}, and the second is a fiber of a trivial Seifert fibration on the atom A directed along γ.

Proposition 8.3. *The following formula holds:*

$$\nu = \exp(2\pi i \rho_0).$$

Proof. Let us note that, in this proposition, we have interchanged the cycles λ and μ in the basis on a Liouville torus (see Lemma 8.5). Therefore, in this case, the rotation number and the period of the Poincaré flow are connected by the relation $\rho^{-1} = \Pi$. Thus, the formula to be proved is actually an assertion on a Hamiltonian system with one degree of freedom given on the transversal section P_{tr}, since the function Π and the multiplier ν characterize the Poincaré flow on P_{tr}. The desired formula is implied by the following fact, which holds for Hamiltonian systems with one degree of freedom.

Let P_0 be a non-degenerate local minimum (or maximum) of a Hamiltonian $F(x,y)$. Set, for definiteness, $F(P_0) = 0$. Let $\Pi(c)$ be the period of the flow $w = \mathrm{sgrad}\, F$ for the closed trajectory of the form $\gamma_c = \{F = c\}$, and let $\Pi_0 = \lim\limits_{c\to 0} \Pi(c)$. Denote by σ the shift by unit time along integral curves of $w = \mathrm{sgrad}\, F$. Consider the linearization $d\sigma$ (i.e., the differential) of σ at the point P_0 and assume that ν and ν^{-1} are its eigenvalues.

Proposition 8.4. *The following formula holds:*

$$\nu = \exp\left(\frac{\pm 2\pi i}{\Pi_0}\right).$$

Proof. According to the Morse lemma, we can choose such local coordinates x, y in a neighborhood of P_0 that $F(x,y) = x^2 + y^2$. Since these coordinates are not necessarily canonical, the symplectic form ω can be written in these coordinates as $\omega = \omega(x,y)\, dx \wedge dy$, where $\omega(x,y)$ is a smooth function. Then the Hamiltonian vector field $w = \mathrm{sgrad}\, F$ becomes

$$\mathrm{sgrad}\, F = \left(\frac{-y}{\omega(x,y)}, \frac{x}{\omega(x,y)}\right).$$

Consider another vector field $\xi = (-y, x)$. It is proportional to the field w, and $w = \dfrac{\xi}{\omega(x,y)}$. The period of ξ is constant and equal to 2π. Therefore, the period $\Pi(c)$ of w can be estimated as follows:

$$2\pi \cdot \min \omega(x,y) \le \Pi(c) \le 2\pi \cdot \max \omega(x,y),$$

where min and max are taken over the circle $\{F = c\}$ centered at P_0. Taking the limit as $c \to 0$, we obtain the equality

$$\Pi(0) = 2\pi\omega(0,0).$$

On the other hand, the linearization of $w = \dfrac{\xi}{\omega(x,y)}$ at the equilibrium point P_0 has the form $\dfrac{\xi}{\omega(0,0)} = \dfrac{(-y,x)}{\omega(0,0)}$. Therefore, the eigenvalues of the linearized mapping σ equal

$$\nu = \exp\left(\frac{\pm i}{\omega(0,0)}\right).$$

Comparing this expression with $\Pi(0)$, we get the desired formula:

$$\nu = \exp\left(\frac{\pm 2\pi i}{\Pi(0)}\right).$$

Thus, Proposition 8.4 is proved.   □

Proposition 8.3 follows immediately from Proposition 8.4.   □

8.5.2. *Realization of a Frame on an Edge of a Molecule*

Here we prove a technical assertion allowing us to realize a system with prescribed rotation function $\rho(f)$ on an edge of a molecule.

Suppose we are given a four-dimensional symplectic manifold

$$M^4 = T^2 \times (a, b) \times (-1, 1),$$

in which the following two open subsets are distinguished (see Fig. 8.5):

$$M_1 = T^2 \times (a, a + \varepsilon) \times (-1, 1) \quad \text{and} \quad M_2 = T^2 \times (b - \varepsilon, b) \times (-1, 1).$$

We assume that two functions H and f are given on M^4, where H is a parameter on the interval $(-1, 1)$, and f is a parameter on the interval (a, b). Let ω_1 and ω_2 be symplectic structures on M_1 and M_2, respectively, such that the natural foliations into 2-tori are Lagrangian foliations. This means that H and f commute on M_i with respect to the given symplectic structure ω_i.

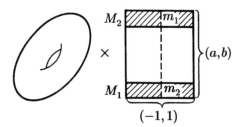

Figure 8.5

On each 2-tori T^2 of the trivial bundle on M^4, we consider a compatible smooth basis (λ, μ) (i.e., this basis depends smoothly on the torus).

Suppose we are now given a smooth function $\rho(f)$ on the interval (a, b). We assume that, for $a < f < a + \varepsilon$ and $b - \varepsilon < f < b$, this function ρ is the rotation function of the integrable Hamiltonian system $v = \operatorname{sgrad} H$ on the isoenergy level $\{H = 0\}$. The parameter for the one-parameter family of tori on $\{H = 0\}$ is the function f.

We wish to realize ρ as the rotation function of some Hamiltonian system with the Hamiltonian H on the level $\{H = 0\}$. That is why we have to impose one more natural restriction on it, which is connected to the following observation. First note that ρ determines the direction of the vector field v on a torus only up to sign. However, if we know the initial position of v and the rotation function, then we can uniquely define its final position by translating v continuously according to the behavior of the rotation function. That is why we shall assume, in addition, that, after formal translation of v from the torus $\{f = a\}$ to the torus $\{f = b\}$ according to the rule defined by the rotation function $\rho(f)$, we obtain again the field $v = \operatorname{sgrad} H$, but not $-v$.

Finally, the last condition is the following: the vector fields sgrad H and sgrad f (considered as a basis on the tangent space to a Liouville torus) define the same orientation of Liouville tori inside M_1 and M_2. (This condition has natural meaning, since we can compare orientations given on different fibers of the trivial T^2-fibration.)

Lemma 8.6. *Under the assumptions stated above, there exists a symplectic structure Ω on the whole manifold M^4 that extends the original structures ω_1 and ω_2 defined on the "boundary collar" $M_1 \cup M_2$ and satisfies the following conditions:*

1) the original trivial fibration of the manifold M^4 into 2-tori is Lagrangian, i.e., the functions H and f commute on M^4.

2) the rotation function of the integrable Hamiltonian system $v = \text{sgrad}\, H$ on the one-parameter family of two-dimensional Liouville tori $\{H = 0\}$ coincides with the prescribed function $\rho(f)$.

Proof. Let us define the action-angle variables inside M_1 and M_2 corresponding to the given symplectic structures ω_1 and ω_2 and the fixed basis cycles (λ, μ).

Then we extend the angle variables φ_1 and φ_2 from the "boundary collar" to all of M^4 in an arbitrary smooth fashion. This can be done, because our T^2-fibration is trivial and its base is contractible. As a result, we obtain two smooth global functions φ_1 and φ_2 on M^4. We now wish to extend the action variables s_1 and s_2 to the whole manifold M^4.

Assume for the moment that we have already extended the symplectic 2-form and action-variables s_1 and s_2 to M^4. Then the rotation function on the level $\{H = 0\}$ can be computed by the formula

$$\rho = \frac{\partial H/\partial s_1}{\partial H/\partial s_2}$$

or, equivalently,

$$\rho = -\frac{\partial s_2/\partial f}{\partial s_1/\partial f}.$$

Actually, the rotation function ρ is known, but we must find the functions $s_1(H, f)$ and $s_2(H, f)$ satisfying this relation, taking into account that the mapping $(H, f) \to (s_1(H, f), s_2(H, f))$ must be an immersion.

Take two smooth functions $a(f)$ and $b(f)$ such that they do not vanish simultaneously and $\rho(f) = \dfrac{a(f)}{b(f)}$. Besides, let $\dfrac{\partial s_1}{\partial f} = -b(f)$ and $\dfrac{\partial s_2}{\partial f} = a(f)$ on the "boundary collar" $M_1 \cup M_2$ for $H = 0$.

Thus, we come to the following problem: to find a smooth curve $\gamma = \gamma(f) = (s_1(f), s_2(f))$ on the (s_1, s_2)-plane such that

$$\frac{d\gamma}{df} = (-b(f), a(f)).$$

This equation has a solution $\gamma = \gamma(f)$ that is uniquely defined up to parallel Euclidean translation.

On each of two-dimensional rectangles m_1 and m_2 in Fig. 8.5 (that represent two "boundary collars" M_1 and M_2), we are given both pairs of functions (H, f) and (s_1, s_2). Therefore, on each rectangle, we can express s_1 and s_2 in terms of H and f. Recall that H and f are in fact Cartesian coordinates on m_1 and m_2. As a result, we obtain a smooth immersion of each rectangle m_1 and m_2 into the (s_1, s_2)-plane. Their images are shown in Fig. 8.6 as two immersed "curvilinear" rectangles $\tilde{m}_1$ and $\tilde{m}_2$.

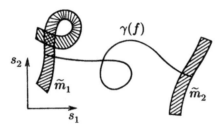

Figure 8.6

Recall that the action variables s_1 and s_2 are defined as functions of H and f up to some additive constants. This means that the immersions are not uniquely defined, but only up to an arbitrary Euclidean translation in the (s_1, s_2)-plane. Thus, we obtain three objects on the (s_1, s_2)-plane: a curve γ and two immersed rectangles $\tilde{m}_1$ and $\tilde{m}_2$. Each of them can be translated in the plane independently of the others. It is clear that, combining these translations in a suitable way, we can obtain the picture shown in Fig. 8.6, where the curve γ starts from one rectangle and ends at the other. Here we have used the following fact: the curve $\gamma(f)$ for $a < f < a + \varepsilon$ and $b - \varepsilon < f < b$ coincides with the images of two intervals $\{H = 0\} \cap m_1$ and $\{H = 0\} \cap m_2$.

All of this means that we are actually given an immersion (into the plane) of two rectangles m_1 and m_2 joined by a straight line segment; our goal is to extend this immersion to the whole rectangle $(a, b) \times (-1, 1)$ (see Fig. 8.5). It is clear that this can be done.

We should point out one subtle point that was left "behind the scenes". If two original immersions of m_1 and m_2 differ from each other in orientation (i.e., if we turned one of them over), then, of course, we would not succeed in extending the immersion to the whole rectangle $(a, b) \times (-1, 1)$. However, we have used the consistency of orientations implied by one of the conditions stated before Lemma 8.6.

Now consider both functions s_1 and s_2 as action variables on the whole M^4.

Then the desired symplectic structure on M^4 can be written in the following canonical form:

$$\Omega = ds_1 \wedge d\varphi_1 + ds_2 \wedge d\varphi_2.$$

Clearly, this form satisfies all the requirements. This completes the proof.   □

8.5.3. Realization of a Frame on the Whole Molecule

Suppose we are given a molecule W and an abstract superfluous t-frame for it:

$$\mathbb{T} = (C_j, R_j^-, R_j^+, \Lambda_c, \Delta_c, Z_c).$$

Here j numbers the edges of the molecule, and c numbers its vertices (atoms). The objects occurring here are not arbitrary. We describe the restrictions that they must satisfy. These restrictions are divided into two types. The first combines natural self-evident conditions satisfied separately by each of these objects. Restrictions of the second type connect different objects and can be called *crossed*.

Restrictions of the first type.

1) The integer 2×2-matrices C_j must have determinant -1.

2) The vectors R_j^- and R_j^+ must be the R-vectors of some smooth functions $\rho_j(t)$ that satisfy the properties described in Section 5.1. This means that these vectors satisfy some obvious and natural conditions implied by the definition of R-vector. For example, between two adjacent minima there must be a maximum or a pole. It is easy to make a formal list of these conditions, but we shall not do it, since their character is completely clear.

Moreover, if an edge exits from a saddle atom (resp. atom A), then the first component of R^- must be infinite (resp. finite). If, on the contrary, the edge enters a saddle atom (resp. atom A), then the last component of R^+ must be infinite (resp. finite). This follows immediately from Lemma 8.5.

3) $\Lambda = (\Lambda_1, \dots, \Lambda_n)$ is a set of n positive real numbers considered up to proportionality (i.e., as a point of projective space), where n is the number of vertices of the atom P_c^2, and all of the numbers Λ_i are in a one-to-one correspondence with the vertices of the atom.

4) $\Delta = (\Delta_1, \dots, \Delta_n)$ is a real 0-chain on the vertices of the atom $V = (P_c^2, K_c)$ belonging to the set $\Delta(V)$ (see Chapter 6).

5) Z is an arbitrary element of the homology group $H_1(\widetilde{P}_c^2, \mathbb{R})$, where $\widetilde{P}_c^2$ is the closed surface that is obtained from P_c^2 by gluing the boundary circles by disks.

Restrictions of the second type (crossed).

1) The rotation functions ρ^+, ρ^- and the components of the gluing matrix on each edge are connected by the relation (indicated in Proposition 1.14)

$$\rho^- = \frac{\alpha \rho^+ + \gamma}{\beta \rho^+ + \delta}.$$

The corresponding relation must also be satisfied by the R-vectors R^- and R^+. This means that R^- can be computed in terms of R^+ and the gluing matrix C.

2) All the edges incident to a saddle atom can be divided into two parts in two different ways. The first way is topological: the edges are partitioned into two types corresponding to the partition of the annuli of the atom into positive and negative. The second is the partition of annuli into positive and negative ones depending on the sign of the rotation function. It is required that these two ways of partitioning give the same result.

3) Let us note that, knowing the first and last components of the R-vectors, we can always define the limit positions $v(0)$ and $v(1)$ of the vector field v, i.e.,

at the beginning and at the end of an edge under consideration. For example, for saddle atoms, the direction of $v(0)$ coincides with that of λ^-, and $v(1)$ indicates the direction of λ^+. For an atom A, we also can uniquely reconstruct the directions of $v(0)$ and $v(1)$. Knowing the first component of R^-, one can reconstruct $v(0)$ up to sign. The final choice of its direction is determined by the condition that the coefficient b in the decomposition $v(0) = a\lambda^- + b\mu^-$ must be positive. The direction of $v(1)$ can be defined similarly, taking into account the decomposition $v(1) = a\lambda^+ + b\mu^+$, where $b > 0$.

Thus, we can always uniquely reconstruct $v(0)$ and $v(1)$ in bases (λ^-, μ^-) and (λ^+, μ^+) respectively. Using the gluing matrix, we can write down these limiting positions in the same basis, for instance, in (λ^-, μ^-).

But on the other hand, starting from the initial position $v(0)$ and moving the vector v according to the information given by the R-vector R^-, we can compute the limit position $v(1)$ in another independent way. The condition imposed on the t-frame is that the limit positions of v at the end of the edge, computed by two different ways, must coincide.

Definition 8.2. An abstract superfluous t-frame of the molecule W that satisfies all the requirements of the first and second types listed above will be called an *admissible* superfluous t-frame of W.

Theorem 8.1. *Suppose we are given an arbitrary admissible superfluous t-frame $\mathbb{T}_0$ of the molecule W. Then there exists a four-dimensional symplectic manifold $M^4 = Q^3 \times D^1$ with an integrable system $v = \operatorname{sgrad} H$ (of the above described type) and a set of transversal sections $\mathbb{P}$ inside $Q^3 = \{H = 0\}$ for all the atoms such that the corresponding superfluous t-frame*

$$\mathbb{T}(v) = (C_j(\mathbb{P}), R_j^-(\mathbb{P}), R_j^+(\mathbb{P}), \Lambda_c(\mathbb{P}), \Delta_c(\mathbb{P}), Z_c(\mathbb{P}))$$

coincides with the original admissible superfluous t-frame $\mathbb{T}_0$.

Proof. We take a molecule W and its gluing matrices C_j; and from this material we first manufacture the 3-manifold Q^3 by gluing it from individual 3-atoms according to the requirements dictated by the gluing matrices (see Chapter 4). At the same time, we choose and fix a transversal section in each atom, which gives us admissible coordinate systems on all the atoms. Then we take the direct product $Q^3 \times (-1, 1)$ and obtain a four-dimensional manifold with a structure of a foliation into 2-tori (and singular leaves).

Following the method already applied once, we take 3-atoms Q_c^3 (as small neighborhoods of singular fibers) and, for each of them, construct the required integrable system on $Q_c^3 \times (-1, 1)$ using Lemma 8.4. This means, in particular, that we have defined a symplectic structure on "4-atoms" $Q_c^3 \times D^1$. All these structures define the same orientation on M^4. This follows from the explicit formula for Ω given in the proof of Lemma 8.4. Note that we have automatically realized all the elements of the superfluous t-frame except for the rotation vectors.

After that, by Lemma 8.6, we can extend the symplectic structure to each edge of the molecule so that, on this edge, we obtain the required rotation vector R^+. (Note that R^- is then uniquely reconstructed from R^+ and the gluing matrix.) Here all conditions on the behavior of the rotation function on the endpoints of the edges required in Lemma 8.6 are automatically fulfilled in view of the above restrictions on the t-frame. It is also necessary to verify the compatibility condition for orientations of the pair of vector fields sgrad H and sgrad f at the endpoints of the edges. But this follows from the following agreement about orientations of M^4, Q^3, and boundary tori of an atom Q_c^3.

The symplectic structure defines such an orientation on M^4 that, for any function f independent of H, the quadruple of vectors

$$\text{sgrad}\, H, \ \text{grad}\, H, \ \text{sgrad}\, f, \ \text{grad}\, f$$

has positive orientation. Since $\text{grad}\, H \neq 0$ everywhere on Q, we can assume by definition that the triple

$$\text{sgrad}\, H, \ \text{sgrad}\, f, \ \text{grad}\, f|_{Q^3}$$

defines positive orientation on Q^3. Here f is an arbitrary integral of the Hamiltonian vector field sgrad H. It is easy to see that the orientation does not depend on the choice of f. Finally, we define the orientation on the boundary tori of an atom by means of the outward normal. As a result, the pair

$$\text{sgrad}\, H, \ \text{sgrad}\, f$$

defines positive orientation on a boundary torus if and only if the vector $\text{grad}\, f|_{Q^3}$ is directed outside of the atom.

Thus, having natural orientation on M^4, we can naturally define orientations on the boundary tori of the atoms. Now it is easily seen that, if f changes monotonically on an edge, then the orientations of the pair sgrad H, sgrad f at the beginning and at the end of the edge will be the same, as required.

This completes the proof of the realization theorem. $\square$

8.6. CONSTRUCTION OF ORBITAL INVARIANTS IN THE TOPOLOGICAL CASE. A t-MOLECULE

In this section, we construct orbital invariants of integrable Hamiltonian systems with two degrees of freedom following the first general principle (see Section 8.4). In other words, the orbital invariants will be constructed as certain functions on the set of superfluous t-frames that are invariant relative to the group of transformations of transversal sections $G\mathbb{P}$.

8.6.1. The R-Invariant and the Index of a System on an Edge

Suppose we are given a molecule W^*, and let e be an arbitrary edge on which two vectors R^+ and R^- are given.

Definition 8.3. If the r-mark on the edge e is finite, then we take the vector $R = \beta R^- - \alpha$ as an invariant associated with e. If $r = \infty$, then we take the rotation vector R^- *modulo* 1 as an invariant on the edge e (sometimes we shall denote it by $R \bmod 1$). In both cases, this invariant will be called the R-*invariant* of the system on the given edge e.

COMMENT. In this definition, the subtraction of a number from a vector is understood as follows: this number is subtracted from each component of the vector (here $\infty - \alpha = \infty$). Speaking about $R \bmod 1$, we have in mind that all the components of the vector are defined up to an integer that is the same for all components.

The fact that the constructed set of numbers R is indeed a well-defined orbital invariant of the system, which does not depend on the choice of transversal sections, follows immediately from the explicit formulas of the action of $G\mathbb{P}$ (Proposition 8.2). Note that, for finite edges (with $r \neq \infty$), the R-invariant has a natural sense. It is just the R-vector for the rotation function ρ written in terms of the "basis" λ^-, λ^+, which is uniquely defined and does not depend on the choice of sections. We use quotation marks for the "basis", since the cycles λ^-, λ^+ are independent, but, in general, do not form a basis in the fundamental group of the Liouville torus.

The rotation vector R actually describes the evolution of the vector $v(t)$ of our system as t changes along an edge of the molecule. We now define another orbital invariant $\operatorname{ind} R$, called the *index of the system on an edge*, that shows the "number of revolutions" completed by the vector $v(t)$ in its motion from the beginning of the edge to the end. This invariant is uniquely computed from the R-vector. Therefore, it is not necessary to include it in the final complete list of independent orbital invariants. However, this invariant (index) is necessary for the statement of restrictions imposed by the system on the rotation vectors.

Since the number of revolutions completed by $v(t)$ is not integer in general, we need an additional construction. We give a precise definition.

Step 1. We reconstruct the rotation function $\rho(t)$ (up to conjugacy) from the R-vector. Then, from $\rho(t)$, we reconstruct the angle function

$$\psi(t) = \operatorname{arccotan} \rho(t) : t \to S^1 .$$

The angles $\psi(0)$ and $\psi(1)$ always satisfy some natural restrictions. The latter are actually formulated in the discussion of restrictions of the second type (crossed restrictions).

Step 2. Note that the system v can be perturbed in a neighborhood of an atom A so that the limit position of v will coincide in direction with any preassigned vector $\nu = a\lambda + b\mu$, where $b > 0$. Here λ and μ form an admissible coordinate system.

The proof follows from the fact that, in a neighborhood of an atom A, the system moves along Liouville tori with a "very small" meridian. Let us write out the vector field in the form $v(t) = a(t)\dfrac{\partial}{\partial \varphi_1} + b(t)\dfrac{\partial}{\partial \varphi_2}$. Here the angle variable φ_1 corresponds to the contractible cycle λ, and, therefore, considering arbitrary (but finite) perturbations of the function $a(t)$ we shall obtain small perturbations of the vector field v. Thus, by perturbing the system, we can arbitrarily change the limit position of v in a half-plane.

Step 3. Let us perturb the system in a neighborhood of an atom A so that the limit position of the angle ψ becomes a multiple of $\pi/2$. This condition defines the limit position uniquely, since $v(t)$ cannot leave the half-plane.

Note that, in a neighborhood of a saddle atom, a perturbation of the system does not affect the limit position of v, but here, even without this perturbation, the limit position of v coincides with one of the basis cycles λ^-, λ^+. Thus, the limit position of the angle is a multiple of $\pi/2$ automatically.

Step 4. After that, we define the index of the system on the edge by setting $\operatorname{ind} R = 2(\psi(1) - \psi(0))/\pi$. The index does not depend on the choice of perturbation of the system. It is clear that it is completely determined only by the rotation vector R and the type of edge.

REMARK. This definition first arose in the bordism theory of integrable Hamiltonian systems and is due to A. V. Bolsinov and Nguyen Tien Zung. In particular, it was used by Nguyen Tien Zung in [259], where an example of an integrable system that is not bordant to zero is constructed.

8.6.2. b-Invariant (on the Radicals of a Molecule)

By analogy with the notion of a family of a molecule, we now introduce the notion of a radical (as a part of a molecule).

Recall that the edges of a molecule that have finite r-marks are termed *finite*. The edges with infinite r-marks are termed *infinite*.

Definition 8.4. An infinite edge whose rotation vector is infinite (i.e., consists only of components equal to $\pm\infty$) will be called *superinfinite*.

In order not to be confused, in what follows, we shall use the term "infinite edge" only for the edges which are infinite, but not superinfinite.

Let us cut the molecule W^* along all the finite and infinite edges (that are not superinfinite by the above convention). As a result, the molecule splits in a disjoint union of subgraphs of two types: the atoms A and pieces that do not contain any atom A. This follows from the fact that the edges incident to atoms A cannot be superinfinite, since the limit of the rotation function on an atom A cannot be infinite.

Definition 8.5. Connected pieces of the second type (i.e., different from A) will be called *radicals*. We shall denote a radical by U.

Note that all the edges that are entirely in a radical are superinfinite. Each family of a molecule splits into a union of a number of radicals. However, there exist radicals that are not contained by any family.

We consider an arbitrary radical U of the molecule W^* and all the edges incident to it, i.e., such that at least one of their endpoints belongs to the radical. The edges entirely contained in a radical (i.e., the superinfinite edges) may naturally be called *interior* edges of the radical. The remaining edges incident to it are termed *exterior* edges (in relation to the given radical). Exterior edges do not belong to the radical.

To each edge e_j incident to the radical U we assign an integer number $[\Theta]_j$ by the following rule:

$$[\Theta]_j = \begin{cases} [\alpha_j/\beta_j] & \text{if } e_j \text{ is finite and exits the radical } U, \\ [-\delta_j/\beta_j] & \text{if } e_j \text{ is finite and enters the radical } U, \\ [MR_j^+] & \text{if } e_j \text{ is infinite and enters the radical } U, \\ -[-MR_j^-] & \text{if } e_j \text{ is infinite and exits the radical } U, \\ -\gamma_j/\alpha_j & \text{if } e_j \text{ is a superinfinite interior edge.} \end{cases}$$

Here $\alpha, \beta, \gamma, \delta$ are the entries of the gluing matrix C_j; MR^+ and MR^- are the arithmetic means of the finite components of the rotation vectors R^+ and R^-; and $[k]$ denotes the integer part of the number k.

It is clear that the set of numbers $[\Theta]$ is a function of a superfluous t-frame.

Definition 8.6. The number

$$b(U) = \sum_j [\Theta]_j .$$

is called the *b-invariant* (of the radical U).

This definition is rather reminiscent of the definition of the n-mark given in Section 4.3. These invariants are in fact closely connected with each other (see below). Here we used the same idea: each term in the above sum changes under a transformation of transversal sections by a certain integer number, but these terms are chosen in such a way that the total sum does not change.

Indeed, let us see what happens to the components of $[\Theta]$ under a transformation of a transversal section for a single atom Q_c belonging to the radical. Proposition 8.2 implies that the change is as follows:

$$[\Theta]_n' = [\Theta]_n + k_n ,$$

where the index n numbers the edges incident to the atom Q_c, and k_n are the coefficients of the difference 2-chain k_c. If we consider a coefficient $[\Theta]_j$ corresponding to an interior edge of the radical, then we must take into account both of the coefficients k_j^-, k_j^+ associated with the beginning and end of the edge respectively, i.e.,

$$[\Theta]_j' = [\Theta]_j + k_j^- + k_j^+ .$$

We shall write this collection of relations as

$$[\Theta]' = [\Theta] + q ,$$

where $q = (k_{c_1}, \ldots, k_{c_p})$ is the set of the difference 2-cochains for all the atoms that belong to U.

No changes happen if we transform a section in the atoms that do not belong to the radical.

Since each 2-cochain k_c is a coboundary, the sum of its coefficients is equal to zero. Hence the sum of all the $[\Theta]_j$'s (over all the edges of the radical) does not vary. Thus, b is invariant with respect to the action of the group $G\mathbb{P}$. According to the first general principle, this means that b is a well-defined topological orbital invariant of a system.

Consider a family in the molecule W^*. As was already remarked, each family splits into a union of radicals. According to Chapter 4, the family is endowed with an integer mark n. At the same time, each radical U is endowed with an integer mark b (the b-invariant).

Proposition 8.5. *The mark n is equal to the sum of the b-invariants of the radicals that are contained in the given family.*

REMARK.   This is the only relation which is satisfied by the b-invariants of the radicals. This actually means that the b-invariants can take arbitrary values independently of each other as well as on the other invariants. In this sense, the b-invariant is a new independent orbital invariant of integrable systems.

Proof (of Proposition 8.5).   We begin with a definition of the mark n. Let S be an arbitrary family. Recall that the radicals that form the given family are obtained from it by cutting along all infinite but not superinfinite edges. To compute the mark n, to each such edge e_k we assign the number $-\dfrac{\gamma_k}{\alpha_k}$. From the point of view of radicals, to the same edge we assign the two numbers

$$[MR_k^+] \quad \text{and} \quad -[-MR_k^-].$$

Now observe that the rotation functions ρ^- and ρ^+ on an infinite (or superinfinite) edge are connected by the relation

$$\rho^- = -\rho^+ - \frac{\gamma}{\alpha}.$$

Therefore, $MR^- = -MR^+ - \dfrac{\gamma}{\alpha}$. Hence

$$[MR_k^+] - [-MR_k^-] = -\frac{\gamma_k}{\alpha_k}.$$

Thus, by passing from the family to the union of the radicals, each number $-\dfrac{\gamma_k}{\alpha_k}$ (assigned with an infinite edge) is decomposed into the sum of the two numbers $[MR_k^+]$ and $-[-MR_k^-]$. No changes happen on the remaining edges of the family. Therefore, the total sum of the numbers $[\Theta]_k$ does not vary at all.   □

8.6.3. $\tilde{\Lambda}$-Invariant

Definition 8.7. As the $\tilde{\Lambda}$-*invariant* for a given saddle atom Q_c, we just take the Λ of the reduced system corresponding to an arbitrary transversal section $P_{\mathrm{tr}} \subset Q_c$.

The fact that Λ does not vary under transformations of the section follows from part 3 of Proposition 8.2.

8.6.4. $\widetilde{\Delta}\,\widetilde{Z}[\widetilde{\Theta}]$-Invariant

We begin the construction of this invariant with the following useful remark. If we look at the formulas for the variation of a superfluous t-frame under the action of the group $G\mathbb{P}$ (Proposition 8.2), then we notice that the difference 2-cochain k_c occurs in almost all these formulas. The only exception is the rule of variation for the Z-invariant. It turns out that, by a little modification of the Z-invariant, we can do so that the variation of Z can be written in terms of the difference 2-cochain k_c, but not 1-chain m_c. This modification will allow us to simplify the group $G\mathbb{P}$ by replacing $\mathbb{M}$ by $\mathbb{K} = \delta\mathbb{M}$. This circumstance has a rather natural interpretation: to compute orbital invariants, we do not need transversal sections themselves, but we need their boundaries only, i.e., admissible coordinate systems on the boundary tori.

Thus, consider the following construction. Let P_{tr} be a transversal section in a 3-atom Q_c. For this section, we can uniquely define the Z-invariant of the system under consideration $Z_c \in H_1(\widetilde{P}_c, \mathbb{R})$. Consider the projection

$$\xi \colon H_1(\widetilde{P}_c, \mathbb{R}) \to H_1(\widetilde{P}_c, S^1) = H_1(\widetilde{P}_c, \mathbb{R})/H_1(\widetilde{P}_c, \mathbb{Z})$$

and the image $\xi(Z_c) \in H_1(\widetilde{P}_c, S^1)$ of the Z-invariant. We assert that, by replacing Z_c by $\xi(Z_c)$, we do not lose any information about the system.

Indeed, consider an arbitrary transformation of a transversal section P_{tr} which does not change its boundary ∂P_{tr}. This means exactly that the difference 2-cochain k_c is equal to zero or, equivalently, the difference 1-cochain m_c is a cocycle from the point of view of the closed surface $\widetilde{P}_c$. What happens to the elements of a superfluous t-frame under such a transformation? According to Proposition 8.2, only the Z-invariant varies:

$$Z_c \to Z_c + \phi_2(m_c)\,.$$

However, we know that the operator ϕ_2 is the Poincaré duality. Therefore, Z_c is changed by an integer cycle, and, consequently, the class $\xi(Z_c)$ remains the same. Moreover, any integer cocycle can be realized by the suitable choice of a cocycle m_c. Thus, we do not lose any information.

Finally, note that, under arbitrary changes of a transversal section, the class $\xi(Z_c)$ changes by the following formula (in which only k_c occurs, but not m_c):

$$\xi(Z_c)' = \xi(Z_c) + \widetilde{\phi}_2(k_c)\,,$$

where the operator $\widetilde{\phi}_2 \colon B^2(\widetilde{P}_c, \mathbb{Z}) \to H_1(\widetilde{P}_c, S^1)$ is uniquely defined by the condition that the diagram

$$
\begin{array}{ccc}
C^1(\widetilde{P}_c, \mathbb{Z}) & \xrightarrow{\ \phi_2\ } & H_1(\widetilde{P}_c, \mathbb{R}) \\
\delta \downarrow & & \downarrow \xi \\
B^2(\widetilde{P}_c, \mathbb{Z}) & \xrightarrow{\ \widetilde{\phi}_2\ } & H_1(\widetilde{P}_c, S^1)
\end{array}
$$

is commutative. This operator is well defined, since, as we have just shown, $\xi\phi_2(\ker(\delta)) = 0$.

Consider again a radical U and an arbitrary set of difference 2-cochains $q = (k_{c_1}, \dots, k_{c_p})$ for the atoms $V_{c_1}, \dots, V_{c_p}$ belonging to the given radical U. Then, for these atoms, we extract the following two subsets of a superfluous t-frame:

$$\Delta = (\Delta_{c_1}, \dots, \Delta_{c_p}) \quad \text{and} \quad Z = (Z_{c_1}, \dots, Z_{c_p}),$$

where Δ_{c_i} and Z_{c_i} are the Δ- and Z-invariants corresponding to the atom V_{c_i}. In addition, we consider the above described set of integer numbers $[\Theta]$ associated with the edges of the radical.

As we already decided, we consider Z-invariants modulo the integer cocycles. For brevity, we shall use the notation

$$\xi(Z) = (\xi(Z_{c_1}), \dots, \xi(Z_{c_p})),$$

and, similarly, denote the sets $\{\tilde{\phi}_2(k_{c_1}), \dots, \tilde{\phi}_2(k_{c_p})\}$ and $\{\phi_1'(k_{c_1}), \dots, \phi_1'(k_{c_p})\}$ by $\tilde{\phi}_2(q)$ and $\phi_1'(q)$ respectively.

Consider the set of all the triples $(\Delta, Z, [\Theta])$.

Definition 8.8. Two sets $(\Delta, Z, [\Theta])$ and $(\Delta', Z', [\Theta]')$ are called *equivalent* if there exists a set of difference 2-cochains q such that
1) $q = [\Theta]' - [\Theta]$,
2) $\phi_1'(q) = \Delta - \Delta'$,
3) $\tilde{\phi}_2(q) = \xi(Z) - \xi(Z)'$.

The equivalence class of the triple $(\Delta, Z, [\Theta])$ is called the $\tilde{\Delta}\tilde{Z}[\tilde{\Theta}]$-*invariant* of the given integrable Hamiltonian system v on the given radical U.

We now prove the invariance of $(\tilde{\Delta}\tilde{Z}[\tilde{\Theta}])$ using the first principle. Let us act to the triple $(\Delta, Z, [\Theta])$ by an element $\mathbb{M}$ of the transformation group $G\mathbb{P}$. We obtain a new triple $(\Delta', Z', [\Theta]')$. It suffices to show that they are equivalent in the sense of Proposition 8.8. To each element $\mathbb{M} \in G\mathbb{P}$ we can assign a set of difference 2-cochains $\mathbb{K} = \{k_c\}$. Then, as the cochain q on the radical U, we take the set of difference 2-cochains corresponding to the atoms of the radical: $q = \{k_{c_1}, \dots, k_{c_p}\}$. Then, for this set, all required relations will obviously be satisfied.

COMMENT. Let us note that the last invariant is rather unwieldy. A natural question arises: is it possible to define these invariants by some simple explicit formulas? It turns out that, in general, such simple formulas do not exist. The point is that the orbit space for the action of $G\mathbb{P}$ is not necessarily a Hausdorff space. This means that there may be no continuous functions that distinguish orbits. However, in some particular cases, it is possible to find explicit formulas for invariants (see Section 8.8 below).

The procedure that we have carried out in this section can be interpreted as an attempt to divide the action of a very big group in a big space into several "smaller" actions on separate pieces of the molecule. These pieces are just the radicals of the molecule. In other words, we tried to decompose the action into irreducible components.

8.6.5. *Final Definition of a t-Molecule for an Integrable System*

Suppose we are given an integrable Hamiltonian system v on an isoenergy 3-manifold Q.

Step 1. Consider the corresponding marked molecule W^*. It contains r-marks, n-marks, and ε-marks.

Step 2. On each edge e_i of the molecule W^* we construct the R-invariant R_i. The set of them is denoted by $\{R\}$.

Step 3. We construct the $\widetilde{\Lambda}$-invariant on each saddle atom V of the molecule W^*.

Step 4. We distinguish the radicals U in the molecule W^*.

Step 5. We construct the $\widetilde{\Delta}\widetilde{Z}[\widetilde{\Theta}]$-invariant on each radical U of W^*.

Definition 8.9. The object

$$W^{*t} = ((W, r, \varepsilon), \{R\}, \{\widetilde{\Lambda}\}, \{\widetilde{\Delta}\widetilde{Z}[\widetilde{\Theta}]\})$$

is called the *t-molecule* of the given integrable system v on the given isoenergy 3-manifold Q.

COMMENT.   Definition 8.9 shows that the classical marked molecule W^* is not completely included in the t-molecule: an important parameter is missing from it, namely, the n-marks. The reason is that the n-invariant actually splits into a "sum" of the b-invariants. And the b-invariants, in turn, are expressed in terms of the $\widetilde{\Delta}\widetilde{Z}[\widetilde{\Theta}]$-invariant.

We have to discuss another important question: what does the coincidence of two t-molecules means? This requires clarification, since the object we have introduced is rather complicated. Thus, suppose we are given two t-molecules W_1^{*t} and W_2^{*t}. First of all, they are graphs with oriented edges and vertices of different types. Under a homeomorphism χ of one molecule onto the other, the orientation of the edges, of course, must be preserved, and each vertex of the first molecule W_1^{*t} must be mapped into a vertex of W_2^{*t} of the same type. Moreover, it is assumed, in advance, that, for each atom, we have a one-to-one correspondence between the edges incident to it and the boundary circles of the corresponding 2-atom from the "canonical list". That is why the homeomorphism χ between the molecules requires the existence of the corresponding homeomorphisms for all of their vertices (considered as 2-atoms, i.e., two-dimensional surfaces with embedded graphs). More precisely, such diffeomorphisms are already defined (by means of χ) on the boundary ∂P_c of each 2-atom (P_c, K_c), and we require that all of them are extendable on the whole 2-atom (P_c, K_c). Finally, the homeomorphism χ between the molecules and the homeomorphisms between the corresponding atoms must preserve all numerical parameters of the t-molecules.

We should also pay attention to the following. The definition of the t-molecule depends on the choice of two orientations: the orientation of the three-dimensional manifold Q^3 and that of the edges of the molecule. (Note that we talked many times about the "beginning" and "end" of an edge and, besides, used the orientation of Q^3 to define admissible coordinate systems.)

In principle, we assume from the very beginning that the orientation of Q^3 must be preserved under orbital isomorphisms. But we can omit this assumption by describing explicitly the transformation of the parameters of the t-molecule under a change of the orientation of Q^3.

In the case of the edges, we may proceed similarly. For example, we can introduce the orientation on edges by choosing the direction of increasing of the additional integral f. Then we may require this direction to be preserved under orbital isomorphisms. After that, the above definition of the t-molecule will be absolutely correct. On the other hand, we can indicate formal rules which show what happens to a t-molecule under changing orientation on its edges; and then t-molecules obtained from one another by such transformations can be considered as equivalent by definition. This approach is quite reasonable, since, from the substantive point of view, changing the direction of the arrow on an edge does not affect anything.

The list of formal transformations that show the influence of the orientation can be found in [53] (see also [62]); in this book we omit it.

8.7. THEOREM ON THE TOPOLOGICAL ORBITAL CLASSIFICATION OF INTEGRABLE SYSTEMS WITH TWO DEGREES OF FREEDOM

We are now ready to formulate and prove one of the main theorems of our book. Let $v = \operatorname{sgrad} H$ be an integrable Hamiltonian system on a symplectic manifold M^4; restrict it onto a compact regular connected isoenergy surface $Q^3 = \{H = h\}$.

Consider the following natural class (v, Q^3) of non-degenerate integrable systems $v = \operatorname{sgrad} H$ on isoenergy 3-manifolds Q^3. Namely, we shall assume the following conditions to be fulfilled.

1) *Topological stability.*
The 3-manifold Q^3 is topologically stable for the given system, i.e., under small variation of the level h of the Hamiltonian H, the type of the Liouville foliation for the system v does not change (in other words, the system $(v, Q^3_{h+\varepsilon})$ remains Liouville equivalent to the initial one (v, Q^3_h)).

2) *Bott property.*
The additional integral f restricted onto Q^3 is a Bott function, i.e., all of its critical submanifolds in Q are non-degenerate. Besides, we assume that all these critical submanifolds are one-dimensional (i.e., are homeomorphic to a circle). In other words, there are neither critical tori nor Klein bottles.

3) *Hyperbolicity of singular trajectories.*
All saddle critical circles of the integral f are hyperbolic trajectories of v. This means that, for each periodic trajectory of v that is a critical saddle circle for f, the differential of the Poincaré map differs from $\pm \operatorname{id}$, where id is the identity mapping.

4) *Non-resonance.*
The system v is not resonant, i.e., irrational Liouville tori are everywhere dense.

5) *Finiteness condition.*
The rotation functions ρ of the system v must have only a finite number of local minima, maxima, and poles.

Our main goal is to classify the systems of the above type up to orbital equivalence (both topological and smooth).

Suppose the system v satisfies conditions (1)–(5) listed above. To each such system, we can assign the corresponding t-molecule W^{*t} introduced in Section 8.6.

Theorem 8.2.

a) *The t-molecule W^{*t} of the integrable Hamiltonian system v is a well-defined orbital topological invariant of the integrable Hamiltonian system v.*

b) *Two integrable Hamiltonian systems v_1 and v_2 of the above type are topologically orbitally equivalent if and only if their t-molecules coincide.*

Proof. Part (a) of this theorem follows from the first general principle and, in fact, has been proved in the previous section.

Let us prove the part (b), i.e., show that the t-molecule is a complete orbital invariant.

In view of the second general principle, it suffices to prove that the parameters of W^{*t}, regarded as functions on the set $\{\mathbb{T}\}$ of all admissible superfluous t-frames, separate the orbits of the action of $G\mathbb{P}$, i.e., for any two distinct orbits of this action, there must be at least one parameter that takes distinct values on them. In other words, two orbits coincide if and only if the values taken by t-molecules on them are the same.

The proof is divided into several steps.

Consider the two elements of the space $\{\mathbb{T}\}$, i.e., two superfluous t-frames corresponding to the given systems v_1 and v_2:

$$\mathbb{T} = (C, R^+, R^-, \Lambda, \Delta, Z) \quad \text{and} \quad \mathbb{T}' = (C', R^{+'}, R^{-'}, \Lambda', \Delta', Z').$$

We know that the values of the t-molecule W^{*t} as a function on $\{\mathbb{T}\}$ coincide on these t-frames. We need to deduce from this that there exists a transformation of transversal sections that makes these two t-frames coincide.

Step 1. We start with saddle atoms, organized into radicals. The partition of W into radicals is uniquely defined by the t-frame, and it can be reconstructed if we know the t-molecule. Indeed, the r-marks and the rotation vectors $\{R\}$ included into the t-molecule allow us to judge which edges of W are finite, infinite, and superinfinite. Therefore, it follows from the equality of the t-molecules W_1^{*t} and W_2^{*t} of the given systems that both these molecules are partitioned into radicals in the same way. To simplify our arguments, we can identify W_1 and W_2, assuming them to be the same molecule W on which two (in general) distinct t-frames $\mathbb{T}$ and $\mathbb{T}'$ are given.

Step 2. Take an arbitrary radical U in the molecule W and two triples (Δ, Z, Θ) and (Δ', Z', Θ'). The sets Θ and Θ' are defined exactly like the set of integer numbers $[\Theta]$ introduced above; but, instead of the integer parts in the definition of $[\Theta]$, we need to take these coefficients themselves. We shall show that, using the r-marks on finite edges, the rotation vectors $R \bmod 1$ on the infinite edges, and the integer parameters $[\Theta]$, we can uniquely reconstruct the real values of Θ. Suppose, for example, that $\Theta = \alpha/\beta$. This number can obviously be reconstructed if we know $[\alpha/\beta]$ and the mark $r = \alpha/\beta \bmod 1$. In the case when $\Theta = MR^-$, this number can be reconstructed if we know $R^- \bmod 1$ and $-[-MR^-]$. Analogous arguments are repeated for other types of values of Θ.

We know that the triples $(\Delta, Z, [\Theta])$ and $(\Delta', Z', [\Theta]')$ are equivalent (since the corresponding values of the $\widetilde{\Delta Z}[\Theta]$-invariants coincide). Their equivalence means that there exists a transformation of transversal sections inside the radical U that maps the first triple into the second one. After making this transformation, we see that now we have the equality

$$(\Delta, Z, [\Theta]) = (\Delta', Z', [\Theta]') .$$

Note that this procedure can be carried out absolutely independently for all the radicals of the molecule. This is a consequence of Proposition 8.2, according to which a transformation of sections in an atom influences only the parameters $(\Delta, Z, [\Theta])$ corresponding to the given atom. Moreover, as we mentioned above, the real values of Θ can be reconstructed uniquely from the integer values of $[\Theta]$. Hence, after making this transformation, we actually match not just integer parts $[\Theta]$ and $[\Theta]'$, but the sets Θ and Θ' themselves. Therefore,

$$(\Delta, Z, \Theta) = (\Delta', Z', \Theta')$$

on all the radicals of the molecule.

Step 3. Now we may assume that all the "saddle atomic parameters", i.e., the parameters (Λ, Δ, Z), coincide in the superfluous t-frames $\mathbb{T}$ and $\mathbb{T}'$. Moreover, the parameters Θ are the same. Recall that Θ is a function of the gluing matrices C and the rotation vectors R^+ and R^-. We claim that, on all the edges of W that join pairs of saddle atoms, the above arguments automatically imply the coincidence of the gluing matrices C and C' as well as the coincidence of the pairs of rotation vectors R^+, R^- and $R^{+'}, R^{-'}$. Indeed, let e be an edge joining two saddle atoms. Three cases are possible:

    a) e is finite,
    b) e is infinite,
    c) e is superinfinite.

We analyze all three cases successively.

Step 4-a. Suppose e is finite, i.e., the r-mark α/β is a finite number ($\beta \neq 0$). Partitioning the molecule W into radicals, we must cut all the finite edges (including e) in the middle. As a result, a cut edge becomes two exterior edges of some radicals. On each of them, we have a number of the form Θ. Thus, corresponding to the edge e are two numbers, which we denote by Θ^+ and Θ^-, where the sign is determined by the orientation of the edge e. It follows from the definition of Θ that

$$\Theta^+ = -\frac{\delta}{\beta} = -\frac{\delta'}{\beta'} \quad \text{and} \quad \Theta^- = \frac{\alpha}{\beta} = \frac{\alpha'}{\beta'} .$$

Here $\alpha, \beta, \gamma, \delta$ are the integer entries of the gluing matrix C corresponding to the given edge.

The numbers α and β are relatively prime (since $\det C = -1$), and, moreover, the signs of β and β' are the same (since $\varepsilon = \varepsilon'$). Hence, the matrices C and C' coincide.

Now we need to prove that the rotation vectors are the same. We can reconstruct them using the following formula (see Definition 8.3):

$$R = \beta R^- - \alpha.$$

The vector R^+ can be reconstructed from R^- and the gluing matrix C, which is the transition matrix from the basis λ^-, μ^- to the basis λ^+, μ^+ (see Proposition 1.14). Thus, we have completely reconstructed the gluing matrix C and the rotation vectors R^+ and R^- on the edge e.

Step 4-b. Suppose e is infinite, i.e., $\beta = 0$ and the rotation vectors R^+ and R^- contain at least one finite element. Again, we have two numbers Θ^+ and Θ^- on the edge e (as in step 4-a). Here $\Theta^+ = MR^+$ (the arithmetic mean of all finite components of R^+) and $\Theta^- = MR^-$. Moreover, we are given the vector $R \bmod 1$, where $R = R^-$ (by definition). It is clear that, if we know Θ^- and $R \bmod 1$, we can uniquely reconstruct the vector R^- itself. Furthermore, using the formula $\rho^+ = -\rho^- + \varepsilon\gamma$, we find that $R^+ = -R^- - \varepsilon\gamma$. Since Θ^+ is the arithmetic mean for the vector R^+, we obtain an analogous formula for Θ^+, i.e., $\Theta^+ = -\Theta^- - \varepsilon\gamma$. Hence we can also reconstruct the number γ uniquely from these formulas, since ε is known. After reconstructing R^-, we can now also reconstruct R^+ using the above formula. In the case of an infinite edge, the gluing matrix C is very simple, namely,

$$C = \begin{pmatrix} \varepsilon & 0 \\ \gamma & -\varepsilon \end{pmatrix}.$$

Hence it is also uniquely reconstructed from γ and ε. Thus, we have uniquely reconstructed C, R^+, and R^-, as required.

Step 4-c. Now suppose that e is superinfinite, i.e., $\beta = 0$ and the rotation vector R does not contain any finite component. In this case, e is an interior edge of some radical U. The number Θ corresponding to e is equal to $-\gamma/\alpha$. But $\alpha = \varepsilon$, and hence (as in step 4-b) the matrix C can be uniquely reconstructed. Moreover, the vector R^- contains no finite elements; so its reduction modulo 2 does not diminish any information. The vector R^+ is expressed in terms of R^- by the formula $R^+ = -R^- - \varepsilon\gamma$, and, consequently, it can differ from R^- only by a "sign of infinities" (which are its components). And γ does not influence the infinite components at all.

Thus, making the transformation of sections inside saddle atoms as described above, we find that all the gluing matrices C and all the rotation vectors R^+ and R^- coincide on the edges joining saddle atoms in the molecule. It remains to analyze the edges for which one of the ends is an atom A (or both ends are atoms A).

Step 5. Suppose the edge e joins a saddle atom with an atom A. Two cases are possible here: e is finite or e is infinite (but e cannot be superinfinite).

Let e be finite. Suppose, for definiteness, that it is directed from the saddle atom to the atom A. The parameter Θ corresponding to the edge e is equal to α/β. As above, since we know the sign of β (i.e., ε), we can uniquely reconstruct the pair of integers α and β, i.e., the first row of the gluing matrix C. The second row cannot be reconstructed uniquely. But we have not yet used the possibility of making a transformation inside the atom A (we have made such transformations above, but only inside the saddle atoms). Now, making a suitable transformation inside A, we achieve that the gluing matrices C and C' are the same. This, of course, can be done. Moreover, as in step 4-a, we can reconstruct the rotation vectors R^+ and R^- using the vector R and the gluing matrix.

Now suppose the edge e is infinite. Suppose, for definiteness, that it is directed from the saddle atom to the atom A. Again, we can make a suitable transformation inside A so that the gluing matrix takes the form $\begin{pmatrix} \varepsilon & 0 \\ 0 & -\varepsilon \end{pmatrix}$. Since the parameters ε are the same, the gluing matrices are now also the same. Furthermore, as in step 4-b, knowing $\Theta^- = MR^-$ and the vector $R^- \bmod 1$, we can reconstruct the vector R^- itself. And the vector R^+ is equal to εR^- (in this case).

Thus, if the molecule W is different from $A\!-\!A$, then part (b) of Theorem 8.2 has been proved.

Step 6. Finally suppose that the molecule W has the form $A\!-\!A$. Then the t-molecule is $W^{*t} = ((W, r, \varepsilon), R)$. Changing "transversal sections" inside the atoms A, we can achieve in a standard way that the gluing matrices C and C' are the same. If e is finite, then, as in step 4-a, the rotation vectors R^+ and R^- are reconstructed from the gluing matrix and the vector R. And if e is infinite, then we may assume that the gluing matrix has the form $\begin{pmatrix} \varepsilon & 0 \\ 0 & -\varepsilon \end{pmatrix}$. Let us make the following changes of coordinates inside both the atoms A:

$$\lambda^+ = \lambda^{+\prime}, \qquad\qquad \lambda^- = \lambda^{-\prime},$$
$$\mu^+ = \mu^{+\prime} + k\lambda^{+\prime}, \qquad \mu^- = \mu^{-\prime} - k\lambda^{-\prime}.$$

It is easy to verify that the gluing matrix does not vary under such a coordinate change (see Proposition 8.2). But the rotation vectors R^+ and R^- vary according to the following rule: $R^+ \to R^+ + k$ and $R^- \to R^- - k$. Using this change, we can arrange that the vectors R^- coincide. Since $R^+ = -R^-$, it follows that the vectors R^+ automatically coincide. As a result, the two superfluous t-frames $\mathbb{T}$ and $\mathbb{T}'$ are the same. This completes the proof of part (b). □

Let us give several remarks on the classification problem. To each integrable Hamiltonian system we have assigned a certain object which allows us to compare systems up to orbital equivalence. However, to complete the classification, we have to answer the question about what abstract t-molecules can be realized as t-molecules of integrable systems. In other words, we must describe the class of admissible t-molecules. In fact, the answer to this question is Theorem 8.1. Let us comment on this in more detail.

Consider the space $\{\mathbb{T}\}$ of all admissible superfluous t-frames for a fixed molecule W (see Section 8.5).

Definition 8.10. The t-molecules corresponding to admissible superfluous t-frames are called *admissible t-molecules*.

It is clear that exactly these molecules can be realized as t-molecules of integrable Hamiltonian systems.

In fact, it is possible to write down the formal conditions on the parameters of a t-molecule W^{*t} that guarantee the admissibility of W^{*t} (see [53]).

8.8. A PARTICULAR CASE: SIMPLE INTEGRABLE SYSTEMS

Definition 8.11. An integrable Hamiltonian system is called *simple* on a given isoenergy 3-manifold Q if each critical level of the additional integral $f: Q \to \mathbb{R}$ contains exactly one critical circle of f. In terms of the molecule W, this means that only three simplest types of atoms are allowed: A, B, and A^* (Fig. 8.7).

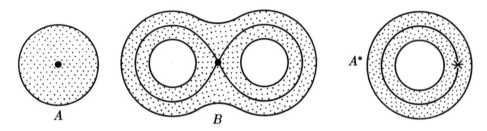

Figure 8.7

It is known (see paper by Nguyen Tien Zung [254]) that an arbitrary smooth non-degenerate integrable system can be made simple by a small smooth perturbation in the class of integrable systems. However, in specific problems, a certain symmetry may occur which leads to the appearance of complicated atoms (see [178], [255], [278]).

Such perturbations, in general, change the type of the Hamiltonian, and, therefore, we obtain another system lying outside the framework of the initial system. That is why above we considered a general case, which includes the theory of complicated atoms and molecules. Nevertheless, the case of simple systems appears quite often. For such systems, the orbital invariants can be simplified; and, in this section, we give a reformulation of the general classification theorem for this special case.

Recall that the t-molecule W^{*t} contains the following orbital invariants:
1) the R-vectors associated to each edge of W^{*t};
2) the Λ-invariants for all saddle atoms;
3) the $\widetilde{\Delta}\widetilde{Z}[\Theta]$-invariants for all radicals.

Let us look at these invariants in the case of simple atoms. The first invariant (R-vector) describes the behavior of the system on an edge of the molecule. Therefore, the simplicity of atoms does not affect it, and the R-invariant "remains the same".

The Λ-invariant becomes trivial, since a simple atom has a single vertex. Thus, we can exclude Λ from the set of orbital invariants.

In the triple $(\Delta, Z, [\Theta])$, the only invariant $[\Theta]$ contains non-trivial information. The two remaining invariants Δ and Z are equal to zero because of simplicity of atoms. Let us analyze Definition 8.8. In the case under consideration, it reduces to the following. Two sets $[\Theta]$ and $[\Theta]'$ are equivalent if and only if its difference $[\Theta] - [\Theta]'$ is the boundary of some integer 1-chain q. This means exactly that the sums of coefficients of $[\Theta]$ and $[\Theta]'$ are the same. But such a sum is, by definition, the b-invariant of the system on the given radical. Thus, the very complicated $\widetilde{\Delta}\widetilde{Z}[\widetilde{\Theta}]$-invariant turns into the b-invariant represented just by one integer number.

As a result, we obtain the following t-molecule W^{*t} in the case of simple systems.

Definition 8.12. In the case of simple systems, the (*simple*) *t-molecule* of a given system is the marked molecule W^* endowed, in addition, with the R-invariants of all edges and the b-invariants of all radicals: $W^{*t} = (W^*, R, b)$.

Taking into account the general classification theorem and the above arguments, we obtain the following result.

Theorem 8.3. *Let v be a simple integrable Hamiltonian system on an isoenergy 3-manifold Q (i.e., its molecule W consists of atoms A, A^*, and B only). Then the corresponding (simple) t-molecule $W^{*t} = (W^*, R, b)$ is a complete topological orbital invariant of the system. This means that two such systems are orbitally topologically equivalent if and only if their (simple) t-molecules coincide.*

As we see, in this important particular case, the orbital invariants can be essentially simplified. Also note that, for simple molecules, all additional information about trajectories of a system is contained in the rotation functions.

8.9. SMOOTH ORBITAL CLASSIFICATION

In this section, we present the theory of smooth orbital classification for integrable Hamiltonian systems with two degrees of freedom [46], [47], [48]. As usual, speaking of smoothness, we mean C^∞-smoothness.

In what follows, we shall assume for simplicity that all 3-atoms under considerations are planar and without stars. In particular, this means that all the transversal sections $P_c = P_{\mathrm{tr}}$ are planar, i.e., admit an embedding into a plane. Note that, in known examples of integrable systems in physics and mechanics, this condition is always fulfilled. In fact, the general method for constructing invariants that was suggested above and will now be applied in the smooth case, can also be used in the case of non-planar atoms as well as atoms with stars. However, as we have already seen, in the general case, some technical problems occur if we wish to describe explicitly the invariants of the action of the transformation group $G\mathbb{P}$. That is why we confine ourselves to the case most important for applications, where, as we shall see soon, the classification theorem has a very natural formulation.

First of all, we construct the so-called superfluous st-frame of the molecule corresponding to a given Hamiltonian system v. Here we shall assume that a first integral f of v is fixed. In particular, the superfluous st-frame and the st-molecule depend on the choice of f.

On the boundary tori of each atom Q_c, we introduce and fix certain admissible coordinate systems (a pair of oriented cycles). One of these cycles is a fiber of the Seifert fibration on Q_c, the other is the intersection of the Liouville torus with a transversal section $P_c \subset Q_c$. Recall that, in the case of planar atoms, fixing a transversal section P_{tr} is equivalent to fixing the corresponding admissible coordinate system, i.e., the boundary of P_{tr}.

As before, we obtain two admissible coordinate systems (λ_j^-, μ_j^-) and (λ_j^+, μ_j^+) on each edge e_j of the molecule W; and we can, therefore, define two rotation functions ρ_j^- and ρ_j^+. Moreover, on each edge, we have an integer gluing matrix C_j that is, by definition, the transition matrix from the basis (λ_j^-, μ_j^-) to the basis (λ_j^+, μ_j^+).

Finally, for each vertex of each saddle atom (P_c, K_c), we define the Λ^*-invariant. As the Hamiltonian of the reduced system on P_c, we consider the additional integral f; the symplectic structure on P_c is chosen according to the Hamiltonian. Thus, having fixed the admissible coordinate systems (or, equivalently, the set of transversal sections $\mathbb{P}$), we can introduce the following object:

$$\mathbb{ST} = \{C_j(\mathbb{P}), \rho_j^-(\mathbb{P}), \rho_j^+(\mathbb{P}), \Lambda_c^*(\mathbb{P})\}.$$

Definition 8.13. The set of all the gluing matrices, rotation functions, and Λ-invariants
$$\mathbb{ST} = \{C_j(\mathbb{P}), \rho_j^-(\mathbb{P}), \rho_j^+(\mathbb{P}), \Lambda_c^*(\mathbb{P})\}$$
is called the *superfluous st-frame* of the molecule W.

Of course, the superfluous st-frame depends essentially on the choice of admissible coordinate systems on Liouville tori.

The next step is the proof of the basic lemma that shows that the information contained in the superfluous st-frame is sufficient for classification. Consider two integrable systems v_1 and v_2 with the same molecule W.

Lemma 8.7 (Basic lemma). *Suppose that, for some choice of admissible coordinate systems, the superfluous st-frames $\mathbb{ST}_1$ and $\mathbb{ST}_2$ of the molecule W corresponding to the systems v_1 and v_2 coincide. Then these systems are smoothly orbitally equivalent.*

Proof. Suppose we are given two integrable Hamiltonian systems restricted on three-dimensional isoenergy manifolds corresponding to the same molecule W (without marks). Suppose further that the superfluous st-frames of the molecule W corresponding to the given systems coincide (for some choice of admissible coordinate systems).

We begin with constructing orbital isomorphisms on atoms; and then we sew them into a single orbital diffeomorphism by extending to the edges in a suitable way.

According to the reduction theorem, to prove the equivalence of v_1 and v_2 on a 3-atom Q^c (i.e., in a small neighborhood of a singular fiber), it suffices to verify that the corresponding Poincaré flows σ_1^t and σ_2^t are conjugate on the transversal sections. We have all necessary tools for such a verification. We mean the theorem on smooth classification of systems on an atom (Theorem 7.1). According to this theorem, in the case of planar atoms, it suffices to check the coincidence of the Λ^*-invariants and period functions. But the Λ^*-invariants coincide as elements of the superfluous st-frames $\mathbb{ST}_1$ and $\mathbb{ST}_2$. The period functions are also the same, since they coincide (see Lemma 8.5) with the rotation functions from the superfluous st-frames. Thus, the systems are smoothly orbitally equivalent on each 3-atom.

They are also equivalent on each edge, since the rotation functions are the same. It remains to sew the existing orbital isomorphisms on the atoms and edges into a single orbital diffeomorphism. But we can do this by using the sewing lemma (Lemma 8.3). This completes the proof of the basic lemma. $\square$

Now, according to the general scheme for constructing invariants, we must consider the action of the group $G\mathbb{P}$ of transformations of transversal sections on the set of superfluous st-frames, and then we must find the complete set of invariants for this action. In the smooth case, this group and its action on the set of st-frames are just the same as those in the topological case (see Sections 8.2 and 8.3). More precisely, these transformations are as follows:

1) $C'_j = \begin{pmatrix} \alpha'_j & \beta'_j \\ \gamma'_j & \delta'_j \end{pmatrix} = \begin{pmatrix} 1 & 0 \\ -k_j^+ & 1 \end{pmatrix}\begin{pmatrix} \alpha_j & \beta_j \\ \gamma_j & \delta_j \end{pmatrix}\begin{pmatrix} 1 & 0 \\ k_j^- & 1 \end{pmatrix} = (A_j^+)^{-1}C_j A_j^-$, where $A_j^{\pm} = \begin{pmatrix} 1 & 0 \\ k_j^{\pm} & 1 \end{pmatrix}$;

2) $(\rho_j^-)' = \rho_j^- + k_j^-$, $(\rho_j^+)' = \rho_j^+ + k_j^+$;

3) $\Lambda_c^{*\prime} = \Lambda_c^*$,

where k_j^+, k_j^- are coefficients of the difference 2-cochains (see Sections 8.2 and 8.3).

We now can easily describe a complete set of invariants of this action. According to the second general principle, such a set is exactly the desired set of parameters for the st-molecule.

Thus, we now turn to the description of these invariants. Actually, they will be very similar to those described in the topological case.

We begin with edge invariants. As we already know, the only invariant on an edge is the rotation function (see Section 5.1). The functions ρ_j^- and ρ_j^+ are not adjusted for our purposes, since they depend on the choice of admissible coordinate systems. Thus, the problem is to choose some natural (uniquely defined) basis on the tori of a one-parameter family. If an edge e_j is finite, then it can easily be done. Indeed, the geometrical condition $r_j \neq \infty$ means that the uniquely defined cycles λ^+ and λ^- are independent on the Liouville tori from the given family. Therefore, we can consider these cycles as a well-defined "basis" for calculating the rotation functions. The fact that these cycles do not form a basis "in the lattice" is not important in the given case.

By rewriting the rotation functions in terms of this "basis", we obtain a new function (see Proposition 1.14)

$$\rho_j = \beta_j \rho_j^- - \alpha_j,$$

where α_j and β_j are coefficients of the gluing matrix C_j. Note that we did just the same when constructing the R-invariant.

If the edge e_j is infinite (i.e., $r_j = \infty$), then λ^- and λ^+ are homologous, and, consequently, we have no unique method for constructing a basis on the tori from the family in question. However, it is possible to determine the rotation function ρ_j^- modulo one. This means that two functions are considered to be equivalent if the difference between them is a constant integer. We shall denote this invariant by $\rho_j \bmod 1$.

The next invariant is just the Λ^*-invariant of each atom. It does not depend on the choice of a section, and, therefore, according to the first general principle, Λ^* is an orbital invariant. It turns out that these two invariants are sufficient for the classification.

Thus, having fixed the additional integral f, we obtain a certain new object $W^{*st} = \{W, r_j, \varepsilon_j, n_k, \rho_j, \Lambda_m\}$, which is called the *st-molecule*. Here

W is the molecule of the system;

r_j, ε_j are the r- and ε-marks on the edges of W (the index j numbers the edges);

n_k are the n-marks on the families of W (the index k numbers the families);

ρ_j is the rotation function on the oriented edge of W (on infinite and superinfinite edges of W, this function is taken modulo the integers, i.e., $\rho \bmod 1$);

Λ_m^* are the Λ^*-invariants of the hyperbolic trajectories of the system associated with the corresponding vertices of saddle atoms (the index m numbers the vertices of the saddle atoms).

Theorem 8.4. *Let (v_1, Q_1) and (v_2, Q_2) be two integrable Hamiltonian systems with two degrees of freedom restricted to isoenergy submanifolds. Let all the atoms from the corresponding molecules be planar and without stars. Then the systems (v_1, Q_1) and (v_2, Q_2) are smoothly orbitally equivalent if and only if there exist Bott integrals f_1 and f_2 of the systems (v_1, Q_1) and (v_2, Q_2) respectively such that the st-molecules W_1^{*st} and W_2^{*st} corresponding to them coincide.*

COMMENT. In other words, the necessary and sufficient conditions for the existence of an orbital smooth isomorphism are as follows:

1) the systems (v_1, Q_1) and (v_2, Q_2) must have the same Liouville foliation;

2) after a suitable change of the first integrals, the rotation functions and Λ^*-invariants of the closed hyperbolic trajectories must coincide.

No other invariants are needed.

COMMENT. The trouble with this theorem is that it is quite impossible at present to understand how the existence (or non-existence) of the pair of integrals required in the theorem can be established. Nevertheless, the first step must always be to compute the *st*-molecules for the systems presented for testing. When these molecules are computed, it is necessary to clarify the following question: is there a change of integral for one of the systems under which its *st*-molecule is transformed into the corresponding *st*-molecule of the second system? In fact, this question may be solved at a formal level (see [46], [47]).

Proof (of Theorem 8.4). We wish to show that the information on the Liouville foliations and the rotation functions ρ and $\rho \bmod 1$ is sufficient for classification.

Thus, suppose that the *st*-molecules W_1^{*st} and W_2^{*st} coincide. We fix some admissible coordinate systems for the first system and take the corresponding superfluous *st*-frame. Then we select a change of admissible coordinate systems to obtain a superfluous *st*-frame coinciding with some fixed *st*-frame corresponding to the second system. In what follows, we shall use indices 1 and 2 to distinguish objects related to the first and second system respectively.

First we make a change of admissible coordinate systems so that all the gluing matrices C_j coincide on all the edges. This is possible by the theory of Liouville classification (Chapter 4). It is easily seen that, after this, the rotation functions ρ_1^- and ρ_2^- on each finite edge coincide, since they can be expressed uniquely in terms of the function ρ from the *st*-molecule and the coefficients of the gluing matrix. The same is obviously true for ρ_1^+ and ρ_2^+. The Λ-invariants also coincide automatically, since they do not change under changes of admissible coordinate systems.

Thus, it remains to equate the rotation functions on the infinite and super-infinite edges. At present, they coincide modulo 1. The next step is to equate the rotation functions on the infinite edges. Let us cut the molecule along all the finite edges. It splits, as a result, into several pieces, which we call *families*. Note that this notion does not completely coincide with that introduced in Chapter 4.

Our first assertion is that, if such a piece is a family in the sense of Definition 4.5 (i.e., does not contain atoms of type A) and, moreover, is a tree, then the rotation functions ρ_{1j}^- and ρ_{2j}^- on the interior edges of the family coincide automatically (not modulo 1, but exactly).

Indeed, consider a vertex V of the family U which is incident to exactly one infinite (or superinfinite) edge e_{j_0} (such a vertex exists, since the family is a tree). Without loss of generality, we shall assume that all the edges incident to V are outgoing edges (with respect to V). The rotation functions ρ_{1j}^- and ρ_{2j}^- coincide on all of these edges e_j except for the only infinite edge e_{j_0}. On this edge, the rotation functions may differ from each other by an integer constant. Let us show that they in fact coincide. To this end, recall that the sum of the so-called finite parts of the rotation functions (or, equivalently, the period functions of the reduced systems) over all the edges incident to a given atom is equal to zero. This immediately implies that the finite parts of $\rho_{1j_0}^-$ and $\rho_{2j_0}^-$ coincide. But then it is obvious that the functions themselves coincide. Indeed, if they had been different by a constant, then their finite parts would have differed by the same constant.

Note that the functions ρ_{1j}^+ and ρ_{2j}^+ also coincide automatically, since they are expressed in terms of ρ_{1j}^- and ρ_{2j}^- and the gluing matrices.

Using this fact, we can now move along the tree-family U proving consecutively that the rotation functions coincide on all the (interior) edges of U.

We now consider the case when the family U either is not a tree or contains atoms A.

If U contains atoms of type A, we produce a new object $\tilde{U}$ from it by gluing all the atoms A into one point (Fig. 8.8). If this new graph $\tilde{U}$ has no cycles, then

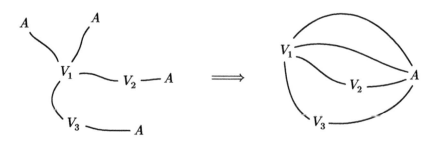

Figure 8.8

the initial family U contained at most one atom A, and the other atoms were saddle. In this case, we can repeat the above argument.

Finally, suppose that $\widetilde{U}$ is not a tree (i.e., contains some cycles). Consider an arbitrary cycle in $\widetilde{U}$. We claim that it is possible to change admissible coordinate systems in such a way that the gluing matrices remain the same, but, on one edge of this cycle, the rotation functions ρ_{1j}^{-} and ρ_{2j}^{-} become equal. Moreover, the transformations of coordinate systems happen only on the edges of this cycle. In particular, the rotation functions do not change on the remaining edges.

Thus, we take an arbitrary cycle in the graph U formed by edges $e_1, \ldots, e_m$. Without loss of generality, we may suppose that the orientation on the edges corresponds to some orientation on the cycle (Fig. 8.9).

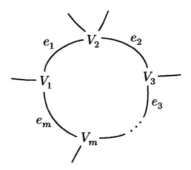

Figure 8.9

Take an arbitrary edge e_{j_0} of this cycle that joins certain atoms V_{j_1} and V_{j_2}. Assume that the rotation functions on e_{j_0} do not coincide; in other words, $\rho_{1j_0}^{-} = \rho_{2j_0}^{-} - k$. Then we make the following admissible coordinate change (on all the edges e_j forming the given cycle):

$$\lambda_j^{-\prime} = \lambda_j^{-}, \qquad\qquad \lambda_j^{+\prime} = \lambda_j^{+},$$
$$\mu_j^{-\prime} = \mu_j^{-} + k\lambda_j^{-}, \qquad \mu_j^{+\prime} = \mu_j^{+} - k\lambda_j^{+}.$$

It is easily seen that such a coordinate change is admissible. Moreover, it does not change the gluing matrices. The only changes to the superfluous st-frame $\mathbb{ST}_1$ are in the rotation functions on the edges forming the cycle. An integer k is added to all those rotation functions ρ_{1j}^{-}. In particular, the rotation functions $\rho_{1j_0}^{-}$ and $\rho_{2j_0}^{-}$ become equal on the edge e_{j_0}.

We now leave this edge alone. If certain cycles remain in U after removing the edge e_1, then we repeat the procedure we have described until the family U becomes a tree (or a disconnected union of trees). For trees, as we have already seen, the rotation functions coincide automatically (provided the rotation functions on all the remaining edges around a given tree have already been equated).

Thus we have equated the rotation functions on all the edges of the molecule, as required. The superfluous st-frames now coincide, and we can apply the basic lemma, according to which the systems will be smoothly orbitally equivalent. This completes the proof. $\square$

Chapter 9

Liouville Classification of Integrable Systems with Two Degrees of Freedom in Four-Dimensional Neighborhoods of Singular Points

In this chapter, we present the results obtained by L. M. Lerman, Ya. L. Umanskiĭ [212], [213], A. V. Bolsinov [44], V. S. Matveev [220], [221], Nguyen Tien Zung [258], [261]. We shall follow the general idea of our book: try to present all facts from the uniform viewpoint of the theory of topological invariants of integrable systems. The preceding chapters were devoted to studying an integrable Hamiltonian system on a three-dimensional isoenergy manifold. Here we wish to discuss its behavior on a four-dimensional symplectic manifold. We shall mainly be interested in the topological structure of the corresponding Liouville foliation.

9.1. l-TYPE OF A FOUR-DIMENSIONAL SINGULARITY

Let $x_0 \in M^4$ be a non-degenerate singular point of the momentum mapping $\mathcal{F} = (H, f) \colon M^4 \to \mathbb{R}^2$ of an integrable Hamiltonian system with a Hamiltonian H and an additional integral f given on a symplectic manifold (M^4, ω). As we have already seen in Chapter 1, non-degenerate singular points can be of the following four types:

    a) center–center,
    b) center–saddle,
    c) saddle–saddle,
    d) focus–focus.

Our aim is to describe the structure of the Liouville foliation in a four-dimensional saturated neighborhood U^4 of the singular leaf L passing through the point x_0. It turns out that singular points of the three first types have some similar invariants which we now describe.

First we impose the following natural assumptions on the integrable system under consideration.

Condition 1. Each leaf of the Liouville foliation is compact.

Condition 2. The singular points that belong to the singular leaf L are all non-degenerate (see Definition 1.23).

Condition 3. The bifurcation diagram in a neighborhood of the point $\mathcal{F}(x_0)$ has the form shown in Fig. 9.1 (a, b, c, d).

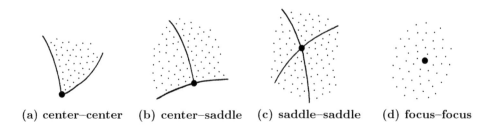

(a) center–center   (b) center–saddle   (c) saddle–saddle   (d) focus–focus

Figure 9.1

COMMENT. One of the prohibited situations is shown in Fig. 9.2, where two curves of the bifurcation diagram touch each other at the singular point $y_0 = \mathcal{F}(x_0)$ with infinite tangency order. This situation is possible for smooth systems, but not analytic ones. Another prohibited situation is shown in Fig. 9.3(a). Here several different points, say of saddle–saddle type, are projected into the same singular point $\mathcal{F}(x_0)$ of the bifurcation diagram Σ. Each of them gives a cross on Σ, but these crosses do not coincide. In the center–saddle case, we forbid the similar situations shown in Fig. 9.3(b).

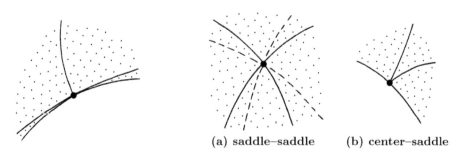

(a) saddle–saddle   (b) center–saddle

Figure 9.2   Figure 9.3

Condition 4. The straight lines given by the equation $\{H = h_0 = \text{const}\}$ intersect the bifurcation diagram Σ transversally (in a neighborhood of y_0).

COMMENT. This condition is not very essential for studying the structure of the foliation into Liouville tori, since, replacing H with the function of the form $\widetilde{H} = H + \lambda f$, we can always satisfy it. However, this condition becomes non-trivial if we want to distinguish the Hamiltonian H among the two-dimensional family of commuting functions. In particular, this condition will guarantee that the Hamiltonian vector field $\operatorname{sgrad} H$ has no equilibrium points except for the singular points lying on the leaf L.

Condition 5. Without loss of generality, we shall assume that the singular leaf L is the preimage of the point y_0 under the momentum mapping $\mathcal{F}$, and its four-dimensional saturated neighborhood U is the preimage of some disc centered at y_0.

COMMENT. This condition simply means that we consider the connected component of the preimage $F^{-1}(y_0)$ of the point $y_0 \in \Sigma$ and the corresponding connected component of the preimage of its neighborhood.

Condition 6. We shall assume that all the objects under consideration (namely, manifolds, symplectic structures, Hamiltonians, integrals, etc.) are real-analytic.

COMMENT. This condition is actually not very essential. All the assertions remain true in the smooth case. But, to prove the corresponding "smooth" statements, we need smooth analogs of Theorems 1.5 and 1.7, which are not proved in our book. As far as we know, a complete proof of these facts has not been published anywhere yet.

Let us study the structure of the set K of critical points of the momentum mapping $\mathcal{F}$ in the neighborhood U^4. For the points of types (a), (b), (c), the set K consists of two parts P_1 and P_2, roughly speaking, of the preimages of the two curves $\gamma_1, \gamma_2 \subset \Sigma$ intersecting at the singular point y_0. More precisely,

$$P_1 = \mathcal{F}^{-1}(\gamma_1) \cap K, \qquad P_2 = \mathcal{F}^{-1}(\gamma_2) \cap K.$$

Proposition 9.1. *Let $z_1, \ldots, z_s$ be the non-degenerate critical points of $\mathcal{F}$ lying on the singular level $L = \mathcal{F}^{-1}(y_0)$. Then:*

1) P_1 and P_2 are two-dimensional symplectic manifolds with boundary which intersect transversally exactly at the points $z_1, \ldots, z_s$;

2) the Hamiltonian H restricted onto submanifolds P_1 and P_2 is a Morse function with the only critical value, and its critical points are exactly $z_1, \ldots, z_s$;

3) the critical points $z_1, \ldots, z_s$ have the same type (in other words, they simultaneously have either type saddle–saddle, or center–saddle, or center–center).

Proof. 1) The set of critical points K consists of zero-dimensional and one-dimensional orbits of the Poisson action of the Abelian group $\mathbb{R}^2$ generated by shifts along integral curves of the vector fields $\operatorname{sgrad} f$ and $\operatorname{sgrad} H$. The points $z_1, \ldots, z_s$ are all non-degenerate by definition and, therefore, are isolated. Clearly, they are zero-dimensional orbits of $\mathbb{R}^2$, and no other zero-dimensional orbits exist in a neighborhood of the singular leaf L. Therefore, we need to analyze the behavior and character of one-dimensional orbits in a neighborhood of L. We claim that these one-dimensional orbits are all non-degenerate. It is easily seen that each of them passes near one of the points $z_1, \ldots, z_s$. Hence, it suffices to verify that every one-dimensional orbit that passes through a neighborhood of z_i is non-degenerate. This easily follows from the local structure of the momentum mapping singularity at z_i. Indeed, according to Theorem 1.5, in a neighborhood of z_i, there are regular local coordinates p_1, p_2, q_1, q_2 in terms of which H and f become

$$H = H(\alpha, \beta), \qquad f = f(\alpha, \beta),$$

where the functions α and β (depending on the type of the singular point) have the form:

a) $\alpha = p_1^2 + q_1^2$, $\beta = p_2^2 + q_2^2$ (center–center case);

b) $\alpha = p_1^2 + q_1^2$, $\beta = p_2 q_2$ (center–saddle case);

c) $\alpha = p_1 q_1$, $\beta = p_2 q_2$ (saddle–saddle case).

Moreover, the smooth change $(H, f) \to (\alpha, \beta)$ is non-degenerate, i.e., $\dfrac{\partial(H, f)}{\partial(\alpha, \beta)} \neq 0$. Therefore, the set of critical points and their properties for the mappings $\mathcal{F} = (H, f) : U \to \mathbb{R}^2$ and $\widetilde{\mathcal{F}} = (\alpha, \beta) : U \to \mathbb{R}^2$ are the same. It remains to observe that the set of critical points for $\widetilde{\mathcal{F}}$ has a very simple local structure: it consists of two surfaces defined by the equations

$$\begin{cases} p_1 = 0 \\ q_1 = 0 \end{cases} \quad \text{and} \quad \begin{cases} p_2 = 0 \\ q_2 = 0 \end{cases}.$$

These surfaces intersect transversally at the non-degenerate singular point z_i and, moreover, are locally symplectic submanifolds in M^4. The non-degeneracy of one-dimensional orbits lying on these surfaces (in the sense of $\widetilde{\mathcal{F}}$) is evident (see Definitions 1.23, 1.25).

Thus, all one-dimensional orbits of the Poisson $\mathbb{R}^2$-action that belong to the neighborhood $U(L)$ turn out to be non-degenerate. As was shown in Proposition 1.16, the set of critical points K near a non-degenerate 1-orbit is a two-dimensional symplectic submanifold. Thus, we have shown that the set $K \cap U(L)$ is a two-dimensional submanifold self-intersecting at the singular points $z_1, \ldots, z_s$. At the same time, it is clear that $K \cap U(L)$ actually consists of two manifolds P_1 and P_2 each of which is two-dimensional and symplectic. They correspond to the curves $\gamma_1, \gamma_2 \subset \Sigma$ and intersect each other at the points $z_1, \ldots, z_s$. Note that the manifolds P_i are not necessarily connected.

2) This assertion follows in essence from Condition 4, i.e., from the fact that the lines $\{H = \text{const}\}$ intersect both γ_1 and γ_2 transversally. According to the corollary of Proposition 1.16, all the points (except for $z_1, \ldots, z_s$) are regular for the function H restricted onto P_i. Although the points $z_1, \ldots, z_s$ themselves are singular, they are non-degenerate (as critical points of $H|_{P_i}$). It follows from the local structure of the singularity. Indeed, the function H has the form $H = H(\alpha, \beta)$; moreover, $\partial H/\partial \alpha \neq 0$ and $\partial H/\partial \beta \neq 0$ (according to Condition 4). In a neighborhood of z_j on the submanifold P_i, we can take p_i, q_i as local coordinates. Hence the restriction of H to P_i in terms of the local coordinates takes the form either $H(p_i^2 + q_i^2)$ or $H(p_i q_i)$. Moreover, $\partial H/\partial \alpha \neq 0$ and $\partial H/\partial \beta \neq 0$. These conditions immediately imply that $H : P_i \to \mathbb{R}$ is a Morse function, as required.

3) The picture of the local structure of a small neighborhood of a non-degenerate singular point is presented in Fig. 9.4. In the first three cases, this neighborhood has a direct product type. It is easy to see that, in the center–center case, the singular leaf L is zero-dimensional, i.e., consists of a single point (since it is assumed to be connected). In the center–saddle case, the leaf L is one-dimensional. The saddle–saddle and focus–focus cases are characterized by the fact that L is two-dimensional. It follows from the non-degeneracy condition that the singular leaf has to have the same dimension at all of its points. Hence center–center and center–saddle points cannot be mixed with any other points.

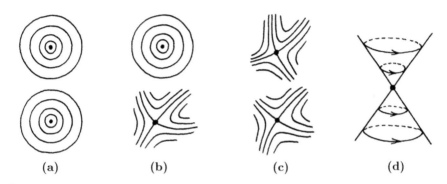

(a)   (b)   (c)   (d)

Figure 9.4

We now prove that a singular leaf cannot contain saddle–saddle and focus–focus points simultaneously. Assume the contrary. Then there exists a two-dimensional orbit O whose closure contains points of different types: saddle–saddle and focus–focus. In the focus–focus case, there exists a linear combination $\lambda H + \mu f$ with constant coefficients such that the trajectories of the vector field $\mathrm{sgrad}(\lambda H + \mu f)$ are all closed on the singular leaf L near the focus–focus point. This follows from the local structure of such a point (see Section 9.8 for detailed description of focus–focus points). The trajectories of this field are closed on the whole two-dimensional orbit O and have the same finite period. On the other hand, in the neighborhood of the saddle–saddle point, the passage time along a trajectory tends to infinity as this trajectory approaches this point. Even if the trajectory had been closed, its period would have tended to infinity. This contradiction proves the assertion. □

Part (3) of Proposition 9.1 implies that we can say not only about the type of a singular point, but also about the type of the singular leaf L itself. The whole leaf L can be related to one of the following types: center–center, center–saddle, saddle–saddle, focus–focus.

REMARK. The submanifolds P_1 and P_2 can be disconnected, but, as we shall see below, the union of them is always connected.

Consider the pairs $V_1 = (H|_{P_1}, P_1)$ and $V_2 = (H|_{P_2}, P_2)$, where H is the Hamiltonian. Part (2) of Proposition 9.1 implies that the function H defines the structure of an atom on each of the surfaces P_1 and P_2. Moreover, the points $z_1, \ldots, z_s$ are the vertices of these atoms. Thus, V_1 and V_2 are atoms naturally connected with the singular leaf L and the point y_0. Note that the atoms V_1 and V_2 are both orientable, since the corresponding surfaces P_1 and P_2 are symplectic and, consequently, oriented.

REMARK. Let us emphasize that, in this context, we allow atoms V_1 and V_2 to be disconnected.

Definition 9.1. The pair (V_1, V_2) is called the *l-type* of the singularity of the momentum mapping $\mathcal{F}$ at the point $y_0 \in \Sigma$.

REMARK. In the focus–focus case, the notion of the l-type is not used.

The types of atoms V_1 and V_2 are completely defined by the type of the given singularity. Namely,

a) if the singularity has center–center type, then the atoms V_1 and V_2 are both of type A,

b) if the singularity has center–saddle type, then one of these atoms is A, whereas the other atom is a saddle one,

c) in the saddle–saddle case, both of the atoms are saddle.

The notion of the l-type can be used for the classification of four-dimensional singularities. The idea is as follows. We can introduce the complexity of a four-dimensional singularity to be the number of singular points $z_1, \ldots, z_s$ on the singular leaf L. Clearly, the number s is the complexity of the atoms V_1 and V_2, i.e., the number of their vertices. Therefore, for singularities of fixed complexity, there is only a finite number of possible l-types, and all of them can easily be listed. By fixing an l-type, we can then try to describe all the singularities corresponding to it. We note, however, that the l-type is not a complete invariant. There may exist several different singularities with the same l-type. As we shall show below, the number of such singularities is always finite.

9.2. THE LOOP MOLECULE
OF A FOUR-DIMENSIONAL SINGULARITY

Let us describe another useful invariant of singularities of the momentum mapping $\mathcal{F}$. Let Σ be the bifurcation diagram located in the plane $\mathbb{R}^2(H, f)$. As we have already seen, the bifurcation diagram is usually represented as a set of smooth curves which may intersect or touch one another at some points. Besides, Σ may contain isolated points. The smooth curves $\gamma_1, \ldots, \gamma_k \subset \Sigma$ correspond, as a rule, to one-parameter families of non-degenerate one-dimensional orbits of the Poisson $\mathbb{R}^2$-action on M^4. Suppose that this condition is fulfilled. The diagram Σ usually has some singular points. A point $y_0 \in \Sigma$ is called *singular* if it belongs to one of the following two types.

Type 1: the point y_0 belongs to the image $\mathcal{F}(K \setminus \widetilde{K})$, where $\widetilde{K}$ is the set of non-degenerate closed one-dimensional orbits of the Poisson action of $\mathbb{R}^2$ on M^4.

Type 2: the point y_0 is an intersection (or self-intersection) point of smooth curves of the diagram Σ.

Denote the set of singular points of Σ by Σ_0. Usually Σ_0 is a finite set of isolated points. If Σ is viewed as a cell complex, then Σ_0 is just the set of its vertices (zero-cells).

We now introduce a notion of an admissible curve.

Definition 9.2. A smooth parameterized curve τ without self-intersections in the plane $\mathbb{R}^2(H, f)$ is called *admissible* if it intersects the bifurcation diagram Σ transversally and does not pass through the singular points of Σ (Fig. 9.5).

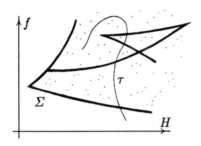

Figure 9.5

Now consider the preimage $Q_\tau = \mathcal{F}^{-1}(\tau)$ of τ in M^4. If τ is an admissible curve, then Q_τ is a smooth 3-manifold. The parameter t of the parametrized curve $\tau(t)$ can be considered as a smooth function on Q_τ. Obviously, this is a Bott function provided τ is admissible. If we wish, we may consider Q_τ as an isoenergy manifold for a certain Hamiltonian system with a Hamiltonian $H_\tau(H, f)$ which satisfies the property that the curve τ on the plane $\mathbb{R}^2(H, f)$ is given by the equation $\{H_\tau = \text{const}\}$. Thus, we have a natural structure of the Liouville foliation on Q_τ and can, therefore, consider its natural invariant, namely, the marked molecule W^*, which we denote by $W^*(\tau)$.

Lemma 9.1. *The marked molecule $W^*(\tau)$ does not change under smooth isotopy τ_s of τ on the plane $\mathbb{R}^2$ in the class of admissible curves.*

Proof. The arguments are standard. $\square$

Suppose now that $y_0 \in \Sigma_0$ is an isolated singular point of the bifurcation diagram. Consider a circle τ of small radius centered at y_0. Suppose that τ is an admissible curve and remains admissible as its radius tends to zero.

Definition 9.3. The marked molecule $W^*(\tau)$ is called the *loop molecule* of the singular point $y_0 \in \Sigma$.

The loop molecule describes the structure of the Liouville foliation on the boundary of the four-dimensional neighborhood $U(L)$ of the singular leaf $L = \mathcal{F}^{-1}(y_0)$. It shows what happens when we move around the singularity. It is important that all the events can be described in terms of the standard marked molecule. As we shall see below, sometimes (and, actually, very often) the loop molecule allows us to describe the structure not only of the boundary of $U(L)$, but also of this neighborhood itself. In other words, it turns out that, in many cases, the loop molecule is a complete topological invariant of a four-dimensional singularity (in the sense of Liouville equivalence).

Conjecture (A. T. Fomenko). If the singular points $z_1, \ldots, z_s$ lying on the singular level L are all non-degenerate, then the loop molecule is a complete topological invariant of the singularity (in the sense of Liouville equivalence). In other words, two Hamiltonian systems are Liouville equivalent in some neighborhoods of non-degenerate singular leaves if and only if their loop molecules coincide.

In what follows, this conjecture will be proved in several important cases. The experience in studying specific examples of integrable systems shows that, as a rule, distinct four-dimensional singularities have distinct loop molecules even if the non-degeneracy condition fails.

As we shall see below, the loop molecules are also useful for computing invariants of integrable systems on arbitrary isoenergy 3-manifolds.

9.3. CENTER–CENTER CASE

Let x be a singular point of center–center type in M^4, let L be the singular leaf passing through it, and let $U(L)$ be its four-dimensional neighborhood in M^4.

Theorem 9.1.   *There exists exactly one singularity of center–center type (up to Liouville equivalence). Its structure is as follows.*

a) *The singular leaf L coincides with the point x itself.*

b) *The neighborhood $U(L)$ is diffeomorphic to a four-dimensional ball.*

c) *The l-type of the singularity is (A, A).*

d) *The loop molecule has the form $A \!-\! A$, and the r-mark is 0 (see Fig. 9.6).*

e) *A canonical model of this singularity is given by the pair of commuting functions $H = \alpha(p_1^2 + q_1^2) + \beta(p_2^2 + q_2^2)$ and $f = p_2^2 + q_2^2$, where the constants α and β are both different from zero.*

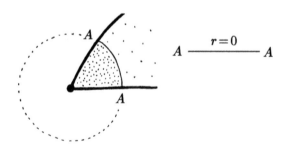

Figure 9.6

COMMENT. This case is the simplest one, and it is possible to prove a stronger result: all center–center singularities are symplectomorphic. Moreover, the same is true for all dimensions. This fact can be found, for example, in the paper by Elliasson [110]. In this case, the Poisson action of $\mathbb{R}^2$ in a neighborhood of a center–center point is factorized to a Poisson action of the torus T^2. Such actions are studied in detail in the book [21] by M. Audin.

Proof (of Theorem 9.1).   The proof immediately follows from the local classification theorem for non-degenerate singular points (Theorem 1.5). According to this theorem, there exists a symplectic coordinate system (p_1, q_1, p_2, q_2) such that $H = H(p_1^2 + q_1^2, p_2^2 + q_2^2)$ and $f = f(p_1^2 + q_1^2, p_2^2 + q_2^2)$. Moreover, the change

$$(H, f) \to (\tilde{H}, \tilde{f}) = (p_1^2 + q_1^2, p_2^2 + q_2^2)$$

is smooth and regular.

Hence the singular leaf $L = \{H = 0,\ f = 0\}$ coincides with the non-degenerate singular point $x = (0,0,0,0)$, and its neighborhood $U(L)$ is a four-dimensional ball. Moreover, we may assume without loss of generality that the coordinates p_1, q_1, p_2, q_2 act on the whole neighborhood. This implies that any two center–center singularities are Liouville equivalent, since every center–center singularity is isomorphic to the singularity corresponding to the Liouville foliation given by two canonical functions $\tilde{H}, \tilde{f}$.

Restricting H to the surfaces $P_1 = \{p_2 = 0,\ q_2 = 0\}$ and $P_2 = \{p_1 = 0,\ q_1 = 0\}$, we obtain functions depending only on $p_1^2 + q_1^2$ and $p_2^2 + q_2^2$, respectively. Such functions correspond to the atoms A. Therefore, the l-type of the center–center singularity is (A, A).

The change $(H, f) \to (p_1^2 + q_1^2, p_2^2 + q_2^2)$ is regular; therefore, studying the topology of the foliation, we can consider the new functions $\tilde{H} = p_1^2 + q_1^2$ and $\tilde{f} = p_2^2 + q_2^2$ instead of H and f. Recall that the loop molecule describes the structure of the Liouville foliation on the 3-manifold $\mathcal{F}^{-1}(\gamma_\varepsilon)$, where γ_ε is the circle of radius ε centered at the point $\mathcal{F}(x)$. In our case, instead of the circle, it is more convenient to take the segment τ_ε given on the plane by the equation $\tilde{H} + \tilde{f} = \varepsilon$ (Fig. 9.7).

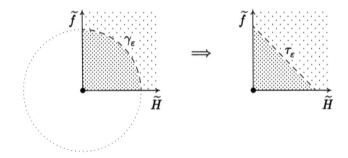

Figure 9.7

Note that the arc of γ_ε that belongs to the image of the momentum mapping can be smoothly deformed into the segment τ_ε in the class of admissible curves. Therefore, the Liouville foliation structures on the preimages of γ_ε and τ_ε are the same. Thus, we need to describe the structure of the foliation given by the function $\tilde{f} = p_2^2 + q_2^2$ on the 3-sphere $p_1^2 + q_1^2 + p_2^2 + q_2^2 = \varepsilon$. This foliation is well known; it is described by the molecule A—A with the mark $r = 0$ (see Proposition 4.3). This completes the proof. $\square$

REMARK. It is worth noticing that a center–center singularity has the type of the direct product $A \times A$ (Fig. 9.8). As we shall see below, the analog of this fact is

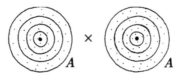

Figure 9.8

valid for all non-degenerate singularities. In more complicated situations, like saddle–saddle, a singularity will be of almost direct product type, i.e., can be presented as a quotient of a direct product of two atoms by a free action of some finite group.

REMARK. In concrete examples of integrable systems in physics and mechanics, center–center singularities correspond to stable non-degenerate equilibrium points. They occur practically in all integrable systems.

9.4. CENTER–SADDLE CASE

Let x be a singular point of center–saddle type in M^4, let L be the singular leaf passing through it, and let $U(L)$ be its four-dimensional neighborhood in M^4.

We start with an example of center–saddle singularity. Consider an arbitrary saddle atom $V = (P, K)$ and the atom A. These atoms are two-dimensional surfaces with symplectic structures and smooth functions f_1 and f_2, which determine the structure of an atom on the corresponding surface (i.e., a one-dimensional Liouville foliation with a single singular leaf). Consider the direct product $A \times V$ and define on it the symplectic structure as the "sum" of the initial symplectic

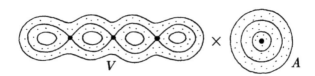

Figure 9.9

structures on A and V (see Fig. 9.9). The functions f_1 and f_2 are naturally lifted onto this direct product and commute with respect to the indicated symplectic structure. Therefore, they determine a Liouville foliation on $A \times V$. It is easy to see that this foliation has exactly one singularity of center–saddle type. Moreover, the direct product $A \times V$ is a regular neighborhood of the singular leaf L. The singular leaf L itself is just the graph K of the atom V. The center–saddle points are the vertices of K. We shall denote such a singularity by $A \times V$.

Theorem 9.2. *Every singularity of center–saddle type is Liouville equivalent to a canonical singularity $A \times V$ for a suitable saddle atom $V = (P, K)$. Moreover, the following assertions hold.*

a) *The singular leaf L coincides with the graph K.*

b) *The neighborhood $U(L)$ is the direct product of the two-dimensional disc and the surface P.*

c) *The l-type of this singularity is (sA, V), where s is the number of vertices of the graph K, and sA denotes a disjoint union of s copies of the atom A.*

d) *The loop molecule of the center–saddle singularity has the form shown in Fig. 9.10; moreover, all the r-marks on its edges are equal to infinity. For all the incoming edges of the atom V (see Fig. 9.10), the ε is the same and, by choosing orientation, can be made equal to $+1$. Then, for all outgoing edges (Fig. 9.10) the ε-mark is -1.*

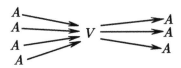

Figure 9.10

e) *The isoenergy 3-manifold Q corresponding to the loop molecule is diffeomorphic to the connected sum of $s + 1$ copies of $S^1 \times S^2$, i.e.,*

$$Q = \underbrace{(S^1 \times S^2) \# \ldots \# (S^1 \times S^2)}_{s+1 \text{ times}} .$$

Proof. Let s be the number of singular points on the singular leaf L. All of them are of center–saddle type. Take any of them. In its neighborhood it is always possible to choose a regular coordinate system (p_1, q_1, p_2, q_2) in terms of which the Liouville foliation is given by the pair of commuting functions $p_1^2 + q_1^2$ and $p_2 q_2$. Therefore, the singular leaf in a neighborhood of each singular point looks like a one-dimensional cross, i.e., two transversally intersecting segments. Therefore, the whole leaf L is obtained by gluing these crosses, i.e., is a graph with s vertices each of which has multiplicity 4. Let us denote it by K. Now we shall see that K is naturally embedded into a certain atom V.

Indeed, as was shown above, the set of critical points of the momentum mapping lying in $U(L)$ represents two intersecting surfaces. Since we consider a center–saddle singularity, one of these surfaces, say P_1, is a disjoint union of s copies of the atom A, and the other surface P_2 has the structure of a saddle atom $V = (P_2, K_2)$. It is easy to see that the singular leaf L consists entirely of critical points and, therefore, belongs to P_2. Besides, $H(L) = \text{const}$ and, consequently, L is noting else but the graph K_2.

Now the idea of the proof is rather natural. First we shall cut the neighborhood $U(L)$ into some standard pieces each of which is a regular neighborhood of a singular center–saddle point $z_i \in L$. The structure of the Liouville foliation near z_i is well known: according to Theorem 1.5, it can be represented as the direct product of the foliated disc and the foliated cross (Fig. 9.12). Then we shall show that, under the above assumptions, the inverse gluing of $U(L)$ from standard pieces can be carried out uniquely up to fiber isotopy. As a result, from local direct products we shall obtain a global one, as required. The same idea will be used in other situations.

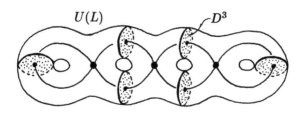

Figure 9.11

Let us indicate the midpoints on all the edges of the graph K (i.e., of the leaf L) embedded into the neighborhood $U(L)$ (Fig. 9.11). In the middle of each edge, we consider a three-dimensional ball transversal to the edge and cut $U(L)$ along all such balls. The neighborhood $U(L)$ splits into a union of 4-bricks each of which is obviously a regular neighborhood of a vertex z_i of the graph $K = L$ and, consequently, has the structure of the above direct product.

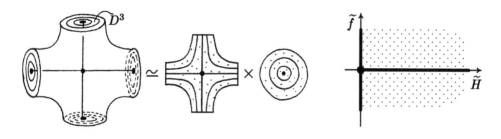

Figure 9.12   Figure 9.13

Consider now the inverse gluing. First we improve the functions H and f in the following way. We make a regular change

$$\widetilde{H} = \widetilde{H}(H, f) \quad \text{and} \quad \widetilde{f} = \widetilde{f}(H, f)$$

so that the bifurcation diagram in the neighborhood of $\mathcal{F}(x)$ straightens, i.e., takes the form shown in Fig. 9.13. As a result, the surfaces P_1 and P_2 in M^4 become critical non-degenerate submanifolds of the new functions $\widetilde{f}, \widetilde{H}$. More precisely, P_1 is a critical submanifold for $\widetilde{f}$, and P_2 is critical for $\widetilde{H}$. Moreover, P_1 is a critical submanifold of index 1, i.e., a saddle one. And P_2 is a critical submanifold of index 0, i.e., corresponds to a minimum. We are in the situation described by the generalized Morse-Bott lemma (Lemma 1.7). Thus, in some neighborhood of P_2, there exist two smooth independent functions x_1, y_1 such that $\widetilde{H} = x_1^2 + y_1^2$. Similarly, in some neighborhood of P_1 (which is a disjoint union of s discs, i.e., atoms A), there exist two smooth independent functions x_2, y_2 such that $\tilde{f} = x_2 y_2$. In fact, this pair of functions x_2, y_2 is defined not on the whole elementary 4-brick, but only on a part of it that is the direct product of the atom A by a small neighborhood of the center of the two-dimensional cross. We need to extend the functions onto the whole cross. This can evidently be done without losing the independence of x_2, y_2 and so that we still have $\tilde{f} = x_2 y_2$. As a result, we obtain four functions x_1, y_1, x_2, y_2 on the whole elementary 4-brick. They are regular coordinates on it (however, not necessarily symplectic) in terms of which we have

$$\tilde{H} = x_1^2 + y_1^2 \quad \text{and} \quad \tilde{f} = x_2 y_2.$$

Thus, on each 4-brick, we have a structure of the foliation shown in Fig. 9.12. Let us look what happens when we glue the boundaries of the 4-bricks. The boundary components are now represented as 3-cylinders $I \times D^2$. Each such cylinder is foliated into concentric circles whose centers are located on the segment I, the axis of the cylinder (Fig. 9.14). These circles are traces of leaves of the Liouville foliation on the boundary of the 4-brick, which are given by the equations $\tilde{H} = \text{const}$ and $\tilde{f} = \text{const}$. Therefore, for each circle on the boundary of one brick, we can always find the corresponding circle on the boundary of the other brick by choosing the circle with the same values of $\tilde{H}$ and $\tilde{f}$ (Fig. 9.14). We now glue these circles. It is clear that, as a result, we define the gluing diffeomorphism for the pair of boundary cylinders $I \times D^2$ uniquely up to fiber isotopy. It remains to observe that the functions $\tilde{H}$ and $\tilde{f}$ are defined globally on the whole of $U(L)$. This gives us a possibility to make all gluing operations by a standard and unambiguous rule. It is clear that, as a result, we obtain the structure of the direct product $A \times V$ on $U(L)$.

Parts (a), (b), (c) of the theorem follow immediately from this statement.

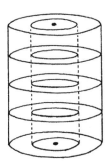

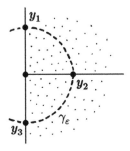

Figure 9.14   **Figure 9.15**

We now turn to computing the loop molecule. In Fig. 9.15, we show the arc γ_ε whose preimage is the 3-manifold Q_ε. The point y_2 corresponds to a saddle bifurcation. Since our singularity has type of the direct product $A \times V$, this bifurcation corresponds to the atom V. Analogously, the points y_1 and y_3 correspond to the bifurcation-atom A. Therefore, the molecule (yet without marks) obviously has the desired form presented in Fig. 9.10. It remains to compute the marks.

Consider a point on an arbitrary edge of the molecule and the Liouville torus corresponding to it. Take two cycles on this torus. The first one is the cycle that shrinks into a point as the torus tends to y_1. The second cycle transforms into the saddle critical circle as the torus tends to y_2. Obviously, this is the same cycle. It is shown in Fig. 9.9 as one of the circles into which the atom A is foliated. In the sense of the direct product structure, this cycle represents a non-singular leaf of the atom A multiplied by a vertex of the atom V. As we know from Chapter 4, this means that the corresponding r-mark is equal to infinity.

Let us compute the ε-marks. They are defined by the mutual orientations on the cycles. Consider the initial Hamiltonian H on $U(L)$. Since the surfaces P_1 and P_2 are invariant under the flow sgrad H, this flow defines the natural orientation on all the critical circles. Consider an arbitrary Liouville torus. It can be naturally identified with the product of two critical circles one of which lies on P_1, and the other lies on P_2. Since the circles are both oriented by the flow, the torus obtains a natural orientation. On the other hand, on this torus (as a leaf of the Liouville foliation on Q_ε), there is another orientation. In our case, the Liouville torus bounds a solid torus and, therefore, can be considered as the boundary torus of the 3-atom A. The orientation of the solid torus induced by the fixed orientation on Q_ε determines the orientation on its boundary torus (for example, by means of the outward normal). Let us compare these two orientations on the same torus. If they coincide, then $\varepsilon = +1$. Otherwise, $\varepsilon = -1$.

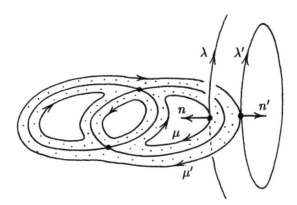

Figure 9.16

In our case, these orientations are illustrated in Fig. 9.16. It is seen that the orientations of the triples (n, λ, μ) and (n', λ', μ') are different on the Liouville tori. Therefore, the ε-marks corresponding to these Liouville tori have different signs. In this figure (Fig. 9.16), we take Liouville tori from two different classes with respect to the atom V (positive and negative in our precedent notation). As is seen from the same figure, for Liouville tori of the same kind (either positive or negative), the ε-mark is also the same. The point is that the orientation of the indicated triples coincide.

It remains to describe the topology of Q_ε. But that has been already done in Proposition 4.5. This completes the proof of Theorem 9.2.   □

9.5. SADDLE–SADDLE CASE

9.5.1. *The Structure of a Singular Leaf*

Let L be a singular leaf of saddle–saddle type and let $y_0 = \mathcal{F}(L)$ be the singular point of the bifurcation diagram Σ corresponding to the given singular leaf. According to the above assumptions, a neighborhood of y_0 on the bifurcation diagram Σ has the form of two transversally intersecting smooth curves γ_1 and γ_2, as shown in Fig. 9.1(c). Let us make a local coordinate change in a neighborhood of y_0 on the plane $\mathbb{R}^2(H, f)$ so that γ_1 and γ_2 transform into segments lying

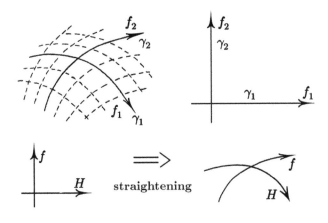

Figure 9.17

on the (new) coordinate axes, as illustrated in Fig. 9.17. We denote these new coordinates by f_1 and f_2. As a result, the curve γ_1 is given by the equation $f_2 = 0$, and γ_2 is given by $f_1 = 0$. We shall assume that $\partial H/\partial f_i > 0$ for $i = 1, 2$. We can always achieve this by changing the sign of f_i if necessary. The functions f_1 and f_2 can obviously be regarded as new integrals of the initial Hamiltonian system instead of H and f.

Proposition 9.2. *Let s be the number of singular points $z_1, \ldots z_s$ of saddle–saddle type on the singular leaf L. Then L is a two-dimensional cell complex glued from $4s$ squares. The interior of each square is a two-dimensional orbit of the Poisson action of the group $\mathbb{R}^2(H, f)$. The edges of the squares (without end-points) are one-dimensional non-closed orbits of the action of $\mathbb{R}^2(H, f)$, and the vertices of the squares are exactly the singular points $z_1, \ldots z_s$, i.e., zero-dimensional orbits. The edges of the squares belong to the submanifolds $P_1 = \mathcal{F}^{-1}(\gamma_1) \cap K$ and $P_2 = \mathcal{F}^{-1}(\gamma_2) \cap K$. Moreover, the opposite edges belong to the same submanifold, and the adjacent edges belong to different ones.*

Proof. We first analyze the structure of L in a neighborhood of each singular point z_i. According to Theorem 1.5, there exists a local coordinate system (p_1, p_2, q_1, q_2) in a neighborhood of z_i such that the Liouville foliation is defined by common level surfaces of functions $p_1 q_1$ and $p_2 q_2$. This means that locally the structure of the Liouville foliation has the type of the direct product of two elementary one-dimensional hyperbolic singularities. In other words, we need to multiply two foliated saddles shown in Fig. 9.18. In particular, the singular leaf L locally (in a neighborhood of z_i) has the form $\Gamma \times \Gamma$, where Γ is the one-dimensional cross, i.e., two transversally intersecting intervals.

It is clear that the points $z_1, \ldots, z_s$ are zero-dimensional orbits of the action of $\mathbb{R}^2$. Moreover, by our assumptions, there are no other zero-dimensional orbits on the leaf L.

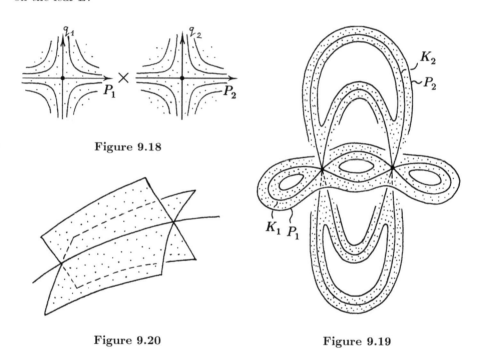

Figure 9.18

Figure 9.20   Figure 9.19

We now study the one-dimensional orbits of the action of $\mathbb{R}^2$ lying on the singular leaf L. All of them, by definition, belong to the set K of singular points of $\mathcal{F}$. On the other hand, in a neighborhood $U(L)$ of L the set of singular points of $\mathcal{F}$ is the union of P_1 and P_2. Therefore, all the one-dimensional orbits (located in $U(L)$) belong either to P_1 or to P_2. Moreover, the one-dimensional orbits lying on L are exactly the edges of the graphs K_1 and K_2 of the atoms $V_1 = (P_1, K_1)$ and $V_2 = (P_2, K_2)$, respectively (Fig. 9.19). In particular, the one-dimensional orbits lying on L are all non-degenerate and non-closed, and the total number of them is $4s$ (namely, $2s$ for each graph K_i). Each orbit has hyperbolic type and, consequently, for an interior point x on it, the singularity of the complex L (in a small neighborhood $U(x)$) looks like a fourfold line (Fig. 9.20), i.e., locally L is homeomorphic to the direct product $\Gamma \times D^1$.

How do the vector fields sgrad H, sgrad f_1, and sgrad f_2 behave on the one-dimensional orbits?

The field $\operatorname{sgrad} f_1$ is non-trivial along 1-orbit lying on P_1 and vanishes on the 1-orbits lying on P_2. In other words, $\operatorname{sgrad} f_1$ flows along the edges of the graph K_1 and is zero on the edges of K_2.

The field $\operatorname{sgrad} f_2$ behaves similarly: it flows along the edges of K_2 and is zero on the edges of K_1.

The field $\operatorname{sgrad} H$ is non-trivial on all the one-dimensional orbits (both in P_1 and P_2). Moreover, $\operatorname{sgrad} H$ flows in the same direction as $\operatorname{sgrad} f_1$ and $\operatorname{sgrad} f_2$ on the corresponding edges. This follows from the condition that $\partial H/\partial f_i > 0$ for $i = 1, 2$.

Turning to the two-dimensional orbits G, we first pay attention to the following general fact: the points that lie in the closure of a two-dimensional orbit, but do not belong to the orbit itself, are necessarily contained in either one-dimensional or two-dimensional orbits. In other words, $\overline{G} \setminus G \subset (K_1 \cup K_2)$.

Now let G be a two-dimensional orbit of the Poisson action of $\mathbb{R}^2 (H, f)$ which belongs to the singular leaf L and contains the point z_1 in its closure. It is known that a two-dimensional orbit is diffeomorphic to either a 2-torus, or a 2-cylinder, or a 2-plane. As we have seen above, the torus and cylinder must be excluded and, therefore, G is diffeomorphic to a plane. Consider its closure $\overline{G}$ and look at its behavior near the boundary points lying in $K_1 \cup K_2$.

The orbit G approaches the vertices $z_1, \ldots, z_s$ of the graph $K_1 \cup K_2$ each time in the same way. As shown above, in a neighborhood of z_i, the singular leaf locally represents the direct product $\Gamma \times \Gamma$. This product evidently admits a stratification into zero-, one-, and two-dimensional strata.

It is easily seen that there are 16 two-dimensional strata each of which is a part of a two-dimensional orbit located in the neighborhood of z_i. On the other hand, the closure of each stratum can evidently be represented as a smooth embedding of the right angle into M^4. Thus, the orbit G near the singular point z_i looks like the usual angle with the vertex z_i, whose sides are edges of two different graphs K_1 and K_2. Notice that the orbit may return to the same singular point several times, so several different angles may correspond to it.

To the one-dimensional orbits, i.e., edges of the graphs K_1 and K_2, the orbit G also approaches in a nice way as a fragment of a half-plane (one of the four smooth sheets shown in Fig. 9.20). Of course, G may return to the same edge several times (in fact, as we shall see soon, at most twice).

Lemma 9.2. *The closure of the orbit $G \subset L$ is a polygon with an even number of edges smoothly immersed into M^4.*

Proof. The fact that the closure $\overline{G}$ is an immersed polygon follows immediately from the local behavior of G near its boundary points.

We only need to prove that the number of edges of G is even. This fact is almost evident. Indeed, let x be a point on the boundary of G lying, for instance, on an edge of K_1. Let us move along this edge (staying on the orbit) in a certain direction. After coming to the nearest vertex of the graph K_1, we shall have to turn to another edge which now belongs to the other graph K_2. This follows from the local behavior of G near a singular point (see above). Having arrived at the end of this edge of K_2, we again turn to an edge of K_1. And so on until we come to the initial point.

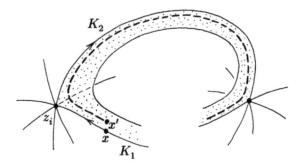

Figure 9.21

Let us emphasize that this sequence (of edges) is completely and uniquely determined by the orbit G which adjoins the graph $K_1 \cup K_2$ by its boundary ∂G (see Fig. 9.21). Since we jumped at each vertex from one graph to the other (from K_1 to K_2, or conversely), the total number of the edges we went along is even.

Lemma 9.2 is proved. □

Thus, $\overline{G}$ can be considered as an embedding into M^4 of a polygon $\widetilde{G}$ with an even number of vertices. Its interior is diffeomorphically mapped onto the orbit G, and the sides are sent to some edges of the graph $K_1 \cup K_2$. In other words, the orbit G can be interpreted as a two-dimensional cell glued to the 1-skeleton $K_1 \cup K_2$ by some mapping of its boundary ∂G.

Lemma 9.3. *The polygon $\widetilde{G}$ is a square.*

REMARK. This means that $\widetilde{G}$ has exactly four sides. It is clear that, in such a case, one pair of the opposite sides is mapped into K_1, and the other is mapped into K_2. Notice that the immersion $\widetilde{G} \to \overline{G} \subset M^4$ is not an embedding in general. Some vertices and even edges may be glued between each other. Nevertheless, the mapping restricted to the interior of $\widetilde{G}$ and to the interior of each edge is an embedding.

Proof (of Lemma 9.3). Consider the commuting fields sgrad f_1 and sgrad f_2 on the closure $\overline{G}$ of the orbit G. Since $\overline{G}$ is the embedding of $\widetilde{G}$, both the fields can be lifted onto the polygon $\widetilde{G}$. Let a be an arbitrary side of $\widetilde{G}$ which belongs to K_1 (after embedding into M^4). Consider the behavior of sgrad f_2 in a neighborhood of the edge a. Since the edge a is a non-degenerate one-dimensional orbit of $\mathbb{R}^2$, integral curves of sgrad f_2 either approach the edge a transversally or go out of it. On the edge a itself, the field sgrad f_2 identically vanishes (Fig. 9.22). The similar argument works for the field sgrad f_1 with respect to the edges of K_2.

Consider an edge of the polygon on which sgrad f_1 is different from zero. Then the field sgrad f_1 vanishes on the two neighboring edges: it flows in one of these edges and flows out of the other. The picture illustrating the behavior of integral curves of sgrad f_1 is shown in Fig. 9.23. As a result, we obtain the complete picture of the field sgrad f_1 in the neighborhood of the boundary of the polygon (Fig. 9.24).

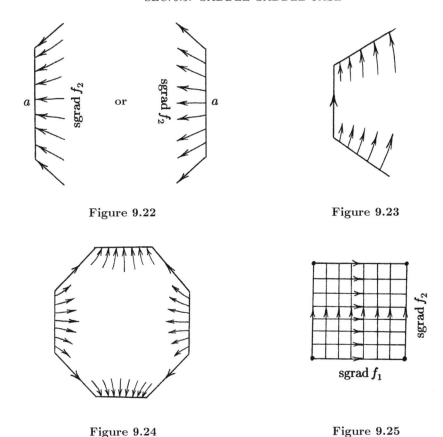

Figure 9.22

Figure 9.23

Figure 9.24

Figure 9.25

We now note that sgrad f_1 has no zeros inside the polygon $\widetilde{G}$. Using standard manipulations with the index of a vector field, we can easily see that such a situation is possible if and only if $\widetilde{G}$ is a square.

Lemma 9.3 is proved.   □

In Fig. 9.25 we present the final picture of the integral curves of the fields sgrad f_1 and sgrad f_2 on the square.

Let us summarize the results obtained. We have described one-dimensional and two-dimensional cells of the complex L. The two-dimensional cells are squares, and the one-dimensional cells are the edges of the graphs K_1 and K_2.

Lemma 9.4. *The number of the squares is 4s, where s is the number of singular points on the singular leaf L.*

Proof. Let k be the number of the squares. Let us compute the total number of the angles of all the squares in two different ways. On the one hand, it is obviously equal to $4k$. On the other hand, at each singular point z_i there are 16 different angles of these squares, and the number of such points is s. Hence the total number of the angles is equal to $16s$. Thus, $16s = 4k$, i.e., $k = 4s$, as was to be proved.   □

Thus, Proposition 9.2 is completely proved.   □

Lemma 9.5. *The singular leaf L is the image of a disjoint union of several tori and Klein bottles under some immersion into M^4.*

Proof. Let us divide the edges of all the squares (from which L is glued) into two classes. The first class consists of the edges of the graphs K_1 and K_2 which integral trajectories of the flows $\operatorname{sgrad} f_2$ and $\operatorname{sgrad} f_1$ enter. The second class, on the contrary, consists of the edges of K_1 and K_2 from which integral curves of $\operatorname{sgrad} f_2$ and $\operatorname{sgrad} f_1$ go out. We now make partial gluing of these squares identifying only those edges which belong to the same class. Note that, making complete gluing, we should identify the edges in quadruples, since each edge of L is a fourfold line. But we have just divided each quadruple into two pairs, and now identify only the edges from the same pair.

Since the edges are glued pairwise, we obtain a certain closed surface $\tilde{L}$ (not necessarily connected). This surface is decomposed into squares. Moreover, it is easily seen that each vertex of this decomposition has degree 4. Computing the Euler characteristic of $\tilde{L}$, we obtain zero. Thus, this surface is a disjoint union of tori or Klein bottles. If we complete this gluing operation, we obtain an immersion of $\tilde{L}$ in M^4, whose image is exactly the singular leaf L. $\quad\square$

As we shall see below, the singular leaf L is a $K(\pi, 1)$-space, i.e., its homotopy groups are all trivial except for the fundamental group $\pi_1(L)$.

It turns out that the topology of the complex L does not determine uniquely the structure of the Liouville foliation in its neighborhood $U(L)$. It turns out that, to construct a complete invariant of a saddle–saddle singularity, it suffices to combine two invariants which have been constructed above: the complex L and l-type. This new invariant is convenient to enumerate possible types of singularities.

9.5.2. Cl-Type of a Singularity

The notion of the Cl-type of a singularity was introduced by A. V. Bolsinov and V. S. Matveev in [65].

Recall that the l-type of a saddle–saddle singularity is a pair of atoms (V_1, V_2), where $V_i = (P_i, K_i)$. Here the union of the graphs K_1 and K_2 is just the 1-skeleton of the complex L, and the surfaces P_1 and P_2 represent the set of critical points of the momentum mapping that belong to the neighborhood $U(L)$.

Definition 9.4. The *Cl-type* of the saddle–saddle singularity is defined as the triple (L, V_1, V_2) together with the two embeddings $\xi_i \colon K_i \to L^{(1)}$ ($i = 1, 2$), where $L^{(1)}$ is the 1-skeleton of L.

In this definition, we ignore orientations on the atoms V_1 and V_2.

COMMENT. Thus, the Cl-type is the union of two objects: the l-type (i.e., the pair of atoms (V_1, V_2)) and the two-dimensional complex L. Moreover, we must remember that the union of the graphs K_1 and K_2 is exactly the 1-skeleton of L. This information is kept in the mappings ξ_1 and ξ_2 (see above). Thus, the elements of the triple (L, V_1, V_2) are not independent, but connected by the 1-skeleton of L.

Theorem 9.3. *The Cl-type is a complete invariant of a saddle–saddle singularity in the sense of Liouville equivalence. This means that, if two saddle–saddle singularities have the same Cl-type, then there exist invariant neighborhoods $U(L)$ and $U'(L')$ of the singular leaves L and L' and there exists a fiber diffeomorphism $U(L) \to U'(L')$ that preserves the direction of the Hamiltonian flows on the one-dimensional orbits.*

The proof is given below (see Section 9.7.1).

This important theorem allows us to classify and enumerate saddle–saddle singularities starting from small complexity. By complexity we mean here the number s of singular points $z_1, \ldots, z_s$ on the singular leaf L. This program of classification will be carried out below: we shall enumerate all saddle–saddle singularities of complexity 1 and 2.

To start the classification of singularities, we need to describe some properties of the Cl-type. The point is that not every triple (L, V_1, V_2) given in an abstract way is admissible, i.e., can be realized as the Cl-type of a certain singularity.

Suppose we are given an abstract triple (L, V_1, V_2), where L is a two-dimensional cell complex, $V_i = (P_i, K_i)$ are atoms of the same complexity s. Moreover, we have also the embeddings $\xi_i \colon K_i \to L^{(1)} \subset L$ ($i = 1, 2$), where $L^{(1)}$ is the 1-skeleton of L.

Definition 9.5. An abstract triple (L, V_1, V_2) with two embeddings ξ_1 and ξ_2 is called an *admissible Cl-type* if the following conditions are fulfilled.

1) All the edges of the 1-skeleton of L can be divided into two classes in such a way that the edges from the i-th class are the images of the edges of the graph K_i. In particular, $L^{(1)} = \xi_1(K_1) \cap \xi_2(K_2)$. (The edges of the first class will be called K_1-edges and denoted by Latin letters, and those of the second class will be called K_2-edges and denoted by Greek letters.)

2) The complex L is glued from $4s$ squares.

3) The opposite edges of each square belong to the same class and have the same orientation. The adjacent edges belong to different classes (Fig. 9.26).

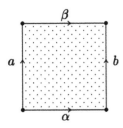
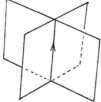
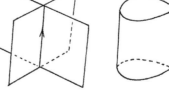

Figure 9.26   **Figure 9.27**

4) Each edge of the complex L is a four-fold line (Fig. 9.20). This means that there are exactly four squares adjacent to it (with multiplicity, i.e., one square may adjoin the edge by two opposite sides (Fig. 9.27)). In other words, if L is cut into the squares, then each letter (denoting an edge) appears exactly four times.

5) All the angles of all the squares are distinct, i.e., either the letters or orientations on their edges are different.

6) Consider the fragment of L consisting of the two squares which are glued along an edge a (Fig. 9.28). The edges are endowed with orientations as shown in Fig. 9.28. The pair of edges α, β can be considered as a basis on the atom V_2 at the beginning of the edge a (Fig. 9.29). The pair of edges γ, δ forms a basis on the same atom V_2 at the end of a. It is required that these bases have the same orientation on V_2.

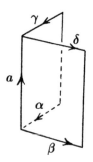

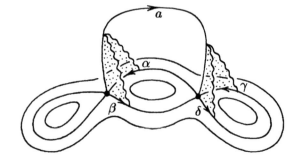

Figure 9.28   Figure 9.29

To clarify this definition, we reformulate it in slightly different terms.

Take a certain l-type, i.e., just a pair of atoms (V_1, V_2). Let us fix some orientation on the surfaces P_1 and P_2 as well as on the edges of the graphs K_1 and K_2. The orientation on the edges must be compatible with the structure of the atoms. There are only two ways to introduce such an orientation. Namely, if K_i is interpreted as a singular level line of a Morse function f_i, then the orientation is given either by the direction of the flow sgrad f_i or by the direction of $-$ sgrad f_i. We denote the edges of K_1 by Latin letters and the edges of K_2 by Greek ones. Besides, we establish a one-to-one correspondence between the vertices of the atom V_1 and those of the atom V_2. (Recall that the surfaces P_1 and P_2 intersect in M^4 exactly at these vertices (Fig. 9.19).) In particular, identifying the corresponding pairs of vertices, we can consider the new graph $K_1 \cup K_2$ (which will play role of the 1-skeleton of the cell complex L).

Having fixed (V_1, V_2), we describe admissible complexes L as sets of squares with oriented and marked (by letters) edges. The conditions to be fulfilled are then as follows.

a) The boundary of each square is a closed path in the union $K_1 \cup K_2$. (Otherwise, this square cannot be glued to the 1-skeleton.)

b) The opposite sides of squares have the same orientation and belong to the same class (i.e., both are either K_1- or K_2-edges). The adjacent sides belong to different classes.

c) The complex L in a neighborhood of each of its vertices has the structure of the direct product (cross) $\times$ (cross). Each such cross can be interpreted as the quadruple of the edges incident to the given vertex in the graphs K_1 and K_2. This gives us the list of all the angles of the squares that meet at this vertex. In particular, at each vertex we have 16 angles, and all of them are different. An example is shown in Fig. 9.30.

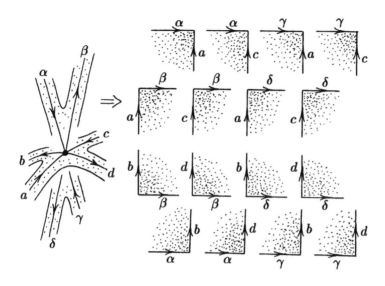

Figure 9.30

d) Compatibility of orientations. Consider an edge a (Fig. 9.29) which starts from some vertex S and comes to some other vertex S'. Since this edge is a fourfold line from the point of view of the complex L (Fig. 9.20), we can represent a neighborhood of a as the direct product $a \times$ (cross). At the starting point S, this cross is just the quadruple of the edges of K_2 that meet at the vertex S. Similarly, at the point S', this cross can be regarded as the quadruple of the edges of K_2 incident to S' (Fig. 9.31). By fixing a cyclic order of these edges, we can define

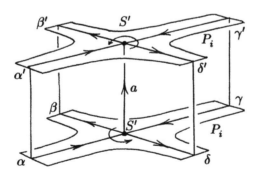

Figure 9.31

some orientation on the surface P_2. Since we can do so both at the beginning S and at the end S' of a, we can compare the two orientations obtained. It is required that these orientations coincide. In other words, each edge $a \in K_1$ allows us to translate the orientation of P_2 from the beginning of a to its end. We require that such a translation does not change the orientation. The similar condition must be fulfilled for each edge $\alpha \in K_2$.

It is possible to suggest one more interpretation of the rules of gluing just described. The complex L satisfies the following conditions.

($*$) L is glued from squares.

($**$) Each edge is adjacent to 4 squares (with multiplicity).

($***$) The fragment of L that is a neighborhood of some edge a looks like one of the two details presented in Fig. 9.32 (the second one is obtained from the first just by the permutation of labels of the edges at the end-point of a).

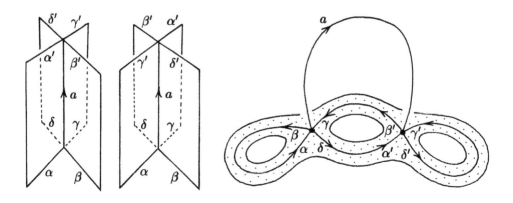

Figure 9.32

Theorem 9.4 (Realization Theorem) (V. S. Matveev). *The admissible Cl-types and only they can be realized as Cl-types of saddle–saddle singularities.*

Therefore, the set of admissible Cl-types gives the complete list of the saddle–saddle singularities up to Liouville equivalence.

The proof of Theorem 9.4 is given below (see Section 9.7.2).

9.5.3. *The List of Saddle–Saddle Singularities of Small Complexity*

In this section, we enumerate all the saddle–saddle singularities of complexity 1 and 2. The idea of the enumeration is as follows. First we describe all the l-types of complexity 1 and 2, and then, using properties ($*$), ($**$), ($***$), for each fixed l-type, we list the complexes L corresponding to it.

In the case of complexity 1, there is only one l-type, namely (B, B). The point is that there exists only one (orientable) saddle atom of complexity 1, this is the atom B.

The squares, from which the complex L (i.e., the singular leaf) is glued, have oriented sides marked by Latin and Greek letters. As usual, moving along the boundary of a square, we can write a word indicating step by step the letters on the edges we meet. Moreover, we endow each letter with power $\varepsilon = \pm 1$ in the standard way taking into account the orientation on the corresponding edge.

Theorem 9.5 (L. M. Lerman, Ya. L. Umanskiĭ [212]). *Suppose that the singular leaf L contains exactly one singular point (i.e., has complexity* 1). *Then L is homeomorphic to one of the four following complexes (Fig.* 9.33):

1) $a\alpha a^{-1}\alpha^{-1}$, $bab^{-1}a^{-1}$, $b\beta b^{-1}\beta^{-1}$, $a\beta a^{-1}\beta^{-1}$.

2) $a\alpha b^{-1}\alpha^{-1}$, $baa^{-1}\alpha^{-1}$, $a\beta b^{-1}\beta^{-1}$, $b\beta a^{-1}\beta^{-1}$.

3) $a\beta a^{-1}\alpha^{-1}$, $bab^{-1}\alpha^{-1}$, $b\beta b^{-1}\beta^{-1}$, $a\alpha a^{-1}\beta^{-1}$.

4) $a\alpha b^{-1}\beta^{-1}$, $a\beta b^{-1}\alpha^{-1}$, $b\beta a^{-1}\alpha^{-1}$, $ba a^{-1}\beta^{-1}$.

Two saddle–saddle singularities of complexity 1 *are Liouville equivalent if and only if their singular leaves are homeomorphic.*

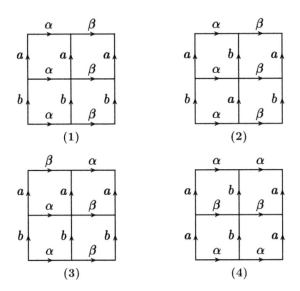

Figure 9.33

Proof. First of all, we note that the four listed complexes are all admissible, i.e., satisfy conditions (∗), (∗∗), (∗∗∗). Let us show that no other complexes exist. Consider the squares adjacent to the edge α. In view of condition (∗∗∗), only two cases presented in Fig. 9.34(a) and Fig. 9.34(b) are possible. Each of them can in turn give three essentially different possibilities. Namely, there may be two, three, or four squares adjacent to the edge α (in both cases (a) and (b)); all these possibilities are shown in Fig. 9.34(a1), Fig. 9.34(a2), Fig. 9.34(a3) and Fig. 9.34(b1), Fig. 9.34(b2), Fig. 9.34(b3) respectively.

In cases (a3) and (b3), we immediately obtain the complete description of the complex L. Both of them are admissible. Case (a3) corresponds to the *second* complex from the above list (after the following changes of notation: $a \leftrightarrow \alpha$, $b \leftrightarrow \beta$), case (b3) corresponds to the *fourth* one.

In cases (a2) and (b2), we need one more square. It can, however, be uniquely defined from the condition that the angles of the squares are all different (see condition (∗∗∗) for the edge β). As a result, we obtain that case (a2) corresponds to the *third* complex from the above list. Case (b2) is not admissible, since condition (∗∗∗) fails for the edges a and b.

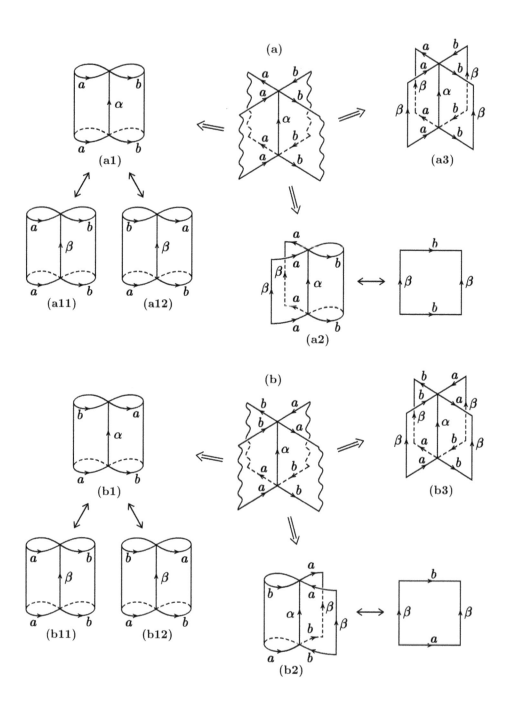

Figure 9.34

Each of cases (a1) and (b1) gives two more possibilities, namely, the cases (a11), (a12) and (b11), (b12).   Additional pairs of squares are shown in Figs. 9.34(a11), Fig. 9.34(a12), Fig. 9.34(b11), and Fig. 9.34(b12). Case (a11) corresponds to the *first* complex from our list. Cases (a12) and (b11) are actually equivalent and can be obtained from each other by interchanging α and β. Both of them correspond to the *third* case. To see this, one needs to make the following changes for notation in case (a12): $a \to \alpha$, $b \to \beta$, $\alpha \to b$, $\beta \to a$.

It remains to consider case (b12). It corresponds to the *second* complex from the above list. $\square$

The next theorem describes the loop molecules corresponding to the saddle–saddle singularities of complexity 1 listed above.

Theorem 9.6 (A. V. Bolsinov [44]).   *The loop molecules of saddle–saddle singularities of complexity* 1 *corresponding to cases* 1–4 *indicated in Theorem* 9.5 *are listed in Fig.* 9.35.

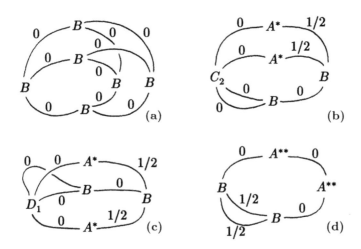

Figure 9.35

Sketch of the proof. Consider the singular point y on the bifurcation diagram that corresponds to the singular leaf L (i.e., $y = \mathcal{F}(L)$). Near the point y, the bifurcation diagram has the form shown in Fig. 9.36, i.e., it represents two smooth curves intersecting transversally at y.

Let x be a regular point close to y. Its preimage $\mathcal{F}^{-1}(x)$ consists of several (or just one) Liouville tori. Since x is close to y, we may assume that these tori are glued from the same squares as the singular leaf L. The difference is that if we glue L, then these squares must be glued in quadruples, whereas, for a regular leaf $\mathcal{F}^{-1}(x)$, the same squares are glued pairwise. In other words, when we move from y to x, the singular leaf splits as illustrated in Fig. 9.37. Each fourfold line splits in two different ways depending on the location of the point x. Namely, x can be located in one of the four quadrants into which the bifurcation diagram divides a neighborhood of y. For each of these quadrants, we can explicitly describe the corresponding family of Liouville tori. Moreover, when we pass from one quadrant to another, we understand how the Liouville tori bifurcate. In other words, we describe all bifurcations of Liouville tori corresponding to the curves of the bifurcation diagram. These bifurcations are just the atoms of the loop molecule.

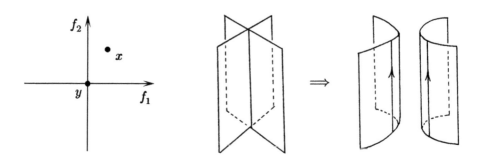

Figure 9.36   **Figure 9.37**

After this, it remains to carry out the computation for each of the four cases separately, by using the above scheme. We omit the details and show the final result in Fig. 9.35. □

We now turn to the classification of saddle–saddle singularities of complexity 2. First of all, we describe all l-types which are *a priori* possible. Since we admit disconnected atoms, we need to add two disconnected atoms BB and BB' (Fig. 9.38) to the standard list of atoms of complexity 2 (they are C_1, C_2, D_1, D_2).

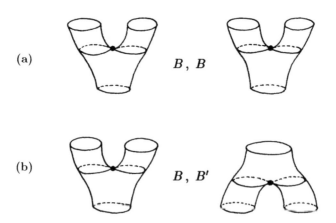

Figure 9.38

Since the singular leaf L is connected, at least one of the two atoms V_1, V_2 from the l-type has to be connected. Therefore, *a priori*, there exist 18 different l-types of complexity 2.

Theorem 9.7 (A. V. Bolsinov [44]). *Suppose the singular leaf L contains exactly two singular points (i.e., has complexity 2). Then the number of different singularities corresponding to a fixed l-type is indicated in the following list:*

$$(BB, C_1) - 7, \quad (BB', C_1) - 1, \quad (C_1, C_1) - 1,$$
$$(BB, C_2) - 7, \quad (BB', C_2) - 1, \quad (C_1, C_2) - 2, \quad (C_2, C_2) - 2,$$
$$(BB, D_1) - 6, \quad (BB', D_1) - 0, \quad (C_1, D_1) - 2, \quad (C_2, D_1) - 2, \quad (D_1, D_1) - 2,$$
$$(BB, D_2) - 6, \quad (BB', D_2) - 0, \quad (C_1, D_2) - 0, \quad (C_2, D_2) - 0, \quad (D_1, D_2) - 0, \quad (D_2, D_2) - 0.$$

Thus, there exist 39 different (up to Liouville equivalence) saddle–saddle singularities of complexity 2. The structure of the corresponding singular leaves L is presented in Table 9.1.

Proof. The proof can be carried out by the same scheme as in Theorem 9.5. We endow each fixed l-type with an additional structure, namely, a bijection between the vertices of the graphs K_1 and K_2 and orientation on their edges. In this case, due to symmetries of atoms of small complexity, this additional structure can be chosen uniquely up to a homeomorphism. Then we enumerate all (admissible) possibilities to glue squares to the 1-skeleton $K_1 \cup K_2$ using conditions $(*)$, $(**)$, $(***)$. □

COMMENT.   Thus, there exist 39 different saddle–saddle singularities of complexity 2. Note that some complexes L from this list are homeomorphic. We have already pointed out this fact: the topology of the singular leaf L does not determine, in general, the topology of its neighborhood $U(L)$. In this case, the different singularities corresponding to the same complex L have different Cl-types.

COMMENT.   The saddle–saddle singularities of complexity 3 can be classified in a similar way. This classification was carried out by N. A. Maksimova by computer analysis. It turns out that the total number of such singularities is equal to 256.

In principle, the above procedure gives the enumeration algorithm for singularities of arbitrary complexity. But the number of singularities grows very fast as complexity increases.

Theorem 9.8 (V. S. Matveev [64], [224]). *The loop molecules of the saddle–saddle singularities of complexity 2 (i.e., with two singular points on the leaf L) that correspond to 39 cases listed in Theorem 9.7 are presented in Table 9.1.*

The proof is carried out by the same scheme as the proof of Theorem 9.6 and actually consists in detailed analysis of each of 39 cases. □

Corollary. *In the case of saddle–saddle singularities of complexity 2, the loop molecule is a complete invariant of the Liouville foliation in a neighborhood of a given singularity.*

Proof. It suffices to note that all loop molecules listed in Fig. 9.35 and in Table 9.1 are different. □

Table 9.1. Saddle–saddle singularities of complexity 2

No	l-TYPE	SINGULAR LEAF L	LOOP MOLECULE
1	(C_1, BB)		
2	(C_1, BB)		
3	(C_1, BB)		
4	(C_1, BB)		
5	(C_1, BB)		

Table 9.1. Saddle–saddle singularities of complexity 2 (continued)

No	l-TYPE	SINGULAR LEAF L	LOOP MOLECULE
6	(C_1, BB)	a c b d; α γ α γ β; a c b d; β δ β δ α; a c b d	0 — M_2 — 0; A^* ... A^*; $1/2$ — C_1 — $1/2$
7	(C_1, BB)	a c b d; α γ β δ α; a c b d; β δ α γ β; a c b d	0 — K_3 — 0; 0 0; B ... B; 0 — C_1 — 0
8	(C_2, BB)	a c b d; α γ α γ α; b d a c; β δ β δ β; a c b d	0 — A^{**} — 0; 0 — A^{**} — 0; C_2 ... C_2; 0 — C_2 — 0
9	(C_2, BB)	b c b c; α γ β δ α; a d a d; α γ β δ α; b c b c	0 — A^{****} — 0; A^{**} ... A^{**}; 0 — C_1 — 0
10	(C_2, BB)	a c b d; α γ α γ α; a c b d; β δ β δ β; a c b d	A^{**}; 0 0 C_2 0 0; 0 0; D_1 ... D_1; 0 — A^{**} — 0

Table 9.1. Saddle–saddle singularities of complexity 2 (continued)

No	l-TYPE	SINGULAR LEAF L	LOOP MOLECULE
11	(C_2, BB)		
12	(C_2, BB)		
13	(C_2, BB)		
14	(C_2, BB)		
15	(D_1, D_1)		

Table 9.1. Saddle–saddle singularities of complexity 2 (continued)

No	l-TYPE	SINGULAR LEAF L	LOOP MOLECULE
16	(D_1, D_1)	$a\ b\ c\ d$ / $\delta\ \alpha\ \beta\ \gamma\ \delta$ / $d\ a\ b\ c$ / $\beta\ \alpha\ \delta\ \gamma\ \beta$ / $a\ b\ c\ d$	0 — B_1^{**} — 0; B — B_1^{**}; $3/4$; $1/4$ — B — 0
17	(C_2, C_2)	$c\ b\ d\ a$ / $\alpha\ \gamma\ \alpha\ \gamma\ \alpha$ / $a\ d\ b\ c$ / $\delta\ \beta\ \delta\ \beta\ \delta$ / $c\ b\ d\ a$	0 — B — $1/2$; $1/2$; B — $1/2$; C_2 0 B B; 0 — C_2 0
18	(C_2, C_2)	$c\ b\ c\ b$ / $\alpha\ \gamma\ \beta\ \delta\ \alpha$ / $a\ d\ a\ d$ / $\gamma\ \alpha\ \delta\ \beta\ \gamma$ / $c\ b\ c\ b$	0 — D_1 — $1/2$; 0 $1/2$; A^{**} B B; $1/2$ $1/2$; 0 — D_1
19	(C_1, BB')	$b\ c\ a\ d$ / $\alpha\ \gamma\ \beta\ \gamma\ \alpha$ / $a\ d\ b\ c$ / $\alpha\ \delta\ \beta\ \delta\ \alpha$ / $b\ c\ a\ d$	$1/2$ — B_1^{**} — 0; $1/2$; B B; 0 $1/2$; B_1^{**} — $1/2$
20	(C_2, BB')	$b\ c\ b\ c$ / $\alpha\ \gamma\ \beta\ \delta\ \alpha$ / $a\ d\ a\ d$ / $\alpha\ \delta\ \beta\ \gamma\ \alpha$ / $b\ c\ b\ c$	0 — B_1^{**} — 0; 0; A^{**} C_2; 0; B_1^{**} — 0

Table 9.1. Saddle–saddle singularities of complexity 2 (continued)

No	l-TYPE	SINGULAR LEAF L	LOOP MOLECULE
21	(D_1, BB)		
22	(D_1, BB)		
23	(D_1, BB)		
24	(D_1, BB)		
25	(D_1, BB)		

Table 9.1. Saddle–saddle singularities of complexity 2 (continued)

No	l-TYPE	SINGULAR LEAF L	LOOP MOLECULE
26	(D_1, BB)	a b c d; β α δ δ β; a b c d; α β γ γ α; a b c d	L_2, D_1; 0 0 0 0 0 0; B A^* A^* B; 0 1/2 1/2 0
27	(C_1, D_1)	a b c d; α γ α γ α; b c d a; β δ β δ β; b c d a	C_1, A^*; 0 0; 3/4 1/4; B B; 1/4 3/4; A^*
28	(C_1, D_1)	a b c d; α γ α γ α; b c d a; β δ β δ β; d a b c	A^{**}, A^*; 1/2 1/2; 3/4 1/4; B B; 1/4 3/4; A^*
29	(D_2, BB)	a b c d; β β γ δ α; a b c d; α α δ γ β; a b c d	V_4', D_2; 0 0 0 0 0 0; B A^* A^* B; 0 1/2 1/2 0
30	(D_2, BB)	a b c d; β β γ γ β; a b c d; α α δ δ α; a b c d	D_2, D_2, D_2; 0 0 0 0 0 0 0 0; B B B B; 0 0

Table 9.1. Saddle–saddle singularities of complexity 2 (continued)

No	l-TYPE	SINGULAR LEAF L	LOOP MOLECULE
31	(D_2, BB)		
32	(D_2, BB)		
33	(D_2, BB)		
34	(D_2, BB)		
35	(C_1, C_2)		

Table 9.1.   Saddle–saddle singularities of complexity 2 (continued)

No	l-TYPE	SINGULAR LEAF L	LOOP MOLECULE
36	(C_1, C_2)	$\begin{array}{cccc} a & b & c & d \\ \alpha \;\gamma\; & \alpha\; \gamma\; & \alpha \\ d & a & b & c \\ \beta\;\delta\; & \beta\;\delta\; & \beta \\ a & b & c & d \end{array}$	$\dfrac{1}{4}\ \overset{A^*}{\frown}\ \dfrac{3}{4}$; A^* ; $\dfrac{3}{4}\quad\dfrac{1}{4}$; $B\quad B$; $0\ \underset{C_1}{\smile}\ 0$
37	(D_1, C_2)	$\begin{array}{cccc} c & d & a & b \\ \alpha\; & \alpha\;\gamma\; & \gamma\; & \alpha \\ a & b & c & d \\ \delta\; & \delta\;\beta\; & \beta\; & \delta \\ c & d & a & b \end{array}$	B ; $\dfrac{1}{2}\ \ \dfrac{1}{2}\ \ 0$; B ; $\dfrac{1}{2}\ \ \dfrac{1}{2}\ \ 0$; $B\quad B\ \ C_2$; $0\ \underset{D_1}{\smile}\ 0\ \ 0$
38	(D_1, C_2)	$\begin{array}{cccc} c & d & a & b \\ \alpha\;\beta\; & \delta\;\gamma\; & \alpha \\ a & b & c & d \\ \gamma\;\delta\; & \beta\;\alpha\; & \gamma \\ c & d & a & b \end{array}$	$\dfrac{1}{2}\ \overset{C_2}{\frown}\ 0$; $\dfrac{1}{2}$; $B\quad B\ \ A^{**}$; $\dfrac{1}{2}\ \ \dfrac{1}{2}\ \ 0$; $0\ \underset{D_1}{\smile}$
39	(C_1, C_1)	$\begin{array}{cccc} c & a & d & b \\ \alpha\;\gamma\; & \beta\;\delta\; & \alpha \\ a & d & c & b \\ \delta\;\alpha\; & \gamma\;\beta\; & \delta \\ d & b & c & a \end{array}$	$\dfrac{1}{2}\ \overset{B}{\frown}\ \dfrac{3}{4}$; $\dfrac{3}{4}$; $B\quad B$; $\dfrac{1}{4}$; $\dfrac{1}{4}\ \underset{B}{\smile}\ \dfrac{1}{2}$

The list of atoms from Table 9.1

A^*

A^{**}

A^{****}

B

$\mathrm{Sym} = \mathbb{Z}_2 = \langle\alpha\rangle$

B_1^{**}

$\mathrm{Sym} = \mathbb{Z}_2 = \langle\alpha\rangle$

The list of atoms from Table 9.1 (continued)

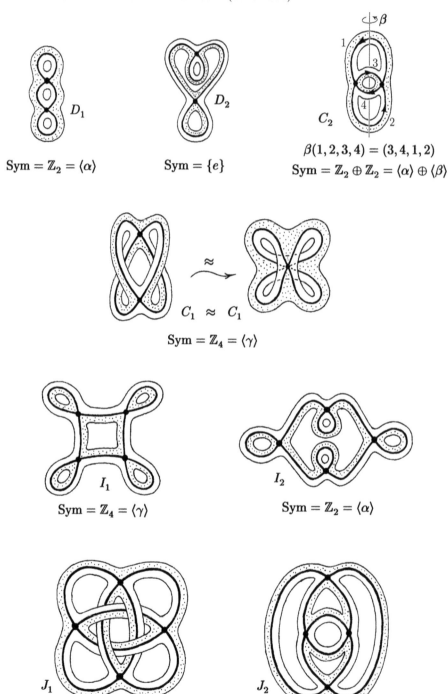

$$D_1$$
$$\text{Sym} = \mathbb{Z}_2 = \langle \alpha \rangle$$

$$D_2$$
$$\text{Sym} = \{e\}$$

$$C_2$$
$$\beta(1,2,3,4) = (3,4,1,2)$$
$$\text{Sym} = \mathbb{Z}_2 \oplus \mathbb{Z}_2 = \langle \alpha \rangle \oplus \langle \beta \rangle$$

$$C_1 \approx C_1$$
$$\text{Sym} = \mathbb{Z}_4 = \langle \gamma \rangle$$

$$I_1$$
$$\text{Sym} = \mathbb{Z}_4 = \langle \gamma \rangle$$

$$I_2$$
$$\text{Sym} = \mathbb{Z}_2 = \langle \alpha \rangle$$

$$J_1$$
$$\text{Sym} = \mathbb{Z}_4 = \langle \gamma \rangle$$

$$J_2$$
$$\text{Sym} = \mathbb{Z}_2 = \langle \alpha \rangle$$

The list of atoms from Table 9.1 (continued)

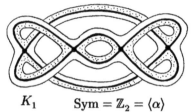

K_1 Sym $= \mathbb{Z}_2 = \langle \alpha \rangle$

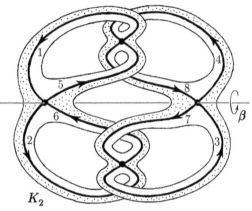

K_2

Sym $= \mathbb{Z}_2 \oplus \mathbb{Z}_2 = \langle \alpha \rangle \oplus \langle \beta \rangle$

$\beta(1, 2, 3, 4, 5, 6, 7, 8) = (6, 5, 8, 7, 2, 1, 4, 3)$

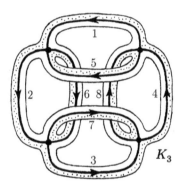

K_3

Sym $= \mathbb{Z}_4 \oplus \mathbb{Z}_2 = \langle \gamma \rangle \oplus \langle \beta \rangle$

$\beta(1, 2, 3, 4, 5, 6, 7, 8) =$
$(7, 8, 5, 6, 3, 4, 1, 2)$

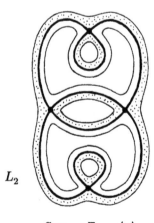

L_2

Sym $= \mathbb{Z}_2 = \langle \alpha \rangle$

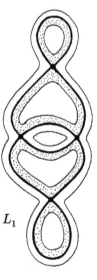

L_1

Sym $= \mathbb{Z}_2 = \langle \alpha \rangle$

The list of atoms from Table 9.1   (continued)

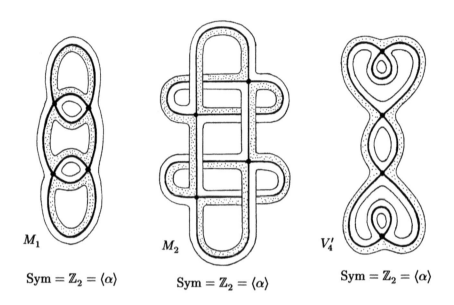

M_1

$\text{Sym} = \mathbb{Z}_2 = \langle \alpha \rangle$

M_2

$\text{Sym} = \mathbb{Z}_2 = \langle \alpha \rangle$

V_4'

$\text{Sym} = \mathbb{Z}_2 = \langle \alpha \rangle$

V_4 $\text{Sym} = \mathbb{Z}_2 = \langle \alpha \rangle$

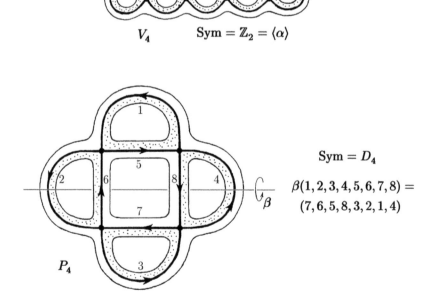

P_4

$\text{Sym} = D_4$

$\beta(1,2,3,4,5,6,7,8) =$
$(7,6,5,8,3,2,1,4)$

9.6. ALMOST DIRECT PRODUCT REPRESENTATION OF A FOUR-DIMENSIONAL SINGULARITY

In Section 9.5, we have obtained the classification of saddle–saddle singularities in terms of their Cl-types. Despite the effectiveness of this description, it is rather unwieldy. We now study the same problem from another point of view. Let us try to imagine what structure these singularities can have. The simplest way to produce a four-dimensional saddle–saddle singularity is just to take a direct product of two two-dimensional singularities, i.e., atoms. Consider the simplest example.

Take two atoms B represented as two-dimensional oriented surfaces P_1, P_2 with Morse functions f_1, f_2. The singular level line of f_i on each atom P_i is the figure eight curve, i.e., the simplest saddle singularity. Let us define a symplectic structure ω_i on P_i and consider the direct product $P_1 \times P_2$ with the natural symplectic structure that is the sum $\omega_1 + \omega_2$. The functions f_1 and f_2 extended on $P_1 \times P_2$ obviously commute with respect to this structure and determine the Liouville foliation on $P_1 \times P_2$. Clearly, this foliation has exactly one singularity of saddle–saddle type. Its singular leaf L is the direct product of two figure eight curves. This singularity is indicated in Theorem 9.5 as the first one.

It is clear that just in the same way we can multiply any saddle atoms V_1 and V_2 to produce more new examples of saddle–saddle singularities.

Let us consider another example. Taking the direct product $C_2 \times B$ of the atoms C_2 and B, we obtain a four-dimensional symplectic manifold (the symplectic structure is defined to be the sum of the symplectic structures on the factors). Consider the central symmetries on the atoms C_2 and B and denote them by τ_1 and τ_2 respectively. It is clear that the involution τ_1 has no fixed points on C_2. We now define an involution τ on $C_2 \times B$ by the formula $\tau(x, y) = (\tau_1(x), \tau_2(y))$. Evidently, it acts freely on $C_2 \times B$, i.e., has no fixed points (Fig. 9.39). Moreover,

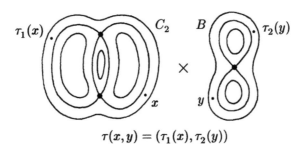

$$\tau(x, y) = (\tau_1(x), \tau_2(y))$$

Figure 9.39

this involution preserves the symplectic structure on $C_2 \times B$. Therefore, we may take the quotient of $C_2 \times B$ with respect to this action of the group $\mathbb{Z}_2$. As a result, we obtain a four-dimensional manifold $(C_2 \times B)/\mathbb{Z}_2$ with an integrable system with a saddle–saddle singularity. Having analyzed the topology of the singular leaf obtained, we see that it corresponds to the second case in Theorem 9.5.

It is easy to see that the third and fourth singularity from Theorem 9.5 can also be included in this construction.

It turns out that this is a general fact: every saddle–saddle singularity can be obtained in two steps: (direct product) + (taking quotient with respect to a free action of a finite group). We now describe this general construction more precisely.

Let V_1 and V_2 be saddle atoms with symplectic structures ω_1 and ω_2 and Morse functions f_1 and f_2 respectively. Suppose that a finite group G acts symplectically on both atoms and, moreover, this action preserves the functions f_1 and f_2. Then we can define the symplectic structure on the direct product $V_1 \times V_2$ to be the sum $\omega_1 + \omega_2$. We also have the structure of the Liouville foliation on $V_1 \times V_2$ given by commuting functions f_1 and f_2, as well as the action of G given by the formula $\varphi(g)(x_1, x_2) = (\varphi_1(g)(x_1), \varphi_2(g)(x_2))$, where φ_i is the action of G on the atom V_i. The action φ is obviously symplectic and preserves the structure of the Liouville foliation. If φ is free, then we can consider the quotient manifold $(V_1 \times V_2)/G$, which is obviously symplectic and has the natural structure of the Liouville foliation induced from $V_1 \times V_2$. Moreover, $(V_1 \times V_2)/G$ is a regular neighborhood of a singular leaf L of saddle–saddle type.

Definition 9.6. The above described four-dimensional saddle–saddle singularity is called a *singularity of almost direct product type*.

Theorem 9.9 (Nguyen Tien Zung [258], [261]). *Every four-dimensional saddle–saddle singularity is a singularity of almost direct product type.*

Proof. In fact, we need to prove that every saddle–saddle singularity admits a finite-sheeted covering which is diffeomorphic to a direct product of two atoms. And, moreover, the group associated with this covering acts freely and component-wise on this direct product.

Consider the universal covering $\pi \colon \widetilde{U} \to U(L)$ over the neighborhood $U(L)$ of the singular leaf L. Clearly, U is a symplectic 4-manifold on which we have two commuting functions $\widetilde{f}_1 = \pi \circ f_1$ and $\widetilde{f}_2 \circ f_2$, where f_1 and f_2 are commuting integrals of the original integrable system on $U(L)$. As a result, we obtain the (induced) Liouville foliation on $\widetilde{U}$. Its leaves will be, however, non-compact unlike the initial foliation on U. Notice that $\widetilde{U}$ can be represented as a four-dimensional regular neighborhood $\widetilde{U}(\widetilde{L})$ of the (non-compact) singular leaf $\widetilde{L}$ given by the equations $\widetilde{f}_1 = 0$, $\widetilde{f}_2 = 0$.

Lemma 9.6.

1) *For any saddle–saddle singularity $U(L)$, the universal covering manifold $\widetilde{U}(\widetilde{L})$ is fiberwise diffeomorphic to the universal covering manifold over the direct product $B \times B$ of two simplest atoms B.*

2) *The manifold $\widetilde{U}(\widetilde{L})$ is diffeomorphic to the direct product $\widetilde{B} \times \widetilde{B}$ of two copies of the universal covering manifold $\widetilde{B}$ over the atom B presented in Fig. 9.40. The manifold $\widetilde{B}$ is a regular two-dimensional neighborhood of the infinite tree all of whose vertices have degree 4.*

Proof. To make the proof clearer, we begin with the two-dimensional case (i.e., with the case of one degree of freedom). Consider an atom V and construct the universal covering over it. We claim that for all the atoms the universal covering space is actually the same (up to a fiber diffeomorphism) and has the form presented in Fig. 9.40. To prove this, we cut the atom V into two-dimensional crosses as shown in Fig. 9.41.

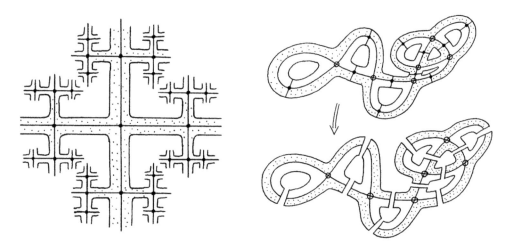

| Figure 9.40 | Figure 9.41 |

The universal covering space $\widetilde{B}$ shown in Fig. 9.40 can also be naturally decomposed into similar crosses. To construct the desired covering $\widetilde{B} \to V$, we take an arbitrary cross from $\widetilde{B}$ and map it homeomorphically onto one of the crosses of V. In addition, we require the orientation and the values of the integral f to be preserved. After this, we extend this mapping to neighboring crosses in $\widetilde{B}$ by mapping them onto the corresponding crosses in V. It is clear that such an extension is always possible and defined uniquely up to fiber isotopy. Since $\widetilde{B}$ does not contain cycles, by continuing this process, we obtain a fiber diffeomorphism $\widetilde{B} \to V$, which is evidently the desired universal covering.

The similar construction works in the four-dimensional case (and even in the multidimensional one). For this, we must consider all the squares which the singular leaf L consists of and cut each of them into four smaller squares, as shown in Fig. 9.42. After this, we need to extend these cuts from the complex L to the whole 4-neighborhood $U(L)$ by analogy with the procedure we have carried out in the two-dimensional case. More formally, we must do the following. Let $S \subset L$ denote the one-dimensional subset along which we cut the singular leaf L. For each point $x \in S$, we consider integral curves of the vector fields

$$\lambda \operatorname{grad} f_1 + \mu \operatorname{grad} f_2$$

for all $\lambda, \mu \in \mathbb{R}$. Taking the union of such curves, we locally obtain a smooth two-dimensional surface which is transversal to the square that contains the point x. As a result, we obtain a three-dimensional cut-surface which is smooth except, perhaps, for the points x lying on the 1-skeleton of L. However, at these points, the cut-surface is also smooth as we shall see below. At the center of each square, four smooth three-dimensional cut-surfaces meet.

Let us now look at the behavior of the cut-surface at those points $x \in S$ which belong to the 1-skeleton of L (i.e., at the midpoints of the edges of $K_1 \cup K_2$). Take one of such points and denote it by P. Since each edge of the 1-skeleton is a fourfold line, four cut-lines coming from the four neighboring squares meet at this point.

These lines are indicated as dotted lines in Fig. 9.42. As we know, in a neighborhood of P, there exists a coordinate system (x, y, z, t) in terms of which the functions f_1 and f_2 take the form

$$f_1 = x^2 - y^2, \qquad f_2 = z,$$

or, conversely,

$$f_2 = x^2 - y^2, \qquad f_1 = z.$$

Now it is easy to see that, choosing a suitable Riemannian metric, we can achieve that the three-dimensional cut-surface is smooth at the point P and, moreover, has the simple form $\{t = t_0 = \mathrm{const}\}$.

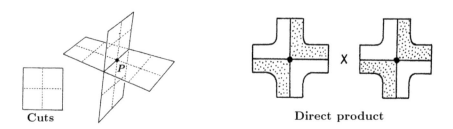

Cuts   Direct product

Figure 9.42   **Figure 9.43**

Furthermore, it is easy to see that, cutting $U(L)$ along these cut-surfaces, we turn $U(L)$ into a disjoint union of a number of four-dimensional elementary bricks each of which is diffeomorphic to the direct product of a cross by a cross (Fig. 9.43). Moreover, on this direct product, we have the natural structure of the two-dimensional foliation given by the functions f_1 and f_2.

Notice that the structure of each elementary bricks is standard. To see this, it suffices to apply the generalized Morse–Bott lemma (Lemma 1.7) to the functions f_1 and f_2 on each brick separately. According to this lemma, in a neighborhood of P_i (more precisely, of that part of P_i which belongs to the given brick), there exist regular coordinates x_i, y_i in terms of which the function f_i takes the form $f_i = x_i y_i$. These functions x_i, y_i can naturally be extended to the whole brick. As a result, the leaves of the foliation will be given by the simple equations $x_1 y_1 = \mathrm{const}$ and $x_2 y_2 = \mathrm{const}$. Thus, on each brick, there are coordinates x_1, y_1, x_2, y_2 in terms of which the initial foliations becomes the standard two-dimensional foliation. In this sense, all the bricks have the same structure.

We now construct the four-dimensional space $\widetilde{U}(\widetilde{L})$ that covers the manifold $U(L)$. We actually wish to show that $\widetilde{U}(\widetilde{L})$ is fiberwise diffeomorphic to the direct product $\widetilde{B} \times \widetilde{B}$.

Indeed, this direct product can also be decomposed into the same elementary bricks. On $\widetilde{B} \times \widetilde{B}$, also there are two functions $\widetilde{f}_1, \widetilde{f}_2$ which determine the structure of a Liouville foliation. Let us now construct a fiber mapping

$$\widetilde{B} \times \widetilde{B} \to U(L).$$

To this end, we take an arbitrary elementary brick from $\widetilde{B} \times \widetilde{B}$ and an arbitrary elementary brick from $U(L)$. Then we map the brick from $\widetilde{B} \times \widetilde{B}$ onto the brick from $U(L)$ by means of a fiber diffeomorphism which transforms the functions $\widetilde{f}_1, \widetilde{f}_2$ into the functions f_1, f_2 and preserves orientation. Now, by analogy with the two-dimensional case, this mapping can easily be extended on all the neighboring bricks (uniquely up to fiber isotopy). By extending this mapping to "all directions", we obtain the desired diffeomorphism $\widetilde{B} \times \widetilde{B} \to U(L)$, which is a universal covering. Moreover, it preserves orientation and the Liouville foliation structure. The point is that we require on each step that the functions $\widetilde{f}_1, \widetilde{f}_2$ transform into functions f_1, f_2. Thus, the universal covering space is the same for all saddle–saddle singularities and diffeomorphic to the direct product $\widetilde{B} \times \widetilde{B}$. Lemma 9.6 is proved. $\quad\square$

Now the mechanism of the appearance of the almost direct product structure is getting clearer. We see that the universal covering over $U(L)$ is the direct product $\widetilde{B} \times \widetilde{B}$. To come back to $U(L)$, we have to factorize this direct product by the action of the fundamental group $\pi_1(U(L))$. As we shall see soon, this factorization can be divided into two steps. First we factorize $\widetilde{B} \times \widetilde{B}$ by some subgroup $H \subset \pi_1(U(L))$ in order to obtain a direct product of two *compact* atoms. After this, we complete the factorization by using the action of the *finite* group $\pi_1(U(L))/H$. This gives the statement of Theorem 9.9.

Consider the universal covering $\widetilde{B} \times \widetilde{B}$ over $U(L)$. This space is simply connected, and the fundamental group $Y = \pi_1(U(L))$ acts naturally on it by translations. It is easy to see that this action is symplectic, i.e., preserves the natural symplectic structure on $\widetilde{B} \times \widetilde{B}$. It also preserves the values of both the integrals $\widetilde{f}_1, \widetilde{f}_2$ and, consequently, maps leaves of the Liouville foliation into leaves. In other words, elements of Y are in fact automorphisms of the Liouville foliation on $\widetilde{B} \times \widetilde{B}$. Consider the total group of automorphisms of this foliation and its quotient with respect to the isotopy subgroup of the foliation. In other words, we shall consider automorphisms up to fiber isotopy. As a result, we obtain a discrete group. This group is obviously isomorphic to the group

$$\mathrm{Aut} \times \mathrm{Aut},$$

where Aut is the discrete group of fiber automorphisms of the space $\widetilde{B}$, i.e., one of the factors in the direct product $\widetilde{B} \times \widetilde{B}$.

It is easy to see that Aut is isomorphic to the free product $\mathbb{Z} * \mathbb{Z}_2$. Let us describe explicitly the action of the generators $a \in \mathbb{Z}$ and $s \in \mathbb{Z}_2$ on $\widetilde{B}$.

The transformation $s: \widetilde{B} \to \widetilde{B}$ is generated by the symmetry of the two-dimensional cross shown in Fig. 9.44. In other words, s is the central symmetry of the space $\widetilde{B}$ with respect to one of the vertices (this symmetry is generated by the central symmetry of the atom B). Clearly, s is an involution.

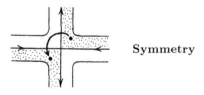

Symmetry

Figure 9.44

The transformation $a : \widetilde{B} \to \widetilde{B}$ is defined as an elementary translation of $\widetilde{B}$ along itself given by Figure 9.45. In other words, we take one of the two natural generators of the fundamental group $\pi_1(B)$ and consider the induced translation on the universal covering $\widetilde{B}$. In Fig. 9.45 we demonstrate the action

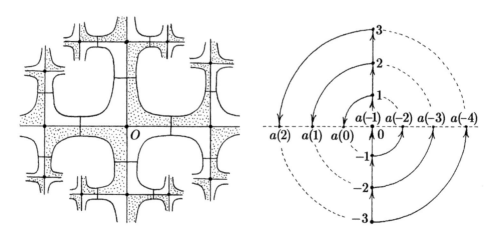

Figure 9.45

of this transformation a. It can be described as follows. We choose an arbitrary vertex O on the tree $\widetilde{B}$ and call it the *center of the tree*. Then we consider the unit shift of $\widetilde{B}$ upward and then rotate the whole tree (as a rigid body) about the point O through the angle $\pi/2$. The result is the transformation a.

Thus, we have described the action of two generators a and s of the group Aut on $\widetilde{B}$. It is clear that the direct product Aut $\times$ Aut acts naturally on the space $\widetilde{B} \times \widetilde{B}$. Moreover, the fundamental group Y can be considered as a subgroup in this group.

Consider the intersection of Y with the factors Aut $\times \{e\}$ and $\{e\} \times$ Aut in the product Aut $\times$ Aut. Set

$$Y_1 = Y \cap (\text{Aut} \times \{e\}), \qquad Y_2 = Y \cap (\{e\} \times \text{Aut}).$$

It is easy to see that each of the subgroups Y_1 and Y_2 is a normal subgroup in Y. Moreover, these subgroups have finite index in Aut $\times \{e\}$ and $\{e\} \times$ Aut respectively.

Now consider $Y_1 \times Y_2$ as a subgroup in Y. Since $Y_1 \times Y_2$ belongs to the group of automorphisms of the covering space $\widetilde{B} \times \widetilde{B}$, we can first take the quotient of $\widetilde{B} \times \widetilde{B}$ by this subgroup. We claim that as a result we obtain a *compact* manifold.

Indeed, the group Y acts freely on $\widetilde{B} \times \widetilde{B}$, since Y is the fundamental group of the base $U(L)$. Therefore, the subgroup $Y_1 \times Y_2$ also acts freely on the covering space and, consequently, the quotient space $(\widetilde{B} \times \widetilde{B})/(Y_1 \times Y_2)$ is a smooth manifold.

Furthermore, since each of the factors Y_1 and Y_2 has finite index in its own group Aut, the subgroup $Y_1 \times Y_2$ has finite index in Aut $\times$ Aut and, consequently, in Y. Hence, the quotient space $(\widetilde{B} \times \widetilde{B})/(Y_1 \times Y_2)$ is compact.

Besides, this space is evidently the direct product of the two-dimensional manifolds

$$V_1 = \tilde{B}/Y_1 \quad \text{and} \quad V_2 = \tilde{B}/Y_2,$$

i.e., has the structure of the direct product of some compact atoms V_1 and V_2.

It remains to notice that Y_1 and Y_2 are both normal subgroups in Y. Therefore, the product $Y_1 \times Y_2$ is also normal in Y. Hence, the factor group $G = Y/(Y_1 \times Y_2)$ is well-defined. Thus, we have

$$V_1 \times V_2 = (\tilde{B} \times \tilde{B})/(Y_1 \times Y_2),$$

but, on the other hand,

$$U(L) = (\tilde{B} \times \tilde{B})/Y.$$

Therefore, $U(L) = (V_1 \times V_2)/G$, where G is a finite group, since $Y_1 \times Y_2$ have finite index in Y. This completes the proof of Theorem 9.9. □

For saddle–saddle singularities of complexity 1 and 2 (all of them are listed in Theorems 9.5 and 9.7), one can produce another representation following Nguyen Tien Zung's theorem (Theorem 9.9), namely, the almost direct product representation.

As an example, we consider the structure of a saddle–saddle singularity in the case when the singular leaf contains only one critical saddle–saddle point. It turns out that, in this case, we have the following four possibilities (corresponding to the four singularities described in Theorem 9.5):

1) the direct product $B \times B$;
2) $(B \times C_2)/\mathbb{Z}_2$, where the group $\mathbb{Z}_2$ acts on each factor as the central symmetry;
3) $(B \times D_1)/\mathbb{Z}_2$, where the group $\mathbb{Z}_2$ acts on each factor as the central symmetry;
4) $(C_2 \times C_2)/(\mathbb{Z}_2 \times \mathbb{Z}_2)$, where the two generators α and β of the group $\mathbb{Z}_2 \times \mathbb{Z}_2$ act in the following way. Consider the model of the atom C_2 shown in Fig. 9.46. Here α acts on the first factor as the symmetry relative to the axis Ox and acts on the second factor as the symmetry relative to Oz. The second generator β, on the contrary, acts on the first factor as the symmetry relative to Oz and on the second factor as the symmetry relative to Ox.

Here we use the standard representations and notation for atoms (see Table 2.1).

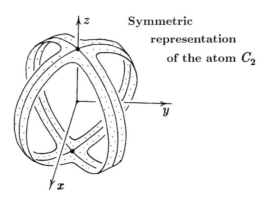

Figure 9.46

It is an interesting fact that the first three singularities from the above list occur in integrable problems of classical mechanics (the first and third occur in the Kovalevskaya case, the second occurs in the Goryachev–Chaplygin–Sretenskiĭ case). We do not know whether the fourth possibility appears in known examples of integrable cases in physics and mechanics. In any case, it does not appear in classical integrable cases in rigid body dynamics, integrable geodesic flows on two-dimensional surfaces, and a number of other examples, for which the topological invariants have been calculated (see Chapters 12, 14).

For all saddle–saddle singularities of complexity 2 (i.e., for 39 cases listed in Theorem 9.7), the almost direct product representation have been described by V. V. Korneev. This result is presented in Table 9.2.

Comments to Table 9.2.

In the second column of Table 9.2, we indicate the numbers that are assigned to the same singularities in Table 9.1. The numbers in Table 9.1 and Table 9.2 turn out to be different because of using two essentially different approaches: in Table 9.1 the singularities are ordered according to their l-types, whereas in Table 9.2 they are ordered according to the types of the groups G acting on direct products.

In the third column of Table 9.2, we indicate the atoms that are the factors in almost direct products. These atoms are presented in the list after Table 9.1. We also indicate there the symmetry groups of the atoms and their generators. Moreover, the atoms themselves are presented in a symmetric form in order to make their symmetry groups more visual. In all the cases, α denotes the central symmetry (i.e., the symmetry relative to the center of an atom). By γ we denote the rotation of an atom through the angle $\pi/2$ about the same center. Finally, β denotes an additional symmetry of an atom which has a different meaning for different atoms. Each time we indicate the action of β explicitly by showing how β acts on the edges of the graph K. From this information, the action of β on the whole atom is uniquely reconstructed.

In the last column of Table 9.2, we indicate the group G acting on the direct product of atoms. All of these groups are commutative with the only exception. In the last case, G is the dihedral group D_4.

In each case, the group G has at most two generators. The component-wise action of these generators is described in the fourth column. For example, in case 32, the first generator e_1 of the group $G = \mathbb{Z}_2 \oplus \mathbb{Z}_2$ acts on the direct product $C_2 \times P_4$ as follows:

$$e_1(C_2 \times P_4) = \alpha(C_2) \times \gamma^2(P_4).$$

This means that, on the first component (i.e., on C_2), the generator e_1 acts as the symmetry α, and, on the second component (i.e., on P_4), it acts as the symmetry γ^2. In this case, α is the central symmetry of C_2, and γ is the rotation of P_4 through the angle $\pi/2$ (in particular, γ^2 is also the central symmetry). Similarly, the second generator e_2 of the group $G = \mathbb{Z}_2 \oplus \mathbb{Z}_2$ acts according to the rule

$$e_1(C_2 \times P_4) = \beta(C_2) \times \beta(P_4).$$

Table 9.2. Singularities of complexity 2 as almost direct products

No	NUMBER FROM TABLE 9.1	FACTORS	ACTION	GROUP
1	4	$B \times C_1$	trivial	$\{e\}$
2	11	$B \times C_2$	trivial	$\{e\}$
3	22	$B \times D_1$	trivial	$\{e\}$
4	30	$B \times D_2$	trivial	$\{e\}$
5	3	$C_1 \times D_1$	$(\gamma^2,\ \alpha)$	$\mathbb{Z}_2$
6	15	$D_1 \times D_1$	$(\alpha,\ \alpha)$	$\mathbb{Z}_2$
7	24	$B \times I_1$	$(\alpha,\ \gamma^2)$	$\mathbb{Z}_2$
8	32	$B \times I_2$	$(\alpha,\ \alpha)$	$\mathbb{Z}_2$
9	31	$B \times K_1$	$(\alpha,\ \alpha)$	$\mathbb{Z}_2$
10	34	$B \times L_1$	$(\alpha,\ \alpha)$	$\mathbb{Z}_2$
11	26	$B \times L_2$	$(\alpha,\ \alpha)$	$\mathbb{Z}_2$
12	25	$B \times J_1$	$(\alpha,\ \gamma^2)$	$\mathbb{Z}_2$
13	33	$B \times J_2$	$(\alpha,\ \alpha)$	$\mathbb{Z}_2$
14	13	$B \times M_1$	$(\alpha,\ \alpha)$	$\mathbb{Z}_2$
15	6	$B \times M_2$	$(\alpha,\ \alpha)$	$\mathbb{Z}_2$
16	21	$B \times V_4$	$(\alpha,\ \alpha)$	$\mathbb{Z}_2$
17	29	$B \times V_4'$	$(\alpha,\ \alpha)$	$\mathbb{Z}_2$
18	1	$C_1 \times C_2$	$(\gamma^2,\ \alpha)$	$\mathbb{Z}_2$
19	17	$C_2 \times C_2$	$(\alpha,\ \alpha)$	$\mathbb{Z}_2$
20	8	$C_2 \times C_2$	$(\alpha,\ \alpha\beta)$	$\mathbb{Z}_2$
21	37	$D_1 \times C_2$	$(\alpha,\ \alpha)$	$\mathbb{Z}_2$
22	10	$D_1 \times C_2$	$(\alpha,\ \alpha\beta)$	$\mathbb{Z}_2$
23	5	$B \times K_2$	$(\alpha,\ \alpha)$	$\mathbb{Z}_2$
24	14	$B \times K_3$	$(\alpha,\ \gamma^2)$	$\mathbb{Z}_2$
25	7	$B \times K_3$	$(\alpha,\ \beta)$	$\mathbb{Z}_2$
26	12	$B \times P_4$	$(\alpha,\ \gamma^2)$	$\mathbb{Z}_2$
27	23	$B \times P_4$	$(\alpha,\ \beta)$	$\mathbb{Z}_2$
28	27	$C_1 \times I_1$	$(\gamma,\ \gamma)$	$\mathbb{Z}_4$
29	35	$C_1 \times K_3$	$(\gamma,\ \gamma)$	$\mathbb{Z}_4$
30	36	$C_1 \times P_4$	$(\gamma,\ \gamma)$	$\mathbb{Z}_4$
31	28	$C_1 \times J_1$	$(\gamma,\ \gamma)$	$\mathbb{Z}_4$
32	38	$C_2 \times P_4$	$(\alpha,\ \gamma^2),\ (\beta,\ \beta)$	$\mathbb{Z}_2 \oplus \mathbb{Z}_2$
33	18	$C_2 \times P_4$	$(\alpha,\ \beta),\ (\beta,\ \beta\gamma^2)$	$\mathbb{Z}_2 \oplus \mathbb{Z}_2$
34	20	$C_2 \times P_4$	$(\alpha,\ \gamma^3\beta),\ (\beta,\ \beta\gamma^3)$	$\mathbb{Z}_2 \oplus \mathbb{Z}_2$
35	19	$C_2 \times K_2$	$(\alpha,\ \beta),\ (\beta,\ \alpha\beta)$	$\mathbb{Z}_2 \oplus \mathbb{Z}_2$
36	9	$C_2 \times K_3$	$(\alpha,\ \beta\gamma^2),\ (\beta,\ \beta)$	$\mathbb{Z}_2 \oplus \mathbb{Z}_2$
37	2	$C_2 \times K_3$	$(\alpha,\ \beta\gamma^2),\ (\beta,\ \gamma^2)$	$\mathbb{Z}_2 \oplus \mathbb{Z}_2$
38	39	$K_3 \times K_3$	$(\gamma,\ \gamma),\ (\beta,\ \beta\gamma^2)$	$\mathbb{Z}_4 \oplus \mathbb{Z}_2$
39	16	$P_4 \times P_4$	$(\gamma,\ \gamma),\ (\gamma^3\beta,\ \beta)$	D_4

9.7. PROOF OF THE CLASSIFICATION THEOREMS

9.7.1. Proof of Theorem 9.3

Consider the singular leaf L of saddle–saddle type and its neighborhood $U(L)$. As we have seen above, this neighborhood can be represented as the result of gluing several standard elementary four-dimensional bricks each of which is the direct product (cross) × (cross). The number of such bricks is exactly the number of singular saddle–saddle points on the leaf L. How $U(L)$ is glued from these bricks? To describe the gluing operation, it suffices to determine the pairs of three-dimensional boundary components of bricks which should be glued together. It is easy to see that each such component is the direct product of the two-dimensional cross by a segment. Moreover, for each pair of boundary components, we need to determine a rule for gluing. If we know the combinatorial structure of the complex L, then we evidently know the structure of decomposition of boundary components into pairs.

It remains to show that each gluing homeomorphism is defined uniquely up to fiber isotopy, provided the Cl-type is fixed. Each boundary component is the direct product (cross) × (interval). Recall that we also have two functions f_1 and f_2 on this component. Moreover, due to the direct product structure, we may assume without loss of generality that f_1 is a function on the cross with one saddle singularity at the center, and f_2 is a function on the interval without singularities. Since we require that the values of f_1 and f_2 coming from different components are the same for points to be glued, we may define the gluing homeomorphism separately on each factor (i.e., on the cross and on the interval). The mapping between two crosses is uniquely determined by the Cl-type. Indeed, each individual gluing operation corresponds to some edge α of the complex L. As we know, α is a fourfold line (Fig. 9.47). The endpoints S_i and S_j of the edge α are exactly the centers of two elementary bricks which must be glued. It is clear that the gluing mapping between

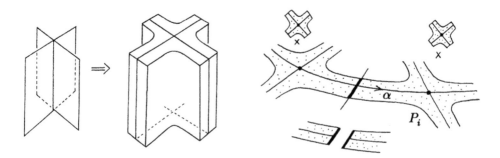

Figure 9.47

the crosses is uniquely defined by the fourfold line α (Fig. 9.47). The mapping between the intervals is also uniquely defined, since on each interval we have the (regular) function f_2 and we only need to identify points with the same values of f_2 (Fig. 9.48).

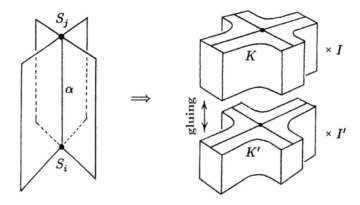

Figure 9.48

Thus, if we know the Cl-type of a saddle–saddle singularity, we can uniquely determine the rule which allows us to glue it from elementary bricks.

The same result can be obtained by using another argument. Consider the universal covering space $\widetilde{U}(\widetilde{L})$ of the singularity $U(L)$. If the Cl-type is fixed, then we know the fundamental group $Y = \pi_1(U(L))$, since $\pi_1(U(L)) = \pi_1(L)$. Then it is not hard to show that the action of $Y = \pi_1 U(L)$ on $\widetilde{U}(\widetilde{L})$ can be uniquely reconstructed from the Cl-type. Since the covering space is the same for all saddle–saddle singularities and we can reconstruct the action of the fundamental group on it, it follows that the manifold $U(L)$ itself (together with the Liouville foliation on it) can also be uniquely reconstructed just by setting $U(L) = (\widetilde{U}(\widetilde{L}))/Y = (\widetilde{B} \times \widetilde{B})/Y$. □

9.7.2. Proof of Theorem 9.4 (Realization Theorem)

Suppose we have an arbitrary admissible Cl-type. We need to realize it as the Cl-type of some saddle–saddle singularity. To this end, it suffices to indicate a subgroup Y in the group Aut × Aut such that the quotient space $(\widetilde{B} \times \widetilde{B})/Y$ has the prescribed Cl-type. Take the complex L and its universal covering space $\widetilde{L}$. We claim that this space is the same for all admissible complexes L. To prove this, it suffices to construct $\widetilde{L}$ explicitly. As $\widetilde{L}$ we take the singular leaf in the standard space $\widetilde{B} \times \widetilde{B}$. This leaf is the direct product of two infinite trees each of which is homeomorphic to the universal covering space for the figure eight curve. It is easy to see that $\widetilde{L}$ satisfies all properties of an admissible complex from Definition 9.5 (except for the fact that $\widetilde{L}$ is infinite). Moreover, the projection $\widetilde{L} \to L$ that preserves "admissibility conditions" is also reconstructed uniquely in the combinatorial sense. The point is that, if we define arbitrarily the image of a certain square, then the projection can be uniquely extended "in all directions".

Now let us notice that the fundamental group $\pi_1(L)$ acts naturally on the universal covering space $\widetilde{L}$ as the group of translations of the covering $\widetilde{L} \to L$.

On the other hand, since $\widetilde{L}$ is embedded into $\widetilde{B} \times \widetilde{B}$, it follows that on $\widetilde{L}$ we have an action of the group $\text{Aut} \times \text{Aut}$. Therefore, the action of $\pi_1(L)$ on $\widetilde{L}$ by translations determine the monomorphism $\pi_1(L) \to \text{Aut} \times \text{Aut}$. Thus, we obtain a subgroup in $\text{Aut} \times \text{Aut}$ corresponding to the given Cl-type. We now take the space $\widetilde{U}(\widetilde{L}) = \widetilde{B} \times \widetilde{B}$, whose 2-skeleton is $\widetilde{L}$, and consider the action of $\pi_1(L)$ on it, where $\pi_1(L)$ is already realized as a subgroup in the total group of automorphisms $\text{Aut} \times \text{Aut}$. Taking the quotient $\widetilde{U}(\widetilde{L}) = \widetilde{B} \times \widetilde{B}$ by this action, we obtain some compact four-dimensional manifold $U(L)$ which gives us the desired realization of the chosen Cl-type.   □

9.8. FOCUS–FOCUS CASE

9.8.1. *The Structure of a Singular Leaf of Focus–Focus Type*

Let L be a singular leaf of a Liouville foliation which contains one or several focus–focus points. We denote these points by $x_1, \ldots, x_n$. Recall that, according to Theorem 1.5, in a neighborhood of the point x_i, there exists a canonical coordinate system p_1, q_1, p_2, q_2 in terms of which the Hamiltonian H and the additional integral f can be written as

$$H = H(f_1, f_2),$$
$$f = f(f_1, f_2),$$

where $f_1 = p_1 q_1 + p_2 q_2$ and $f_2 = p_1 q_2 - p_2 q_1$. Note that the change

$$(H, f) \to (f_1, f_2)$$

is regular and, therefore, the functions f_1, f_2 can be locally expressed as smooth functions of H and f. In particular, the foliation given by H and f coincides with the one given by the simpler functions f_1 and f_2.

To understand the local structure of the foliation in a neighborhood of a focus–focus point, it is convenient to pass to complex variables

$$z = q_1 + i q_2,$$
$$w = p_1 - i p_2.$$

Then the functions f_1 and f_2 can be represented as the real and imaginary part of one complex function $F = zw$. In particular, it follows from this that locally the singular leaf L is represented as a pair of Lagrangian discs transversally intersecting at the focus–focus points and given by the equations $z = 0$ and $w = 0$.

REMARK. Note that the real singularity of focus–focus type coincides with the simplest non-degenerate complex singularity of double point type.

It is easy to see that the function f_2 defines a Poisson S^1-action in a neighborhood of the focus–focus point x_i. This follows from the fact that the integral curves of the vector field sgrad f are all closed with period 2π. Moreover, this action is free everywhere except for the only fixed point x_i. This follows from the explicit form of sgrad f_2 in terms of the complex coordinates w and z:

$$\dot{w} = iw, \qquad \dot{z} = -iz.$$

Hence the action of the circle S^1 is given by the simple formula:

$$(z, w) \to (e^{-i\varphi}z, e^{i\varphi}w).$$

Consider the neighborhood of x_i given by the inequality $|z|^2 + |w|^2 <$ const. The boundary of this neighborhood is the 3-sphere, on which the above S^1-action defines a fibration which is topologically equivalent to the well-known Hopf fibration on S^3.

We now study the structure of the singular leaf L as a whole. By our assumption, L contains n focus–focus points. In a neighborhood of each of them, the singular leaf is the union of two transversally intersecting discs. They can be presented as a neighborhood of the vertex of a cone in $\mathbb{R}^3$ (Fig. 9.49). We claim that the whole singular leaf L is obtained by gluing them in a chain as shown in Fig. 9.50. As a result, we obtain a consequence of n two-dimensional spheres

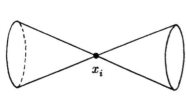

Figure 9.49   Figure 9.50

which are glued with each other at the focus–focus points (Fig. 9.50). Let us emphasize that two neighboring spheres intersect at the point x_i transversally (inside the four-dimensional symplectic manifold M^4). The same leaf L can be described in another way. Recall that the singular leaf L appears from the two-dimensional Liouville torus as the latter tends to L. This process can be imagined as follows: we distinguish n parallel non-trivial cycles on the torus and then shrink each of them into a point. As a result, we obtain the torus with n neckings (Fig. 9.50).

Lemma 9.7. *The singular leaf L is homeomorphic to the two-dimensional torus with n neckings, where n is the number of focus–focus points on L.*

Proof. The singular leaf L consists of orbits of the Poisson $\mathbb{R}^2$-action. Each orbit can be homeomorphic to one of the following manifolds: a point (zero-dimensional orbit); a line or a circle (one-dimensional orbit); a disc, a 2-torus, or an annulus. Clearly, the zero-dimensional orbits (in L) are exactly the points $x_1, \ldots, x_n$. Furthermore, we claim that L contains no one-dimensional orbits. Recall that, by our assumption, the orbits of the $\mathbb{R}^2$-action are non-degenerate. Indeed, if we admit the existence of such orbits, then these orbits are always organized into one-parameter families (Proposition 1.16). Under the momentum mapping, such a family is projected onto some one-dimensional curve on the bifurcation diagram. But, in our case, the bifurcation diagram is just one isolated point.

Now consider two-dimensional orbits in L. We assert that neither tori nor discs can occur. Indeed, since there are no one-dimensional orbits, each two-dimensional orbit must be adjacent at least to one singular focus–focus point (this condition excludes the torus, since the torus is closed and not adjacent to anything). In a neighborhood of this point, we have a free S^1-action (see above). This action is naturally extended to the orbits which pass through this neighborhood. Thus, on each two-dimensional orbit from L, there is a free S^1-action. Clearly, this is possible provided each orbit is homeomorphic to an annulus. Thus, L is glued from zero-dimensional orbits and two-dimensional annuli. Moreover, the annuli are adjacent to the focus–focus point as was shown above: in other words, each boundary component of the annulus shrinks into a point. As a result, we see that the singular focus–focus points $x_1, \ldots, x_n$ are connected by a sequence of annuli (Fig. 9.50). This is obviously a reformulation of the fact to be proved. $\square$

Lemma 9.8. *On the neighborhood $U(L)$, there is a smooth Poisson S^1-action which is free everywhere except for the singular points $x_1, \ldots, x_n$. Moreover, this action leaves each leaf of the Liouville foliation invariant. Such an action is uniquely defined up to reversing orientation on the acting circle S^1.*

Proof. As was already shown, the S^1-action with the desired properties is defined in a neighborhood of each singular point x_i separately. Take one of them. In its neighborhood, in a suitable canonical coordinate system (p_1, q_1, p_2, q_2), the Hamiltonian of the action has the form $f_2 = p_1 q_2 - p_2 q_1$. In fact, this function is a certain function of f and H, i.e., $f_2 = f_2(f, H)$. Since f and H are defined globally on the whole $U(L)$, the function f_2 is also globally defined. On the other hand, integral trajectories of sgrad f_2 are closed with period 2π in the neighborhood of x_i. In particular, they are closed on each Liouville torus T^2 passing through this neighborhood. But, in such a case, all the integral trajectories of sgrad f_2 will be closed on the whole Liouville torus with the same period (i.e., not only near x_1, but everywhere on T^2). Thus, the integral trajectories of sgrad f_2 are closed with period 2π on each Liouville torus in $U(L)$. Hence, by continuity argument, they are also closed on the whole neighborhood $U(L)$. But this exactly means that we have a smooth Poisson S^1-action on $U(L)$ with the desired properties. The uniqueness of this action on $U(L)$ (up to reversing of orientation) follows from the local uniqueness of such an action in a neighborhood of just one singular point x_i. $\square$

9.8.2. *Classification of Focus–Focus Singularities*

It is convenient to represent the singular leaf L as a union of n elementary bricks L_i each of which is homeomorphic to a neighborhood of the vertex of a cone (Fig. 9.49). In other words, L_i is a neighborhood of the point x_i in the singular leaf.

We now extend this decomposition of L to its neighborhood $U(L)$ to represent $U(L)$ as the union of n elementary four-dimensional bricks U_i. To this end, we cut $U(L)$ along three-dimensional manifolds homeomorphic to $S^1 \times D^2$, where S^1's are the circles located in the middle of annuli (two-dimensional orbits) connecting the focus–focus points. Cutting the leaf L along these circles, we decompose it into the elementary 2-bricks L_i. To cut $U(L)$, we need to extend the cut from the singular leaf L to the neighboring Liouville tori. On each neighboring torus, we need to take a circle which is close to the selected circle on L. Since the Liouville tori form a two-parameter family, all such circles form a 3-manifold homeomorphic to a solid torus $S^1 \times D^2$ (see Fig. 9.51). By cutting $U(L)$ along these 3-manifold, we obtain the desired decomposition of $U(L)$ into elementary bricks U_i.

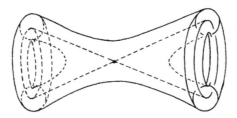

Figure 9.51

From the topological point of view, U_i can be considered as a regular neighborhood of the focus–focus point x_i in $U(L)$. In other words, U_i is the intersection of $U(L)$ with the four-dimensional ball D^4 centered at x_i.

The intersection of U_i with the boundary sphere $S^3 = \partial D^4$ consists of two solid tori located in the sphere S^3 in the following way. Without loss of generality, we shall use the canonical complex coordinates z and w in a neighborhood of x_i and assume that S^3 is embedded into $\mathbb{C}^2(z, w)$ in the standard way and is given by the equation $|z|^2 + |w|^2 = 1$. Then the elementary brick U_i can be represented as $\{|zw| < \varepsilon\} \cap D^4$. Recall that the equation $zw = 0$ defines the singular level L. It is easy to see that the set $\{|zw| < \varepsilon\}$ cut out two solid tori in the boundary sphere $S^3 = \partial D^4$. The boundary tori of these solid tori are given in S^3 by the following equations:

$$|z| = \alpha, \quad |w| = \beta \quad \text{(the first torus)};$$
$$|z| = \beta, \quad |w| = \alpha \quad \text{(the second torus)}.$$

Here α and β satisfy the relations: $\alpha^2 + \beta^2 = 1$, $\alpha\beta = \varepsilon$, $\alpha > \beta > 0$. Notice that these two solid tori are linked inside the sphere with linking number 1.

Thus, all the elementary 4-bricks U_i have the same structure. The neighborhood $U(L)$ is obtained from them by gluing neighboring 4-bricks U_i and U_{i+1} (and also U_n and U_1) by some diffeomorphism of boundary solid tori.

Theorem 9.10. *The number n of singular focus–focus points on the singular leaf L is a complete topological invariant of the focus–focus singularity (up to Liouville equivalence). In other words, two focus–focus singularities are Liouville equivalent (topologically, but, in general, not smoothly) if and only if their singular leaves contain the same number of singular points.*

Proof.   It suffices to prove that the four-dimensional neighborhood $U(L)$ of the singular leaf L is uniquely defined (up to fiber homeomorphism) if we know in advance the number n of singular points on L. In other words, it suffices to verify that $U(L)$ can uniquely be glued from n elementary 4-bricks.

To this end, we first analyze the gluing of two neighboring boundary solid tori. Each of them is foliated into circles parallel to the axis of the solid torus. These circles are intersections of the Liouville tori with the boundary of the given 4-brick. Therefore, in each brick they are defined by the equations $f = \mathrm{const}$, $H = \mathrm{const}$, and $|z|^2 + |w|^2 = 1$.

The solid tori are glued in such a way that these two foliations are fiberwise identified.   In how many ways can this be done?   To answer this question, it is convenient to represent each solid torus as the direct product $D^2 \times S^1$. Recall that, on each elementary brick U_i, we have standard canonical coordinates p_1, p_2, q_1, q_2 such that

$$H = H(f_1, f_2), \quad f = f(f_1, f_2),$$

where $f_1 = p_1 q_1 + p_2 q_2$, and $f_2 = p_1 q_2 - p_2 q_1$. Moreover, the change

$$(H, f) \to (f_1, f_2)$$

is regular and, consequently, f_1, f_2 can locally be expressed in terms of H and f. Without loss of generality, we may assume that the Jacobian of the change $(H, f) \to (f_1, f_2)$ is positive.

On the boundary solid torus $D^2 \times S^1$ of the i-th brick U_i, the functions f_1, f_2 can be considered as local coordinates on D^2. We know that, under gluing operations, the functions H and f must be preserved. On the other hand, the expressions for H and f in terms of f_1 and f_2 may be different on neighboring bricks U_i and U_{i+1}. Therefore, the gluing mapping written in terms of f_1 and f_2 becomes, in general, a non-trivial diffeomorphism with positive Jacobian.

The next technical assertion shows that we may assume, nevertheless, such a diffeomorphism to be simply identical.

Lemma 9.9. *Consider a neighborhood $U(x)$ of a singular point $x \in M^4$ of focus–focus type (a standard 4-brick) and the corresponding momentum mapping $\mathcal{F}: U(x) \to \mathbb{R}^2$, $\mathcal{F}(x) = (0,0)$. Let ξ be an arbitrary local orientation preserving diffeomorphism in the image (i.e., $\xi: \mathbb{R}^2 \to \mathbb{R}^2$) such that $(0,0)$ is a fixed point. Then there exists a fiber homeomorphism $\widehat{\xi}: U(x) \to U(x)$ such that $\mathcal{F}\widehat{\xi} = \xi\mathcal{F}$. Moreover, there exists a fiber isotopy $\widehat{\xi}_t$, $t \in [0,1]$ such that $\widehat{\xi}_1 = \widehat{\xi}$ and $\widehat{\xi}_0 = \mathrm{id}$.*

REMARK. This lemma actually says that $U(x)$ has a rather large group of fiber automorphisms. Using such homeomorphisms, we can rearrange the leaves of the Liouville foliation in an arbitrary way.

Proof (of Lemma 9.9). It will be convenient to prove this lemma in the complex form. That is why we shall consider $U(x)$ as a domain in complex space $\mathbb{C}^2\,(z,w)$, and the momentum mapping $\mathcal{F}$ as a mapping from $U(x)$ into $\mathbb{C}$. Moreover, in view of Theorem 1.5, we suppose that $\mathcal{F}(z,w) = zw$.

First consider a real analog of the statement. Let $\zeta\colon \mathbb{R} \to \mathbb{R}$ be an arbitrary orientation preserving diffeomorphism such that $\zeta(0) = 0$. We now construct a homeomorphism (in fact, it is not hard to construct a diffeomorphism) $\widehat{\zeta}\colon \mathbb{R}^2 \to \mathbb{R}^2$ such that $F\widehat{\zeta} = \zeta F$, where $F\colon \mathbb{R}^2 \to \mathbb{R}$ is the mapping of the form $F(x,y) = xy$. In other words, we construct a fiber homeomorphism for the foliation given on $\mathbb{R}^2$ by level lines of F. This homeomorphism must mix the leaves according to the diffeomorphism ζ (given on the base).

Consider the vector field $\operatorname{grad} F = (y,x)$ and define the mapping $\varphi(x,y,\alpha)$ which moves the point (x,y) along the integral trajectory of $\operatorname{grad} F$ to the point (x',y') such that $F(x',y') = x'y' = \alpha$. We assume here that the sign of α coincides with the sign of $F(x,y) = xy$.

Now the desired homeomorphism $\widehat{\zeta}$ can be given by the following formula:

$$\widehat{\zeta}(x,y) = \varphi(x,y,\zeta(xy)).$$

It is easy to verify that $\widehat{\zeta}$ satisfies the required properties. Note, by the way, that $\widehat{\zeta}$ is fixed on the singular leaf $\{F = 0\}$.

Similarly, we construct now the homeomorphism $\widehat{\xi}$ in the complex case. Writing a complex number z as a pair $(|z|, \arg(z))$, we can represent the momentum mapping $\mathcal{F}$ in the form

$$(|z|, \arg(z), |w|, \arg(w)) \to (|zw|, \arg(z) + \arg(w)).$$

We now consider the mapping $\widehat{\xi}\colon \mathbb{C}^2 \to \mathbb{C}^2$ given in the following way:

$$(|z|, |w|, \arg(z), \arg(w)) \to (|z'|, |w'|, \arg(z'), \arg(w')),$$

where

$$(|z'|, |w'|) = \varphi(|z|, |w|, |\xi(zw)|),$$
$$\arg(z') = \arg(z) + \lambda(\arg(\xi(zw)) - \arg(zw)),$$
$$\arg(w') = \arg(w) + (1 - \lambda)(\arg(\xi(zw)) - \arg(zw)).$$

The mapping φ has been defined above, and $\lambda = \dfrac{2}{\pi}\arctan\left(\dfrac{|w|}{|z|}\right)$.

It is easily verified that this mapping satisfies the required property $\mathcal{F}\widehat{\xi} = \xi\mathcal{F}$ and is a homeomorphism. Moreover, $\widehat{\xi}$ is identical on the singular leaf $\{\mathcal{F} = 0\}$.

To construct the isotopy, it suffices to consider an isotopy ξ_t in the image such that $\xi_0 = \operatorname{id}$, $\xi_1 = \xi$, and then apply the indicated formulas.

Lemma 9.9 is proved. $\square$

By means of this lemma, all the gluing diffeomorphisms can be made identical (in terms of f_1, f_2) on the discs D^2. It remains to understand how to glue the S^1-fibers.

It is clear that *a priori* there are only two ways to identify the fibers of the solid tori (up to fiber isotopy): preserving or reversing orientation. In fact, no ambiguity occurs. To see this, it suffices to use the existence of a global Poisson S^1-action on $U(L)$. Using this action, we can introduce orientation on all the fibers. After this, the gluing operations are all uniquely defined (up to fiber isotopy). Thus, the obtained Liouville foliation on $U(L)$ is also uniquely defined in the above sense. Therefore, if the number n of singular points on the singular level L is prescribed, then the topology of $U(L)$ is uniquely reconstructed as well as the Liouville foliation on it. □

COMMENT. It is important that Theorem 9.10 gives the classification of focus–focus singularities up to fiber homeomorphisms (but not diffeomorphisms). The point is that, if the singular leaf L contains several singular points, then the smooth classification becomes more complicated. It turns out that topologically equivalent singularities may be different from the smooth point of view (i.e., may not be transformed into each other by a fiber diffeomorphism). In other words, there exist non-trivial smooth invariants that distinguish focus–focus singularities up to fiber diffeomorphisms. The reason is that in the smooth case, Lemma 9.9 fails. In particular, for the existence of a fiber diffeomorphism $\hat{\xi}$ (that covers ξ acting on the base), it is necessary that the differential $d\xi$ at the fixed point is *complex*.

9.8.3. Model Example of a Focus–Focus Singularity and the Realization Theorem

Consider the two-dimensional complex space $\mathbb{C}^2 (z, w)$ with the symplectic structure defined by the formula $\operatorname{Re}(dw \wedge dz) = dp \wedge dq = dp_1 \wedge dq_1 + dp_2 \wedge dq_2$. Let U be an open domain in $\mathbb{C}^2 (z, w)$ given by the following inequalities:

$$|zw| < \varepsilon, \quad |z| < 1 + \delta, \quad |w| < 1 + \delta.$$

This domain is a regular neighborhood of two discs intersecting transversally at the point $(0,0)$ (Fig. 9.52):

$$\{|z| \leq 1, \, w = 0\} \quad \text{and} \quad \{z = 0, \, |w| \leq 1\}.$$

Consider regular neighborhoods of the boundary circles of these discs in U and denote them by

$$U_z = U \cap \{(1 + \delta)^{-1} < z < 1 + \delta\},$$
$$U_w = U \cap \{(1 + \delta)^{-1} < w < 1 + \delta\}.$$

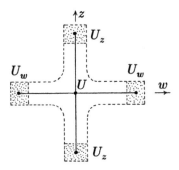

Figure 9.52

In Fig. 9.52 they are shown as shaded regions. Topologically, each of them is homeomorphic to the direct product $S^1 \times D^3$. Let us glue these neighborhoods U_z and U_w by the mapping $\xi \colon U_w \to U_z$, where

$$\xi \colon (z, w) \to (w^{-1}, zw^2) \,.$$

Clearly, ξ is a complex mapping. Moreover, it is easily seen that ξ is symplectic (since $dw \wedge dz = d(zw^2) \wedge d(w^{-1})$) and preserves the function zw.

Hence, as a result, from the domain U we obtain a complex symplectic 4-manifold U_1 with a holomorphic function F on it which has the form zw in terms of local coordinates z, w. This function has exactly one critical point $(0,0)$. The singular level $\{F = 0\}$ is obtained as a result of gluing two transversally intersecting discs along their boundaries, i.e., is homeomorphic to the sphere with one self-intersection point. The manifold U_1 is a regular neighborhood of this level (i.e., $U_1 = \{|F| < \varepsilon\}$) foliated into compact non-singular levels of F each of which is diffeomorphic to a 2-torus.

From the real point of view, we have obtained two commuting functions $f_1 = \operatorname{Re} F$ and $f_2 = \operatorname{Im} F$, which determine the structure of a Liouville foliation on U_1 with the only singular leaf of focus–focus type, which contains one singular point.

Thus, U_1 can be considered as the simplest model example of a focus–focus singularity of complexity 1.

In the similar way, we can construct a model example of a singularity whose singular leaf contains n focus–focus points. To this end, we only need to glue successively n copies of the domain U as was described above. As a result, we obtain a manifold U_n with the desired properties.

The same manifold can be obtained in another way. Namely, we can just consider an n-sheeted covering over U_1. Here we use the fact that the fundamental group of a singular leaf in U_1 (as well as that of U_1 itself) is isomorphic to $\mathbb{Z}$. So we can produce such a covering in the standard way by taking the subgroup $n\mathbb{Z} \subset \mathbb{Z}$.

Notice that the universal covering U_∞ over U_1 coincides with the universal covering over U_n. Moreover, the structure of the (non-compact) Liouville foliation on U_∞ lifted from the base of the covering will be the same for all U_m, $m \in \mathbb{N}$ (Fig. 9.53). On this universal covering, there is a natural action of $\mathbb{Z}$, and every focus–focus singularity (considered up to a fiber homeomorphism) can be obtained

Figure 9.53

from this universal model by taking a quotient with respect to the subgroup in $\mathbb{Z}$ of index n. Let us emphasize that this statement holds only in topological sense. From the symplectic point of view, it is not the case. The point is that there exist non-trivial symplectic invariants which distinguish focus–focus singularities with the same number of singular points on the singular leaf L.

9.8.4. *The Loop Molecule and Monodromy Group of a Focus–Focus Singularity*

Let, as above, L be a singular leaf of focus–focus type, and let $y = \mathcal{F}(L) \subset \mathbb{R}^2\,(H, f)$ be its image under the momentum mapping. By our assumption, y is an isolated singular point of the bifurcation diagram. Consider a circle γ_ε of small radius ε centered at the point y and its preimage $Q_{\gamma_\varepsilon} = \mathcal{F}^{-1}(\gamma_\varepsilon)$ (Fig. 9.54). It is clear that the 3-manifold Q_{γ_ε} is a fiber bundle over the circle γ_ε whose fibers are Liouville tori T^2. This fiber bundle is completely determined by its monodromy group, i.e., the group of the automorphisms of the fundamental group of a fiber $\pi_1(T^2)$ corresponding to closed loops on the base. Since $\pi_1(T^2) = \mathbb{Z} \oplus \mathbb{Z}$ and the base is the circle γ_ε, the monodromy group is just a cyclic subgroup in $\mathbb{Z} \oplus \mathbb{Z}$.

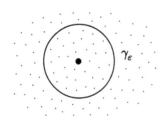

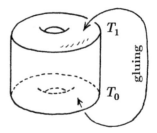

Figure 9.54   **Figure 9.55**

Note that Q_{γ_ε} can be represented as the result of identification of the boundary tori T_0 and T_1 of the 3-cylinder $T^2 \times [0,1]$ by some diffeomorphism $\psi: T_0 \to T_1$ (Fig. 9.55). The gluing diffeomorphism ψ induces an automorphism ψ_* of the fundamental group of the torus $\mathbb{Z} \oplus \mathbb{Z}$. The automorphism ψ_* is uniquely defined by an integer unimodular matrix. This matrix depends, of course, on the choice of a basis on the torus. But its conjugacy class is a well-defined complete invariant of the fiber bundle $Q_{\gamma_\varepsilon} \xrightarrow{T^2} \gamma_\varepsilon$. This matrix is called the *monodromy matrix*.

Recall that the loop molecule of the singularity in question is the marked molecule that describes the Liouville foliation on Q_{γ_ε}. In our case, this foliation has no singular leaves (i.e., the loop molecule has no atoms). Thus, the loop molecule is represented as a circle (whose points parametrize the Liouville tori in Q^{γ_ε}) endowed with a mark, namely, the monodromy matrix (or, more precisely, its conjugacy class).

Theorem 9.11. *The loop molecule is a complete invariant (in the sense of Liouville equivalence) of a focus–focus singularity. If the singular leaf L contains n focus–focus singular points, then the monodromy matrix has the form*

$$\begin{pmatrix} 1 & n \\ 0 & 1 \end{pmatrix}.$$

Proof. We shall give two different proofs. The first is by indicating a basis on the Liouville torus and the result of its translation along the base γ_ε. The second one is by analyzing some relations in the fundamental group $\pi_1(Q_{\gamma_\varepsilon})$.

We begin with the case when $n = 1$, i.e., L contains just one focus–focus point. To prove the theorem in this case, it suffices to use the model example of a focus–focus singularity and to compute explicitly the monodromy matrix. Recall that, on the model 4-manifold U_1, we have complex coordinates (z, w). Although they cover the whole manifold U_1, they are not uniquely defined.

We also have a holomorphic function $F = zw$ on U_1, which maps U_1 into the complex plane $\mathbb{R}^2 = \mathbb{C}$. Clearly, F is just the momentum mapping corresponding to the commuting real-valued functions $f_1 = \operatorname{Re} F$ and $f_2 = \operatorname{Im} F$. Therefore, the 3-manifold Q_{γ_ε} is given by the equation $|F(z, w)| = |zw| = \varepsilon$. Let us describe a useful representation of the manifold Q_{γ_ε}. Consider the 3-manifold with boundary in $\mathbb{C}^2$ given by

$$|zw| = \varepsilon, \qquad |z| \leq 1, \qquad |w| \leq 1.$$

Its boundary consists of two tori given as follows:

$$T_w = \{|z| = \varepsilon, |w| = 1\}, \qquad T_z = \{|w| = \varepsilon, |z| = 1\}.$$

It is easy to see that Q_{γ_ε} is obtained from this manifold by gluing its boundary tori by the diffeomorphism

$$\xi : T_w \to T_z, \qquad \text{where} \quad \xi : (z, w) \to (w^{-1}, zw^2).$$

The 3-manifold Q_{γ_ε} obtained is foliated into two-dimensional Liouville tori T_φ, where φ is the natural angle-parameter on γ_ε, i.e.,

$$T_\varphi = \{zw = \varepsilon e^{i\varphi}\}.$$

On each of these tori T_φ, we now construct a basis $(\lambda_\varphi, \mu_\varphi)$ which smoothly depends on the parameter φ. Recall that our aim is to compute the monodromy matrix. In terms of the above bases, this matrix is the transition matrix between the bases (λ_0, μ_0) and $(\lambda_{2\pi}, \mu_{2\pi})$, which actually correspond to the same torus $T_0 = T_{2\pi}$.

By definition, we set

$$\lambda_\varphi(t) = (\varepsilon e^{i\varphi} e^{2\pi it}, e^{-2\pi it}), \qquad t \in [0,1],$$
$$\mu_\varphi(s) = (\varepsilon s^{-1} e^{i\varphi\tau(s)}, s e^{i\varphi(1-\tau(s))}), \quad s \in [\varepsilon, 1].$$

Here $\tau(s) = (s-\varepsilon)(1-\varepsilon)^{-1}$, and s and t are parameters on the cycles (Fig. 9.56).

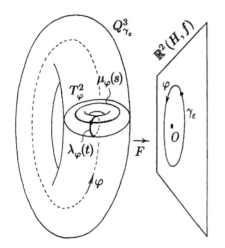

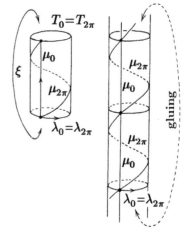

Figure 9.56   Figure 9.57

It is easily seen that both cycles belong to the torus $T_\varphi = \{zw = \varepsilon e^{i\varphi}\}$. Moreover, $\lambda_\varphi(t)$ and $\mu_\varphi(s)$ are non-trivial cycles without self-intersections and form a basis on T_φ.

Let us compare the bases (λ_0, μ_0) and $(\lambda_{2\pi}, \mu_{2\pi})$ on the torus $T_0 = T_{2\pi}$. We have

$$\lambda_0(t) = (\varepsilon e^{2\pi it}, e^{-2\pi it}), \qquad \mu_0(s) = (\varepsilon s^{-1}, s).$$
$$\lambda_{2\pi}(t) = (\varepsilon e^{2\pi it}, e^{-2\pi it}), \qquad \mu_{2\pi}(s) = (\varepsilon s^{-1} e^{2\pi i\tau(s)}, s e^{-2\pi i\tau(s)}).$$

It is clear that $\lambda_0 = \lambda_{2\pi}$ and $\mu_{2\pi} = \lambda_0 + \mu_0$. Hence the monodromy matrix has the desired form. This proves the theorem for $n = 1$.

If n is arbitrary, then the above formulas take the following form: $\lambda_0 = \lambda_{2\pi}$ and $\mu_{2\pi} = n\lambda_0 + \mu_0$. Indeed, to obtain the monodromy matrix in the case of complexity n, we just need to raise the monodromy matrix for $n = 1$ to power n. This fact can easily be observed by using the n-sheeted covering $U_n \to U_1$ (Fig. 9.57). □

COMMENT. The boundary torus T_z (i.e., the torus T_w after gluing) intersects with all the tori of the form T_φ along the cycles λ_φ. All these cycles on T_z are homologous to the same cycle λ_0. Moreover, the cycles λ_φ are orbits of the S^1-action on Q_{γ_ε}.

We now briefly describe another proof of the same theorem. Note that the 3-manifold Q_{γ_ε} admits two different foliations into two-dimensional tori. The first one is the Liouville foliation into tori T_φ. The base of this foliation is the circle γ_ε. The second is the foliation into the tori T'_s of the form

$$T'_s = \{|z| = \varepsilon s^{-1}, \ |w| = s\}, \qquad \text{where} \quad s \in [\varepsilon, 1].$$

The torus T_z (coinciding with T_w after gluing) is naturally included in this family. Namely, $T'_1 = T_w$ and $T'_\varepsilon = T_z$.

The monodromy matrix of the fiber bundle $Q_{\gamma_\varepsilon} \to \gamma_\varepsilon$ can be interpreted in terms of the fundamental group $\pi_1(Q_{\gamma_\varepsilon})$.

To explain this, we use the following general construction. Let $\pi \colon Q_{\gamma_\varepsilon} \to S^1$ be a T^2-fiber bundle over the circle. Consider three natural generators in $\pi_1(Q_{\gamma_\varepsilon})$. Let γ be the generator corresponding to a loop which is projected onto the base S^1

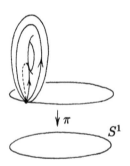

Figure 9.58

homeomorphically, and let α and β be two generators on the fiber T^2 (Fig. 9.58). Then the fundamental group $\pi_1(Q_{\gamma_\varepsilon})$ is generated by α, β, and γ with the following relations:

1) α and β commute, i.e., $\alpha\beta\alpha^{-1}\beta^{-1} = e$;

2) $\begin{pmatrix} \gamma\alpha\gamma^{-1} \\ \gamma\beta\gamma^{-1} \end{pmatrix} = C\begin{pmatrix} \alpha \\ \beta \end{pmatrix}$, where C is some integer matrix, which is just the monodromy matrix of the given fiber bundle. Note that the latter relation is written in additive form, but we may do so, since the generators α and β commute.

Let us compute the fundamental group of Q_{γ_ε} by using first the structure of the foliation into tori T'_s. As γ we take the cycle $\gamma = \mu_0(s) = (\varepsilon s^{-1}, s)$ (see above). As α and β we take the basis cycles on the torus $T_w = \{|w|=1, |z|=\varepsilon\}$ given by the following explicit formulas in coordinates (z, w):

$$\alpha = (\varepsilon e^{2\pi i t}, 1), \qquad \beta = (\varepsilon, e^{-2\pi i t}), \qquad t \in [0, 1].$$

Using the gluing mapping $\xi\colon (z,w) \to (w^{-1}, zw^2)$ between the tori $T_1' = T_w$ and $T_\varepsilon' = T_z$, we find immediately the monodromy matrix in the basis α, β. This matrix defines the induced automorphism ξ_* of the fundamental group of the torus, and it has the form

$$\begin{pmatrix} 0 & -1 \\ 1 & 2 \end{pmatrix}.$$

Therefore, the relations in the fundamental group $\pi_1(Q_{\gamma_\varepsilon})$ are as follows:

$$\gamma \alpha \gamma^{-1} = \beta^{-1}, \qquad \gamma \beta \gamma^{-1} = \alpha \beta^2\,.$$

Now we pass to other generators in $\pi_1(Q_{\gamma_\varepsilon})$ which correspond to the foliation into Liouville tori: $Q_{\gamma_\varepsilon} \to \gamma_\varepsilon$. On the torus T_φ that is the fiber of this foliation for $\varphi = 0$, we take the generators $\tilde\alpha = \lambda_0$ and $\tilde\beta = \mu_0$ (see above). It follows from the explicit formulas for these cycles that

$$\tilde\alpha = \alpha\beta\,, \qquad \tilde\beta = \gamma\,.$$

As the third generator $\tilde\gamma$ (which maps homeomorphically onto γ_ε), we take, for instance, the cycle α.

Let us rewrite the relations in the fundamental group $\pi_1(Q_{\gamma_\varepsilon})$ in terms of new generators $\tilde\alpha, \tilde\beta, \tilde\gamma$. Notice that $\tilde\alpha$ and $\tilde\beta$ commute as basis cycles on the torus $T_{\varphi=0}$. However, this can be seen formally using the above relations for α, β, and γ.

The desired monodromy matrix $\tilde C$ in terms of the new generators is defined from the standard relation:

$$\begin{pmatrix} \tilde\gamma\tilde\alpha\tilde\gamma^{-1} \\ \tilde\gamma\tilde\beta\tilde\gamma^{-1} \end{pmatrix} = \tilde C \begin{pmatrix} \tilde\alpha \\ \tilde\beta \end{pmatrix}.$$

Let us compute the elements $\tilde\gamma\tilde\alpha\tilde\gamma^{-1}$ and $\tilde\gamma\tilde\beta\tilde\gamma^{-1}$. We get

$$\tilde\gamma\tilde\alpha\tilde\gamma^{-1} = \alpha\alpha\beta\alpha^{-1} = \alpha\beta = \tilde\alpha\,,$$
$$\tilde\gamma\tilde\beta\tilde\gamma^{-1} = \alpha\gamma\alpha^{-1} = \alpha(\gamma\alpha^{-1}\gamma^{-1})\gamma = \alpha\beta\gamma = \tilde\alpha\tilde\beta\,.$$

Thus, the monodromy matrix $\tilde C$ is

$$\begin{pmatrix} 1 & 0 \\ 1 & 1 \end{pmatrix},$$

as was to be proved.

REMARK. Consider the manifold $U(L) \setminus L$ which is obviously a fiber bundle whose fibers are Liouville tori over the disc with one point removed. According to J. J. Duistermaat's theorem [104], there always exists a natural affine structure on the base of this bundle. Consider the holonomy group of this affine structure. It is easy to see that, in our case, this group just coincides with the monodromy group of the fiber bundle $Q_{\gamma_\varepsilon} \to \gamma_\varepsilon$.

REMARK. In the next chapters of our book we give some examples of integrable systems from physics and classical mechanics, where focus–focus singularities appear. Among these systems, in particular, we can see the well-known Lagrange and Clebsch integrable cases in rigid body dynamics, as well as the spherical pendulum and four-dimensional rigid body systems.

REMARK. The focus–focus singularities have been studied in detail in papers [89], [90], [212], [221], [258].

REMARK. Notice that locally a focus–focus singularity can be considered as a singularity of one complex functions of two complex variables. One can apply (locally) the classical Picard–Lefschetz theory to such foliations. In particular, the character of the monodromy also has a natural classical interpretation. From the point of view of Picard–Lefschetz theory, this case is the simplest one. However, from the global viewpoint, a focus–focus singularity does not have to be complex (it is complex only locally). Therefore, to study its global properties, we need some additional arguments.

9.9. ALMOST DIRECT PRODUCT REPRESENTATION FOR MULTIDIMENSIONAL NON-DEGENERATE SINGULARITIES OF LIOUVILLE FOLIATIONS

In this section, we briefly discuss Nguyen Tien Zung's theorem [258] on the topological structure of multidimensional singularities of integrable systems. This theorem generalizes the above proved result on the decomposition of a four-dimensional singularity into an almost direct product of 2-atoms. It turns out that a similar result holds for a wide class of multidimensional singularities of Liouville foliations.

Recall that by a singularity of a Liouville foliation we mean a small neighborhood of a singular leaf considered up to fiber equivalence. In other terms, we can speak about the germ of the foliation on the singular leaf.

A singularity of a Liouville foliation is called *non-degenerate* if the singular points of the momentum mapping $\mathcal{F}$ lying on the singular leaf of the foliation are all non-degenerate in the sense of Definition 1.24. Consider the singular points of minimal rank i on the singular leaf L. Then, for each of them, we can define its *type* to be the triple of integers (m_1, m_2, m_3) described in Section 1.8.4. It can be shown that the type (m_1, m_2, m_3) is the same for all singular points of minimal rank on L. Therefore, one can speak of the type (m_1, m_2, m_3) of the singular leaf L itself as well as of its rank i.

Before formulating the theorem, we need one more additional assumption to the singularities of Liouville foliations. This assumption (the so-called *non-splitting condition*) distinguishes a wide and natural class of singularities. For all specific examples of integrable systems, which we know, this condition is fulfilled.

We begin with the simplest example. Suppose we have an integrable system with two degrees of freedom. Consider the restriction of it onto a regular energy level $Q^3 = \{H = h\}$. We have above introduced the topological stability condition for a system on Q^3. Recall that the topological stability means that, under small variation of the energy level h, the topological type of the Liouville foliation does not change. More precisely, the Liouville foliation on an energy level sufficiently close to h is fiberwise equivalent to the initial foliation on Q^3.

In fact, this condition is local and can be reformulated in terms of the bifurcation diagram of the momentum mapping $\mathcal{F}$ in a neighborhood of each single 3-atom in Q.

The only case when a system with non-degenerate singularities is non-stable is the situation of disintegration of a complicated 3-atom into a union of several simpler atoms. This disintegration can be seen in the bifurcation diagram. Indeed, consider non-degenerate critical circles of the given 3-atom. All of them are located on the same level of the additional integral f. Under the momentum mapping $\mathcal{F}$, they are mapped into the same point. On the other hand, each of these circles is included in a one-parameter family. This family of circles is mapped into a smooth curve on the plane $\mathbb{R}^2 (H, f)$. The number of such curves is equal to the number of critical circles of the given 3-atom.

If the disintegration of the 3-atom does not happen, then all these curves obviously coincide on $\mathbb{R}^2 (H, f)$. On the contrary, if the 3-atom splits, then there are circles which move to different levels of the integral f. As a result, the corresponding curves become different (Fig. 9.59).

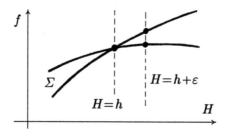

Figure 9.59

The same condition can also be reformulated in the following way. Consider the singular leaf L and those points $x \in L$ where the momentum mapping $\mathcal{F}$ has minimal rank. In our case, these are the points x lying on the critical circles. Note that, under the momentum mapping $\mathcal{F}$, the whole singular leaf L is mapped into a single point $y = \mathcal{F}(L)$. Choose a point x_1 on each of the critical circle. For each point x_i, consider a small neighborhood in M^4, its image in $\mathbb{R}^2 (H, f)$, and the local bifurcation diagram Σ_{x_i}. It is easy to see that Σ_{x_i} does not depend on the choice of a representative x_i on the given critical circle. As a result, we obtain a number of local bifurcation diagrams Σ_{x_i} on the plane $\mathbb{R}^2 (H, f)$.

Now the condition that the 3-atom does not split can be formulated in the following way: all the local bifurcation diagrams Σ_{x_i} must coincide with each other and, consequently, with the whole bifurcation diagram Σ in a neighborhood of the point $y = \mathcal{F}(L)$. In other words, the bifurcation diagram Σ does not split into several pieces.

Just similarly the non-splitting condition can be formulated in the case of multidimensional non-degenerate singularities.

For analytic multidimensional systems, it suffices to repeat almost literally the above definition. Consider a non-degenerate singularity of the Liouville foliation on M^{2n}. Let L be the corresponding singular leaf, and Σ_L denote the local bifurcation diagram of the momentum mapping $\mathcal{F}$ restricted to sufficiently small neighborhood of L in M^{2n}. Now consider the points of minimal rank i on L. It is possible to show that such points form one or several critical tori of dimension i. Choose a point-representative x_j on each of them and construct the local bifurcation diagram $\Sigma_{x_j} \subset \mathbb{R}^n$ for $\mathcal{F}$ restricted to a sufficiently small neighborhood of x_j. It is required that, for all the points-representatives $\{x_j\}$, these local bifurcation diagrams $\{\Sigma_{x_j}\}$ are the same. Moreover, we must also require that Σ_L does not contain anything else, i.e., $\Sigma_L = \Sigma_{x_j}$ for each x_j.

Now recall that the form of local bifurcation diagrams Σ_{x_j} has been already described (see Section 1.8.4). This is the canonical bifurcation diagrams of the model singularities (up to a diffeomorphism). These bifurcation diagrams are in fact piecewise linear and consist of pieces of linear subspaces (see examples in Fig. 1.12).

Definition 9.7. We shall say that a non-degenerate singularity of the Liouville foliation in M^{2n} satisfies the *non-splitting condition* if its bifurcation diagram Σ in $\mathbb{R}^n$ can be reduced by some diffeomorphism to the canonical diagram corresponding to the type of the given singularity (see Section 1.8.4).

COMMENT. Definition 9.7 is based on some properties of the bifurcation diagram of the given system. The point is that, if we wish to analyze a certain system, first we need to verify the non-splitting condition. Such a verification can be based on the analysis of the bifurcation diagram which is usually known.

COMMENT. Non-degenerate singularities satisfying the non-splitting condition are, in some sense, the simplest ones. Their bifurcation diagrams do not contain "anything additional". We mean the following. The bifurcation diagram Σ of a non-degenerate singularity of general type (i.e., perhaps without non-splitting condition) always contains the canonical bifurcation diagram as a subset. The non-splitting condition means that, besides this canonical diagram, Σ contains nothing else.

COMMENT. Consider an example of a non-degenerate, but splittable singularity. In Fig. 9.60 we show a singular leaf L of the Liouville foliation which contains one focus–focus point and a saddle circle. The circle is the line along which the 2-torus touches the sphere with two points identified. These points just give a focus–focus singularity. It is clear that the bifurcation diagram of this singularity is a smooth curve passing through the point that is the projection both of the focus–focus point and the saddle circle. In the sense of our definition, this singularity is splittable. It would have satisfied the non-splitting condition if it had consisted of the only point. But, in the case under consideration, there is an additional piece, the curve passing through this point.

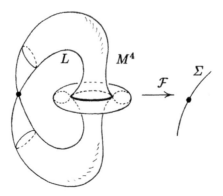

Figure 9.60

Here we see one of the mechanisms generating splittable singularities. The appearance of the additional curve is caused by the fact that, together with a focus–focus point, the singular leaf contains a closed orbit of the Poisson $\mathbb{R}^2$-action. In general, the situation is similar: our assumption forbids the existence of closed orbits of rank $> i$ on a singular leaf L of rank i. Note that, in the above example, the focus–focus point and one-dimensional closed orbit can be moved onto different leaves by small perturbation of the Poisson action of $\mathbb{R}^2$. As a result, the singularity splits into two simpler singularities each of which satisfies the non-splitting condition.

COMMENT. Consider another example of a non-degenerate splittable singularity. Take a two-dimensional surface P^2 in $\mathbb{R}^3(x, y, z)$ shown in Fig. 9.61.

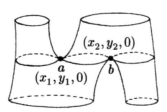

Figure 9.61

The height function $f(x, y, z) = z$ has the only critical value $z = 0$ on this surface, and the corresponding singular level contains two saddle critical points $a = (x_1, y_1, 0)$ and $b = (x_2, y_2, 0)$. Let P be a regular level surface of some smooth function H, i.e., $P^2 = \{H(x, y, z) = 0\}$. Consider the four-dimensional Euclidean space as a symplectic manifold $M^4 = \mathbb{R}^4(x, y, u, v)$ with the symplectic structure $\omega = dx \wedge dy + du \wedge dv$. Take two more functions $\tilde{H}$ and $\tilde{f}$ on M^4 by setting

$$\tilde{H} = H(x, y, u^2 + v^2), \qquad \tilde{f} = f(x, y, u^2 + v^2).$$

The functions $\tilde{H}$ and $\tilde{f}$ commute on M^4 and define the momentum mapping $\mathcal{F}: M^4 \to \mathbb{R}^2(\tilde{H}, \tilde{f})$. The points $\tilde{a} = (x_1, y_1, 0, 0)$ and $\tilde{b} = (x_2, y_2, 0, 0)$ are isolated non-degenerate critical points of center–saddle type for the Liouville foliation on M^4 defined by $\widetilde{H}, \widetilde{f}$. Both points belong to the same singular leaf L of the Liouville foliation.

The bifurcation diagram of $\mathcal{F}$ is presented in Fig. 9.62(a). The set of critical points of the momentum mapping consists of three components. The first one is the 2-plane that consists of the points $(x, y, 0, 0)$. All such points are critical for the function $\widetilde{f}$. The second and third components are also two-dimensional and are generated by the points a and b. Under the momentum mapping $\mathcal{F}$, the first component is mapped onto the horizontal straight line (Fig. 9.62), the boundary of the upper half-plane. The second and third components are mapped onto the left and right rays respectively. Local bifurcation diagrams for each of the points $\widetilde{a}$ and $\widetilde{b}$ are shown in Fig. 9.62(b). It is seen that none of them coincides with the whole bifurcation diagram, which consists of the horizontal straight line with both rays. Therefore, this singularity does not satisfy the non-splitting condition.

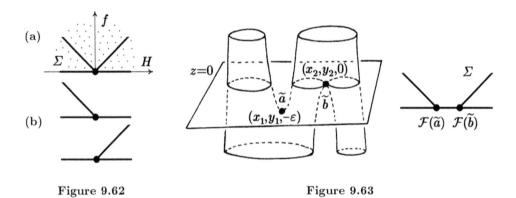

(a)

(b)

Figure 9.62   **Figure 9.63**

As in the preceding example, the described singularity of the Liouville foliation can be split into two non-splittable singularities by small perturbation of the Poisson action of $\mathbb{R}^2$. Such a perturbation is illustrated in Fig. 9.63.

COMMENT. Another mechanism for generating splittable singularities occurs in the smooth case. This is splitting of the bifurcation diagram at a singular point, which was already discussed in Chapter 1. Here we can see the difference from the analytic case, where such a mechanism does not work.

It turns out that there may be situations when, under an analytic perturbation, a non-degenerate singularity does not change its topological type, whereas, under a suitable smooth perturbation, the topological type changes. This is connected (see Chapter 1) with the fact that a smooth perturbation can split one of the curve of Σ into two curves which touch each other at a singular point of Σ (with infinite tangency order). Analytic perturbations do not allow us to produce such an effect.

COMMENT. It should be noted that we have partially changed the terminology used in the original work [258] by Nguyen Tien Zung. He called the above singularities *stable*. We say instead that they satisfy the *non-splitting condition*. The point is that, speaking of stability, one usually means the stability under certain perturbations. In fact, the stability in this sense and the non-splitting condition are closely connected, as the above examples show. Nevertheless, these conditions are not equivalent.

We now describe a simple and natural method for constructing multidimensional singularities of Liouville foliations.

First we take a direct product of a number of "elementary singularities" $U^{2n} = V_1 \times V_2 \times \ldots \times V_p$, where each factor has one of the following four types:

1) an elliptic singularity (i.e., an atom A) represented as a 2-disc foliated into concentric circles;

2) a hyperbolic singularity (i.e., a saddle atom) represented as a two-dimensional surface P with an embedded graph K;

3) a focus–focus singularity represented as one of the model four-dimensional singularities described in Section 9.8.3;

4) a trivial S^1-foliation without singularities $S^1 \times D^1$.

Note that each factor V_i has the natural Liouville foliation structure. This means that V_i is endowed with a symplectic structure ω_i and a function f_i (or two commuting functions f_i, f_i' in the focus–focus case) whose level lines determine the leaves of the foliation.

It is clear that, on the direct product U^{2n}, we can define a natural symplectic structure to be the product of the symplectic structures from each factor, i.e, $\omega = \sum \omega_i$. Then the functions f_i extended from the factors V_i to the whole direct product commute with respect to ω and determine the structure of the Liouville foliation on U^{2n} whose leaves are obviously direct products of leaves of the elementary foliations defined on each factor.

The multidimensional singularities of this kind are said to be singularities of *direct product type*.

This construction can naturally be generalized to the class of singularities which are called *almost direct products*. To this end, we consider a model singularity $U^{2n} = V_1 \times V_2 \times \ldots \times V_p$ of direct product type. Suppose that some finite group G acts on U^{2n}, and this action φ satisfies the following conditions.

1) The action φ is free.

2) The action φ is component-wise, i.e., G transforms each factor V_k into itself. More precisely, this means that the action of G commutes with the projections to each of the factors. In other words, the action on $U^{2n} = V_1 \times V_2 \times \ldots \times V_p$ can be written in the following form:

$$\varphi(g)(x_1, \ldots, x_p) = (\varphi_1(g)(x_1), \ldots, \varphi_p(g)(x_p)),$$

where $g \in G$, and φ_k is a certain action of G on the component V_k.

3) On each component, the action φ_k is symplectic and preserves the structure of the Liouville foliation. Moreover, we shall assume that φ_k preserves the corresponding functions that define the Liouville foliation on V_k.

4) On each elliptic component V_s (i.e., on the atoms of type A), the action φ_s is trivial.

Consider now the quotient space of $U^{2n} = V_1 \times V_2 \times \ldots \times V_p$ with respect to the action of G. Since the action φ is free, U^{2n}/G is a smooth manifold.

Moreover, the symplectic structure and Liouville foliation can also be transferred to U^{2n}/G from U^{2n}. As a result, we obtain a symplectic $2n$-dimensional manifold U^{2n}/G with the Liouville foliation defined by n commuting independent functions. Clearly, U^{2n}/G can be considered as a neighborhood of a singular leaf all of whose singular points are non-degenerate. In other words, U^{2n}/G represents a model non-degenerate singularity of an integrable Hamiltonian system with n degrees of freedom.

Definition 9.8. The singularities of type U^{2n}/G (as well as those which are Liouville equivalent to them) are called *singularities of almost direct product type*.

Theorem 9.12 (Nguyen Tien Zung). *Each non-degenerate multidimensional singularity that satisfies the non-splitting condition is a singularity of almost direct product type.*

For systems with two degrees of freedom, the proof of Theorem 9.12 has been obtained in Sections 9.3–9.8; for the general case, see [258].

REMARK. It should be emphasized that Theorem 9.12 has the topological, but not symplectic character. More precisely, it states that every non-degenerate singularity satisfying the non-splitting condition is smoothly fiberwise equivalent to a certain model singularity of almost direct product type. But the corresponding fiber mapping does not have to be a symplectomorphism. That is, the symplectic structure on the direct product is not necessarily the product of symplectic forms of the factors.

As an application of this theorem, we give the classification of six-dimensional saddle–saddle–saddle singularities of complexity 1, which has been recently obtained by V. V. Kalashnikov (Jr.).

Theorem 9.13 (V. V. Kalashnikov (Jr.) [172]). *Suppose that the singular leaf L of saddle–saddle–saddle type (in a six-dimensional symplectic manifold) contains exactly one non-degenerate singular point. Then there exist 32 types of such singularities. They are listed in Table 9.3.*

COMMENT. It is worth noticing that all these singularities are produced from 2-atoms of four types only: B, D_1, C_2, P_4. These atoms together with their symmetries are presented in the list after Table 9.1. No other more complicated components appear in this case.

The classification is given in terms of almost direct products of 2-atoms. In the second column of Table 9.3, we indicate the three factors that determine the given singularity. In the last column of the table, we show the group G acting on the direct product of these 2-atoms. The singularity itself is obtained by taking a quotient with respect to the action of G. This action is described in the third column of Table 9.3. In all the cases (except for case 19), G is the direct sum of the groups $\mathbb{Z}_2$ and, therefore, has k natural generators (where k is the number of factors). For each generator, we indicate its action on the direct product of the three 2-atoms. Since this action is component-wise, we indicate the corresponding symmetries for each component.

Table 9.3. Saddle–saddle–saddle singularities as almost direct products

No	FACTORS			ACTION	GROUP
1	B	B	B	trivial	$\{e\}$
2	C_2	B	C_2	$(\alpha, \mathrm{id}, \alpha\beta)$, $(\alpha\beta, \mathrm{id}, \alpha)$	$\mathbb{Z}_2 \oplus \mathbb{Z}_2$
3	B	B	D_1	$(\alpha, \mathrm{id}, \alpha)$	$\mathbb{Z}_2$
4	B	B	C_2	$(\alpha, \mathrm{id}, \alpha)$	$\mathbb{Z}_2$
5	C_2	C_2	C_2	$(\alpha, \mathrm{id}, \alpha\beta)$, $(\mathrm{id}, \alpha, \alpha\beta)$, $(\alpha\beta, \alpha\beta, \alpha)$	$\mathbb{Z}_2 \oplus \mathbb{Z}_2 \oplus \mathbb{Z}_2$
6	D_1	C_2	C_2	$(\alpha, \mathrm{id}, \alpha\beta)$, $(\mathrm{id}, \alpha, \alpha\beta)$, $(\mathrm{id}, \alpha\beta, \alpha)$	$\mathbb{Z}_2 \oplus \mathbb{Z}_2 \oplus \mathbb{Z}_2$
7	B	C_2	C_2	$(\mathrm{id}, \alpha, \alpha\beta)$, $(\alpha, \alpha\beta, \alpha)$	$\mathbb{Z}_2 \oplus \mathbb{Z}_2$
8	C_2	D_1	C_2	$(\alpha, \mathrm{id}, \alpha\beta)$, $(\mathrm{id}, \alpha, \alpha\beta)$, $(\alpha\beta, \mathrm{id}, \alpha)$	$\mathbb{Z}_2 \oplus \mathbb{Z}_2 \oplus \mathbb{Z}_2$
9	D_1	D_1	B	$(\alpha, \mathrm{id}, \alpha)$, $(\mathrm{id}, \alpha, \alpha)$	$\mathbb{Z}_2 \oplus \mathbb{Z}_2$
10	B	D_1	C_2	$(\mathrm{id}, \alpha, \alpha\beta)$, $(\alpha, \mathrm{id}, \alpha)$	$\mathbb{Z}_2 \oplus \mathbb{Z}_2$
11	C_2	D_1	B	$(\alpha, \mathrm{id}, \alpha)$, $(\mathrm{id}, \alpha, \alpha)$	$\mathbb{Z}_2 \oplus \mathbb{Z}_2$
12	C_2	C_2	C_2	$(\alpha, \mathrm{id}, \alpha\beta)$, $(\mathrm{id}, \alpha, \alpha\beta)$, $(\alpha\beta, \mathrm{id}, \alpha)$	$\mathbb{Z}_2 \oplus \mathbb{Z}_2 \oplus \mathbb{Z}_2$
13	D_1	C_2	B	$(\alpha, \mathrm{id}, \alpha)$, $(\mathrm{id}, \alpha, \alpha)$	$\mathbb{Z}_2 \oplus \mathbb{Z}_2$
14	B	C_2	C_2	$(\mathrm{id}, \alpha, \alpha\beta)$, $(\alpha, \mathrm{id}, \alpha)$	$\mathbb{Z}_2 \oplus \mathbb{Z}_2$
15	C_2	C_2	B	$(\alpha, \mathrm{id}, \alpha)$, $(\mathrm{id}, \alpha, \alpha)$	$\mathbb{Z}_2 \oplus \mathbb{Z}_2$
16	D_1	B	C_2	$(\alpha, \mathrm{id}, \alpha\beta)$, $(\mathrm{id}, \alpha, \alpha)$	$\mathbb{Z}_2 \oplus \mathbb{Z}_2$
17	C_2	C_2	C_2	$(\alpha, \alpha\beta, \alpha\beta)$, $(\alpha\beta, \alpha, \alpha\beta)$, $(\alpha\beta, \alpha\beta, \alpha)$	$\mathbb{Z}_2 \oplus \mathbb{Z}_2 \oplus \mathbb{Z}_2$
18	C_2	C_2	C_2	$(\alpha, \alpha\beta, \mathrm{id})$, $(\alpha\beta, \alpha, \alpha\beta)$, $(\alpha\beta, \alpha\beta, \alpha)$	$\mathbb{Z}_2 \oplus \mathbb{Z}_2 \oplus \mathbb{Z}_2$
19	P_4	P_4	B	$(\beta, \gamma\beta, \alpha)$, $(\gamma^2\beta, \gamma\beta, \mathrm{id})$, $(\gamma\beta, \beta, \alpha)$, $(\gamma\beta, \gamma^2\beta, \mathrm{id})$	G_{16}
20	C_2	C_2	P_4	$(\alpha, \alpha\beta, \mathrm{id})$, $(\alpha\beta, \alpha, \mathrm{id})$, $(\alpha\beta, \alpha\beta, \beta)$, $(\alpha\beta, \mathrm{id}, \gamma^2\beta)$	$\mathbb{Z}_2 \oplus \mathbb{Z}_2 \oplus \mathbb{Z}_2 \oplus \mathbb{Z}_2$
21	C_2	C_2	C_2	$(\alpha, \alpha\beta, \mathrm{id})$, $(\alpha\beta, \alpha, \alpha\beta)$, $(\alpha\beta, \mathrm{id}, \alpha)$	$\mathbb{Z}_2 \oplus \mathbb{Z}_2 \oplus \mathbb{Z}_2$
22	C_2	C_2	B	$(\alpha, \alpha\beta, \alpha)$, $(\alpha\beta, \alpha, \alpha)$	$\mathbb{Z}_2 \oplus \mathbb{Z}_2$
23	C_2	C_2	P_4	$(\alpha, \alpha\beta, \mathrm{id})$, $(\alpha\beta, \alpha, \mathrm{id})$, $(\alpha\beta, \alpha\beta, \beta)$, $(\mathrm{id}, \alpha\beta, \gamma^2\beta)$	$\mathbb{Z}_2 \oplus \mathbb{Z}_2 \oplus \mathbb{Z}_2 \oplus \mathbb{Z}_2$
24	D_1	C_2	B	$(\alpha, \alpha\beta, \alpha)$, $(\mathrm{id}, \alpha, \alpha)$	$\mathbb{Z}_2 \oplus \mathbb{Z}_2$
25	P_4	C_2	B	$(\beta, \alpha\beta, \mathrm{id})$, $(\gamma^2\beta, \mathrm{id}, \alpha)$, $(\mathrm{id}, \alpha, \alpha)$	$\mathbb{Z}_2 \oplus \mathbb{Z}_2 \oplus \mathbb{Z}_2$
26	P_4	C_2	B	$(\beta, \alpha\beta, \alpha)$, $(\gamma^2\beta, \mathrm{id}, \alpha)$, $(\mathrm{id}, \alpha, \alpha)$	$\mathbb{Z}_2 \oplus \mathbb{Z}_2 \oplus \mathbb{Z}_2$
27	D_1	B	C_2	$(\alpha, \alpha, \alpha\beta)$, $(\mathrm{id}, \alpha, \alpha)$	$\mathbb{Z}_2 \oplus \mathbb{Z}_2$
28	P_4	B	C_2	$(\beta, \alpha, \mathrm{id})$, $(\gamma^2\beta, \mathrm{id}, \alpha\beta)$, $(\mathrm{id}, \alpha, \alpha)$	$\mathbb{Z}_2 \oplus \mathbb{Z}_2 \oplus \mathbb{Z}_2$
29	P_4	B	C_2	$(\beta, \alpha, \alpha\beta)$, $(\gamma^2\beta, \mathrm{id}, \alpha\beta)$, $(\mathrm{id}, \alpha, \alpha)$	$\mathbb{Z}_2 \oplus \mathbb{Z}_2 \oplus \mathbb{Z}_2$
30	C_2	P_4	B	$(\alpha, \mathrm{id}, \alpha)$, $(\alpha\beta, \beta, \alpha)$, $(\mathrm{id}, \gamma^2\beta, \alpha)$	$\mathbb{Z}_2 \oplus \mathbb{Z}_2 \oplus \mathbb{Z}_2$
31	C_2	C_2	C_2	$(\alpha, \alpha\beta, \mathrm{id})$, $(\mathrm{id}, \alpha, \alpha\beta)$, $(\alpha\beta, \mathrm{id}, \alpha)$	$\mathbb{Z}_2 \oplus \mathbb{Z}_2 \oplus \mathbb{Z}_2$
32	C_2	C_2	B	$(\alpha, \alpha\beta, \alpha)$, $(\mathrm{id}, \alpha, \alpha)$	$\mathbb{Z}_2 \oplus \mathbb{Z}_2$

For example, the singularity number 8 is obtained as follows. First, we take the direct product $C_2 \times D_1 \times C_2$ of three 2-atoms and then consider the action of $\mathbb{Z}_2 \oplus \mathbb{Z}_2 \oplus \mathbb{Z}_2$ on it. This group acts on the direct product in the following way. The first generator e_1 of the group $\mathbb{Z}_2 \oplus \mathbb{Z}_2 \oplus \mathbb{Z}_2$ acts as follows:

$$e_1(C_2 \times D_1 \times C_2) = \alpha(C_2) \times \mathrm{id}(D_1) \times \alpha\beta(C_2).$$

The second generator e_2 of the group $\mathbb{Z}_2 \oplus \mathbb{Z}_2 \oplus \mathbb{Z}_2$ acts as follows:

$$e_2(C_2 \times D_1 \times C_2) = \mathrm{id}(C_2) \times \alpha(D_1) \times \alpha\beta(C_2).$$

And, finally, the third generator e_3 of the group $\mathbb{Z}_2 \oplus \mathbb{Z}_2 \oplus \mathbb{Z}_2$ acts as follows:

$$e_3(C_2 \times D_1 \times C_2) = \alpha\beta(C_2) \times \mathrm{id}(D_1) \times \alpha(C_2).$$

Here α and β denote the same symmetries of 2-atoms as in Table 9.1. In case 19, G is a non-commutative group of order 16. In Table 9.3, we indicate the action of four generators $e_1, e_2, e_3, e_4 \in G$.

Chapter 10

Methods of Calculation of Topological Invariants of Integrable Hamiltonian Systems

10.1. GENERAL SCHEME FOR TOPOLOGICAL ANALYSIS OF THE LIOUVILLE FOLIATION

10.1.1. *Momentum Mapping*

Let $v = \operatorname{sgrad} H$ be an integrable Hamiltonian system on a four-dimensional symplectic manifold M^4. We assume that the Hamiltonian H is given in an explicit way, as well as the additional integral f. It should be noted that the integral f is not uniquely defined and can be replaced by an arbitrary function of f and H. We recall that, in the non-resonant and non-degenerate case, the topology of the Liouville foliation does not depend on the specific choice of the integral f. In this case, the molecules corresponding to two different Bott integrals f and f' will coincide. We may use this fact by choosing possibly simplest integral f. The recommendation is that, among different possible integrals, we choose a function f which has the least number of critical points. For example, sometimes, to make f better, it is useful to extract the root: $\sqrt{f}$.

Having chosen f, we should consider the momentum mapping $\mathcal{F}: M^4 \to \mathbb{R}^2$, i.e., $\mathcal{F}(x) = (H(x), f(x))$. Here H and f are taken as Cartesian coordinates in $\mathbb{R}^2$. In the most specific problems, the image of M^4 is a closed subset $\mathcal{F}(M^4)$ in the plane.

In the most interesting cases, all isoenergy 3-manifolds $Q^3 = \{H = \operatorname{const}\}$ are compact. Moreover, the preimage $\mathcal{F}^{-1}(A)$ of any compact subset $A \subset \mathbb{R}^2$ is compact. In what follows, we assume this condition to be fulfilled.

10.1.2. Construction of the Bifurcation Diagram

Consider the set K of critical points of the momentum mapping $\mathcal{F}$:

$$K = \{x \in M^4 : \operatorname{rank} d\mathcal{F}(x) < 2\}.$$

This set is closed in M^4, and its image $\mathcal{F}(K) \subset \mathbb{R}^2$ is denoted by Σ and called the *bifurcation diagram* of the momentum mapping. The bifurcation diagram Σ is usually closed in $\mathbb{R}^2$ and consists of a number of smooth curves, which may intersect or be tangent to each other; Σ may also contain isolated points.

The construction of Σ is reduced to the description of the set where $\operatorname{grad} H$ and $\operatorname{grad} f$ are linearly dependent. This problem is purely analytical provided H and f are given by explicit formulas. It may turn out to be rather complicated. But for many series of integrable systems which we discuss below, this problem can be effectively solved.

The set Σ separates the image of the momentum mapping $\mathcal{F}(M)$ into open connected subsets which we shall call *cameras*. Usually there is a finite number of them. Some of them are bounded, some go to infinity (Fig. 10.1). The boundary of a camera consists of several smooth curves $\gamma_i \subset \Sigma$ (called *walls* of the camera)

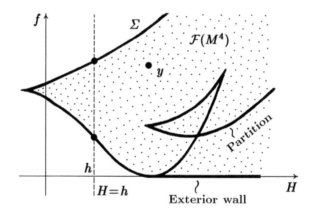

Figure 10.1

and some singular points of Σ (*corners* of the camera). Walls can be of two types: *partitions* and *exterior* ones. A wall is called a partition if it separates two cameras; a wall is called exterior if it belongs to the boundary of the image of the momentum mapping $\mathcal{F}(M^4)$ (see Fig. 10.1).

10.1.3. Verification of the Non-degeneracy Condition

The next step is to verify the condition that the additional integral is a Bott function on each isoenergy manifold $Q_h^3 = \{H = h\}$. In other words, we should prove that the singularities of $\mathcal{F}$ corresponding to each smooth piece γ of the bifurcation diagram Σ are non-degenerate in the sense of Definition 1.21. In most cases, this condition is indeed fulfilled, and the set $K(\gamma)$ represents a family of one-dimensional non-degenerate orbits of the Poisson action of $\mathbb{R}^2(H, f)$.

The same can be reformulated in terms of the isoenergy 3-manifold Q_h^3, which is the preimage of the vertical line $\{H = h\} \subset \mathbb{R}^2$ in M^4 under the momentum mapping $\mathcal{F}$ (Fig. 10.1). Suppose that the line $\{H = h\}$ intersects smooth pieces of the diagram Σ transversally and does not pass through the singular points of Σ. In such a case, the non-degeneracy of singularities in M^4 is equivalent to the fact that the restriction of f onto Q_h^3 is a Bott function. Varying h, i.e., moving the line $\{H = h\}$ to the left and right, we obtain the set of all isoenergy 3-manifolds of the given integrable system. Moreover, under this variation, the structure of the Liouville foliation remains the same unless we pass through the singular points of Σ or break the transversality of intersection of the line $\{H = h\}$ with the bifurcation diagram Σ. On the other hand, this approach allows us to indicate those bifurcational energy levels at which a bifurcation of the topological type of the Liouville foliation on Q_h^3 happens.

Concrete examples showing how to verify the non-degeneracy condition are presented in Chapter 12 (for integrable geodesic flows on two-dimensional surfaces) and in Chapter 14 (for integrable cases in rigid body dynamics).

10.1.4. Description of the Atoms of the System

If a point $y = \mathcal{F}(x)$ lies inside a camera, then its preimage $\mathcal{F}^{-1}(y)$ consists of a number of two-dimensional Liouville tori in M^4. Varying y inside the camera, we make these tori move isotopically in M^4. Therefore, for any two points y and y' from the same camera, the number of Liouville tori over them is the same. In other words, no bifurcations happen inside the camera.

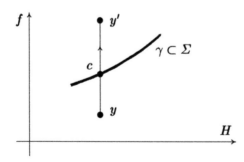

Figure 10.2

On the contrary, if y approaches a wall of the camera along a certain smooth path τ, pierces the wall at some point c (Fig. 10.2), and passes to the neighboring camera, then a bifurcation of the Liouville tori happens. We shall always assume that the path τ is transversal to the wall. Since we have already verified the non-degeneracy of the system on the wall of the camera, this bifurcation is coded by a certain atom (or a disjoint union of several atoms). The present step is the description of these atoms. In general, this problem can be rather non-trivial, and the list of atoms presented in Chapter 3 turns out to be very useful. Let us illustrate this by an example.

Suppose that we have just one Liouville torus before and after a bifurcation. Let the point c (i.e., the point where τ pierces the wall of the camera) correspond to exactly one critical circle. Then the appearing atom has to be of type A^*. This follows immediately from Table 3.1, since there is the only atom satisfying these properties.

Thus, it is useful to find out in advance how many tori there were before bifurcation, and how many tori appear after bifurcation, as well as how many critical circles correspond to the critical level c. In many cases, this allows us to reject most possibilities and to reduce the problem to the analysis of a comparatively small number of atoms.

Consider one more example. If a camera has an exterior wall, then the atoms corresponding to it are all of type A. Partitions may correspond both to atoms of type A and to saddle ones. One of such possibilities is shown in Fig. 10.3.

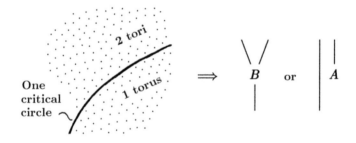

Figure 10.3

In one camera, there is one Liouville torus; in the other, there are two. The wall between them corresponds to one critical circle. Two cases are possible: either the bifurcation of type B or one Liouville torus passes through the wall without any bifurcation and the other undergoes the bifurcation of type A and as a result disappears.

10.1.5. Construction of the Molecule of the System on a Given Energy Level

On the previous step, we have found out how many Liouville tori "hang" over each camera and what are the atoms corresponding to the walls of cameras. Now we can collect all this information in the form of a single molecule W. In other words, we should take all the atoms that code the appearing bifurcations (Fig. 10.4), consider their "ends" and glue them in the order determined by the topology of the isoenergy manifold Q_h^3. This procedure is not automatic, since the "ends"

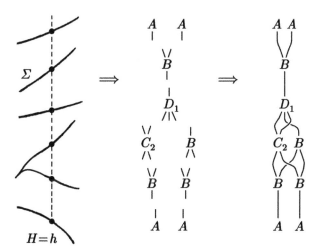

Figure 10.4

of atoms can be joined in different ways which lead, generally speaking, to different molecules. We must choose the only possibility that is realized in the given system. This usually requires some additional efforts and arguments.

10.1.6. *Computation of Marks*

To complete the construction of the marked molecule W^*, it remains to compute the marks r_i, ε_i, and n_k. It is rather hard to suggest a general algorithm for this. In different situations one has to use different reasons and methods. Some of them are discussed in the next section.

This is the last step. As a result, we obtain the marked molecule W^* that is, as we already know, a complete invariant of the system on the given energy level in the sense of Liouville equivalence.

10.2. METHODS FOR COMPUTING MARKS

As we shall see below, in real physical and geometrical problems, the following 3-manifolds often appear to be isoenergy surfaces of Hamiltonian systems with two degrees of freedom:

1) the sphere S^3,
2) the projective space $\mathbb{R}P^3$,
3) the three-dimensional torus T^3,
4) the direct product $S^1 \times S^2$,
5) the unit (co)tangent vector bundle over a two-dimensional surface $Q^3 \xrightarrow{S^1} X^2$,
6) connected sums of several copies of $S^1 \times S^2$.

The first important example is the simplest molecule $A\!-\!A$ with marks r and ε on the edge (there is no mark n in this case). It turns out that, in this case, the topology of Q^3 determines the mark r almost uniquely.

Proposition 10.1.

a) *The manifold Q^3 corresponding to the molecule $A\!-\!A$ is obtained by gluing two solid tori along their boundaries and is homeomorphic to one of the following manifolds: $S^3, \mathbb{R}P^3, S^1 \times S^2$, and lens spaces $L_{p,q}$.*

b) *If $Q^3 = S^3$, then $r = 0$ and the mark $\varepsilon = \pm 1$ is defined by the orientation chosen on S^3.*

c) *If $Q^3 = \mathbb{R}P^3$, then $r = 1/2$, and the mark $\varepsilon = \pm 1$ is defined by the orientation chosen on $\mathbb{R}P^3$.*

d) *If $Q^3 = S^1 \times S^2$, then $r = \infty$, and the mark $\varepsilon = \pm 1$ is defined as follows. The Hamiltonian flow* sgrad H *corresponding to the molecule $A\!-\!A$ has exactly two singular periodic trajectories (the axes of the atoms A) which are fibers of the trivial S^1-fibration on Q^3. If* sgrad H *flows along these trajectories in the same direction, then $\varepsilon = -1$. If these directions are opposite, then $\varepsilon = +1$.*

The proof immediately follows from Proposition 4.3. We only recall that the molecule $A\!-\!A$ means exactly that Q^3 is glued of two solid tori. By fixing bases on their boundary tori (i.e., admissible coordinate systems) as described in Chapter 4, we obtain the following gluing matrix

$$\begin{pmatrix} \alpha & \beta \\ \gamma & \delta \end{pmatrix},$$

in terms of which one can uniquely describe the topology of Q^3. In Chapter 4 we showed that the topology of Q^3 was defined by the r-mark in a one-to-one manner. In particular, in the cases of the sphere, projective space, and direct product $S^1 \times S^2$, the r-mark is equal to 0, $1/2$, and ∞ respectively, as required. □

10.3.   THE LOOP MOLECULE METHOD

Let us recall the notion of a loop molecule introduced in Chapter 9. Let Σ be the bifurcation diagram of the momentum mapping $\mathcal{F}$ of an integrable system $v = $ sgrad H. The diagram Σ is located in the plane $\mathbb{R}^2(H, f)$. Suppose that Σ represents a collection of smooth curves $\gamma_1, \ldots, \gamma_k$, which may intersect or be tangent at certain points. We also assume that each curve γ_i satisfies the non-degeneracy condition (see Section 10.1.3). Note that Σ may contain isolated points.

Consider the singular points of the bifurcation diagram Σ, i.e., the intersection points, points of tangency, cusps, isolated points, etc. (see Chapter 9 for details).

Denote the set of singular points of Σ by Σ_0. We shall assume that Σ_0 is a finite set. If Σ is regarded as a one-dimensional cell complex, then Σ_0 is just the set of its vertices (zero-cells). The edges of this complex are $\gamma_1, \ldots, \gamma_k$.

Recall that a smooth curve τ without self-intersections in the plane $\mathbb{R}^2 (H, f)$ is called *admissible* if it intersects the bifurcation diagram transversally and does not pass through the singular points of Σ.

As was already explained, the preimage $Q_\tau = \mathcal{F}^{-1}(\tau)$ is a three-dimensional smooth submanifold in M^4 with the structure of a Liouville foliation, all of whose singularities are non-degenerate. Thus, we can consider the invariant of this foliation on Q_τ, namely, the marked molecule, which we denote by $W^*(\tau)$. We showed in Chapter 9 that this molecule does not change under isotopy of τ in the class of admissible curves.

Now let $y_0 \in \Sigma_0$ be one of isolated singular points of the bifurcation diagram. Consider a circle τ of small radius ε centered at the point y_0. Suppose that τ is an admissible curve for any sufficiently small ε. Then the molecule $W^*(\tau)$ is well defined and is called the *loop molecule* of the singular point $y_0 \in \Sigma$. We shall denote it by $W^*(y_0)$.

The idea to use loop molecules to describe the global structure of the Liouville foliation on isoenergy surfaces can be explained as follows. The loop molecule is a local invariant of a singularity. That is why, it is usually more readily identified than the molecule for an isoenergy surface $W^*(Q^3)$, which is a global invariant. If the type of the singularity corresponding to a singular point $y_0 \in \Sigma_0$ is understood and described, then calculating the corresponding loop molecule is just a formal procedure and does not lead to any serious difficulties. Moreover, in specific problems, we usually meet singularities from some finite list. Therefore, one can collect such typical (standard) singularities and their loop molecules in advance, once and forever, and then use this list to study different systems.

Suppose that we have computed all the loop molecules for all singular points of the bifurcation diagram. Then, for any admissible curve τ on the plane $\mathbb{R}^2 (H, f)$, the marked molecule $W^*(\tau)$ can be glued from pieces of the loop molecules. It turns out that this procedure allows us to get much information about the desired molecule $W^*(\tau)$, in particular, about the marks. Sometimes, it is possible to compute $W^*(\tau)$ completely.

Let us illustrate this idea by Fig. 10.5. In Fig. 10.5(a) we show a bifurcation diagram Σ with three singular points y_1, y_2, y_3. An admissible curve $\tau = \{H = h_0\}$ intersects four smooth parts γ_1, γ_2, γ_3, γ_4 of this bifurcation diagram. Each part γ_i corresponds to a certain atom V_i. Let us note that these atoms are the same as those which occur in the loop molecules $W^*(y_1)$, $W^*(y_2)$, $W^*(y_3)$. Moreover, the r-marks on the edges between the atoms also coincide with the corresponding r-marks for the loop molecules. For example, the molecules $W^*(\tau)$ and $W^*(y_1)$ have the common fragment of the form $V_1 - V_2$ (see Fig. 10.6).

In Fig. 10.5(b) we show deformations of the edges of the loop molecules which drag these edges to the corresponding segments of the admissible curve τ. The same idea is illustrated in Fig. 10.5(c). It is seen from Fig. 10.6 that, under such an isotopy, the r-marks on the deforming edges of the loop molecules do not change. As a result, the desired molecule $W^*(\tau)$ with r-marks can be composed from the corresponding fragments of the loop molecules $W^*(y_1)$, $W^*(y_2)$, and $W^*(y_3)$.

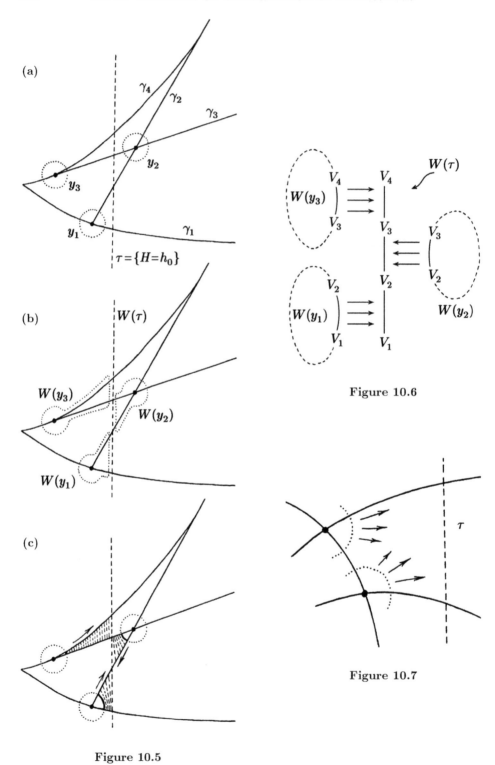

Figure 10.5

Figure 10.6

Figure 10.7

We have actually described a rather simple case.   A real situation may, of course, be more complicated. For example, it may happen that one fragment of τ corresponds to several edges of loop molecules (see Fig. 10.7).   In this case, one needs to define the rule for summation of the marks coming from different loop molecules.   We discuss one of such rules below.   Some other difficulties may also occur. Nevertheless, this scheme works successfully in many situations.

We now consider the situation presented in Fig. 10.7 in more detail. Suppose that we know the mark $r = p/q$ for the fragment of the loop molecule of the singular point y_1, as well as the mark $r = s/t$ for the fragment of the loop molecule of the singular point y_2 (Fig. 10.8). Is it possible to find the r-mark on the segment of τ corresponding to these fragments of the two loop molecules?

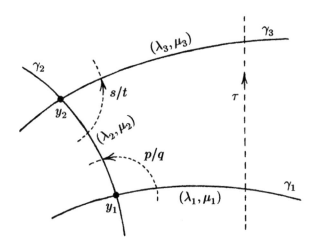

Figure 10.8

Consider the atoms corresponding to the parts $\gamma_1, \gamma_2, \gamma_3$ of Σ (Fig. 10.8). Let us distinguish the region bounded by $\gamma_1, \gamma_2, \gamma_3$.   Each of its points corresponds to a certain Liouville torus.   Moving along the vertical segment τ, this torus generates an edge of the desired molecule $W^*(\tau)$.   Moving along small circles centered at y_1 and y_2 (Fig. 10.8), the Liouville torus generates two edges of the corresponding loop molecules.   When the Liouville torus approaches the bifurcation diagram, it can be considered as a boundary torus of the corresponding atom. Therefore, on this torus, we can define three admissible coordinate systems (λ_1, μ_1), (λ_2, μ_2), and (λ_3, μ_3) corresponding to γ_1, γ_2, and γ_3 respectively.   Then the transition matrix from γ_1 to γ_3 (i.e., the gluing matrix for the corresponding atoms) is the product of the two other transition matrices, namely, from γ_1 to γ_2 and from γ_2 to γ_3. This allows us to obtain the following formula that connects the marks on the edges of the loop molecules with the mark on τ.

Proposition 10.2. *Suppose that the r-marks on the indicated edges of the loop molecules (Fig. 10.8) are equal to p/q and s/t; then the r-mark on the segment τ is*

$$\frac{(s+mt)p+t\gamma}{(s+mt)q+t\delta}\,,$$

where m is some integer, and $\gamma, \delta \in \mathbb{Z}$ are uniquely defined from the conditions $p\delta - q\gamma = -1$ and $0 \le \delta < |q|$.

The proof easily follows from Definition 4.3 and the rule of multiplication for gluing matrices.   □

It is seen that in general the r-mark on the edge τ cannot be uniquely defined from p/q and s/t, since the answer contains an arbitrary parameter m. However, in some particular cases the formula from Proposition 10.2 gives the exact answer.

Corollary. *If one of the r-marks in Fig. 10.8 is equal to infinity, then the resulting r-mark on the edge τ coincides with the other r-mark. For example, if $q = 0$, then the resulting r-mark is equal to s/t.*

10.4.   LIST OF TYPICAL LOOP MOLECULES

10.4.1. *Loop Molecules of Regular Points of the Bifurcation Diagram*

We begin with the simplest case and indicate the loop molecules for non-singular points of the bifurcation diagram. This information sometimes may also be useful.

Let $y \in \Sigma$ be a non-singular point of the bifurcation diagram lying on a smooth curve $\gamma \subset \Sigma$. We assume that all the points of γ (including y) correspond to a certain non-degenerate singularity, i.e., to a certain atom V. Consider a circle τ of small radius centered at the point y (Fig. 10.9). This circle is

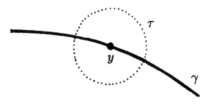

Figure 10.9

obviously admissible. Therefore, we can consider its preimage $Q_y^3 = \mathcal{F}^{-1}(\tau)$ and the corresponding loop molecule $W^*(y)$ that describes the Liouville foliation on Q^3.

We distinguish two cases:
1) the case, when V is the atom of type A;
2) the case of a saddle atom V.

Proposition 10.3 (The case of the atom A).

Suppose that $y \in \Sigma$ is a non-singular point of the bifurcation diagram corresponding to the atom A (Fig. 10.10(a)). Then the loop molecule $W^*(y)$ has the form

$$A - A,$$

where $r = \infty$, $\varepsilon = -1$. Moreover, the manifold $Q_y^3 = \mathcal{F}^{-1}(\tau)$ is diffeomorphic to the direct product $S^1 \times S^2$.

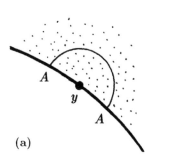

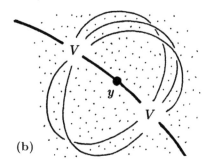

(a)   (b)

Figure 10.10

Proposition 10.4 (The case of a saddle atom V).

Suppose that $y \in \Sigma$ is a non-singular point of the bifurcation diagram corresponding to a saddle atom V. Then the loop molecule $W^*(y)$ has the form shown in Fig. 10.10(b). Moreover, all the r-marks are equal to infinity, all the ε-marks are equal to $+1$, and the mark n is equal to zero. If V is an atom without stars, then $Q_y^3 = \mathcal{F}^{-1}(\tau)$ is diffeomorphic to the direct product

$$P_k^2 \times S^1,$$

where P_k^2 is the closed oriented two-dimensional surface of genus

$$k = 2g(V) + (\text{the number of the edges of } V) - 1.$$

If V is an atom with stars, then $Q_y^3 = \mathcal{F}^{-1}(\tau)$ is diffeomorphic to the Seifert S^1-fibration over the same surface P_k^2, whose singular fibers are all of type $(2,1)$ and are in one-to-one correspondence with the star-vertices.

Proof. We prove both Propositions 10.3 and 10.4 simultaneously.

Let us deform the circle τ (see Fig. 10.9) into the curve shown in Fig. 10.11, which can be regarded as two segments whose endpoints are glued. Moreover, both segments are transversal to the bifurcation diagram. Therefore, the preimage of each segment is a 3-atom. Thus, we obtain two copies of the same atom V which

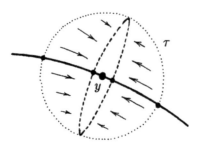

Figure 10.11

should be glued along the boundary tori by the identity mapping. More precisely, each boundary torus is glued with its duplicate by the identity mapping. Hence the loop molecule $W(y)$ has the desired form shown in Fig. 10.10.

The explicit form of the marks r, ε, and n indicated in Propositions 10.3 and 10.4 follows immediately from the fact that the gluing is carried out by the identity mapping. We only note that, in the case of the atom A, we have $\varepsilon = -1$ (by Proposition 10.1), since the Hamiltonian flow sgrad H has the same directions on both the critical circles (axes of the atoms A).

We now describe the topology of the 3-manifold $Q_y^3 = \mathcal{F}^{-1}(\tau)$. Clearly, Q_y^3 is obtained by gluing two copies of the 3-atom V along their boundaries: $Q_y^3 = V + V$. At the same time, the 3-atom V is either the direct product $Y \times S^1$ or the Seifert fibration over Y. Here Y is a two-dimensional surface with boundary. Hence Q_y^3 is either the direct product $(Y+Y) \times S^1$ or the Seifert fibration over $Y + Y$. It remains to compute the genus of the surface $Y + Y$. In the case of the atom A, the base Y is a 2-disc and, consequently, $Y + Y$ is homeomorphic to S^2. Therefore, in this case, $Q_y^3 = S^2 \times S^1$. If V is a saddle atom, then the genus of $Y + Y$ is expressed via the genus of Y and the number of its boundary components according to the above formula.   □

10.4.2. Loop Molecules of Non-degenerate Singularities

For each type of non-degenerate singularities (i.e., center–center, center–saddle, saddle–saddle, and focus–focus), the loop molecules have been described in Chapter 9. On the other hand, it is interesting to look at those of them which really occur in specific problems of geometry and classical mechanics. We have completed such a list on the basis of a wide analysis of concrete examples of integrable Hamiltonian systems.

The result is presented in Table 10.1.

For each loop molecule included in this table, we indicate those integrable cases where the molecule occurs.

This list is, probably, not complete; some new cases may appear in integrable problems which we have not examined yet from the point of view of the topology of their Liouville foliations. However, our list includes a number of the most interesting cases of integrability and, in this sense, is rather representative.

Table 10.1. Typical loop molecules for non-degenerate singularities

center–center	center–saddle
$A \overset{r=0}{\text{———}} A$ almost all cases	$A \overset{r=\infty}{\diagdown} \; A \overset{r=\infty}{\diagup} \quad B \overset{r=\infty}{\text{———}} A$ Kovalevskaya, Sretenskiĭ, Steklov cases

focus–focus	
$\begin{pmatrix} 1 & 0 \\ 1 & 1 \end{pmatrix}$ Lagrange case	$A \overset{r=\infty}{\diagdown} \quad \overset{r=\infty}{\diagup} A$ $\qquad C_2 \qquad$ $A \overset{r=\infty}{\diagup} \quad \overset{r=\infty}{\diagdown} A$ Euler, Clebsch, Steklov cases, 4-dimensional rigid body
$\begin{pmatrix} 1 & 0 \\ 2 & 1 \end{pmatrix}$ Clebsch case, 4-dimensional rigid body	$A \overset{r=\infty}{\text{———}} D_1 \begin{array}{l} \overset{r=\infty}{\diagup} A \\ \overset{r=\infty}{\text{—}} A \\ \overset{r=\infty}{\diagdown} A \end{array}$ Sretenskiĭ case

saddle–saddle

Kovalevskaya case

Kovalevskaya case

Sretenskiĭ case

Clebsch case, 4-dimensional rigid body

Steklov case

10.5. THE STRUCTURE OF THE LIOUVILLE FOLIATION FOR TYPICAL DEGENERATE SINGULARITIES

In this section, we examine the case of one-dimensional degenerated orbits and corresponding singularities of the Liouville foliation. Consider the momentum mapping $\mathcal{F}: M^4 \to \mathbb{R}^2 (H, f)$ of a given integrable system $v = \operatorname{sgrad} H$ and the corresponding Poisson action $\Phi: \mathbb{R}^2 \to \operatorname{Diff}(M^4)$ generated by translation along integral curves of the Hamiltonian vector fields $v = \operatorname{sgrad} H$ and $w = \operatorname{sgrad} f$.

Let $O(x_0)$ be a periodic one-dimensional orbit of this action. If this orbit is degenerate (see Definition 1.20), then its image $y_0 = \mathcal{F}(O(x_0))$ is usually an isolated singular point of the bifurcation diagram Σ. For definiteness, we assume that $dH(x_0) \neq 0$; in particular, $O(x_0)$ is a periodic trajectory of $v = \operatorname{sgrad} H$.

Suppose that any circle τ of sufficiently small radius centered at $y_0 = \mathcal{F}(x_0)$ is an admissible curve on the plane $\mathbb{R}^2 (H, f)$. It will be convenient to introduce the function $f_\tau: Q_\tau \to S^1 = \tau$, where $f_\tau(x) = \mathcal{F}(x) \in \tau$. Then we obtain a mapping of Q_τ into the circle $\tau \subset \mathbb{R}^2(H, f)$. This function f_τ has non-degenerate singularities only and determines the structure of a Liouville foliation on Q_τ to be described.

Without loss of generality, we may assume that the preimage $\mathcal{F}^{-1}(y_0)$ is connected and coincides, consequently, with a leaf of the Liouville foliation. Note that the orbit $O(x_0)$ belongs to this leaf, but two cases are possible. The first is that the orbit $O(x_0)$ coincides with the leaf. In the second case, the orbit is "less" than the leaf, i.e., the leaf $\mathcal{F}^{-1}(y_0)$ also contains some other orbits.

We begin with the first case.

We first describe a certain model Liouville foliation. Consider an arbitrary Morse function g on the 2-sphere S^2 and extend it to the cylinder $D^1 \times S^2$ in a natural way. After that, we glue the bases of the cylinder, i.e., the 2-spheres $\{0\} \times S^2$ and $\{1\} \times S^2$, by a diffeomorphism σ which preserves the function g. As a result, the cylinder $D^1 \times S^2$ turns into a 3-manifold $Q^3 = S^1 \times S^2$, and the function g becomes a smooth Bott function on Q^3. The foliation on $S^1 \times S^2$ defined by this function g is just our model Liouville foliation.

Theorem 10.1 [44]. *Suppose that the orbit $O(x_0)$ coincides with the leaf $\mathcal{F}^{-1}(y_0)$. Then*

a) *the "isoenergy" 3-manifold Q_τ is diffeomorphic to the direct product $S^1 \times S^2$,*

b) *the Liouville foliation on Q_τ is diffeomorphic to one of the model Liouville foliations on $S^1 \times S^2$ for a suitable Morse function $g: S^2 \to \mathbb{R}$ and a diffeomorphism $\sigma: S^2 \to S^2$.*

Proof. We begin with part (a). Consider the following smooth function on M^4:

$$h_\tau(x) = (H(x) - H(x_0))^2 + (f(x) - f(x_0))^2 .$$

Clearly, the 3-manifold Q_τ is a level surface of this function: $Q_\tau = \{h_\tau(x) = \varepsilon\}$, where $\varepsilon > 0$ is a small number. Note that h_τ is an integral of the Hamiltonian system $v = \operatorname{sgrad} H$. Consider a two-dimensional transversal (local) section L^3 to the orbit $O(x_0)$ in M^4 (see Fig. 10.12). It is evident that the intersection of it with the level surface $Q_\tau = \{h_\tau(x) = \varepsilon\}$ is homeomorphic to the two-dimensional sphere of "small" radius. This follows from the explicit form of h_τ and the fact that the point x_0 is an isolated point of local minimum of h_τ on the transversal section L^3.

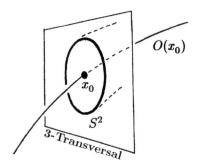

Figure 10.12

Consider the Poincaré map $\sigma: L^3 \to L^3$ generated by the Hamiltonian vector field $v = \operatorname{sgrad} H$. Clearly, the sphere $\{h_\tau = \varepsilon\} \cap L^3$ is invariant under the Poincaré map, since h_τ is an integral of v. The isoenergy 3-manifold Q_τ is obviously the union of all integral curves of v starting from the 2-sphere $\{h_\tau = \varepsilon\} \cap L^3$. By taking the restriction of σ to this sphere, we obtain a diffeomorphism of the sphere $S^2 = \{h_\tau = \varepsilon\} \cap L^3$ onto itself. Hence Q_τ can be viewed as the result of gluing the bases of the cylinder $D^1 \times S^2$ by some diffeomorphism. This diffeomorphism preserves orientation and, consequently, is homotopic to the identity mapping. Therefore, Q_τ is diffeomorphic to the direct product $S^1 \times S^2$. The first part of Theorem 10.1 is proved.

We now turn to part (b). Consider the function $f_\tau(x) = \mathcal{F}(x) = (H(x), f(x))$ on Q_τ which determines the Liouville foliation structure on Q_τ. By our assumption, f_τ is a Bott function on Q_τ. Therefore, its restriction to the 2-sphere $S^2 = \{h_\tau = \varepsilon\} \cap L^3$ is a Morse function, since this sphere is transversal to the Hamiltonian flow $v = \operatorname{sgrad} H$. We denote this restriction by $g(x)$. Note that the function f_τ is constant along integral curves of $\operatorname{sgrad} H$ and, therefore, can be considered as an extension of g from the sphere to the whole cylinder. In other words, the function f_τ has the required structure (see the description of the model Liouville foliation above). This completes the proof. □

In fact, not every model foliation can be realized as a Liouville foliation on Q_τ for a suitable degenerate orbit. The point is that the Morse function g and the diffeomorphism σ must satisfy some natural restrictions, which we now describe. As is seen from the proof of Theorem 10.1, the sphere S^2 lies in the transversal section L^3. Since $dH(x_0) \neq 0$, there exist local canonical coordinates p_1, q_1, p_2, q_2 such that $H = p_1$ and $f = f(p_1, p_2, q_2)$. Clearly, p_1, p_2, q_2 are local coordinates on L^3. Hence H is a height function on the sphere S^2. But the foliations on S^2 given by the Hamiltonian H and the function $g = f_\tau|_{S^2}$ coincide. Thus, without loss of generality, the function g can be considered as a height function on S^2. The condition on the diffeomorphism σ is that σ must be extendable to the interior of the ball in L^3 bounded by the sphere S^2. Moreover, for each ε, the diffeomorphism σ of the (contracting) sphere $S^2_\varepsilon = \{h_\tau = \varepsilon\}$ onto itself preserves the function g, and the topology of the one-dimensional foliation defined on $S^2_\varepsilon = \{h_\tau = \varepsilon\}$ by the function g does not depend on ε.

The analysis of these restrictions leads to the fact that, among the model foliations on $S^1 \times S^2$ described above, it suffices to consider only the following (see [269] for details).

Let g be a height function on the sphere S^2 smoothly embedded into $\mathbb{R}^3$. Suppose that the sphere S^2 and the function g on it are invariant under the rotation about the vertical axis Oz through the angle $2\pi p/q$ for some integers p and q. This rotation is taken as the diffeomorphism σ.   After that, we construct the Liouville foliation on $S^1 \times S^2$ corresponding to the pair (g, σ) as described above. The Liouville foliations constructed in this way are exactly those which are realized on Q_τ in the case when the degenerate orbit $O(x_0)$ coincide with the leaf $\mathcal{F}^{-1}(y_0)$.

All possible loop molecules $W^*(y_0)$ that appear in the above situation are described in [269].

It follows from the description of model foliations that the isoenergy manifold Q_τ has natural structure of a Seifert fibration compatible with the Liouville foliation (i.e., each fiber of the Seifert fibration lies on a certain leaf of the Liouville foliation). Moreover, this Seifert fibration either has no singular fibers (and is then trivial) or has exactly two singular fibers of the same type (p, q).

We now turn to the analysis of the second case when together with $O(x_0)$ the leaf $\mathcal{F}^{-1}(y_0)$ contains some other orbits. The most typical situation is that $\mathcal{F}^{-1}(y_0)$ contains a one-dimensional orbit N homeomorphic to an annulus $S^1 \times \mathbb{R}^1$, and $O(x_0)$ belongs to the closure of N, i.e., lies on the boundary of the annulus. In what follows, we shall assume this condition to be fulfilled. In this case, the situation is similar to the first case. However, the Seifert fibration on Q_τ may have a more complicated structure.

We also assume that all the objects are real-analytic.

Theorem 10.2 [44]. *Under the above assumptions, the isoenergy manifold Q_τ has natural structure of a Seifert fibration compatible with the Liouville foliation.*

Proof. Consider the two-dimensional orbit N such that $O(x_0) \subset \bar{N}$. By our assumption, the orbit N is homeomorphic to an annulus $S^1 \times \mathbb{R}^1$. Then there exists a linear combination

$$w = \lambda \operatorname{sgrad} H + \mu \operatorname{sgrad} f$$

such that the integral trajectories of w are all closed on the annulus $N = S^1 \times \mathbb{R}^1$. Moreover, one can construct such a function $F(H, f)$ that $w = \operatorname{sgrad} F$ and trajectories of w are closed not only on N, but also on all the orbits $O(x)$ passing near N. The construction of this function F was described in Chapter 3, where F was called a periodic integral. Recall that, to construct F, we consider a closed trajectory $\gamma_0 \subset N$ of w and move it isotopically to the nearest orbits $O(x)$. Denoting the cycles obtained on these orbits by γ, we can define F by the explicit formula:

$$F = \oint_\gamma \alpha,$$

where $d\alpha = \omega$. Here we use the standard method for constructing action variables in a neighborhood of a Liouville torus (see Section 1.4). As a result, we obtain a function F which is defined on all the orbits $O(x)$, where x is sufficiently close to γ_0. It is clear that the trajectories of $w = \operatorname{sgrad} F$ are closed on all such orbits.

Now we observe that F is a real-analytic function of H and f in a neighborhood V of the point $(H(x_0), f(x_0)) = y_0$. Therefore, it can be naturally extended to the whole neighborhood U of the singular leaf $\mathcal{F}^{-1}(y_0)$ of the form $U = \mathcal{F}^{-1}(V)$ by setting $F(x) = F(H(x), f(x))$. Being closed on some open subset in M^4, the trajectories of $w = \operatorname{sgrad} F$ will be closed on the whole neighborhood U (due to the analyticity of F) and, in particular, on the isoenergy manifold $Q_\tau \subset U$. Thus, Q_τ is foliated into closed trajectories each of which lies on a certain leaf of the Liouville foliation. As a result, we obtain on Q_τ the required structure of a Seifert fibration. $\square$

10.6. TYPICAL LOOP MOLECULES CORRESPONDING TO DEGENERATE ONE-DIMENSIONAL ORBITS

We shall list the main examples of singularities of the above type (as well as the corresponding loop molecules) appearing in classical mechanics and mathematical physics.

Example 1 (Parabolic orbits). Two types of non-degenerate periodic orbits are well known: elliptic (stable) and parabolic (unstable) ones. One can often meet a situation when, under the change of some parameters, a trajectory changes its type, for example, from elliptic to hyperbolic. At the moment of transition, the trajectory becomes degenerate, and the character of transition can be, as a rule, described by the following model example (such a case is called parabolic). Consider two functions $H = p_2$ and $f = f(p_1, q_1, p_2)$ in $\mathbb{R}^3(p_1, q_1, p_2)$ such that the level lines of f on horizontal planes $\{H = \text{const}\}$ have the form presented in Fig. 10.13. It is seen that, when the energy H increases, two non-degenerate critical points appear "out of nothing" (one hyperbolic, the other elliptic). The functions H and f define a one-dimensional foliation in $\mathbb{R}^3$. Starting from there, one can naturally construct a two-dimensional foliation on $\mathbb{R}^3 \times S^1$ by taking the Cartesian product with the circle S^1. We shall assume that S^1 is endowed with a periodic coordinate q_2 so that we have on $\mathbb{R}^3 \times S^1$ four coordinates p_1, q_1, p_2, q_2 which we put by definition to be symplectic (i.e., $\omega = dp_1 \wedge dq_1 + dp_2 \wedge dq_2$). It is clear that $H(p, q)$ and $f(p, q)$ commute and define a Liouville foliation structure on the four-dimensional symplectic manifold $\mathbb{R}^3 \times S^1$. (One can naturally consider $\mathbb{R}^3$ as a cross-section.)

The corresponding bifurcation diagram of the momentum mapping $\mathcal{F}: \mathbb{R}^3 \times S^1 \to \mathbb{R}^2(H, f)$ is shown in Fig. 10.13. It has a cusp which is entirely located inside the image of $\mathcal{F}$. The corresponding loop molecule W_τ (obtained by going around the cusp) is also shown in Fig. 10.13.

As an additional integral f, one can take a function which locally (near the origin) has the form

$$f = p_1^2 + q_1^3 - p_2 q_1.$$

Such a function is exactly the canonical form of a parabolic singularity [63]. In this case, the set of critical points of $\mathcal{F}$ on the three-dimensional cross-section is a parabola in $\mathbb{R}^3$ lying on the plane $\{p_1 = 0\}$ and given by the equation $q_1^2 = p_2/3$. For $q_1 > 0$, the singular points are elliptic; for $q_1 < 0$, they are hyperbolic. For $q_1 = 0$, we obtain the so-called parabolic singular point. The bifurcation diagram Σ is given on the plane $\mathbb{R}^2 (H, f)$ by the equation

$$f = \pm \frac{2}{3\sqrt{3}} H^{3/2}.$$

In the case when the system under consideration satisfies additional conditions, for example, if it admits a $\mathbb{Z}_2$-symmetry, two parabolic orbits may appear

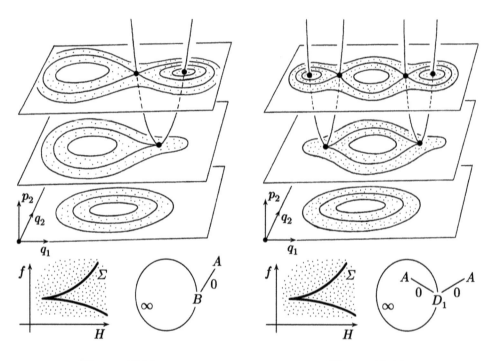

Figure 10.13   Figure 10.14

simultaneously on the same leaf of the Liouville foliation. Such a case is illustrated in Fig. 10.14, where we also show the corresponding loop molecule.

Example 2. This is an integrable variant of the well-known pitch-fork and period-doubling bifurcations.

The qualitative picture is again provided by two functions $H = p_2$ and $f = f(p_1, q_1, p_2)$ on $\mathbb{R}^3 (p_1, q_1, p_2)$. The level lines of f on horizontal planes $\{H = \mathrm{const}\}$ are shown in Fig. 10.15. When H increases, the elliptic singularity transforms into three non-degenerate singularities: two elliptic and one hyperbolic. The functions H and f define a one-dimensional foliation in $\mathbb{R}^3$, which can be transformed into a Liouville foliation on $\mathbb{R}^3 \times S^1$. But, unlike the parabolic case, there are now two possibilities due to the symmetry.

The first possibility is to take the Cartesian product with the circle S^1. For a moment, we call this case *orientable*. As before, we take a 2π-periodic coordinate q_2 on S^1 and define the symplectic structure to be $\omega = dp_1 \wedge dq_1 + dp_2 \wedge dq_2$. It is clear that $H(p,q)$ and $f(p,q)$ commute and define a Liouville foliation on the symplectic 4-manifold $\mathbb{R}^3 \times S^1$.

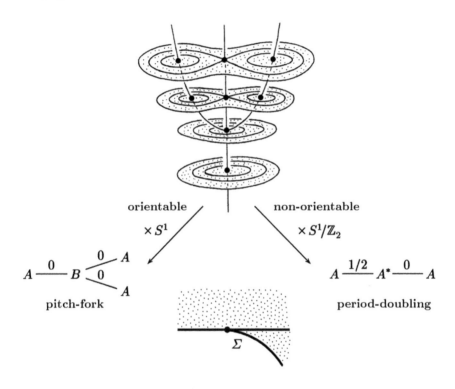

Figure 10.15

The second case, which we call *non-orientable*, can be described as follows. One needs to multiply the above picture in $\mathbb{R}^3$ by the circle S^1 and to take then the quotient with respect to the $\mathbb{Z}_2$-action on $\mathbb{R}^3 \times S^1$ given by

$$(p_1, q_1, p_2, q_2) \to (-p_1, -q_1, p_2, q_2 + \pi).$$

One can obtain the same result by multiplying the picture shown in Fig. 10.15 with the segment $[0, \pi]$ and then gluing the basements of the cylinder so obtained with a π-twist about the vertical line p_2.

The corresponding bifurcation diagram of the momentum mapping $\mathcal{F} \colon \mathbb{R}^3 \times S^1 \to \mathbb{R}^2 (H, f)$ is shown in Fig. 10.15. The two loop molecules W_τ^{orient} and $W_\tau^{\text{non-orient}}$ obtained by going around the singular point are shown in the same Fig. 10.15.

As an additional integral f, one can take, for example, a function which locally has the following form:

$$f = p_1^2 + q_1^4 - p_2 q_1^2.$$

The set of critical points on the cross-section $\mathbb{R}^3$ is given by the following equations (and looks like a fork; see Fig. 10.15):

$$q_1 = 0, \quad p_1 = 0 \qquad \text{(a vertical line)},$$
$$2q_1^2 = p_2, \quad p_1 = 0 \qquad \text{(a parabola in the vertical plane)}.$$

The bifurcation diagram Σ in the (H, f)-plane consists of two pieces. The first one is the straight line $\{f = 0\}$. The second is a half of the parabola $\{f = -H^2/2,\ H > 0\}$ (Fig. 10.15).

Example 3. This case is analogous to the previous one but with elliptic and hyperbolic singularities interchanged. As a result, we obtain the picture in $\mathbb{R}^3$ presented in Fig. 10.16. One hyperbolic singularity transforms into one elliptic and

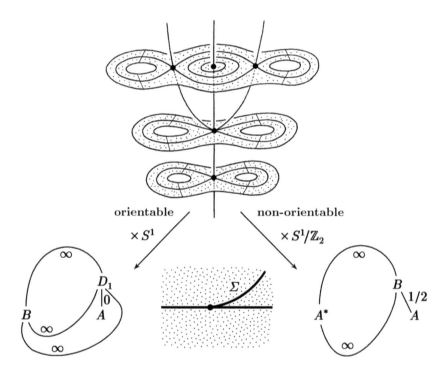

Figure 10.16

two hyperbolic singularities. Again we can distinguish two different possibilities: orientable and non-orientable. As before, we put $H = p_2$ and take f with the following local behavior:

$$f = p_1^2 - q_1^4 + p_2 q_1^2.$$

The set of critical points is given then by the same equations as in Example 2. The bifurcation diagram is also similar to the previous one. It consists of two pieces: the straight line $\{f = 0\}$ and a half of the parabola $\{f = H^2/2,\ H > 0\}$ (Fig. 10.16).

Example 4. This case is locally the same as the previous example. The formulas for H and f, the set of critical points, and the bifurcation diagram are identical. But the global picture is different. By connecting the separatrix in the other way, we obtain the atom C_2 (on the cross-section) instead of the atom D_1. The corresponding loop molecules are shown in Fig. 10.17.

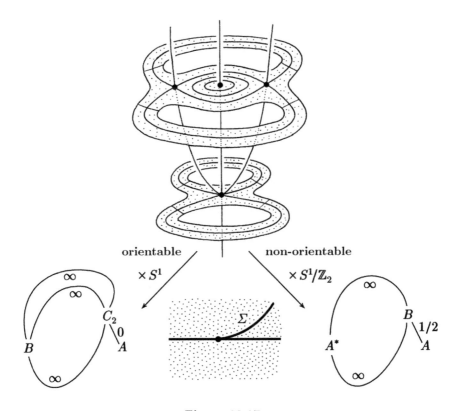

Figure 10.17

It is worth mentioning that, in the non-orientable case, we have obtained the same molecule as in Example 3. The point is that the corresponding Liouville foliations are actually diffeomorphic (although this is not obvious).

In conclusion we collect all the listed molecules in Table 10.2 (with referring to those integrable cases where they occur).

We confined ourselves to the most frequently occurring degenerate singularities. Among them, we can distinguish the class of the so-called topologically stable singularities (which preserve their topological structure under small integrable perturbations of the system). The classification of such singularities has recently been obtained by V. V. Kalashnikov (Jr.) [63]. In Table 10.2, the singularities with numbers 1, 4, 6 are stable. Unstable singularities (numbers 2, 3, 7 in Table 10.2) usually occur due to some additional $\mathbb{Z}_2$-symmetry which obstructs their destruction.

Table 10.2. Typical loop molecules for degenerate singularities

1 A 0 B ∞ Kovalevskaya, Zhukovskiĭ, Clebsch, Steklov cases, 4-dimensional rigid body	**5** ∞ D_1 0 B A ∞ ∞ No examples
2 A 0 D_1 0 A ∞ Goryachev–Chaplygin–Sretenskiĭ case	**6** ∞ B $1/2$ A^* A ∞ Goryachev–Chaplygin–Sretenskiĭ, Kovalevskaya cases
3 $A \overset{0}{\longrightarrow} B \overset{0}{\underset{0}{\diagup}} \begin{matrix} A \\ A \end{matrix}$ Goryachev–Chaplygin–Sretenskiĭ, Kovalevskaya, Steklov cases, Neumann problem	**7** ∞ ∞ C_2 0 B A ∞ Kovalevskaya, Steklov cases
4 $A \overset{1/2}{\longrightarrow} A^* \overset{0}{\longrightarrow} A$ Goryachev–Chaplygin–Sretenskiĭ case	

10.7. COMPUTATION OF r- AND ε-MARKS BY MEANS OF ROTATION FUNCTIONS

We now describe a method which allows us to calculate the gluing matrix on an edge of the molecule in the case when the rotation function is known. The idea is as follows. Suppose that we know the rotation function $\rho(t)$ on the given edge in terms of some basis (λ, μ). As we know, under a change of basis, the rotation function ρ is transformed according to the rule given by Proposition 1.13. On the other hand, the rotation function ρ written in an admissible coordinate system has very specific properties. This gives us a possibility to find the transition matrix from (λ, μ) to the admissible coordinate system. We now explain this idea more precisely.

Consider an edge e adjacent to a saddle atom V. Recall that e represents a one-parameter family of Liouville tori $T^2(t)$. Let (λ^*, μ^*) be a basis on the torus $T^2(t)$.

Proposition 10.5 (Admissibility criterion). *The basis (λ^*, μ^*) is admissible with respect to the atom V (without taking into account orientations on the cycles) if and only if the rotation function $\rho(t)$ written in this basis tends to infinity as the Liouville torus approaches the atom V.*

Proof. The necessity of this condition follows from the Corollary to Lemma 8.5. The sufficiency follows from the transformation formula for the rotation function:

$$\rho = \frac{a\rho^* + c}{b\rho^* + d}.$$

Let us assume that ρ is written in an admissible coordinate system. Then $\lim \rho = \infty$. On the other hand, we have $\lim \rho^* = \infty$. It is easy to see that such a situation is possible if and only if $b = 0$. Therefore, the transition matrix between (λ^*, μ^*) and (λ, μ) has the form

$$\begin{pmatrix} \pm 1 & 0 \\ c & \pm 1 \end{pmatrix}.$$

This transformation is obviously admissible and, consequently, the basis (λ^*, μ^*) is admissible, as was to be proved. $\square$

Using this assertion, we can now construct an admissible coordinate system if we know the rotation function in a certain basis (λ, μ). First suppose that the edge e joins two saddle atoms V^- and V^+.

For definiteness, we assume that the basis (λ, μ) is positively oriented in the sense of the atom V^-. Let (λ^-, μ^-) be an admissible coordinate system on V^- and let

$$\begin{pmatrix} \lambda^- \\ \mu^- \end{pmatrix} = \begin{pmatrix} c_1 & c_2 \\ c_3 & c_4 \end{pmatrix} \begin{pmatrix} \lambda \\ \mu \end{pmatrix}$$

be the desired coordinate change. Then we have

$$\rho = \frac{c_1 \rho^- + c_3}{c_2 \rho^- + c_4}.$$

Passing to the limit as the torus tends to the atom V^-, we see that $\lim \rho = c_1/c_2$. Since the rotation function ρ is known, we easily find the first row (c_1, c_2)

of the transition matrix up to sign (here we use the fact that the transition matrix is unimodular and $c_i \in \mathbb{Z}$). The second row (c_3, c_4) can be chosen arbitrarily. The only condition is $\det C = 1$.

Just in the same way we can find the transition matrix B from the basis (λ, μ) to the basis (λ^+, μ^+):

$$\begin{pmatrix} \lambda^+ \\ \mu^+ \end{pmatrix} = \begin{pmatrix} b_1 & b_2 \\ b_3 & b_4 \end{pmatrix} \begin{pmatrix} \lambda \\ \mu \end{pmatrix}$$

Note that the second row (b_3, b_4) should be chosen so that $\det B = -1$. This is necessary in order for the gluing matrix (i.e., the transition matrix from (λ^-, μ^-) to (λ^+, μ^+)) to have determinant -1.

Thus the gluing matrix is the product of B and C^{-1}:

$$\begin{pmatrix} b_1 & b_2 \\ b_3 & b_4 \end{pmatrix} \begin{pmatrix} c_1 & c_2 \\ c_3 & c_4 \end{pmatrix}^{-1}.$$

It is an important fact that the ambiguity in the choice of the second rows (b_3, b_4) and (c_3, c_4) of the transition matrices C and B has no influence on the r-mark.

However, the constructed admissible bases (λ^-, μ^-) and (λ^+, μ^+) are defined up to the simultaneous change of orientation on the basis cycles. Therefore, the resulting gluing matrix is also defined up to multiplication by -1. This choice determines the mark ε. So, we need to avoid this ambiguity.

Let us turn to the rotation function ρ again. We now use its global behavior along the whole edge.

We distinguish two cases: finite and infinite edges.

The condition that an edge e is infinite (i.e., $r = \infty$) can be reformulated in terms of ρ as follows. The edge e is infinite if and only if the limits of ρ at the beginning and end of this edge are the same. Indeed, as the Liouville torus approaches a saddle atom, the Hamiltonian vector field $v(t)$ tends to the direction of the first basis cycle, i.e., tends to λ^- at the beginning and to λ^+ at the end. The coincidence of the limits means exactly that these two cycles are parallel, or equivalently, $r = \infty$.

We begin with the case of a finite edge. The cycles λ^- and λ^+ are independent; therefore, we can write ρ with respect to this pair of cycles and compute the corresponding rotation index (see Section 8.6.1). If the orientations on λ^- and λ^+ have been chosen correctly, then $\mathrm{ind} = 1 \bmod 4$. Otherwise, $\mathrm{ind} = 3 \bmod 4$. To correct such a situation, we only need to multiply the gluing matrix by -1. Let us formulate this statement as a lemma.

Lemma 10.1. *Suppose that the edge e is finite. If the matrix BC^{-1} coincides with the real gluing matrix, then the rotation index for the function ρ written in terms of (λ^-, λ^+) is equal to $1 \bmod 4$. If BC^{-1} differs from the gluing matrix by sign, then this index is equal to $3 \bmod 4$.*

Proof. In the case of a finite edge, we may consider the pair of cycles λ^-, λ^+ to be a basis in the tangent plane to the Liouville torus. The Hamiltonian vector field v can be written in terms of this basis so that the coordinates of v will be constant on each fixed Liouville torus. Let us draw the curve $v(t)$ on the plane (λ^-, λ^+), where t is a parameter of the family of tori (i.e., t is a parameter on the whole edge e from V^- to V^+). The qualitative picture is shown in Fig. 10.18.

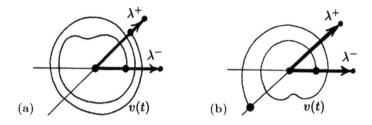

Figure 10.18

Case (a) corresponds to the correct choice of orientations on the basis cycles, case (b) corresponds to the wrong one. Here we use the following fact: if the orientations are correct, then the initial point of $v(t)$ belongs to the ray generated by λ^-, and the end-point of $v(t)$ must belong to the ray generated by λ^+. It is seen from Fig. 10.18(a) that, in this case, the index of ρ is indeed $1 \bmod 4$. In fact, the same situation happens if we make the double mistake, i.e., the orientations on λ^+ and λ^1 are both wrong. But, in this situation, the gluing matrix is correct anyway. Finally, if only one of the orientations is wrong, then ind $= 3 \bmod 4$ (Fig. 10.18(b)). $\square$

Now we consider the case of an infinite edge. This case is simpler. We need to compute the index of the rotation function ρ^- written in terms of the basis (λ^-, μ^-).

Lemma 10.2. *If the index is equal to $0 \bmod 4$, then $\varepsilon = 1$. If the index is equal to $2 \bmod 4$, then $\varepsilon = -1$.*

Proof. It is seen from Fig. 10.19 that the end-points of the curve $v(t)$ in both cases belong to the line generated by λ^-. On the other hand, the curve $v(t)$ ends on the ray generated by λ^+. Figure 10.19 shows that the cycles λ^+ and λ^- have

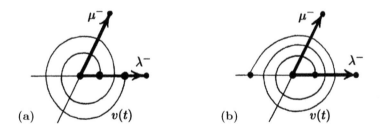

Figure 10.19

the same direction if and only if the index of ρ is equal to $0 \bmod 4$. Conversely, the cycles λ^+ and λ^- have opposite orientations if and only if the index of ρ is equal to $2 \bmod 4$. In the first case, by definition we have $\varepsilon = +1$, whereas the second case corresponds to $\varepsilon = -1$. $\square$

Thus, we see that the rotation function ρ allows us to compute the invariants r and ε on an edge joining two saddle atoms.

In the case when one of the atoms has type A, or both atoms have type A, we can apply a similar construction. However, here we need some additional information on the admissible coordinate system on the atom A. The point is that the limit of the rotation function can be arbitrary. But in fact, in specific examples, one often succeeds in finding an admissible coordinate system on the atom A directly, since usually the contractible cycles are easily detected.

10.8. COMPUTATION OF THE n-MARK BY MEANS OF ROTATION FUNCTIONS

In the preceding section, we have explained how to find an admissible coordinate system on an individual boundary torus of a 3-atom. The natural question is how to make these admissible coordinate systems compatible in order to obtain a uniform admissible coordinate system for the whole atom $U(L)$. We recall that, although the second basis cycle μ_i on an individual torus $T_i \subset \partial U(L)$ can be chosen more or less arbitrarily, all these cycles together must satisfy one important additional property: there must exist a global section $P \subset U(L)$ such that $\mu_i = P \cap T_i$ or, equivalently, $\partial P = \bigcup \mu_i$. Is it possible to find whether such a section exists for a given collection of cycles μ_i? The answer is positive and given by the following construction. For simplicity, we restrict ourselves to the case of atoms without stars.

Let $V = U(L)$ be a saddle atom (without stars), and $e_1, \ldots, e_m$ be the edges incident to it. Each edge in fact denotes a one-parameter family of Liouville tori. Suppose that, on each edge e_i, we have a basis (λ_i, μ_i) and the corresponding rotation function $\rho_i(t)$. We assume here that the parameter t on each edge is taken to be the value of the additional integral f. This is necessary in order for the parameters on different edges to be compatible. Without loss of generality, we shall assume that $f(L) = 0$, where L is the singular leaf. As usual, we separate the edges into two classes: positive and negative ones depending on the sign of f.

Proposition 10.6 (Admissibility criterion for a coordinate system on an atom). *The collection of bases $\{(\lambda_i, \mu_i)\}$ form an admissible coordinate system for the atom $V = U(L)$ if and only if the rotation functions $\{\rho_i\}$ satisfy the following conditions.*

1) The limit $\lim_{t \to 0} \rho_i(t)$ is infinite; in this case, the asymptotic behavior of ρ_i has the form

$$\rho_i(t) = a_i \ln |t| + b_i(t),$$

where $b_i(t)$ is a continuous function including the point $t = 0$.

2) Either

$$\rho_i(t) \to +\infty \quad as \quad t \to 0 \qquad for\ positive\ edges\ e_i,$$
$$\rho_i(t) \to -\infty \quad as \quad t \to 0 \qquad for\ negative\ edges\ e_i$$

or, on the contrary,

$$\rho_i(t) \to -\infty \quad as \quad t \to 0 \qquad for\ positive\ edges\ e_i,$$
$$\rho_i(t) \to +\infty \quad as \quad t \to 0 \qquad for\ negative\ edges\ e_i.$$

3) The equality $\sum_i b_i(0) = 0$ holds, where the sum is taken over all the edges (both positive and negative).

Proof. In fact, the necessity of the listed conditions has been already proved. Namely, parts (1) and (2) follow from Proposition 10.5 and Lemma 8.5; part (3) is, in essence, a reformulation of the remark to Lemma 6.7.

Let us prove the sufficiency. The fact that, on each individual edge, the basis $\{(\lambda_i, \mu_i)\}$ is admissible follows from Proposition 10.5. In particular, each λ_i is a fiber of the trivial S^1-fibration on $U(L)$. Thus, it remains to show that there exists a global section P (of this fibration) such that $\partial P = \bigcup \mu_i$.

Consider an arbitrary admissible coordinate system $\{(\lambda_i^0, \mu_i^0)\}$ on V and extract the collection of the second basis cycles $\{\mu_i^0\}$. Since $\lambda_i^0 = \lambda_i$, the second basis cycles are connected by the relation

$$\mu_i^0 = \mu_i + k_i \lambda_i \,,$$

where k_i are some integers. Then the rotation functions ρ_i and ρ_i^0 (computed in terms of the bases $\{(\lambda_i, \mu_i)\}$ and $\{(\lambda_i^0, \mu_i^0)\}$ respectively) are connected as follows:

$$\rho_i = \rho_i^0 + k_i \,.$$

Similar relations hold for the "finite parts" of ρ_i and ρ_i^0, i.e.,

$$b_i(0) = b_i^0(0) + k_i \,.$$

Taking the sum over all i's, we obtain

$$\sum b_i(0) = \sum b_i^0(0) + \sum k_i \,.$$

By our assumption, $\sum b_i(0) = 0$. On the other hand, $\sum b_i^0(0) = 0$ because of the admissibility of $\{(\lambda_i^0, \mu_i^0)\}$. Therefore, $\sum k_i = 0$. But this implies that the change $\mu_i^0 \to \mu_i$ is admissible and, consequently, the coordinate system $\{(\lambda_i, \mu_i)\}$ is admissible as required. $\square$

Sometimes, the condition $\sum b_i(0) = 0$ is hard to verify, since it is not clear how to extract the "finite parts" of rotation functions. The following reformulation allows us to avoid this procedure.

Proposition 10.7. *The condition $\sum b_i(0) = 0$ is equivalent to the following relation:*

$$\lim_{t \to 0} \left(\underbrace{\sum \rho_i(t)}_{\substack{\text{positive} \\ \text{edges}}} + \underbrace{\sum \rho_i(-t)}_{\substack{\text{negative} \\ \text{edges}}} \right) = 0 \,.$$

Proof. We should verify that all the logarithmic terms in the above sum are canceled. This follows from the fact that the coefficient of a logarithmic term is equal (up to sign) to the sum of Λ-invariants of those critical circles which are adjacent to the corresponding one-parameter family e_i of Liouville tori. Note that each circle is adjacent to four families of tori: two of them are positive, and the other two are negative. Therefore, taking the sum over all families, we see that the sum of the logarithmic terms vanishes, since each Λ-invariant enters the sum twice with sign "$+$" and twice with sign "$-$". This completes the proof. $\square$

We now discuss the question of how to compute n-marks by means of the rotation functions ρ_i.

Consider an arbitrary edge e of the molecule. We showed in the preceding section how to find admissible bases $\{(\lambda^-, \mu^-)$ and $(\lambda^+, \mu^+)\}$ on e. In particular, on each finite edge, we can rewrite the rotation function with respect to the pair of cycles (λ^-, λ^+). On an infinite edge, we consider the function ρ^-, which is defined up to an additive integer constant. Thus, we assume that exactly these rotation functions are determined on all the edges.

Consider an arbitrary family in the molecule. Suppose, for definiteness, that all the exterior edges are outgoing from the family. Recall that each rotation function ρ has the following asymptotic behavior as the torus approaches a singular leaf: $\rho(f) = a \ln |f - f_0| + b(f)$, where f_0 is the value of the additional integral f on the singular leaf. Let us define the following number:

$$-\tilde{n} = \sum_{\substack{\text{exterior} \\ \text{edges } e_i}} \left(\lim_{T^2 \to e_i^-} b(f) \right) + \sum_{\substack{\text{interior} \\ \text{edges } e_j}} \left(\lim_{T^2 \to e_j^-} b(f) - \lim_{T^2 \to e_j^+} b(f) \right),$$

where e^- and e^+ denote the beginning and end of the edge e, respectively.

In a slightly different way, the same number can be defined as follows (without using the "finite parts" of rotation functions). Consider an arbitrary atom and the edges incident to it. These edges are divided into two parts: positive and negative ones; and, on each of them, we have a rotation function $\rho(f)$. Assume that f is equal to zero on the singular leaf of the atom and consider the function

$$N(f) = \sum_{\substack{\text{positive} \\ \text{edges}}} \pm \rho(f) + \sum_{\substack{\text{negative} \\ \text{edges}}} \pm \rho(-f), \qquad \text{for } f > 0.$$

Here the sign $\pm$ is chosen in the following way. If an edge is incoming, we take "$-$"; if an edge is outgoing, we take "$+$".

We assert that, although each function in this sum tends to infinity, the function $N(f)$ is continuous on $[0, \varepsilon)$ and has a finite limit at zero. This is equivalent to the fact that all the logarithmic terms are canceled, which can be proved just by the same method as in Proposition 10.7.

Let $N(\rho, V)$ denote the limit of $N(f)$ as $f \to 0$; here ρ denotes the collection of the rotation functions on the edges incident to the atom V. If the value of f on the singular leaf is not zero, but f_0, then the formula for $N(\rho, V)$ can be rewritten as

$$N(\rho, V) = \lim_{\varepsilon \to 0} \left(\sum_{\substack{\text{positive} \\ \text{edges}}} \pm \rho(f_0 + \varepsilon) + \sum_{\substack{\text{negative} \\ \text{edges}}} \pm \rho(f_0 - \varepsilon) \right).$$

It is easy to see that, as a result, the expression for $\tilde{n}$ can be rewritten as follows:

$$\tilde{n}(\text{family}) = - \sum_{V \in \text{family}} N(\rho, V).$$

Let us note that, in general, $\tilde{n}$ is not integer. But, as we shall see soon, it is always rational.

We now recall the definition of the n-mark. To each edge e_i incident to the given family, we assign an integer number Θ_i according to the following rule:

$$\Theta_i = \begin{cases} [\alpha_i/\beta_i] & \text{if } e_i \text{ is an outgoing edge,} \\ [-\delta_i/\beta_i] & \text{if } e_i \text{ is an incoming edge,} \\ -\gamma_i/\alpha_i & \text{if } e_i \text{ is an interior edge,} \end{cases}$$

where $\alpha_i, \beta_i, \gamma_i, \delta_i$ are the elements of the gluing matrix C_i on the edge e_i.

Then n is defined to be the sum $\sum \Theta_i$. Since we have oriented the exterior edges in such a way that all of them are outgoing, this formula becomes simpler, and finally we have

$$n = \sum_{\substack{\text{exterior} \\ \text{edges}}} [\alpha_i/\beta_i] - \sum_{\substack{\text{interior} \\ \text{edges}}} \gamma_i/\alpha_i.$$

Proposition 10.8. *The number $\tilde{n}$ depends neither on the rotation functions $\rho_i(f)$ nor on the choice of the additional integral f. In fact, $\tilde{n}$ is an invariant of the Liouville foliation, which can be defined by the following explicit formula:*

$$\tilde{n} = \sum_{\substack{\text{exterior} \\ \text{edges}}} \alpha_i/\beta_i - \sum_{\substack{\text{interior} \\ \text{edges}}} \gamma_i/\alpha_i.$$

In particular, the rational number $\tilde{n}$ and the integer mark n (both defined for the given family) are connected by the simple formula:

$$n = \tilde{n} - \sum_{\substack{\text{exterior} \\ \text{edges}}} r_i,$$

where $r_i = \{\alpha_i/\beta_i\}$ is the r-mark on the edge e_i.

Proof. By definition, we have

$$\tilde{n}(\text{family}) = -\sum_{V \in \text{family}} N(\rho, V) = -\sum_{V \in \text{family}} \lim_{\varepsilon \to 0} \left(\sum_{\substack{\text{positive} \\ \text{edges}}} \pm \rho(f_0 + \varepsilon) + \sum_{\substack{\text{negative} \\ \text{edges}}} \pm \rho(f_0 - \varepsilon) \right).$$

Let us rewrite this sum in a different way by decomposing it into two sums: over all exterior edges and over all interior edges. Recall that, on the exterior edges, we have $\rho_i = \rho_i^- - \alpha_i/\beta_i$; and, on the interior edges, $\rho_i = \rho_i^-$. (Here $\rho^-(t)$ is the rotation function in terms of the basis λ^-, μ^-.) Each interior edge enters this sum twice: for its beginning, we take ρ_i^-; and, for its end, the corresponding term can be represented as $-\rho_i^- = \rho_i^+ + \gamma_i/\alpha_i$. Taking the sum over all the atoms $V \in$ family, we can distinguish two groups of terms. The first group is the sum of all the rotation functions ρ_i^- and ρ_i^+. The second group consists of the terms of the form $-\alpha_i/\beta_i$ and γ_i/α_i. Thus, the second group takes the form

$$\sum_{\substack{\text{exterior} \\ \text{edges}}} (-\alpha_i/\beta_i) + \sum_{\substack{\text{interior} \\ \text{edges}}} \gamma_i/\alpha_i.$$

It remains to observe that, according to Proposition 10.7, the sum of the first group is zero. This gives the desired formula for $\tilde{n}$, which immediately implies the independence of $\tilde{n}$ of $\rho_i(f)$ and f. $\quad\square$

In what follows, we shall sometimes call $\tilde{n}$ the *energy of the family*, following P. Topalov [198], who used this invariant to describe the relationship between the marks of the molecule W^* and the topology of the corresponding isoenergy manifold Q^3.

10.9. RELATIONSHIP BETWEEN THE MARKS OF THE MOLECULE AND THE TOPOLOGY OF Q^3

In this section, we briefly describe a construction suggested by P. Topalov. Let W^* be the marked molecule of a certain Liouville foliation on a 3-manifold Q^3. The idea is that the topology of Q^3 is completely determined by the marked molecule W^*. This means that topological invariants of Q^3 (the fundamental group, homology groups, etc.) can be expressed as some functions of the marked molecule W^*. In many cases, these functions can be written explicitly. On the other hand, for specific integrable problems in physics, mechanics, and geometry, the topology of the isoenergy 3-manifold Q^3 is often known in advance. Moreover, in real examples, Q^3 is rather simple (like S^3, $\mathbb{R}P^3$, $S^1 \times S^2$, T^3). Therefore, if we know the topology of Q^3 in advance, then we obtain some relations among numerical marks of the molecule W^*. Sometimes, these relations allow us to compute the marks r, ε, n. In fact, we have already demonstrated how this idea works in the case of the simplest molecules $A - A$ (see Proposition 4.3).

We begin with some preliminary constructions. Let us cut the molecule W^* along all the finite edges. As a result, W^* splits into pieces of three types. The pieces of the first type are *families*, i.e., the subgraphs that contain saddle atoms only. The pieces of the second type are isolated atoms A. The third type consists of the pieces that contain both saddle atoms and atoms A.

Each piece of the third type necessarily contains at least one infinite edge joining a saddle atom with an atom A. We choose and fix one of such edges in each piece (they will be called *fixed*).

An edge of W^* is said to be *essential* if it is not fixed and none of its end-points belongs to any family. Note that essential edges can be of two kinds. They are either interior edges of pieces of the third type or edges between pieces of the second and third types (all variants are possible: second with second, second with third, and third with third).

Let $g(V)$ denote the genus of an atom V, i.e., the genus of the closed surface $\widetilde{P}$ obtained from the base P of the Seifert fibration on $V = U(L)$ by gluing discs to each component of the boundary ∂P. The valency of the atom V is denoted by $q(V)$. Recall that the valency is the number of edges incident to V, if V is considered as a vertex of the graph W. In other words, the valency is the number of families of Liouville tori incident to the singular leaf $L \subset U(L) = V$. Similarly, we introduce the *genus* $g(F)$ and the *valency* $q(F)$ of a family F. Recall that each family F carries the structure of a Seifert fibration over some two-dimensional base. In what follows, we shall assume for simplicity that this base is orientable.

By W' we shall denote the graph obtained from the molecule W^* by shrinking each of its families into a point (i.e., a vertex of W'). Thus, the graph W' has vertices of three kinds:

1) families,
2) atoms A,
3) saddle atoms which do not belong to any family.

We denote the vertices of W' by U.

We now introduce the notion of the *energy* of the marked molecule. By definition, we set

$$N(W^*) = \begin{cases} 0 & \text{if} \quad \operatorname{rank} H_1(Q^3) > \operatorname{rank} H_1(W') + 2 \sum_{U \neq A} g(U), \\ |\operatorname{Tor} H_1(Q^3)| & \text{otherwise}. \end{cases}$$

Here $|\operatorname{Tor} H_1(Q^3)|$ denotes the order of the subgroup of elements of finite order in the one-dimensional homology group $H_1(Q^3)$ (i.e., the order of the torsion subgroup $\operatorname{Tor} H_1(Q^3)$). If $\operatorname{Tor} H_1(Q^3)$ is trivial, then we set its order to be 1.

It turns out that the energy of the marked molecule can be expressed in terms of the marks r_i, ε_i, n_k by a rather simple formula. A general algorithm which gives such a formula for an arbitrary molecule W^* has been obtained by P. Topalov in [344]. We do not discuss this algorithm here in detail, restricting ourselves to several examples.

Example 1. Suppose that the molecule W has the form shown in Fig. 10.20, where $e_1, e_2, \ldots, e_m$ are edges of the molecule, and F is a family. The edges $e_1, e_2, \ldots, e_m$ are all incident to the family F. Moreover, the edges $e_1, e_2, \ldots, e_m$ are all finite, i.e., the corresponding r-marks are finite. Then the energy of the molecule is given by the following formula:

$$N(W^*) = \pm \beta_1 \beta_2 \ldots \beta_m \tilde{n}(F),$$

where $\beta_1, \ldots, \beta_m$ are the denominators of the r-marks $r_1 = \alpha_1 / \beta_1, \ldots, r_m = \alpha_m / \beta_m$ corresponding to the edges $e_1, \ldots, e_m$, and $\tilde{n}(F)$ is the energy of the family F introduced in the preceding section.

Thus, if we know all the r-marks and the topology of Q^3, then we can uniquely determine the n-mark for the family F. It is expressed in terms of $\tilde{n}$ by the formula from Proposition 10.8.

Example 2. Suppose that the molecule W has the form shown in Fig. 10.21, where F_1 and F_2 denote some families. The edges $e_0, e_1, e_2, \ldots, e_m$ incident to them are all finite. In this case, we have

$$N(W^*) = \pm \beta_0 \beta_1 \beta_2 \ldots \beta_m \left(\tilde{n}(F_1) \tilde{n}(F_2) - \frac{1}{\beta_0^2} \right).$$

Here β_i are the denominators of the marks r_i on the edges e_i, and $\tilde{n}(F_j)$ is the energy of the family F_j.

Example 3. Suppose that the molecule W has the form shown in Fig. 10.22, where F_0, F_1, F_2 are families and, consequently, the edges $e_0, e_1, \ldots, e_m$ incident to them are all finite. The family F_2 may, in general, have no interior edges except for e_0 and e_1. Then

$$N(W^*) = \pm\beta_0\beta_1\beta_2 \ldots \beta_m \left(\tilde{n}(F_0)\tilde{n}(F_1)\tilde{n}(F_2) - \frac{\tilde{n}(F_0)}{\beta_0^2} - \frac{\tilde{n}(F_1)}{\beta_1^2} \right).$$

Here β_i are the denominators of the marks r_i on the edges e_i, and $\tilde{n}(F_j)$ is the energy of F_j.

Example 4. Suppose that the molecule W has the form shown in Fig. 10.23, where F is a family and the edges $e_0, e_1, \ldots, e_m$ are all finite. One of these edges, namely e_0, is a loop. In this case, the energy of W^* is given by the formula

$$N(W^*) = \pm\beta_0\beta_1 \ldots \beta_m \left(\tilde{n}(F) - \frac{2}{\beta_0} \right).$$

Here β_i are the denominators of the marks r_i on the edges e_i, and $\tilde{n}(F)$ is the energy of the family F.

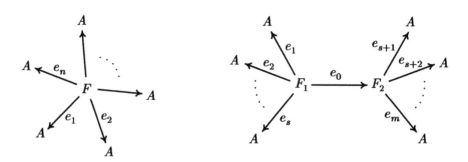

Figure 10.20   Figure 10.21

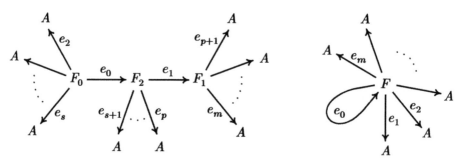

Figure 10.22   Figure 10.23

In these examples, we assumed that the molecule W had no atoms with stars. If such atoms exist, then this situation is reduced to the previous one in the following way. If a certain atom V contains a star-vertex, then, without changing the manifold Q^3, we can change its decomposition into 3-atoms. For this, we consider the Liouville foliation in a neighborhood of the singular fiber S of type $(2,1)$ of the Seifert fibration on V, which corresponds to the star-vertex. Let us formally cut Q^3 along an additional torus which is the boundary of a tubular neighborhood $U(S)$ of S. As a result, a new 3-atom of type A appears together with a new edge incident to it. This atom is just the neighborhood of S. By cutting out the tubular neighborhood of all singular fibers from the atom V, we obtain the structure of a trivial S^1-fibration on the remaining part of V. This remaining part $V \setminus \bigcup U(S_i)$ can be considered as a new atom without stars. As a result, we transform each star-vertex into an additional edge of the form $\longrightarrow A$, on which we should put the marks $r = 1/2$ and $\varepsilon = 1$. The structure of families does not change.

After such a reduction, it is not hard to obtain the explicit formulas for the energies of the molecules shown in Figs. 10.20, 10.21, 10.22, 10.23 in the case of atoms with stars.

If the molecule W from Example 1 (Fig. 10.20) contains p star-vertices, then the formula for its energy becomes

$$N(W^*) = \pm 2^p \beta_1 \beta_2 \ldots \beta_m \tilde{n}(F),$$

where the "new" energy $\tilde{n}(F)$ of a family containing star-vertices is computed as follows:

$$(\tilde{n}(F))_{\text{new}} = (\tilde{n}(F))_{\text{old}} + \frac{p}{2},$$

where $(\tilde{n}(F))_{\text{old}}$ is given by Proposition 10.8.

Just the same happens in all the remaining Examples 2, 3, and 4. As a result, the energies of the molecules shown in Figs. 10.21, 10.22, and 10.23, become

$$N(W^*) = \pm 2^p \beta_0 \beta_1 \beta_2 \ldots \beta_m \left(\tilde{n}(F_1)\tilde{n}(F_2) - \frac{1}{\beta_0^2} \right),$$

$$N(W^*) = \pm 2^p \beta_0 \beta_1 \beta_2 \ldots \beta_m \left(\tilde{n}(F_0)\tilde{n}(F_1)\tilde{n}(F_2) - \frac{\tilde{n}(F_0)}{\beta_0^2} - \frac{\tilde{n}(F_1)}{\beta_1^2} \right),$$

$$N(W^*) = \pm 2^p \beta_0 \beta_1 \beta_2 \ldots \beta_m \left(\tilde{n}(F) - \frac{2}{\beta_0} \right),$$

respectively, where p is the total number of star-vertices in all the families F_i, and $\tilde{n}(F)$ is the "new" energy of the family F.

The listed examples help to calculate the numerical marks of W^* in many situations.

In conclusion we formulate some more results by P. Topalov which demonstrate links between the topology of Q^3 and the marks of the molecule W^*.

Theorem 10.3 (P. Topalov [344]). *Let W^* be the marked molecule of a certain integrable Hamiltonian system on an isoenergy 3-manifold Q^3, and let W' be the graph obtained from W by contracting each family into a point (see above).*

1) *The following relations hold:*

$$\operatorname{rank} H_1(Q^3) \geq \operatorname{rank} H_1(W') + 2 \sum_{U \neq A} g(U),$$

$$\operatorname{rank} H_1(Q^3) \leq \operatorname{rank} H_1(W') + 2 \sum_{U \neq A} g(U) + \sum_{U \neq A} q(U),$$

where each sum is taken over all vertices of W' different from an atom A.

2) *Any essential edge e of the molecule W^* gives one term of the form $\mathbb{Z}_{\beta(e)}$ in the decomposition of the one-dimensional homology group $H_1(Q^3, \mathbb{Z})$ into cyclic subgroups, where $\beta(e)$ is the denominator of the r-mark α/β on the edge e. (Here we formally assume that for $\beta = 0$ the group $\mathbb{Z}_{\beta(e)}$ is isomorphic to $\mathbb{Z}$, and for $\beta = 1$ the group $\mathbb{Z}_{\beta(e)}$ is trivial.) If an atom with stars V does not belong to any family, then each star-vertex gives one term of the form $\mathbb{Z}_2$ in this decomposition.*

Corollary. *If $\operatorname{rank} H_1(Q^3) < 2$ (in particular, if $Q^3 \simeq S^3$ or $Q^3 \simeq \mathbb{R}P^3$), then all the atoms V in the molecule W^* are planar, i.e., $g(V) = 0$.*

Corollary.

1) *If Q^3 is homeomorphic to the sphere S^3 or the torus T^3, then the r-marks on the essential edges of W^* are all zeros.*

2) *If Q^3 is homeomorphic to $\mathbb{R}P^3$, then the r-marks on the essential edges of W^* are either zeros or $1/2$. Moreover, the mark $r = 1/2$ may appear only once.*

3) *If Q^3 is homeomorphic to $S^1 \times S^2$ (or the connected sum of several copies of $S^1 \times S^2$), then the r-marks on the essential edges of W^* are either zeros or ∞.*

Corollary. *If the molecule W^* contains no families, then the product of the denominators β_i of all the finite r-marks is less than or equal to the order of the torsion group $\operatorname{Tor} H_1(Q^3)$. If the molecule W^* contains p star-vertices, then the following stronger estimation holds:*

$$2^p \prod \beta_i \leq |\operatorname{Tor} H_1(Q^3)|.$$

As an application of the above construction, we consider an example of the molecule W^* of the form

$$A \longleftarrow A^* \longrightarrow A$$

with the r-marks $r_1 = \{\alpha_1/\beta_1\}$, $r_2 = \{\alpha_2/\beta_2\}$, and one mark n on the atom A^*. The molecule of this kind occurs, in particular, in the integrable Goryachev–Chaplygin case (see Chapter 14). In this case, the corresponding isoenergy 3-manifold is homeomorphic to the 3-sphere. Applying the above results (see Example 1), we obtain the following relation between the r-marks and the mark $n = [\alpha_1/\beta_1] + [\alpha_2/\beta_2]$ (where $[x]$ is the integer part of x):

$$2\beta_1\beta_2\left(\frac{\alpha_1}{\beta_1} + \frac{\alpha_2}{\beta_2} + \frac{1}{2}\right) = 1.$$

In the Goryachev–Chaplygin case, the marked molecule $A \longleftarrow A^* \longrightarrow A$ is symmetric with respect to the atom A^* (this easily follows from some additional symmetry of the system). Hence the r-marks coincide. This implies that $\{\alpha_1/\beta_1\} = \{\alpha_2/\beta_2\}$ and, consequently, $\beta_1 = \beta_2$. Therefore, the above relation can be rewritten as

$$2\beta^2 \left(\frac{\alpha_1}{\beta} + \frac{\alpha_2}{\beta} + \frac{1}{2} \right) = 1 .$$

Thus, $\beta(2(\alpha_1 + \alpha_2) + \beta) = 1$, and we obtain two possibilities: either $\beta = 1$, $\alpha_1 + \alpha_2 = 0$ (Fig. 10.24(a)), or $\beta = -1$, $\alpha_1 + \alpha_2 = 1$ (Fig. 10.24(b)).

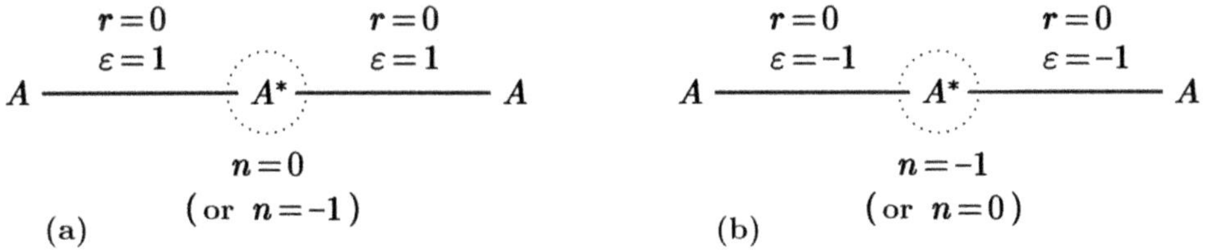

Figure 10.24

REMARK. Recall that n depends on the choice of orientation on Q^3 (see Chapter 4). In our situation, under change of the orientation, the mark $n = 0$ is transformed into $n = -1$ and vice versa. In this sense, "$n = 0$" and "$n = -1$" mean actually the same.

This example shows that the above method allows us to compute non-trivial topological invariants of integrable Hamiltonian systems for concrete examples or, at least, to reduce the problem to the analysis of several possibilities.

Chapter 11

Integrable Geodesic Flows
on Two-dimensional Surfaces

11.1. STATEMENT OF THE PROBLEM

Let M^n be a smooth Riemannian manifold with a Riemannian metric $g_{ij}(x)$. Recall that geodesics of the given metric are defined as smooth parameterized curves

$$\gamma(t) = (x^1(t), \ldots, x^n(t))$$

that are solutions to the system of differential equations

$$\nabla_{\dot\gamma}\dot\gamma = 0,$$

where $\dot\gamma = \dfrac{d\gamma}{dt}$ is the velocity vector of the curve γ, and ∇ is the covariant derivation operator related to the symmetric connection associated with the metric g_{ij}. In local coordinates, these equations can be rewritten in the form

$$\frac{d^2 x^i}{dt^2} + \sum \Gamma^i_{jk} \frac{dx^j}{dt} \frac{dx^k}{dt} = 0,$$

where $\Gamma^i_{jk}(x)$ are smooth functions called the Christoffel symbols of the connection ∇ and defined by the following explicit formulas:

$$\Gamma^i_{jk}(x) = \frac{1}{2} \sum g^{is} \left(\frac{\partial g_{sj}}{\partial x^k} + \frac{\partial g_{ks}}{\partial x^j} - \frac{\partial g_{kj}}{\partial x^s} \right).$$

The geodesics can be interpreted as trajectories of a single mass point that moves on the manifold without any external action, i.e., by inertia. Indeed, the equation of geodesics means exactly that the acceleration of the point equals zero.

Recall the main properties of geodesics (see [31], [75], [101], [148], and [183]).

1) They are locally shortest, i.e., for two sufficiently close points lying on a geodesic, the length of the geodesic segment between them is strictly less than the length of any other smooth curve connecting these points.

2) For every point $x \in M$ and for every tangent vector $\xi \in T_x M$, there exists a unique geodesic γ such that $\gamma(0) = x$ and $\dot{\gamma}(0) = \xi$.

3) If M is compact, then any geodesic can be infinitely extended in the sense of its parameter. In other words, every solution $\gamma(t)$ to the equation of geodesics is defined for any $t \in \mathbb{R}$.

4) If M is compact, then any two points $x, y \in M$ can be connected by a geodesic (there can be many such geodesics in general).

5) If M is compact, then any homotopy class of closed curves (i.e., mappings of the circle into the manifold) contains a closed geodesic. Such a geodesic may have self-intersections.

6) If M is compact, then for any point $x \in M$ and for any element of the fundamental group $\pi_1(M, x)$, there exists a geodesic starting from this point and returning to it which realizes the chosen element of the fundamental group. This geodesic is not necessarily closed. In other words, the initial and terminal velocity vectors can be different, i.e., x can be a transversal self intersection point of this geodesic.

One can consider the equation of geodesics as a Hamiltonian system on the cotangent bundle $T^* M$, and the geodesics themselves can be regarded as the projections of trajectories of this Hamiltonian system onto M. To this end, consider natural coordinates x and p on the cotangent bundle $T^* M$, where $x = (x^1, \ldots, x^n)$ are the coordinates of a point on M and $p = (p_1, \ldots, p_n)$ are the coordinates of a covector from the cotangent space $T_x^* M$ in the basis $dx^1, \ldots, dx^n$. Take the standard symplectic structure $\omega = dx \wedge dp$ on $T^* M$ and consider the following function as a Hamiltonian:

$$H(x, p) = \frac{1}{2} \sum g^{ij}(x) p_i p_j = \frac{1}{2} |p|^2 .$$

Proposition 11.1.

a) *Let $\gamma(t) = (x(t), p(t))$ be an integral trajectory of the Hamiltonian system $v = \operatorname{sgrad} H$ on $T^* M$. Then the curve $x(t)$ is a geodesic, and its velocity vector $\dot{x}(t)$ is connected with $p(t)$ by the following relation*:

$$\frac{dx^i(t)}{dt} = \sum g^{ij}(x) p_j(t) .$$

b) *Conversely, if a curve $x(t)$ is a geodesic on M, then the curve $(x(t), p(t))$, where $p_i(t) = \sum g_{ij}(x) \dfrac{dx^j}{dt}$, is an integral trajectory of the Hamiltonian system $v = \operatorname{sgrad} H$.*

Proof. Consider the Hamiltonian equations related to the Hamiltonian H:

$$\frac{dp_i}{dt} = -\frac{\partial H}{\partial x^i}, \qquad \frac{dx^i}{dt} = \frac{\partial H}{\partial p_i} .$$

In local coordinates, we get

$$\frac{dx^i}{dt} = \frac{\partial H}{\partial p_i} = \sum g^{ij} p_j \,, \qquad \frac{dp_i}{dt} = -\frac{\partial H}{\partial x^i} = -\frac{1}{2}\sum \frac{\partial g^{\alpha\beta}}{\partial x^i} p_\alpha p_\beta \,.$$

Identifying vectors and covectors by means of the Riemannian metric g, i.e., by substituting $p_i = \sum g_{ij}\dot{x}^j$, we obtain

$$\dot{x}^i = \frac{dx^i}{dt} \qquad\qquad\qquad \text{(the first equation)},$$

$$\frac{d}{dt}\left(g_{is}\frac{dx^s}{dt}\right) = -\frac{1}{2}\sum \frac{\partial g^{\alpha\beta}}{\partial x^i} g_{\alpha k}\frac{dx^k}{dt} g_{\beta j}\frac{dx^j}{dt} \qquad \text{(the second equation)}.$$

By transforming the second equation, we get

$$\frac{\partial g_{ik}}{\partial x^j}\frac{dx^k}{dt}\frac{dx^j}{dt} + g_{is}\frac{d^2 x^s}{dt^2} = \frac{1}{2}\sum \frac{\partial g_{kj}}{\partial x^i}\frac{dx^k}{dt}\frac{dx^j}{dt}\,.$$

Using the evident identity

$$\frac{\partial g_{ik}}{\partial x^j}\frac{dx^k}{dt}\frac{dx^j}{dt} = \frac{1}{2}\left(\frac{\partial g_{ki}}{\partial x^j} + \frac{\partial g_{ij}}{\partial x^k}\right)\frac{dx^k}{dt}\frac{dx^j}{dt}\,,$$

we rewrite the equation obtained in the form

$$g_{is}\frac{d^2 x^s}{dt^2} + \frac{1}{2}\sum\left(\frac{\partial g_{ij}}{\partial x^k} + \frac{\partial g_{ki}}{\partial x^j} - \frac{\partial g_{kj}}{\partial x^i}\right)\frac{dx^k}{dt}\frac{dx^j}{dt} = 0\,.$$

Hence, we have

$$\frac{d^2 x^s}{dt^2} + \frac{1}{2}\sum g^{is}\left(\frac{\partial g_{ij}}{\partial x^k} + \frac{\partial g_{ki}}{\partial x^j} - \frac{\partial g_{kj}}{\partial x^i}\right)\frac{dx^k}{dt}\frac{dx^j}{dt} = 0\,.$$

That is

$$\frac{d^2 x^s(t)}{dt^2} + \sum \Gamma^s_{jk}\frac{dx^j(t)}{dt}\frac{dx^k(t)}{dt} = 0\,. \qquad \square$$

A natural and important question arises: in what cases it is possible to solve the equations of geodesics explicitly, for example, in quadratures? How to describe the behavior of geodesics in a qualitative way?   Since we study integrable Hamiltonian systems in our book, the most interesting problem for us would be the description of those cases where the geodesic flow is completely Liouville integrable. In fact, the following two substantive questions arise.

a) On which manifolds do there exist Riemannian metrics whose geodesic flows are integrable?

b) Given a manifold on which such metrics exist, how does one describe them?

A large number of papers, both classical and contemporary, are devoted to studying these problems. See, for example, G. Darboux [92], [93], M. L. Raffy [303], G. Birkhoff [32], S. I. Pidkuiko, A. M. Stepin [290], A. M. Stepin [327], A. Thimm [337], M. Adler, P. van Moerbeke [1], [2], V. N. Kolokol'tsov [187], [189], A. S. Mishchenko [243], M. L. Bialy [76], I. A. Taimanov [329], [330], [331], V. V. Kozlov, D. V. Treshchev [201], [202], V. V. Kozlov, N. V. Denisova [198], [199], G. Paternain [283], [285], R. Spatzier [321], G. Paternain, R. Spatzier [284], I. K. Babenko, N. N. Nekhoroshev [26]. In our book, we restrict ourselves to the case of two-dimensional surfaces, which has been studied most explicitly.

11.2. TOPOLOGICAL OBSTRUCTIONS TO INTEGRABILITY OF GEODESIC FLOWS ON TWO-DIMENSIONAL SURFACES

Theorem 11.1 (V. V. Kozlov [192], [195]).   *Let a two-dimensional compact real-analytic manifold with a negative Euler characteristic be endowed with a real-analytic Riemannian metric. Then the geodesic flow of this metric does not admit any non-trivial real-analytic integral.*

COMMENT.   Recall that any two-dimensional compact manifold M is represented either as the sphere with handles (in the orientable case) or as the sphere with Möbius strips (in the non-orientable case). The Euler characteristic of M is then computed as $\chi = 2 - 2g$, where g is the number of handles, or $\chi = 2 - m$, where m is the number of Möbius strips. The condition $\chi \geq 0$ means that, in the orientable case, the number of handles is not greater than 1 (then the manifold is either the sphere or the torus), and, in the non-orientable case, the number of Möbius strips is not greater than 2 (such a manifold is either the projective space or the Klein bottle). Therefore, real-analytic Riemannian metrics with integrable geodesic flows cannot exist on any 2-manifold, except for the sphere, the torus, the projective plane, and the Klein bottle. We shall show below that on these manifolds, they really exist.

COMMENT.   We shall give a proof of Theorem 11.1 that is different from the original proof by V. V. Kozlov and based on some general properties of Liouville foliations. In our opinion, such a new approach clarifies the topological nature of that kind of obstructions to integrability which was discovered by V. V. Kozlov. The analyticity condition is needed, in fact, only in order for the Liouville foliation singularities not to be pathologically bad. Roughly speaking, there should not be too many of them. For example, instead of the analyticity condition, one can require the integral of the flow to be tame (see [236] for definition) or geometrically simple (see [329]).

COMMENT.   Note that in Kozlov's theorem, it is sufficient to require that an additional integral of the flow is analytic only on the isoenergy surface (the integral is allowed to have singularities on the zero section). If we assume the integral to be analytic on the whole phase space, then the problem is reduced to the study of flows that admit polynomial (in momenta) integrals. To see this, one can expand the integral into a power series in momenta and take its homogeneous polynomial components as new integrals. In this case, the theorem can be proved by the method suggested by V. N. Kolokol'tsov [187]. The analyticity (in coordinates x) is no longer important. One only assume that the integral is a polynomial in momenta, but the coefficients of the polynomial can be arbitrarily smooth functions.

COMMENT. In fact, the statement of Kozlov's theorem remains valid for Finsler metrics [41]. This circumstance shows that the main cause that obstructs integrability is actually the topology of a manifold. In other words, the property of the Hamiltonian to be a quadratic form is not important here.

Proof (of Theorem 11.1). Let M be a two-dimensional compact smooth manifold with an integrable geodesic flow. First, for simplicity, we assume that the geodesic flow possesses a Bott integral F. We shall explain afterwards how to pass from the situation with non-degenerate singularities (i.e., Bott singularities) to the analytic case.

Consider a regular isoenergy 3-surface $Q = \{H = \text{const}\}$. It has the structure of an S^1-fiber bundle over M^2 (i.e., the space of unit (co)tangent vectors). As we know from the general theory, the structure of the Liouville foliation on Q^3 is described by a marked molecule W^*. Since F is a Bott function, the molecule W^* consists of a finite number of atoms and a finite number of edges. In other words, the manifold Q^3 can be divided into a finite number of one-parameter families of Liouville tori and a finite number of singular leaves.

Consider the natural projection p of Q^3 onto the base M^2 and the induced homomorphism $p_*: H_1(Q^3, \mathbb{Z}) \to H_1(M^2, \mathbb{Z})$ between the one-dimensional homology groups with integer coefficients. It follows from property (5) of geodesics (see above) that each element $\alpha \in H_1(M^2, \mathbb{Z})$ can be realized by a closed geodesic γ on M^2. In turn, this geodesic γ is the projection of a closed trajectory γ' of the geodesic flow on Q^3. Since the geodesic flow is integrable, γ' lies on some leaf of the Liouville foliation (either singular or regular one).

1) Let γ' lie on a Liouville torus T^2. Then α is contained in the image of the group $H_1(T^2, \mathbb{Z}) \simeq \mathbb{Z} \oplus \mathbb{Z}$. Note that all Liouville tori that belong to the same one-parameter family as the torus T^2 itself are isotopic to each other, and therefore, their one-dimensional homology groups are mapped into the same group $p_* H_1(T^2, \mathbb{Z})$.

2) Let γ' lie on some singular leaf of the Liouville foliation. As we know, this leaf consists of a finite number of one-dimensional orbits of the Poisson $\mathbb{R}^2$-action (generated by commuting functions H and f) and a finite number of two-dimensional orbits each of which is homeomorphic to the annulus $S^1 \times D^1$. In each of these cases, the one-dimensional homology group of the orbit is isomorphic to $\mathbb{Z}$. Since γ' lies in a certain orbit O_i, the element α is contained in the image of the group $H_1(O_i, \mathbb{Z}) \simeq \mathbb{Z}$.

Thus, taking into account that the Liouville foliation on Q^3 has a finite number of one-parameter families of Liouville tori and a finite number of singular orbits O_i (i.e., different from tori), we see that the whole group $H_1(M^2, \mathbb{Z})$ (as a set) is the union of a finite number of its subgroups each of which has at most two generators. Since $H_1(M^2, \mathbb{Z})$ is isomorphic either to $\mathbb{Z}^{2g}$ (in the orientable case) or to $\mathbb{Z}^k \oplus \mathbb{Z}_2$ (in the non-orientable case), it follows immediately from this that the one-dimensional homology group of the manifold M^2 should be one of the following four groups $\{e\}$, $\mathbb{Z} \oplus \mathbb{Z}_2$, $\mathbb{Z}_2$, $\mathbb{Z} \oplus \mathbb{Z}$. All the other groups are too large.

Therefore, M^2 is diffeomorphic to one of the following four manifolds: the sphere S^2, the projective plane $\mathbb{R}P^2$, the torus T^2, and the Klein bottle K^2. Thus, Theorem 11.1 is proved for the case where the geodesic flow admits a Bott integral.

It is easily seen from our proof that the Bott condition for F is too strong. The same arguments can literally be repeated in the case where the integral is not a Bott function, but the foliation of Q^3 into orbits of the Poisson action contains only finite number of singular orbits and one-parameter families of regular Liouville tori. For example, the arguments remain valid for the so-called *tame integrals* studied in [236].

To complete the proof of Theorem 11.1, it remains to observe that the analytic integrals are tame. $\square$

The proof of Theorem 11.1 implies the following consequence.

Corollary. *For any integrable geodesic flow on a two-dimensional compact surface that admits a non-trivial tame integral, there exists a Liouville torus T^2 such that $p_* H_1(T^2, \mathbb{Q}) = H_1(M, \mathbb{Q})$, i.e., under the natural projection $p : Q^3 \to M^2$, the one-dimensional rational homology group of the torus covers completely the one-dimensional rational homology group of the surface M.*

The similar statement is also true in the multidimensional case.

Theorem 11.2 (I. A. Taimanov [329]). *Let a real-analytic geodesic flow on a real-analytic n-dimensional compact manifold M be Liouville integrable in the class of real-analytic integrals. Then the one-dimensional Betti number $b_1(M)$ is not greater than n.*

Proof. The scheme of the proof is, in essence, the same as that in the two-dimensional case. As before, each element α of the homology group $H_1(M^n, \mathbb{Z})$ can be realized by a closed trajectory γ' of the geodesic flow on Q^{2n-1} lying either on a regular Liouville torus or on a singular leaf of the Liouville foliation. If γ' lies on a singular leaf, then there exists a small isotopy of this trajectory (perhaps, taken with some multiplicity) which sends (pushes) the trajectory onto a close Liouville torus T^n.

To show this, consider a transversal section P^{2n-2} at an arbitrary point $x \in \gamma'$ and define the Poincaré map σ on it.

Let $f_1, \ldots, f_{n-1}$ be independent first integrals of the geodesic flow restricted to the isoenergy surface $Q^{2n-1} = \{H = 1\}$. We assume that the Hamiltonian H is f_n. Next consider a regular Liouville torus T^n passing near the point x. We can suppose that this torus is given as a common level surface of the integrals: $\{f_1 = c_1, \ldots, f_{n-1} = c_{n-1}\}$. Consider the intersection of the torus T^n and the transversal section P^{2n-2}. In the analytic case, without loss of generality, we can assume that the intersection $T^n \cap P^{2n-2}$ consists (locally) of a finite number of components, and moreover, this number is bounded below by the same constant for all Liouville tori. This fact easily follows from the general properties of real-analytic functions.

Let y belong to one of such components sufficiently closed to x. The Poincaré map σ sends the set $T^n \cap P^{2n-2}$ into itself, permuting somehow its connected components (Fig. 11.1). Since the number of connected components is finite, there is a number q such that σ^q sends each connected component of $T^n \cap P^{2n-2}$ into itself.

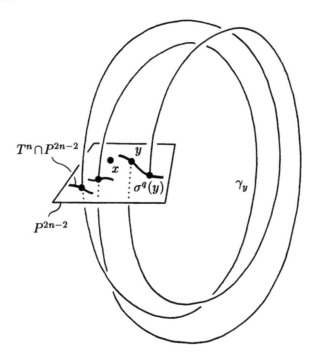

Figure 11.1

Now the points y and $\sigma^q(y)$ lie in the same component and, therefore, can be connected by a curve β inside this component (Fig. 11.1). Consider the closed curve γ^* that consists of the geodesic segment connecting the points y and $\sigma^q(y)$, and the curve β. By construction, γ^* lies entirely on the Liouville torus and, besides, is isotopic to the trajectory γ taken with multiplicity q.

Thus, we have proved that any closed trajectory of the geodesic flow (taken, perhaps, with some multiplicity) can be isotopically deformed onto a regular Liouville torus. It follows immediately from this that for any $\alpha \in H_1(M, \mathbb{Q})$ there exists a Liouville torus T^n and a one-dimensional homology class $\alpha' \in H_1(T^n, \mathbb{Q})$ such that $\alpha = p_*(\alpha')$, where $p_*: H_1(T^n, \mathbb{Q}) \to H_1(M, \mathbb{Q})$ is the homeomorphism induced by the natural projection $p: T^n \to M$.

Thus, the whole rational homology group $H_1(M, \mathbb{Q})$ is the set-theoretical union of the images of the one-dimensional homology groups $\mathbb{Q}^n = H_1(T^n, \mathbb{Q})$ of Liouville tori T^n. Notice that the number of families of (homologically) different Liouville tori is finite. (This follows again from the properties of real-analytic functions.) Therefore, the group $H_1(M, \mathbb{Q})$ is the union of a finite number of its subgroups of the form $p_* H_1(T^n, \mathbb{Q})$ and, consequently, has to coincide with one of them. This claim repeats exactly the corollary of Theorem 11.1. Hence

$$b_1(M) = \dim H_1(M, \mathbb{Q}) = \dim p_* H_1(T^n, \mathbb{Q}) \leq \dim H_1(T^n, \mathbb{Q}) = n,$$

as was to be proved. $\square$

There are some other generalizations of Theorems 11.1 and 11.2. Here are, for example, several statements proved by I. A. Taimanov [330].

Theorem 11.3. *Suppose a real-analytic geodesic flow on M^n is integrable in the class of real-analytic integrals.*

1) The fundamental group $\pi_1(M^n)$ is almost commutative, i.e., contains a commutative subgroup of a finite index.

2) If $\dim H_1(M^n, \mathbb{Q}) = d$, then the cohomology ring $H^*(M^n, \mathbb{Q})$ contains a subring isomorphic to the rational cohomology ring of the d-dimensional torus T^d.

3) If $\dim H_1(M^n, \mathbb{Q}) = n$, then the rational cohomology rings of the manifold M^n and n-dimensional torus T^n are isomorphic: $H^*(M^n, \mathbb{Q}) \simeq H^*(T^n, \mathbb{Q})$.

4) There always exists an n-dimensional Liouville torus $T^n \subset T^* M^n$ such that the image of its fundamental group in $\pi_1(M^n)$ under the natural projection has a finite index.

REMARK. As we have already noticed, the analyticity condition in the above statements is needed only in order for the structure of momentum mapping singularities not to be too complicated. In particular, Theorems 11.2 and 11.3 are still valid under a weaker assumption on the geometrical simplicity of the geodesic flow [330].

There are a number of results by I. A. Taimanov [330], G. Paternain [285], E. I. Dinaburg [96], and others which are devoted to the investigation of links between the integrability, the topological entropy of a geodesic flow, and the topology of the manifold.

The obstructions to the integrability of a rather different kind have been discovered, in particular, by S. V. Bolotin [42], [40]. Side by side with the papers on topological obstructions to integrability, there are many works connected with the construction of explicit examples of integrable geodesic flows on Riemannian manifolds. See, in particular, A. S. Mishchenko and A. T. Fomenko [245], A. V. Brailov [72], A. S. Mishchenko [243], A. Thimm [337], G. P. Paternain and R. J. Spatzier [284].

11.3. TWO EXAMPLES OF INTEGRABLE GEODESIC FLOWS

11.3.1. *Surfaces of Revolution*

Consider a two-dimensional surface of revolution M^2 in $\mathbb{R}^3$ given by the equation $r = r(z)$ in the standard cylindrical coordinates r, φ, z. As local coordinates on M^2, we take z and φ. Then the Riemannian metric induced on M^2 by the Euclidean metric has the form

$$ds^2 = (1 + r'^2)\, dz^2 + r^2(z)\, d\varphi^2 .$$

Let $z = z(t)$, $\varphi = \varphi(t)$ be a geodesic on M^2, and let ψ be the angle between the velocity vector of the geodesic and a parallel on the surface of revolution.

Theorem 11.4 (A. Clairaut).

1) *The geodesic flow of the surface of revolution in R^3 is completely integrable. The function $r \cos\psi$ is constant along each geodesic, i.e., is its first integral.*

2) *The equation of geodesics on the surfaces of revolution has the form*

$$\frac{d\varphi}{dz} = \frac{c\sqrt{1 + r'^2}}{r\sqrt{r^2 - c^2}} ,$$

where c is an arbitrary constant and the geodesics themselves (without taking into account their parameterization) are defined by the following explicit formula:

$$\varphi(z) = \int \frac{d\varphi}{dz}\, dz = \int \frac{c\sqrt{1 + r'^2}}{r\sqrt{r^2 - c^2}}\, dz .$$

Proof. Write down the Hamiltonian of the geodesic flow in the natural coordinates $z, \varphi, p_z, p_\varphi$ on the cotangent bundle:

$$H(z, \varphi, p_z, p_\varphi) = \frac{p_z^2}{1 + r'^2} + \frac{p_\varphi^2}{r^2} .$$

Since $r = r(z)$, the Hamiltonian H does not depend on φ, and the function p_φ is an additional integral of the flow. To clarify the geometrical meaning of this integral, we pass to coordinates on the tangent bundle. Recall (see Proposition 11.1) that momenta and velocities are connected by the following relations:

$$p_z = (1 + r'^2)\dot{z} \quad \text{and} \quad p_\varphi = r^2 \dot{\varphi}.$$

Let $e_\varphi = (0, 1)$ be a tangent vector to a parallel on the surface of revolution M^2. Computing the angle ψ between this vector and the velocity vector $\dot{\gamma} = (\dot{z}, \dot{\varphi})$ of a geodesic $\gamma(t) = (z(t), \varphi(t))$, we obtain

$$\cos\psi = \frac{\langle \dot{\gamma}, e_\varphi \rangle}{\sqrt{\langle \dot{\gamma}, \dot{\gamma} \rangle \langle e_\varphi, e_\varphi \rangle}} = \frac{r^2 \dot{\varphi}}{r\sqrt{E}} = \frac{p_\varphi}{r\sqrt{E}} .$$

Hence $p_\varphi = r \cos\psi \sqrt{E}$.

The value of $\sqrt{E}$ coincides with the length of the velocity vector and, consequently, is constant along any geodesic. Therefore, the function $r \cos\psi$ is an additional integral of the geodesic flow on the surface of revolution. This integral is usually called the Clairaut integral.

We now write the equations of geodesics on the surface of revolution in an explicit form. Using the condition $r \cos \psi = c = \text{const}$, we obtain

$$r \cos \psi = c = \frac{r^2 \dot{\varphi}}{\sqrt{(1 + r'^2)\dot{z}^2 + r^2 \dot{\varphi}^2}}.$$

In this equation, the parameter t on a geodesic can be arbitrary, i.e., it is not necessary the arclength. If we take the coordinate z as t, then the equation can easily be rewritten in the desired form. By integrating its right-hand side, we obtain explicit formulas for geodesics on the surface of revolution. The theorem is proved.   □

11.3.2. *Liouville Metrics*

Definition 11.1. A Riemannian metric on a two-dimensional surface M^2 is called a *Liouville metric* if, in appropriate local coordinates x and y, it has the form

$$ds^2 = (f(x) + g(y))(dx^2 + dy^2),$$

where $f(x)$ and $g(y)$ are some smooth positive functions.

Theorem 11.5.

1) *The geodesic flow of the Liouville metric on a two-dimensional surface is completely integrable.*

2) *The equation of geodesics of the Liouville metric can be written as follows:*

$$\frac{dx}{dy} = \pm \frac{\sqrt{f(x) + a}}{\sqrt{g(y) - a}},$$

and the geodesics themselves are defined by the relation

$$\int \frac{dx}{\sqrt{f(x) + a}} \pm \int \frac{dy}{\sqrt{g(y) - a}} = c,$$

where a and c are arbitrary constants.

Proof. Writing the Hamiltonian of the geodesic flow on the cotangent bundle in standard coordinates x, y, p_x, p_y, we have

$$H = \frac{p_x^2 + p_y^2}{f + g}.$$

It is easy to verify that the function

$$F = p_x^2 - fH$$

is an integral of this flow.

Since the function H is preserved by the geodesic flow, the function $\dfrac{F}{H} = \dfrac{p_x^2}{H} - f$ is an integral, too. Using the standard change of coordinates

$$p_x = (f+g)\dot{x}, \qquad p_y = (f+g)\dot{y},$$

we pass to the tangent bundle.

By rewriting the integral F/H in these coordinates, we obtain

$$\frac{(f+g)^2 \dot{x}^2}{(f+g)(\dot{x}^2 + \dot{y}^2)} - f = a,$$

or

$$\frac{g\dot{x}^2 - f\dot{y}^2}{\dot{x}^2 + \dot{y}^2} = a = \text{const}.$$

In this equation, the parameterization is not important because of the homogeneity of the above expression with respect to derivatives. After simple transformations, we obtain the following equation in total differentials:

$$\frac{dx}{\sqrt{f(x)+a}} \pm \frac{dy}{\sqrt{g(y)-a}} = 0,$$

which can easily be solved. The theorem is proved. ☐

REMARK. The integral $\dfrac{g\dot{x}^2 - f\dot{y}^2}{\dot{x}^2 + \dot{y}^2}$ can be rewritten in the following form, which allows us to reveal its geometrical meaning:

$$g\frac{\dot{x}^2}{\dot{x}^2 + \dot{y}^2} - f\frac{\dot{y}^2}{\dot{x}^2 + \dot{y}^2} = g\cos^2 \psi - f\sin^2 \psi,$$

where ψ is the angle between the velocity vector of a geodesic and the level lines of coordinate x on the surface. This expression resembles an analogous formula from the Clairaut theorem for a surface of revolution. It is not by chance, but can be explained by the observation that the metrics on a surface of revolution is a particular case of a Liouville metric. To see this, one needs to make the change of coordinates $z = z(y)$, $\varphi = x$, which reduces the metric on a surface of revolution to the conformal form. This form will be obviously of the Liouville type. It is sufficient to set $g = r^2$ and $f = 0$. Then the integral of the Liouville metric turns into the square of the Clairaut integral.

REMARK. Sometimes in the literature, by the Liouville form of a metric, one means its representation in the form

$$ds^2 = (f(x) + g(y))(\alpha(x)dx^2 + \beta(y)dy^2),$$

where f, g, α, and β are smooth functions. Such metrics are called *almost Liouville*. Of course, it is not difficult to reduce this metric to the Liouville form (in the above sense). It can be done just by setting

$$x' = \int \sqrt{\alpha(x)}\, dx, \qquad y' = \int \sqrt{\beta(y)}\, dy.$$

However, even without such a reduction, it is possible to write explicit formulas for the geodesics of almost Liouville metrics. The formulas are as follows:

$$\int \frac{\sqrt{\alpha(x)}\,dx}{\sqrt{f(x)+a}} \pm \int \frac{\sqrt{\beta(y)}\,dy}{\sqrt{g(y)-a}} = c\,,$$

where a and c are some constants.

Let us notice that in the two considered examples of Riemannian metrics, the integrals of the geodesic flows are either linear or quadratic in momenta. It turns out (see the next section) that these two types of metrics, in essence, exhaust all the cases where the geodesic flow (on a two-dimensional surface) admits a linear or quadratic integral.

11.4. RIEMANNIAN METRICS WHOSE GEODESIC FLOWS ARE INTEGRABLE BY MEANS OF LINEAR OR QUADRATIC INTEGRALS. LOCAL THEORY

Definition 11.2. Geodesic flows that are integrable by means of an integral which is linear or quadratic in momenta (but not reducible to a linear one) are said to be *linearly* or *quadratically integrable* respectively.

In this section, we describe local properties of Riemannian metrics with such geodesic flows. This description was in fact obtained in classical papers by G. Darboux [92], [93], U. Dini [97], M. L. Raffy [303], G. Birkhoff [32]. In modern terms, this theory was then developed by V. N. Kolokol'tsov [187], [189]; see also the book by V. V. Kozlov [196] and papers by K. Kiyohara [181], [182].

11.4.1. *Some General Properties of Polynomial Integrals of Geodesic Flows. Local Theory*

Let a Riemannian metric be locally (i.e., in a neighborhood of some point on a surface) written as $ds^2 = \lambda(x,y)(dx^2+dy^2)$. In other words, x and y are local conformal (isothermal) coordinates for the given metric. Suppose that its geodesic flow is integrable by means of some polynomial (in momenta) integral

$$F = \sum b_i(x,y)p_x^{n-i}p_y^i$$

of degree n. Consider the function

$$R(z) = (b_0 - b_2 + b_4 - b_6 + \ldots) + i(b_1 - b_3 + b_5 - b_7 + \ldots)\,,$$

where $z = x + iy$. Note that the function $R(z)$ and the integral $F(x,y,p_x,p_y)$ are connected by a simple relation. It is sufficient to set $p_x = 1$ and $p_y = i$ (the imaginary unit) in the expression for F. In other words,

$$R(z) = F(x,y,1,i)\,.$$

The following two statements (Propositions 11.2 and 11.3) were proved by G. Birkhoff for $n = 1, 2$ [32]. For arbitrary n, this result was obtained by V. N. Kolokol'tsov [187].

Proposition 11.2. $R(z)$ *is a holomorphic function of z.*

Proof. Let us compute the Poisson bracket of the integral F and the Hamiltonian $H = (p_x^2 + p_y^2)/\lambda(x, y)$:

$$\{F, H\} = 2\lambda^{-1} \sum_k \left(\frac{\partial b_k}{\partial x} p_x^{n-k+1} p_y^k + \frac{\partial b_k}{\partial y} p_x^{n-k} p_y^{k+1} \right)$$

$$- \sum_k b_k (n-k) \frac{\partial \lambda^{-1}}{\partial x} (p_x^2 + p_y^2) p_x^{n-k-1} p_y^k$$

$$- \sum_k b_k k \frac{\partial \lambda^{-1}}{\partial y} (p_x^2 + p_y^2) p_x^{n-k} p_y^{k-1} = 0.$$

If we put $p_x = 1$ and $p_y = i$ in this identity, then we obtain $p_x^2 + p_y^2 = 0$ and, therefore,

$$\frac{\partial}{\partial x} \sum_k b_k i^k + i \frac{\partial}{\partial y} \sum_k b_k i^k = 0.$$

This equality coincides with the Cauchy–Riemann equations for the complex function $R(z)$. Indeed, since $\sum_k b_k i^k = R(z)$, the condition obtained means

$$\frac{\partial R}{\partial x} + i \frac{\partial R}{\partial y} = 0,$$

i.e., $\dfrac{\partial}{\partial \bar{z}} (R(z)) = 0$. □

Thus, the polynomial integral F allows us to assign a certain holomorphic function $R(z)$ to each isothermal coordinate system x, y. We wish now to find out how this function changes under transition to another isothermal coordinates u, v. For definiteness, we shall consider only orientation preserving changes of coordinates. Recall that, under this condition, the new coordinates u, v are isothermal if and only if the transformation $w = w(z)$, where $w = u + iv$, is a holomorphic function.

Denote by $S(w)$ the holomorphic function constructed from the integral F and related to the new coordinate system u, v.

Proposition 11.3. *Let $w = w(z)$ be a holomorphic change of isothermal coordinates on the surface. Then the functions $R(z)$ and $S(w)$ are connected by the relation $R(z) = S(w(z))(w'(z))^{-n}$, where $w'(z)$ is the complex derivative.*

Proof. Let $w = u + iv$, and p_u, p_v be the new canonical momenta. Then

$$\begin{pmatrix} p_x \\ p_y \end{pmatrix} = \begin{pmatrix} \partial u/\partial x & \partial v/\partial x \\ \partial u/\partial y & \partial v/\partial y \end{pmatrix} \begin{pmatrix} p_u \\ p_v \end{pmatrix}.$$

Taking into account the Cauchy–Riemann equations, we transform this matrix to the following form:

$$\begin{pmatrix} \partial u/\partial x & \partial v/\partial x \\ -\partial v/\partial x & \partial u/\partial x \end{pmatrix}.$$

We have

$$F = \sum_{k=0}^{n} b_k \left(\frac{\partial u}{\partial x} p_u + \frac{\partial v}{\partial x} p_v \right)^{n-k} \left(-\frac{\partial v}{\partial x} p_u + \frac{\partial u}{\partial x} p_v \right)^{k}.$$

To compute the function $S(w(z))$, we substitute $p_u = 1$ and $p_v = i$ into the above expression. We obtain

$$S(w(z)) = \sum_{k=0}^{n} b_k \left(\frac{\partial u}{\partial x} + i\frac{\partial v}{\partial x} \right)^{n-k} \left(-\frac{\partial v}{\partial x} + i\frac{\partial u}{\partial x} \right)^{k}$$

$$= \sum_{k=0}^{n} b_k i^k \left(\frac{\partial u}{\partial x} + i\frac{\partial v}{\partial x} \right)^{n} = R(z)(w'(z))^{n}.$$

Proposition 11.3 is proved.   □

Propositions 11.2 and 11.3 can be reformulated as follows: an additional polynomial integral F of degree n in momenta defines a "differential form"

$$\frac{(dz)^n}{R(z)}$$

on M which is invariant with respect to holomorphic coordinate changes. Such a form is usually called an n-differential.   As a complex structure on M we consider the structure that is uniquely defined by the Riemannian metric. Namely, as a local complex coordinate we take $z = x + iy$, where x, y are isothermal coordinates.

Corollary (V. N. Kolokol'tsov). *On a closed two-dimensional surface M^2 with negative Euler characteristic there exist no polynomially integrable geodesic flows.*

Proof. Without loss of generality, we may assume that the surface M^2 is orientable.   Otherwise one should consider its two-sheeted covering.   Suppose the polynomial integral has the least possible degree.   Note that the form $\frac{(dz)^n}{R(z)}$ has no zeros, since the function $R(z)$ is locally holomorphic (Proposition 11.2).   But there are no such n-differentials on the surfaces different from the sphere and torus, since their zeros and poles must satisfy the following relation (see, for example, [112]):

$$\text{(the number of zeros)} - \text{(the number of poles)} = 2n(\chi - 1),$$

where χ is the genus of the surface.   The only possibility is that $R(z) \equiv 0$. In this case, as it is easy to verify, the integral F is divisible (without a remainder) by the Hamiltonian H (see [150] for details). But this contradicts the fact that F has the least possible degree.   □

Let us note that Kozlov's theorem (Theorem 11.1) does not follow from this statement, because in this theorem the polynomial integrability is not assumed. Moreover, in Kozlov's theorem, the additional assumptions on the character of the integral F must be fulfilled for a single isoenergy surface only.

11.4.2. *Riemannian Metrics Whose Geodesic Flows Admit a Linear Integral. Local Theory*

Consider a smooth Riemannian metric $ds^2 = E\,dx^2 + 2F\,dx\,dy + G\,dy^2$ on a two-dimensional surface M^2. Suppose that in a neighborhood of some point $P \in M^2$ the geodesic flow of this metric possesses an integral $F(x, y, p_x, p_y) = a(x, y)p_x + b(x, y)p_y$, linear in momenta. The following theorem yields a local description of such metrics at a generic point.

Theorem 11.6. *Let the geodesic flow of the metric ds^2 possess a linear (in momenta) integral F in a neighborhood of a point $P \in M^2$. Suppose that F is not identically zero on the (co)tangent plane at the point P. Then, in some neighborhood of this point, there exist local coordinates x and y in which the metric has the form*

$$ds^2 = \lambda(x)(dx^2 + dy^2).$$

Proof. The linear integral F can be considered as a smooth vector field w_F on the surface M^2, since at each point, it is represented as a linear function on the cotangent plane of M^2. The additional condition of the theorem means that the field w_F does not vanish at the point P. It is well known that in this case, in some neighborhood of P, there exist local regular coordinates u and v such that $w_F = \partial/\partial v$, or which is the same, $F = p_v$. Since the Hamiltonian H commutes with the integral F, this means exactly that the function H does not depend on the variable v. Therefore, the metric takes the form

$$ds^2 = E(u)\,du^2 + 2F(u)\,du\,dv + G(u)\,dv^2.$$

Now consider the following coordinate change which transform the metric to the diagonal form:

$$u = u', \qquad v = v' - \int \frac{F(u')}{G(u')}\,du'.$$

Then, in the new coordinate system we have

$$ds^2 = A(u')\,du'^2 + B(u')\,dv'^2.$$

It remains to make one more coordinate change to reduce the metric to the conformal form. Let us set

$$u' = u'(x), \qquad v' = y,$$

where $u'(x)$ is the solution of the differential equation

$$\frac{du'}{dx} = \sqrt{\frac{A(u')}{B(u')}}.$$

After this change, the metric takes the desired form

$$ds^2 = \lambda(x)(dx^2 + dy^2),$$

where $\lambda(x) = A(u'(x))$. $\square$

Let us indicate one more important property of the metrics with linearly integrable geodesic flows. It is easy to verify that the commutativity condition for the Hamiltonian H and linear integral F can be expressed in terms of w_F as $L_{w_F}(g^{ij}) = 0$, where L_{w_F} is the Lie derivative along w_F and $\{g^{ij}\}$ is the tensor inverse to the metric. This immediately implies

$$L_{w_F}(g_{ij}) = 0\,,$$

i.e., the vector field w_F generates a one-parameter group of local diffeomorphisms each of which is an isometry of the metric. The converse statement is also true: every one-parameter isometry group generates a linear integral of the geodesic flow. In other words, the linear integrability is equivalent to the existence of a one-parameter isometry group. This statement is a particular case of the well-known Noether theorem.

REMARK. We assume in Theorem 11.6 that the vector field w_F that determines the linear integral F does not vanish at the point P. In a certain sense, this is the non-degeneracy condition for F. What happens if F has singularity at the given point, i.e., if $w_F(P) = 0$? To what canonical form can the metric be reduced in the neighborhood of such a point? Which singularities of the vector field w_F are admissible? We show below that all such singularities have the simplest form, and the field w_F in the neighborhood of a singular point can be reduced to the form $w_F = c(-y, x) = c\,\partial/\partial\varphi$, where c is some constant.

11.4.3. Riemannian Metrics Whose Geodesic Flows Admit a Quadratic Integral. Local Theory

Theorem 11.7. *Let the geodesic flow of a metric ds^2 possess a quadratic integral F in a neighborhood of a point $P \in M$. Suppose that this integral is not proportional (as a quadratic form) to the Hamiltonian H on the (co)tangent space at the point P. Then, in some neighborhood of this point, the metric is a Liouville one, i.e., there exist local coordinates u and v in which the metric takes the form*

$$ds^2 = (f(u) + g(v))(du^2 + dv^2)\,,$$

where $f(u)$ and $g(v)$ are smooth positive functions. Moreover,

$$F(u, v, p_u, p_v) = \frac{-(f(u) - C)p_v^2 + (g(v) + C)p_u^2}{f(u) + g(v)}\,,$$

where C is some constant.

Proof. From the very beginning, we can assume that local coordinates x, y are chosen to be conformal and the metric has already the form $ds^2 = \lambda(x, y)(dx^2 + dy^2)$. Then

$$H = \frac{p_x^2 + p_y^2}{2\lambda}\,,$$

where p_x and p_y are momenta.

Write the integral $F(x, y, p_x, p_y)$ in the form

$$F = b_1 p_x^2 + 2b_2 p_x p_y + b_3 p_y^2,$$

where $b_i(x, y)$ are some smooth functions. Consider the following function: $R(z) = (b_1 - b_3) + 2ib_2$, where $z = x + iy$. According to Proposition 11.2, this function is holomorphic. Note that the non-proportionality condition for the Hamiltonian H and the integral F on the tangent space implies that there exists a change of coordinates under which the function R becomes constant. Indeed, consider the holomorphic transformation $w = w(z)$ by taking $w(z)$ to be a solution to the differential equation $w'(z) = (R(z))^{-1/2}$. Then, by virtue of Proposition 11.3, a new function $S(w)$ takes the form $S(w) = (w'(z))^2 R(z) \equiv 1$.

We now consider the quadratic integral F in the coordinate system u, v, where $w = u + iv$. We get

$$F = a(u, v)p_u^2 + 2b(u, v)p_u p_v + c(u, v)p_v^2.$$

However, $(a - c) + 2ib = S(w) \equiv 1$; therefore,

$$F = (c + 1)p_u^2 + cp_v^2,$$

where c is some smooth function. The Hamiltonian of the geodesic flow preserves its conformal form in the new coordinate system:

$$H = \frac{p_u^2 + p_v^2}{2\Lambda},$$

where $\Lambda(u, v) = \lambda(x, y)|w'(z)|^{-2}$.

Let us write explicitly the condition that H and F commute. After standard transformations, we obtain

$$\{H, F\} = \left\{ \frac{p_u^2 + p_v^2}{2\Lambda}, (c + 1)p_u^2 + cp_v^2 \right\}$$

$$= \frac{p_u^2 + p_v^2}{\Lambda^2} \left(p_u \frac{\partial}{\partial u}(\Lambda(c + 1)) + p_v \frac{\partial}{\partial v}(\Lambda c) \right) = 0.$$

Finally, we have the following simple system of two equations:

$$\frac{\partial}{\partial u}(\Lambda(c + 1)) = 0, \qquad \frac{\partial}{\partial v}(\Lambda c) = 0.$$

Differentiating the first equation in v and the second one in u, and then subtracting one from the other, we obtain the following equation for the conformal multiplier:

$$\frac{\partial^2 \Lambda}{\partial u \, \partial v} = 0.$$

This means exactly that

$$\Lambda(u, v) = f(u) + g(v),$$

where $f(u)$ and $g(v)$ are some smooth functions, which can be supposed to be positive, since $\Lambda(u, v) > 0$.

Now it is easy to find a general form for the function $c(u, v)$:

$$c = \frac{-f(u) + C}{f(u) + g(v)},$$

where C is an arbitrary constant. Thus,

$$ds^2 = (f(u) + g(v))(du^2 + dv^2),$$

$$F = \frac{(g(v) + C)p_u^2 - (f(u) - C)p_v^2}{f(u) + g(v)}.$$

The theorem is proved.   □

In Theorem 11.7, we have studied the canonical form of a metric with the quadratically integrable geodesic flow in a neighborhood of a generic point P at which the Hamiltonian H and integral F are not proportional (as quadratic forms on the (co)tangent plane). What happens if, however, they turn out to be proportional? What is the local classification of such metrics in this case? The answer is given by the following theorem, which also includes the case of linearly integrable geodesic flows.

Theorem 11.8. *Let the geodesic flow of a Riemannian metric ds^2 possess a quadratic integral F in the neighborhood of a point $P \in M^2$. Suppose that on the (co)tangent plane at this point, the Hamiltonian H and integral F are proportional as quadratic forms. Then the following two cases are possible.*

1) In some neighborhood of the point P, there exist local regular coordinates u, v in which the metric has the form

$$ds^2 = f(u^2 + v^2)(du^2 + dv^2),$$

where $f(t)$ is a positive smooth function. In this case, the geodesic flow possesses one more integral that is linear and has the form $\widetilde{F} = vp_u - up_v$.

2) In some neighborhood of the point P, there exist local regular coordinates u, v in which the metric has the form

$$ds^2 = \frac{h(r + u) - h(u - r)}{2r}(du^2 + dv^2),$$

where $r = \sqrt{u^2 + v^2}$ and h is a smooth (in a neighborhood of zero) function satisfying the condition $h'(0) > 0$. The quadratic integral F of the geodesic flow (up to a linear combination with the Hamiltonian) has the form

$$F(u, v, p_u, p_v) = -\frac{r(h(u + r) + h(u - r))}{h(u + r) - h(u - r)}(p_u^2 + p_v^2) + (up_u^2 + 2vp_up_v - up_v^2).$$

REMARK. Obviously, the converse statement is also true: any metric given by the above explicit formulas is smooth and its geodesic flow admits a smooth linear or quadratic integral of the type described.

Proof (of Theorem 11.8). The condition that the integral F and Hamiltonian H are proportional at the point P is equivalent to the fact that the function $R(z)$ vanishes at P, where $z = x + iy$ (see the proof of Theorem 11.7). Let us study the character of its zero.

Lemma 11.1. *Let $R(z)$ be given and different from zero in a neighborhood of a point z_0. Then the local Liouville coordinates x', y' corresponding to this function (i.e., those in which $R(z) \equiv 1$) are uniquely defined in this neighborhood up to a parallel translation on the plane and the reflection $(x', y') \to (-x', -y')$. In particular, the coordinate lines $x' = $ const and $y' = $ const are uniquely defined. These two families of coordinate lines are determined as solutions to the following two differential equations on the plane (x, y):*

$$\frac{dz}{dt} = \sqrt{R(z)}, \qquad \frac{dz}{dt} = i\sqrt{R(z)}.$$

Proof. Proposition 11.3, which gives the transformation rule for the function R under holomorphic coordinate changes, implies that $dz' = \dfrac{dz}{\pm\sqrt{R(z)}}$ where x', y' are Liouville coordinates, and $z' = x' + iy'$.

Therefore, $z' = z'(z)$ is the primitive of the function $\dfrac{1}{\pm\sqrt{R(z)}}$ in a neighborhood of the point z_0, where $R(z_0) \neq 0$, and, consequently, is uniquely defined up to transformation $z' \to \pm z' + $ const, as required.

Further, the coordinate lines $x' = $ const and $y' = $ const of the Liouville coordinate system are given by the equations $\dfrac{dz'}{dt} = 1$ and $\dfrac{dz'}{dt} = i$, which can be rewritten in the desired form

$$\frac{dz}{dt} = \sqrt{R(z)}, \qquad \frac{dz}{dt} = i\sqrt{R(z)}.$$

The lemma is proved. □

REMARK. The above differential equations determine the coordinate lines only as geometrical curves, i.e., without orientation on them. The point is that the right-hand side of the equations does not determine a vector field but only a direction field.

Let $z = 0$ be a singular point, i.e., $R(0) = 0$.

Lemma 11.2. *Let $R(z)$ have a first order zero at the point $z = 0$. Then, in a neighborhood of $z = 0$, there exists a holomorphic transformation $w = w(z)$ that transforms the function R to a linear function $S(w) = w$.*

Proof. According to Proposition 11.2, the desired function $w(z)$ must be the solution of the differential equation

$$\left(\frac{dw}{dz}\right)^2 = \frac{w}{R(z)}.$$

It suffices, therefore, to show that in a neighborhood of zero this equation has a holomorphic solution satisfying the condition $w'(0) \neq 0$.

We indicate this solution explicitly. Let us represent the function $R(z)$ in the form $R(z) = z\,b(z)$, where $b(0) \neq 0$. Then the function $\dfrac{1}{\sqrt{b(z^2)}}$ is an even holomorphic function in the neighborhood of zero. Consider its primitive $a(z)$ satisfying the condition $a(0) = 0$. The function $a(z)$ will be an odd function (due to the evenness of its derivative). This implies that the function $w(z) = \left(a\sqrt{z}\right)^2$ is holomorphic. It is not hard to verify that $w(z)$ is a solution of the equation in question and, moreover, $w'(0) \neq 0$, since $a'(0) = \dfrac{1}{\sqrt{b(0)}} \neq 0$. $\square$

Lemma 11.3. *Let $R(z)$ have a second order zero at the point $z = 0$. Then, in a neighborhood of $z = 0$, there exists a holomorphic transformation $w = w(z)$ which makes the function R into the function $S(w) = A^2 w^2$, where A is a complex number.*

Proof. As in Lemma 11.2, we must find a holomorphic (in a neighborhood of zero) solution of the differential equation

$$\left(\frac{dw}{dz}\right)^2 = \frac{A^2 w^2}{R(z)}$$

for an appropriate complex number $A \neq 0$. It can be done explicitly again.

In the neighborhood of zero, the function $R(z)$ can be represented in the form

$$R(z) = A^2 z^2 (1 + c_1 z + c_2 z^2 + \ldots).$$

That is why we can take a root of it, that is, there exists a holomorphic function $q(z) = Az\,b(z)$ such that $q(z)^2 = R(z)$ and, moreover, $b(0) = 1$. Thus, it suffices to solve the following equation:

$$\frac{dw}{w} = \frac{dz}{z\,b(z)}.$$

We shall search the solution in the form $w(z) = z\,c(z)$. By substituting this in the equation, we obtain

$$\frac{c\,dz + z\,dc}{cz} = \frac{dz}{zb}, \quad \text{i.e.,} \quad \frac{dc}{dz} = \frac{c\,(1 - b)}{zb}.$$

By virtue of the fact that $b(0) = 1$, the function $m(z) = \dfrac{1 - b}{zb}$ is holomorphic. Therefore, the equation has the holomorphic solution of the form

$$c(z) = \exp\!\left(\textstyle\int m(z)\,dz\right).$$

Thus, $w(z) = z\exp\!\left(\int m(z)\,dz\right)$ is the desired change. $\square$

COMMENT. Lemmas 11.2 and 11.3 can be obtained as a consequence of general theorems on the normal forms of analytic differential equations at singular points.

Lemma 11.4. *The order of a zero of $R(z)$ is at most 2.*

Proof.   Consider the coordinate lines of the Liouville coordinate system in a punctured neighborhood of the point $z = 0$, where $R(0) = 0$. As we have already shown, these lines are uniquely defined (since the function R is given and fixed). They are solutions to the equations

$$\frac{dz}{dt} = \sqrt{R(z)} \quad \text{and} \quad \frac{dz}{dt} = i\sqrt{R(z)}\,.$$

We assert that, if the order of zero of $R(z)$ at $z = 0$ is greater than 2, then $z = 0$ is a limit point for each of these coordinate lines (i.e., each line enters zero). Indeed, consider, first, the case where $R(z)$ has the form z^k, where $k > 2$. Then one can easily see that these lines have the form shown in Fig. 11.2 (for even and odd k). It is seen that each line enters zero. In a general case, we have

$$R(z) = z^k + \ldots = z^k(1 + n(z))\,,$$

where $n(z)$ is a holomorphic function. It is clear that, for small z, the picture shown in Fig. 11.2 does not change from the qualitative point of view. That is, each coordinate line still enters zero.

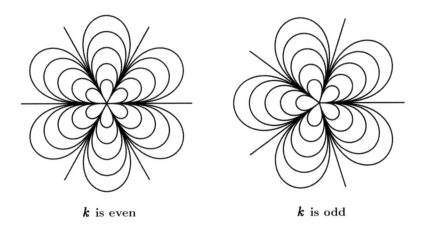

k is even   **k** is odd

Figure 11.2

Consider now the metric. By writing it in the Liouville coordinates and returning to the initial coordinate system $z = x + iy$, we see that it can be represented as

$$\frac{(f + g)\, dz\, d\bar{z}}{|R(z)|}\,,$$

where f is a function that is constant on coordinate lines of the first family, and g is a function that is constant on coordinate lines of the second family.

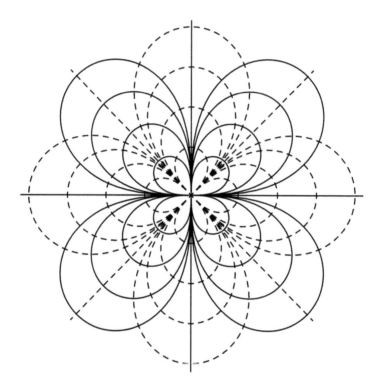

Figure 11.3

Since all these lines, as just shown, enter zero, i.e., become arbitrarily close to one another (see Fig. 11.3), the functions f and g are both constant in a neighborhood of the point $z = 0$. Moreover, the function $\dfrac{f + g}{|R(z)|}$ is smooth in this neighborhood. Taking into account that $R(0) = 0$, we conclude that $f(0)+g(0) = 0$. Hence $f + g \equiv 0$, but this is impossible by virtue of non-degeneracy of the metric. We arrive at a contradiction, which completes the proof. □

These three lemmas in fact give us a complete list of possible singularities of Liouville coordinates.

Let us turn to the proof of Theorem 11.8.

Thus, we have shown that the function R can be reduced by a holomorphic transformation in a neighborhood of its zero to one of the following canonical forms: it can be either w or $A^2 w^2$, where A is a complex number. Let us examine both these cases. This function written in a canonical coordinate system w will be still denoted by $R = R(w)$ (but not by $S(w)$ as in Lemmas 11.2 and 11.3).

Case $R(w) = w$. It is convenient to take $2w$ here instead of w, i.e., we set $R(w) = 2w$. Then, to find the change of coordinates that reduces the metric to the Liouville form, we need to solve the equation

$$dz = \frac{dw}{\sqrt{2w}}.$$

This equation has an obvious solution $w = z^2/2$. The coordinate lines of the Liouville coordinates x, y in a neighborhood of P are shown in Fig. 11.4 (here, as above, $z = x + iy$ and $w = u + iv$).

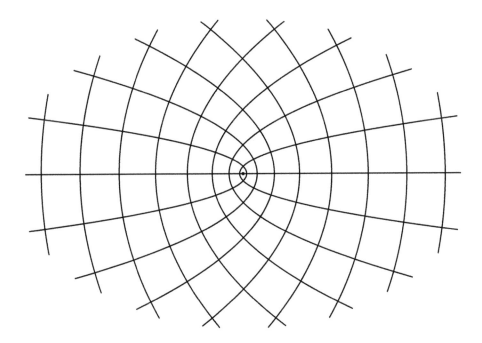

Figure 11.4

Being written in coordinates x, y, the metric becomes $(f(x) + g(y))(dx^2 + dy^2)$. Doing the inverse change $x = \sqrt{r + u}$, $y = \sqrt{r - u}$, where $r = \sqrt{u^2 + v^2}$, we obtain the following form of this metric in coordinates u, v:

$$\frac{f(\sqrt{r + u}) + g(\sqrt{r - u})}{2r}(du^2 + dv^2).$$

Let us show that the functions f and g are closely related and, in fact, both arise from some smooth function h.

Let $ds^2 = \Lambda(u, v)(du^2 + dv^2)$. Set $h(t) = t\Lambda(t/2, 0)$. We assert that

$$\Lambda(u, v) = \frac{h(u + r) - h(u - r)}{2r}.$$

To verify this, notice that the denominator in the right-hand side has the desired form, i.e., is represented as a sum of two functions, one of which depends only on x and the other depends only on y. Hence it is sufficient to verify our identity only on two coordinate lines, for example, $\{x = 0\}$ and $\{y = 0\}$. From the viewpoint of coordinates $w = u + iv$, this means that the verification should be done on the real axis $\{v = 0\}$ only. Let us do it.

For $u \geq 0$ and $v = 0$, we have $r = u$. Then

$$\frac{h(u+r) - h(u-r)}{2r} = \frac{h(2u) - h(0)}{2u} = \frac{2u\Lambda(u,0) - 0}{2u} = \Lambda(u,0).$$

Analogously, if $u < 0$ and $v = 0$, then $r = -u$ and

$$\frac{h(u+r) - h(u-r)}{2r} = \frac{h(0) - h(2u)}{-2u} = \frac{0 - 2u\Lambda(u,0)}{-2u} = \Lambda(u,0).$$

Thus, in a neighborhood of P, the Riemannian metric has the form

$$ds^2 = \frac{h(u+r) - h(u-r)}{2r}(du^2 + dv^2),$$

where h is a smooth function (and even analytic in the case of an analytic metric); moreover, $h'(0) = \Lambda(0,0) \neq 0$, as was to be proved.

It is useful to point out the following relation between the function h and the functions f, g:

$$f(t) = h(t^2), \qquad g(t) = -h(-t^2).$$

It remains to write the explicit form for the integral

$$F = \frac{g(y)p_x^2 - f(x)p_y^2}{f(x) + g(y)}$$

after the transformation $w = z^2/2$. Substituting the explicit expressions for the momenta and coordinates

$$\begin{pmatrix} p_x \\ p_y \end{pmatrix} = \begin{pmatrix} x & y \\ -y & x \end{pmatrix} \begin{pmatrix} p_u \\ p_v \end{pmatrix},$$

$$x = \sqrt{r+u}, \qquad y = \sqrt{r-u},$$

into the above formula for F, we obtain the required expression for the integral, that is,

$$F(u,v,p_u,p_v) = -r\frac{h(u+r) + h(u-r)}{h(u+r) - h(u-r)}(p_u^2 + p_v^2) + (up_u^2 + 2vp_up_v - up_v^2).$$

Thus, the case where P is a first order zero of F is examined.

Case $R(w) = A^2w^2$. We show first that the number A^2 has to be real. The proof of this fact is similar to that of Lemma 11.4. Consider the level lines of the Liouville coordinate system in a neighborhood of a zero of R. They are defined by equations

$$\frac{dw}{dt} = Aw, \qquad \frac{dw}{dt} = iAw.$$

If $A = a + ib$, $a \neq 0$, and $b \neq 0$, then all these level lines enter zero as shown in Fig. 11.5. Two orthogonal families of infinite spirals wind around zero. The arguments completely similar to those which have been used in the proof of Lemma 11.4 show us that the function $f + g$ becomes identically zero, which is impossible. Therefore, the number A^2 is necessarily real.

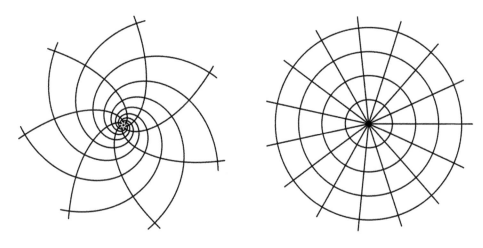

Figure 11.5   Figure 11.6

Suppose, for definiteness, $A^2 > 0$, i.e., $A \in \mathbb{R}$ (the case $A \in i\mathbb{R}$ can be examined analogously). Here the transformation has the form $w = \exp(Az)$, and the coordinate lines of Liouville coordinates x, y are radial rays going out from the origin and the concentric circles (Fig. 11.6). The metric can be written again as

$$(\widetilde{f}(x) + \widetilde{g}(y))(dx^2 + dy^2) = \frac{\widetilde{f} + \widetilde{g}}{A^2 r^2}(du^2 + dv^2),$$

where the first function $\widetilde{f}$ is constant on the circles and the second $\widetilde{g}$ is constant on the rays going out from the point P. But a smooth function, being constant on the rays, has to be constant identically, since all rays meet at the point P. The first function should depend on the sum of squares $r^2 = u^2 + v^2$ only. As a result, we conclude that the conformal multiplier Λ in coordinates u and v becomes $\frac{\widetilde{f}(r^2)}{A^2 r^2}$. Denoting this expression by $f(r^2)$, we have

$$ds^2 = f(r^2)(du^2 + dv^2).$$

The smoothness of the function $f(t)$ is implied by the smoothness of the conformal multiplier $f(r^2) = f(u^2 + v^2)$ in the variables u and v. Theorem 11.8 is proved. □

Thus, Liouville coordinates can have only two types of singularities shown in Figs. 11.4 and 11.6. This observation will essentially be used for the global description of Riemannian metrics with quadratically integrable geodesic flows on the sphere.

11.5. LINEARLY AND QUADRATICALLY INTEGRABLE GEODESIC FLOWS ON CLOSED SURFACES

A number of papers by several authors have been devoted to the classification of linearly and quadratically integrable geodesic flows on two-dimensional closed surfaces. The examples of such flows have, of course, been well known since Jacobi and Liouville. However, the first results on their complete description have been obtained quite recently by V. N. Kolokol'tsov. Then his results have been developed and completed in the papers by K. Kiyohara [182], I. K. Babenko and N. N. Nekhoroshev [26], I. K. Babenko [25], V. S. Matveev [224]. In our book we shall follow the terminology and notation suggested by V. S. Matveev, because they are most convenient for the further study of integrable geodesic flows in the framework of the Liouville and orbital classification theory.

11.5.1. *The Torus*

Consider the standard two-dimensional torus T^2. Every point is given as a pair (x, y), where $x \in \mathbb{R} \bmod T_x$, $y \in \mathbb{R} \bmod T_y$. In other words, we consider the torus as the quotient space $\mathbb{R}^2 / \Gamma$, where Γ is the lattice generated by vectors $e_1 = (T_x, 0)$ and $e_2 = (0, T_y)$. Two real numbers x and y defined modulo T_x and T_y respectively are called *global periodic coordinates* on the torus, T_x and T_y are called their periods. Note that on the same torus there exist infinitely many different global periodic coordinates.

Theorem 11.9 (The case of a linear integral).

1) *Let the geodesic flow of a Riemannian metric ds^2 on the torus T^2 be linearly integrable. Then there exist global periodic coordinates x, y, where $x = x$ (mod 2π) and $y = y$ (mod 2π), on the torus in which the metric has the form*

$$ds^2 = h(y)(a\, dx^2 + c\, dx\, dy + b\, dy^2),$$

where $h(y)$ is some positive 2π-periodic smooth function, a, b, c are real numbers such that the form $a\, dx^2 + c\, dx\, dy + b\, dy^2$ is positively defined.

2) *And conversely, the geodesic flow of such a metric on the torus T^2 is linearly integrable.*

This theorem can be reformulated as follows.

Recall that, by virtue of the uniformization theorem, for any Riemannian metric ds^2 on the torus $T^2 = \mathbb{R}^2 / \Gamma$ there exist global isothermal coordinates. More precisely, this means that on the covering plane $\mathbb{R}^2$ there exist global coordinates x, y in which the metric has the form $ds^2 = \lambda(x, y)(dx^2 + dy^2)$, where λ is a doubly periodic function (i.e., invariant with respect to transitions by the elements of the lattice). Here, of course, we suppose that the action of the lattice Γ in coordinates x, y is the standard linear action: each element $g \in \Gamma$ is an integer linear combination $mf_1 + nf_2$, where $f_1, f_2 \in \mathbb{R}^2$ is a basis of Γ, and its action to a point $X = (x, y)$ is simply $g(X) = X + g$. Note that from the point of view of global isothermal coordinates x, y the lattice Γ can be distorted. In particular, the coordinate lines $\{x = \text{const}\}$ and $\{y = \text{const}\}$ do not have to be closed (in contrast to the case of global periodic coordinates).

It is worth mentioning that global isothermal coordinates on the torus are defined uniquely up to transformation $w = az + b$ or $w = a\bar{z} + b$. We shall use this fact below.

The following theorem indicates the form of the conformal multiplier λ in global isothermal coordinates for linearly integrable geodesic flows.

Theorem 11.10. *Let the geodesic flow of a Riemannian metric ds^2 on the torus $T^2 = \mathbb{R}^2/\Gamma$ be linearly integrable and x, y be global isothermal coordinates. Then*

$$ds^2 = f(-\delta x + \gamma y)(dx^2 + dy^2),$$

where (γ, δ) are the coordinates of some vector of the lattice Γ.

Proof (of Theorem 11.9).   Let u, v be arbitrary global isothermal coordinates on the torus T^2.

Consider a linear integral of the geodesic flow $F = b_0 p_u + b_1 p_v$ lifted to the covering plane and written in coordinates u, v, p_u, p_v. It is clear that F is a doubly periodic function with respect to variables u, v. Therefore, the function $R(z) = b_0 + i b_1$ constructed above (see Proposition 11.2) has the same property. Since $R(z)$ is holomorphic on the whole torus, it has to be constant. Then $R(z)$ can be made equal to 1 identically by means of a linear complex transformation (note that after such transformations the coordinates remain global isothermal). Therefore, without loss of generality we can assume that $b_0 + i b_1 = 1$, hence $b_0 = 1$, $b_1 = 0$, i.e., $F = p_u$. Thus, the conformal multiplier $\lambda(u, v)$ of the Riemannian metric does not depend on u, since H commutes with $F = p_u$.

Thus, $\lambda(u, v) = q(v)$ and $ds^2 = q(v)(du^2 + dv^2)$, where q is some smooth function which is periodic with respect to the lattice Γ.

Consider two cases.

a) Let $q(v) = \text{const}$. Then we obtain a flat torus with the metric $ds^2 = \text{const}(du^2 + dv^2)$. Passing from the global isothermal coordinates u, v to the global periodic coordinates x, y on the covering plane, we turn the metric $ds^2 = \text{const}(du^2 + dv^2)$ into the metric $ds^2 = \text{const}(a\,dx^2 + c\,dx\,dy + b\,dy^2)$, as was to be proved.

b) Suppose that the function $q(v)$ is not constant. Then there exists a basis f_1, f_2 of the lattice Γ such that $f_1 = (\alpha, 0)$. Indeed, if it is not so, then on the straight line $\{v = 0\}$ there is no element of the lattice, and its image on the torus (under the natural projection $\mathbb{R}^2 \to T^2 = \mathbb{R}^2/\Gamma$) is everywhere dense. But in this case the function $q(v)$, being constant on this line, has to be constant on the whole torus, which contradicts our assumption.

Take such a vector $f_1 = (\alpha, 0)$ and complement it up to a basis by another vector $f_2 \in \Gamma$. Consider a linear change of coordinates $(u, v) \to (x, y)$ such that in new coordinates x, y the basis vectors of the lattice become $f_1 = (2\pi, 0)$ and $f_2 = (0, 2\pi)$. The coordinates x, y will be obviously global 2π-periodic coordinates on the torus.   It is easy to see that they are connected with the isothermal coordinates u, v by the relations $u = \dfrac{\alpha}{2\pi} x + \beta y$, $v = \gamma y$. Substituting these relations into the above expression for the metric in isothermal coordinates, we obtain

$$ds^2 = h(y)(a\,dx^2 + c\,dx\,dy + b\,dy^2),$$

where $h(y) = q(\gamma y)$. The theorem is proved. $\square$

Proof (of Theorem 11.10).   We have already proved (Theorem 11.9) that there exist global isothermal coordinates u, v in which the metric has the form $q(v)(du^2 + dv^2)$. Moreover, we have shown that one of vectors of the lattice Γ is collinear to the basic vector of the isothermal coordinate system. Thus, this form of the metric satisfies the assertion of Theorem 11.9. If we take another isothermal coordinate system on the torus, then it suffices to remember that any two such systems are obtained one from the other by some transformation of the form $w = az + b$, where $a, b \in \mathbb{C}$ (the case $w = a\bar{z} + b$ can be considered by analogy). It is easy to see that such transformation preserves the desired form of the metric. Indeed, by writing this transformation as

$$u = a_1 x - a_2 y + b_1 ,$$
$$v = a_2 x + a_1 y + b_2 ,$$

where $a = a_1 + ia_2$, $b = b_1 + ib_2$, we obtain

$$ds^2 = (a_1^2 + a_2^2)q(a_2 x + a_1 y + b_2)(dx^2 + dy^2) .$$

On the other hand, the basic vector f_1 of the lattice Γ, which had coordinates $(\alpha, 0)$, after the transformation will have coordinates $(\gamma, \delta) = \dfrac{\alpha}{a_1^2 + a_2^2}(a_1, -a_2)$. That is why the metric can be rewritten in the desired form

$$ds^2 = f(-\delta x + \gamma y)(dx^2 + dy^2) ,$$

where $f(t) = (a_1^2 + a_2^2)\, q\left(\dfrac{a_1^2 + a_2^2}{\alpha} t + b_2 \right)$. $\square$

COMMENT. As we know, a linear integral of the geodesic flow can be viewed as a vector field on the surface, i.e., on the torus in our case. It follows from the explicit form of the integral F (see the proof of Theorem 11.9) that this vector field has no singular points on the torus. Its integral curves are just level lines

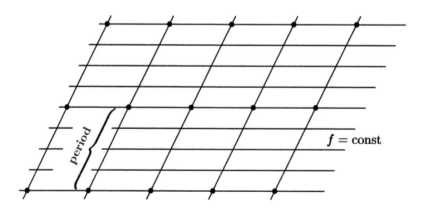

Figure 11.7

of the function f (Fig. 11.7), which are parallel to one of the vectors of Γ. Therefore, a Riemannian metric whose geodesic flow is integrable admits a one-parameter isometry group $\mathcal{G}$, which is isomorphic to $S^1 \simeq SO(2)$ and acts freely by translations along one of basis vectors of the lattice. The converse is also true: if a non-flat metric on the torus admits a one-parameter isometry group $\mathcal{G}$, then its geodesic flow is linearly integrable (E. Noether's theorem), and the group $\mathcal{G}$ is isomorphic to the circle which acts on the torus just in the same way as above.

Definition 11.3. A Riemannian metric on the torus is called a *global Liouville metric* if there exist global periodic coordinates x and y on the torus in which the metric has the form

$$ds^2 = (f(x) + g(y))(dx^2 + dy^2),$$

where $f(x)$ and $g(y)$ are some smooth positive functions with periods T_x and T_y, respectively, different from constants.

Note that this property of a metric implies, in particular, that the lattice Γ has an orthogonal basis. In general, it is not the case.

It is clear that the geodesic flow of a global Liouville metric is integrable. However, a Riemannian metric on the torus with quadratically integrable geodesic flow, as we shall see now, does not have to be a global Liouville one.

Theorem 11.11 (The case of a quadratic integral [26]).

1) *The geodesic flow of a Riemannian metric ds^2 on the torus T^2 is integrable by means of a quadratic integral (irreducible to a linear one) if and only if there exists finite-sheeted covering of T^2 by another torus $\widetilde{T}^2$*

$$\pi : \widetilde{T}^2 \to T^2$$

such that the metric $d\tilde{s}^2 = \pi^* ds^2$ lifted from T^2 to $\widetilde{T}^2$ is a global Liouville metric on the torus $\widetilde{T}^2$.

2) *There exist Riemannian metrics on the torus which are not global Liouville ones, but, nevertheless, whose geodesic flows are quadratically integrable.*

Such metrics ds^2 are said to be *finite-sheeted Liouville metrics* on the torus (unless they are global Liouville ones themselves).

Proof. The beginning of the proof is quite similar to that of Theorem 11.9. It follows from the uniformization theorem that on the torus there exist global isothermal coordinates x, y in which the metric becomes $\lambda(x, y)(dx^2 + dy^2)$. From the local theory we already know that every isothermal coordinate system leads to a holomorphic function $R(z)$, which will be globally holomorphic on the torus due to globality of the coordinates x, y. Since the torus is compact, $R(z)$ has to be constant. It is clear that $R(z) \neq 0$; otherwise the Hamiltonian and the integral are linearly dependent. By means of a complex linear change of coordinates, we can make the function $R(z) = R_0$ equal 1 identically on the whole torus.

As was shown above, in this case $\lambda(x, y) = f(x) + g(y)$. Since we assume the quadratic integral not to be reducible to a linear one, neither $f(x)$ nor $g(y)$ is a constant.

Thus, there exist global isothermal coordinates on the torus in which the given metric takes the form

$$ds^2 = (f(x) + g(y))(dx^2 + dy^2).$$

Such coordinates are said to be *global Liouville coordinates*.

A priori it is not clear how these global Liouville coordinates x, y are connected with the lattice Γ related to the torus. Let us examine this question.

Denote by e_1 and e_2 the basis vectors of the global Liouville coordinate system x, y on the covering plane, i.e., the vectors with coordinates $(1, 0)$ and $(0, 1)$. Let us show that the periodicity of the function $f(x) + g(y)$ (as a function on the covering plane) with respect to the lattice Γ implies the following statement.

Lemma 11.5. *There exist elements $\tilde{f}_1$ and $\tilde{f}_2$ of the lattice Γ such that*

$$\tilde{f}_i = \alpha_i e_i, \qquad \text{where} \quad i = 1, 2.$$

Proof. Note, first of all, that it follows from the boundedness of the function $f(x) + g(y)$ on the torus that each of the functions $f(x)$ and $g(y)$ is also bounded on the covering plane. Besides, the fact that the function $f(x) + g(y)$ is doubly periodic with respect to Γ implies that each of the functions $f(x)$ and $g(y)$ is individually periodic with respect to the same lattice Γ. Indeed, by writing down the periodicity condition for the function $f(x) + g(y)$, we obtain

$$f(x + \omega_1) + g(y + \omega_2) = f(x) + g(y),$$

where (ω_1, ω_2) is an arbitrary element of the lattice Γ. Rewrite this identity in the following way:

$$f(x + \omega_1) - f(x) = g(y + \omega_2) - g(y).$$

Since the left-hand side of this identity depends only on x, and the right-hand side depends only on y, then they are both equal to a certain constant C. Hence

$$f(x + \omega_1) = f(x) + C.$$

But it follows immediately from the boundedness of f that the constant C is actually zero. Therefore, f is a periodic function. Analogously we obtain that g is also periodic. Thus each of the functions $f(x)$ and $g(y)$ is in fact a function on the torus (not only on the covering plane).

Take the function $f(x)$ and consider one of its level lines on the torus given by the equation $x = 0$. This level line is necessarily closed. Indeed, if it is not the case, then the line $\{x = 0\}$, being the projection of a straight line under the standard projection $\mathbb{R}^2 \to T^2$, has to be everywhere dense on the torus. Then f is identically constant, which contradicts the irreducibility of the quadratic integral to a linear one. In other words, if we consider the line $\{x = 0\}$ on the covering plane $\mathbb{R}^2$, then it necessarily contains one of the vectors of Γ. Therefore, there exists an element $\tilde{f}_2 \in \Gamma$ such that $\tilde{f}_2 = \alpha_2 e_2$. By analogy, we prove the existence of an element $\tilde{f}_1$ such that $\tilde{f}_1 = \alpha_1 e_1$. Lemma 11.5 is proved. $\square$

Consider the sublattice $\tilde{\Gamma} \subset \Gamma$ generated by elements $\tilde{f}_1 = (\alpha_1, 0)$ and $\tilde{f}_2 = (0, \alpha_2)$ and the torus $\tilde{T}^2 = \mathbb{R}^2 / \tilde{\Gamma}$ related to the sublattice $\tilde{\Gamma}$. It is clear that the initial torus T^2 is covered by $\tilde{T}^2$. On the new torus $\tilde{T}^2$, consider the coordinates x and y taken from the covering plane $\mathbb{R}^2$. It is clear that they are global periodic coordinates on the torus $\tilde{T}^2$ with periods $T_x = \alpha_1$ and $T_y = \alpha_2$. At the same time the metric in these coordinates has the Liouville form and is, therefore, a global Liouville metric on the torus $\tilde{T}^2$.

The first statement of Theorem 11.11 is proved.

It remains to show that on the two-dimensional torus there exist Riemannian metrics with quadratically integrable geodesic flows which are not global Liouville ones. In other words, the finite-sheeted coverings that appear in Theorem 11.11 are actually essential for the complete description of integrable geodesic flows. We give an example.

Consider the metric on the plane $\mathbb{R}^2$

$$ds^2 = (f(x) + g(y))(dx^2 + dy^2),$$

where $f(x)$ is periodic with period 1, and $g(y)$ is periodic with period $\dfrac{\sqrt{3}}{2}$. Take the lattice Γ on the plane generated by the vectors $f_1 = (2,0)$ and $f_2 = (1, \sqrt{3})$ (Fig. 11.8). Consider ds^2 as a Riemannian metric on the corresponding torus $T^2 = \mathbb{R}^2/\Gamma$. Its geodesic flow is obviously quadratically integrable. But the metric

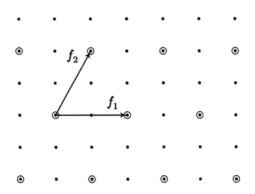

Figure 11.8

itself is not a global Liouville metric on the torus. Indeed, if it had been a global Liouville metric, then the lattice Γ would have admitted an orthogonal basis. However, it is quite clear that such an orthogonal basis does not exist.

The theorem is proved.   □

Note that Theorem 11.11 can be easily derived from the paper by I. K. Babenko and N. N. Nekhoroshev [26], as well as the next result, where the tori T^2 and $\tilde{T}^2$ are interchanged in some sense.

Theorem 11.12. *The geodesic flow of a Riemannian metric ds^2 on the torus T^2 is integrable by means of a quadratic integral (irreducible to a linear one) if and only if there exists another torus $\tilde{T}^2$ with a global Liouville metric $d\tilde{s}^2$ and the covering*

$$\rho: T^2 \to \tilde{T}^2$$

such that $ds^2 = \rho^* d\tilde{s}^2$.

In other words, all Riemannian metrics on the torus with quadratically integrable geodesic flows can be obtained from global Liouville metrics by taking finite-sheeted covering.

Proof. As was proved above, there exist global isothermal coordinates x, y on the covering plane in which the metric ds^2 takes the form $(f(x)+g(y))(dx^2+dy^2)$. Besides, we have proved that each of the functions $f(x)$ and $g(y)$ is periodic. Consider the smallest periods of these functions. Let a be the smallest period for $f(x)$, and b be the smallest period for $g(y)$. Consider the new orthogonal lattice $\widetilde{\Gamma}$ generated by vectors $(a, 0)$ and $(0, b)$ (Fig. 11.9). Recall that on the plane

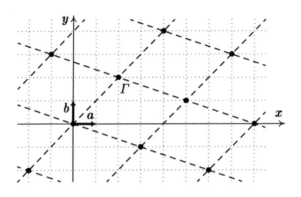

Figure 11.9

there is the initial lattice Γ that determines the given torus. This lattice Γ does not have to be orthogonal, and its basis vectors do not have to go along the directions of x and y. We assert that the lattice Γ is in fact a sublattice of the new lattice $\widetilde{\Gamma}$. Indeed, if (ω_1, ω_2) is an arbitrary element of Γ, then, as was proved above, $f(x + \omega_1) = f(x)$ and $g(y + \omega_2) = g(y)$, i.e., ω_1 is a period of the function $f(x)$, and ω_2 is a period of the function $g(y)$. Since a and b are the smallest periods of these functions, ω_1 and ω_2 must be multiples of them, i.e., $\omega_1 = k_1 a$ and $\omega_2 = k_2 b$ for some integers k_1 and k_2. Therefore,

$$(\omega_1, \omega_2) = k_1(a, 0) + k_2(0, b).$$

This just means that (ω_1, ω_2) is an element of the lattice $\widetilde{\Gamma}$ as claimed.

Consider the new torus $\widetilde{T}^2 = \mathbb{R}^2 / \widetilde{\Gamma}$ that corresponds to the lattice $\widetilde{\Gamma}$. On this torus we define the metric given formally by the same formula as above, i.e., $(f(x) + g(y))(dx^2 + dy^2)$. The point is that the conformal multiplier $f(x) + g(y)$ is invariant with respect to the lattice $\widetilde{\Gamma}$. The obtained metric on the torus $\widetilde{T}^2$ is obviously a global Liouville metric. On the other hand, the torus T^2 covers the torus $\widetilde{T}^2$, as was to be proved. $\square$

We now prove one more useful statement.

Proposition 11.4.

a) *Let the geodesic flow of a metric ds^2 on the torus be linearly integrable. If the metric ds^2 is not flat, then the linear integral of the flow is defined uniquely up to a constant multiplier.*

b) *Let the geodesic flow of a metric ds^2 on the torus be quadratically integrable. Then its quadratic integral F is defined uniquely up to an arbitrary linear combination with the Hamiltonian H.*

Proof. a) As we showed, the linear integral must have the form $F = b_0 p_u + b_1 p_v$, where b_0 and b_1 are constants. By taking the Poisson bracket of this expression with the Hamiltonian $H = \dfrac{p_u^2 + p_v^2}{q(v)}$, we obtain the following necessary condition: $b_1 \dfrac{\partial}{\partial v}(q(v)) \equiv 0$. Hence we see that, for a not flat metric, $b_1 \equiv 0$, i.e., $F = \text{const} \cdot p_u$.

b) Suppose F and F' are two quadratic integrals of the geodesic flow. Each of them corresponds to a certain global Liouville coordinate system. Denote these coordinates by x, y and u, v and consider the corresponding conformal factors $\lambda(x, y) = f(x) + g(y)$ and $\lambda'(u, v) = f'(u) + g'(v)$. Recall that then, according to Theorem 11.7, we have

$$F(x, y, p_x, p_y) = \frac{-(f(x) - C)p_y^2 + (g(y) + C)p_x^2}{f(x) + g(y)},$$

$$F'(u, v, p_u, p_v) = \frac{-(f'(u) - C')p_v^2 + (g'(v) + C')p_u^2}{f'(u) + g'(v)},$$

where C and C' are certain constants.

Since the coordinate systems are both isothermal, they are connected by a linear transformation either complex or anti-complex. Therefore, the functions λ and λ' differ from each other by a constant factor, i.e.,

$$f(x) + g(y) = \text{const}\,(f'(u) + g'(v)).$$

When is such a situation possible? If the axes of the new and old coordinates are not parallel, it is easy to verify that $f(x) + g(y) = a(x^2 + y^2) + bx + cy + d$. But this is impossible due to the periodicity of the conformal factor.

Thus, the axes of the Liouville coordinate systems under consideration are parallel. This exactly means that the coordinates $z = x + iy$ and $w = u + iv$ are connected by one of the following transformations:

a) $w = az + b$,
b) $w = i(az + b)$,
c) $w = a\bar{z} + b$,
d) $w = i(a\bar{z} + b)$,

where a is real, and b is a complex number.

Applying any of them to the integral F', we can rewrite it in the same coordinate system x, y as that related to the integral F, and verify the claim.   □

Above we have described linearly and quadratically integrable geodesic flows on the torus. However, if we want to deal with the classification problem in a more formal way, we must produce the complete list of canonical forms for the corresponding metrics, and indicate afterwards which metrics from the list are isometric. As a result, the problem of isometry classification for the metrics in question will be solved in the strict sense. Such an approach was carried out by V. S. Matveev [224].

We begin with the case of linearly integrable geodesic flows. As was seen above, for every such metric ds^2 on the torus $T^2 = \mathbb{R}^2/\Gamma$ there exist global isothermal coordinates u, v on the covering plane in which $ds^2 = q(v)(du^2 + dv^2)$; here the function $q(v)$ is invariant under translations by the elements of the lattice Γ. Besides, we have shown that the first basis element of Γ can be taken in the form

$f_1 = (\alpha, 0)$ (unless ds^2 is flat). After scaling $u' = u/\alpha$, $v' = v/\alpha$, we can assume that $f_1 = (1, 0)$. Then the second basis vector f_2 can be chosen as $f_2 = (t, L)$, where $t \in [0, 1)$, $L > 0$. Since $q(v)$ is invariant with respect to the lattice, L is a period of the function q. So, the metric ds^2 can be determined (coded) by means of the triple (q, t, L), where $t \in [0, 1)$, $L > 0$, $q(v)$ is a function with period L.

Conversely, if we are given an arbitrary triple (q, t, L), then, using it, we can construct a natural metric on the torus. Namely, we need to consider the metric $ds^2 = q(v)(du^2 + dv^2)$ on the plane $\mathbb{R}^2$ with Cartesian coordinates u, v and then take the quotient space $\mathbb{R}^2/\Gamma$, where Γ is the lattice generated by vectors $f_1 = (1, 0)$ and $f_2 = (t, L)$. For brevity, we shall call such metrics (q, t, L)-metrics.

Theorem 11.13 (V. S. Matveev). *If the geodesic flow of a metric on the torus admits a linear integral, then the metric is either flat or isometric to a (q, t, L)-metric. Two metrics corresponding to the triples (q, t, L) and $(\widehat{q}, \widehat{t}, \widehat{L})$ are isometric if and only if for some real number c one of the following four relations is fulfilled:*

1) $(\widehat{q}(v), \widehat{t}, \widehat{L}) = (q(v + c), t, L)$,
2) $(\widehat{q}(v), \widehat{t}, \widehat{L}) = (q(v + c), 1 - t, L)$,
3) $(\widehat{q}(v), \widehat{t}, \widehat{L}) = (q(-v + c), t, L)$,
4) $(\widehat{q}(v), \widehat{t}, \widehat{L}) = (q(-v + c), 1 - t, L)$.

Proof. For every linearly integrable geodesic flow on the torus, we have described the procedure of constructing the corresponding (q, t, L)-model. This task is reduced to searching global isothermal coordinates u, v satisfying two conditions:

a) in terms of these coordinates the metric takes the form $q(v)(du^2 + dv^2)$,

b) the first basic vector of the lattice has coordinates $(1, 0)$.

Therefore, the only ambiguity in coding a given metric by its (q, t, L)-models is that one can choose such coordinates u, v on the covering plane in different ways.

Let us describe the transformations preserving conditions (a) and (b). We know that any transformation between two global isothermal coordinate systems must have the form $w = az + b$ (or $w = a\bar{z} + b$), where a, b are some complex numbers. But in our case, these transformations have, in addition, to preserve the form of the function q in the sense that it must remain a function of the second variable v only. That is why the indicated transformations have to preserve the level lines $\{v = \text{const}\}$. Hence a is a real number. Condition (b) (the norming condition) then implies $a = \pm 1$. The number b can be arbitrary. As a result, we obtain four types of transformations illustrated in Fig. 11.10. By dotted lines we show the new coordinates on the covering plane after the corresponding transformation.

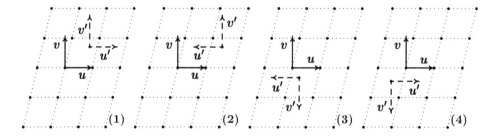

Figure 11.10

Every such transformation changes the global isothermal coordinate system and, consequently, changes the (q, t, L)-model, i.e., the conformal factor of the metric and the coordinates of the basic vectors of the lattice.

The cases (1), (2), (3), (4) presented in Fig. 11.10 exactly correspond to the transformations (1), (2), (3), (4) indicated in the theorem.   □

Consider now the case of Riemannian metrics on the torus whose geodesic flows admit quadratic integrals (irreducible to linear ones). To begin with we shall construct a canonical form for every such metric by using Theorem 11.12. According to this theorem, such a metric is presented as a triple $(\rho, (\widetilde{T}^2, d\widetilde{s}^2))$, where $d\widetilde{s}^2$ is a global Liouville metric on a torus $\widetilde{T}^2$, and $\rho \colon T^2 \to \widetilde{T}^2$ is a finite-sheeted covering. Such a representation is very useful, but, generally speaking, not uniquely defined. Our goal now is to choose one canonical representation among these triples. To this end we need to choose a canonical covering

$$\rho_0 \colon T^2 \to \widetilde{T}_0^2 \,.$$

To explain the main point, we shall say at once that ρ_0 must be chosen to be the covering with the least possible number of sheets.

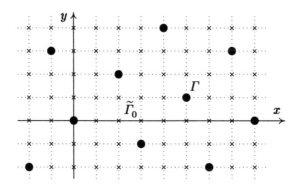

Figure 11.11

Formally we proceed as follows. Consider the Liouville coordinate system x, y on the covering plane of the torus T^2 and the lattice Γ that determines the torus. As we noted above, the lattice Γ can be distorted with respect to the orthogonal coordinate system x, y (Fig. 11.11). We construct a new lattice $\widetilde{\Gamma}_0$ in the following way. Through each node of the lattice Γ we draw vertical and horizontal straight lines (Fig. 11.11). As a result, we obtain an orthogonal lattice on the plane. It is just that lattice which we shall take as $\widetilde{\Gamma}_0$.

Let us indicate the following important property of the lattice $\widetilde{\Gamma}_0$.

Lemma 11.6. *Let $\widetilde{\Gamma}$ be any other orthogonal lattice on the covering plane, which contains Γ and whose basis vectors are directed along the axes x and y. Then $\widetilde{\Gamma}_0$ is a sublattice of $\widetilde{\Gamma}$. In particular, $\widetilde{\Gamma}_0$ does not contain non-trivial sublattices satisfying the above properties.*

Proof. The proof follows from the following obvious remark. If $\tilde{\Gamma}$ is an arbitrary orthogonal lattice with the indicated properties and P and Q are two its nodes, then the intersection point of two straight lines passing through P and Q in parallel to the axes x and y respectively also belongs to the lattice $\tilde{\Gamma}$. Lemma 11.6 is proved. $\square$

This statement implies the following important property of the lattice $\tilde{\Gamma}_0$ and the corresponding covering $\rho_0 : T^2 \to \tilde{T}_0^2$. Consider an arbitrary covering $\rho : T^2 \to \tilde{T}^2$ such that $ds^2 = \rho^* d\tilde{s}^2$, and let $d\tilde{s}^2$ be a global Liouville metric on the torus $\tilde{T}^2$. Then the lattice $\tilde{\Gamma}$ corresponding to the torus $\tilde{T}^2$ satisfies the assumptions of Lemma 11.6, and, therefore, the covering ρ factors through ρ_0. In other words, there exists a covering $\pi : \tilde{T}_0^2 \to \tilde{T}^2$ such that $\rho = \pi \circ \rho_0$.

It follows from this, in particular, that $ds^2 = (\rho_0)^* d\tilde{s}_0^2$, and the metric $d\tilde{s}_0^2$ on the torus $\tilde{T}_0^2$ is a global Liouville one. Moreover, the covering ρ_0 has the least number of sheets among all the coverings over the tori with global Liouville metrics.

The next step is to choose a canonical basis in the lattice Γ. Denote by e_1, e_2 an orthogonal basis in the lattice $\tilde{\Gamma}_0$. Here we assume that e_1 is parallel to the axis x, and e_2 is parallel to the axis y, where x, y are global Liouville coordinates on the covering plane. It is clear that such a basis is defined uniquely up to multiplication of its vectors by -1. As the basis f_1, f_2 in the lattice Γ, we take two following vectors: $f_1 = me_1$, where $m > 0$ is some integer, and $f_2 = ke_1 + ne_2$, where $0 \le k < m$ and $n > 0$. These two properties determine the basis vectors $f_1 = (m, 0)$ and $f_2 = (k, n)$ uniquely.

Lemma 11.7. *In fact, $n = 1$, and k and m are relatively prime.*

Proof. Assume the contrary, that is, n is a positive integer number greater than 1. Consider the lattice Γ' generated by vectors e_1 and ne_2. Then $\Gamma \subset \Gamma' \subset \tilde{\Gamma}_0$ that contradicts Lemma 11.6.

Let the integers k and m be not relatively prime. Then $k = sk'$ and $m = sm'$, where $s \ne 1$. Considering the lattice Γ' generated by vectors se_1 and e_2, we see again that it satisfies the property $\Gamma \subset \Gamma' \subset \tilde{\Gamma}_0$, which contradicts Lemma 11.6. Lemma 11.7 is proved. $\square$

Thus, the canonical covering $\rho_0 : T^2 \to \tilde{T}_0^2$ constructed above is determined by the matrix

$$\begin{pmatrix} m & k \\ 0 & 1 \end{pmatrix},$$

composed from the coordinates of vectors f_1, f_2 with respect to the basis of the lattice $\tilde{\Gamma}_0$.

Let (x, y) be global Liouville coordinates on the torus $\tilde{T}_0^2$. In these coordinates, $e_1 = (T_x, 0)$ and $e_2 = (0, T_y)$. Here T_x and T_y are the periods of Liouville coordinates. Passing to other Liouville coordinates

$$x' = \frac{x}{T_x}, \qquad y' = \frac{y}{T_x},$$

we may always assume that the period T_x equals 1.

Thus, each Riemannian metric with quadratically integrable geodesic flow on the torus can be determined (coded) by means of the following quadruple:

$$(L, f, g, k/m),$$

where $L = T_y$ (after rescaling so that $T_x = 1$) is an arbitrary positive number, f and g are two periodic smooth function with periods 1 and L respectively, and k/m is a rational number from $[0, 1)$.

Given such a quadruple, the torus T^2 with a quadratically integrable geodesic flow is constructed as follows.

First consider a global Liouville metric on the Euclidean plane (x, y):

$$(f(x) + g(y))(dx^2 + dy^2).$$

Then, by taking the quotient with respect to the lattice Γ generated by vectors $f_1 = (m, 0)$ and $f_2 = (k, L)$, one gets the torus T^2 with the desired quadratically integrable metric g_{ij}, which will be, for brevity, called an $(L, f, g, k/m)$-*metric* or $(L, f, g, k/m)$-*model*. Thus, we have proved the following statement.

Proposition 11.5. *Any Riemannian metric on the torus with quadratically integrable geodesic flow can be represented as an* $(L, f, g, k/m)$-*metric.*

As we already pointed out, the coding of a quadratically integrable metric by means of the quadruple $(L, f, g, k/m)$ is not, generally speaking, uniquely defined. We examine the character of this ambiguity below.

Proposition 11.6. *Let* (x, y) *and* (u, v) *be two global Liouville coordinate systems on the covering plane of the torus for a given Liouville metric (global or finite-sheeted). Then the coordinates* (x, y) *and* (u, v) *are connected by a sequence of the following four transformations:*

1) $z \to z + b$, *where* b *is a real number (translation along the axis* x),
2) $z \to \bar{z}$ *(complex conjugation),*
3) $z \to -i\bar{z}$ *(interchanging* x *and* y),
4) $z \to az$, *where* a *is a positive real number (homothety).*

REMARK. The compositions of these transformations form a group consisting of eight connected components:
   a) $w = az + b$,
   b) $w = i(az + b)$,
   c) $w = a\bar{z} + b$,
   d) $w = i(a\bar{z} + b)$,
where $a \in \mathbb{R} \setminus \{0\}$, and $b \in \mathbb{C}$.

Observe that the real number a can be either positive or negative. Changing the sign changes the connected component in the group.

Proof (of Proposition 11.6).   We have already shown (see the proof of Proposition 11.4) that two global Liouville coordinate systems must be connected by one of the above transformations (a), (b), (c), (d). Each of these transformations can, in turn, be represented as a superposition of the transformations (1), (2), (3), (4).   □

Let us describe four elementary operations, which we shall use to transform the codes $(L, f, g, k/m)$.

Operation α_b: $\alpha_b(L, f, g, k/m) = (L, \widehat{f}, g, k/m)$, where $\widehat{f}(x) = f(x - b)$. The meaning of this operation is that we make the transformation $z \to z + b$ (see Proposition 11.6).

Operation β: $\beta(L, f, g, k/m) = (L, f, \widehat{g}, (m - k)/m)$, where $\widehat{g}(y) = g(-y)$. The meaning of this operation is that we make the conjugation $z \to \bar{z}$ (see Proposition 11.6).

Operation γ: $\gamma(L, f, g, k/m) = (1/L, \widehat{f}, \widehat{g}, \widehat{k}/m)$, where $\widehat{f}(x) = L^2 g(Lx)$, $\widehat{g}(y) = L^2 f(Ly)$, and $m > \widehat{k} \geq 0$, $k\widehat{k} = 1 \pmod{m}$. The meaning of this operation is interchanging variables x and y and rescaling. More precisely, in terms of Proposition 11.6 the operation γ corresponds to the complex transformation $w(z) = -i\bar{z}/L$. One has to rescale the complex coordinate in order for the first real coordinate x to have period 1.

Operation δ_c: $\delta_c(L, f, g, k/m) = (L, \widehat{f}, \widehat{g}, k/m)$, where $\widehat{f}(x) = f(x) + c$, $\widehat{g}(y) = g(y) - c$, and c is an arbitrary constant. The meaning of this operation is that the conformal multiplier λ can be decomposed into a sum of two functions $f(x)$ and $g(y)$ in two different ways, namely, $\widehat{f}(x) + \widehat{g}(y) = f(x) + g(y)$.

Theorem 11.14 (V. S. Matveev). *Two Riemannian metrics on the torus given by the quadruples $(L, f, g, k/m)$ and $(\widehat{L}, \widehat{f}, \widehat{g}, \widehat{k}/\widehat{m})$ are isometric if and only if these sets of parameters can be transformed to each other by a composition of the four elementary operations $\alpha_b, \beta, \gamma, \delta_c$.*

Proof.   As we already explained, after fixing global Liouville coordinates on the covering plane, the construction of the quadruple $(L, f, g, k/m)$ related to the given metric is unambiguous (modulo the representation of the conformal factor λ as the sum $f + g$, i.e., up to the operation δ_c). Thus, the only ambiguity of the $(L, f, g, k/m)$-model consists in the choice of global Liouville coordinates on the covering plane of the torus. But we already know that any two global Liouville coordinate systems can be obtained from each other by means of elementary transformations described in Proposition 11.4. These elementary transformations induce some transformations of quadruples $(L, f, g, k/m)$. It remains to observe that these induced transformations are exactly the operations α_b, β, γ. The operation δ has been already taken into account. It should be noted that we do not discuss here the homothety transformation $z \to az$, where a is real and positive. The point is that we have the additional norming condition for the Liouville coordinate system: we choose the coordinates x, y in such a way that the first basic vector of the lattice $\widetilde{\Gamma}_0$ has the form $(1, 0)$.

Let us prove the converse statement. It is required to verify that the transformations $\alpha_b, \beta, \gamma, \delta_c$ do not change the metric, i.e., more precisely, transfer it into an isometric one. But this is evident, because each of these transformations can be considered as just a change of global Liouville coordinates without changing the metric and lattice. $\square$

11.5.2. *The Klein Bottle*

Consider integrable geodesic flows on the Klein bottle K^2. As we shall see, among them there exist flows of different types, in particular, linearly and quadratically integrable. Moreover, on the Klein bottle there are metrics whose geodesic flows admit a polynomial integral of degree 4, which cannot be reduced to linear and quadratic ones.

Recall that the Klein bottle admits a two-sheeted covering by the torus. Such a covering is uniquely defined (up to the natural equivalence of coverings). It is clear that by lifting an integrable geodesic flow from the Klein bottle K^2 to the torus, we obtain an integrable geodesic flow on the torus. However, as we shall see, the degree of the integral can become lower. For example, from a quadratically integrable flow we may obtain a linearly integrable geodesic flow on the torus. And from the geodesic flow with an integral of degree 4 we may obtain a quadratically integrable geodesic flow. See I. K. Babenko [25] and V. S. Matveev [222], [223].

Definition 11.4. Let the geodesic flow of a Riemannian metric on the Klein bottle be quadratically integrable. This metric is called a *global Liouville metric* if after lifting to the covering torus it becomes a global Liouville metric on the torus.

Definition 11.5. Let the geodesic flow of a Riemannian metric on the Klein bottle be quadratically integrable. This flow is called *quasi-linearly integrable* if after lifting to the covering torus it becomes linearly integrable.

A priori we could expect that on the Klein bottle there exist finite-sheeted Liouville metrics by analogy with the case of the torus. However, on the Klein bottle no such metrics occur. It turns out that the lattice of the covering torus is always orthogonal. Thus, unlike the torus, there are no distorted lattices here.

We begin with the complete description of quadratically integrable geodesic flows on the Klein bottle. Consider two positive functions $f(x)$ and $g(y)$ with the properties

1) $f(x)$ is periodic with period $1/2$ and is not constant,
2) $g(y)$ is an even periodic function with period L,

and take the Liouville metric $ds^2 = (f(x) + g(y))(dx^2 + dy^2)$ on the plane $\mathbb{R}^2$. Consider the torus T^2 as a quotient space $\mathbb{R}^2/\Gamma$, where the orthogonal lattice Γ is generated by the vectors

$$f_1 = (1,0) \quad \text{and} \quad f_2 = (0, L).$$

Due to the periodicity of the functions f and g under the translations by the elements of Γ, the metric ds^2 can be viewed as a metric on the torus T^2.

Consider an involution ξ on this torus defined by

$$\xi(x, y) = (x + 1/2, -y).$$

This is the standard involution which has no fixed points and changes orientation. Therefore, the Klein bottle can be represented as $K^2 = T^2/\xi$. It is easy to see that the metric ds^2 is invariant with respect to ξ and, consequently, can be dropped from the torus down to the Klein bottle. We obtain a Riemannian metric on K^2. By construction, such metrics are determined by three parameters L, f, g, where

L is one of the periods of the lattice, f and g are two functions satisfying the above properties.

Definition 11.6. Such a metric on the Klein bottle is called an (L, f, g)-*metric*.

Theorem 11.15.

a) *If the geodesic flow of a Riemannian metric on the Klein bottle is quadratically integrable, then the metric is isometric to some (L, f, g)-metric with appropriate parameters L, f, g.*

b) *An (L, f, g)-metric is a Liouville metric on the Klein bottle if and only if the function $g(y)$ is not constant. Otherwise, the geodesic flow of the (L, f, g)-metric is quasi-linearly integrable.*

Proof. Consider a quadratically integrable geodesic flow on K^2 and cover K^2 by the torus T^2 and, then, by the plane $\mathbb{R}^2$. On the plane $\mathbb{R}^2$ there appears a metric whose geodesic flow is either quadratically integrable or linearly integrable (if the degree of the integral is reduced after unfolding the Klein bottle). In this case, as was shown above, there exist global Liouville coordinates x, y on the plane in terms of which the metric has the form $(f(x) + g(y))(dx^2 + dy^2)$. Here at least one of the functions $f(x)$ and $g(y)$ is not constant, because otherwise the metric on K^2 is flat and admits, consequently, a linear integral. Let, for definiteness, $f(x)$ be non-constant. Denote the lattice of the torus by Γ, as before.

Let us lift the involution ξ from the torus onto the plane $\mathbb{R}^2$. It can always be done (however, not uniquely). We denote the new involution again by ξ. Let us find the explicit formula for its action in terms of coordinates x, y. First notice that ξ preserves the metric and, in particular, its Liouville form. That is why, as we saw above, ξ must have one of the following forms:

$$z \to az + b,$$
$$z \to a\bar{z} + b,$$
$$z \to iaz + b,$$
$$z \to ia\bar{z} + b,$$

where a is a real number, and b is a complex one.

Since ξ^2 is identical on the torus, ξ^2 on the plane $\mathbb{R}^2$ preserves the lattice Γ, i.e., maps it into itself. Hence $a^2 = 1$, i.e., $a = \pm 1$ and no dilatations exist.

Besides, the mapping ξ changes orientation and, therefore, only the following possibilities of the list remain:

1) $z \to \bar{z} + b$,
2) $z \to -\bar{z} + b$,
3) $z \to i\bar{z} + b$,
4) $z \to -i\bar{z} + b$.

Let us show that the cases 3 and 4 are not allowed.

Lemma 11.8. *The involution ξ can be neither of type 3 nor of type 4.*

Proof. Consider the case 3. Since the involution ξ has the form $z \to i\bar{z} + b$, in terms of coordinates x and y it becomes

$$x \to y + b_1, \qquad y \to x + b_2.$$

As ξ preserves the metric on the torus, we obtain the following conditions on the functions f and g:

$$f(x) + g(y) = f(y + b_1) + g(x + b_2).$$

Evidently, it follows from this that

$$f(x) = g(x + b_2) + C_0, \qquad g(y) = f(y + b_1) - C_0.$$

Since the involution ξ is defined on the torus, it also preserves the integral F which is the pull-back of the initial integral from the Klein bottle. Now remember that, according to Proposition 11.4, the form of this quadratic integral on the torus is defined uniquely up to a linear combination with the Hamiltonian. This form is as follows:

$$F(x, y, p_x, p_y) = \frac{-(f(x) - C)p_y^2 + (g(y) + C)p_x^2}{f(x) + g(y)},$$

Let us act on this form by the involution ξ, taking into account the relations for f and g obtained above. We get

$$
\begin{aligned}
\xi^* F &= \frac{-(f(y + b_1) - C)p_x^2 + (g(x + b_2) + C)p_y^2}{f(y + b_1) + g(x + b_2)} \\
&= \frac{-(g(y) + C_0 - C)p_x^2 + (f(x) - C_0 + C)p_y^2}{f(x) + g(y)}.
\end{aligned}
$$

By equating $\xi^* F$ and F, we obtain

$$-g(y) - C_0 + C = g(y) + C, \qquad f(x) - C_0 + C = -f(x) + C.$$

Hence the functions $f(x)$ and $g(y)$ are both constants whose sum is zero. But this is impossible because the conformal factor $\lambda = f(x) + g(y)$ must be a positive function. Thus, case 3 is completely treated.

Case 4 is examined just in the same way. It suffices to make the change $x \to -x$, after which we get exactly the previous case 3.   □

Let us now analyze the involution of type 2.   First suppose that none of the functions $f(x)$ and $g(y)$ is constant. Then, by interchanging x and y, we turn type 2 into type 1. In other words, in this case the variables x and y have equal status. Case 1 itself will be analyzed later.

If in the case 2 the function $g(y)$ turns out to be a constant equal to some number g, and $f(x)$ is not constant, then the metric has the form

$$(f(x) + g)(dx^2 + dy^2)$$

and evidently admits the linear integral p_y. At the same time, the involution ξ acts as follows: $(x, y) \to (-x + b_1, y + b_2)$. Clearly, the integral p_y is invariant under ξ, and, therefore, can be descended down, i.e., to the Klein bottle. As a result we obtain a *linear* integral on the Klein bottle. This contradicts our assumption that the initial quadratic integral is not reduced to a linear one.

Thus, we can assume that the involution ξ has the form $z \to \bar{z} + b$, i.e., $\xi(x, y) = (x + b_1, -y + b_2)$ (case 1). By a translation of the coordinate system along the axis y one can achieve that $b_2 = 0$, i.e., $\xi(x, y) = (x + b_1, -y)$.

Consider the lattice Γ of the torus $T^2 = \mathbb{R}^2/\Gamma$.

Lemma 11.9. *There exists an orthogonal basis* f_1, f_2 *of the lattice* Γ *such that* $f_1 = (a, 0)$, $f_2 = (0, b)$, *and, moreover,* $b_1 = (n + 1/2)a$.

Proof. Consider the involution ξ on the torus. Then ξ^2 is identical on the torus and, consequently, by considering ξ^2 on the covering plane, we obtain that ξ^2 maps the lattice Γ into itself and has the form $\xi^2(x, y) = (x + 2b_1, y)$, i.e., ξ is a translation along the axis x. Hence, the vector $(2b_1, 0)$ is an element of the lattice Γ. Therefore, there exists a basic vector f_1 of the lattice Γ such that $kf_1 = (2b_1, 0)$ for some natural k, i.e., $f_1 = (a, 0)$. Let us show that k is in fact odd. Assume the contrary, then the vector $(b_1, 0)$ belongs to the lattice Γ. But in this case, the point $(0,0)$ is mapped to the point $(b_1, 0)$ under the action of ξ. In particular, this point remains in the lattice and, consequently, from the point of view of the torus, is a fixed point of the involution ξ. But this contradicts the definition of ξ. As a result, it follows from this that $b_1 = (n + 1/2)a$.

By ξ_0 we denote the linear part of the affine mapping ξ on the plane, i.e., $\xi_0(x, y) = (x, -y)$. We claim that ξ_0 maps the lattice Γ into itself. Indeed, since ξ is well-defined on the torus, it follows that $\xi(P + \omega) = \xi(P) + \omega'$, where P is an arbitrary point in the plane, ω and ω' belong to the lattice. On the other hand, $\xi(P + \omega) = \xi(P) + \xi_0(\omega)$. Thus, $\xi_0(\omega) = \omega'$, i.e., the lattice is actually mapped into itself.

Observe that ξ_0 is the symmetry of the plane with respect to the axis x (see the above formula), which preserves the lattice Γ. There exist only two types of such lattices. They are shown in Fig. 11.12. The first type is an orthogonal lattice, i.e., just that we need. Lattices of the second type are generated by isosceles triangles (Fig. 11.12(b)). We now give the formal proof of this fact.

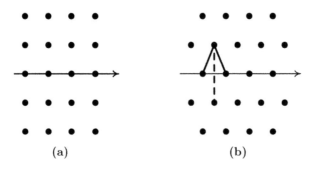

(a)   (b)

Figure 11.12

Choose the second basic vector of the lattice in the form $f_2 = (c, b)$, where $b > 0$ and $0 \leq c < a$. Then $\xi_0(f_2) = (c, -b)$. Consider the element $\xi_0(f_2) + f_2 = (2c, 0) \in \Gamma$. Recall that the first basic vector is $f_1 = (a, 0)$. Consequently, c is equal to either zero or $a/2$. This exactly corresponds to the two types of lattices presented in Fig. 11.12.

Actually, the lattices of the second type cannot appear. The point is that, in this case, the involution ξ would have a fixed point on the torus, but this is forbidden. Indeed, consider the point $(0, -b/2)$. Its image under ξ has the form $((n + 1/2)a, b/2)$. These two points differ by the vector

$$((n + 1/2)a, b) = n(a, 0) + (a/2, b) = nf_1 + f_2.$$

But this vector belongs to the lattice, i.e., the point $(0, -b/2)$ is fixed under the action of ξ on the torus.

Thus, the only admissible lattice is the orthogonal lattice with basis $f_1 = (a, 0)$ and $f_2 = (0, b)$. The lemma is proved. $\square$

We now turn to the proof of the theorem. Consider the conformal coordinate change $x \rightarrow x/a$, $y \rightarrow y/a$. Then the basis of the lattice becomes $f_1 = (1, 0)$ and $f_2 = (0, L)$, where $L = b/a$. The involution ξ in terms of the new coordinate system (which we still denote by x, y) is written as

$$\xi(x, y) = (x + n + 1/2, -y).$$

However, this involution and the involution of the form $(x, y) \rightarrow (x + 1/2, -y)$ coincide on the torus, because n is even. Therefore, one can suppose that $\xi(x, y) = (x + 1/2, -y)$.

Summarizing, we can say that we have represented the metric on the Klein bottle as an (L, f, g)-metric. This completes the proof of item (a).

We now discuss the question in what case this metric is a global Liouville metric on the Klein bottle. This means that it must be such after lifting from the Klein bottle to the torus. Clearly, in our case, this is equivalent to the fact that none of the functions $f(x)$ and $g(y)$ is constant. This is exactly the first part of (b).

It remains to analyze the case when $g(y)$ is constant. In this case, the degree of the integral can be reduced after lifting to the torus, because one can take p_y as a new linear integral. It needs to show that the initial integral on the Klein bottle remains quadratic (that is, cannot be reduced to a linear one). This will mean the quasi-linearity of the given (L, f, g)-metric on the Klein bottle. In order for the linear integral p_y to determine a linear integral on the Klein bottle, it is necessary for p_y to be invariant under the involution ξ on the torus. But p_y evidently changes sign under ξ and, consequently, cannot be descended onto K^2. Only by squaring it, we obtain the quadratic integral p_y^2 which can be successfully descended onto K^2 and determines a quadratic integral there.

It remains to notice that such a linear integral on the torus is defined uniquely up to a constant factor. This completes the proof of (b). $\square$

To complete the classification of quadratically integrable geodesic flows on the Klein bottle, we need to answer the question: which triples (L, f, g) and $(\hat{L}, \hat{f}, \hat{g})$ correspond, in fact, to the same metric on the Klein bottle? Or, more precisely: which (L, f, g)-metrics are isometric among themselves?

Consider the following four operations $\alpha_v, \beta, \gamma, \delta_c$ on the set of triples $\{(L, f, g)\}$:

$$\alpha_v(L, f(x), g(y)) = (L, f(x + v), g(y)),$$
$$\beta\,(L, f(x), g(y)) = (L, f(x), g(y + L/2)),$$
$$\gamma\,(L, f(x), g(y)) = (L, f(-x), g(y)),$$
$$\delta_c\,(L, f(x), g(y)) = (L, f(x) + c, g(y) - c).$$

Theorem 11.16 (V. S. Matveev). *The (L, f, g)-metric and $(\widehat{L}, \widehat{f}, \widehat{g})$-metric on the Klein bottle are isometric if and only if the triples (L, f, g) and $(\widehat{L}, \widehat{f}, \widehat{g})$ can be obtained from each other by compositions of the operations $\alpha_v, \beta, \gamma, \delta_c$.*

Proof. The scheme of reasoning repeats, of course, the proof of the analogous theorem for the torus case. We only note some distinctions. The difference is that in this case we should take into account the involution ξ, which has the special form in global Liouville coordinates:

$$\xi(x, y) = (x + 1/2, -y).$$

It is seen that, unlike the torus case, here the coordinates x and y are not equivalent. In particular, among the transformations of triples (L, f, g) there is not the permutation of x and y which appears in the torus case. The transformations generated by translations along the axis y are also absent. It is explained by the fact that the axis $y = 0$ is an invariant line with respect to ξ. This line generates an invariant cycle on the torus. There is one more invariant cycle corresponding to the straight lines on the plane given by $y = L/2 + kL$, where k is integer. The set of such lines is invariant under translation by L along the axis y. There are no other ξ-invariant cycles on the torus. That is why y can be shifted only by $L/2$, as reflected in the operation β.

Finally, the symmetry with respect to the axis x does not change anything due to the evenness of $g(y)$; therefore, this operation is also excluded from the list of elementary operations. $\square$

We now describe the class of linearly integrable geodesic flows on the Klein bottle. Consider an orthogonal lattice with the basis $f_1 = (1, 0)$ and $f_2 = (0, L)$ on the plane $\mathbb{R}^2$ with Cartesian coordinates x and y, where L is an arbitrary positive number. Consider the metric

$$g(y)(dx^2 + dy^2),$$

where $g(y)$ is an even function with period L. As we already know, this metric gives us a linearly integrable geodesic flow on the torus. Consider an involution ξ on the torus, defined by the following formula (in coordinates on the covering plane):

$$\xi(x, y) = (x + 1/2, -y).$$

It is easy to see that the function $g(y)$ is invariant with respect to ξ, and, therefore, taking the quotient T^2/ξ, we obtain the Klein bottle with the Riemannian metric $g(y)(dx^2 + dy^2)$.

We shall call such metrics on the Klein bottle (L, g)-*metrics*. It is easy to see that their geodesic flows are all linearly integrable. Indeed, the linear integral of the metric lifted back to the torus is p_x. But since this function p_x is ξ-invariant, we can drop this integral down and obtain, as a result, a linear integral of the geodesic flow of the (L, g)-metric on the Klein bottle. It turns out that no other linearly integrable geodesic flows on the Klein bottle exist.

Theorem 11.17 (V. S. Matveev). *If the geodesic flow of a Riemannian metric on the Klein bottle admits a linear integral, then the metric is either flat or isometric to some (L, g)-metric with appropriate parameters L and g. Moreover, the (L, g)-metric and $(\widehat{L}, \widehat{g})$-metric corresponding to different parameters are isometric if and only if $L = \widehat{L}$ and $g(y) = \widehat{g}(y + L/2)$.*

Proof. In essence, the proof repeats the arguments for the case of a quadratic integral. Moreover, we may formally square p_x and repeat all the preceding arguments. We assume here, of course, that the metric is not flat.

Choose orthogonal coordinates x, y on the covering plane in such a way that the metric takes the form $g(y)(dx^2 + dy^2)$ and then analyze the admissible form of the involution ξ written in terms of these coordinates. There are four possible cases only (see above):

1) $z \to \bar{z} + b$,
2) $z \to -\bar{z} + b$,
3) $z \to i\bar{z} + b$,
4) $z \to -i\bar{z} + b$.

The same argument as in the case of a quadratic integral shows that cases 3 and 4 cannot appear. Under the involution of type 2, i.e., the symmetry with respect to the axis y, the linear integral p_x of the system on the covering plane (or on the torus) turns into $-p_x$, i.e., is not invariant under ξ. Therefore, it cannot be obtained by lifting a linear integral from the Klein bottle. Thus, the case 2 is also impossible.

It remains to consider case 1. But for this case we have already shown (see the proof of Theorem 11.15) that the lattice has to be orthogonal, and the involution must have the desired form.

Thus, it is always possible to find a coordinate system x, y on the covering plane in terms of which the lattice Γ is orthogonal and generated by the vectors $f_1 = (1, 0)$ and $f_2 = (0, L)$, the metric becomes $g(y)(dx^2 + dy^2)$, and the involution ξ has the form $\xi(x, y) = (x + 1/2, -y)$. Thus, it is proved that any Riemannian metric on the Klein bottle whose geodesic flow admits a linear integral is either flat or isometric to some (L, g)-metric.

To prove the second statement of the theorem, it suffices to turn to the description of operations $\alpha_v, \beta, \gamma, \delta_c$ in the proof of Theorem 11.16 and to take only those of them which correspond, from the formal viewpoint, to the case $f(x) = 0$. Such an operation is β. The others act trivially.   □

Finally, consider one more family of metrics on the Klein bottle, whose geodesic flows can be naturally called *quasi-quadratically integrable*. They possess a polynomial integral F of degree 4, which can be locally (but not globally) presented in the form $F = \widetilde{F}^2$, where $\widetilde{F}$ is a quadratic polynomial. In particular, being lifted to the covering torus this geodesic flow becomes quadratically integrable.

Let m and n be natural relatively prime numbers, and f be a positive periodic function of one variable with period 1 (different from a constant). Consider the standard Cartesian coordinates x, y on the plane and define a metric by

$$ds^2 = (f(x) + f(y + n/2))(dx^2 + dy^2).$$

Let Γ be an orthogonal lattice on the plane generated by vectors $f_1 = (m, -m)$ and $f_2 = (n, n)$ (Fig. 11.13). It is obtained from a standard lattice by rotating through the angle $\pi/4$. By a standard lattice we mean here a lattice with basis vectors directed along the axes x and y. It is easy to see that the described metric is invariant with respect to Γ and, therefore, defines some metric on the torus $T^2 = \mathbb{R}^2/\Gamma$. By the way, the latter is a finite-sheeted Liouville metric on the torus (since Γ is not standard).

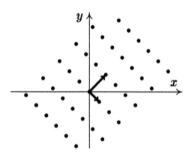

Figure 11.13

We now define an involution $\xi: (x, y) \to (y + n/2, x + n/2)$ on the torus. It is easy to see that the metric on the torus is invariant with respect to ξ and, consequently, determines some metric on the Klein bottle $K^2 = T^2/\xi$. We denote this metric by $ds^2_{m,n,f}$.

Theorem 11.18 (V. S. Matveev).

1) *The geodesic flow of the metric $ds^2_{m,n,f}$ on the Klein bottle has a polynomial (in momenta) integral of degree 4. This integral is not reduced to integrals of degree less than 4.*

2) *The metrics of type $ds^2_{m,n,f}$ exhaust (up to isometries) all metrics on the Klein bottle whose geodesic flows do not have an additional quadratic integral, but obtain such an integral after lifting to the covering torus. (It is natural to call such geodesic flows quasi-quadratically integrable.)*

3) *Two such metrics $ds^2_{m,n,f}$ and $ds^2_{\widehat{m},\widehat{n},\widehat{f}}$ are isometric if and only if their parameters are connected by the relations $m = \widehat{m}$, $n = \widehat{n}$, $f(x + t) = \widehat{f}(x)$ for some real number t.*

Proof. We begin with constructing the integral of degree 4 on K^2. Since $ds^2_{m,n,f}$ is a finite-sheeted Liouville metric, its geodesic flow on the torus admits the quadratic integral

$$F(x, y, p_x, p_y) = \frac{p_x^2 f(y + n/2) - p_y^2 f(x)}{f(y + n/2) + f(x)}.$$

This function is not invariant under the involution ξ and, consequently, cannot be descended onto K^2 as a single-valued integral. More precisely, $\xi^* F = -F$. Therefore, this integral can be considered as a two-valued function on $T^* K^2$. Clearly, to construct a well-defined single-valued integral on the Klein bottle it suffices to square F. Thus, as the desired integral of degree 4 on K^2 we take F^2.

Let us prove that this metric does not admit any non-trivial quadratic integrals. Note that at the same time we shall show that it has no linear integrals either (because, by squaring a linear integral, we would get a quadratic one).

As we showed in Proposition 11.4, the quadratic integral of the above metric on the torus is defined uniquely up to a linear combination with the Hamiltonian, i.e., has the form $c_1 F + c_2 H$, where H is the Hamiltonian, and c_1, c_2 are some constants. Acting on this integral by the involution ξ we see that $\xi(c_1 F + c_2 H) = -c_1 F + c_2 H$. In order for this integral to descend correctly down to the Klein bottle, it is necessary that $c_1 = 0$. Therefore, any quadratic integral of our metric on the Klein bottle is proportional to H, that is, trivial. Thus we have shown that in this case neither linear nor quadratic integrals exist.

We now prove that the metric $ds^2_{m,n,f}$ admits no integrals of degree 3. It turns out that the more general statement holds.

Lemma 11.10. *The geodesic flows of global Liouville metrics on the torus do not admit any cubic integrals.*

REMARK. It follows immediately from this that the same property is fulfilled for finite-sheeted Liouville metrics on the torus.

Corollary. *The geodesic flow of the metric $ds^2_{m,n,f}$ on the Klein bottle admits no cubic integrals.*

The proof of the corollary follows from the fact that the metric $ds^2_{m,n,f}$ is covered by a finite-sheeted Liouville metric on the torus. It remains to apply Lemma 11.10 and to use the above remark. □

Proof (of Lemma 11.10). Let the Liouville metric have the form

$$(f(x) + g(y))(dx^2 + dy^2).$$

Consider the quadratic integral F of the corresponding geodesic flow:

$$F(x, y, p_x, p_y) = \frac{-p_x^2 g(y) + p_y^2 f(x)}{g(y) + f(x)}.$$

To prove the lemma we need to analyze the structure of the Liouville tori. The common level surface $\{H = 1, F = a = \text{const}\}$ is given by the following equations:

$$p_x^2 = f(x) - a, \qquad p_y^2 = g(y) + a.$$

Indeed,

$$\frac{p_x^2 + p_y^2}{f + g} = 1, \qquad \frac{f p_y^2 - g p_x^2}{f + g} = a.$$

By multiplying the first equation by f and subtracting the second equation from it, we obtain

$$\frac{f p_x^2}{f + g} + \frac{g p_x^2}{f + g} = f - a,$$

that is, $p_x^2 = f - a$. Analogously, we get $p_y^2 = g(y) + a$.

For regular values of a this level surface consists, generally speaking, of several Liouville tori. Consider a positive number a closed to the absolute minimum f_0 of the function $f(x)$. More precisely, set $a = f_0 + \varepsilon$, where ε is a small positive number (Fig. 11.14). Select that part of the graph of $f(x)$ which lies above $a = f_0 + \varepsilon$. This part of the graph is projected onto the axis x. Consider

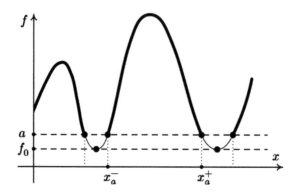

Figure 11.14

one of the connected components of this projection, i.e., the arc on the circle parameterized by x. Denote the ends of this arc by x_a^- and x_a^+. It is easy to see that the set of points

$$\{p_x = \pm\sqrt{f - a}, \quad p_y = \sqrt{g + a}, \quad x \in [x_a^-, x_a^+], \quad y \text{ is arbitrary}\}$$

is a Liouville torus. Denote this torus by T_a. By varying a, we obtain a family of Liouville tori. The rotation function ρ of the geodesic flow cannot be constant on all of these tori. This follows from the fact that the rotation function ρ tends to infinity as $a \to f_0$, i.e., as $\varepsilon \to 0$. In particular, our geodesic flow is non-resonant at least for small ε. We shall show this below in Chapter 13, where the orbital classification of integrable geodesic flows on the torus is discussed. That is why, without loss of generality, we can assume that every integral of the geodesic flow is constant on each Liouville torus from the family T_a.

Suppose now that there exists an integral K of degree 3:

$$K(x, y, p_x, p_y) = A_1(x, y)p_x^3 + B_1(x, y)p_x^2 p_y + A_2(x, y)p_x p_y^2 + B_2(x, y)p_y^3.$$

Let us prove that all the coefficients of this integral are identically zero. Let us write down the conditions to the coefficients which appear from the non-resonance condition. It is clear that the Liouville torus T_a is projected onto the annulus lying on the torus T^2 and having the form

$$[x_a^-, x_a^+] \times S^1(y).$$

Consider an arbitrary interior point (x, y) of this annulus. This point has two different preimages under the natural projection $T_a \to T^2$. It is important for us that these two preimages (i.e., two covectors) are symmetric with respect to the axis y in the sense that their coordinates have the form

$$(p_x(a), p_y(a)) \quad \text{and} \quad (-p_x(a), p_y(a)).$$

Since both the points belong to the same Liouville torus T_a, the values of the polynomial K must coincide at these points. Hence we obtain the relation to the coefficients

$$A_1(x, y)p_x^3(a) + A_2(x, y)p_x^2(a)p_y(a) = 0,$$

which evidently remains valid under small perturbations of the point (x, y) and the parameter a. This implies that the coefficients of the polynomial K vanish in the neighborhood of (x, y). Now it remains to note that, by our construction, (x, y) can be an arbitrary point on the torus T^2 provided x is not the global minimum point for the function f. Therefore, A_1 and A_2 vanish identically.

It is analogously shown that $B_1 \equiv B_2 \equiv 0$. This completes the proof of Lemma 11.10 (as well as the first statement of Theorem 11.18). ☐

The proof of the two remaining statements of Theorem 11.18 is carried out by the scheme that we already applied several times, and we omit it. We only notice that the metrics $ds^2_{m,n,f}$ are, in essence, just those metrics which were rejected when we studied quadratically integrable geodesic flows on the Klein bottle (Theorem 11.15).

REMARK. The fact that the described geodesic flow on the Klein bottle does not have one-valued quadratic integral (but possesses only an integral of degree 4) is reflected on the qualitative properties of the metric $ds^2_{m,n,f}$. For example, this metric does not admit any non-trivial geodesic equivalence. This means that on the Klein bottle there are no other metrics with the same geodesics as the metric $ds^2_{m,n,f}$ has (here geodesics are considered geometrically, i.e, as curves without parameterization). That is one of the distinctions between the metric $ds^2_{m,n,f}$ and the metrics with linearly and quadratically integrable geodesic flows. This result follows from the global Dini theorem, which we discuss in Chapter 15.

11.5.3. The Sphere

In the case of the sphere the description of linearly and quadratically integrable geodesic flows was obtained by V. N. Kolokol'tsov [187], [189]. Later Nguyen Tien Zung, E. N. Selivanova, L. S. Polyakova [263], [264], and V. S. Matveev [224] suggested an apt geometrical formulation of these results. This geometrical approach turned out to be more convenient for studying the global topology of integrable geodesic flows on the sphere.

We begin with the case of linearly integrable geodesic flows. Let the standard unit sphere S^2 be given in $\mathbb{R}^3\,(x,y,z)$ by equation

$$x^2 + y^2 + z^2 = 1\,,$$

where z is the vertical coordinate. Denote by ds_0^2 the standard metric on it induced by the ambient Euclidean metric in $\mathbb{R}^3$. In the usual spherical coordinates θ, φ this metric takes the form

$$ds_0^2 = d\theta^2 + \sin^2\theta d\varphi^2\,.$$

Let $f(z)$ be an arbitrary positive smooth function on the segment $[-1,1]$.

Theorem 11.19.

a) *The geodesic flow of a Riemannian metric ds^2 on the sphere is linearly integrable if and only if ds^2 is isometric to a metric of the form $f(z)ds_0^2$, where $f(z)$ is a smooth positive function on $[-1,1]$.*

b) *Two metrics of this kind $f(z)ds_0^2$ and $\widehat{f}(z)ds_0^2$ are isometric if and only if the functions $f(z)$ and $\widehat{f}(z)$ are connected by one of the two following relations:*

$$\text{either}\quad \alpha\widehat{f}(z)\ =\ f\left(\frac{(\alpha+1)z - (1-\alpha)}{(\alpha-1)z + (1+\alpha)}\right)$$

$$\text{or}\quad \alpha\widehat{f}(-z) = f\left(\frac{(\alpha+1)z - (1-\alpha)}{(\alpha-1)z + (1+\alpha)}\right)\,,$$

where α is some real number.

This theorem can be reformulated in other terms, by using global conformal coordinates on the sphere.

Theorem 11.20.

a) *The geodesic flow of a Riemannian metric on the sphere is linearly integrable if and only if there exist global conformal coordinates x,y in which the metric has the form*

$$ds^2 = f(x^2 + y^2)(dx^2 + dy^2)\,,$$

where $f(t)$ is a positive smooth function on the semi-axis $[0,+\infty)$ such that the function $\dfrac{f(1/t)}{t^2}$ is positive and smooth on the semi-axis $[0,+\infty)$ too.

b) *Two metrics of this kind $f(x^2 + y^2)(dx^2 + dy^2)$ and $\widehat{f}(x^2 + y^2)(dx^2 + dy^2)$ are isometric if and only if*

$$\text{either}\quad f(t/\alpha) = \alpha\widehat{f}(t)\qquad \text{or}\qquad \frac{f(\alpha/t)}{t^2} = \frac{\widehat{f}(t)}{\alpha}$$

for some positive number α.

REMARK. The indicated conditions for the function f actually mean that ds^2 is a smooth metric on the whole sphere.

It is seen from Theorem 11.19 that the metrics on the sphere with linearly integrable geodesic flows are invariant under rotations of the sphere around the vertical axis z. In other words, they admit a one-parameter isometry group.

Corollary. *The geodesic flow of a metric on the sphere is linearly integrable if and only if the metric admits a smooth action of the circle $S^1 = \mathrm{SO}(2)$ by isometries. Moreover, such an action is conjugate to the standard one (i.e., to the rotation of the sphere about axis z).*

In particular, such an action has exactly two fixed points (the poles of the sphere). The complement to the poles is foliated into smooth circles, which are one-dimensional orbits of the action (Fig. 11.15).

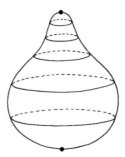

Figure 11.15

Note that the metrics on surfaces of revolution in $\mathbb{R}^3$ are particular cases of metrics with linearly integrable geodesic flows. They, however, do not exhaust all linearly integrable metrics described above. In other words, not every metric with linearly integrable geodesic flow admits an isometrical imbedding of the sphere into $\mathbb{R}^3$ as a surface of revolution.

The proofs of Theorems 11.19 and 11.20 shall be obtained below in the framework of a general construction which works both in linear and quadratic cases.

We now describe the list of all Riemannian metrics on the sphere with quadratically integrable geodesic flows.

We begin with constructing a series of model examples of such metrics. Then we shall show that the list obtained is complete, i.e., includes all the metrics with quadratically integrable geodesic flows.

Consider the two-dimensional torus T^2 as a quotient space of the plane $\mathbb{R}^2$ with Cartesian coordinates x, y with respect to an orthogonal lattice Γ whose basis consists of two orthogonal vectors $f_1 = (1,0)$ and $f_2 = (0, L)$, where L is an arbitrary positive number. Consider the involution σ on the torus given (on the covering plane) by the following symmetry:

$$\sigma(x, y) = (-x, -y),$$

i.e., central symmetry with respect to the origin.

It is clear that the lattice Γ sustains this symmetry, hence σ is actually an involution on the torus. Consider the natural projection $\varkappa: T^2 \to T^2/\sigma$.

Lemma 11.11. *The quotient space T^2/σ is homeomorphic to the two-dimensional sphere S^2. The projection $\varkappa: T^2 \to S^2 = T^2/\sigma$ is a two-sheeted branching covering over the sphere with four branch points, each of which has exactly one preimage on the torus.*

Proof. Divide the square in halves by its middle horizontal line (Fig. 11.16). One of the rectangles obtained (for example, the upper one) is glued according to the rule of identification for its edges as shown in Fig. 11.16. As a result we get

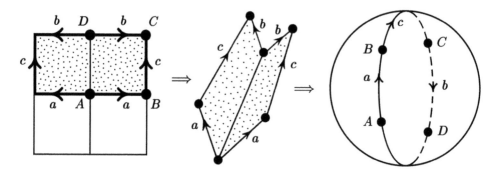

Figure 11.16

the sphere. The four branch points of this covering are the points with coordinates $(0,0)$, $(0, L/2)$, $(1/2, 0)$, $(1/2, L/2)$ depicted as filled points A, B, C, D in Fig. 11.16. These points are projected onto four different points of the sphere. To simplify notation, we denote their images on the sphere by the same letters A, B, C, D. Under the projection onto the sphere, the boundary of the rectangle becomes the arc connecting points A and D. The points B and C are interior points of the arc.

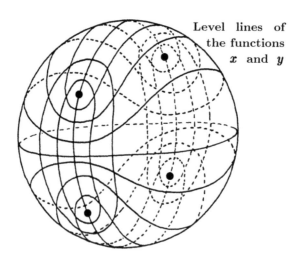

Level lines of
the functions
x and y

Figure 11.17

By cutting the sphere along this arc we obtain the initial rectangle. It is shown in Fig. 11.17 how the orthogonal coordinate net on the torus is projected onto the sphere. As a result, on the sphere we obtain two families of closed curves. Lemma 11.11 is proved. □

It turns out that the projection $\varkappa$ can be described in terms of complex structures on the torus and on the sphere. Note that, in fact, we have already defined a complex structure on the torus by having represented the torus as the quotient space $\mathbb{R}^2/\Gamma$, where $\mathbb{R}^2$ is identified with the complex plane $\mathbb{C}^1$, and $z = x + iy$, where x and y are the Cartesian coordinates. Fix this complex structure on the torus.

Lemma 11.12.

a) *One can choose a complex structure on the sphere S^2 in such a way that the projection $\varkappa$ becomes a holomorphic mapping of the torus T^2 onto the sphere S^2.*

b) *Each of the four branch points of $\varkappa$ is a branch point of order two. In other words, in a neighborhood of such a point and in a neighborhood of its image there exist local complex coordinates $\tilde{z}$ and $\tilde{w}$ in which $\varkappa$ takes the form $\tilde{w} = \tilde{z}^2/2$.*

REMARK.   The holomorphic mapping $\varkappa\colon T^2 \to S^2$ in fact presents the well-known meromorphic Weierstrass function $w = \wp(z)$. Recall that the Weierstrass function can be defined by the following formula:

$$\wp(z) = \frac{1}{z^2} + {\sum_{w\in\Gamma}}' \left(\frac{1}{(z-w)^2} - \frac{1}{w^2} \right),$$

where the sign $'$ means that w runs over all elements of the lattice Γ except zero. This series converges to a doubly periodic holomorphic function for all z that do not belong to the lattice Γ. At all the nodes of Γ this function has second order poles. Thus, $\wp$ is a meromorphic function on the whole torus. For more detailed description of the properties of the Weierstrass function see [113], [160], [302], [308].

Proof (of Lemma 11.12). In fact, the formulation of the statement itself contains the rule for defining a complex structure on the sphere. If the point $(x,y) \in T^2$ is not a singular point of the projection $\varkappa$, then in a neighborhood of its image $\varkappa(x,y)$ on the sphere we introduce (locally) a complex structure just by taking $\tilde{w} = x + iy$ as a local complex coordinate. If $P_0 = \varkappa(z_0)$ is a branch point on the sphere, then as a local coordinate in its neighborhood we take $\tilde{w}(P) = (\varkappa^{-1}(P) - z_0)^2$, where P is an arbitrary point from the neighborhood of P_0 (Fig. 11.18). This coordinate is well-defined in spite of the fact that the point $P \neq P_0$ has two preimages z and z'. Indeed, the definition of $\varkappa$ immediately implies that $z - z_0 = -(z' - z_0)$ and, consequently, after squaring, the ambiguity disappears.

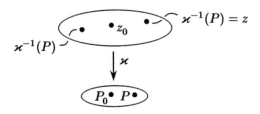

Figure 11.18

Thus, locally, in a neighborhood of each point, we have introduced a complex coordinate. It is easy to see that all the transition functions are complex-analytic. This means that the complex structure is well-defined on the whole sphere.   □

It is important that as a result we have constructed a *smooth* two-sheeted mapping of the torus onto the sphere which has exactly that topological structure which we need, i.e., is the gluing of the torus by the involution σ.

Let us now describe the class of Riemannian metrics on the sphere with quadratically integrable geodesic flows. We shall use the mapping $\varkappa$ and, instead of describing the desired metric in terms of the sphere, we shall describe its pull-back on the torus. It is clear that the metric on the sphere is uniquely reconstructed from its pull-back on the covering torus. Such a description of a metric on the sphere turns out to be much simpler than the one in terms of isothermal coordinates on the sphere itself.

On the covering plane of the torus we define two periodic smooth functions $f(x)$ and $g(y)$ satisfying the following conditions.

a) $f(x)$ is non-negative smooth even function with period 1.

b) $g(y)$ is non-negative smooth even function with period L.

c) This condition describes the asymptotic behavior of f and g near their zeros. The function $f(x)$ vanishes at the points $x = m/2$, where $m \in \mathbb{Z}$. The function $g(y)$ vanishes at the points $y = Lk/2$, where $k \in \mathbb{Z}$. For every point $(m/2, kL/2)$ there exists a smooth (in a neighborhood of zero) function $h(t)$ such that $h(0) = 0$, $h'(0) \neq 0$, and

$$f(m/2 + t) = h(t^2), \qquad g(kL/2 + t) = -h(-t^2).$$

If we want to consider real-analytic metrics on the sphere, then the functions f and g (as well as the function h in the third condition) have to be analytic.

Concerning the third condition, the two following remarks seem to be useful.

COMMENT. Due to the periodicity of f and g, the condition (c) needs to be checked only in four points $(0,0)$, $(1/2, 0)$, $(0, L/2)$, $(1/2, L, 2)$. It can be reformulated in terms of the Taylor expansions of these functions. Denote the Taylor expansion of the function f at a point t_0 by $f_{t_0}^{\#}(t) = \sum c_k (t - t_0)^k$. The similar notation is used for the Taylor expansion of g. Then the condition (c) can be rewritten as follows:

$$f_0^{\#}(t) = -g_0^{\#}(it) = f_{1/2}^{\#}(1/2 + t) = -g_{L/2}^{\#}(L/2 + it).$$

Hence these Taylor expansions in fact have the form

$$f_0^{\#} = \sum a_{2k} t^{2k}, \qquad\qquad g_0^{\#} = -\sum (-1)^k a_{2k} t^{2k},$$
$$f_{1/2}^{\#} = \sum a_{2k}(t - 1/2)^{2k}, \qquad g_{L/2}^{\#} = -\sum (-1)^k a_{2k}(t - L/2)^{2k},$$

where the first coefficient a_2 of the series is positive.

COMMENT. In the case of a real-analytic metric the functions f and g have to be analytic. That is why the above conditions to their power series imply several useful corollaries. First, since the Taylor expansions of f at the points $t = 0$ and $t = 1/2$ coincide, the function f in fact has period $1/2$, but not 1. By the same reason the function g has period $L/2$. Besides, the condition $f_0^{\#}(t) = -g_0^{\#}(it)$ turns now into a relation between the functions f and g themselves, namely:

$$f(t) = -g(it).$$

This means that the functions f and g appear from the same complex-analytic function. In other words, on the complex plane $z = x + iy$ there exists a doubly periodic (with periods $1/2$ and $iL/2$) even analytic function $\mathcal{R}$ which takes real values on the real and imaginary axes and satisfies the relations

$$f(x) = \mathcal{R}(x) \quad \text{and} \quad g(y) = -\mathcal{R}(iy).$$

Let us explain that the domain of definition of this function, generally speaking, is not the whole complex plane $\mathbb{C}$, but some neighborhood of the orthogonal net generated by lines $\{x = m/2\}$, $\{y = kL/2\}$. In other words, if we consider $\mathcal{R}$ as a function on the complex torus, then it is defined, generally speaking, only in a neighborhood of the parallel and meridian.

Thus, in the real-analytic case the parameter of the metric is the only complex function $\mathcal{R}$, but not two functions f and g. In particular, the conformal multiplier λ takes the form

$$\lambda = f(x) + g(y) = \mathcal{R}(x) - \mathcal{R}(iy).$$

We now return to constructing model quadratically integrable geodesic flows on the sphere. Consider a 2-form $(f(x) + g(y))(dx^2 + dy^2)$ on the torus with the above functions f and g. We use the word "2-form" instead of "a metric", because it has zeros at the branch points of $\varkappa$ and, consequently, is not a Riemannian metric in the strict sense. It is easily seen that, by construction, this 2-form is invariant on the torus with respect to σ. Therefore, by pushing it down, we get some well-defined 2-form on the sphere except for the four branch points. It turns out that this form can be defined at these points in such a way that we obtain a smooth Riemannian metric on the whole sphere.

Proposition 11.7.

1) *Let* $(f(x)+g(y))(dx^2+dy^2)$ *be a 2-form on the torus* T^2 *satisfying properties* (a), (b), (c) *and* $\varkappa\colon T^2 \to T^2/\sigma = S^2$ *be the two-sheeted covering described above. Then on the sphere* S^2 *there exists the only smooth Riemannian metric* ds^2 *such that* $\varkappa^*(ds^2) = (f(x)+g(y))(dx^2+dy^2)$. *If the functions* f *and* g *are real-analytic, then the metric* ds^2 *is real-analytic too.*

2) *Conversely, consider the 2-form* $\varkappa^*(ds^2)$ *on the torus* T^2, *where* ds^2 *is a smooth metric on the sphere* S^2. *If it has the form* $(f(x)+g(y))(dx^2+dy^2)$, *then the functions* f *and* g *satisfy the properties* (a), (b), (c).

Proof. Conditions (a) and (b) exactly mean that the form $(f(x)+g(y))(dx^2+dy^2)$ is well-defined on the torus and is invariant under the involution σ. Condition (c) says that the zeros of this form coincide with the singular points of the mapping $\varkappa$, i.e., with the branch points A, B, C, D, and describes the character of these zeros. The necessity and sufficiency of this condition follows immediately from the local Theorem 11.8. $\square$

As a result, to each triple (L, f, g) satisfying properties (a), (b), (c), we have assigned a smooth Riemannian metric ds^2 on the sphere S^2.

Definition 11.7. By analogy with the case of the torus we call this metric an (L, f, g)-*metric* on the sphere.

REMARK.   In the analytic case, however, such a metric could be called an $\mathcal{R}$-metric, since here the complex function $\mathcal{R}$ (see the above commentary) contains all information about the corresponding triple (L, f, g). Nevertheless, in order not to complicate our notation, we shall use the term "an (L, f, g)-metric" both in the smooth and analytic case.

Theorem 11.21. *The geodesic flow of a smooth (or real-analytic) Riemannian metric ds^2 on the 2-sphere is quadratically integrable if and only if the metric ds^2 is isometric to an (L, f, g)-metric with appropriate parameters L, f, g.*

Proof.   The proof of the fact that the geodesic flow of a (L, f, g)-metric on the sphere admits a quadratic integral is rather simple.   This integral can be written explicitly on the torus and, then, descended to the sphere. Let us do it.

Given an (L, f, g)-metric on the sphere, it is required to prove that its geodesic flow possesses a non-trivial quadratic integral, which is smooth or analytic depending on the type of the metric itself.

Lift this metric onto the covering torus and take the obtained 2-form $(f(x) + g(y))(dx^2 + dy^2)$ with the four degenerate points A, B, C, D. This form corresponds to some geodesic flow with singularities given by the Hamiltonian $\tilde{H} = \dfrac{p_x^2 + p_y^2}{f(x) + g(y)}$. However, outside the singular points A, B, C, D this is a usual geodesic flow which possesses a quadratic integral

$$\tilde{F} = \frac{f(x)p_y^2 - g(y)p_x^2}{f(x) + g(y)}.$$

Evidently, this integral sustains the involution σ and, consequently, can be descended down to the sphere with four points removed. It remains to verify that on the sphere this integral can be extended to these singular points (smoothly or analytically respectively). But this follows immediately from local Theorem 11.8, since the conditions imposed on the functions f and g correspond exactly to the situation described in this theorem. The explicit form of the quadratic integral is also indicated there.

The proof of the converse is not trivial and requires certain efforts. Here we shall follow the arguments by V. N. Kolokol'tsov. At the same time we shall get the proof of Theorems 11.19 and 11.20.

The idea of the proof of Theorems 11.19, 11.20, and 11.21 is the following. A geodesic flow with a quadratic integral uniquely determines a holomorphic function $R(z)$. This function, in turn, defines Liouville coordinates on the whole sphere, except for those points, where R becomes zero. As was proved above, the corresponding coordinate net (with singularities) is uniquely determined by the function R. Moreover, according to the local theorem, the singularities of the net can be only of two types shown in Figs. 11.4 and 11.6. It is natural to try to draw all possible coordinate nets of this kind on the sphere. It can be done, and it turns out that there are three possibilities. They are presented in Fig. 11.19.

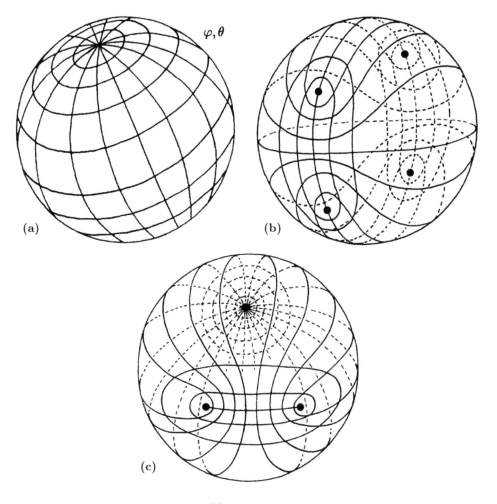

Figure 11.19

The first net just coincides with the net of the standard spherical coordinates θ, φ (Fig. 11.19(a)). The second net looks like the net of elliptic coordinates on the sphere (Fig. 11.19(b)). The third one is shown in Fig. 11.19(c).

The third net is in fact impossible. The point is that in Liouville coordinates the conformal multiplier λ of the metric is presented as the sum of two functions f and g, each of which is constant on the corresponding family of coordinate lines. But the third net has a point (pole) which all the lines of one family enter. This leads to the fact that the corresponding function (f or g) is constant. On the other hand, the same net has a singular point of the second type for which none of the functions f and g can be constant. In other words, the singularities of two different types, shown in Figs. 11.4 and 11.6, cannot occur simultaneously.

Thus, only two possibilities remain.

It follows from the local theory that in the first case (Fig. 11.19(a)) the geodesic flow admits a linear integral, and the metric has the form $\lambda(dx^2 + dy^2)$, where λ

is a function which is constant on the circles of the coordinate net, i.e., λ depends on $x^2 + y^2$ only. It is just the statement of Theorem 11.19 (or Theorem 11.20), which describes the linearly integrable geodesic flows on the sphere.

In the second case (Fig. 11.19(b)), it is easily seen that the net is the image of the standard orthogonal coordinate net on the torus under a two-sheeted covering $\varkappa\colon T^2 \to S^2$ with four branch points. In other words, after lifting from the sphere to the torus, the metric becomes a Liouville one. That is, in fact, what Theorem 11.21 claims.

We now give a formally strong proof of the steps listed above.

Consider global isothermal coordinates x, y on the sphere and the corresponding function $R(z)$, where $z = x + iy$ (see Proposition 11.2).

Lemma 11.13. *The function $R(z)$ is a polynomial of degree at most 4.*

Proof. Since the integral F is well-defined and smooth on the whole sphere, the function $R(z)$ and the function $S(w)$ obtained from $R(z)$ by transformation $z = 1/w$ are both holomorphic functions on the whole complex plane. At the same time, we have shown that R and S are connected by the relation $R(z)(w')^2 = S(w)$. Hence the function $\dfrac{R(z)}{z^4}$ is bounded as $z \to \infty$, since $\dfrac{R(z)}{z^4}$ tends to $S(0)$, i.e., to a finite number. Therefore, according to the well-known theorem of complex analysis, $R(z)$ is a polynomial, and its degree is at most 4, as was to be proved.   □

Note that the function $R(z)$ cannot vanish identically. Indeed, if $R(z) \equiv 0$, then the integral F is proportional to the Hamiltonian H at each point of the sphere, i.e., $F = kH$, where $k(z)$ is some function on the sphere itself. Hence $k(z)$ is an integral of the geodesic flow and, consequently, is constant on the sphere. Then F and H are functionally depended, which contradicts the assumptions of the theorem.

Without loss of generality we may assume that, being written in an appropriate coordinate system, the polynomial $R(z)$ has exactly degree 4. To check this, it is sufficient to make a transformation $w = \dfrac{1}{z - z_0}$, where z_0 is chosen in such a way that $R(z_0) \neq 0$.

Let us list all possibilities for the roots of such a polynomial:
1) $1 + 1 + 1 + 1$, i.e, all the roots are simple,
2) $2 + 1 + 1$, i.e., one root is double, the other two are simple,
3) $2 + 2$, i.e., two double roots,
4) $3 + 1$, i.e., one root is triple, the other is simple,
5) 4, i.e., one root of multiplicity four.

Lemma 11.14. *Only the cases 1 and 3 are possible.*

Proof. The cases 4 and 5 are impossible, since, as was proved above, the function $R(z)$ cannot have zeros of order more than 2.

The case 2 is mixed, here there is one point of the type shown in Fig. 11.6 (second order zero) and two points of the type shown in Fig. 11.4 (first order zeros). Consider two families of coordinate lines on the sphere corresponding to the given function $R(z)$. In a neighborhood of the second order zero these lines behave as shown in Fig. 11.6. Therefore, one of the functions (for example, f; recall that $\lambda = (f + g)/|R|$) has to be the same constant on the radial rays entering this singular point. These radial rays, while moving off from this point, must cover almost the whole sphere. Therefore, at some moment they will reach one of the other singular points, namely, first order zeros of R. But in this case, in a neighborhood of this point the function f becomes locally constant; this is impossible by virtue of the local classification theorem. The behavior of the lines of both the families is shown in Fig. 11.19(c). It can be easily seen by drawing these lines for the polynomial $A^2(1-z^2)$. We can reduce a polynomial of degree 4 to such a form in the case $2+1+1$. This picture appears, of course, only for real A. If A is a complex number, then in a neighborhood of the second order zero we shall see the picture presented in Fig. 11.5. It is prohibited by the local theory (see above). Lemma is proved. □

Consider the case 3 in more detail. Here we have two double roots. Using a linear-fractional transformation, we may always assume that one of these roots is zero, and the other is infinity. Then the polynomial $R(z)$ becomes A^2z^2 on the whole sphere. As we know from the local theory, the coefficient A has to be real. By repeating local arguments, we see that in the given coordinate system on the sphere the metric have the form

$$f(x^2+y^2)(dx^2+dy^2),$$

where $f(t)$ is a smooth function. The function $\dfrac{f(1/t)}{t^2}$ also must be smooth. The point is that after the transformation $w = 1/z$ the metric becomes $\widetilde{f}(|w|^2)\,dw\,d\overline{w}$, where $\widetilde{f}(|w|^2) = \dfrac{f(1/|w|^2)}{|w|^4}$.

This completes the proof of the first part of Theorem 11.20.

We now pass to the second part of the theorem. Which of the described metrics are isometric among themselves? This question can be reformulated as follows: which global transformations of the sphere preserve the form of a metric, more precisely, the form of its conformal multiplier? Consider such a transformation of the sphere into itself: $w = w(z)$. It must be linear-fractional, since it is global and preserves the conformal form of the metric. Let the metric be different from the constant curvature metric. We need now to describe the transformations $w = w(z)$ which make the metric $f(|z|^2)\,dz\,d\overline{z}$ into a metric of the same kind $\widehat{f}(|w|^2)\,dw\,d\overline{w}$, where f and $\widehat{f}$ are smooth functions. We assert that, in the new coordinate system w, the circles $\{|z| = \text{const}\}$ are defined by the similar equation $|w| = \text{const}$. To check this, we should characterize these circles intrinsically, i.e., independently of the choice of coordinate system. It can be easily done if we observe that these circles are orbits of the isometry group action. This group is isomorphic to $S^1 = SO(2)$ and uniquely defined (unless the metric has constant curvature). So, the conformal transformation $w = w(z)$ may have only the following form: either az or a/z, where a is a constant. If we also consider the transformations changing orientation, then two more transformations occur: $a\overline{z}$ and $a/\overline{z}$.

It remains to watch what happens to the function $f(|z|^2)$ under such transformations. By straightforward calculation we check that this function is changed just according to the formulas indicated in Theorem 11.20. This completes the proof of Theorem 11.20. □

It remains to consider the case $1 + 1 + 1 + 1$, i.e., when the roots of $R(z)$ are all simple.

We apply again a linear-fractional transformation which sends one of the roots to infinity. As a result, we lower the degree of the polynomial R and make it into a cubic polynomial. Moreover, this polynomial can be always reduced to the form

$$R(z) = 4z^3 - g_2 z - g_3, \qquad \text{where} \quad g_2^3 - 27g_3^2 \neq 0.$$

The condition $g_2^3 - 27g_3^2 \neq 0$ is equivalent to the fact that all the roots of $R(z)$ are simple. Let us try now to find a holomorphic transformation $z = z(w)$ after which the function $R(z)$ becomes identically equal to 1 on the whole sphere without four points (zeros of $R(z)$). In other words, we want to find global Liouville coordinates u, v (where $w = u + iv$). For this we need to solve the differential equation

$$z'^2 = 4z^3 - g_2 z - g_3.$$

The solution of such an equation is well known. It is the Weierstrass function $z = \wp(w)$ (see the formula above). Note that the parameters g_2, g_3 are connected with the lattice Γ related to the Weierstrass function in the following way:

$$g_2 = 60 \sideset{}{'}\sum_{w \in \Gamma} \frac{1}{w^4},$$

$$g_3 = 140 \sideset{}{'}\sum_{w \in \Gamma} \frac{1}{w^6}.$$

Without loss of generality, we assume here that the lattice is spanned on the vectors $f_1 = (1, 0)$ and $f_2 = (b, L)$, where $b \in [0, 1)$ and $L > 0$.

The Weierstrass function is a doubly periodic (i.e., it is periodic with respect to the lattice Γ) meromorphic function on $\mathbb{C}$, and, consequently, can be considered as a meromorphic function on the torus $\mathbb{C}/\Gamma$:

$$\wp \colon T^2 = \mathbb{C}/\Gamma \to S^2 = \overline{\mathbb{C}}.$$

It is well known that this mapping is a two-sheeted covering over the sphere with four branch points $0, \frac{1}{2}f_1, \frac{1}{2}f_2, \frac{1}{2}(f_1 + f_2)$. The first branch point $z = 0$ goes to infinity under the mapping $\wp$, the others go to the roots of the polynomial $R(z)$. The topology of this mapping is very simple. As was already shown above, it can be considered as factorization of the torus by the involution $w \to -w$ (indeed, it is easy to see that $\wp$ is an even function).

Lemma 11.15. *The lattice Γ is orthogonal, i.e., $f_2 = (0, L)$.*

Proof. Clearly, in the general case, the lattice related to the Weierstrass function (for an arbitrary cubic polynomial $R(z)$) may be arbitrary. However, the presence of the Riemannian metric in our reasoning imposes some additional restrictions to the mutual disposition of the roots and, consequently, to the lattice Γ itself.

To see this, we lift the Riemannian metric from the sphere onto the torus. We obtain some quadratic form, which defines a Riemannian metric everywhere except for the branch points, where it becomes zero. On the other hand, in coordinates $w = u + iv$ the lifted metric takes the form

$$(f(u) + g(v))(du^2 + dv^2),$$

where $f(u)$ and $g(v)$ are periodic functions. Since the function $f(u) + g(v)$ is non-negative on the whole plane, each of the functions f and g can be assumed to be non-negative. Indeed, consider the functions $\tilde{f} = f - f_0$ and $\tilde{g} = g + f_0$, where f_0 is the minimum of f. It is a finite number due to the smoothness and periodicity of f. The function $\tilde{f}$ is obviously non-negative, so is $\tilde{g}$. Otherwise there exists a point v_0 at which $\tilde{g}(v_0) < 0$, but then the sum $f + g = \tilde{f} + \tilde{g}$ is negative at the point (u_0, v_0), where $f(u_0) = f_0$, which contradicts the condition $f + g \geq 0$.

Thus, we assume the functions f and g both to be non-negative. Consider the set of zeros of the sum $f(u) + g(v)$. This set is the solution of two equations $f(u) = 0$, $g(v) = 0$ (due to the non-negativity of both the functions). From the geometrical point of view, the set of zeros has a very simple structure. These zeros are obtained as intersections of two orthogonal families of straight lines: vertical and horizontal ones. This set is illustrated in Fig. 11.20.

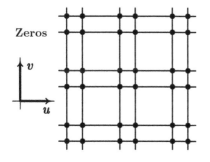

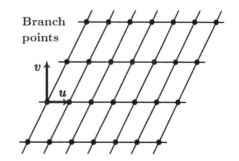

Figure 11.20   Figure 11.21

On the other hand, these zeros are exactly the branch points of the Weierstrass function.

It is well known that the set of branch points of the Weierstrass function is the lattice of its half-periods (Fig. 11.21). Thus, two pictures presented in Figs. 11.20 and 11.21 must coincide. Clearly, this happens if and only if the lattice of periods Γ is orthogonal. The lemma is proved. □

Thus, the lattice Γ has a basis consisting of two vectors $f_1 = (1, 0)$ and $f_2 = (0, L)$ (or, in complex notation, $f_1 = 1$ and $f_2 = iL$). Besides, it follows from the proof of the lemma that $f(0) = f(1/2) = 0$ and $g(0) = g(L/2) = 0$, and there are no other zeros of f and g.

The evenness of functions f and g follows immediately from the fact that the form $(f(u) + g(v))(du^2 + dv^2)$ is obtained from the initial Riemannian metric by lifting under the mapping that glues the pairs of points (u, v) and $(-u, -v)$.

The condition (c) for f and g follows from the local theory (see Theorem 11.8). Theorem 11.21 is proved. □

The following useful statement about the uniqueness of a quadratic integral will be used below to prove Theorem 11.22.

Proposition 11.8 (V. N. Kolokol'tsov [189]). *Let ds^2 be a metric on the sphere with quadratically integrable geodesic flow. If this metric is different from the constant curvature metric, then the quadratic integral F of its geodesic flow is defined uniquely up to a linear combination with the Hamiltonian.*

Corollary. *If the geodesic flow of a Riemannian metric ds^2 on the sphere possesses two quadratic integrals F_1 and F_2 such that F_1, F_2, H are linearly independent (here H is the Hamiltonian of the geodesic flow), then ds^2 is the constant positive curvature metric.*

COMMENT. Recall that, for non-resonant integrable Hamiltonian systems, any integral F_2 must be functionally dependent with H and F_1. This fact, however, cannot be directly used to prove the above proposition. There are two reasons for this. First, in our statement, it is said about the uniqueness up to linear (but not functional) dependence of functions. Besides, as we shall show below, there exist resonant quadratically integrable geodesic flows on the sphere whose geodesics are all closed.

As before, a natural question arises: which (L, f, g)-metrics are isometric among themselves? In other words, how ambiguous is the (L, f, g)-representation of a given metric?

Introduce the following operations $\alpha_{1/2}$ and γ on (L, f, g)-metrics:

$$\alpha_{1/2}(L, f(x), g(y)) = (L, f(x + 1/2), g(y)),$$
$$\gamma(L, f(x), g(y)) = (1/L, L^2 g(Lx), L^2 f(Ly)).$$

Each of these operations is an involution. However, they do not commute, but generate the dihedral group.

Theorem 11.22 (V. S. Matveev). *Let (L, f, g)-metric and $(\widehat{L}, \widehat{f}, \widehat{g})$-metric be both different from the constant curvature metric on the sphere. Then they are isometric if and only if they can be transformed to each other by the compositions of the transformations $\alpha_{1/2}$ and γ.*

Proof. The statement will be proved if we give the answer to the following question. Given a Riemannian metric on the sphere whose geodesic flow is quadratically integrable, how many different ways exist to represent it as an (L, f, g)-metric? This means that we must find all the coverings $\varkappa$ of the sphere by the torus branched at four points such that there exist coordinates u and v on the torus in terms of which the lifted metric on the torus (with four degenerate points) has the form $(f(u) + g(v))(du^2 + dv^2)$?

If such a covering is given and fixed, then the triple (L, f, g) is uniquely defined. Thus the non-uniqueness of an (L, f, g)-representation is reduced to the ambiguity of the covering $\varkappa$ and that of the choice of Liouville coordinates u, v on the torus. However, no ambiguity of the choice of $\varkappa$ exists. The point is that the branched two-sheeted covering over the sphere with four branch points is unique from the topological point of view. Namely, any two such coverings with fixed branch points on the sphere are topologically equivalent in the sense that they are transferred one onto the other by a homeomorphism of the torus. Moreover, taking into account that at branch points such a covering looks like raising to the second power $w = z^2$, it is easy to see that such coverings are equivalent in the smooth sense, i.e., the above homeomorphism of the torus is actually a diffeomorphism. It remains to explain why the branch points can be considered as fixed, i.e., uniquely defined for a given metric. According to Proposition 11.8, the quadratic integral is uniquely defined. But the branch points are just those points where the integral (as a quadratic form on the tangent plane to the sphere) is proportional to the Hamiltonian. Therefore, if the Hamiltonian (or the metric) is fixed, then the branch points are uniquely defined.

Thus, the ambiguity of an (L, f, g)-representation takes place just because of the non-uniqueness of global Liouville coordinates on the covering torus, and our problem is reduced to the case of the torus. In Theorem 11.14, when we studied this case, we indicated four operations $\alpha, \beta, \gamma, \delta$ acting on the set of possible $(L, f, g, k/m)$-representations. It remains to select only those of them which correspond to the situation in question.

The first operation has the form α_b and is just the shift of argument of f by an arbitrary real number b. In other words, in the torus case we could change the origin for x. But in our case, there is the additional condition on f, namely, $f(0) = 0$. Therefore, we can shift the origin only by b equal to the distance up to the next zero of f, that is, $b = 1/2$. Thus, the first operation α has exactly the form announced in Theorem 11.22.

The next operation β is not interesting for us: it acts trivially, since g is an even function.

The operation γ corresponds exactly to the operation γ described in Theorem 11.22. It interchanges the coordinates u and v.

The operation δ was connected with the non-uniqueness of the decomposition of $f(u) + g(v)$ into the sum of $f(u) + c$ and $g(v) - c$. But in our case, there is no ambiguity (see above), since there is one more additional condition: the minimum of f (as well as the minimum of g) is zero. Therefore, in our case, $f(u) + g(v)$ is uniquely represented as the sum of $f(u)$ and $g(v)$. This completes the proof of Theorem 11.22. $\square$

This theorem can be also proved in another way.

We have explained above the method to reduce a given Riemannian metric on the sphere with quadratically integrable geodesic flow to the form of an (L, f, g)-metric. This method is not quite uniquely defined. Namely, the choice of the integral was ambiguous (we might take any linear combination with the Hamiltonian), consequently, the function $R(z)$ was not uniquely defined (only up to a real factor). Besides, we sent to infinity one of the four roots of $R(z)$. There was no arbitrariness in the rest. How does this ambiguity affect the (L, f, g)-representation of a given metric? It is easy to check that multiplying $R(z)$ by -1 leads to the permutation of coordinates u and v in the (L, f, g)-model, i.e., just to the transformation γ (see above). The ambiguity in the choice of a root to be sent to infinity leads to the translation of Γ by one of three following vectors: $(1/2, 0)$, $(0, L/2)$, $(1/2, L/2)$. The first translation can be simulated by the operation $\alpha_{1/2}$, the other two can be realized by combinations of operations α and γ. Namely, the translation by $(0, L/2)$ is the composition $\gamma \circ \alpha_{1/2} \circ \gamma$, and the translation by $(1/2, L/2)$ corresponds to $\gamma \circ \alpha_{1/2} \circ \gamma \circ \alpha_{1/2}$.

Corollary. *The set of all linearly (quadratically) integrable geodesic flows on the sphere is arcwise connected. In other words, for any two metric of this kind G_1 and G_2 there exists a deformation of the metric G_1 in the class of metrics with linearly (resp. quadratically) integrable geodesic flows such that, as a result, we obtain a metric which is isometric to G_2.*

Proof. According to Theorem 11.21 all metrics with quadratically integrable geodesic flows are coded up to isometry by three parameters (L, f, g). It is clear that the space of parameters is arcwise connected. The metrics with linearly integrable geodesic flows are coded by one parameter f. This space is evidently also arcwise connected. Moreover, any metric whose geodesic flow is quadratically integrable can be smoothly deformed into the standard metric induced on the triaxial ellipsoid (with preserving the quadratic integrability). On the other hand, any metric whose geodesic flow is linearly integrable can be deformed into the metric on the ellipsoid of revolution (with preserving the linear integrability). It is clear that by deforming the triaxial ellipsoid (in the class of all ellipsoids) into the ellipsoid of revolution, we connect the metrics on them without losing the integrability. Thus, the set of metrics with quadratically and linearly integrable geodesic flows is also arcwise connected. □

REMARK. In the case of the torus the analogous statement is not true. The point is that for the torus one of the parameters, which determine metrics with quadratically integrable geodesic flows, is discrete and cannot be changed by a continuous deformation.

11.5.4. The Projective Plane

The classification of linearly and quadratically integrable geodesic flows on the projective plane was obtained by V. S. Matveev [224].

First we describe the class of metrics on $\mathbb{R}P^2$ whose geodesic flows admit a linear integral. Consider $\mathbb{R}P^2$ as a quotient space $S^2/\mathbb{Z}_2$, where the sphere S^2 is presented as the extended complex plane $\overline{\mathbb{C}} = \mathbb{C} + \{\infty\}$ with the involution $\tau(z) = -1/\bar{z}$. It is clear that this involution has no fixed points on the sphere and changes orientation, so $\overline{\mathbb{C}}/\tau = \mathbb{R}P^2$. We now take the following metric on the sphere $S^2 = \overline{\mathbb{C}}$:

$$ds^2 = f(|z|^2)\, dz\, d\bar{z},$$

where $f(t)$ is a smooth positive function on the semi-axis $[0, \infty)$ satisfying an additional property $f(t) = \dfrac{f(1/t)}{t^2}$. This in fact means that the above metric is invariant under τ and, consequently, determines a metric on the projective plane.

Definition 11.8. The metrics of this kind on the projective plane are called *f-metrics.*

Theorem 11.23 (V. S. Matveev).

a) *The geodesic flow of a Riemannian metric ds^2 on $\mathbb{R}P^2$ is linearly integrable if and only if ds^2 is isometric to an f-metric.*

b) *Two such metrics, i.e., an f-metric and an $\hat{f}$-metric, are isometric if and only if the functions f and $\hat{f}$ coincide identically.*

Proof. The above f-metrics are indeed linearly integrable. The linear integral on the sphere can be presented (see above) as the vector field $\partial/\partial\varphi$, where φ is the standard angle coordinate (longitude) on the sphere.

It is seen that this field is invariant with respect to the involution τ and, therefore, can be considered as a well-defined vector field on $\mathbb{RP}^2$. As a result, we obtain a linear integral of the f-metric on $\mathbb{RP}^2$.

Conversely, let us be given a smooth metric on $\mathbb{RP}^2$ whose geodesic flow is linearly integrable. By lifting it to the sphere, we obtain a smooth metric on the sphere with the same property. The structure of such metrics is completely described by Theorem 11.20. That is, there exists a global conformal coordinate $z = x + iy$ such that the metric has the form $f(x^2 + y^2)(dx^2 + dy^2)$. We now need to understand how the involution is arranged in terms of these coordinates.

It must preserve the above metric and, in particular, the form of its conformal factor. But all such transformations have been described in Theorem 11.20. They must have one of the following forms: $z \to az$, $z \to a\bar{z}$, $z \to a/z$, $z \to a/\bar{z}$. But since we are interested only in orientation-reversing involutions without fixed points, it follows that the only admissible involution is $\tau(z) = -a/\bar{z}$, where $a > 0$ is a real constant. If $a \neq 1$, then we make the coordinate change $z = a^{1/2}w$. As a result, the involution $\tau(z) = -a/\bar{z}$ turns into the involution $\tau(w) = -1/\bar{w}$. The desired additional property of the function f follows then from the fact that f is preserved under this involution. This completes the proof of the first statement (a).

Let us prove statement (b). Suppose we are given an f-metric and an $\hat{f}$-metric on $\mathbb{RP}^2$, and they are isometric. By lifting them onto the covering sphere, we obtain two isometric metrics with linearly integrable geodesic flows. According to Theorem 11.20, two such metrics corresponding to the functions f and $\hat{f}$ are isometric if and only if

$$\text{either} \quad f(t/\alpha) = \alpha\hat{f}(t) \qquad \text{or} \qquad \frac{f(\alpha/t)}{t^2} = \frac{\hat{f}(t)}{\alpha}$$

for some positive number α. Consider the first case. Then f and $\hat{f}$ must satisfy the following relations:

$$f(t/\alpha) = \alpha\hat{f}(t), \quad f(t) = \frac{f(1/t)}{t^2}, \quad \hat{f}(t) = \frac{\hat{f}(1/t)}{t^2}.$$

But these three conditions imply one more property of the function f (the analogous property is, of course, fulfilled also for $\hat{f}$):

$$\alpha^2 f(t) = f(t/\alpha^2).$$

This is possible if and only if $\alpha = 1$. But then the first of the above conditions $f(t/\alpha) = \alpha\hat{f}(t)$ becomes the identity $f = \hat{f}$, as required.

The second case $\dfrac{f(\alpha/t)}{t^2} = \dfrac{\hat{f}(t)}{\alpha}$ can be studied analogously. It is reduced to the first one, because of the additional relation $\dfrac{f(1/t)}{t^2} = f(t)$. These two relations imply that $f(t/\alpha) = \alpha\hat{f}(t)$. This case has already been examined. Thus, we obtain again $f = \hat{f}$. This completes the proof of the second statement (b). $\square$

REMARK. In fact, the proof is based on a simple fact which explains why, in the case of the sphere, the functions f and $\hat{f}$ may be different, whereas in the case of $\mathbb{R}P^2$ they must coincide. The point is that here we have an additional condition: the involution must have the form $\tau(z) = -1/\bar{z}$. This condition uniquely determines that coordinate system in terms of which we write the metric in conformal form. No possibilities for its deformation by means of dilatation remain.

We now describe quadratically integrable geodesic flows on $\mathbb{R}P^2$. In Section 11.5.3 we presented the sphere as the quotient space of the torus by an involution σ and introduced a class of (L, f, g)-metrics as

$$ds^2 = (f(x) + g(y))(dx^2 + dy^2),$$

where f and g satisfy conditions (a), (b) (c) (see above). Now we present $\mathbb{R}P^2$ as a quotient space of the sphere by defining an involution τ on it, namely

$$\tau((x, y), (-x, -y)) = ((-x + 1/2, y + L/2), (x - 1/2, -y - L/2)).$$

Here a point of the sphere is understood as a pair of equivalent points on the torus. It is easy to see that τ has no fixed points on the sphere and changes orientation. This involution, however, can be lifted up to an involution on the covering torus which commutes with σ. Clearly, $S^2/\tau \simeq \mathbb{R}P^2$. Select those metrics from the set of (L, f, g)-metrics which are invariant under the involution τ. This is equivalent to the relation

$$f(x) + g(y) = f(1/2 - x) + g(L/2 + y).$$

Hence $f(x) = f(1/2 - x)$ and $g(y) = g(L/2 + y)$. Taking into account the evenness of f, one gets $f(x) = f(x - 1/2)$. Thus, the condition that an (L, f, g)-metric is τ-invariant exactly means that f and g have periods $1/2$ and $L/2$ respectively.

REMARK. In the analytic case, the periodicity condition for f (with period $1/2$) and g (with period $L/2$) is automatically fulfilled. We already indicated this property while discussing condition (c) for the sphere. Therefore, in the analytic case, no additional conditions on f and g appear. In particular, if the geodesic flow on the sphere is analytic and quadratically integrable, then the metric admits an isometry $\tau\colon S^2 \to S^2$ which is an involution without fixed points.

Definition 11.9. A Riemannian metric on the projective plane is called an (L, f, g)-metric if, after being lifted to the sphere, it becomes an (L, f, g)-metric on the sphere.

Theorem 11.24 (V. S. Matveev).

a) *The geodesic flow of a Riemannian metric ds^2 on the projective plane is quadratically integrable if and only if ds^2 is isometric to an (L, f, g)-metric on $\mathbb{R}P^2$ for appropriate parameters L, f, g.*

b) *Two such metrics, i.e., an (L, f, g)-metric and an $(\widehat{L}, \widehat{f}, \widehat{g})$-metric, are isometric if and only if*

$$\text{either} \quad \widehat{L} = L, \quad \widehat{f} = f, \quad \widehat{g} = g$$

$$\text{or} \quad \widehat{L} = 1/L, \quad \widehat{f}(x) = L^2 g(Lx), \quad \widehat{g}(y) = L^2 f(Ly).$$

Proof. We begin with the first statement (a). Let ds^2 be a Riemannian metric on $\mathbb{R}P^2$ whose geodesic flow admits a quadratic integral. By lifting ds^2 from $\mathbb{R}P^2$ onto the covering sphere, we obtain a metric on the sphere, whose geodesic flow is quadratically integrable. The degree of the integral cannot be reduced, since otherwise it would turn into the linear integral $\partial/\partial\varphi$. But this integral is preserved by the involution τ, and, consequently, the initial integral would be reduced to a liner one, but it is not the case.

We have already described the metrics which admit quadratic integrals on the sphere. They are represented as (L, f, g)-metrics with some additional assumptions on the functions f and g. We need to show that the involution τ' on the torus covering the sphere has exactly the same form as was indicated above, namely

$$\tau'(x, y) = (1/2 - x, L/2 + y).$$

Such a form of the involution follows from the fact that τ must satisfy three conditions at the same time. The first one is that it reverses orientation. The second is that it has no fixed points. The third is that it preserves the metric and, in particular, its (L, f, g)-representation. Therefore, the explicit form of the involution can easily be obtained from Theorem 11.22.

The periodicity of f and g, with periods $1/2$ and $L/2$ respectively, follows from the fact that the (L, f, g)-metric is τ-invariant.

We now prove statement (b). Let an (L, f, g)-metric and an $(\widehat{L}, \widehat{f}, \widehat{g})$-metric be isometric on $\mathbb{R}P^2$. Then the corresponding metrics obtained by lifting onto the covering sphere are also isometric. But on the sphere, such metrics can be obtained one from the other by the operations of two types: $\alpha_{1/2}$ and γ (Theorem 11.22). The first of them, i.e., the operation $\alpha_{1/2}$, becomes trivial in our case, since f is periodic exactly with period $1/2$. The only remaining operation is γ, as statement (b) claims. $\square$

Chapter 12

Liouville Classification of Integrable Geodesic Flows on Two-Dimensional Surfaces

In this chapter, we discuss the results by E. N. Selivanova [311], V. V. Kalash-nikov (Jr.) [175], Nguyen Tien Zung, L. S. Polyakova [263], [264], V. S. Matveev [224] devoted to the topology of Liouville foliations of integrable geodesic flows on two-dimensional surfaces. We begin with the simplest case of global Liouville metrics on the two-dimensional torus, where the structure of the Liouville foliation, on the one hand, is the most natural and, on the other hand, serves a good model for the description of all other cases. As we shall see shortly, integrable geodesic flows on a two-dimensional surface are similar in many respects. However, each class of such geodesic flows is distinguished among the others by some specific properties of Liouville foliations. Following the general idea of our book, we shall formulate the final answer in terms of marked molecules.

12.1. THE TORUS

Consider a Riemannian metric ds^2 with linearly or quadratically integrable geodesic flow on the two-dimensional torus T^2. First of all, we want to find out under which conditions the singularities of the corresponding Liouville foliation are non-degenerate or, which is the same, when the integral of the geodesic flow is a Bott function on the isoenergy surface.

For the sake of simplicity, we consider a global Liouville metric. This means that there exist global periodic coordinates (x, y) (with periods 1 and L, respectively) in which the metric takes the form

$$ds^2 = (f(x) + g(y))(dx^2 + dy^2),$$

where $f(x)$ and $g(y)$ are smooth strictly positive periodic functions with periods 1 and L respectively.

The Hamiltonian of the geodesic flow related to the metric ds^2 is

$$H = \frac{p_x^2 + p_y^2}{f(x) + g(y)},$$

where (x, y, p_x, p_y) are the standard coordinates on the cotangent bundle T^*T^2, and the quadratic integral is given by

$$F(x, y, p_x, p_y) = \frac{g(y)p_x^2 - f(x)p_y^2}{f(x) + g(y)}.$$

If one of the functions f and g is constant, then the geodesic flow possesses a linear integral ($F = p_x$ or $F = p_y$ respectively).

Proposition 12.1. *The integral F is a Bott function on the isoenergy surface $Q^3 = \{H = \text{const}\}$ if and only if each of the functions $f(x)$ and $g(y)$ is either a Morse function or a constant.*

Proof. First suppose that neither f nor g are constant. If we fix the energy level $\{H = h\}$, then we obtain the isoenergy surface Q_h^3 which is diffeomorphic to the three-dimensional torus T^3. As local coordinates on Q_h^3, we can choose either (x, y, p_x) or (x, y, p_y). Restricting the integral F to the isoenergy surface, we obtain either the function

$$F|_Q = p_x^2 - f(x)$$

in terms of coordinates (x, y, p_x) or the function

$$F|_Q = -p_y^2 + g(y)$$

in terms of coordinates (x, y, p_y).

Therefore, the critical points of $F|_Q$ are exactly those of the form

(x is a critical point of $f(x)$, y is arbitrary, $p_x = 0$)

or

(x is arbitrary, y is a critical point of $g(y)$, $p_y = 0$).

It is easy to see from this that the singularities of the integral F on Q are of Bott type if and only if f and g are Morse functions.

Consider now the case when one of the functions, say $f(x)$, is constant. Without loss of generality we shall assume that $f(x) \equiv 0$. In this case, the linear integral F

has the form p_x. After restricting F onto the isoenergy surface $Q^3 = \{H = 1\}$, we obtain

$$F|_Q = p_x \quad \text{in terms of coordinates } (x, y, p_x),$$

or

$$F|_Q = \pm\sqrt{g(y) - p_y^2} \quad \text{in terms of coordinates } (x, y, p_y).$$

The point is that $p_x^2 + p_y^2 = g(y)$ on Q^3.

In this case, the critical points of the linear integral F are evidently the points (x, y, p_y) such that

$$(\; x \text{ is arbitrary, } \quad y \text{ is a critical point of } g(y), \quad p_y = 0 \;).$$

Such a point is non-degenerate for the integral F on Q if and only if y is a non-degenerate critical point for the function $g(y)$.

Finally, in the exceptional case, when both the functions f and g are constant, we obtain a flat metric. It is easy to see that the set of critical points of any its linear integral F (restricted on Q^3) is the disjoint union of two non-degenerate critical two-dimensional tori which, however, cease to be critical after replacing F by another linear integral. As a result, the Liouville foliation represents the trivial T^2-fibration without singularities over a circle. $\square$

REMARK. If only one of the functions f and g is constant, then we obtain a linearly integrable geodesic flow. If f and g are both constant, then the metric ds^2 is flat (its geodesic flow is clearly linearly integrable). It is easy to check that, in this case, the linear integral has two non-degenerate critical submanifolds, each of which is diffeomorphic to a torus.

REMARK. The statement remains valid for arbitrary linearly and quadratically integrable geodesic flows on the torus. Recall that the only difference from the simplest case is that the basis of the lattice Γ related to the torus T^2 is distorted with respect to the Liouville coordinate system (x, y).

Let us give the formal construction of the molecule W in the case of a global Liouville metric, assuming that both f and g are Morse functions.

It is clear that the topology of the Liouville foliation is completely determined by these functions. We begin with assigning to each of them some model foliation into two-dimensional tori, from which we shall then glue the global Liouville foliation on the whole isoenergy surface.

Consider the isoenergy surface $Q^3 = \{H = 1\}$, which is, from the topological point of view, the trivial S^1-fibration over the torus T^2. Its fiber is a circle which lies in the cotangent plane. Then, as a local coordinate on the fiber, we can take either p_x or p_y. As a result, local coordinates on Q^3 can be either (x, y, p_x) or (x, y, p_y). It is easy to check that the integral F written in these coordinates takes one of the following forms:

$$F|_Q = \quad p_x^2 - f(x) \quad \text{in coordinates } (x, y, p_x);$$
$$F|_Q = -p_y^2 + g(y) \quad \text{in coordinates } (x, y, p_y).$$

Take, for definiteness, the second case: $F = -p_y^2 + g(y)$. Consider F as a function on the flat annulus, where y and p_y are thought as the angle and radial coordinates on it (see Fig. 12.1). We shall assume that the middle circle of the annulus

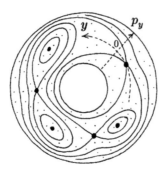

Figure 12.1

corresponds to $p_y = 0$. As a result, the graph of the function $-p_y^2 + g(y)$ can be interpreted as a mountain relief on the annulus whose critical points are exactly critical points of F (see Fig. 12.2(a)). It can also be described as follows. Take the graph of the function $g(y)$ on the segment $[0, L]$ and make it into a linear mountain relief as illustrated in Fig. 12.2(b). After this, we should bend this segment into a circle. The linear relief is transformed as a result into an annulus mountain relief, i.e., given on the annulus (see Fig. 12.2(a)).

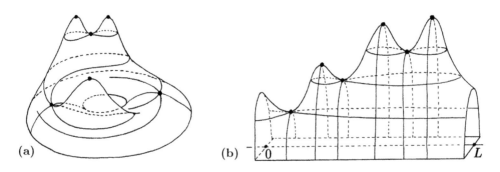

(a)   (b)

Figure 12.2

Note that we are interested only in the domain that corresponds to positive values of F, i.e., is located above the "sea-level". In other words, we consider an annulus of variable width $P = \{g(y) - p_y^2 \geq 0\}$.

The level lines of the mountain relief (i.e., of the function $F = -p_y^2 + g(y)$) determine a foliation with Morse singularities on P, which is described by a certain molecule $W(g)$ with two free edges related to the boundary of the annulus. Let us now consider the direct product of this one-dimensional foliation by a circle. As a result, we obtain a Liouville foliation on the three-dimensional manifold $P \times S^1$, which would be naturally called a *direct product type foliation*. The corresponding molecule obviously coincides with molecule $W(g)$ related to the Morse function of two variables $F = -p_y^2 + g(y)$.

Suppose that all the critical circles of this Liouville foliation have the same orientation. Then the corresponding marks r and ε on the edges of the molecules can be easily described: all the ε-marks equal 1; all r-marks on the edges between saddle atoms equal infinity, and those between saddle atoms and atoms A equal zero. Thus, starting from the function g, we have constructed a Liouville foliation of direct product type and completely described the corresponding molecule with marks.

In the same way, we deal with the function $f(x)$. As a result, we obtain, generally speaking, another molecule $W(f)$, but again with two free ends.

The structure of molecules $W(f)$ and $W(g)$ can be described in greater detail. Let us describe, for example, the construction of $W(f)$. Take the graph of $f(x)$ on the segment $[0, 1]$, and consider the domain U that lies beneath the graph and is bounded from below by the axis Ox. Let us foliate it by horizontal segments, that is, by lines $\{f = \text{const}\}$ (see Fig. 12.3). Each such segment $\{f = \text{const}\} \cap U$ corresponds to a connected component of a level line of the function $p_x^2 - f(x)$ on the annulus. The only exception is the family of horizontal segments located below the global minimum of $f(x)$. Let us note that such segments always exist, since f and g are both positive and, therefore, their global minima are strictly greater than zero. Every such level-segment $\{f = c\}$ is related to two different connected components of the level line $\{p_x^2 - f(x) = -c\}$ on the annulus.

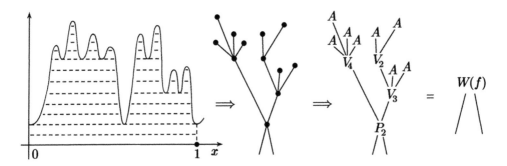

Figure 12.3

Next we shrink each of the above segments into a point. To each of the segments located below the global minimum of $f(x)$, we assign not one, but two points. It is clear that the domain U as a result turns into a graph which coincides with the Reeb graph of the function $p_x^2 - f(x)$ given on the annulus (see Fig. 12.3). This graph is obviously a tree.

The interior vertices of the graph correspond to local minima of $f(x)$ considered as a function on the circle, and its end-vertices correspond to the local maxima. To each vertex, we now put a certain atom. Namely, the end-vertices of the graph are replaced with atoms A. The interior vertices (except for that related to the global minimum of the function f) correspond to atoms of type V_k. Here V_k is a planar atom illustrated in Fig. 12.4, and k denotes the number of its vertices which coincides with the number of local minima of f that lie on the same horizontal segment $\{f = \text{const}\} \cap U$. The planar atom P_m, which is assigned to the global minimum of f, has the form shown in Fig. 12.5; here m is the number of local minimum points of f (there may be several of them). In Fig. 12.3, as an example, we show the molecule $W(f)$ for the function f whose graph is illustrated on the same figure. Note that each of the graphs $W(f)$ and $W(g)$ is a tree.

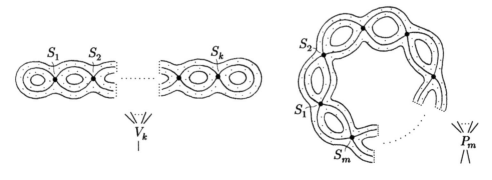

Figure 12.4   Figure 12.5

We now construct a new molecule W by taking two copies of $W(f)$ and two copies of $W(g)$ and gluing their free ends crosswise as shown in Fig. 12.6(a).

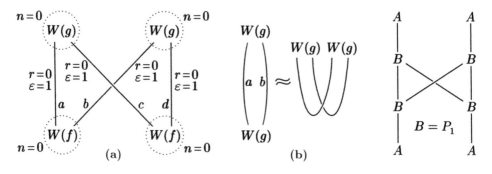

Figure 12.6   Figure 12.7

In the simplest case, when each of the functions f and g has exactly one minimum and one maximum, the complete molecule W is of the form shown in Fig. 12.7. Here the atom P_1 coincides with the atom B.

In the case of linearly integrable geodesic flows on the torus, we proceed as follows. For definiteness, assume that $f(x)$ is identically zero. Then the corresponding molecule $W(f)$ can be thought as trivial, i.e., consisting just of a segment with two free ends (without any atoms). After this, the procedure is repeated, and the construction of W is reduced to gluing two copies of $W(g)$ between themselves. It is clear that W will have the form illustrated in Fig. 12.6(b).

Theorem 12.1 (E. N. Selivanova).

a) *Let the metric ds^2 on the torus T^2 be a global Liouville one (i.e., let ds^2 be an $(L, f, g, 0)$-metric in the notation of Chapter 11). Then the marked molecule W^* corresponding to its geodesic flow has the form shown in Fig. 12.6(a), and the marks are given in the following way. The four edges a, b, c, d, as well as all the edges incident to atoms A, have the mark r equal to zero. The other edges have the mark $r = \infty$. All the marks n_k are equal to zero; and all the marks ε_i are equal to 1.*

b) *Let the geodesic flow of a Riemannian metric ds^2 on the torus T^2 be linearly integrable (i.e., let ds^2 be an (g, t, L)-metric in the notation of Chapter 11). Then the marked molecule W^* corresponding to its geodesic flow has the form shown in Fig. 12.6(b), and the marks are given in the following way. All the edges that do not contain atoms A have the mark $r = \infty$. The other edges (i.e., incident to atoms A) have the mark $r = 0$. The only family has the mark n equal to zero. The marks ε on the edges a and b are equal to -1; on the other edges $\varepsilon = +1$.*

A similar statement holds in the remaining case of finite-sheeted Liouville metrics. From the point of view of the molecule, only the marks on the edges a, b, c, d are changed.

Theorem 12.2 (V. V. Kalashnikov (Jr.)). *Let ds^2 be a finite-sheeted Liouville metric on the torus (i.e., an $(L, f, g, k/m)$-metric with $k \neq 0$ in the notation of Chapter 11). Then the marked molecule W^* corresponding to its geodesic flow has the form shown in Fig. 12.8, and the marks are given in the following way. All the edges that contain an atom A are endowed with the r-mark equal to zero. The edges b, c are endowed with the r-mark equal to k/m, and the edges a, d are endowed with the r-mark equal to $-k/m$. All the other edges have the mark $r = \infty$. The marks n_k are equal to -1; and the marks ε_i are equal to 1.*

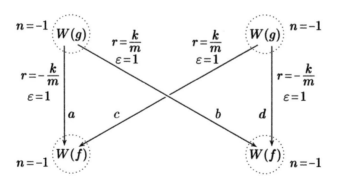

Figure 12.8

Theorems 12.1, 12.2 give, as a result, the complete Liouville classification of all quadratically and linearly integrable geodesic flows on the two-dimensional torus.

Proof (of Theorem 12.1). We begin with the case of a quadratically integrable geodesic flow. Consider the 3-manifold $Q = \{H = 1\}$. It is diffeomorphic to the three-dimensional torus and has the natural structure of a trivial S^1-fibration over T^2. Consider the level surface $\{F = 0\}$ in Q. It is easily checked that all the critical points of $F|_Q$ are located out of the level $\{F = 0\}$. Therefore, $\{F = 0\}$ is non-singular and hence consists of several regular Liouville tori. The number of these tori is 4 and they are given by the following explicit expressions:

$$T_a^2 = \{p_x = +\sqrt{f(x)},\, p_y = +\sqrt{g(y)}\},$$
$$T_b^2 = \{p_x = +\sqrt{f(x)},\, p_y = -\sqrt{g(y)}\},$$
$$T_c^2 = \{p_x = -\sqrt{f(x)},\, p_y = +\sqrt{g(y)}\},$$
$$T_d^2 = \{p_x = -\sqrt{f(x)},\, p_y = -\sqrt{g(y)}\}.$$

Each of these tori is diffeomorphically projected onto the base T^2, since the values of p_x and p_y for any point $(x,y) \in T^2$ can be uniquely reconstructed from the above formulas. It is convenient to picture the level lines of the integral $F(x,y,p_x,p_y)$ in each tangent plane of the torus (Fig. 12.9). It is seen from the explicit formula for F that the zero level line is a pair of intersecting straight lines, and any other level is a hyperbola. The four tori that compose the zero level surface of F are presented in Fig. 12.9 as four points denoted by a, b, c, d on the circle $\{H = 1\}$ (that lies in the (co)tangent plane). We preserve the same notation for the corresponding edges of the molecule to which these tori belong.

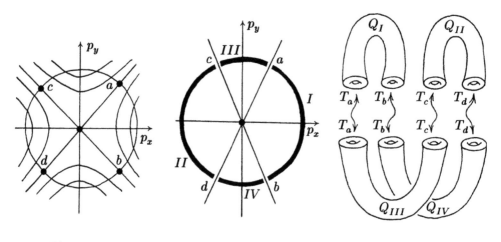

Figure 12.9   Figure 12.10   Figure 12.11

The four tori $T_a^2, T_b^2, T_c^2, T_d^2$ divide the manifold Q into four connected pieces $Q_I, Q_{II}, Q_{III}, Q_{IV}$ (Fig. 12.10) that correspond to the four arcs of the circle $\{H = 1\}$ into which it is divided by points a, b, c, d. Each of these pieces has a very simple structure being the direct product of a torus by a segment. The boundary of each of them consists of two tori. The corresponding mutual gluing operations are illustrated in Fig. 12.11. The rule of gluing is extracted from Fig. 12.10.

Each of pieces $Q_I, Q_{II}, Q_{III}, Q_{IV}$ is foliated into Liouville tori. Note that the Liouville foliations on Q_I and Q_{II} have the same structure. Indeed, to see this, one only needs to consider the diffeomorphism

$$(x,y,p_x,p_y) \to (x,y,-p_x,-p_y),$$

which sends one foliation into the other. This diffeomorphism just changes the direction of every geodesic. The same statement is also valid for Q_{III} and Q_{IV}. We now look at the Liouville foliation on each piece, for instance, Q_I. It is seen from Fig. 12.10 that (x, y, p_y) can be taken as global coordinates on this 3-manifold. Then the integral F on Q_I becomes $F = -p_y^2 + g(y)$. Since the coordinate x does not take part in the explicit expression for F, the foliation into Liouville tori that appears on Q_I is of direct product type and is determined by the molecule $W(g)$ described above.

Using an analogous argument, we see that the same molecule $W(g)$ corresponds to the piece Q_{II}, and two copies of $W(f)$ correspond to Q_{III} and Q_{IV}. Thus, we have proved that the molecule W is glued from two copies of $W(f)$ and two copies of $W(g)$ as shown in Fig. 12.6. Therefore, W (without marks yet) has the desired form.

The next step is computing the marks of the molecule. We begin with the pieces Q_I, Q_{II}, Q_{III}, Q_{IV}, that is, with the molecules $W(f)$, $W(g)$. The statement about r-marks assigned to their interior edges follows immediately from the fact that the Liouville foliation has the direct product type. The condition $\varepsilon = 1$ is equivalent to the fact that all critical circles of the foliations have the same orientation. To show this, in the case Q_I, one needs to check that the derivative of the angle coordinate x along the flow sgrad H has always the same sign. This derivative is easily computed:

$$\frac{dx(t)}{dt} = \frac{\partial H}{\partial p_x} = \frac{2p_x}{f(x) + g(y)}.$$

It remains to notice that the value of p_x is positive everywhere in Q_I.

Turn to the edges a, b, c, d of the molecule W. To compute the related marks, we should write down gluing matrices with respect to some admissible coordinate systems on the tori $T_a^2, T_b^2, T_c^2, T_d^2$. Let us describe these admissible coordinate systems. Consider, for example, the piece Q_I. It has the type of direct product $P_I \times S^1$. Here S^1-fibers are given by the relations $p_y = \text{const}$, $y = \text{const}$, $x = t$; and P_I is an annulus which can be realized as the global section $P_I \subset Q_I$ given by the equation $x = 0$. According to the definition, in the capacity of the first basis cycle on the boundary tori T_a^2 and T_b^2 of Q_I we should take the S^1-fiber, and in the capacity of the second cycle we should take the boundary of the section P_I (taking into account the orientation). It is easy to see that these cycles are obtained by lifting the standard basis cycles $\lambda = \{y = \text{const}, x = t\}$ and $\mu = \{x = \text{const}, y = t\}$ from the torus T^2 (recall that T_a^2 and T_b^2 are diffeomorphically projected to T^2). Thus, on T_a^2 and T_b^2, admissible coordinate systems are given by the pairs of cycles (λ, μ) and $(\lambda, -\mu)$, respectively. Changing the sign for the second basis cycle is dictated by the standard rule of orientation for ∂P_I. In a similar way, we can introduce admissible coordinate systems on the boundary tori of the remaining pieces and check that the gluing matrices on the edges a, b, c, d have the form

$$\begin{pmatrix} 0 & 1 \\ 1 & 0 \end{pmatrix}.$$

Hence all the r-marks on these edges equal zero, and the ε-marks equal 1.

It is easy to see that the molecule W has four families related exactly to the pieces Q_I, Q_{II}, Q_{III}, Q_{IV} (speaking more precisely, in order to obtain families from these pieces, we should remove out of them the solid tori that are neighborhoods of stable closed geodesics, i.e., all the atoms A in our terms). The marks n corresponding to these families equal zero. To check this, we may, for example, write down the gluing matrices on all edges of the molecule. It can be easily done by considering admissible coordinate system on each edge, related to the global sections P_I, P_{II}, P_{III}, P_{IV}. As a result, all gluing matrices between saddle atoms become

$$\begin{pmatrix} 1 & 0 \\ 0 & -1 \end{pmatrix},$$

and, on the edges connecting saddle atoms with atoms A, gluing matrices are

$$\begin{pmatrix} 0 & 1 \\ 1 & 0 \end{pmatrix}.$$

It remains to apply the formula from the definition of the n-mark.

In fact, we can manage without any computations if we take into account the topological meaning of n as an obstruction to extending the section. The condition $n = 0$, for example, for the family Q_I, actually means that one can find a global section of the S^1-fibration on Q_I spanned on the cycles μ and $-\mu$ that lie on the boundary tori T_a and T_b. But this is evident: such a section is the annulus $P_I = \{x = 0\} \subset Q_I$.

The first part of Theorem 12.1 is proved.

Now turn to the case of a linear integral. Consider the metric

$$ds^2 = g(y)(dx^2 + dy^2),$$

where x and y are standard angle coordinates on the torus, $x \in [0, 1]$, $y \in [0, L]$; and suppose that the torus is obtained from the plane $\mathbb{R}^2(x, y)$ by factorization with respect to the lattice generated by vectors $f_1 = (1, 0)$ and $f_2 = (0, L)$. Recall that $g(y)$ is an L-periodic function.

The Hamiltonian H and integral F take the form

$$H = \frac{p_x^2 + p_y^2}{g(y)}, \qquad F = p_x.$$

The main difference from the quadratic case is that the zero level line $\{F = 0\}$ turns here into a straight line in each cotangent plane (p_x, p_y) (instead of two intersecting lines in the quadratic case). In other words, two (of four) arcs of the circle are shrunk into a point. As a result, Fig. 12.9 turns into Fig. 12.12. Therefore, two lower pieces Q_{III} and Q_{IV} disappear (turn into a Liouville torus). In terms of the molecule W, this means that the subgraphs $W(f)$ disappear (see Fig. 12.13).

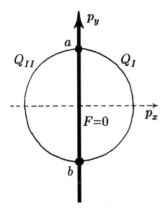

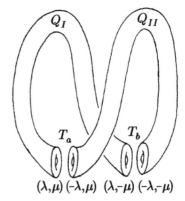

(λ,μ) $(-\lambda,\mu)$ $(\lambda,-\mu)$ $(-\lambda,-\mu)$

Figure 12.12   Figure 12.13

The pieces Q_I and Q_{II} are determined in Q by conditions

$$Q_I = \{p_x \geq 0\}, \qquad Q_{II} = \{p_x \leq 0\}.$$

(see Fig. 12.12). They border on themselves along two Liouville tori, which we denote by T_a, T_b. Namely,

$$T_a^2 = \{p_x = 0, p_y = +\sqrt{g(y)}\}, \qquad T_b^2 = \{p_x = 0, p_y = -\sqrt{g(y)}\}.$$

The structure of the Liouville foliation on Q_I and Q_{II} is the same as in the case of a quadratic integral. The argument is completely the same as above.

Thus, we have described the molecule W in the case of a linear integral. It is illustrated in Fig. 12.6(b). Turn to computing the marks.

On the subgraph $W(g)$, all the marks remain the same as in the quadratic case, and admissible coordinate systems on Liouville tori are constructed just in the same way (see Fig. 12.13). The only difference occurs for the gluing matrices on the edges a and b. They now take the form

$$\begin{pmatrix} -1 & 0 \\ 0 & 1 \end{pmatrix}.$$

Therefore, the mark r is equal to ∞, and ε is -1 on both the edges. Since a and b carry the infinite r-mark, in this case there is only one family, which coincides with the whole molecule W except for the atoms A. The mark n is easily computed by using the above frame, i.e., the collection of all gluing matrices assigned to the edges of the molecule. It equals zero.

However, we should also consider the case of distorted lattices generated by vectors $f_1 = (1,0)$ and $f_2 = (t, L)$, where $t \in [0, 1]$. The corresponding metrics were called (g, t, L)-metrics. For $t = 0$, we obtain the situation just examined.

In the case of distorted lattices, all the above arguments can be applied without changing. The point is that a (g, t, L)-metric depends on the real parameter t continuously. At the same time, neither the Hamiltonian H nor linear integral F depends on t. On the other hand, the marked molecule W^* is a discrete object.

Therefore, under continuous change of the parameter t, the molecule W^* cannot be changed. Hence the marked molecule of the integrable geodesic flow related to the (g, t, L)-metric has to coincide with that related to the $(g, 0, L)$-metric (which is obtained from (g, t, L)-metric as t tends to zero).

Theorem 12.1 is completely proved. $\square$

Proof (of Theorem 12.2).   Let us say a few words about the proof of Theorem 12.2. Let ds^2 be a finite-sheeted Liouville metric on the torus T^2. As we already know, this metric admits the following description. Consider the metric

$$ds^2 = (f(x) + g(y))(dx^2 + dy^2),$$

on the plane $\mathbb{R}^2(x, y)$, where $f(x)$ and $g(y)$ are smooth periodic functions with periods 1 and L respectively. Then the torus T^2 is considered as the quotient space $\mathbb{R}^2/\Gamma$ for the lattice Γ generated by vectors $e_1 = (m, 0)$, $e_2 = (k, L)$, where m and k are relatively prime natural numbers.

Since the metric is invariant under translation by elements of the lattice, it induces a metric ds^2 on the torus T^2 (such a metric was called above an $(L, f, g, k/m)$-metric).

To describe the topology of the corresponding Liouville foliation, it will be convenient to use Theorem 11.12, according to which there exists a canonical covering $\rho : (T^2, ds^2) \to (\widetilde{T}^2, \widetilde{ds}^2)$ such that $ds^2 = \rho^*(\widetilde{ds}^2)$ and $\widetilde{ds}^2$ is a global Liouville metric on the torus $\widetilde{T}^2$. In our case, the torus $\widetilde{T}^2$ is taken as the quotient space $\mathbb{R}^2(x, y)$ with respect to the lattice $\widetilde{\Gamma}$ generated by $\widetilde{e}_1 = (1, 0)$ and $\widetilde{e}_2 = (0, L)$. The metric $\widetilde{ds}^2$ on the torus $\widetilde{T}^2$ is induced in a natural way by the periodic metric on the plane.

The covering $\rho : T^2 \to \widetilde{T}^2$ induces a natural covering of isoenergy 3-surfaces $Q \to \widetilde{Q}$, which obviously sends the leaves of one Liouville foliation to those of the other, i.e., is a fiber mapping. This covering has a very simple structure. The isoenergy 3-surfaces Q and $\widetilde{Q}$, respectively, can be presented as

$$Q = T^2 \times S^1, \qquad \widetilde{Q} = \widetilde{T}^2 \times S^1.$$

The covering $\rho : Q \to \widetilde{Q}$ preserves this decomposition. In other words, on the first factor it coincides with the initial covering, on the other one (i.e., on the circle) it can be thought as the identity mapping.

Let us examine the structure of the covering $\rho : Q \to \widetilde{Q}$.

Recall that the structure of the Liouville foliation on $\widetilde{Q}$ is completely determined by Theorem 12.1. The manifold $\widetilde{Q}$ is divided into four connected pieces $\widetilde{Q}_I, \widetilde{Q}_{II}, \widetilde{Q}_{III}, \widetilde{Q}_{IV}$. It is easy to see that we get a similar picture inside Q. This means that Q is divided by the level $\{F = 0\}$ into four connected pieces $Q_I, Q_{II}, Q_{III}, Q_{IV}$, each of which covers the corresponding 3-piece $\widetilde{Q}_I, \widetilde{Q}_{II}, \widetilde{Q}_{III}$, or $\widetilde{Q}_{IV}$.

Consider the covering $\rho : Q_I \to \widetilde{Q}_I$. We have already shown (see the proof of Theorem 12.1) that $\widetilde{Q}_I = \widetilde{P}_I^2 \times S^1$, i.e., is of direct product type. It is not hard to see that Q_I has the similar structure, i.e., $Q_I = P_I^2 \times S^1$. However, in this case, it is more convenient to define the section P_I^2 by the equation $x - ky = 0$.

It is easy to check that the covering $\rho\colon Q_I \to \tilde{Q}_I$ is compatible with the direct product structure. Namely, ρ is a diffeomorphism on the first factor:

$$\rho\colon P_I^2 \to \tilde{P}_I^2 \, ;$$

and it is the m-multiple winding mapping on the second one:

$$\rho\colon S^1 \to S^1, \qquad \rho(z) = z^k \, .$$

It follows immediately from this that the Liouville foliation structures on Q_I^3 and $\tilde{Q}_I^3$ are the same and determined by the molecule $W(g)$ described above in detail.

The same argument works, of course, for the other 3-pieces Q_{II}, Q_{III}, Q_{IV}. Thus, the Liouville foliation on the isoenergy surface Q is glued from four pieces on each of which it has the same structure as before. Therefore, the whole molecule W (without marks) has the required form. Moreover, all the marks related to interior edges of the subgraphs $W(f)$ and $W(g)$ are just the same as in Theorem 12.1 (i.e., in the case of a global Liouville metric). The difference appears only for r- and ε-marks on the four edges a, b, c, d, as well as for marks n.

To define the remaining marks, we need to fix admissible coordinate systems on the tori $T_a^2, T_b^2, T_c^2, T_d^2$, along which the pieces $Q_I, Q_{II}, Q_{III}, Q_{IV}$ are glued among them, and to compute the corresponding gluing matrices.

Take the piece Q_I^3 and consider its boundary tori T_a^2 and T_b^2. Just by analogy with the proof of Theorem 12.1, as the basic cycles on these tori we must take the fiber of the trivial S^1-fibration on Q_I^3 and the boundary of a global section P_I^2. It is easy to see that, in our case, these cycles can be obtained by lifting the cycles, corresponding to the basic elements e_1, e_2 of the lattice Γ, from the base T^2 onto T_a^2 and T_b^2. We suppose that λ denotes the pull-back of e_1, and μ denotes that of e_2. Then our construction implies that the admissible coordinate systems are the following pairs: on the torus T_a^2 these are (λ, μ); and on the torus T_b^2 these are $(\lambda, -\mu)$. Changing the sign of μ has been explained above (see the proof of Theorem 12.1).

Analogously, the admissible coordinate systems on the boundary tori T_c^2 and T_d^2 of Q_{II}^3 are $(-\lambda, \mu)$ and $(-\lambda, -\mu)$, respectively.

The arguments for the two remaining pieces Q_{III}^3 and Q_{IV}^3 are similar; but here, instead of (e_1, e_2), we should consider another basis of the lattice, namely,

$$e_1' = m\tilde{e}_2, \qquad e_2' = \tilde{e}_1 + k'\tilde{e}_2 \, ,$$

where k' is the integer number uniquely defined from the relations $k'k = 1 \bmod m$ and $0 < k' < m$. Then, on the tori T_a^2 and T_c^2 regarded as boundary tori of Q_{III}^3, the admissible coordinate systems are (λ', μ') and $(\lambda', -\mu')$, respectively. Analogously, on the tori T_b^2 and T_d^2 regarded as boundary tori of Q_{IV}^3, we can take $(-\lambda', \mu')$ and $(-\lambda', -\mu')$ as admissible coordinate systems (see Fig. 12.14). All the cycles $\lambda, \mu, \lambda', \mu'$ are explicitly expressed from the basis $\tilde{e}_1, \tilde{e}_2$:

$$\begin{cases} \lambda = m\tilde{e}_1 \\ \mu = \tilde{e}_2 + k\tilde{e}_1 \end{cases} \quad \text{and} \quad \begin{cases} \lambda' = m\tilde{e}_2 \\ \mu' = \tilde{e}_1 + k'\tilde{e}_2 \end{cases} .$$

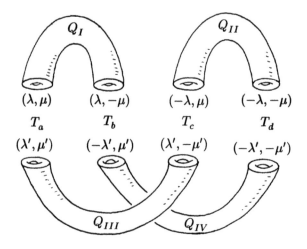

Figure 12.14

We now can compute the gluing matrices for the Liouville foliation on Q^3. They can be easily obtained by using the above formulas for cycles. The result is as follows. On the tori T_a^2, T_d^2, the gluing matrices have the form

$$\begin{pmatrix} -k & m \\ p & k' \end{pmatrix}.$$

On the tori T_b^2, T_c^2, they are

$$\begin{pmatrix} k & m \\ p & -k' \end{pmatrix}.$$

It follows from this that the r-marks and ε-marks on the edges a, b, c, d are as shown in Fig. 12.8.

It remains to find the n-marks. In our case, there are exactly four families related to two subgraphs $W(f)$ and two subgraphs $W(g)$. They correspond to the four pieces $Q_I, Q_{II}, Q_{III}, Q_{IV}$. Consider, for example, the piece Q_I. On all the edges except for a and b, the gluing matrices are very simple. On the edges between two saddle atoms, they are

$$\begin{pmatrix} 1 & 0 \\ 0 & -1 \end{pmatrix}.$$

On the edges connecting saddles atoms with atoms A, the gluing matrices are

$$\begin{pmatrix} 0 & 1 \\ 1 & 0 \end{pmatrix}.$$

It is seen that such gluing matrices give no contribution to the n-mark (see Definition 4.6). The contribution comes only from the gluing matrices related to the edges a and b, which are indicated above. Therefore, n can be computed as follows:

$$n = [k/m] + [-k/m] = -1.$$

The same argument remains correct for the other three families. As a result, we see that n equals -1 on each of the four families. This completes the proof of Theorem 12.2. $\square$

COMMENT. We encounter here a very interesting class of integrable systems, whose topological invariants may have r-marks of arbitrary type k/m. Let us note that usually in integrable systems one can meet only very simple r-marks like 0, $1/2$, ∞ (see examples below).

12.2. THE KLEIN BOTTLE

12.2.1. Quadratically Integrable Geodesic Flow on the Klein Bottle

Recall the description of the metrics on K^2 whose geodesic flows are quadratically integrable. Consider two positive functions $f(x)$ and $g(y)$ with the following properties:

1) $f(x)$ is a non-constant function with period $1/2$;
2) $g(y)$ is a non-constant even function with period L.

Consider the Liouville metric $(f(x)+g(y))(dx^2+dy^2)$ on the covering plane $\mathbb{R}^2$ and represent the torus T^2 as the quotient space $\mathbb{R}^2/\Gamma$, where the lattice Γ is generated by the vectors

$$f_1 = (1,0) \quad \text{and} \quad f_2 = (0,L).$$

Consider the involution ξ on the torus given by

$$\xi(x,y) = (x+1/2, -y)$$

in coordinates x, y.

Then the Klein bottle can obviously be represented as $K^2 = T^2/\xi$. Due to the choice of f and g, the involution ξ preserves the metric on the torus. Therefore, this metric can be descended from the torus down to the Klein bottle. The obtained metric on K^2 is determined by three parameters L, f, g and is called an (L, f, g)-metric.

Consider the two-sheeted covering $\pi: T^2 \to K^2$ related to the involution ξ. It induces a two-sheeted covering (for simplicity, it will be denoted by the same letter) $\pi: Q_T^3 \to Q_K^3$ between the corresponding isoenergy 3-surfaces related to the torus and Klein bottle. It is clear that π is a fiber mapping in the sense of the Liouville foliations on Q_T^3 and Q_K^3. In particular, it follows from this that the molecule W_T (that describes the foliation Q_T^3) covers somehow the molecule W_K (that describes the foliation on Q_K^3). The involution ξ naturally defines an involution on Q_T^3. We denote it also by ξ. It is clear that the projection $\pi: Q_T^3 \to Q_K^3$ is actually the factorization of Q_T^3 by ξ. Hence ξ can also be considered as an involution on the molecule W_T, and, moreover, W_K is obtained from W_T by factorization with respect to ξ.

It is easy to see that the involution ξ preserves the orientation on the isoenergy surface $Q_T^3 \simeq T^3$. Therefore, the quotient manifold Q_T^3/ξ is again diffeomorphic to the 3-torus T^3.

Let us describe the action of this involution on the foliation on Q_T in more detail. The structure of this foliation was described in the previous section, and the corresponding molecule is shown in Fig. 12.6(a). The scheme of the foliation is also shown in Fig. 12.11. As we explained above in detail, the manifold Q_T is naturally divided into four connected pieces $Q_I, Q_{II}, Q_{III}, Q_{IV}$. The Liouville foliations on the pieces Q_I and Q_{II} are isomorphic and have the direct product type $P^2 \times S^1$, where P^2 is an annulus with coordinates (y, p_y), and y is a periodic coordinate. The foliation on $P^2 \times S^1$ is determined by the function $-p_y^2 + g(y)$. It is easily seen that the involution ξ maps $Q_I = P^2 \times S^1$ into itself, namely,

$$\xi \colon (y, p_y, x) \to (-y, -p_y, x + 1/2)\,.$$

Here $x = x \bmod 1$ is a coordinate on the circle S^1. Since the function $g(y)$ is even, the involution ξ preserves the level lines of the function $-p_y^2 + g(y)$ on the annulus and, consequently, preserves the foliation on Q_I. The action of ξ on the annulus P^2 is shown in Fig. 12.15. The annulus is rotated around the vertical axis through the angle π. Clearly, ξ is a fiber mapping of Q_I onto itself. Therefore, it generates a natural foliation on Q_I/ξ. Denote the molecule corresponding to this foliation by $\widetilde{W}(g)$. The similar events happen on Q_{II}. We obtain the foliation on Q_{II}/ξ, which is described by the same molecule $\widetilde{W}(g)$.

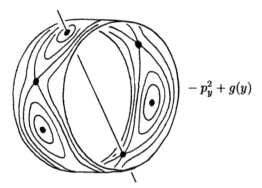

$$-p_y^2 + g(y)$$

Figure 12.15

To describe $\widetilde{W}(g)$, we consider the action of ξ on the atoms of the original molecule $W(g)$. Recall that, in this case, there are only two types of saddle atoms: V_k and P_m. Consider the annulus in Fig. 12.15 foliated into level lines of the function $-p_y^2 + g(y)$ and list all possible cases of the action of ξ. Note that a fixed point of the involution is a critical point of $-p_y^2 + g(y)$; consequently, it is a vertex of some atom.

Case (a). The axis of involution goes past the atom V_{2k}, which is mapped into itself under ξ (Fig. 12.16(a)). It is clear that here the atom V_{2k} after factorization turns into a new atom V_k. The number of its vertices becomes twice less.

Case (b). The axis of involution pierces through the atom V_{2k+1} at one of its vertices. Here the atom goes into itself under ξ (Fig. 12.16(b)). It is clear that, in this case, the atom V_{2k+1} after factorization turns into a new atom V_k^*, where the star indicates that one special star-vertex occurs. This is exactly that vertex, which is the only fixed point of ξ.

Case (c). The axis of involution goes past the atom P_{2k} (Fig. 12.16(c)). Here P_{2k} is mapped into itself under ξ and, after factorization, turns into a new atom V_k.

Case (d). The axis of involution pierces through the atom P_{2k+1} at one of its vertices (Fig. 12.16(d)). Then, after factorization, P_{2k+1} turns into a new atom V_k^*, where the star means that one star-vertex occurs.

Case (e). The axis of involution pierces through the atom P_{2k} at two opposite vertices (Fig. 12.16(e)). Then, after factorization, P_{2k} turns into a new atom V_{k-1}^{**}, which has two star-vertices.

Case (f). The involution interchanges two isomorphic saddle atoms of type V_k (Fig. 12.16(f)). As a result, they are transformed into one atom V_k.

Case (g). The involution maps the atom A into itself (Fig. 12.16(g)). As a result of factorization, A turns again into an atom of the same type A.

Case (h). The involution interchanges two isomorphic atoms A (Fig. 12.16(h)). As a result of factorization, the two atoms A are glued into one atom A.

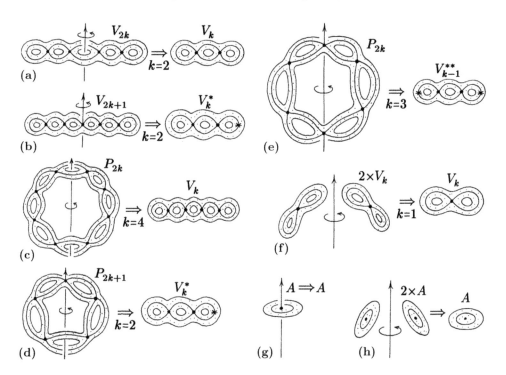

Figure 12.16

We now can explain how to construct the molecule $\widetilde{W}(g)$ from $W(g)$. Take the molecule $W(g)$ and consider the involution ξ on it. Then, according to the above list of rules (a)–(h), look after the result of the action of ξ on all of its atoms. Replacing each atom (or pair of atoms) by its image under factorization, we obtain the molecule $\widetilde{W}(g)$.

Let us illustrate this algorithm by several simplest examples. In Fig. 12.17(1–4), we list all the cases where the number of critical points of the function $-p_y^2 + g(y)$ is less than 5. We show the initial foliation on the annulus P^2, the corresponding molecules $W(g)$, and the molecules $\widetilde{W}(g)$ obtained by factorization. The stars denote the fixed points of ξ.

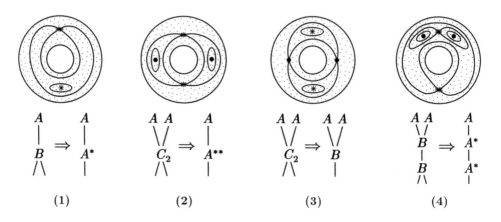

Figure 12.17

We have examined the Liouville foliations on Q_I and Q_{II}. Let us turn to the remaining pieces Q_{III} and Q_{IV}.

The Liouville foliations on Q_{III} and Q_{IV} are isomorphic to each other and have the direct product type $P^2 \times S^1$, where P^2 is an annulus with coordinates (x, p_x), and x is a periodic coordinate. The function $p_x^2 - f(x)$ defines a foliation on $P^2 \times S^1$. It is easy to see that ξ interchanges Q_{III} and Q_{IV}, and the mapping $\xi\colon Q_{III} \to Q_{IV}$ can be written in coordinates as follows:

$$\xi\colon (y, p_x, x) \to (-y, p_x, x + 1/2)\,.$$

Here $y = y \bmod L$ is a coordinate on the circle S^1. Since $f(x)$ has period $1/2$, the involution ξ preserves the foliation and, moreover, is a fiber diffeomorphism between Q_{III} and Q_{IV}. Hence two copies of $W(f)$ (corresponding to the pieces Q_{III} and Q_{IV}) should be identified under factorization and as a result turn into one molecule $W(f)$. Note that $W(f)$ is not arbitrary here, but has some additional symmetries, since f has period $1/2$, but not 1 as usual. Thus, the function $p_x^2 - f(x)$ on P^2 is invariant under the rotation through the angle π and, therefore, $W(f)$ is Z_2-symmetric.

Summarizing all this information about the action of ξ on $Q_I, Q_{II}, Q_{III}, Q_{IV}$, we see that, after factorization, the molecule W turns into the molecule $\widetilde{W}$ shown in Fig. 12.18. The action of ξ on the initial molecule W could be seen from Fig. 12.18. One needs to rotate W around the axis shown in Fig. 12.18 by a dotted line. Two graphs $W(f)$ are identified, and each of two graphs $W(g)$ remains on its place, but undergoes the factorization turning as a result into $\widetilde{W}(g)$.

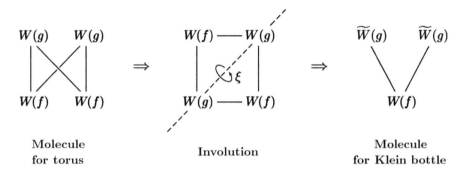

| Molecule for torus | Involution | Molecule for Klein bottle |

Figure 12.18

Thus, we have described the molecule $\widetilde{W}$ corresponding to a given quadratically integrable geodesic flow on the Klein bottle (both functions f and g differ from a constant). To compute the numerical marks r_i, ε_i, n_k, we can apply the same algorithm as in the case of T^2, by using admissible coordinate systems and gluing matrices. As a result, we obtain the following statement.

Theorem 12.3 (V. S. Matveev). *Consider the (L, f, g)-metric on the Klein bottle, where both f and g are not constant. Then the Liouville foliation of the geodesic flow on the isoenergy surface $Q = T^3$ is determined by the molecule $\widetilde{W}$ constructed above and shown in Fig. 12.18. The marks on the graph should be put in the following way.*

a) *On the edges of the graph $W(f)$, we put the same marks as in the case of the torus. Namely, on the edges between saddle atoms, the r-marks are equal to infinity; on the edges between saddle atoms and atoms of type A, the r-marks are equal to zero. On the two edges connecting the graph $W(f)$ with the two copies of the graph $\widetilde{W}(g)$, the r-marks are equal to zero.*

b) *In the graph $\widetilde{W}(g)$, the marks are as follows. On the edges between saddle atoms, the r-marks are equal to infinity. Between saddle atoms and atoms of type A, the r-marks are equal to zero, except for those cases where the atom A corresponds to a fixed point of the involution ξ (see Fig. 12.16(g)); in the latter case, the r-mark equals $1/2$.*

c) *All the marks ε in the molecule $\widetilde{W}$ are equal to $+1$.*

d) *There are exactly three families in the molecule $\widetilde{W}$. One of them is the graph $W(f)$, the other two families are the graphs $\widetilde{W}(g)$. On the family $W(f)$, the mark n is equal to zero. On each of the families $\widetilde{W}(g)$, the mark n equals -1.*

The proof follows from the reason that the Liouville foliation for the Klein bottle is obtained from the above described Liouville foliation for the torus by factorization with respect to the involution ξ. We have explicitly described the structure of this covering. To find the marks, one should again consider the admissible coordinate systems and compute the gluing matrices following the algorithm demonstrated above. We omit these technical details.   □

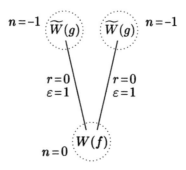

Figure 12.19

As a result, we obtain a complete Liouville classification of quadratically integrable geodesic flows on the Klein bottle (see Fig. 12.19).

12.2.2. *Linearly Integrable Geodesic Flows on the Klein Bottle*

In this case, the metric on the Klein bottle has the same type as above; the only difference is that here $f(x)$ is constant. Then its geodesic flow admits the linear integral $F = p_x$. By analogy with the previous case, the Liouville foliation on the isoenergy surface Q_K can be obtained by factorization of the Liouville foliation on Q_T with respect to the same involution ξ.

In this case, the Liouville foliation for the torus has the form illustrated in Fig. 12.20 by means of the corresponding molecule. The action of ξ is just the rotation of the molecule about the vertical axis through π. So, the algorithm that transforms the molecule W into $\widetilde{W}$ is the same as in the quadratic case. As a result, we obtain the molecule shown in Fig. 12.20.

Theorem 12.4 (V. S. Matveev). *Consider an (L, g)-metric $g(y)(dx^2 + dy^2)$ on the Klein bottle, where g is not a constant. Then the Liouville foliation of the geodesic flow on the isoenergy surface Q is determined by the molecule $\widetilde{W}$ described above and shown in Fig. 12.20. The marks on the graph should be put in the following way. On the edge connecting two copies of $\widetilde{W}(g)$, the r-mark is infinity. The mark ε on this edge equals -1. Inside the graphs $\widetilde{W}(g)$, the marks r are the same as in Theorem 12.3. There is only one family here, and the mark n corresponding to it is -2.*

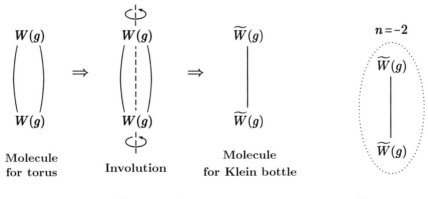

Figure 12.20

Figure 12.21

Thus, we have obtained a complete Liouville classification of linearly integrable geodesic flows on the Klein bottle (see Fig. 12.21).

12.2.3. Quasi-Linearly Integrable Geodesic Flows on the Klein Bottle

In this case, the Riemannian metric has the same form as in Section 12.2.1, but $g(y)$ is identically zero. Thus, we obtain the metric on the Klein bottle of the form $f(x)(dx^2 + dy^2)$, where $f(x)$ is a positive smooth periodic function on $[0,1]$ that admits an additional period $1/2$. The integral F is quadratic, namely, $F = p_y^2$ (recall that we cannot take just p_y, since this function is not uniquely defined on K^2).

Following the same procedure as before, we arrive at the classification of quasi-linearly integrable geodesic flows on K^2 up to the Liouville equivalence. The scheme for factorization of the molecule W is illustrated in Fig. 12.22. The involution ξ leaves two Liouville tori to be fixed. These tori are the connected components of the level $\{p_y = 0\}$. Each of them is invariant under the action of ξ. It turns out that these tori become Klein bottles after factorization. It follows from the fact that, under the natural projection $Q_T \to T^2$, each of these Liouville tori is diffeomorphically projected onto T^2. The involution ξ on the Liouville torus is just the pull-back of ξ on T^2. Since the quotient space T^2/ξ is homeomorphic to K^2, the same is true for these two Liouville tori. As a result, we obtain the molecule illustrated in Fig. 12.22: the molecule $W(f)$ is transformed into a molecule with two special atoms K related to critical Klein bottles. The structure of the subgraph $W(f)$ is not changed and remains the same as in Theorem 12.3.

Theorem 12.5 (V. S. Matveev). *Consider the $(L, f, 0)$-metric $f(x)(dx^2 + dy^2)$ on the Klein bottle, where $f \neq$ const. Then the Liouville foliation of its geodesic flow on the isoenergy surface $Q_K \simeq T^3$ is determined by the molecule $\widetilde{W}$ described above and shown in Fig. 12.22. The numerical marks on $\widetilde{W}$ should be put in the following way.*

a) *The marks on the graph* $W(f)$ *are the same as in Theorem 12.3. Namely, on the edges between saddle atoms, r-marks equal infinity; on the edges between saddle atoms and atoms* A, *the r-marks equal zero.*

b) *On the two edges connecting* $W(f)$ *with the atoms* K, *the r-marks equal zero.*

c) *All marks* ε *in the molecule* $\widetilde{W}$ *equal* $+1$.

d) *There are three families in the molecule* $\widetilde{W}$. *One of them is the graph* $W(f)$, *and the other two are the atoms* K. *On each family, the mark* n *is zero.*

The proof of the theorem is obtained by considering the double covering over the molecule $\widetilde{W}$. We omit the details.   □

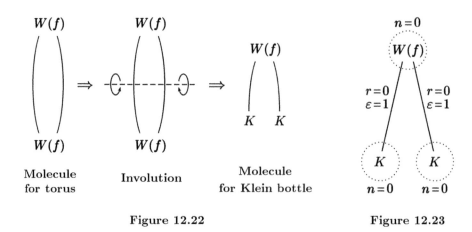

Figure 12.22   Figure 12.23

So, we obtain a complete Liouville classification of all quasi-linearly integrable geodesic flows on the Klein bottle (see Fig. 12.23).

12.2.4. *Quasi-Quadratically Integrable Geodesic Flows on the Klein Bottle*

We begin with recalling the canonical representation of the metrics on the Klein bottle whose geodesic flows are quasi-quadratically integrable (see Chapter 11).

Let a and b be natural relatively prime numbers, and let f be a non-constant positive function of one variable with period 1. Consider the standard Cartesian coordinates x, y on the plane and define a metric by

$$ds^2 = (f(x) + f(y + b/2))(dx^2 + dy^2).$$

Let Γ be an orthogonal lattice on the plane generated by the vectors $f_1 = (a, -a)$ and $f_2 = (b, b)$ (Fig. 12.24). It is obtained from a standard lattice via rotation through the angle $\pi/4$. Here by a standard lattice we mean an orthogonal lattice with basis vectors directed along the coordinate axes x and y. It is easy to see that the above metric is invariant with respect to translations by elements of Γ. Therefore, ds^2 determines some metric on the torus. By the way, it is a finite-sheeted Liouville metric (since Γ is not standard).

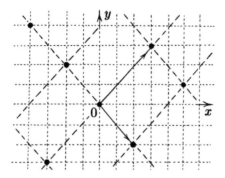

Figure 12.24

We now define the involution $\xi: (x, y) \rightarrow (y + b/2, x + b/2)$ on the torus. Factorizing the torus by the action of ξ, we obviously obtain the Klein bottle. It is easy to see that the metric on the torus is ξ-invariant and, therefore, defines some metric on the Klein bottle K^2, which was denoted above by $ds^2_{a,b,f}$.

Let us cover the Klein bottle by the torus T^2 and consider again the pull-back of $ds^2_{a,b,f}$ on T^2. We obtain a finite-sheeted Liouville metric on the torus whose molecule W was already described. Following the above scheme, we observe that ξ generates an involution of the Liouville foliation on Q_T. Therefore, factorizing Q_T by the action of ξ, we see that the molecule W is also factorized. This process is illustrated in Fig. 12.25. The molecule W is bent in the middle, and, as a result, two copies of $W(f)$ are identified with two other copies of $W(f)$; and, in the middle of the two vertical edges, two new atoms K (critical Klein bottles) appear. The new molecule $\widetilde{W}$ obtained is presented in Fig. 12.25.

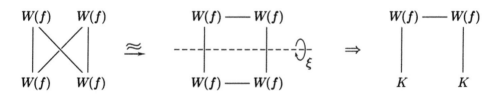

Figure 12.25

Denote by $k = k(a, b)$ the integer that is defined by the following rule. The number k belongs to the interval $(0, 2ab)$, $k - 1$ is divisible by $2a$, and $k + 1$ is divisible by $2b$. It is easy to verify that such a number k exists and is uniquely defined.

Theorem 12.6 (V. S. Matveev). *Let $ds^2_{a,b,f}$ be a Riemannian metric on the Klein bottle whose geodesic flow is quasi-quadratically integrable. Then the Liouville foliation of its geodesic flow on the isoenergy surface Q is defined by the molecule $\widetilde{W}$ described above and shown in Fig. 12.26. Moreover, the marks on the graph $\widetilde{W}$ should be put as follows.*

a) *The marks on the graph $W(f)$ are the same as in Theorem 12.3. Namely, on the edges between saddle atoms, the r-marks are equal to infinity, and on the edges between saddle atoms and those of type A, the r-marks are equal to zero.*

b) *$r = -k/(2ab) \bmod 1$ on the edge connecting the two copies of the graph $W(f)$.*

c) *$r = -a/b \bmod 1$ on the two edges connecting the graph $W(f)$ with the two copies of the K.*

d) *All the marks ε in the molecule $\widetilde{W}$ are equal to $+1$.*

e) *There are exactly four families in the molecule $\widetilde{W}$. Two of them are the graphs $W(f)$, the other two families are the atoms K. On each of the families $W(f)$, the mark n is equal to zero. On each of the families K, the mark n is equal to $[-a/b]$.*

The proof is similar to that of Theorems 12.3, 12.4, and 12.5. We omit the details.　□

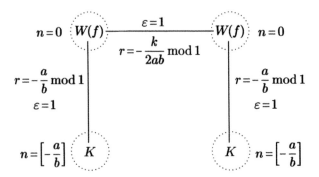

Figure 12.26

Thus, we obtain a complete Liouville classification of all quasi-quadratically integrable geodesic flows on the Klein bottle (see Fig. 12.26).

12.3. THE SPHERE

12.3.1. *Quadratically Integrable Geodesic Flows on the Sphere*

Recall that every metric on the sphere whose geodesic flow is quadratically integrable can be obtained from an appropriate degenerate metric on the torus by factorization with respect to some involution σ. The metric on the torus has the Liouville form:

$$ds^2 = (f(x) + g(y))(dx^2 + dy^2),$$

where f and g are smooth non-negative even functions with periods 1 and L respectively. Hence, by the way, $f(x)$ is symmetric with respect to $x = 1/2$, and $g(y)$ is symmetric with respect to $y = L/2$. These functions vanish at the points $k/2$ and $Lk/2$ respectively, where $k \in \mathbb{Z}$. The torus is considered here as the quotient space $\mathbb{R}^2/\Gamma$, where the lattice is generated by the vectors $(1,0)$ and $(0,L)$. The involution σ that transforms the torus to the sphere is given on the plane $\mathbb{R}^2$ by

$$\sigma(x,y) = (-x,-y).$$

The projection $\varkappa:T^2 \to T^2/\sigma = S^2$ can be considered as the Weierstrass function $\mathcal{P}(z)$, where z is the complex coordinate $z = x + iy$ on the plane. This mapping is a two-sheeted branching covering with four branch points. Since the metric on the torus is invariant under the involution σ, its "projection" yields a metric on the sphere S^2. This metric will be smooth provided the additional condition (c) is fulfilled (see Section 11.5.3).

Our goal is to describe the topology of the Liouville foliation associated with the geodesic flow of this metric. We will use the same method that we applied when examining integrable geodesic flows on the Klein bottle. In other words, we will examine the topology of the Liouville foliation "above" on the torus and then look what happens to it under the projection "down" onto the sphere.

The integral of the geodesic flow on the covering torus has the form

$$F(x,y,p_x,p_y) = \frac{g(y)p_x^2 - f(x)p_y^2}{f(x) + g(y)}.$$

This integral has singularities at the four branch points of $\varkappa$ on the torus. On the sphere, the integral of the geodesic flow will be actually the same. One only needs to project the indicated function down by means of $\varkappa$. All the singularities of the integral F disappear, and F becomes a smooth function, which we denote by $\widetilde{F}$. The Hamiltonian of the geodesic flow on the torus has the form

$$H(x,y,p_x,p_y) = \frac{p_x^2 + p_y^2}{f(x) + g(y)}.$$

The corresponding Hamiltonian $\widetilde{H}$ is also a smooth function on the cotangent bundle of S^2. Recall that we always assume F to be a Bott function.

Proposition 12.2. *The integral $\widetilde{F}$ is a Bott function on the isoenergy surface $\widetilde{Q} = \{\widetilde{H} = \text{const}\}$ if and only if $f(x)$ and $g(y)$ are both Morse functions.*

This statement follows easily from its analog in the case of the torus (Proposition 12.1). $\square$

Note that the isoenergy 3-surface $Q \subset T^*T^2$ is not compact, but goes to infinity because of the degeneracy of ds^2 at the four branch points on the torus. Consider a sufficiently small positive number δ and the two following subsets in Q:

$$Q_{+\delta} = \{F \geq +\delta,\ H = 1\}, \qquad Q_{-\delta} = \{F \leq -\delta,\ H = 1\}.$$

Consider the analogous subsets for the 2-sphere, that is, subsets in T^*S^2 given by

$$\widetilde{Q}_{+\delta} = \{\widetilde{F} \geq +\delta,\ \widetilde{H} = 1\}, \qquad \widetilde{Q}_{-\delta} = \{\widetilde{F} \leq -\delta,\ \widetilde{H} = 1\}.$$

It is clear that $Q_{\pm\delta}$ covers $\widetilde{Q}_{\pm\delta}$ in a two-sheeted way under the projection $\varkappa$. Moreover, the Liouville foliation on $Q_{\pm\delta}$ is invariant with respect to the involution σ; therefore, $\varkappa$ maps fiberwise the Liouville foliation on $Q_{\pm\delta}$ to that on $\widetilde{Q}_{\pm\delta}$. Recall that $Q_{\pm\delta}$ was studied above. In particular, we have described the structure of the Liouville foliation on $Q_{\pm\delta}$. Each of the pieces $Q_{+\delta}$ and $Q_{-\delta}$ consists of four connected components. The Liouville foliation on each component is just the direct

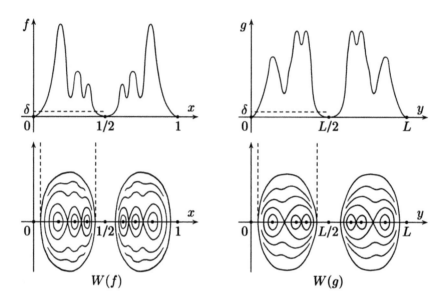

Figure 12.27

product of a foliated 2-disc and a circle. In Fig. 12.27, for each of the functions f and g, we have drawn two discs, since the functions are symmetric with respect to $x = 1/2$ and $y = L/2$. Each of these 2-discs is foliated into level lines of its own function, namely, $p_x^2 - f(x)$ and $p_y^2 - g(y)$. By $P_{+\delta}$ and $P_{-\delta}$ we denote these pairs of discs. It is convenient to consider them as subsets in the plane (y, p_y) and in the plane (x, p_x), respectively; namely,

$$P_{+\delta} = \{g(y) - p_y^2 \geq \delta,\, y \in [0, L]\},$$
$$P_{-\delta} = \{f(x) - p_x^2 \geq \delta,\, x \in [0, 1]\}.$$

Denote the molecules related to the obtained foliations on 2-discs by $W(f)$ and $W(g)$ respectively. It is clear that the corresponding Liouville foliation on $Q_{\pm\delta}$ are described by the same molecules $W(f)$ and $W(g)$. It follows from the direct product structure that

$Q_{+\delta} = \{$(two components of $P_{+\delta}$) $\times\, S^1$, where $p_x > 0\}$

$\cup\, \{$(two components of $P_{+\delta}$) $\times\, S^1$, where $p_x < 0\}$,

$Q_{-\delta} = \{$(two components of $P_{-\delta}$) $\times\, S^1$, where $p_y > 0\}$

$\cup\, \{$(two components of $P_{-\delta}$) $\times\, S^1$, where $p_y < 0\}$.

The involution σ acts on each of the 3-manifolds $Q_{\pm\delta}$ as follows. In the case of $Q_{+\delta}$, two components of $P_{+\delta} \times S^1$, where $p_x > 0$, are interchanged with two components of $P_{+\delta} \times S^1$, where $p_x < 0$. Analogously, in the case $Q_{-\delta}$, two components of $P_{-\delta} \times S^1$, where $p_y > 0$ are interchanged with two components of $\{P_{-\delta} \times S^1$, where $p_y < 0\}$. Therefore, from the four copies of $W(f)$, we obtain two copies of $W(f)$. Similarly, from four copies of $W(g)$, we obtain two copies

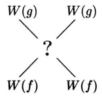

Figure 12.28

of $W(g)$ (see Fig. 12.28). It remains to find out what atom (?) should be put at the center of the molecule in Fig. 12.28. This atom lies in the 3-manifold Q on the level $\{\widetilde{F} = 0\}$. Let us describe the properties of this 3-atom.

1) The atom (?) is connected. Indeed, the leaf $\{F = 0, H = 1\}$ of the geodesic flow of the degenerate metric on the torus is connected.

2) The atom (?) has complexity 2, i.e., contains exactly two critical circles of the integral F. Indeed, in Fig. 12.29, one can see the torus with 16 geodesics of the degenerate metric lifted from the sphere. Eight of them can be obtained

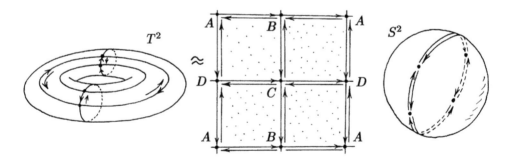

Figure 12.29

from the other eight ones by reversing the orientation. Therefore, there are only 8 geometrically different geodesics. After projection onto the sphere, they become segments of the closed geodesic on the sphere, which pass through four branch points. Taking this curve with two different orientations, we obtain two different geodesics lying on the critical leaf $\{\widetilde{F} = 0\}$. They are the critical circles of the atom (?).

3) The atom (?) has exactly four ends (two upper and two lower). This means that, under this bifurcation, two Liouville tori are transformed into two.

4) The atom (?) is symmetric with respect to the involution

$$(x, y, p_x, p_y) \rightarrow (x, y, -p_x, -p_y) \,.$$

The point is that integral $F(x, y, p_x, p_y)$ is obviously invariant under this involution (see the formula for F). In particular, the singular leaf (i.e., the atom (?)) goes into itself. Moreover, this symmetry is not trivial in the sense that it interchanges the two critical circles of (?).

According to the classification theorem for atoms of lower complexity (see Chapter 3), there exist the only atom that satisfies all these conditions. This is the atom C_2.

Thus, we have proved the following statement.

Proposition 12.3. *The molecule W related to a quadratically integrable geodesic flow on the sphere has the form shown in Fig. 12.30.*

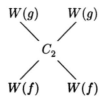

Figure 12.30

It remains to compute the numerical marks on W and thus obtain a complete Liouville classification of such flows. Consider an arbitrary Liouville torus and two basis cycles λ and μ on it given by the equations $\lambda = \{x = \text{const}\}$ and $\mu = \{y = \text{const}\}$ respectively. Fix an orientation on them in such a way that the integrals of the action form $\alpha = p_x \, dx + p_y \, dy$ along λ and μ are positive.

Let a Liouville torus be given by the equations $F = \text{const}$, $H = 1$; consider its projection on the base T^2. We obtain an annulus on T^2 given by one of the following two conditions:

$$y \in [y_0, y_1], \quad x \text{ is arbitrary};$$

or

$$x \in [x_0, x_1], \quad y \text{ is arbitrary}.$$

In other words, the projection of each Liouville torus is an annulus shown in Fig. 12.31 as a rectangle.

Then the first integral has the form

$$\int_\lambda \alpha = 2 \int \sqrt{g(y) + F_0} \, dy \,,$$

where the latter integral is taken over $[0, L/2]$ if $F_0 > 0$, and is taken over the admissible segment $[y_0, y_1]$ related to the projection of the Liouville torus on T^2

(see Fig. 12.31) if $F_0 < 0$. Here y_0 and y_1 are roots of the equation $g(y) + F_0 = 0$; and $g(y) + F_0 > 0$ on the interval (y_0, y_1).

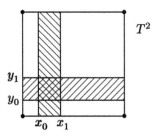

Figure 12.31

The second integral is

$$\int_\mu \alpha = 2 \int \sqrt{f(x) - F_0} \, dx \,,$$

where the latter integral is taken over $[0, 1/2]$ if $F_0 < 0$, and is taken over the segment $[x_0, x_1]$ if $F_0 > 0$. The numbers x_0 and x_1 are roots of the equation $f(x) - F_0 = 0$; and $f(x) - F_0 > 0$ on the interval (x_0, x_1).

Thus, the orientation of cycles λ and μ is fixed. Note that λ and μ give us an admissible coordinate system on each of the trees $W(g)$ and $W(f)$ (see Fig. 12.30). The Liouville foliation on each of these pieces has the direct product type $P^2 \times S^1$ (see above). Consider two upper pieces, that is, two copies of $W(g)$. Here μ is an oriented fiber of this direct product and λ is its section. In the two lower pieces $W(f)$, the situation is opposite. Here μ is a section of the direct product and λ is a fiber. Hence, on the interior edges of $W(g)$ and $W(f)$, the gluing matrices are very simple. Between two saddle atoms, they are

$$\begin{pmatrix} 1 & 0 \\ 0 & -1 \end{pmatrix}.$$

Between saddle atoms and atoms A, the gluing matrices look like

$$\begin{pmatrix} 0 & 1 \\ 1 & 0 \end{pmatrix}.$$

In the case of the torus, we observed the same situation. In particular, the numerical marks on $W(g)$ and $W(f)$ in the case of the sphere are the same as in the case of the torus.

To compute the numerical marks on the four central edges of W that are incident to the central atom C_2, we construct an admissible coordinate system on the boundary Liouville tori of C_2. The first cycle $\varkappa$ of this system should be isotopic to the critical circle of the atom C_2. It can be expressed from the basis cycles λ and μ on a neighboring Liouville torus.

Lemma 12.1. $\varkappa = \lambda + \mu$.

Proof. It is clear that we only need to verify the following relation:

$$\int_{\varkappa} \alpha = \int_{\lambda} \alpha + \int_{\mu} \alpha .$$

It will immediately follow from this that $\varkappa = \lambda + \mu$, since the integrals in the right-hand side are, in general, independent over $\mathbb{Z}$. The desired equation $\int_{\varkappa} \alpha = \int_{\lambda} \alpha + \int_{\mu} \alpha$ follows from Fig. 12.32. Indeed, the critical circle-fiber $\varkappa$ corresponds to a geodesic on the sphere that belongs to the singular leaf and pass through the four branch points A, B, C, D on the sphere. The cycle $\varkappa$ is

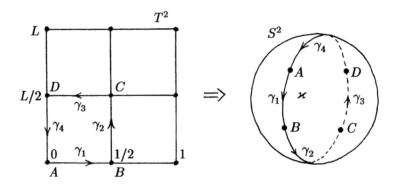

Figure 12.32

the image (under the projection $T^2 \to S^2$) of four geodesic segments γ_1, γ_2, γ_3, γ_4 located on the torus as shown in Fig. 12.32. These segments are projected onto four geodesic segments connecting the branch points A, B, C, D and forming the closed geodesic $\varkappa$. Therefore,

$$\int_{\varkappa} \alpha = \left(\int_{\gamma_1} \alpha + \int_{\gamma_3} \alpha \right) + \left(\int_{\gamma_2} \alpha + \int_{\gamma_4} \alpha \right) = 2\int_0^{1/2} \sqrt{f(x)}\, dx + 2\int_0^{L/2} \sqrt{g(y)}\, dy = \lim_{F_0 \to 0} \int_{\lambda+\mu} \alpha$$

Thus, $\int_{\varkappa} \alpha = \int_{\lambda} \alpha + \int_{\mu} \alpha$, as was to be proved. $\square$

Now we need to indicate the second basis cycle of the admissible coordinate system. Take the cycle $\lambda = \{x = \text{const}\}$ which lies on a Liouville torus close to C_2. Being projected down onto the sphere, it will look like a circle shown in Fig. 12.33. On the covering torus, this circle corresponds to a vertical segment (actually, a cycle) given by $x = $ const (Fig. 12.33). On the isoenergy 3-manifold $\tilde{Q} \simeq \mathbb{R}P^3$, this circle corresponds to a two-dimensional section of the 3-atom C_2 transversal to the geodesic flow. Denote this section by P_{tr}. The transversality is easily seen from Fig. 12.33, since all the geodesics starting from the four branched points A, B, C, D meet this circle $\{x = \text{const}\}$ transversely. Therefore, in Q^3, the geodesic flow is also transversal to the corresponding 2-section. Note that the circle $\{x = \text{const}\}$ defines "above" two connected transversal sections (each of them is homeomorphic to the 2-atom C_2). The corresponding picture on the covering torus is also shown in Fig. 12.33.

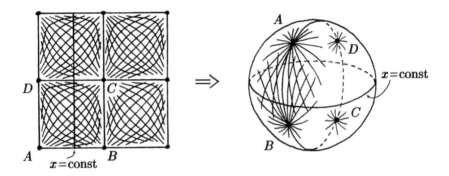

Figure 12.33

Thus, we have constructed a transversal two-dimensional section in Q^3 whose boundary circles coincide with the cycles λ. It remains to look after the correct orientation of these cycles (from the point of view of the admissible coordinate system).

Take y and p_y as regular coordinates on P_{tr}. Then the section P_{tr} can be defined as follows:

$$P_{\text{tr}} = \{|F| \leq \varepsilon\}.$$

Since the integral F has the form $F = g(y) - p_y^2$ in coordinates (y, p_y), the indicated inequality defines the domain on the plane (y, p_y) shown in Fig. 12.34. By the way, it is seen that this domain is homeomorphic to the 2-atom C_2. Consider the four boundary circles of this section that geometrically coincide with the cycles λ. The orientation on λ is defined by the condition that the integral of α along them is positive. In coordinates (y, p_y) on P_{tr}, this integral becomes $\int p_y \, dy$. It is seen from Fig. 12.34 that, in order for this integral to be positive, it is necessary and sufficient that the cycles λ are oriented as shown in Fig. 12.34(a).

We now produce the desired basis cycles of the admissible coordinate system from the cycles λ by defining the right orientation on them (namely, the orientation induced by that of P_{tr}; see Fig. 12.34(b)). Comparing the right orientation with the initial orientation on λ, we see that we need to change the orientation in two cases, and to preserve it in the other two. Namely, as the second basis cycle, we take $-\lambda$ on the two edges incident to the subgraphs $W(f)$, and λ itself on the edges incident to $W(g)$.

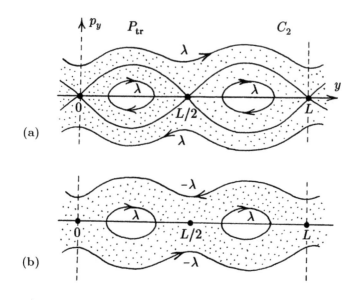

Figure 12.34

Thus, we have completely described the admissible coordinate systems on Liouville tori related to the edges incident to C_2.

The calculation of gluing matrices and all numerical marks is presented in Fig. 12.35. Note that here we have to distinguish three different cases depending on the structure of $W(f)$ and $W(g)$. In other words, we have to consider separately the case where one or both of these subgraphs consist of the only atom A. The point is that the definitions of admissible coordinate systems are different for saddle atoms and atoms of type A (see Chapter 4). According to this, the molecule can contain either one, or three, or five families.

The final result can be formulated as follows.

Theorem 12.7 (Nguyen Tien Zung, L. S. Polyakova, V. S. Matveev). *Consider the (L, f, g)-metric on the 2-sphere. Then the Liouville foliation of its geodesic flow on the isoenergy surface $Q = \mathbb{R}P^3$ is determined by the molecule W described above. The marks on the graph W are presented in Fig. 12.35. The marks inside each of the trees $W(f)$ and $W(g)$ are the same as in Theorem 12.3, that is, the marks r between saddle atoms are equal to ∞, and those between saddle atoms and atoms of type A are equal to zero. All the marks ε are equal to $+1$ (inside $W(f)$ and $W(g)$).*

Thus, we have obtained a complete Liouville classification of all quadratically integrable (L, f, g)-metrics on the sphere.

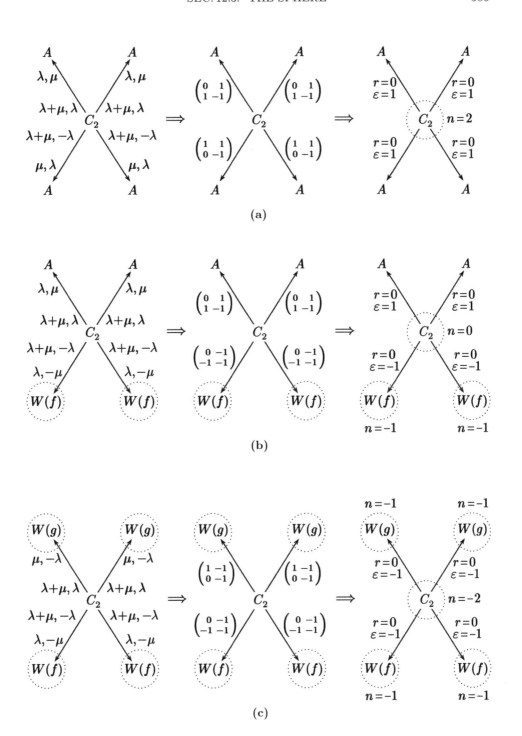

Figure 12.35

12.3.2. *Linearly Integrable Geodesic Flows on the Sphere*

Recall that a geodesic flow on the sphere is linearly integrable if and only if there exist global conformal coordinates x, y on the sphere in which the metric takes the form

$$ds^2 = f(x^2 + y^2)(dx^2 + dy^2),$$

where $f(t)$ is a positive smooth function on the semi-axis $[0, +\infty)$ such that $\dfrac{f(1/t)}{t^2}$ is also smooth on $[0, +\infty)$.

It will be convenient for us to reformulate this statement as follows.

Theorem 12.8.   *The geodesic flow of a Riemannian metric on the sphere is linearly integrable if and only if there exist smooth global coordinates (τ, φ) on the sphere with two singular points (similar to the standard spherical coordinates (θ, φ)), where τ varies from 0 to some τ_0, and φ is a periodic coordinate defined modulo 2π, such that the metric in these coordinates has the form*

$$ds^2 = d\tau^2 + f(\tau)\, d\varphi^2.$$

REMARK.   The condition that the metric is smooth means automatically that $f(\tau)$ smoothly depends on τ on the closed interval $[0, \tau_0]$, is positive inside the interval $(0, \tau_0)$, and vanishes at its ends. Moreover, in neighborhoods of 0 and τ_0, the function $f(\tau)$ can be represented as $f(\tau) = \widetilde{h}(\tau^2)$ and $f(\tau) = \widetilde{g}((\tau - \tau_0)^2)$ respectively, where $\widetilde{h}$ and $\widetilde{g}$ are smooth functions.

It should be noted that the described metrics naturally generalize the usual metrics on the spheres of revolution in $\mathbb{R}^3$. In this case, φ is the usual angle that defines the rotation about the axis of revolution (that is, the standard angle on parallels), and s is the arclength parameter on meridians. The singular points of such a coordinate system are the poles of the sphere. In this case, $f(\tau)$ yields the square of the distance between a point and the axis of revolution (Fig. 12.36).

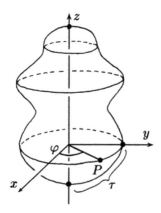

Figure 12.36

However, we repeat once more that not every metric with the linearly integrable geodesic flow on the sphere can be realized as the metric on a surface of revolution.

Note that the Hamiltonian H of the geodesic flow of the above metric has the form

$$H = p_\tau^2 + f(\tau)^{-1} p_\varphi^2 ,$$

and the integral F is

$$F = p_\varphi .$$

As before, we assign to f the graph $W(f)$ that describes the foliation of a two-dimensional disc into level lines of the function $f(\tau) - p_\tau^2$. This disc itself is given by the inequality $f(\tau) - p_\tau^2 \geq 0$ on the plane (τ, p_τ) (see Fig. 12.27).

Theorem 12.9 (Nguyen Tien Zung, L. S. Polyakova). *Consider the Riemannian metric*

$$ds^2 = d\tau^2 + f(\tau) \, d\varphi^2$$

on the sphere.

a) *The molecule W corresponding to its geodesic flow has the form shown in Fig. 12.37, where the graph $W(f)$ is constructed from the function $f(\tau)$ as described above. The numerical marks inside each of the two molecules $W(f)$ are standard, that is, on the edges between saddle atoms, the r-marks are equal to ∞, and on the edges between saddle atoms and atoms of type A, the r-marks are equal to zero; all the ε-marks are $+1$.*

b) *Suppose the molecule $W(f)$ is different from the atom A (i.e., contains at least one saddle atom). Then, on the single central edge connecting two copies of the molecule $W(f)$, the mark r is equal to ∞, and $\varepsilon = -1$. Here there is one family only (it is the molecule W from which all the end-atoms A are removed). The mark n on this family is equal to 2.*

c) *If the molecule $W(f)$ is reduced to a single atom A, then the whole molecule W has the simplest form A——A. In this case, we have $r = 1/2$ and $\varepsilon = 1$. There are no families here.*

Thus, we have obtained a complete Liouville classification of all linearly integrable metrics on the two-dimensional sphere.

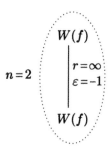

Figure 12.37

Proof. As we already did several times, we cut the isoenergy manifold Q into several pieces by the level surface $\{F = 0\}$, where F is an additional integral.

In our case, F has a very simple form $F = p_\varphi$. Here the level surface $\{p_\varphi = 0\}$ is a regular Liouville torus consisting of all meridians given by $\varphi = \text{const}$ (since $p_\varphi = 0$ is equivalent to $\varphi = \text{const}$). We obtain the family of all meridians passing through the north and south poles of the sphere. The union of them is just a Liouville torus in Q.

Thus, after being cut along $\{p_\varphi = 0\}$, the isoenergy surface Q is divided into two pieces Q_+ and Q_-. Both of them are of direct product type. This can easily be seen by choosing an appropriate global transversal section P_{tr}. In our case, P_{tr} can be given by the equation $\varphi = \text{const}$. Let us examine the structure of F restricted to P_{tr}. As local coordinates on P_{tr}, we consider (τ, p_τ). It is easy to see that, if $H = p_\tau^2 + f(\tau)^{-1} p_\varphi^2 = 1$, then

$$p_\varphi = \pm((1 - p_\tau^2) f(\tau))^{1/2} .$$

Here we take the sign "+" for the piece Q_+, and "−" for Q_-. The fibers of the Seifert fibration are given here as follows:

$$\{\tau = \text{const}, \ p_\varphi = \text{const}, \ p_\tau = \text{const}, \ \varphi = \pm t\},$$

where t is a parameter on the fiber-circle. Therefore, the Liouville foliation is the direct product of the circle (a fiber of the Seifert fibration) and the one-dimensional foliation on P_{tr} given by level lines of $p_\varphi = \pm((1 - p_\tau^2) f(\tau))^{1/2}$. Note that the exponent $1/2$ can be removed, since it does not affect the topology of the foliation. By making the transformation $\tilde{p}_\tau = p_\tau (f(\tau))^{1/2}$, we obtain that the foliation on the disc with coordinates $\tau, \tilde{p}_\tau$ can be defined by level lines of the following function:

$$f(\tau) - \tilde{p}_\tau^2 .$$

But the topology of this foliation is described just by the molecule $W(f)$ that was defined above. Thus, the whole molecule W has the required form, namely, it is obtained by gluing two copies of $W(f)$.

The numerical marks inside the graphs $W(f)$ correspond to the direct product topology, and consequently, have just the same values as in Theorem 12.9.

It remains to find the gluing matrix on the central edge e_0 that connects two copies of $W(f)$. Consider the zero level surface of F in $Q^3 = \mathbb{R}P^3$. It is clear that this is a Liouville torus T_0 that corresponds to the middle of e_0. Here Q_+ corresponds to the upper subgraph $W(f)$, and Q_- corresponds to the lower one. Consider two Liouville tori $T_{+\varepsilon} = \{F = +\varepsilon\}$ and $T_{-\varepsilon} = \{F = -\varepsilon\}$ close to T_0. Projecting each of them down onto the sphere S^2 (using the natural projection $T^* S^2 \to S^2$), we obtain the result shown in Fig. 12.38. The torus $T_{+\varepsilon}$ is projected onto an annulus that covers the whole sphere except for two small discs around the poles, so is $T_{-\varepsilon}$. The projection of T_0 is the whole sphere S^2.

Suppose first that $W(f)$ contains at least one saddle atom. Let us draw the projections of basis cycles λ^+, μ^+ and λ^-, μ^- on this annulus. The cycles

λ^+ and λ^- can be thought of as equators of the sphere (Fig. 12.38). The cycles μ^+ and μ^- are projected onto vertical arcs of meridians, which should be counted twice (up and down). Consider the cycle γ on the central torus T_0 that is projected onto a meridian on the sphere passing through the south and north poles. Let us push γ isotopically to the Liouville tori $T_{\pm\varepsilon}$. Then its projection takes the form

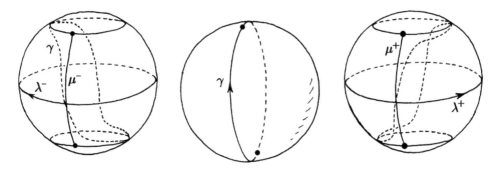

Figure 12.38

shown in Fig. 12.38. On $T_{-\varepsilon}$, the perturbed cycle γ moves along the segment μ^+, but near the north and south poles, it turns and goes around them. The same events happen on the torus $T_{+\varepsilon}$, but γ goes around the poles from the other side. Analytically, this means that

$$\gamma = \mu^- + \lambda^- , \qquad \gamma = \mu^+ + \lambda^+ .$$

Therefore, the basis cycles $\mu^-, \lambda^-, \mu^+, \lambda^+$ are connected by the relation $\mu^- + \lambda^- = \mu^+ + \lambda^+$. Moreover, as is seen from Fig. 12.38, one more relation holds, i.e., $\lambda^+ = -\lambda^-$. Hence the gluing matrix on e_0 is

$$\begin{pmatrix} \lambda^+ \\ \mu^+ \end{pmatrix} = \begin{pmatrix} -1 & 0 \\ 2 & 1 \end{pmatrix} \begin{pmatrix} \lambda^- \\ \mu^- \end{pmatrix} .$$

It follows from this that, if $W(f)$ contains at least one saddle atom, then the numerical marks on the central edge of W have the form indicated in the theorem. Namely, $r = \infty$, $\varepsilon = -1$, and $n = 2$.

Let W be of the form $A\!-\!\!-\!A$. Then, according to the definition of admissible coordinate system (for the case of the atom A), we need only to interchange λ and μ. The gluing matrix becomes

$$\begin{pmatrix} \mu^+ \\ \lambda^+ \end{pmatrix} = \begin{pmatrix} 1 & 2 \\ 0 & -1 \end{pmatrix} \begin{pmatrix} \mu^- \\ \lambda^- \end{pmatrix} .$$

Therefore, $r = 1/2$ and $\varepsilon = 1$. The theorem is proved. □

REMARK. It is useful to give another proof by using the fact that the gluing of two pieces Q_+ and Q_- along the central torus T_0 leads us to the 3-manifold $\mathbb{R}P^3$. We see from the above discussion that $\lambda^+ = -\lambda^-$. Therefore, the gluing matrix for two solid tori Q_+ and Q_- should have the following form:

$$\begin{pmatrix} -1 & 0 \\ k & 1 \end{pmatrix},$$

where k is some integer number. But we know that k completely determines the fundamental group of the 3-manifold Q glued from two solid tori, and vice versa. Namely, $\pi_1(Q) = Z_k$. Since $Q = \mathbb{R}P^3$ in our case, we have $\pi_1(Q) = Z_2$ and, consequently, $k = 2$ (up to a sign).

Consider an important particular case of the metric on a surface of revolution in $\mathbb{R}^3$ (homeomorphic to the sphere). As we know, the geodesic flows of such metrics are linearly integrable. Let us compute the marked molecule in the case of a surface of revolution that is obtained by rotating the graph of a function $y = f(x)$ (see Fig. 12.39).

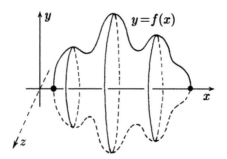

Figure 12.39

Corollary. *The marked molecule of the geodesic flow on the surface of revolution corresponding to the function $f(x)$ coincides with the molecule described in Theorem 12.9 and shown in Fig. 12.37.*

Proof. Note that the meaning of the function f in Theorem 12.9 and in this corollary is different. Here f defines the meridian of the surface of revolution, whereas in Theorem 12.9, f is a parameter of a Riemannian metric. If we reduce the metric on the surface of revolution to the form indicated in Theorem 12.9, i.e.,

$$ds^2 = d\tau^2 + \tilde{f}(\tau)\, d\varphi^2 ,$$

then f and $\tilde{f}$ will be, in general, different. We need to verify that the molecules $W(f)$ and $W(\tilde{f})$ related to f and $\tilde{f}$ coincide. However, it is easy to see that the functions $f(x)$ and $\tilde{f}(\tau)$ are connected by a monotone change of the parameter. More precisely, we have $\tilde{f}(\tau(x)) = f(x)$, where $\tau(x)$ is the arclength parameter on the graph of $f(x)$. Hence the molecules coincide, as was claimed. $\square$

We said above that not every Riemannian metric with the linearly integrable geodesic flow could be realized as the induced metric on an appropriate surface of revolution (sphere) in $\mathbb{R}^3$. Nevertheless, the following interesting statement holds.

Corollary. *Let ds^2 be a smooth metric on the two-dimensional sphere with the linearly integrable geodesic flow. Then there exists a surface of revolution in $\mathbb{R}^3$ such that the geodesic flow on it is Liouville equivalent to the geodesic flow of ds^2.*

Proof. Consider an arbitrary metric $ds^2 = d\tau^2 + f(\tau) \, d\varphi^2$ on the sphere with the linearly integrable geodesic flow. Construct the surface of revolution in R^3 by taking the graph of $f(\tau)$ as its generatrix. Here, of course, we should smooth out the function $f(\tau)$ at the poles of the sphere, i.e., at the ends of the segment $[0, \tau_0]$ (Fig. 12.40). The point is that the graph of $f(\tau)$ is tangent to the axis τ at the ends of the segment. As a result, two singular points occur on the surface of revolution under consideration. However, this problem

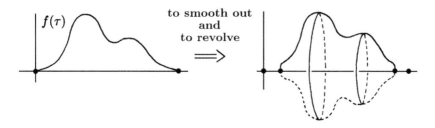

Figure 12.40

can be avoided as shown in Fig. 12.40. Since the bifurcations of the Liouville foliation happen far from the ends of the segment, such a smoothing operation has no influence on the Liouville foliation type of the metric obtained. The previous corollary implies that the geodesic flow of this metric (of revolution) is Liouville equivalent to the initial one. $\square$

12.4. THE PROJECTIVE PLANE

12.4.1. *Quadratically Integrable Geodesic Flows on the Projective Plane*

Recall the description of the metrics on the projective plane with quadratically integrable geodesic flows. By lifting such a metric to the sphere, we obviously obtain a metric whose geodesic flow is quadratically integrable. To describe such metrics on the sphere S^2, we represent S^2 as the quotient space T^2/σ and introduce the class of the following metrics on the torus:

$$ds^2 = (f(x) + g(y))(dx^2 + dy^2),$$

where f and g satisfies the conditions (a), (b), (c) described above in Section 11.5.3. Now we represent $\mathbb{R}P^2$ as the quotient space of the sphere by the involution ξ

defined as follows:

$$\xi((x,y),(-x,-y)) = ((-x+1/2, y+L/2),(x-1/2, -y-L/2)).$$

It is easy to verify that ξ has no fixed points on the sphere and can be lifted up to an involution on the covering torus $\xi'\colon (x,y) \to (1/2 - x, L/2 + y)$. The involutions σ and ξ obviously commute.

Now choose those metrics from the set of all (L, f, g)-metrics which are ξ-invariant. Consider f and g such that

$$f(x) + g(y) = f(1/2 - x) + g(L/2 + y).$$

Clearly, $f(x) = f(1/2 - x)$ and $g(y) = g(L/2 + y)$. Taking into account the evenness of f, we obtain $f(x) = f(x - 1/2)$, i.e., f is just periodic with period $1/2$. Analogously, g is periodic with period $L/2$.

We have proved above that any metric on RP^2 with a quadratically integrable geodesic flow can be obtained by this procedure. So, it is given by the following three parameters: L, f, and g. We now want to produce the molecule for the metric on the projective plane from the molecule for the corresponding metric on the sphere. To do this, we need to understand how ξ acts on the Liouville foliation in the isoenergy manifold $Q_{S^2} = \mathbb{R}P^3$.

Note first that ξ has no fixed points on $Q_{S^2} = \mathbb{R}P^3$. This property implies the following statement.

Proposition 12.4.   *The isoenergy manifold $Q_{\mathbb{R}P^2}$ of the geodesic flow on the projective plane $\mathbb{R}P^2$ is diffeomorphic to the lens space $L_{4,1}$.*

Since the involution ξ acts on $\mathbb{R}P^3$, it also acts on the molecule W corresponding to the case of the sphere (it is shown in Fig. 12.30). It is easy to verify that ξ maps the central atom C_2 and the edges incident to it into themselves. The involution on the 3-atom C_2 (represented as the direct product of the 2-atom C_2 by the circle) is just the translation along the circle by the angle π. On the pieces $W(f)$ and $W(g)$, the action of the involution is described as follows. Note that $f(x)$ satisfies an additional symmetry condition on the closed interval $[0, 1/2]$. It is symmetric with respect to $x = 1/4$. It follows from the fact that f is even and has the period $1/2$. Hence the function $f(x) - p_x^2$ is invariant with respect to the central symmetry of the disc given by the inequality $f(x) - p_x^2 > 0$. The center of symmetry is $(1/4, 0)$.

The Liouville foliation can be represented as the direct product of the one-dimensional foliation on the disc by the circle. The involution acts on this direct product in the following way. It rotates the S^1-fiber through π, and it is the central symmetry on the disc-base with the center $(1/4, 0)$. By analogy with the case of the Klein bottle, we denote by $\widetilde{W}(f)$ the molecule that corresponds to the Liouville foliation obtained from the direct product by factorization with respect to ξ. Here we will distinguish two cases.

The first is the case where the fixed point of the involution on the disc-base is a point of local maximum of the function $f(x) - p_x^2$. This point corresponds to an atom A that preserves its own structure after factorization.

In the second case, the fixed point is a saddle critical point of $f(x) - p_x^2$ and belongs to an atom of the type V_{2k-1}. After factorization, this atom turns into the atom V_k^* with one star-vertex (see Fig. 12.41).

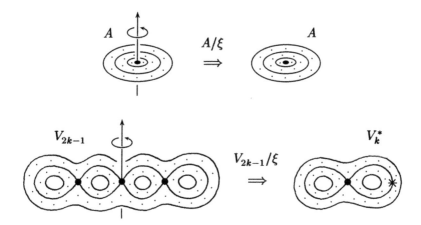

Figure 12.41

The same construction should be done for $g(y)$. As a result, we obtain the molecule $\widetilde{W}(g)$.

The similar procedure is carried out in the case of the Klein bottle. The difference is that there the base of the direct product is an annulus, but here it is a disc. So, the structure of $\widetilde{W}(f)$ and $\widetilde{W}(g)$ is easily extracted from the analysis of the Klein bottle.

Finally, we obtain the molecule $\widetilde{W}$ shown in Fig. 12.42.

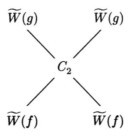

Figure 12.42

Since the action of ξ on the molecule W corresponding to the sphere is explicitly described, it is easy to compute the numerical marks on the quotient-molecule $\widetilde{W}$. Omitting the details, we formulate the final result.

Theorem 12.10 (V. S. Matveev). *Consider a Riemannian (L, f, g)-metric*

$$ds^2 = (f(x) + g(y))(dx^2 + dy^2)$$

on the projective plane and the geodesic flow corresponding to it. The isoenergy 3-manifold Q^3 is diffeomorphic to the lens space $L_{4,1}$.

a) *The molecule $\widetilde{W}$ corresponding to this geodesic flow has the form shown in Fig. 12.42, where the molecules $\widetilde{W}(f)$ and $\widetilde{W}(g)$ are constructed in the way described above. The numerical marks inside each of the molecules $\widetilde{W}(f)$ and $\widetilde{W}(g)$ are similar to those described in the classification theorem 12.3 for the Klein bottle.*

b) *Namely, on the graphs $\widetilde{W}(f)$ and $\widetilde{W}(g)$, the marks are as follows. On the edges between saddle atoms the r-marks are equal to infinity. Between saddle atoms and atoms of type A, the r-marks are equal to zero, except for those cases, where the atom A (i.e., the local maximum of the function) corresponds to the fixed point of the involution ξ (see Fig. 12.16(g)). In this case, the r-mark is equal to $1/2$.*

c) *On the remaining four edges of the molecule $\widetilde{W}$ (i.e., on the central edges incident to the atom C_2), the marks are as in Fig. 12.43. Here three different cases are distinguished depending on how many critical points the functions $f(x)$ and $g(y)$ have. In accordance with this, the molecule has either one, or three, or five families.*

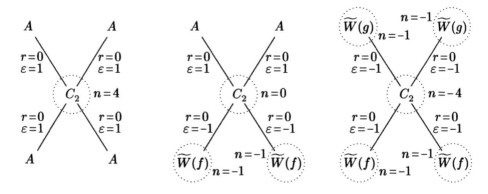

Figure 12.43

Thus, we have obtained a complete Liouville classification of all quadratically integrable geodesic flows on the projective plane (see Fig. 12.43).

12.4.2. Linearly Integrable Geodesic Flows on the Projective Plane

As we already know, if we have a Riemannian metric on the projective plane with the linearly integrable geodesic flow, then, by taking the standard two-sheeted covering over $\mathbb{R}P^2$, we can lift this metric to the sphere S^2. The geodesic flow of the lifted metric will be again linearly integrable. The Liouville classification for all such geodesic flows on the sphere was obtained in Theorem 12.9. Therefore, to obtain the Liouville classification of linearly integrable geodesic flows on the projective plane, one needs to consider the action of the involution ξ on the molecule W related to the covering sphere and described in Theorem 12.9. By taking "the quotient" W/ξ, we will obtain the desired molecule $\widetilde{W}$ that corresponds to the geodesic flow on $\mathbb{R}P^2$.

As before, the lifted metric on the sphere can be written as follows:

$$ds^2 = d\tau^2 + f(\tau)\,d\varphi^2 .$$

Note that the function f cannot be arbitrary here and ds^2 should be ξ-invariant. Without loss of generality, we can assume that τ is defined on the closed interval $[-\tau_0, \tau_0]$ and the involution ξ has the form

$$\xi: (\tau, \varphi) \to (-\tau, -\varphi).$$

Then the additional condition is that f should be even.

Theorem 12.11 (V. S. Matveev). *Consider the Riemannian metric*

$$ds^2 = d\tau^2 + f(\tau)\,d\varphi^2$$

on the projective plane. Then the molecule $\widetilde{W}$ corresponding to its geodesic flow has the form shown in Fig. 12.44, where the molecule $\widetilde{W}(f)$ is constructed from the function $f(\tau)$ in the way described above in Theorem 12.9.

a) In both the graphs $\widetilde{W}(f)$, the marks are as follows. On the edges between saddle atoms, the r-marks are equal to infinity. Between saddle atoms and atoms of type A, the r-marks are equal to zero, except for the case where the atom A (i.e., the local maximum of the function) corresponds to a fixed point of the involution ξ (see Fig. 12.16(g)). In this case, the r-mark is equal to $1/2$.

b) Suppose that the molecule $\widetilde{W}(f)$ is different from the atom A (i.e., contains at least one saddle atom). Then, on the single central edge connecting two copies of the $\widetilde{W}(f)$, the mark r is equal to infinity and $\varepsilon = -1$. Here there is only one family (it is the molecule W from which all the end-atoms A are removed). The mark n on this family is equal to -1.

c) If the molecule $\widetilde{W}(f)$ is reduced to a single atom A, then the whole molecule W has the simplest form A——A. In this case, we have $r = 1/4$ and $\varepsilon = 1$. There are no families here.

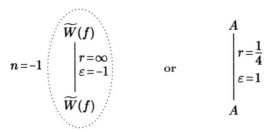

Figure 12.44

Thus, we have obtained a complete Liouville classification of linearly integrable geodesic flows on the projective plane.

Chapter 13

Orbital Classification
of Integrable Geodesic Flows
on Two-Dimensional Surfaces

13.1. CASE OF THE TORUS

13.1.1. *Flows with Simple Bifurcations (Atoms)*

An orbital classification of linearly and quadratically integrable geodesic flows on the torus was obtained by E. N. Selivanova [312].

As was shown in Chapters 5 and 8, for orbital classification of integrable systems, first of all, it is necessary to calculate the rotation function on the edges of the marked molecule W^*. Consider a torus with an integrable geodesic flow. We restrict ourselves to the case of a global Liouville metric.

As usual, we first consider a global Liouville metric

$$(f(x) + g(y))(dx^2 + dy^2)$$

on the Euclidean plane $\mathbb{R}(x, y)$, and then factor it with respect to an orthogonal lattice Γ generated by two vectors $f_1 = (1, 0)$ and $f_2 = (0, L)$. Here we assume that f and g are both strictly positive non-constant functions with the periods 1 and L respectively. We obtain the torus $T^2 = \mathbb{R}^2/\Gamma$ with the so-called (L, f, g)-metric having the quadratically integrable geodesic flow. In the similar way, we can construct a linearly integrable geodesic flow if we set $g(y) \equiv 0$.

First, we assume that both functions f and g differ from a constant. As before, we denote by F an additional quadratic integral.

Recall that Liouville tori in the isoenergy manifold $Q = \{H = 1\} \subset T^*T^2$ are given in the standard coordinates (x, y, p_x, p_y) by the following equations:

$$p_x^2 = f(x) + F, \qquad p_y^2 = g(y) - F.$$

These equations can actually define several Liouville tori that lie on the same level of F in Q.

Recall that, in our case, the molecule W has the form shown in Fig. 12.6(a) from Chapter 12. It is seen from its structure that the Liouville tori are naturally divided into three groups depending on the value of F:

1) for $F > \min(g(y))$, the Liouville torus gets into one of the upper subgraphs $W(g)$ in W;

2) for $F \in (-\min(f), +\min(g))$, the Liouville torus belongs to one of the four central edges a, b, c, d;

3) for $F < -\min(f)$, the Liouville torus gets into one of the lower subgraphs $W(g)$ in W.

Having fixed the value of F, we obtain several segments on the y-axis on each of which the function $p_y^2 = g(y) - F$ is non-negative. Every such segment corresponds to a pair of Liouville tori that differ from each other only by the direction of geodesics.

If, for example, $F > \min(g(y))$, then the parameter y can vary inside several segments (see Fig. 13.1). We choose one of them and denote it by $[y_1, y_2]$. The other variable x varies in its domain of definition, i.e., in the whole closed interval $[0, 1]$.

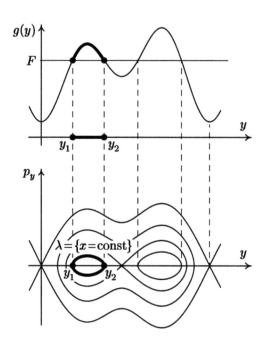

Figure 13.1

Analogously, if $F \in (-\min(f), +\min(g))$, then x and y can independently take any admissible values, i.e., $x \in [0, 1]$ and $y \in [0, L]$.

Finally, if $F < -\min(f)$, then x takes values in some closed interval $[x_1, x_2]$, and y runs over its domain of definition, i.e., the whole closed interval $[0, L]$.

It is useful to draw the projection of the described Liouville tori on the base T^2. According to the above three cases, these projections will look as shown in Fig. 13.2: an annulus, the whole torus, and again an annulus.

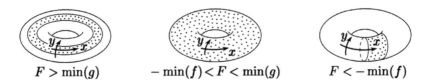

$$F > \min(g) \qquad\qquad -\min(f) < F < \min(g) \qquad\qquad F < -\min(f)$$

Figure 13.2

On each of these Liouville tori, we fix the basis cycles λ and μ given by the equations

$$\lambda = \{x = \text{const}\} \quad \text{and} \quad \mu = \{y = \text{const}\}.$$

Then the value of the rotation function on this Liouville torus with respect to the fixed basis is defined by the following statement.

Proposition 13.1. *Consider the quadratically integrable geodesic flow of the (L, f, g)-metric on the torus.*

a) *If $F > \min(g(y))$, then the rotation function ρ on the Liouville torus (lying on the level $\{F = \text{const}\}$) is given by*

$$\rho_e(F) = \frac{\displaystyle\int_0^1 \frac{dx}{\sqrt{f(x)+F}}}{\displaystyle\int_{y_1}^{y_2} \frac{2\,dy}{\sqrt{g(y)-F}}}.$$

Here e belongs to one of two subgraphs $W(g)$ (see Fig. 12.6(a) in Chapter 12).

b) *If $F \in (-\min(f), +\min(g))$, then the rotation function ρ on the Liouville torus (lying on the level $F = \text{const}$) is given by*

$$\rho_e(F) = \frac{\displaystyle\int_0^1 \frac{dx}{\sqrt{f(x)+F}}}{\displaystyle\int_L^0 \frac{dy}{\sqrt{g(y)-F}}}.$$

Here e is one of the four central edges a, b, c, d of the molecule W.

c) *If $F < -\min(f)$, then the rotation function ρ on the Liouville torus (lying on the level $F = \text{const}$) is given by*

$$\rho_e(F) = \frac{\displaystyle\int_{x_1}^{x_2} \frac{2\,dx}{\sqrt{f(x)+F}}}{\displaystyle\int_L^0 \frac{dy}{\sqrt{g(y)-F}}}.$$

Here e belongs to one of two subgraphs $W(f)$ (see Fig. 12.6(a) in Chapter 12).

Proof. To compute the rotation function, we use the action variables, which can be explicitly calculated. Recall the following formula for the rotation function in the action variables:

$$\rho(F) = \frac{\partial H/\partial I_1}{\partial H/\partial I_2} = -\frac{\partial I_2/\partial F}{\partial I_1/\partial F}.$$

Here I_1, I_2 are the action variables corresponding to the cycles λ and μ. Namely,

$$I_1 = \frac{1}{2\pi} \int_\lambda \alpha \quad \text{and} \quad I_2 = \frac{1}{2\pi} \int_\mu \alpha,$$

where $\alpha = p_x \, dx + p_y \, dy$. On the Liouville torus, we can express p_x and p_y through x, y, and F. For definiteness, we assume that $F > \min(g(y))$. Then

$$I_1 = \frac{1}{2\pi} \int_{y_1}^{y_2} 2\sqrt{g(y) - F} \, dy, \qquad I_2 = \frac{1}{2\pi} \int_0^1 \sqrt{f(x) + F} \, dx.$$

The coefficient 2 occurs in the formula for I_1, since the integration over λ (Fig. 13.1) is equivalent to the double integration over the closed interval $[y_1, y_2]$.

Differentiating the actions by F and substituting the result into the above formula for ρ, we obtain the desired statement.

In the cases where $F \in (-\min(f), +\min(g))$ or $F < -\min(f)$, the argument is similar. The proposition is proved.   □

Note that the values of ρ turned out to be positive. It can be explained by the choice of the orientation on the cycles λ and μ. The orientation is chosen in such a way that the values of actions are positive.

Let us describe some properties of the rotation function.

Proposition 13.2.   *Consider a quadratically integrable geodesic flow on the torus. Then, on each of the four central edges a, b, c, d of the molecule W, the rotation function is strictly monotone and varies from zero to infinity. In particular, on these edges, the rotation function gives no contribution to the orbital invariant of the geodesic flow.*

REMARK. Unlike the four central edges of W, on the other edges, the behavior of the rotation function can be different. For example, ρ may be non-monotone. Therefore, the R-vectors on these edges may be non-trivial.

Proof. The case under consideration corresponds to the case (b) of Proposition 13.1. It is seen from the explicit formula for ρ that, in this case, the numerator of the ratio, i.e., the integral

$$\int_0^1 \frac{1}{\sqrt{f(x) + F}} \, dx$$

decreases (as a function of F) from $+\infty$ up to some finite limit. The denominator, i.e., the integral

$$\int_0^L \frac{1}{\sqrt{g(y) - F}} \, dy$$

increases from a finite value up to $+\infty$. Hence, ρ goes monotonically from $+\infty$ up to zero.   □

Let us turn to the case of a Riemannian metric on the torus whose geodesic flow is integrable by means of a linear integral. We consider the simplest case, which can be obtained from the previous one by setting $g(y) \equiv 0$. The function $f(x)$ is assumed to be strictly positive. Recall that the corresponding metric on the torus is called an $(f, 0, L)$-metric. The molecule of the geodesic flow has the form shown in Fig. 12.6(b). The edges of this molecule can naturally be divided into two groups according to the value of an additional (linear) integral F, namely, the pair of central edges and the pair of subgraphs $W(f)$. The following statement yields the explicit formulas for the rotation function in each of these cases.

Proposition 13.3.

a) *If* $|F| > \min(f(x))$, *then the rotation function* ρ *on the Liouville torus* (*lying on the level* $\{F = \text{const}\}$) *is given by*

$$\rho_e(F) = \frac{1}{L} \int_{x_1}^{x_2} \frac{2F \, dx}{\sqrt{f(x) - F^2}} \, .$$

Here e *belongs to one of two subgraphs* $W(f)$ (*see Fig. 12.6(b) in Chapter 12*).

b) *If* $|F| < \min(f)$, *then the rotation function* ρ *on the Liouville torus* (*lying on the level* $\{F = \text{const}\}$) *is given by*

$$\rho_e(F) = \frac{1}{L} \int_{0}^{1} \frac{F \, dx}{\sqrt{f(x) - F^2}} \, .$$

Here e *is one of two central edges* a *and* b *of the molecule* W.

Proof. We begin with the case (a). Using the same method as before, we compute the action variables I_1 and I_2. In this case, the integral F has the simple form $F = p_y$. Hence a Liouville torus is given by two simple equations:

$$F = p_y, \qquad p_x^2 = f(x) - F^2 \, .$$

Therefore,

$$I_1 = \frac{1}{2\pi} \int_{0}^{L} F \, dy, \qquad I_2 = \frac{1}{2\pi} \int_{x_1}^{x_2} 2\sqrt{f(x) - F^2} \, dx \, .$$

By substituting these expressions into the general formula for ρ, we obtain

$$\rho(F) = -\frac{\partial I_2 / \partial F}{\partial I_1 / \partial F} = \frac{1}{L} \int_{x_1}^{x_2} \frac{2F \, dx}{\sqrt{f(x) - F^2}} \, .$$

The case (b) is analogously examined. $\square$

Proposition 13.4. *Consider a linearly integrable geodesic flow on the torus. Then, on each of two central edges a and b of the molecule W (see Fig. 12.6(b)), the rotation function is strictly monotone and varies from $-\infty$ to $+\infty$. In particular, on these four edges, the rotation function gives no contribution to the orbital invariant of the geodesic flow.*

Proof. The statement immediately follows from the explicit form of the rotation function on both central edges (see Proposition 13.3(b)).   □

REMARK. As in the quadratic case, the rotation function may be non-monotone on the molecule edges lying inside the subgraphs $W(f)$. Therefore, non-trivial R-vectors may occur.

Theorem 13.1. *Let ds^2 and ds'^2 be two Riemannian metrics on the torus whose geodesic flows are linearly or quadratically integrable. Suppose that the corresponding molecules W^* and W'^* are simple, that is, all their atoms are simple (i.e., either A or B). Then the geodesic flows of these metrics are topologically orbitally equivalent if and only if the marked molecules W^* and W'^* coincide (in the exact sense described above), and in addition, the rotation functions on the corresponding edges of the molecules are continuously conjugate.*

Proof. Since all the singularities of the Liouville foliation (i.e., atoms) are assumed to be simple, no atomic invariants (like Λ, Δ, and Z) appear in this case. This means that the only orbital invariant is the marked molecule W^* endowed with the collection of R-vectors, which are uniquely determined by the rotation functions. The coincidence of the R-vectors is equivalent to the conjugacy of the corresponding rotation functions. Hence the statement of the theorem follows directly from the general classification theory (see Chapter 8).   □

Using this theorem, one can construct examples of topologically orbitally equivalent geodesic flows on the torus. Consider, for example, two global Liouville metrics

$$ds^2 = (f(x) + g(y))^2(dx^2 + dy^2) \quad \text{and} \quad d\widetilde{s}^2 = (\widetilde{f}(x) + \widetilde{g}(y))^2(dx^2 + dy^2)$$

on the torus, where each of the periodic functions f, g, $\widetilde{f}$, $\widetilde{g}$ has exactly one minimum and one maximum on the interval of periodicity (see Fig. 13.3). Then the corresponding molecules W^* and $\widetilde{W}^*$ have the simplest form shown in Fig. 13.4. We do not indicate the marks r, ε, and n, since they are not needed here.

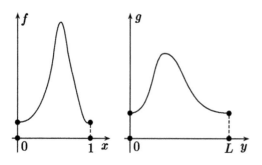

Figure 13.3

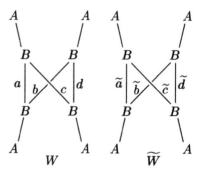

Figure 13.4

For simplicity, assume that the rotation functions are monotone on every edge. On each of the four central edges, the rotation function varies from zero to $+\infty$, and consequently, does not contribute to the orbital invariant. On the other edges (incident to atoms A), the rotation function ρ has the following structure. Its limit on the saddle atom B is either zero or infinity, and therefore, does not affect the invariant. But, as a Liouville torus tends to the atom A, the limit of ρ is a finite number, and its value (according to the general theory) is an orbital topological invariant of the geodesic flow. This limit is obviously the same on both upper atoms A. We denote it by p. The analogous limit of ρ on both lower atoms A is denoted by q.

In the case of the geodesic flow of $\widetilde{ds}^2$, we denote the corresponding limits by $\widetilde{p}$ and $\widetilde{q}$. Thus, to each of the flows, we have assigned two orbital invariants, namely, p and q (and respectively $\widetilde{p}$ and $\widetilde{q}$). All the other invariants are the same in both cases, and therefore, we can just forget them. Therefore, we conclude that the geodesic flows of the metrics ds^2 and $\widetilde{ds}^2$ are topologically orbitally equivalent if and only if either $(p,q) = (\widetilde{p}, \widetilde{q})$ or $(p,q) = (\widetilde{q}^{-1}, \widetilde{p}^{-1})$.

The point is that the molecules presented in Fig. 13.4 are symmetric with respect to the horizontal axis passing through the middle of the molecule. So we have two different possibilities for a homeomorphism $W \to \widetilde{W}$. In fact, it just means that we can interchange the coordinates x and y. This leads us to interchanging basis cycles λ and μ and, finally, to the transformation of the rotation function according to the rule $\rho \to 1/\rho$.

This result can be reformulated as follows. Under the above assumptions on f and g, the geodesic flow has exactly two stable closed geodesics. One of them corresponds to the upper pair of atoms A, and the other corresponds to the lower one. Two atoms A from each pair denote the same geodesic but with two different orientations. Every closed geodesic is characterized by its multiplier, which is completely determined by the limit of the rotation function ρ. More precisely, the following statement holds.

Consider the rotation number ρ on the Liouville torus T^2 neighboring to a stable closed geodesic γ; let ρ_0 be the limit of ρ as $T^2 \to \gamma$. The closed geodesic γ is also characterized by its index (usually called the Morse index). Recall that $\operatorname{ind}(\gamma)$ is defined to be the number of points that are conjugate to an initial point P along γ; each conjugate point is taken into account only once. Note that here we mean a simple geodesic (i.e., taken without multiplicity). If the initial point P is conjugate to itself along γ, then we also take it into account.

Recall that the *multiplier* of γ is defined to be an eigenvalue $\nu = \exp(2\pi i\varphi)$ of the linearization of the Poincaré map along γ.

Proposition 13.5. *The limit of the rotation number ρ_0 on the closed stable geodesic (on a two-dimensional surface), its Morse index* $\operatorname{ind}(\gamma)$*, and its multiplier ν are connected by the following relations:*

$$\operatorname{ind}(\gamma) = [2\rho_0] \quad (\textit{i.e., the integer part of } 2\rho_0), \qquad \nu = \exp(2\pi i\rho_0).$$

Proof. First note that the equality $\nu = \exp(2\pi i\rho_0)$ has been already proved in Chapter 8 (Proposition 8.3). And we only need to verify the first relation between the index and the rotation function.

Let γ be a closed geodesic, and let P be a point on it. Consider another geodesic $\widetilde{\gamma}$ starting from P in the direction of a vector a close to the velocity vector $\dot{\gamma}(0)$ of the initial geodesic γ (see Fig. 13.5). Since γ is stable, $\widetilde{\gamma}$ goes near γ intersecting it at some points.

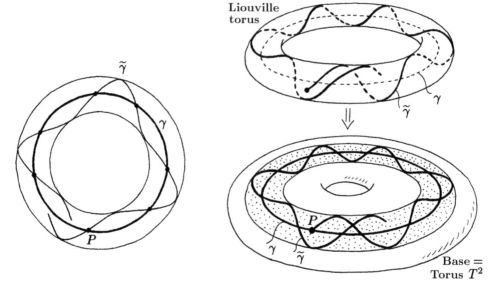

Figure 13.5   **Figure 13.6**

Consider the Liouville torus in Q which contains the geodesic $\widetilde{\gamma}$. Since γ is stable, we may assume that the whole Liouville torus is located in some narrow tubular neighborhood of γ in Q. Therefore, by projecting this Liouville torus down to the 2-torus T^2, we obtain a narrow annulus on T^2 with the axis γ inside of which $\widetilde{\gamma}$ moves (see Fig. 13.5). It turns out that each pair of sequential intersections of $\widetilde{\gamma}$ and γ corresponds to one complete turn of the geodesic $\widetilde{\gamma}$ along the generatrix of the Liouville torus (see Fig. 13.6). This follows easily from the fact that the coordinate net of the angle variables on the Liouville torus is projected into the standard coordinate net $\{x = \text{const}\}$, $\{y = \text{const}\}$ on the torus T^2.

By $n_{\widetilde{\gamma}}(k)$ we denote the number of intersections of $\widetilde{\gamma}$ and γ on the torus T^2 after k complete turns along the closed geodesic γ. By $n_{\widetilde{\gamma}}(1 + \varepsilon)$ we denote the number of intersections of $\widetilde{\gamma}$ and γ on the segment from 0 to $1 + \varepsilon$. Here we assume that 1 is the period of γ, that is, its length after an appropriate norming. Further, let ε be an arbitrary sufficiently small positive number. Then

$$\text{ind}(\gamma) = \lim n_{\widetilde{\gamma}}(1 + \varepsilon), \qquad \text{as} \quad \widetilde{\gamma} \to \gamma.$$

First assume that P is not conjugate to itself along the geodesic γ. Then it is clear that in the above formula we can set $\varepsilon = 0$, i.e.,

$$\text{ind}(\gamma) = \lim n_{\widetilde{\gamma}}(1), \qquad \text{as} \quad \widetilde{\gamma} \to \gamma.$$

This means that the index of γ is exactly equal to the number s of intersection points of γ with the close geodesic $\widetilde{\gamma}$ after one complete turn (Fig. 13.6).

On the other hand, for the limit $\rho_0 = \lim \rho(\tilde{\gamma})$ of the rotation function $\rho(\tilde{\gamma})$ as $\tilde{\gamma} \to \gamma$, the following formula takes place:

$$\rho_0 = \lim_{\tilde{\gamma} \to \gamma} \lim_{k \to \infty} \frac{n_{\tilde{\gamma}}(k)}{2k}.$$

It suffices to show that

$$s \le \frac{n_{\tilde{\gamma}}(k)}{k} < s + 1.$$

Let us prove the left inequality. Assume the contrary, i.e., $s > n_{\tilde{\gamma}}(k)/k$. Divide the segment $[0, k]$ into unit intervals of the form $[m, m + 1)$. Then there exists at least one interval inside of which the number of intersection points of γ and $\tilde{\gamma}$ is strictly less than s. Consider the geodesic τ starting from m (i.e., from the left end-point of the interval) and close to γ (see Fig. 13.7). Since the index of γ on the interval $[m, m + 1)$ is s, it follows that the number of intersection points of τ and γ on this interval also must be equal to s.

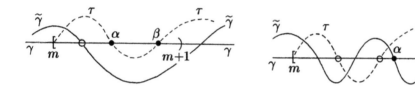

Figure 13.7   Figure 13.8

In this case, we obtain a pair of neighboring points α and β on $[m, m + 1)$ at which τ intersects γ. Moreover, as is seen from Fig. 13.7, there are no other intersection points of γ and $\tilde{\gamma}$ between α and β. Without loss of generality we may assume that both geodesics τ and $\tilde{\gamma}$ lie on the same Liouville torus. But then we obtain a contradiction, since the intersection points of the geodesics τ and $\tilde{\gamma}$ with γ must alternate. This observation follows from the Liouville theorem and the fact that all Liouville tori close to γ are projected (down to the base T^2) onto concentric annuli with the same axis γ. Thus, the first inequality is proved.

We now turn to the second part of the inequality. Assume the contrary, that is, $n_{\tilde{\gamma}}(k)/k \ge s + 1$. By analogy with the previous case, we obtain that there exists at least one interval $[m, m + 1)$ on the segment $[0, k]$ which contains at least $s + 2$ points of intersection of the geodesics γ and $\tilde{\gamma}$ (see Fig. 13.8). Consider the geodesic τ starting from m and lying on the same Liouville torus as $\tilde{\gamma}$. Then it is clear that there exist two points α and β of intersection of the geodesics γ and $\tilde{\gamma}$ between which there is no intersection points of γ and τ. But this is impossible for the same reason as before. This contradiction gives, consequently, the second inequality.

By setting $k \to \infty$, we obtain

$$s \le 2\rho < s + 1.$$

Therefore, $[2\rho] = s$. Since s, in turn, is equal to the index of γ, we get the required equality:

$$[2\rho] = \text{ind}(\gamma).$$

Thus, in the case when P is not conjugate to itself along γ, the assertion has been proved.

Now let P be conjugate to itself along γ. This means that the geodesic $\tilde{\gamma}$ returns after one complete turn to a point P' close to the initial point P up to infinitesimal of the second order. In other words, there exists a Jacobi field along γ which vanishes both at the point $t = 0$ and at the point $t = 1$, where t is the parameter along the closed geodesic γ, and 1 is its length (i.e., period). In this case, the multiplier ν is equal to ± 1. It follows from the above formula $\nu = \exp(2\pi i \rho_0)$ that $2\rho_0$ is an integer number. Consequently, it remains to verify that $\operatorname{ind}(\gamma) = 2\rho_0$. But this follows from the above definition of the index of a closed geodesic, where (in the case when P is conjugate to itself) we must take into account the end-point of the geodesic as a conjugate one. $\square$

This statement allows us to formulate a simple orbital equivalence criterion for "simple" integrable geodesic flows on the torus.

Corollary. *Consider two quadratically integrable geodesic flow on the torus with "simple" molecules W and W' shown in Fig. 13.4. Suppose the rotation functions are all monotone. Then these geodesic flows are topologically orbitally equivalent if and only if the corresponding closed stable geodesics have the same Morse indices and the same multipliers.*

13.1.2. *Flows with Complicated Bifurcations (Atoms)*

In this section, we do not formulate general theorems, but restrict ourselves to several comments.

If a singularity of the Liouville foliation (i.e., an atom) is complicated, then, in addition to rotation functions, new orbital invariants Λ and Δ appear. (Note that the Z-invariant does not appear here, since all the atoms are planar.)

According to the general theory, the orbital invariants of the initial Hamiltonian system are conjugacy invariants of the reduced system on a two-dimensional transversal section. This system is called the *Poincaré flow*. Its Hamiltonian (the so-called *Poincaré Hamiltonian*) can be written explicitly. After that, the invariants Λ and Δ can be easily computed.

As we have shown above, in the case of torus, the isoenergy 3-surface Q^3 is separated into 4 parts for each of which there exists a global transversal 2-section. In Fig. 12.11 (Chapter 12), we illustrate these four pieces Q_I, Q_{II}, Q_{III}, and Q_{IV}. Consider, for example, Q_I. Its transversal section (as shown in Chapter 12) can be the surface P_I given by $x = 0$ in Q^3. The local coordinates on P_I are y and p_y. The symplectic structure is obviously standard, i.e., takes the form $dp_y \wedge dy$. An additional integral of the geodesic flow, being restricted to P_I, is given by

$$F = -p_y^2 + g(y)\,.$$

It is clear that the Poincaré Hamiltonian F^* is some function of F, i.e., can be written as $F^* = F^*(F)$. To compute the conjugacy invariants Λ and Δ of the flow $\operatorname{sgrad} F^*$, we can use the following statement, which holds not only for the torus, but also for any Hamiltonian systems on 2-atoms.

Lemma 13.1. *Consider the Hamiltonian flows* sgrad F *and* sgrad $F^*(F)$ *on a 2-atom* (P, K). *Then*

1) $\Lambda(F^*) = \Lambda(F)$,

2) $\Delta(F^*) = \Delta(F) \cdot \left(\dfrac{dF^*}{dF} \right)^{-1}$,

3) $Z(F^*) = Z(F) \cdot \left(\dfrac{dF^*}{dF} \right)^{-1}$.

Here the derivative dF^*/dF is computed on the singular fiber.

Proof. The proof easily follows from the definition of these invariants. □

In the case of the torus, the function $F^*(F)$ can easily be computed by using the above explicit formulas for the action variables (Proposition 13.1) and the Topalov formula for the Poincaré Hamiltonian (Proposition 5.6). The result is

$$F^*(F) = \int_0^1 \sqrt{f(x) + F}\, dx .$$

Then

$$\frac{dF^*}{dF} = \int_0^1 \frac{dx}{\sqrt{2(f(x) + F)}} .$$

Thus, in the case of the torus, the problem is reduced to the calculation of the invariants Λ and Δ on the section P_I for a very simple Hamiltonian, which is equal to

$$F = -p_y^2 + g(y) .$$

Lemma 13.2. *Let a singular fiber K on P_I contain singular points $S_1, \ldots, S_m$. In coordinates (y, p_y), these points are $(y_i, 0)$, where y_i is a critical point of $g(y)$. (See Fig. 13.9.) Then the Λ-invariant on K is given by*

$$\Lambda = (\lambda_1 : \ldots : \lambda_m), \qquad where \quad \lambda_i = \sqrt{\frac{d^2 g(y_i)}{dy^2}} .$$

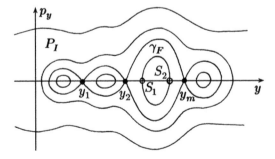

Figure 13.9

Proof. One needs just to apply the formula, which expresses Λ from the symplectic structure and second derivatives of the Hamiltonian (see Chapter 6). □

To compute the Δ-invariant, we also can use the formula, which expresses it from the "finite parts" of the period functions (see Chapter 6). The period functions themselves are

$$\Pi(F) = \int \frac{dy}{\sqrt{g(y) - F}},$$

where, for each closed integral curve of $\operatorname{sgrad} F$, the integral is taken over the corresponding closed interval $[s_1, s_2]$ (see Fig. 13.9). $\quad\square$

13.2. CASE OF THE SPHERE

The case of the sphere is examined more or less in the same way as that of the torus.

As usual, for orbital classification of integrable geodesic flows, we need to compute the rotation functions.

Consider an (L, f, g)-metric on the sphere. As we know, it is obtained from the global Liouville "metric" on the torus by means of a two-sheeted covering with four branch points. Therefore, we can define such a metric on the sphere as a metric on the covering torus that is invariant under the corresponding involution ξ.

As before, we consider a global Liouville metric

$$(f(x) + g(y))(dx^2 + dy^2)$$

on the Euclidean plane $\mathbb{R}^2 (x, y)$, and then factor the plane with respect to the orthogonal lattice generated by the vectors $f_1 = (1, 0)$ and $f_2 = (0, L)$. Here we assume that f and g are periodic even non-negative functions such that $f(k/2) = 0$ and $g(kL/2) = 0$ for $k \in \mathbb{Z}$. We obtain the torus T^2 with the metric ds^2, which can be pushed forward to the sphere by using the covering $T^2 \to T^2/\xi = S^2$, where ξ is the involution given by $\xi(x, y) = (-x, -y)$. The obtained metric on the sphere is an (L, f, g)-metric. Denote an additional integral of the geodesic flow by F.

The Liouville tori in the isoenergy manifold $Q = \{H = 1\} \subset T^*S^2$, being lifted to the cotangent bundle T^*T^2, are given in coordinates (x, y, p_x, p_y) by

$$p_x^2 = f(x) + F, \qquad p_y^2 = g(y) - F.$$

These equations can determine several Liouville tori lying on the same level of F. Recall that the molecule W for the geodesic flow on S^2 has the form shown in Fig. 12.30. As we see from the structure of the molecule, the Liouville tori are divided into two groups depending on the value of F:

1) for $F > 0$, the Liouville torus gets into one of the upper subgraphs $W(g)$ in W;

2) for $F < 0$, the Liouville torus is located inside one of two lower subgraphs $W(f)$ in W.

Having fixed the value of F, we obtain several closed intervals on the y-axis in each of which the function $p_y^2 = g(y) - F$ is non-negative. Choose one of them and denote it by $[y_1, y_2]$. This closed interval corresponds to a pair of Liouville tori, which differ from each other only by the direction of geodesics. The coordinate x for these tori can be arbitrary and varies on the whole closed interval $[0, 1]$.

If $F < 0$, then x takes values in some closed interval $[x_1, x_2]$ (one of the several possible ones), and y varies on the whole closed interval $[0, L]$.

According to the above two cases, the projections of these tori look as shown in Fig. 13.10. In both cases, we obtain an annulus.

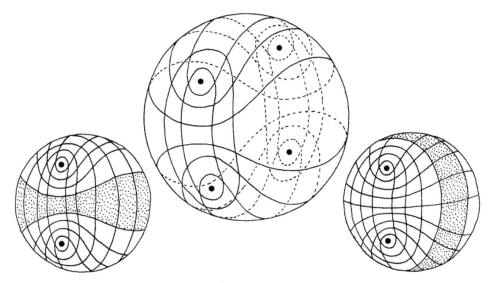

Figure 13.10

On each Liouville torus, we consider the basis cycles λ and μ given by

$$\lambda = \{x = \text{const}\} \quad \text{and} \quad \mu = \{y = \text{const}\}.$$

The next statement gives us an explicit formula for the rotation function ρ with respect to this basis.

Proposition 13.6.

a) *If $F > 0$, then the rotation number ρ on the Liouville torus (lying on the level $\{F = \text{const}\}$) is given by*

$$\rho_e(F) = \frac{\displaystyle\int_0^1 \frac{dx}{\sqrt{f(x)+F}}}{\displaystyle\int_{y_1}^{y_2} \frac{2\,dy}{\sqrt{g(y)-F}}}.$$

b) *If $F < 0$, then the rotation number ρ on the Liouville torus (lying on the level $\{F = \text{const}\}$) is given by*

$$\rho_e(F) = \frac{\displaystyle\int_{x_1}^{x_2} \frac{2\,dx}{\sqrt{f(x)+F}}}{\displaystyle\int_0^1 \frac{dy}{\sqrt{g(y)-F}}}.$$

The proof immediately follows from that of Proposition 13.3. $\quad\square$

Theorem 13.2. *Let ds^2 and ds'^2 be two metrics on the sphere whose geodesic flows are quadratically integrable. Suppose that the corresponding molecules W^* and W'^* are simple in the sense that all their atoms, except for the central atom C_2, are simple (i.e., either A or B). Then their geodesic flows are topologically orbitally equivalent if and only if the marked molecules W^* and W'^* coincide, and in addition, their rotation functions on the corresponding edges of the molecules are continuously conjugate.*

The prove is just the same as in the case of the torus.   □

REMARK. Of course, Theorem 13.2 can also be formulated in the general case where the singularities of the Liouville foliation (i.e., atoms) are not supposed to be simple. But, in this case, the formulation becomes unwieldy and we omit it. Here we have to take into account other orbital invariants such as Λ and Δ.

Using this theorem, one can construct examples of orbitally equivalent geodesic flows on the sphere. Consider, for example, two quadratically integrable geodesic flows on the sphere corresponding to the triples (L, f, g) and $(\widetilde{L}, \widetilde{f}, \widetilde{g})$. Suppose, in addition, that each of the functions f, g, $\widetilde{f}$, and $\widetilde{g}$ has exactly one local minimum and one local maximum (on its half-period). Then the corresponding molecules W^* and $\widetilde{W}^*$ have the simplest form shown in Fig. 12.35(a).

Suppose for simplicity that the rotation functions are monotone. Then every rotation function is as follows. Its limit on the saddle singular leaf C_2 equals 1. This follows from the above formulas for ρ and from the asymptotics of f and g at their zeros.

So, this limit does not depend on the choice of f and g, and, consequently, we do not pay any attention to it. On the contrary, as the Liouville torus tends to the atom A (i.e., shrinks into the stable closed geodesic), the limit of ρ is a finite number, which will be one of the orbital invariants of the geodesic flow ds^2. The limits on the upper atoms A obviously coincide, and we denote them by p. For the lower atoms A, we denote the limit of ρ by q.

For other geodesic flows, we consider analogous limits $\widetilde{p}$ and $\widetilde{q}$ of the rotation function. So, for each geodesic flow, we have described two orbital invariants p and q. Using the general classification theory (see Chapter 8), it is easily seen that no other essential invariants occur in this case (more precisely, all other invariants coincide automatically). Thus, by analogy to the case of the torus, such geodesic flows on the sphere are topologically orbitally equivalent if and only if either $(p, q) = (\widetilde{p}, \widetilde{q})$ or $(p, q) = (\widetilde{q}^{-1}, \widetilde{p}^{-1})$. We discuss one of such examples in Chapter 10.

In the same way as for the torus, this result can be reformulated as follows. Under the above assumptions, the geodesic flow has exactly two stable closed geodesics (or four if we take into account their orientation), presented in the molecule W by atoms A.

Using Proposition 13.5, we can formulate the following simple criterion for the orbital equivalence of "simple" integrable geodesic flows on the sphere.

Corollary. *Consider two quadratically integrable geodesic flows on the sphere. Suppose that the corresponding molecules W and W' have the simplest structure (Fig. 12.35(a)) and the rotation functions are monotone. Then these geodesic flows are topologically orbitally equivalent if and only if the corresponding closed stable geodesics have the same Morse indices and the same multipliers.*

13.3. EXAMPLES OF INTEGRABLE GEODESIC FLOWS ON THE SPHERE

13.3.1. *The Triaxial Ellipsoid*

The geodesic flow on the standard ellipsoid (both two- and multi-dimensional) was investigated by many authors. See, for example, [23], [137], [167], [184], [185], [353].

Consider the usual ellipsoid in $\mathbb{R}^3$ given in the Cartesian coordinates by

$$\frac{x^2}{a} + \frac{y^2}{b} + \frac{z^2}{c} = 1\,,$$

where $a < b < c$. Consider the Riemannian metric induced on it by the Euclidean metric from $\mathbb{R}^3$. The corresponding geodesic flow turns out to be integrable. C. Jacobi was first who discovered this fact in his famous "Vorlesungen über Dynamik" [167]. To show this, it is convenient, following C. Jacobi, to use elliptic coordinates in $\mathbb{R}^3$.

The elliptic coordinates of a point $P = (x, y, z) \in \mathbb{R}^3$ are defined to be three real roots $\lambda_1 > \lambda_2 > \lambda_3$ of the cubic equation

$$\frac{x^2}{a + \lambda} + \frac{y^2}{b + \lambda} + \frac{z^2}{c + \lambda} = 1\,.$$

If P does not belong to any coordinate plane in $\mathbb{R}^3$, then the above equation has three real roots $\lambda_1 > \lambda_2 > \lambda_3$; moreover, $\lambda_1 \in (-a, \infty)$, $\lambda_2 \in (-b, -a)$, and $\lambda_3 \in (-c, -b)$. The corresponding coordinate surfaces $\lambda_i = \text{const}$ in $\mathbb{R}^3$ are ellipsoids, and one- and two-sheeted hyperboloids for $i = 1, 2, 3$, respectively (see Fig. 13.11). The initial ellipsoid itself is given then by $\lambda_1 = 0$. The other two coordinates λ_2 and λ_3 can be considered as local regular coordinates on it. Their level lines are shown in Fig. 13.12. (Notice, by the way, that they are curvature lines on the ellipsoid.)

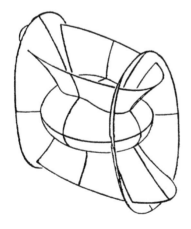

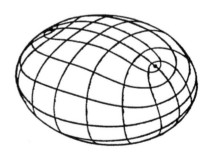

Figure 13.11   Figure 13.12

It is useful to point out explicit formulas, which express the Cartesian coordinates from elliptic ones:

$$x^2 = \frac{(a + \lambda_1)(a + \lambda_2)(a + \lambda_3)}{(a - b)(a - c)},$$

$$y^2 = \frac{(b + \lambda_1)(b + \lambda_2)(b + \lambda_3)}{(b - a)(b - c)},$$

$$z^2 = \frac{(c + \lambda_1)(c + \lambda_2)(c + \lambda_3)}{(c - a)(c - b)}.$$

The Euclidean metric $ds^2 = dx^2 + dy^2 + dz^2$ in $\mathbb{R}^3$, being written in elliptic coordinates, takes the following form:

$$ds^2 = \frac{1}{4}\left(\frac{(\lambda_1 - \lambda_2)(\lambda_1 - \lambda_3)}{(a + \lambda_1)(b + \lambda_1)(c + \lambda_1)}\, d\lambda_1^2 + \frac{(\lambda_2 - \lambda_1)(\lambda_2 - \lambda_3)}{(a + \lambda_2)(b + \lambda_2)(c + \lambda_2)}\, d\lambda_2^2 \right.$$
$$\left. + \frac{(\lambda_3 - \lambda_1)(\lambda_3 - \lambda_2)}{(a + \lambda_3)(b + \lambda_3)(c + \lambda_3)}\, d\lambda_3^2 \right).$$

Restricting this metric to the ellipsoid, i.e., setting $\lambda_1 = 0$, we obtain the following formula for the metric on the ellipsoid in elliptic coordinates:

$$ds^2 = \frac{1}{4}(\lambda_2 - \lambda_3)\left(\frac{\lambda_2}{(a + \lambda_2)(b + \lambda_2)(c + \lambda_2)}\, d\lambda_2^2 - \frac{\lambda_3}{(a + \lambda_3)(b + \lambda_3)(c + \lambda_3)}\, d\lambda_3^2 \right).$$

Denoting the polynomial $(a + \lambda)(b + \lambda)(c + \lambda)$ by $P(\lambda)$, one can write this metric as

$$ds^2 = \frac{1}{4}(\lambda_2 - \lambda_3)\left(\frac{\lambda_2}{P(\lambda_2)}\, d\lambda_2^2 - \frac{\lambda_3}{P(\lambda_3)}\, d\lambda_3^2 \right).$$

We see that this metric is similar to the Liouville one.   Such metrics are sometimes called *almost Liouville metrics*, since, under the simple transformation

$$\sqrt{\frac{\lambda_2}{P(\lambda_2)}} = du, \qquad \sqrt{\frac{\lambda_3}{P(\lambda_3)}} = dv,$$

such a metric is reduced to the Liouville form:

$$ds^2 = \frac{1}{4}(\lambda_2(u) - \lambda_3(v))(du^2 + dv^2).$$

Based on the previous results, one can easily find the molecule W^* of the geodesic flow on the ellipsoid. It is sufficient to note that the functions $\lambda_2(u)$ and $\lambda_3(v)$ have exactly one local minimum and one local maximum as λ_2 varies from $-b$ to $-a$ and λ_3 varies from $-b$ to $-c$. This easily follows from the fact that the derivatives $d\lambda_2/du$ and $d\lambda_3/dv$ do not change their sign on these intervals.

Thus, we obtain the situation of Theorem 7.7, which describes the molecule W^* for arbitrary metrics of the form

$$ds^2 = \frac{1}{4}(\lambda_2(u) - \lambda_3(v))(du^2 + dv^2).$$

Since each of the functions $\lambda_2(u)$ and $\lambda_3(v)$ has only one local maximum as we have already verified, the molecule W^* of the geodesic flow on the ellipsoid has the form shown in Fig. 12.35(a).

It remains to find the rotation functions on the four edges of W^*. We just use Proposition 13.6. Note that, in Proposition 13.6, the explicit formula for ρ is written in the Liouville coordinates u, v. However, one can carry out all calculations in the initial elliptic coordinates λ_2, λ_3. To this end, we first need to describe Liouville tori and introduce the parameterization on them. Consider an annulus on the ellipsoid given by $-c \leq \lambda_3 \leq -t$ (see Fig. 13.13). The annulus is the projection of a Liouville torus, and we can regard t as a parameter on the corresponding family of tori. It is well known that the geodesics lying on the torus behave as is shown in Fig. 13.13, i.e., move along the annulus touching its boundaries from time to time. If we consider these geodesics with the opposite orientation, we obtain a Liouville torus from another family.

Figure 13.13

The other two families of Liouville tori can be defined in the similar way. They correspond to the annuli $-t \leq \lambda_2 \leq -a$. Thus, the parameter t on Liouville tori varies from a to c. Varying from a to b, the parameter t describes the Liouville tori that belong to the lower edges of W^* shown in Fig. 12.35(a). When t changes from b to c, we obtain Liouville tori from two upper edges. The value $t = b$ is, consequently, a bifurcation. At this instant, Liouville tori undergo the bifurcation, which is presented as the atom C_2 in the molecule W^*.

We now can find the rotation function for the Liouville torus that corresponds to a given value of t.

The Hamiltonian H of the geodesic flow on the ellipsoid is

$$H = \frac{2}{\lambda_2 - \lambda_3}\left(\frac{P(\lambda_2)p_2^2}{\lambda_2} - \frac{P(\lambda_3)p_3^2}{\lambda_3}\right).$$

Without loss of generality, we can assume that $H = 1$ on the given Liouville torus. Then this torus is given by the equations

$$\frac{2P(\lambda_2)p_2^2}{\lambda_2} - \lambda_2 = t, \qquad \frac{2P(\lambda_3)p_3^2}{\lambda_3} - \lambda_3 = t.$$

First, we calculate the action variables I_λ and I_μ on the given Liouville torus related to the cycles λ and μ, where $\lambda = \{\lambda_2 = \text{const}\}$ and $\mu = \{\lambda_3 = \text{const}\}$. For definiteness we assume that $t \in [a, b]$. We have

$$I_\lambda = \frac{1}{2\pi} \int_\lambda (p_2 \, d\lambda_2 + p_3 \, d\lambda_3) \,,$$

where the integral is taken over $\lambda = \{\lambda_2 = \text{const}\}$. Taking this fact into account, we obtain

$$I_\lambda = \frac{1}{2\pi} \int_\lambda p_3 \, d\lambda_3 \,.$$

Substituting the expression for p_3 as a function of λ_3 and taking into account that the cycle λ is obtained as λ_3 runs through the closed interval $[-c, -b]$ four times, we obtain

$$I_\lambda = \frac{2}{\pi} \int_{-c}^{-b} \sqrt{\frac{(t + \lambda_3)\lambda_3}{2P(\lambda_3)}} \, d\lambda_3 \,.$$

The action variable I_μ is calculated analogously; it is equal to

$$I_\mu = \frac{2}{\pi} \int_{-t}^{-a} \sqrt{\frac{(t + \lambda_2)\lambda_2}{2P(\lambda_2)}} \, d\lambda_2 \,.$$

Now we can compute the rotation function (on the lower edges of W) by using the following standard formula for ρ:

$$\rho(t) = \frac{-\partial I_\mu / \partial t}{\partial I_\lambda / \partial t} = \frac{\int_{-t}^{-a} \Phi(u, t) \, du}{\int_{-c}^{-b} \Phi(u, t) \, du} \,,$$

where $\Phi(u, t) = \dfrac{\sqrt{u}}{\sqrt{(u + a)(u + b)(u + c)(u + t)}}$ and $t \in [a, b]$.

For the upper edges of W (i.e., for other two families of Liouville tori), the rotation function ρ is

$$\rho(t) = \frac{\int_{-b}^{-a} \Phi(u, t) \, du}{\int_{-c}^{-t} \Phi(u, t) \, du} \,.$$

Here $t \in [b, c]$.

Thus, we have completely described the molecule W^* and rotation functions for the geodesic flow on the ellipsoid.

Note that we used several times the assumption that the semi-axes of the ellipsoid are pairwise different. If some of them coincide, then the ellipsoid turns into an ellipsoid of revolution, and its geodesic flow becomes linearly integrable.

13.3.2. *The Standard Sphere*

The integration of the geodesic flow on the standard sphere is free of any difficulties. However, it would be interesting to look at the metric on the sphere written in the Liouville form. For example, this representation will be used for constructing a large family of smooth metrics with closed geodesics.

First, we introduce specific coordinates in R^3, the so-called *sphero-conical* ones. They can be regarded as a limit case of the elliptic coordinates. To show this, consider the behavior of elliptic coordinates at infinity, more precisely, as $\lambda_1 \to \infty$. This means that the ellipsoids (given as the level surface $\{\lambda_1 = \text{const}\}$) inflate and transform into spheres in the limit. The level surfaces of the second and third families (namely, one-sheeted and two-sheeted hyperboloids) are transformed into two families of elliptic cones at infinity. By applying a contracting homothety to this asymptotic picture, we can transfer the elliptic coordinates "at infinity" into a bounded region in R^3. As a result, we obtain just the sphero-conical coordinates in R^3. The coordinate surfaces of the first family are concentric spheres, and the coordinate surfaces of two others are elliptic cones.

More precisely, the sphero-conical coordinates ν_2, ν_3 are defined to be the roots of the (quadratic) equation

$$\frac{x^2}{a+\nu} + \frac{y^2}{b+\nu} + \frac{z^2}{c+\nu} = 0,$$

which can be thought as the limit of the (cubic) equation

$$\frac{x^2}{a+\lambda} + \frac{y^2}{b+\lambda} + \frac{z^2}{c+\lambda} = 1$$

as $(x, y, z) \to \infty$. The first sphero-conical coordinate ν_1 is just the sum of the squares:

$$\nu_1 = x^2 + y^2 + z^2.$$

Thus, consider the sphero-conical coordinates (ν_1, ν_2, ν_3) in R^3. The explicit formulas that express the Cartesian coordinates x, y, z through ν_1, ν_2, ν_3 are

$$x^2 = \frac{\nu_1(a+\nu_2)(a+\nu_3)}{(a-b)(a-c)},$$

$$y^2 = \frac{\nu_1(b+\nu_2)(b+\nu_3)}{(b-a)(b-c)},$$

$$z^2 = \frac{\nu_1(c+\nu_2)(c+\nu_3)}{(c-a)(c-b)}.$$

Then the Euclidean metric $ds^2 = dx^2 + dy^2 + dz^2$ in R^3 can be written in sphero-conical coordinates in the following way:

$$ds^2 = \frac{1}{4}\left(\frac{1}{\nu_1} d\nu_1^2 - \frac{\nu_1(\nu_2-\nu_3)}{(a+\nu_2)(b+\nu_2)(c+\nu_2)} d\nu_2^2 - \frac{\nu_1(\nu_3-\nu_2)}{(a+\nu_3)(b+\nu_3)(c+\nu_3)} d\nu_3^2\right).$$

Restrict this metric to the standard sphere S^2, which is given in sphero-conical coordinates by a simple equation: $\nu_1 = 1$. We obtain the following form of the standard metric on S^2 in sphero-conical coordinates:

$$ds^2 = \frac{1}{4}(\nu_2 - \nu_3)\left(-\frac{d\nu_2^2}{(a+\nu_2)(b+\nu_2)(c+\nu_2)} + \frac{d\nu_3^2}{(a+\nu_3)(b+\nu_3)(c+\nu_3)}\right).$$

Using the previous notation, we have

$$ds^2 = \frac{1}{4}(\nu_2 - \nu_3)\left(-\frac{d\nu_2^2}{P(\nu_2)} + \frac{d\nu_3^2}{P(\nu_3)}\right),$$

where $P(\nu) = (a+\nu)(b+\nu)(c+\nu)$.

Thus, the standard metric on the sphere turns out to be written in an almost Liouville form. It is interesting to point out that the above representations for the metrics on the sphere and on the ellipsoid turn out to be quite similar.

As is seen from the above construction, the Liouville coordinates on the standard 2-sphere are not uniquely defined. For example, one can choose the values of a, b, and c in different ways.

Note that, having constructed the Liouville representation for the metric, we immediately obtain a quadratic integral of the geodesic flow. This integral determines the structure of a Liouville foliation on $Q^3 \subset T^*S^2$; and we can distinguish one-parameter families of Liouville tori, examine their bifurcations, and compute the marked molecule with the rotation functions. By repeating the arguments related to the ellipsoid, we can see that the molecule W^* for the standard sphere is the same as for the ellipsoid. Liouville tori are projected again onto two families of annuli on the sphere given by the equations $-c \leq \nu_3 \leq -t$ and $-t \leq \nu_2 \leq -a$ (which are obviously similar to the ellipsoid case).

In conclusion, following our general scheme, we want to write the rotation function ρ for the standard metric on the sphere and to see that $\rho \equiv 1$. (Of course, this fact is known in advance and is obvious.) By repeating literally the above arguments (see Section 13.3.1), we obtain the following answer:

$$\rho(t) = \frac{\int\limits_{-t}^{-a} N(u,t)\,du}{\int\limits_{-c}^{-b} N(u,t)\,du} \qquad \text{if} \quad t \in [a,b]\,,$$

$$\rho(t) = \frac{\int\limits_{-b}^{-a} N(u,t)\,du}{\int\limits_{-c}^{-t} N(u,t)\,du} \qquad \text{if} \quad t \in [b,c]\,.$$

Here

$$N(u,t) = \frac{1}{\sqrt{-(u+a)(u+b)(u+c)(u+t)}}\,.$$

Note that $N(u,t)$ can be rewritten as

$$N(u,t) = \frac{1}{-P(u)(u+t)},$$

where $P(u) = (u+a)(u+b)(u+c)$.

The fact that both the expressions for $\rho(t)$ identically equal 1 follows from the well-known formulas in the theory of elliptic integrals.

13.3.3. The Poisson Sphere

The Poisson sphere is defined to be the two-dimensional sphere endowed with the Riemannian metric

$$ds^2 = \frac{a\,dx^2 + b\,dy^2 + c\,dz^2}{\frac{x^2}{a} + \frac{y^2}{b} + \frac{z^2}{c}},$$

where x,y,z are the Cartesian coordinates in $\mathbb{R}^3$, and $a < b < c$ are arbitrary positive numbers. We assume that the above metric is restricted to the sphere S^2 which is imbedded in $\mathbb{R}^3(x,y,z)$ in the standard way: $S^2 = \{x^2 + y^2 + z^2 = 1\}$. The same metric on the Poisson sphere can be defined in another way. Consider the group $SO(3)$ endowed with a left-invariant Riemannian metric defined by the diagonal matrix $\operatorname{diag}(a,b,c)$. This matrix determines the inner product on the Lie algebra $so(3)$. By extending this product with left shifts along the group, we obtain a left-invariant metric on $SO(3)$. Then we take the left action of the circle $S^1 \simeq SO(2)$ on $SO(3)$ and consider the corresponding quotient space $SO(3)/S^1$. We obtain the two-dimensional sphere S^2. The initial left-invariant metric on $SO(3)$ naturally induces some metric on the base S^2 (submersion metric), which is just the metric on the Poisson sphere.

Let us write this metric in sphero-conical coordinates ν_2, ν_3 on the sphere imbedded into $\mathbb{R}^3$. The Cartesian coordinates x,y,z on the sphere $\{x^2 + y^2 + z^2 = 1\}$ are expressed through ν_2, ν_3 in the following way:

$$x^2 = \frac{(a+\nu_2)(a+\nu_3)}{(a-b)(a-c)},$$

$$y^2 = \frac{(b+\nu_2)(b+\nu_3)}{(b-a)(b-c)},$$

$$z^2 = \frac{(c+\nu_2)(c+\nu_3)}{(c-a)(c-b)}.$$

Substituting these expressions onto the explicit formula of the Poisson metric, we have

$$ds^2 = \frac{1}{4}abc\left(\frac{1}{\nu_3} - \frac{1}{\nu_2}\right)\left(\frac{\nu_2\,d\nu_2^2}{P(\nu_2)} - \frac{\nu_3\,d\nu_3^2}{P(\nu_3)}\right),$$

where $P(\nu) = (a+\nu)(b+\nu)(c+\nu)$.

Note that the numerator in the formula for the Poisson metric $a\,dx^2 + b\,dy^2 + c\,dz^2$ defines the ellipsoid metric on the imbedded sphere, and the conformal multiplier

$$\frac{1}{\frac{x^2}{a} + \frac{y^2}{b} + \frac{z^2}{c}}$$

in sphero-conical coordinates becomes

$$\frac{abc}{\nu_2 \nu_3}.$$

As a result, we see that the metric on the Poisson sphere is connected with that on the ellipsoid as follows:

$$ds^2_{\text{Poisson sphere}} = \frac{abc}{\nu_2 \nu_3}\, ds^2_{\text{ellipsoid}}.$$

It is easily verified that the marked molecule W^* for the metric on the Poisson sphere has the same form as for the ellipsoid (see Fig. 12.35(a) in Chapter 12).

It remains to write the rotation function ρ for the Poisson sphere. The calculation is similar to that in the case of the ellipsoid. As a result, we obtain the following explicit formulas:

$$\rho(t) = \frac{\int\limits_{-t}^{-a} S(u,t)\,du}{\int\limits_{-c}^{-b} S(u,t)\,du} \qquad \text{if} \quad t \in [a,b],$$

$$\rho(t) = \frac{\int\limits_{-b}^{-a} S(u,t)\,du}{\int\limits_{-c}^{-t} S(u,t)\,du} \qquad \text{if} \quad t \in [b,c].$$

Here $S(u,t) = \dfrac{-u}{\sqrt{-(u+a)(u+b)(u+c)(u+t)}}.$

13.4. NON-TRIVIALITY OF ORBITAL EQUIVALENCE CLASSES AND METRICS WITH CLOSED GEODESICS

Consider all Riemannian metrics whose geodesic flows admit non-trivial linear or quadratic integrals and divide this set into classes by including into the same class those metrics whose geodesic flows are smoothly orbitally equivalent (on isoenergy surfaces). The following theorem by E. N. Selivanova shows that these classes are not trivial. In other words, for any metric ds^2, there are many other metrics which are not isometric to ds^2, but have orbitally equivalent geodesic flows.

Theorem 13.3 (E. N. Selivanova [311]). *Let ds^2 be a smooth Riemannian metric on the sphere or torus whose geodesic flow admits a non-trivial linear or quadratic Bott integral. In addition, let ds^2 be different from the flat metric on the torus. Then, on this surface, there exists a family of Riemannian metrics depending on a functional parameter (and pairwise non-isometric) whose geodesic flows are smoothly orbitally equivalent to the geodesic flow of ds^2.*

Proof. The scheme for the proof of this theorem is quite natural. Let, for example, ds^2 be a global Liouville metric on the two-dimensional torus, i.e., $ds^2 = (f(x) + g(y))(dx^2 + dy^2)$ in some global periodic coordinates (x, y). The main orbital invariant of its geodesic flow is the molecule W^* to each of whose edges the corresponding rotation function is assigned. The formulas for them in terms of the functions f and g are explicitly written. The idea is to perturb the functions f and g in such a way that the rotation functions do not change. The structure of critical points of f and g should certainly be preserved in order for the molecule W^* not to change itself. It turns out that it is possible to choose such a perturbation almost explicitly, and the parameter of the perturbation can be chosen to be a function in one variable. We demonstrate this in one important particular case below. Let us notice that the case of the flat metric on the torus has to be excluded from this construction, since the functions f and g are constant in this case, and any perturbation of them (i.e., transformation to a non-constant function) leads us immediately to the bifurcation of the molecule W^*. However, for a fixed flat metric on the torus, the corresponding equivalence class is non-trivial anyway, since it includes all flat metrics, which are not necessarily isometric to each other. Their parameters are the coordinates of basis vectors of the corresponding lattice.

We now apply this scheme to the case of the metric

$$(f(x) + g(y))(dx^2 + dy^2),$$

on the torus, where f and g are periodic, and at least one of them is not constant. First consider the simplest case when each of these functions has exactly two critical points: one minimum and one maximum. In essence, the only orbital invariant in this case is the rotation function ρ on the edges of the molecule W. Thus, we need to construct the perturbation $f \to \tilde{f}$, $g \to \tilde{g}$ which does not change the rotation function ρ. Since ρ is actually expressed by means of the derivatives of the action variables, it suffices to construct a functional family of perturbations which do not change the action variables. As we know the action variables have the form

$$I_1(F) = \frac{1}{2\pi} \int_{y_1}^{y_2} 2\sqrt{g(y) - F}\, dy \qquad \text{if} \quad F > \min(g),$$

$$I_1(F) = \frac{1}{2\pi} \int_0^L \sqrt{g(y) - F}\, dy \qquad \text{if} \quad F < \min(g).$$

We want to perturb the function g in such a way that $I_1(F)$ is not changed. From the geometrical point of view, the function $I_1(F)$ is proportional to the area of the domain U_F lying on the plane with coordinates (y, p_y) and given by the following inequalities (see Fig 13.14):

$$0 < y < L, \qquad g(y) - p_y^2 > F.$$

The perturbation of f can be constructed by the same method as for g, and we restrict ourselves to the function g. Let us divide the segment $[0, L]$ into two parts where g is monotone: from 0 to y_0 and from y_0 to L. Here we assume,

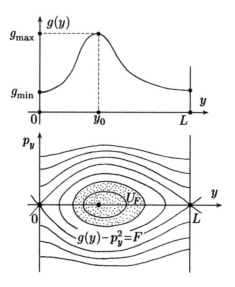

Figure 13.14

for definiteness, that g has its maximum at the points 0 and L (see Fig. 13.14). Let $\alpha_1(s)$ and $\alpha_2(s)$ denote the functions inverse to g on the two intervals of monotonicity. Then the formulas for the action variable $I_1(F)$ can be rewritten as follows:

$$I_1(F) = \frac{1}{2\pi} \int_F^{g_{\max}} 2\sqrt{s - F}(\alpha_1'(s) - \alpha_2'(s))\, ds \qquad \text{if} \quad F > \min(g),$$

$$I_1(F) = \frac{1}{2\pi} \int_{g_{\min}}^{g_{\max}} \sqrt{s - F}(\alpha_1'(s) - \alpha_2'(s))\, ds \qquad \text{if} \quad F < \min(g).$$

These formulas show how one should perturb the function g. The perturbation must not change the difference $\alpha_1'(s) - \alpha_2'(s)$. Consider the perturbed function $\tilde{g}(y)$ on the interval $[0, y_0]$ (Fig. 13.15) such that $\tilde{g}(y)$ remains monotone and coinciding with $g(y)$ in the neighborhoods of 0 and y_0. On the other interval $[y_0, L]$, the perturbed function $\tilde{g}(y)$ is uniquely reconstructed from the condition

$$\tilde{\alpha}_1'(s) - \tilde{\alpha}_2'(s) = \alpha_1'(s) - \alpha_2'(s).$$

In other words, the perturbed function $\tilde{g}$ on $[y_0, L]$ is inverse to the function $\tilde{\alpha}_2 = \alpha_2 - \alpha_1 + \tilde{\alpha}_1$. It is clear that such a perturbation always exists in the class of smooth functions. Thus, in the simplest case when f and g have only two critical points, the statement is proved.

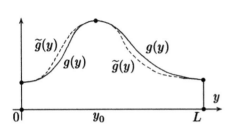

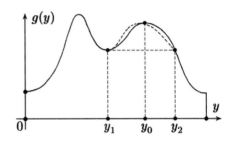

| Figure 13.15 | Figure 13.16 |

If f and g are Morse functions of general type, that is, have several local minima and maxima, the arguments remain, in essence, the same. It suffices to consider a neighborhood of an arbitrary local maximum of g (or f) and to repeat the above arguments (see Fig. 13.16).

The case of the sphere in fact does not differ from that of the torus.   □

The above theorem has interesting corollaries if it is applied in the case of the standard constant curvature metric on the two-dimensional sphere. Namely, we obtain a large family of metrics whose geodesic flows are orbitally equivalent to the geodesic flow of the constant curvature metric. In particular, all their geodesics are closed and, as can easily be shown, have the same length. Such metrics are usually called *Zoll metrics*.

In the case of the so-called revolution metrics (that is, metrics admitting an S^1-action by isometries), this fact is well known (see, for instance, the book by A. Besse [31]). Nevertheless, we suppose that it would be very interesting to look at it in the context of the theory of integrable geodesic flows.

Consider the constant curvature sphere as a surface of revolution. Then its metric can be written in the following standard form (we use here the spherical coordinates):

$$ds_0^2 = d\theta^2 + \sin^2 \theta \, d\varphi^2 .$$

The geodesic flow admits a linear integral $F = p_\varphi$. The molecule has the simplest form $A\!-\!\!-\!\!A$, i.e., there exists a single one-parameter family of Liouville tori. The rotation function on this family is

$$\rho_0(F) = \frac{1}{\pi} \int\limits_{x}^{\pi - x} \frac{F \, d\theta}{\sin \theta \sqrt{\sin^2 \theta - F^2}} ,$$

where $\sin x = F$, $x \in (0, 2\pi)$. It is easy to verify that $\rho_0(F) \equiv 1$; this is just the condition for all geodesics to be closed.

REMARK. In fact, the necessary and sufficient condition for all geodesics to be closed can be written as follows: $\rho(F) \equiv \text{const} = p/q \in \mathbb{Q}$. G. Darboux was the first who obtained it. One of the examples of the surface satisfying this condition for $p = 2$ and $q = 1$ is the well-known Tannery pear (see [31]). However, it has a singularity at one of the poles. This circumstance is essential: the revolution metrics without singularities whose geodesics are all closed exist for $p = 1$ and $q = 1$ only.

Let us look now at how to perturb the constant curvature metric in such a way that the rotation function would not change. Following A. L. Besse's book [31], we will search for the perturbation in the form

$$ds_h^2 = (1 + h \cos \theta)^2 \, d\theta^2 + \sin^2 \theta \, d\varphi^2 \,,$$

where h is some smooth function. Then the rotation function can be written as follows:

$$\rho_h(F) = \frac{1}{\pi} \int_x^{\pi - x} \frac{(1 + h \cos \theta) F \, d\theta}{\sin \theta \sqrt{\sin^2 \theta - F^2}} \,.$$

The closedness condition $\rho_h(F) \equiv 1$ for geodesics can be rewritten in the following way:

$$\rho_h(F) - \rho_0(F) = \frac{1}{\pi} \int_x^{\pi - x} \frac{h(\cos \theta) F \, d\theta}{\sin \theta \sqrt{\sin^2 \theta - F^2}} \equiv 0 \,.$$

It is easy to see that this condition is satisfied for any odd function h. The same condition is necessary. Namely, the following result holds.

Theorem 13.4 [31]. *The revolution metric ds^2 on the two-dimensional sphere is a Zoll metric if and only if it can be written in the following form*:

$$ds_h^2 = (1 + h \cos \theta)^2 d\theta^2 + \sin^2 \theta \, d\varphi^2 \,,$$

where h is an odd function.

COMMENT. Here, in order to write the metric, we use the following analog of spherical coordinates:

$$\varphi \in \mathbb{R} \bmod 2\pi \,, \qquad \theta \in (-\pi/2, \pi/2) \,;$$

they act on the whole sphere, except for its two poles. The metric ds_h^2 can be extended to a smooth metric on the whole sphere if and only if h is a smooth function on $[-1, 1]$ and, moreover, $h(1) = h(-1) = 0$. In addition, choosing h in an appropriate way, it is possible to make this metric real-analytic.

A more interesting series of Zoll metrics can be obtained by considering the constant curvature metric on the sphere as an (L, f, g)-metric. In this case, the perturbation of the standard metric should be done in another way.

Theorem 13.5. *There exists a family of pairwise non-isometric C^∞-smooth metrics $ds^2_{\alpha,\beta}$ on S^2 depending continuously on two functional parameters α and β such that*

1) *the geodesic flows of these metrics are quadratically integrable,*

2) *their geodesics are all closed and have the same length,*

3) *all these metrics are perturbations of the constant curvature metric on the sphere in the sense that, for $\alpha = 0$ and $\beta = 0$, the metric $ds^2_{\alpha,\beta}$ becomes the standard metric ds^2_0,*

4) *all metrics $ds^2_{\alpha,\beta}$ are given by explicit formulas.*

Proof. First, we represent the constant curvature metric ds^2_0 on the two-dimensional sphere as an (L, f, g)-metric. In fact, it has been done in Section 13.3.2, where we have pointed out the following Liouville representation for ds^2_0 in sphero-conical coordinates:

$$ds^2_0 = \frac{1}{4}(\nu_2 - \nu_3)\left(-\frac{d\nu_2^2}{P(\nu_2)} + \frac{d\nu_3^2}{P(\nu_3)}\right),$$

where $P(\nu) = (a+\nu)(b+\nu)(c+\nu)$.

Let us comment this formula. In fact, it can be considered as four separate formulas on the four domains defined as follows:

$U_{++} = \{x > 0, z > 0\}$,
$U_{-+} = \{x < 0, z > 0\}$,
$U_{+-} = \{x > 0, z < 0\}$,
$U_{--} = \{x < 0, z < 0\}$.

It is important that on each of these four domains, the functions $P(\nu_2)$ and $P(\nu_3)$ can be defined independently of each other (by its own formula). The only condition is that they should be compatible on the boundaries of the domains for the smoothness of the metric. For the original metric on the sphere, all four functions are given by the same formula. However, it is not necessary at all. Moreover, two functions $P(\nu_2)$ and $P(\nu_3)$ are given by the same function P (just depending on different variables). It is not necessary, too. We can take two distinct functions, for example, Q and R. The described observations open a way to construct appropriate perturbations of the standard metric on the sphere in such a way that the rotation function would not change. To this end, let us define a metric on each of the four domains as follows:

$$ds^2 = \frac{1}{4}(\nu_2 - \nu_3)\left(-\frac{d\nu_2^2}{Q_+(\nu_2)} + \frac{d\nu_3^2}{R_+(\nu_3)}\right) \qquad \text{on } Q_{++},$$

$$ds^2 = \frac{1}{4}(\nu_2 - \nu_3)\left(-\frac{d\nu_2^2}{Q_+(\nu_2)} + \frac{d\nu_3^2}{R_-(\nu_3)}\right) \qquad \text{on } Q_{-+},$$

$$ds^2 = \frac{1}{4}(\nu_2 - \nu_3)\left(-\frac{d\nu_2^2}{Q_-(\nu_2)} + \frac{d\nu_3^2}{R_+(\nu_3)}\right) \qquad \text{on } Q_{+-},$$

$$ds^2 = \frac{1}{4}(\nu_2 - \nu_3)\left(-\frac{d\nu_2^2}{Q_-(\nu_2)} + \frac{d\nu_3^2}{R_-(\nu_3)}\right) \qquad \text{on } Q_{--}.$$

Here we have to choose the functions in such a way that the obtained metric on the whole sphere becomes smooth. For example, we can just assume that new functions Q_+ and Q_- (defined on $[-b, -a]$) coincide with P near the boundary points $-b$ and $-a$; and analogously, R_+ and R_- (defined on $[-c, -b]$) coincide with P near the boundary points $-b$ and $-c$.

Changing the functions Q_+, Q_-, R_+, R_-, we obtain a functional family of smooth Liouville metrics $ds^2(Q_+, Q_-, R_+, R_-)$ on the sphere.

Recall that, having represented the metric ds_0^2 in the Liouville form, we automatically obtain a quadratic integral of its geodesic flow and, consequently, the structure of a Liouville foliation on the isoenergy surface (i.e., on the unit (co)tangent vector bundle). This foliation has four one-parameter families of Liouville tori, for each of which one can write the formula for the rotation function. As was shown above (see Section 13.3.2), this function (on one of four families of tori) looks like

$$\rho(t)_{\text{standard sphere}} = \frac{\int\limits_{-t}^{-a} N(u,t)\,du}{\int\limits_{-c}^{-b} N(u,t)\,du},$$

where $N(u,t) = \dfrac{1}{\sqrt{-P(u)(u+t)}}$, $t \in (a, b)$. It can easily be verified that

$$\rho(t)_{\text{standard sphere}} \equiv 1.$$

This property is equivalent to the closedness of all the geodesics lying on the Liouville tori from the family under consideration.

The Liouville foliation structure for the metrics $ds^2(Q_+, Q_-, R_+, R_-)$ is just the same as for ds_0^2, and the rotation function on the corresponding family of tori can be written as follows:

$$\rho(t)_{\text{perturbed sphere}} = \frac{\int\limits_{-t}^{-a} N_2(u,t)\,du}{\int\limits_{-c}^{-b} N_3(u,t)\,du},$$

where

$$N_2(u,t) = \frac{1}{\sqrt{u+t}}\left(\frac{1}{\sqrt{-Q_+(u)}} + \frac{1}{\sqrt{-Q_-(u)}}\right),$$

$$N_3(u,t) = \frac{1}{\sqrt{-u-t}}\left(\frac{1}{\sqrt{R_+(u)}} + \frac{1}{\sqrt{R_-(u)}}\right).$$

Now let us choose the functions Q_+, Q_-, R_+, and R_- so that the following condition holds:

$$\rho(t)_{\text{perturbed sphere}} \equiv 1.$$

It will exactly guarantee the closedness of the geodesics. Let us choose the functions in such a way that the integrals do not change at all, i.e.,

$$\frac{1}{\sqrt{-Q_+(u)}} + \frac{1}{\sqrt{-Q_-(u)}} = \frac{2}{\sqrt{-P(u)}},$$

$$\frac{1}{\sqrt{R_+(u)}} + \frac{1}{\sqrt{R_-(u)}} = \frac{2}{\sqrt{P(u)}}.$$

As a result, we obtain

$$\rho(t)_{\text{standard sphere}} = \rho(t)_{\text{perturbed sphere}}.$$

In other words, perturbing the standard metric, we preserve the rotation function.

It is easy to write an explicit formula for these perturbations of the standard metric on the sphere. It is sufficient to set

$$Q_+(\nu_2) = \left(\frac{1}{\sqrt{-P(\nu_2)}} + \alpha(\nu_2) \right)^{-2},$$

$$Q_-(\nu_2) = \left(\frac{1}{\sqrt{-P(\nu_2)}} - \alpha(\nu_2) \right)^{-2},$$

$$R_+(\nu_3) = \left(\frac{1}{\sqrt{P(\nu_3)}} + \beta(\nu_3) \right)^{-2},$$

$$R_-(\nu_3) = \left(\frac{1}{\sqrt{P(\nu_3)}} - \beta(\nu_3) \right)^{-2}.$$

In this representation, α and β are almost arbitrary smooth functions. We have only to require the asymptotics of the function P in its poles not to change after adding α and β, and the expression in brackets to remain always positive. No other assumptions on α and β are needed. One can see from this that the constant curvature metric admits a lot of perturbations leaving the geodesic flow quadratically integrable with the same rotation function, and, consequently, with closed geodesics. Notice that the above perturbations need not be small. The listed conditions can easily be satisfied even under an appropriate choice of large perturbations.

It is worth emphasizing that the constructed perturbations are smooth but not analytic. Moreover, the described method does not allow one to make them analytic. The point is that, as we showed above when classifying quadratically integrable geodesic flows on the sphere, the functions Q_+ and Q_- should coincide in the analytic case, since each real-analytic (L, f, g)-metric admits an additional $\mathbb{Z}_2$-symmetry.

It remains to prove that the perturbed metrics $ds^2(Q_+, Q_-, R_+, R_-)$ are not isometric to the standard metric on the sphere and are not isometric to each other. Non-isometricity to the standard metric easily follows from the fact that the metrics defined by the above formulas have no restrictions imposed on their curvature.

The fact that the perturbed metrics are not isometric to each other follows from the uniqueness of a Liouville representation for the metrics different from the standard one.

The theorem is proved.   □

COMMENT. In a slightly different formulation, this theorem was obtained by K. Kiyohara in [182]. He also generalized this result to the multidimensional case [181]. In our book we follow the construction suggested by E. N. Selivanova.

COMMENT. The Riemannian metrics on the sphere constructed in Theorem 13.5 are new in the sense that they differ from the Zoll metrics of revolution described in Theorem 13.4. In particular, they do not admit any smooth S^1-action by isometries. Moreover, the isometry group of such metrics is discrete. Namely, in the general case (that is, for non-zero α and β), this group is isomorphic to the group $\mathbb{Z}_2$ generated by the reflection with respect to the plane Oxz under which the point (x, y, z) goes to $(x, -y, z)$.

COMMENT. It should be mentioned that, in the papers by V. N. Kolokol'tsov [188], [189], the Zoll metrics in the class of all metrics with quadratically integrable geodesic flows are investigated from the general point of view. In particular, he found the necessary and sufficient conditions for the functions f and g (see the definition of an (L, f, g)-metric) that guarantee the closedness of all geodesics. It was proved that, under these conditions, all the geodesics have the same length and do not have self-intersections. In other words, the surface with such a metric is an *SC-manifold*. Moreover, he, in fact, showed that the structure of the Liouville foliation for their geodesic flows has the simplest form. The topology of this Liouville foliation is shown in Fig. 12.35(a) in the form of a molecule. V. N. Kolokol'tsov also constructed explicit examples of *SC*-metrics on the sphere; however, these metrics are not smooth but have singularities in four points.

In conclusion, let us notice some links between Theorem 13.5 and the paper by V. Guillemin [149] (see also A. L. Besse's book [31]). V. Guillemin proved the existence theorem for a large set of smooth perturbations of the constant curvature metric on the sphere which give metrics with closed geodesics and trivial isometry group. The difference (and some advantage) of our approach is that the perturbed smooth metrics are produced by explicit formulas.

Note that geodesic flows of the metrics with closed geodesics are certainly integrable. It would be very interesting to realize the nature of the integrals appearing here. Are they polynomials in momenta? If yes, then what degree can they have? If no, then how do they look like in general? In fact, at present, there is no explicit example of integrable geodesic flows with non-polynomial integrals.

Geodesic flows of metrics with closed geodesics form a rather special and interesting class in the set of all integrable systems. There are many papers devoted to their investigation. See, for example, [70], [138], [188], [368]. The most complete presentation of this subject can be found in the famous book by A. L. Besse [31].

Chapter 14

The Topology of Liouville Foliations in Classical Integrable Cases in Rigid Body Dynamics

14.1. INTEGRABLE CASES IN RIGID BODY DYNAMICS

In this chapter, we discuss the results on computing of topological invariants for the main integrable cases in rigid body dynamics. The bifurcations of Liouville tori, bifurcation diagrams, and molecules W for these cases were first calculated by M. P. Kharlamov [178] and A. A. Oshemkov [277], [278], [280]. Then the complete invariants of the Liouville foliations (marked molecules W^*) were computed in a series of papers by several authors (A. V. Bolsinov [44], P. Topalov [344], A. V. Bolsinov, A. T. Fomenko [55], [59], O. E. Orel [270], O. E. Orel, S. Takahashi [275]). As a result, a complete classification of the main integrable cases in rigid body dynamics has been obtained up to Liouville equivalence. Just this classification will be presented in this chapter.

The classical *Euler–Poisson equations* [18], [44] that describe the motion of a rigid body with a fixed point in the gravity field have the following form (in the coordinate system whose axes are directed along the principal moments of inertia of the body):

$$A\dot{\omega} = A\omega \times \omega - P \cdot r \times \nu \,,$$
$$\dot{\nu} = \nu \times \omega \,. \tag{14.1}$$

Here ω and ν are phase variables of the system, where ω is the angular velocity vector, ν is the unit vector for the vertical line. The parameters of (14.1) are the diagonal matrix $A = \mathrm{diag}(A_1, A_2, A_3)$ that determines the tensor of inertia of the body, the vector r joining the fixed point with the center of mass, and the weight P of the body. Notation $a \times b$ is used for the vector product in $\mathbb{R}^3$. The vector $A\omega$ has the meaning of the angular momentum of the rigid body with respect to the fixed point.

N. E. Zhukovskiĭ studied the problem on the motion of a rigid body having cavities entirely filled by an ideal incompressible fluid performing irrotational motion [367]. In this case, the angular momentum is equal to $A\omega + \lambda$, where λ is a constant vector characterizing the cyclic motion of the fluid in cavities. The angular momentum has a similar form in the case when a flywheel is fixed in the body such that its axis is directed along the vector λ. Such a mechanical system is called a gyrostat. The motion of a gyrostat in the gravity field, as well as some other problems in mechanics (see, for instance, [180]), are described by the system of equations

$$A\dot{\omega} = (A\omega + \lambda) \times \omega - P \cdot r \times \nu \,,$$
$$\dot{\nu} = \nu \times \omega \,, \tag{14.2}$$

whose particular case for $\lambda = 0$ is system (14.1).

Another generalization of equations (14.1) can be obtained by replacing the homogeneous gravity field with a more complicated one. The equations of motion of a rigid body with a fixed point in an arbitrary potential force field were obtained by Lagrange. If this field has an axis of symmetry, then this axis can be assumed to be vertical, and the equations become

$$A\dot{\omega} = A\omega \times \omega + \nu \times \frac{\partial U}{\partial \nu} \,,$$
$$\dot{\nu} = \nu \times \omega \,, \tag{14.3}$$

where $U(\nu)$ is the potential function, and $\dfrac{\partial U}{\partial \nu}$ denotes the vector with coordinates $\left(\dfrac{\partial U}{\partial \nu_1}, \dfrac{\partial U}{\partial \nu_2}, \dfrac{\partial U}{\partial \nu_3} \right)$. For $U = P\langle r, \nu \rangle$, we obtain system (14.1). By $\langle a, b \rangle$ we denote the standard Euclidean inner product in $\mathbb{R}^3$.

The generalized equations (14.2) and (14.3) can be combined by considering the motion of a gyrostat in an axially symmetric force field. The most general equations that describe various problems in rigid body dynamics have the following form (see, for example, Kharlamov's book [178]):

$$A\dot{\omega} = (A\omega + \varkappa) \times \omega + \nu \times \frac{\partial U}{\partial \nu} \,,$$
$$\dot{\nu} = \nu \times \omega \,, \tag{14.4}$$

where $\varkappa(\nu)$ is the vector function whose components are the coefficients of a certain closed 2-form on the rotation group $SO(3)$, the so-called form of gyroscopic forces. Moreover, $\varkappa(\nu)$ is not arbitrary, but has the form

$$\varkappa = \lambda + (\Lambda - \operatorname{div} \lambda \cdot E)\nu \,, \tag{14.5}$$

where $\lambda(\nu)$ is an arbitrary vector function, $\operatorname{div} \lambda = \dfrac{\partial \lambda_1}{\partial \nu_1} + \dfrac{\partial \lambda_2}{\partial \nu_2} + \dfrac{\partial \lambda_3}{\partial \nu_3}$, and $\Lambda = \left(\dfrac{\partial \lambda_i}{\partial \nu_j} \right)^{\mathsf{T}}$ is the transposed Jacobi matrix. Obviously, systems (14.1)–(14.3) are particular cases of (14.4).

System (14.4) always possesses the geometrical integral

$$F = \langle \nu, \nu \rangle = 1$$

and the energy integral

$$E = \frac{1}{2}\langle A\omega, \omega \rangle + U(\nu).$$

If the vector function $\varkappa(\nu)$ has the form (14.5), then there exists another integral, the so-called area integral

$$G = \langle A\omega + \lambda, \nu \rangle.$$

We now show that equations (14.4), (14.5) are Hamiltonian on common four-dimensional levels of the geometrical and area integrals. Moreover, (14.4) and (14.5) can be represented as the Euler equations for the six-dimensional Lie algebra e(3) of the group of transformations of three-dimensional Euclidean space.

On the dual space e(3)*, there is the standard *Lie–Poisson bracket* defined for arbitrary smooth functions f and g:

$$\{f, g\}(x) = x([d_x f, d_x g]),$$

where $x \in$ e(3)*, $[\cdot, \cdot]$ denotes the commutator in the Lie algebra e(3), and $d_x f$ and $d_x g$ are the differentials of f and g at the point x. These differentials in fact belong to the Lie algebra e(3) after standard identification of e(3)** with e(3). In terms of the natural coordinates

$$S_1, \; S_2, \; S_3, \; R_1, \; R_2, \; R_3$$

on the space e(3)* this bracket takes the form

$$\{S_i, S_j\} = \varepsilon_{ijk} S_k, \qquad \{R_i, S_j\} = \varepsilon_{ijk} R_k, \qquad \{R_i, R_j\} = 0, \qquad (14.6)$$

where $\{i, j, k\} = \{1, 2, 3\}$, and $\varepsilon_{ijk} = \frac{1}{2}(i - j)(j - k)(k - i)$.

A Hamiltonian system on e(3)* relative to the bracket (14.6), i.e., the so-called *Euler equations*, by definition has the form:

$$\dot{S}_i = \{S_i, H\}, \qquad \dot{R}_i = \{R_i, H\},$$

where H is a function on e(3)* called the Hamiltonian of the system. By introducing the vectors

$$S = (S_1, S_2, S_3) \quad \text{and} \quad R = (R_1, R_2, R_3),$$

these equations can be rewritten in the form of the generalized *Kirchhoff equations*:

$$\dot{S} = \frac{\partial H}{\partial S} \times S + \frac{\partial H}{\partial R} \times R, \qquad \dot{R} = \frac{\partial H}{\partial S} \times R. \qquad (14.7)$$

Proposition 14.1. *The mapping* $\varphi \colon \mathbb{R}^6\,(\omega, \nu) \to \mathbb{R}^6\,(S, R)$ *given by the formulas*

$$S = -(A\omega + \lambda)\,, \qquad R = \nu \tag{14.8}$$

establishes an isomorphism between system (14.4), (14.5) *and system* (14.7) *with the Hamiltonian*

$$H = \frac{(S_1 + \lambda_1)^2}{2A_1} + \frac{(S_2 + \lambda_2)^2}{2A_2} + \frac{(S_3 + \lambda_3)^2}{2A_3} + U\,, \tag{14.9}$$

where the parameters A_1, A_2, A_3 *and the functions* $\lambda_1, \lambda_2, \lambda_3,$ *and* U *are taken from* (14.4), (14.5), *but the functions are defined not on the space* $\mathbb{R}^3\,(\nu)$, *but on* $\mathbb{R}^3\,(R)$.

Proof. It suffices to prove that the differential $d\varphi$ of (14.8) sends the vector field defined by (14.4), (14.5) to the vector field defined by (14.7), (14.9). In terms of the chosen coordinates, the differential is given by the following 6×6-matrix:

$$d\varphi = \begin{pmatrix} -A & -\Lambda^\top \\ 0 & E \end{pmatrix}.$$

Thus, we need to show that, for any point $P \in \mathbb{R}^6\,(\omega, \nu)$, the following equality holds:

$$\begin{pmatrix} -A & -\Lambda^\top \\ 0 & E \end{pmatrix}_P \begin{pmatrix} \dot{\omega} \\ \dot{\nu} \end{pmatrix}_P = \begin{pmatrix} \dot{S} \\ \dot{R} \end{pmatrix}_{\varphi(P)}.$$

By computing, we obtain

$$\dot{S} = \frac{\partial H}{\partial S} \times S + \frac{\partial H}{\partial R} \times R = (A^{-1}(S + \lambda)) \times S + (\Lambda A^{-1}(S + \lambda)) \times R + \frac{\partial U}{\partial R} \times R\,,$$

$$-A\dot{\omega} - \Lambda^\top \dot{\nu} = -(A\omega + \lambda + (\Lambda - \operatorname{div}\lambda \cdot E)\nu) \times \omega - \nu \times \frac{\partial U}{\partial \nu} - \Lambda^\top(\nu \times \omega)$$

$$= \omega \times (A\omega + \lambda) - ((\Lambda\nu) \times \omega + \Lambda^\top(\nu \times \omega) - \operatorname{div}\lambda \cdot (\nu \times \omega)) + \frac{\partial U}{\partial \nu} \times \nu$$

$$= \omega \times (A\omega + \lambda) - (\Lambda\omega) \times \nu + \frac{\partial U}{\partial \nu} \times \nu\,.$$

Here we used the formula

$$(Ca) \times b + a \times (Cb) + C^\top(a \times b) = \operatorname{trace} C \cdot (a \times b)\,,$$

which holds for any matrix C and any vectors $a, b \in \mathbb{R}^3$. Comparing the expressions obtained and taking (14.8) into account, we have

$$\dot{S}(\varphi(P)) = (-A\dot{\omega} - \Lambda^\top \dot{\nu})(P)\,.$$

The equality $\dot{R}(\varphi(P)) = \dot{\nu}(P)$ can be verified in a similar way.   □

Corollary. *Condition* (14.5) *imposed on the vector function* $\varkappa(\nu)$ *is equivalent to the fact that* (14.4) *is isomorphic to the Euler equations* (14.7) *on* $e(3)^*$ *with the quadratic (in variables S) Hamiltonian of the form*

$$H = \langle CS, S \rangle + \langle W, S \rangle + V \,, \tag{14.10}$$

where C is a constant symmetric 3×3-matrix, $W(R)$ is a vector function, and $V(R)$ is a smooth scalar function.

Proof. The sufficiency of (14.5) has already been proved. Suppose that system (14.4) with parameters $A, \varkappa, U$ is equivalent to system (14.7) with some Hamiltonian of the form (14.10). Notice that, by a suitable rotation in the S-space, we can reduce Hamiltonian (14.10) to the form (14.9) with some parameters $\widetilde{A}, \widetilde{\lambda}, \widetilde{U}$. Thus, we obtain that two systems of the form (14.4) with parameters $(A, \varkappa, U)$ and $(\widetilde{A}, \widetilde{\varkappa}, \widetilde{U})$ define the same vector field in $\mathbb{R}^6 \,(w, \nu)$. It is easy to verify that such a situation is possible if and only if the following relations are fulfilled:

$$A = k\widetilde{A}\,, \qquad \varkappa(\nu) = k\widetilde{\varkappa}(\nu)\,, \qquad U(\nu) = k\widetilde{U}(\nu) + \psi(\nu_1^2 + \nu_2^2 + \nu_3^2)\,,$$

where k is constant, and $\psi \colon \mathbb{R} \to \mathbb{R}$ is a certain function. Since $\widetilde{\varkappa}$ has the form (14.5), it follows that

$$\varkappa = \lambda + (\Lambda - \operatorname{div} \lambda)\nu\,,$$

where $\lambda = k\widetilde{\lambda}$. $\square$

Under mapping (14.8), the integrals $F = \langle \nu, \nu \rangle$ and $G = \langle A w + \lambda, \nu \rangle$ transform into the invariants of the Lie algebra $e(3)$

$$f_1 = R_1^2 + R_2^2 + R_3^2\,, \qquad f_2 = S_1 R_1 + S_2 R_2 + S_3 R_3\,;$$

and the energy integral $E = \dfrac{1}{2} \langle A w, w \rangle + U(\nu)$ transforms into Hamiltonian (14.9). System (14.7) is Hamiltonian on common four-dimensional level surfaces of the two invariants f_1 and f_2

$$M_{c,g}^4 = \{ f_1 = R_1^2 + R_2^2 + R_3^2 = c,\ f_2 = S_1 R_1 + S_2 R_2 + S_3 R_3 = g \}\,. \tag{14.11}$$

For almost all values of c and g, these common levels are non-singular smooth submanifolds in $e(3)^*$. In what follows, we shall assume that c and g are such regular values.

It is easily seen that these symplectic 4-manifolds $M_{c,g}^4$ are diffeomorphic (for $c > 0$) to the cotangent bundle TS^2 of the 2-sphere S^2. The symplectic structure on $M_{c,g}^4$ is given by the restriction of the Lie–Poisson bracket onto $TS^2 = M_{c,g}^4$ from the ambient six-dimensional space $e(3)^*$. Since the linear transformation $S' = S$, $R' = \gamma R$, where $\gamma = \mathrm{const}$, preserves bracket (14.6), we shall assume in what follows that $c = 1$.

Thus, from now on, we shall consider equations (14.7) with Hamiltonian (14.9) on symplectic four-dimensional manifolds $M_{1,g}^4 = \{ f_1 = 1, f_2 = g \}$ in the six-dimensional space $e(3)^*$. In each specific problem, the phase variables and parameters of the system obtain a concrete physical meaning.

Now we give the list of main integrable cases of equations (14.7), (14.9) with necessary comments. For each case we indicate explicitly the Hamiltonian H and the additional integral K independent of H. Note that sometimes the additional integral K may exist only for exceptional values of the area constant g.

The *Euler case* (1750). The motion of a rigid body about a fixed point that coincides with its center of mass.

$$H = \frac{S_1^2}{2A_1} + \frac{S_2^2}{2A_2} + \frac{S_3^2}{2A_3}, \qquad K = S_1^2 + S_2^2 + S_3^2. \qquad (14.12)$$

The *Lagrange case* (1788). The motion of an axially symmetric rigid body about a fixed point located at the symmetry axis.

$$H = \frac{S_1^2}{2A} + \frac{S_2^2}{2A} + \frac{S_3^2}{2B} + aR_3, \qquad K = S_3. \qquad (14.13)$$

The *Kovalevskaya case* (1899). The motion of a rigid body about a fixed point with the special symmetry conditions indicated below.

$$H = \frac{S_1^2}{2A} + \frac{S_2^2}{2A} + \frac{S_3^2}{A} + a_1 R_1 + a_2 R_2,$$
$$K = \left(\frac{S_1^2 - S_2^2}{2A} + a_2 R_2 - a_1 R_1\right)^2 + \left(\frac{S_1 S_2}{A} - a_1 R_2 - a_2 R_1\right)^2. \qquad (14.14)$$

The integral K has degree 4. In this case, $A_1 = A_2 = 2A_3$ (in particular, the body is axially symmetric), and the center of mass is located in the equatorial plane related to the coinciding axes of the inertia ellipsoid.

The *Goryachev–Chaplygin case* (1899). The motion of a rigid body about a fixed point with the special symmetry conditions indicated below.

$$H = \frac{S_1^2}{2A} + \frac{S_2^2}{2A} + \frac{2S_3^2}{A} + a_1 R_1 + a_2 R_2,$$
$$K = S_3(S_1^2 + S_2^2) - AR_3(a_1 S_1 + a_2 S_2). \qquad (14.15)$$

The integral K has degree 3. In this case, $A_1 = A_2 = 4A_3$, and the center of mass is located in the equatorial plane related to the coinciding axes of the inertia ellipsoid.

In this case, the Poisson bracket of H and K is

$$\{H, K\} = (S_1 R_1 + S_2 R_2 + S_3 R_3)(a_2 S_1 - a_1 S_2).$$

Hence the functions H and K do not commute on all the manifolds $M_{1,g}^4$. Therefore, the system is integrable only on the single special manifold $M_{1,0}^4 = \{f_1 = 1, f_2 = 0\}$. This is a case of partial integrability corresponding to the zero value of the area constant f_2.

Each of these four cases admits an integrable generalization to the case of gyroscopic forces.

The *Zhukovskiĭ case* (1885). The motion of a gyrostat in the gravity field when the body is fixed at its center of mass.

$$H = \frac{(S_1 + \lambda_1)^2}{2A_1} + \frac{(S_2 + \lambda_2)^2}{2A_2} + \frac{(S_3 + \lambda_3)^2}{2A_3},$$

$$K = S_1^2 + S_2^2 + S_3^2.$$

(14.16)

This case is a generalization of the classical Euler case (the Euler case is obtained for $\lambda_1 = \lambda_2 = \lambda_3 = 0$).

The *Lagrange case with gyrostat.*

$$H = \frac{S_1^2}{2A} + \frac{S_2^2}{2A} + \frac{(S_3 + \lambda)^2}{2B} + aR_3, \qquad K = S_3.$$

(14.17)

The classical Lagrange case corresponds to $\lambda = 0$.

The *Kovalevskaya–Yahia case* (1986). The Kovalevskaya case with gyrostat.

$$H = \frac{S_1^2}{2A} + \frac{S_2^2}{2A} + \frac{(S_3 + \lambda)^2}{A} + a_1 R_1 + a_2 R_2,$$

$$K = \left(\frac{S_1^2 - S_2^2}{2A} + a_2 R_2 - a_1 R_1 \right)^2 + \left(\frac{S_1 S_2}{A} - a_1 R_2 - a_2 R_1 \right)^2$$

$$- \frac{2\lambda}{A^2}(S_3 + 2\lambda)(S_1^2 + S_2^2) + \frac{4\lambda R_3}{A}(a_1 S_1 + a_2 S_2).$$

(14.18)

The classical Kovalevskaya case is obtained for $\lambda = 0$.

The *Sretenskiĭ case* (1963). The Goryachev–Chaplygin case with gyrostat.

$$H = \frac{S_1^2}{2A} + \frac{S_2^2}{2A} + \frac{2(S_3 + \lambda)^2}{A} + a_1 R_1 + a_2 R_2,$$

$$K = (S_3 + 2\lambda)(S_1^2 + S_2^2) - AR_3(a_1 S_1 + a_2 S_2).$$

(14.19)

If $\lambda = 0$, then we obtain the classical Goryachev–Chaplygin case. The system is integrable on the zero level of the area integral.

The *Clebsch case* (1871). Motion of a rigid body in a fluid.

$$H = \frac{S_1^2}{2A_1} + \frac{S_2^2}{2A_2} + \frac{S_3^2}{2A_3} + \frac{\varepsilon}{2}(A_1 R_1^2 + A_2 R_2^2 + A_3 R_3^2),$$

$$K = \frac{1}{2}(S_1^2 + S_2^2 + S_3^2) - \frac{\varepsilon}{2}(A_2 A_3 R_1^2 + A_3 A_1 R_2^2 + A_1 A_2 R_3^2).$$

(14.20)

The *Steklov–Lyapunov case* (1893). Motion of a rigid body in a fluid.

$$H = \frac{S_1^2}{2A_1} + \frac{S_2^2}{2A_2} + \frac{S_3^2}{2A_3} + \varepsilon(A_1 S_1 R_1 + A_2 S_2 R_2 + A_3 S_3 R_3)$$

$$+ \frac{\varepsilon^2}{2}(A_1(A_2^2 + A_3^2)R_1^2 + A_2(A_3^2 + A_1^2)R_2^2 + A_3(A_1^2 + A_2^2)R_3^2),$$

$$K = (S_1^2 + S_2^2 + S_3^2) - 2\varepsilon(A_2 A_3 S_1 R_1 + A_3 A_1 S_2 R_2 + A_1 A_2 S_3 R_3)$$

$$+ \varepsilon^2(A_1^2(A_2 - A_3)^2 R_1^2 + A_2^2(A_3 - A_1)^2 R_2^2 + A_3^2(A_1 - A_2)^2 R_3^2).$$

(14.21)

When studying the topology of the listed integrable cases, we often need to find critical points of a function given on a common level surface of other functions, and to calculate their indices. Let us describe one of the possible methods of calculation to be used in what follows.

Suppose we are given smooth functions $h_0, h_1, \ldots, h_m$ in $\mathbb{R}^n$ and let

$$M^{n-m} = \{h_1 = p_1, \ldots, h_m = p_m\}$$

be their common non-singular level surface. Then the vectors

$$\operatorname{grad} h_1, \ldots, \operatorname{grad} h_m$$

are linearly independent at every point $x \in M^{n-m} \subset \mathbb{R}^n$. The point $x_0 \in M^{n-m}$ is a critical point for the function $\widetilde{h}_0 = h_0|_M$ if and only if there is a collection of coefficients $(\lambda_1, \ldots, \lambda_m)$ such that

$$\operatorname{grad} h_0(x_0) = \sum_{i=1}^{m} \lambda_i \operatorname{grad} h_i(x_0). \tag{14.22}$$

Consider the matrix

$$G = G_0 - \sum_{i=1}^{m} \lambda_i G_i, \tag{14.23}$$

where G_i is the Hesse matrix of the function h_i at the point x_0, and the coefficients $\lambda_1, \ldots, \lambda_m$ are taken from (14.22). The Hesse matrix at a critical point is the symmetric matrix whose elements are the second derivatives of the function. This matrix determines a certain quadratic form on tangent vectors at the critical point. The number of negative eigenvalues is called the *index* of the critical point, and the number of zero eigenvalues is the *degeneracy index* of the critical point.

Lemma 14.1. *Let condition (14.22) be fulfilled for the point $x_0 \in M^{n-m} \subset \mathbb{R}^n$. Then the quadratic form determined by the Hesse matrix of the function $\widetilde{h}_0 = h_0|_M$ at the point x_0 is the restriction of the form (14.23) to the tangent space $T_{x_0} M^{n-m}$.*

Proof. Let $\gamma(t) \subset M^{n-m}$ be an arbitrary smooth curve such that $\gamma(0) = x_0$. Then the value of the quadratic form, determined by the Hesse matrix of the function $\widetilde{h}_0$, on the vector $a = \dot\gamma(0) \in T_{x_0} M^{n-m}$ is equal to

$$\widetilde{G}_0(a, a) = \left.\frac{d^2}{dt^2}\right|_{t=0} \widetilde{h}_0(\gamma(t)) = \left.\frac{d^2}{dt^2}\right|_{t=0} h_0(\gamma(t))$$

$$= G_0(a, a) + \langle \operatorname{grad} h_0(x_0), \ddot\gamma(0) \rangle$$

$$= G_0(a, a) + \sum_{i=1}^{m} \lambda_i \langle \operatorname{grad} h_i(x_0), \ddot\gamma(0) \rangle$$

$$= G_0(a, a) + \sum_{i=1}^{m} \lambda_i \left(\left.\frac{d^2}{dt^2}\right|_{t=0} h_i(\gamma(t)) - G_i(a, a) \right) = G(a, a),$$

because $h_i(\gamma(t)) \equiv p_i$. Here $G_i(a, a)$ is the value of the quadratic form G_i on the vector a, and $\langle \operatorname{grad} h_i(x_0), \ddot\gamma(0) \rangle$ denotes the standard pairing in $\mathbb{R}^n$ of the vector $\ddot\gamma(0)$ and covector $\operatorname{grad} h_i(x_0)$. $\square$

The given method of calculation is convenient, because one need not introduce any local coordinate systems on the common level surfaces of the functions in order to find the indices of critical points. It is sufficient to choose some basis in the tangent space of this surface and calculate the value of the form (14.23) on the basis vectors.

The next simple assertion, which often simplifies the computation, consists in the following.

Suppose, as before, that we are given some functions $h_1, \ldots, h_m, f, g$ in $\mathbb{R}^n$, and let $M = \{h_1 = p_1, \ldots, h_m = p_m\}$ be a non-singular common level surface of the functions $h_1, \ldots, h_m$. Consider the mapping

$$\tilde{f} \times \tilde{g} \colon M \to \mathbb{R}^2 \,,$$

where $\tilde{f} = f|_M$, $\tilde{g} = g|_M$, $(\tilde{f} \times \tilde{g})(x) = (\tilde{f}(x), \tilde{g}(x)) \in \mathbb{R}^2$. Let K be the set of critical points of the mapping $\tilde{f} \times \tilde{g}$, and let $\gamma(t) \subset K$ be a smooth curve. Moreover, let $\operatorname{grad} \tilde{f} \neq 0$ at each point $y \in \gamma(t)$. The mapping $\tilde{f} \times \tilde{g}$ takes the curve $\gamma(t)$ to a certain curve $(a(t), b(t))$ in $\mathbb{R}^2$, where $a(t) = f(\gamma(t))$, $b(t) = g(\gamma(t))$. Since each point of $y \in \gamma(t)$ is critical for the mapping $\tilde{f} \times \tilde{g}$ and $\operatorname{grad} \tilde{f} \neq 0$ at points of $\gamma(t)$, it follows that there exist uniquely defined functions $\lambda(t), \lambda_1(t), \ldots, \lambda_m(t)$ such that

$$\operatorname{grad} g(\gamma(t)) = \lambda(t) \operatorname{grad} f(\gamma(t)) + \sum_{i=1}^{m} \lambda_i(t) \operatorname{grad} h_i(\gamma(t)) \,. \qquad (14.24)$$

Lemma 14.2. *For each value t, we have the following relation:*

$$\frac{db}{dt} = \lambda(t) \frac{da}{dt} \,,$$

where $\lambda(t)$ is the coefficient at $\operatorname{grad} f$ in expansion (14.24) for $\operatorname{grad} g$.

Proof. By straightforward calculation, we obtain

$$\frac{db}{dt} = \frac{d}{dt}(g(\gamma(t)) = \langle \operatorname{grad} g(\gamma(t)), \dot{\gamma}(t) \rangle$$

$$= \lambda(t) \langle \operatorname{grad} f(\gamma(t)), \dot{\gamma}(t) \rangle + \sum_{i=1}^{m} \lambda_i(t) \langle \operatorname{grad} h_i(\gamma(t)), \dot{\gamma}(t) \rangle$$

$$= \lambda(t) \frac{d}{dt}(f(\gamma(t))) + \sum_{i=1}^{m} \lambda_i(t) \frac{d}{dt}(h_i(\gamma(t)))$$

$$= \lambda(t) \frac{da}{dt} \,,$$

since $h_i(\gamma(t)) \equiv p_i$. $\square$

14.2. TOPOLOGICAL TYPE OF ISOENERGY 3-SURFACES

14.2.1. The Topology of the Isoenergy Surface and the Bifurcation Diagram

In this section, we describe the topological types of isoenergy three-dimensional surfaces for Hamiltonians (14.12)–(14.21).

In this case, an isoenergy surface Q^3 is defined to be a common level surface of the functions f_1, f_2, H given on the Euclidean space $\mathbb{R}^6 (S, R)$, where

$$f_1 = R_1^2 + R_2^2 + R_3^2, \qquad f_2 = S_1 R_1 + S_2 R_2 + S_3 R_3,$$

and H is the Hamiltonian. Since we assume that $f_1 = 1$, different surfaces Q^3 are determined by two parameters g and h (the values of the functions f_2 and H). The description of the topological types of Q^3 will be given in the following way. We consider the bifurcation diagram for the integrals f_2 and H. As a result, we obtain certain curves on the plane $\mathbb{R}^2 (g, h)$ separating it into regions so that, for all points (g, h) from the same region, the topological type of the corresponding isoenergy surfaces

$$Q^3 = \{f_1 = 1,\ f_2 = g,\ H = h\}$$

is also the same.

Let us emphasize that these bifurcation diagrams have nothing to do with the integrability of a given system and can be constructed for arbitrary Hamiltonians.

Thus, consider the mapping

$$F = f_2 \times H \colon S^2 \times \mathbb{R}^3 \to \mathbb{R}^2 (g, h)$$

given by $F(P) = (f_2(P), H(P)) \in \mathbb{R}^2 (g, h)$, where $P \in S^2 \times \mathbb{R}^3$. The image of the set of critical points of F is the bifurcation diagram $\Sigma \in \mathbb{R}^2 (g, h)$. The preimage of an arbitrary point $(g, h) \notin \Sigma$ is a non-singular isoenergy surface $Q^3 = \{f_1 = 1,\ f_2 = g,\ H = h\}$. For the Hamiltonians of the form (14.9), F is a proper mapping, i.e., the preimage of any compact subset in $\mathbb{R}^2 (g, h)$ is compact. Hence, for all the points (g, h) that belong to the same connected component of the set $\mathbb{R}^2 (g, h) \setminus \Sigma$, the topological type of Q^3 will be also the same.

How to find the type of the isoenergy surface Q for each connected component of the set $\mathbb{R}^2 (g, h) \setminus \Sigma$? The answer can be obtained by using the following result by S. Smale [316].

Proposition 14.2. Let $H = \langle CS, S \rangle + \langle W, S \rangle + V$, where C is a constant positively defined 3×3-matrix, $W(R)$ is an arbitrary vector function, and $V(R)$ is an arbitrary smooth function. Consider the projection $\pi(Q^3)$ of the isoenergy surface Q^3 to the two-dimensional Poisson sphere given in $\mathbb{R}^3 (R)$ by the equation $R_1^2 + R_2^2 + R_3^2 = 1$, where $\pi(S, R) = R$. Let $\pi(Q^3)$ be homeomorphic to the sphere with m holes. Then:

1) if $m = 0$, then Q^3 is diffeomorphic to $\mathbb{R}P^3$;
2) if $m = 1$, then Q^3 is diffeomorphic to S^3;
3) if $m > 1$, then Q^3 is diffeomorphic to $\underbrace{(S^1 \times S^2) \# \ldots \# (S^1 \times S^2)}_{m-1 \text{ times}}$.

Proof. Consider the projection of Q^3 onto the Poisson sphere S^2. Take an arbitrary point $R = (R_1, R_2, R_3) \in \pi(Q^3)$ and the fiber $\pi^{-1}(R)$ over it. This fiber is the intersection of the plane $\{R_1 S_1 + R_2 S_2 + R_3 S_3 = g\}$ with the ellipsoid $\{H = \langle CS, S \rangle + \langle W, S \rangle + V = \text{const}\}$. Therefore, it is homeomorphic to either a circle or a point, or is empty. More specifically, if R lies inside the image $\pi(Q^3)$, then it is easy to see that the fiber is homeomorphic to a circle. If R belongs to the boundary of $\pi(Q^3)$, then the fiber is a point. If $R \notin \pi(Q^3)$, then the fiber is empty. Therefore, topologically the 3-manifold Q^3 can be characterized in the following way.

If the projection $\pi(Q^3)$ coincides with the whole Poisson sphere, then Q^3 represents an S^1-fibration over the sphere topologically equivalent to the unit tangent vector bundle. Therefore, $Q^3 = \mathbb{RP}^3$.

If the projection $\pi(Q^3)$ does not coincide with the whole sphere, i.e., has some holes, then Q^3 can be obtained as follows. First we consider the direct product $\pi(Q^3) \times S^1$, and then we shrink a fiber S^1 into a point over each boundary point $R \in \partial(\pi(Q^3))$. We already met the three-dimensional manifolds obtained in this way in Chapter 4, and their topology was described in Proposition 4.5. The two last assertions (2) and (3) are just a reformulation of Proposition 4.5. □

Thus, having constructed the bifurcation diagram of F and indicated the topological type of Q^3 for each region into which Σ separates the plane $\mathbb{R}^2(g, h)$, we obtain a complete description of all topological types of non-singular isoenergy surfaces of a given Hamiltonian system.

In the papers by S. B. Katok [177] and Ya. V. Tatarinov [333], [334], the above approach was applied to study the topology of surfaces Q^3 for the problem on the motion of a rigid body with a fixed point. In these papers, the bifurcation diagrams of the mapping $F = f_2 \times H$ were constructed for Hamiltonians of sufficiently general types (not necessarily integrable). In particular, the cases by Euler, Lagrange, and Kovalevskaya were studied. We now describe the bifurcation diagrams for integrable cases in detail.

To this end, we use the following general idea. As before, we represent the system of equations that describe the motion of a rigid body (or, more general, a gyrostat) as the Euler equations on the coalgebra $e(3)^*$ with coordinates $(S_1, S_2, S_3, R_1, R_2, R_3)$ and the Poisson bracket

$$\{S_i, S_j\} = \varepsilon_{ijk} S_k, \qquad \{S_i, R_j\} = \varepsilon_{ijk} R_k, \qquad \{R_i, R_j\} = 0.$$

The kernel of this bracket is generated by the functions $f_1 = R_1^2 + R_2^2 + R_3^2$ (geometrical integral) and $f_2 = S_1 R_1 + S_2 R_2 + S_3 R_3$ (area integral). The regular common level surfaces $\{f_1 = 1, f_2 = g\}$ are all diffeomorphic to the tangent bundle of S^2. The Hamiltonian is taken in the form

$$H = \frac{(S_1 + \lambda_1)^2}{2A_1} + \frac{(S_2 + \lambda_2)^2}{2A_2} + \frac{(S_3 + \lambda_3)^2}{2A_3} + U(R_1, R_2, R_3),$$

where A_1, A_2, A_3 are the principal moments of inertia of the body, $\lambda = (\lambda_1, \lambda_2, \lambda_3)$ is the gyrostatic momentum, $U(R_1, R_2, R_3)$ is the potential, which in our case has the form $a_1 R_1 + a_2 R_2 + a_3 R_3$. Thus, we need to describe the topology of isoenergy surfaces $Q^3_{h,g} = \{f_1 = 1, \; f_2 = g, \; H = h\}$ for different constants $A_1, A_2, A_3, \lambda_1, \lambda_2, \lambda_3, g, h$ and functions $U(R_1, R_2, R_3)$.

Let us introduce the following notation: $S = (S_1, S_2, S_3)$, $R = (R_1, R_2, R_3)$, $\lambda = (\lambda_1, \lambda_2, \lambda_3)$, $a = (a_1, a_2, a_3)$, $A = \mathrm{diag}(A_1, A_2, A_3)$, where S, R, λ, a are three-dimensional vectors and A is a diagonal matrix. We will assume that all these vectors belong to the same Euclidean space $\mathbb{R}^3$ with scalar product $\langle \cdot, \cdot \rangle$. Then

$$f_1 = \langle R, R \rangle, \qquad f_2 = \langle S, R \rangle, \qquad H = \frac{1}{2} \langle A^{-1}(S + \lambda), S + \lambda \rangle + U(R).$$

Consider the projection $\pi \colon (S, R) \to R$. Since we have $f_1(R) \equiv 1$, the image of the isoenergy surface $Q^3_{h,g}$ under this projection is some subset of the Poisson sphere $S^2 = \{R_1^2 + R_2^2 + R_3^2 = 1\} \subset \mathbb{R}^3 (R_1, R_2, R_3)$. Obviously, the point $R \in S^2$ belongs to the image $\pi(Q^3_{h,g})$ if and only if there exists a solution of the equations

$$\langle S, R \rangle = g, \qquad \langle A^{-1}(S + \lambda), S + \lambda \rangle = 2(h - U(R)),$$

where $(S_1, S_2, S_3) = S$ is unknown. The first equation determines a plane, and the second determines an ellipsoid in $\mathbb{R}^3 (S_1, S_2, S_3)$ provided $h - U(R) > 0$. It is easily seen that these two surfaces intersect if and only if

$$\frac{(g + \langle \lambda, R \rangle)^2}{2 \langle AR, R \rangle} + U(R) \geq h.$$

Moreover, the intersection is a circle in the case of the strong inequality, and a point in the case of the equality.

Consider the function

$$\varphi_g(R) = \frac{(g + \langle \lambda, R \rangle)^2}{2 \langle AR, R \rangle} + U(R),$$

given on the Poisson sphere and called the *reduced potential*.

It turns out that the topology of isoenergy surfaces $Q^3_{h,g}$ and their bifurcations under varying the energy level h is completely determined by the function $\varphi_g(R)$. Namely, combining Proposition 14.2 with the above arguments, we obtain the following result.

Theorem 14.1. *Let $\varphi_g(R)$ be the reduced potential.*

1) If $h < \min \varphi_g(R)$, then $Q^3_{g,h}$ is empty.

2) If h is a regular value of $\varphi_g(R)$, where $\min \varphi_g(R) < h < \max \varphi_g(R)$, then the set $\{\varphi_g(R) \leq h\}$ is a disjoint union of two-dimensional manifolds with boundary $P_{i_1}, \ldots, P_{i_m}$ embedded into the Poisson sphere, where P_k is a 2-disc with k holes. In this case, the isoenergy surface $Q^3_{h,g}$ is a smooth three-dimensional manifold which is homeomorphic to a disjoint union of three-dimensional manifolds $N_{i_1}, \ldots, N_{i_m}$, where N_0 is the three-dimensional sphere, and N_k ($k \geq 1$) is the connected sum of k copies of $S^1 \times S^2$.

3) If $h > \max \varphi_g(R)$, then $Q^3_{g,h} \simeq \mathbb{R}P^3$.

Using this theorem, we can suggest the following algorithm for description of the topological type of $Q_{g,h}^3$.

Step 1. We construct the Reeb graph of the reduced potential $\varphi_g(R)$. If the value of h is less than the minimum (or greater than the maximum) of $\varphi_g(R)$, then $Q_{h,g}^3$ is empty (or homeomorphic to $\mathbb{R}P^3$ respectively).

Step 2. Let h belong to the image of the function $\varphi_g(R)$. Then we cut the Reeb graph along the level h and consider its lower part. The number of connected components of the lower part coincides with the number of connected components of $Q_{g,h}^3$. Moreover, if a connected component of the graph has k boundary points (here we mean those points along which the graph was cut, but not vertices of the graph), then the corresponding connected component of $Q_{g,h}^3$ is homeomorphic to the connected sum of $k-1$ copies of $S^1 \times S^2$ for $k \geq 1$, or to the 3-sphere S^3 for $k=1$.

We now analyze each integrable case separately.

14.2.2. Euler Case

In this case, $\lambda = 0$ and $U(R) \equiv 0$. The reduced potential has the form

$$\varphi_g(R) = \frac{g^2}{2\langle AR, R\rangle}.$$

For $g \neq 0$, the critical points of this function on the Poisson sphere are the same as those of the quadric $\langle AR, R\rangle$, i.e., those points where the corresponding ellipsoid $\{\langle AR, R\rangle = \text{const}\}$ is tangent to the sphere $\{\langle R, R\rangle = 1\}$. For each $g \neq 0$, there are six critical points

$$R = (\pm 1, 0, 0,), \ (0, \pm 1, 0), \ (0, 0, \pm 1).$$

They correspond to three critical values: $h_i = \dfrac{g^2}{2A_i}$. For $g = 0$, the reduced potential has the only critical value equal to zero. Thus, the bifurcation diagram in the Euler case consists of three parabolas (Fig. 14.1).

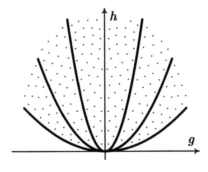

Figure 14.1

The topological type of isoenergy surfaces is reconstructed by means of the corresponding Reeb graphs presented in Fig. 14.2. In the case when some

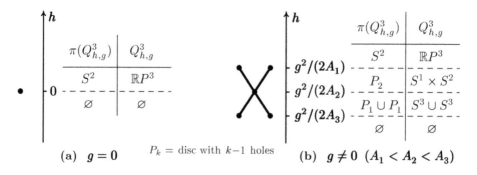

(a) $g = 0$ P_k = disc with $k-1$ holes   (b) $g \neq 0$ ($A_1 < A_2 < A_3$)

Figure 14.2

of principal moments of inertia coincide, some vertices of the Reeb graph shrink into one point. In the bifurcation diagram, the corresponding parabolas also become the same.

Let us summarize the results. The bifurcation diagram of the mapping $f_2 \times H$ for the Hamiltonian (corresponding to the Euler case)

$$H = \frac{S_1^2}{2A_1} + \frac{S_2^2}{2A_2} + \frac{S_3^2}{2A_3}$$

has the simple form shown in Fig. 14.3. It consists of 3 parabolas $\left\{ h = \frac{g^2}{2A_i} \right\}$, $i = 1, 2, 3$, which divide the plane $\mathbb{R}^2 (g, h)$ into 6 regions. In each region in Fig. 14.3,

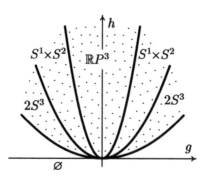

Figure 14.3

we indicate the topological type of the 3-manifold Q^3. The sign $\varnothing$ means that, for each point (g, h) from this region, its preimage under the mapping $f_2 \times H$ is empty, i.e., $(f_2 \times H)^{-1}(g, h) = \varnothing$. Figure 14.3 represents the case when the parameters A_1, A_2, A_3 of a rigid body are connected by the relation $0 < A_3 < A_2 < A_1$. If $0 < A_3 = A_2 < A_1$, then two upper parabolas coincide.

If $0 < A_3 < A_2 = A_1$, then two lower parabolas coincide. For $A_1 = A_2 = A_3$, the Hamiltonian H of the Euler case becomes resonant, since the corresponding system has two functionally independent integrals (for example, S_1 and S_2). In fact, this case is very simple (it is actually equivalent to the geodesic flow on the standard constant curvature sphere) and we shall not consider it here.

14.2.3. Lagrange Case

For the classical Hamiltonian of the Lagrange case

$$ H = \frac{1}{2}\left(S_1^2 + S_2^2 + \frac{S_3^2}{\beta} \right) + R_3, \qquad \text{where} \quad \beta > 0, $$

the bifurcation diagrams are essentially different in the following cases:
 a) $0 < \beta < 1$,
 b) $\beta = 1$,
 c) $1 < \beta \le 4/3$,
 d) $\beta > 4/3$.
They are presented in Fig. 14.4.

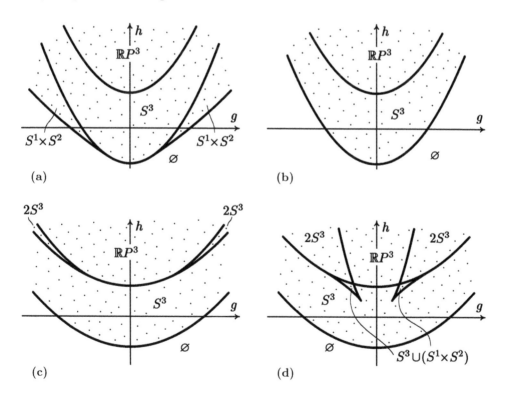

Figure 14.4

The bifurcation diagrams consist of two parabolas $\left\{ h = \dfrac{g^2}{2\beta} \pm 1 \right\}$ and the curve (for $\beta \neq 1$) that can be given parametrically as follows:

$$g = t + \frac{1}{(\beta - 1)t^3}, \qquad h = \frac{t^2}{2} + \frac{3}{2(\beta - 1)t^2}, \qquad t^2 \geq \frac{1}{|\beta - 1|}.$$

This curves touches one of the parabolas at the points

$$\left(\pm \frac{\beta}{\sqrt{|1 - \beta|}}, \frac{3\beta - 2}{2|1 - \beta|} \right) \in \mathbb{R}^2 (g, h)$$

and, for $\beta > 4/3$, has two cusp points with coordinates

$$\left(\pm \frac{4}{3} \sqrt[4]{\frac{3}{\beta - 1}}, \sqrt{\frac{3}{\beta - 1}} \right).$$

The topological type of Q^3 is indicated in Fig. 14.4. Here $S^3 \cup (S^1 \times S^2)$ denotes the disjoint union of S^3 and $S^1 \times S^2$.

We now comment on these results.

First consider a Hamiltonian of a more general kind given by the following parameters:

$$A_1 = A_2 = B, \quad \lambda_1 = \lambda_2 = 0, \quad U(R) = U(R_3).$$

The reduced potential has then the form

$$\varphi_g(R) = \frac{(g + \lambda_3 R_3)^2}{2(BR_1^2 + BR_2^2 + A_3 R_3^2)} + U(R_3) = \frac{(g + \lambda_3 R_3)^2}{2(B + (A_3 - B)R_3^2)} + U(R_3).$$

It depends on the third coordinate R_3 only. Therefore, for any values of g and h, the region defined on the Poisson sphere by the condition $\varphi_g(R) \leq h$ consists of several annuli and, perhaps, one or two discs. Therefore, the isoenergy surface $Q^3_{g,h}$ is homeomorphic to a disjoint union of $S^1 \times S^2$ and, perhaps, one or two 3-spheres.

Consider now the standard Lagrange case with a liner potential in detail. Without loss of generality, we shall assume that

$$A_1 = A_2 = 1, \quad A_3 = \beta, \quad \lambda = (0, 0, 0), \quad U(R) = R_3.$$

Then the reduced potential becomes

$$\varphi_g(R) = f(R_3) = \frac{g^2}{2(1 + (\beta - 1)R_3^2)} + R_3.$$

Let $0 < \beta < 1$. It is easy to see that $f(-1) = \dfrac{g^2}{2\beta} - 1$ and $f(1) = \dfrac{g^2}{2\beta} + 1$; moreover, $f''(R_3) > 0$ on the whole segment $[-1, 1]$. Therefore, from the qualitative point of view, there are two possibilities for the graph of the function f shown in Figs. 14.5(a) and 14.5(b). The choice of one of these possibilities is determined by the sign of the derivative $f'(-1) = 1 - \dfrac{g^2(1-\beta)}{\beta^2}$. Thus, cases (a) and (b) are obtained for $g^2 < \dfrac{\beta^2}{1-\beta}$ and for $g^2 > \dfrac{\beta^2}{1-\beta}$ respectively. The corresponding Reeb graphs are presented in Fig. 14.6.

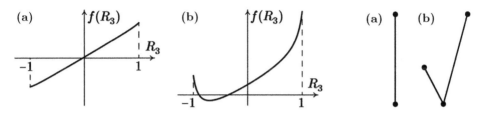

Figure 14.5   Figure 14.6

At the critical points corresponding to the degree 1 vertices of the Reeb graph, the function f takes values $\pm 1 + \dfrac{g^2}{2\beta}$. The minimal value in case (b) can be calculated in the following way. The condition that the derivative

$$f'(R_3) = 1 - \frac{g^2(\beta-1)R_3}{(1+(\beta-1)R_3^2)^2}$$

is equal to zero is fulfilled only for $(\beta-1)R_3 > 0$. Denote $(\beta-1)R_3$ by t^{-2}. Then $|t|^2 \geq (1-\beta)^{-1}$, since $|R_3| \leq 1$. By rewriting the condition $f'(R_3) = 0$ in terms of t and substituting $(\beta-1)^{-1}t^{-2}$ instead of R_3, we obtain

$$g = t + \frac{1}{t^3(\beta-1)}, \qquad h = \frac{t^2}{2} + \frac{3}{2(\beta-1)t^2}, \qquad |t| \geq \frac{1}{\sqrt{|1-\beta|}}.$$

Regarding these equations as a parametrized curve on the plane $\mathbb{R}^2(g,h)$, we see that, in our case, the bifurcation diagram consists of this curve and two parabolas $\left\{ h = \pm 1 + \dfrac{g^2}{2\beta} \right\}$, as shown in Fig. 14.7.

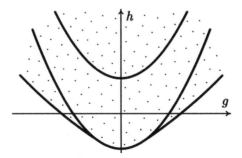

Figure 14.7

In the case $\beta = 1$ the reduced potential is linear: $f(R_3) = R_3 + \dfrac{g^2}{2}$. The bifurcation diagram consists of two parabolas $\{h = \pm 1 + g^2\}$ only.

Suppose that $\beta > 1$. Then four types of graphs of $f(R_3)$ are possible (see Fig. 14.8). The corresponding Reeb graphs are presented in Fig. 14.9.

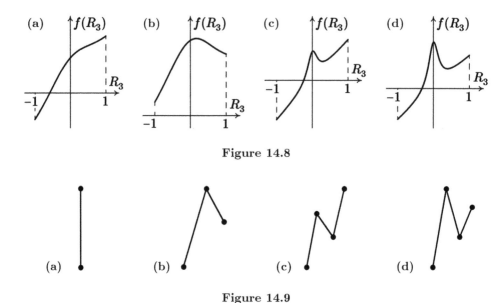

Figure 14.8

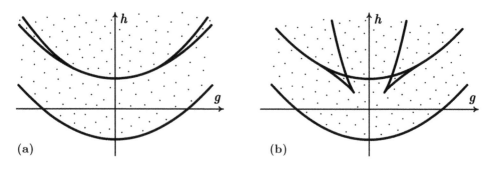

Figure 14.9

The specific type for various values of β and g is defined by the number of zeros of $f'(R_3)$ on the interval $(-1, 1)$ (or, equivalently, by the number of zeros of the polynomial $P(x) = (1 + (\beta - 1)x^2)^2 - g^2(\beta - 1)x$ on $(-1, 1)$) and by the values of f at these zeros.

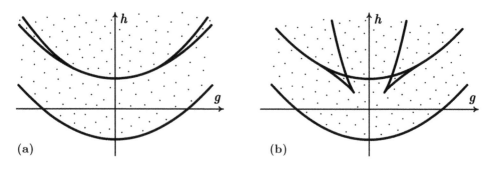

Figure 14.10

Having analyzed each case separately, we obtain the answer. The bifurcation diagram consists of two parabolas and the curve given by the same equation as in the case $\beta < 1$. It has different type for $1 < \beta \le 4/3$ and $\beta > 4/3$ (see Fig. 14.10(a) and Fig. 14.10(b), respectively).

The tangency points of this curve with the parabola on the plane $\mathbb{R}^2(g, h)$ (as well as in the case $\beta < 1$) have the coordinates

$$\left(\pm \frac{\beta}{\sqrt{|1 - \beta|}}, \frac{3\beta - 2}{2|1 - \beta|} \right).$$

In the case $\beta > 4/3$, there are two cusp points with the coordinates

$$\left(\pm \frac{4}{3} \sqrt[4]{\frac{3}{\beta - 1}}, \sqrt{\frac{3}{\beta - 1}} \right).$$

14.2.4. Kovalevskaya Case

We first consider a more general Hamiltonian

$$H = \frac{1}{2}(S_1^2 + S_2^2 + \beta S_3^2) + R_1 \tag{14.25}$$

than the one corresponding to the Kovalevskaya case. The bifurcation diagrams for Hamiltonian (14.25) have exactly the same form as those in the Lagrange case. They are obtained from the bifurcation diagrams presented in Fig. 14.4 by the following contraction (along the g-axis): $(h, g) \rightarrow (h, \beta g)$. However, the topological type of isoenergy surfaces Q^3 will be different. Hamiltonian (14.25) corresponds to the Kovalevskaya case when $\beta = 2$. That is why we present in Fig. 14.11 the bifurcation diagram for $\beta > 4/3$ only.

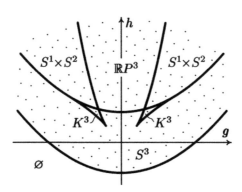

Figure 14.11

If $\beta = 2$, then the points where the curve is tangent to the parabola have coordinates $(\pm\sqrt{2}, 2)$. The coordinates of the transversal intersection points are $(\pm 2\sqrt{\sqrt{2} - 1}, 2(\sqrt{2} - 1))$. The topological type of Q^3 can be described by Smale's method (see Section 14.2); the result is presented in Fig. 14.11. The 3-manifold denoted here by K^3 is the connected sum of two copies of $S^1 \times S^2$.

We now describe the computation of the bifurcation diagram for the Kovalev-skaya case in more detail.

For the Kovalevskaya parameters

$$A_1 = A_2 = 1, \quad A_3 = 1/2, \quad \lambda = (0,0,0), \quad U(R) = R_1,$$

the reduced potential has the form

$$\varphi_g(R) = \frac{g^2}{2 - R_3^2} + R_1.$$

Consider this function in coordinates (R_3, R_1) on the hemi-sphere $\{R_2 > 0\}$. Note that, for $\{R_2 < 0\}$, the picture is symmetric.

Let us draw the level lines of $\varphi_g(R)$ in projection to the (R_3, R_1)-plane. The equation $\varphi_g(R) = c$ is equivalent to the following:

$$R_1 = c - \frac{g^2}{2 - R_3^2}.$$

Three qualitatively different pictures of level lines on the disc $\{R_1^2 + R_3^2 \leq 1\}$ are presented in Fig. 14.12. In this figure, R_3 is the horizontal axis, and R_1 is the horizontal one. In case (c), there are two possibilities depending on the value of $\varphi_g(R)$ at the maximum points. The corresponding Reeb graphs are shown in Fig. 14.13.

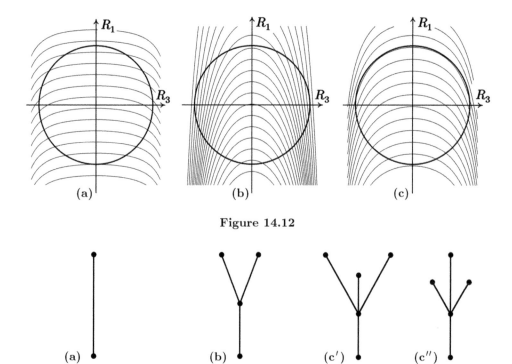

Figure 14.12

Figure 14.13

The values of $\varphi_g(R)$ at points $(\pm 1, 0, 0)$ are equal to $\dfrac{g^2}{2} \pm 1$. This defines two parabolas $\{h = g^2/2 \pm 1\}$ of the bifurcation diagram. By computing the values for

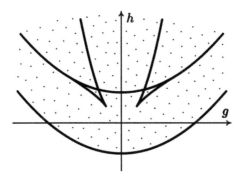

Figure 14.14

the remaining points, we obtain one more curve (Fig. 14.14). The computations are similar to those in the Lagrange case. Moreover, the curve obtained differs from that in the Lagrange case for $\beta = 2$ only by contracting along the g-axis with the coefficient $\sqrt{2}$.

14.2.5. *Zhukovskiĭ Case*

Suppose that the parameters of the Hamiltonian in the Zhukovskiĭ case

$$H = \frac{(S_1 + \lambda_1)^2}{2A_1} + \frac{(S_2 + \lambda_2)^2}{2A_2} + \frac{(S_3 + \lambda_3)^2}{2A_3} \tag{14.26}$$

satisfy the conditions $0 < A_1 < A_2 < A_3$ and $\lambda_1 \lambda_2 \lambda_3 \neq 0$.

Since $U(R) = 0$, the reduced potential has the form

$$\varphi_g(R) = \frac{(g + \langle R, \lambda \rangle)^2}{2 \langle AR, R \rangle} .$$

The critical points of the mapping

$$f_2 \times H \colon S^2 \times \mathbb{R}^3 \to \mathbb{R}^2$$

are defined by the condition

$$\operatorname{grad} H = \mu_1 \operatorname{grad} f_1 + \mu_2 \operatorname{grad} f_2 , \qquad f_1 = 1, \tag{14.27}$$

where μ_1 and μ_2 are some real numbers. Introducing the vectors $S = (S_1, S_2, S_3)$, $R = (R_1, R_2, R_3)$, $\lambda = (\lambda_1, \lambda_2, \lambda_3)$ and the matrix $A = \operatorname{diag}(A_1, A_2, A_3)$, we rewrite equations (14.27) in a vector form:

$$A^{-1}(S + \lambda) = \mu_2 R , \quad 2\mu_1 R + \mu_2 S = 0 , \quad (R, R) = 1 . \tag{14.28}$$

If $\mu_2 = 0$, then $\mu_1 = 0$, since $R \neq 0$. Hence we obtain the following solution of (14.28):

$$S = -\lambda, \quad \langle R, R \rangle = 1, \quad \mu_1 = \mu_2 = 0. \qquad (14.29)$$

The set defined by (14.29) is a two-dimensional sphere in $\mathbb{R}^6 (S, R)$. The value of the Hamiltonian (14.26) at these points is equal to zero, and the value of f_2 is $g = -\langle \lambda, R \rangle$. Therefore, the image of set (14.29) under the mapping $f_2 \times H$ is the segment $\{h = 0, |g| \leq \langle \lambda, \lambda \rangle^{1/2}\}$ on the coordinate axis $\{h = 0\}$ in $\mathbb{R}^2 (g, h)$. The preimage of each interior point of the segment is a circle on which the function

$$\widetilde{H} = H|_{\{f_1 = 1, f_2 = g\}}, \qquad \text{where} \quad |g| \leq \sqrt{\langle \lambda, \lambda \rangle},$$

has a minimum.

Now suppose that $\mu_2 \neq 0$ in (14.28). Then

$$S = -2\mu_1 \mu_2^{-1} R. \qquad (14.30)$$

Since $\langle S, R \rangle = g$, condition (14.30) implies that

$$g = \langle S, R \rangle = \langle -2\mu_1 \mu_2^{-1} R, R \rangle = -2\mu_1 \mu_2^{-1},$$

and, therefore, $S = gR$. By substituting the expression $S = gR$ into the first equation of (14.28), we obtain

$$(\mu_2 A - g)R = \lambda. \qquad (14.31)$$

Taking into account (14.30) and (14.31), we can write the solution of (14.28) in the following way:

$$S_i = \frac{\lambda_i g}{A_i t - g}, \quad R_i = \frac{\lambda_i}{A_i t - g} \quad (i = 1, 2, 3),$$
$$\mu_1 = -gt/2, \qquad \mu_2 = t, \qquad (14.32)$$

where t is a certain parameter, and the function $g(t)$ is defined by the implicit equation

$$\frac{\lambda_1^2}{(A_1 t - g)^2} + \frac{\lambda_2^2}{(A_2 t - g)^2} + \frac{\lambda_3^2}{(A_3 t - g)^2} = 1. \qquad (14.33)$$

At the points given by (14.32), the Hamiltonian H takes the following values:

$$h = \frac{t^2}{2} \left(\frac{\lambda_1^2 A_1}{(A_1 t - g)^2} + \frac{\lambda_2^2 A_2}{(A_2 t - g)^2} + \frac{\lambda_3^3 A_3}{(A_3 t - g)^2} \right). \qquad (14.34)$$

Thus, for any t and g satisfying (14.33), we obtain the only point (14.32) at which the gradients of f_2 and H are dependent. The image of this point under the mapping $f_2 \times H$ is the point with coordinates (g, h) in the plane $\mathbb{R}^2$, where h is defined by (14.34). Thus, we need to construct the curve $(g(t), h(t))$ on $\mathbb{R}^2 (g, h)$ given by (14.33) and (14.34).

It follows from Lemma 14.2 that $\dfrac{dh}{dt} = t \cdot \dfrac{dg}{dt}$. Therefore, it suffices to construct curve (14.33) on the plane $\mathbb{R}^2\,(t, g)$. Using polar coordinates $t = r\cos\varphi$, $g = r\sin\varphi$, we obtain the expression of r in terms of φ:

$$r = \left(\frac{\lambda_1^2}{(A_1^2 + 1)\sin^2(\varphi-\varphi_1)} + \frac{\lambda_2^2}{(A_2^2 + 1)\sin^2(\varphi-\varphi_2)} + \frac{\lambda_3^2}{(A_3^2 + 1)\sin^2(\varphi-\varphi_3)} \right)^{1/2},$$

where $\tan\varphi_i = A_i$ ($i = 1, 2, 3$). The curve defined by this relation is shown in Fig. 14.15. This curve is symmetric with respect to the origin and has asymptotes $\{g = A_i t \pm \lambda_i\}$ indicated in Fig. 14.15 by dotted lines.

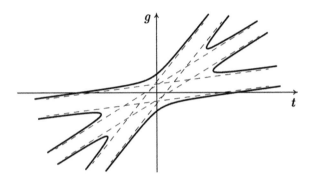

Figure 14.15

Formula (14.34) defines a mapping of this curve into $\mathbb{R}^2\,(g, h)$. Taking into account that $\dfrac{dh}{dt} = t \cdot \dfrac{dg}{dt}$, it is easy to see that the local minima and maxima of $g(t)$ (see Fig. 14.15) correspond to the cusp points of the bifurcation curve, and the local minima and maxima of the inverse function $t(g)$ correspond to the points of inflection. As a point lying on the curve shown in Fig. 14.15 tends asymptotically to the straight line $\{g = A_i t \pm \lambda_i\}$ (i.e., $t \to \infty$), the image of this point on the bifurcation diagram tends asymptotically to the parabola

$$\left\{ h = \frac{(g \mp \lambda_i)^2}{2A_i} + \frac{1}{2}\left(\frac{\lambda_j^2 A_j}{(A_j - A_i)^2} + \frac{\lambda_k^2 A_k}{(A_k - A_i)^2} \right) \right\},$$

where $\{i, j, k\} = \{1, 2, 3\}$.

The bifurcation diagram of the mapping $f_2 \times H$ for the Zhukovskiĭ case is the union of the constructed curve with the segment

$$\{h = 0,\ |g| \le \langle \lambda, \lambda \rangle^{1/2}\},$$

as shown in Fig. 14.16. The bifurcation diagram is symmetric with respect to the axis $\{g = 0\}$. The points where the segment is tangent to the curve have the coordinates

$$\left(\pm\sqrt{\lambda_1^2 + \lambda_2^2 + \lambda_3^2},\, 0 \right) \in \mathbb{R}^2\,(g, h),$$

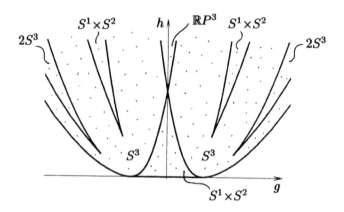

Figure 14.16

and the intersection point of the two branches of the curve has the coordinates

$$\left(0, \frac{\lambda_1^2}{2A_1} + \frac{\lambda_2^2}{2A_2} + \frac{\lambda_3^2}{2A_3}\right) \in \mathbb{R}^2(g, h).$$

The topological type of Q_h^3 in each region can be determined by means of Proposition 14.2, i.e., by considering the projection of Q_h^3 to the Poisson sphere. The same result can be obtained in another way. Let us replace λ_i with $\lambda_i' = \alpha\lambda_i$. As $\alpha \to 0$, the bifurcation diagram shown in Fig. 14.16 turns into the bifurcation diagram corresponding to the Euler case (Fig. 14.3). Moreover, the regions in Fig. 14.16 for which Q_h^3 is homeomorphic to $2S^3$, $S^1 \times S^2$, $\mathbb{R}P^3$ turns into the corresponding regions in Fig. 14.3. It is clear that the topological type of Q_h^3 does not change under passing to the limit. The manifolds Q_h^3 corresponding to the curvilinear triangle adjacent to the horizontal segment in Fig. 14.16 are homeomorphic to $S^1 \times S^2$. For the neighboring regions, Q^3 is homeomorphic to S^3. These assertions easily follows from the Morse–Bott theory, since the preimage of each point from the segment is a critical circle (on which the function $\widetilde{H} = H|_{\{f_1=1, f_2=g\}}$ has a minimum), whereas the preimage of any other points of the lower boundary is an isolated singular point (global minimum of $\widetilde{H}$).

14.2.6. Goryachev–Chaplygin–Sretenskiĭ Case

Now consider the Hamiltonian

$$H = \frac{1}{2}(S_1^2 + S_2^2 + 4(S_3 + \lambda)^2) + R_1. \tag{14.35}$$

In this case, the additional integral exists only on a single 4-manifold $\{f_1 = 1, \ f_2 = 0\} \subset \mathbb{R}^6(S, R)$. Therefore, we need not consider the mapping $f_2 \times H$. However, the topological type of Q^3 (as well as the corresponding marked molecule W^*) will depend on the parameter λ. That is why, in this case, we construct the separating curves on the plane $\mathbb{R}^2(\lambda, h)$ and define the topological type of Q^3 for each region.

For Hamiltonian (14.35), the reduced potential has the form

$$\varphi_0(R) = \frac{2\lambda^2 R_3^2}{4 - 3R_3^2} + R_1 \,. \tag{14.36}$$

To determine the type of the region

$$\{\varphi_0(R) \le h\} \,, \tag{14.37}$$

we find the critical points of function (14.36) on the two-dimensional sphere $\{R_1^2 + R_2^2 + R_3^2 = 1\} \subset \mathbb{R}^3(R)$. They are determined by the following system of equation:

$$\frac{\partial \varphi_0}{\partial R_i} = 2\mu R_i \qquad (i = 1, 2, 3) \,,$$

$$R_1^2 + R_2^2 + R_3^2 = 1 \,.$$

Its solutions are two critical points existing for any λ

$$R_1 = \pm 1, \quad R_2 = R_3 = 0, \qquad \mu = \pm 1/2 \,, \tag{14.38}$$

as well as two critical points depending on λ

$$R_1 = \frac{1}{2\xi}, \quad R_2 = 0, \quad R_3 = \pm\sqrt{1 - \frac{1}{4\xi^2}}, \qquad \mu = \xi \,, \tag{14.39}$$

where $\lambda^2 = \dfrac{(4\xi^2 + 3)^2}{128\xi^3}$ and $\xi \ge \dfrac{1}{2}$.

The value of function (14.36) at points (14.38) is equal to $\varphi_0(R) = \pm 1$. Since λ is arbitrary, we obtain two straight lines $\{h = \pm 1\}$ on the plane $\mathbb{R}^2(\lambda, h)$, which separate regions corresponding to different topological types of Q^3. At points (14.39), we have

$$\varphi_0(R) = \frac{16\xi^4 + 40\xi^2 - 3}{64\xi^3} \,.$$

Thus, these points determine a separating curve on the plane $\mathbb{R}^2(\lambda, h)$ given parametrically by

$$h = \frac{16\xi^4 + 40\xi^2 - 3}{64\xi^3}, \quad \lambda^2 = \frac{(4\xi^2 + 3)^2}{128\xi^3}, \quad \xi \ge \frac{1}{2} \,. \tag{14.40}$$

Taking the union of the straight lines $\{h = \pm 1\}$ and curve (14.40), we obtain the bifurcation diagram in $\mathbb{R}^2(\lambda, h)$ presented in Fig. 14.17. It is symmetric with respect to the axis $\{\lambda = 0\}$. The cusps of curve (14.40) have the coordinates

$$\left(\pm \frac{1}{\sqrt{3}}, \frac{7}{9} \right) \in \mathbb{R}^2(\lambda, h) \,.$$

The points where the line $\{h = 1\}$ is tangent to curve (14.40) have the coordinates

$$(\pm 1, 1) \in \mathbb{R}^2 (\lambda, h).$$

Curve (14.40) asymptotically approaches the parabola $\{h = \lambda^2\}$.

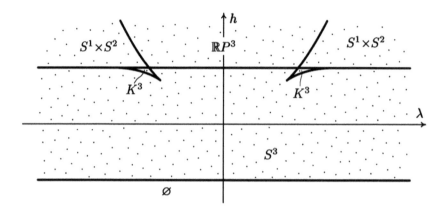

Figure 14.17

Each point of the straight line $\{h = \pm 1\}$ corresponds to one critical point of function (14.36), and each point of curve (14.40) corresponds to two critical points. Moreover, computing the indices of these critical points, we obtain the following. For the straight line $\{h = -1\}$, the index is zero. For the straight line $\{h = 1\}$, the index is equal to 2 on the segment between two tangency points, and is equal to 1 on the remaining part of the line. For curve (14.40), the index is 1 on the arc joining the tangency point with the cusp, and is 2 on the remaining part of the curve. If we know the indices of the critical points, we can easily describe the form of regions (14.37) on the Poisson sphere. They are, respectively,

the empty set $\varnothing$,
a disc D^2,
an annulus $S^1 \times \mathbb{R}^1$,
a disc D^2 with two holes,
and, finally, the whole sphere S^2.

The corresponding 3-manifolds Q^3 are indicated in Fig. 14.17 (as before, $K^3 \simeq S^1 \times S^2 \# S^1 \times S^2$).

14.2.7. Clebsch Case

Let us construct the bifurcation diagram of the mapping $f_2 \times H$ for

$$H = \frac{S_1^2}{2A_1} + \frac{S_2^2}{2A_2} + \frac{S_3^2}{2A_3} + \frac{\varepsilon}{2}(A_1 R_1^2 + A_2 R_2^2 + A_3 R_3^2),$$

where $0 < A_1 < A_2 < A_3$ and $\varepsilon = \pm 1$.

The condition $\operatorname{grad} H = \mu_1 \operatorname{grad} f_1 + \mu_2 \operatorname{grad} f_2$ that determines the critical points of the mapping $f_2 \times H \colon S^2 \times \mathbb{R}^3 \to \mathbb{R}^2$ can be rewritten in the following way:

$$
\begin{pmatrix}
A_1^{-1} & 0 & 0 & -\mu_2 & 0 & 0 \\
0 & A_2^{-1} & 0 & 0 & -\mu_2 & 0 \\
0 & 0 & A_3^{-1} & 0 & 0 & -\mu_2 \\
-\mu_2 & 0 & 0 & \varepsilon A_1 - 2\mu_1 & 0 & 0 \\
0 & -\mu_2 & 0 & 0 & \varepsilon A_2 - 2\mu_1 & 0 \\
0 & 0 & -\mu_2 & 0 & 0 & \varepsilon A_3 - 2\mu_1
\end{pmatrix}
\begin{pmatrix}
S_1 \\ S_2 \\ S_3 \\ R_1 \\ R_2 \\ R_3
\end{pmatrix}
= 0 . \quad (14.41)
$$

Since $R_1^2 + R_2^2 + R_3^2 = 1$, equality (14.41) is fulfilled only in the case when the determinant of the above matrix is zero. The matrix is decomposed into three 2×2-blocks whose determinants are equal to

$$
D_i = \varepsilon - 2\mu_1 A_i^{-1} - \mu_2^2 \qquad (i = 1, 2, 3) . \tag{14.42}
$$

If $D_i \neq 0$, then $S_i = R_i = 0$. Therefore, if only one of the conditions $D_i = 0$ is fulfilled, then, taking into account that $R_1^2 + R_2^2 + R_3^2 = 1$, we obtain the following critical points:

$$
(t_1, 0, 0, \pm 1, 0, 0) , \quad (0, t_2, 0, 0, \pm 1, 0) , \quad (0, 0, t_3, 0, 0, \pm 1) \quad \in \mathbb{R}^6 \, (S, R) ,
$$

where t_1, t_2, t_3 are some parameters.

By computing the values of f_2 and H at these points, we obtain three parabolas in $\mathbb{R}^2 \, (g, h)$:

$$
\left\{ h = \frac{g^2}{2A_i} + \varepsilon \frac{A_i}{2} \right\} \qquad (i = 1, 2, 3) . \tag{14.43}
$$

The preimage of each point lying on parabolas (14.43) contains exactly two critical points of the mapping $f_2 \times H$. Since $A_i \neq A_j$ for $i \neq j$, none of determinants (14.42) vanish simultaneously in the case $\varepsilon = -1$. Therefore, for $\varepsilon = -1$, we obtain the bifurcation diagram consisting of three non-intersecting parabolas (Fig. 14.18).

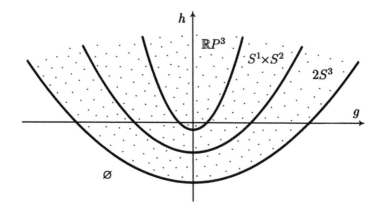

Figure 14.18

Consider now the case $\varepsilon = 1$. If any two of determinants (14.42) vanish, then $\mu_1 = 0$ and $\mu_2^2 = 1$. Therefore, the third determinant is also zero. From (14.41), we obtain the following system of equations:

$$S_i = \pm A_i R_i \qquad (i = 1, 2, 3),$$
$$R_1^2 + R_2^2 + R_3^2 = 1. \tag{14.44}$$

Equations (14.44) define two spheres in $\mathbb{R}^6 (S, R)$ filled by critical points of the mapping $f_2 \times H$. The values of f_2 and H on set (14.44) are equal to

$$g = \pm(A_1 R_1^2 + A_2 R_2^2 + A_3 R_3^2),$$
$$h = A_1 R_1^2 + A_2 R_2^2 + A_3 R_3^2.$$

Thus, under the mapping $f_2 \times H$, the spheres (14.44) are mapped onto two segments

$$\{h = |g|, A_1 \leq |g| \leq A_3\} \subset \mathbb{R}^2(g, h). \tag{14.45}$$

The preimage of each interior point of the segment consists of two circles filled by critical points.

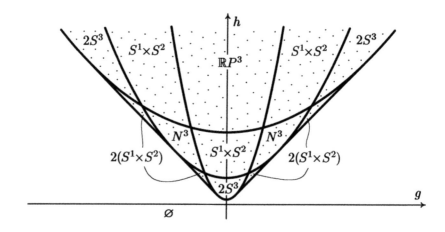

Figure 14.19

Taking the union of parabola (14.43) (for $\varepsilon = 1$) and segments (14.45), we obtain the bifurcation diagram presented in Fig. 14.19. Segments (14.45) are tangent to each of the three parabolas at points $(\pm A_i, A_i) \in \mathbb{R}^2 (g, h)$. Parabolas (14.43) intersect at points

$$\left(\pm\sqrt{A_i A_j}, \ \frac{A_i + A_j}{2}\right) \in \mathbb{R}^2 (g, h).$$

The topological type of $Q_h^3 = \{f_1 = 1,\; f_2 = 0,\; H = h\}$ for each region in Fig. 14.18 and Fig. 14.19 can be determined by analyzing the projection of Q_h^3 on the Poisson sphere.

The image of Q_h^3 under the projection is defined by the condition

$$g^2 z^{-1} + \varepsilon z \le 2h, \qquad A_1 \le z \le A_3, \tag{14.46}$$

where z denotes the expression $A_1 R_1^2 + A_2 R_2^2 + A_3 R_3^2$. The graph of the function $\varphi(z) = g^2 z^{-1} + \varepsilon z$ is shown in Fig. 14.20(a) for $\varepsilon = -1$, and in Fig. 14.20(b) for $\varepsilon = 1$.

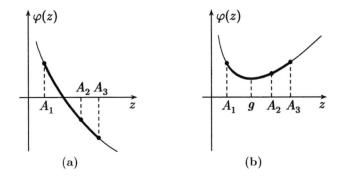

(a)   (b)

Figure 14.20

In the case $\varepsilon = 1$, the function $\varphi(z)$ has a minimum at the point $z = g$. We present one of possible positions of g on the z-axis with respect to A_1, A_2, A_3 in Fig. 14.20(b).

It is not hard to analyze all possibilities and find out the form of domain (14.46) on the Poisson sphere in each case. After this, the topological type of Q^3 is determined in the standard way using Proposition 14.2. The list of all isoenergy manifolds Q^3 is presented in Fig. 14.19. By N^3 we denote the connected sum $(S^1 \times S^2) \# (S^1 \times S^2) \# (S^1 \times S^2)$. Its projection to the Poisson sphere is a disc with three holes.

14.2.8. *Steklov Case*

The construction of the bifurcation diagrams of the mapping $f_2 \times H$ for the Steklov case can be carried out similarly to the Clebsch case.

The bifurcation diagram is, as before, a union of three parabolas and segments tangent to each parabola. But, unlike the Clebsch case, the axes of these parabolas do not coincide any more. As a result, we obtain a number of different possibilities depending on the relations between the parameters A_1, A_2, A_3 of the Steklov case. The complete description of this case becomes rather awkward, and we omit it in our book. The possible molecules W for the Steklov case are calculated in Section 14.9.

14.3. LIOUVILLE CLASSIFICATION OF SYSTEMS IN THE EULER CASE

From now on, we shall consider another pair of integrals, namely, the Hamiltonian H and additional integral K on an invariant symplectic 4-manifold $M_{1,g}^4$. Therefore, the bifurcation diagrams which we are going to study now have another meaning. They correspond to bifurcations of two-dimensional Liouville tori, while the bifurcation diagrams of the mapping $f_2 \times H$ studied in the previous section corresponded to topological bifurcations of isoenergy manifolds Q^3.

We begin with constructing the bifurcation diagram for the Euler case. The Hamiltonian H and the integral K have the following form

$$H = \frac{S_1^2}{2A_1} + \frac{S_2^2}{2A_2} + \frac{S_3^2}{2A_3}, \qquad K = S_1^2 + S_2^2 + S_3^2.$$

Recall that in the Euler case we assume that $A_1 > A_2 > A_3 > 0$.

The functions H and K commute on the symplectic 4-manifold $M_{1,g}^4 = \{f_1 = 1, f_2 = g\}$ for any g. Consider the corresponding momentum mapping $\mathcal{F} = (H, K) \colon M_{1,g}^4 \to \mathbb{R}^2$ given by these two functions and construct its bifurcation diagram.

The diagram depends on the parameter g, i.e., on the area constant. For each value of g, the critical points of the momentum mapping $\mathcal{F}$ are those points where the gradients of H and K (restricted to $M_{1,g}^4 = \{f_1 = 1, f_2 = g\}$) are linearly dependent. Therefore, a critical point of $\mathcal{F}$ is either a critical point of the function H restricted onto $M_{1,g}^4$, or the following relation holds at this point:

$$\lambda_1 \operatorname{grad} f_1 + \lambda_2 \operatorname{grad} f_2 + \lambda_3 \operatorname{grad} H = \operatorname{grad} K, \qquad (14.47)$$

for some real numbers $\lambda_1, \lambda_2, \lambda_3$.

The critical points of H on $M_{1,g}^4$ were already found when we studied the topological type of isoenergy surfaces in the Euler case. Recall (see Section 14.2.2) that, for any g, there are 6 critical points of H

$$(\pm g, 0, 0, \pm 1, 0, 0), \quad (0, \pm g, 0, 0, \pm 1, 0), \quad (0, 0, \pm g, 0, 0, \pm 1), \qquad (14.48)$$

at which the Hamiltonian H takes values $\dfrac{g^2}{2A_i}$, and the integral K has the same value g^2.

For the remaining critical points of the momentum mapping $\mathcal{F}$, we have relation (14.47). Thus, in terms of variables $(S, R) = (S_1, S_2, S_3, R_1, R_2, R_3)$, the system of equations that defines critical points of the momentum mapping can be written in the following way:

$$2\lambda_2 R + \lambda_3 A^{-1} S = 2S, \qquad 2\lambda_1 R + \lambda_2 S = 0,$$
$$\langle R, R \rangle = 1, \qquad \langle S, R \rangle = g, \qquad (14.49)$$

where A denotes the diagonal matrix with elements A_1, A_2, A_3, and $\langle\, ,\rangle$ is the inner product in $\mathbb{R}^3$.

First we find the solutions of (14.49) for $\lambda_2 \neq 0$.

From the second equation of system (14.49), we obtain $S = -\dfrac{2\lambda_1}{\lambda_2}R$. Taking into account that $\langle R, R \rangle = 1$, and $\langle S, R \rangle = g$, we have

$$S = gR, \qquad \lambda_1 = -\frac{\lambda_2 g}{2}.$$

Substituting gR instead of S into the first equation of system (14.49), after some simple transformations, we obtain

$$(2(g - \lambda_2)A - \lambda_3 gE)R = 0,$$

where E denotes the identity 3×3-matrix.

Since the parameters A_1, A_2, A_3 are mutually different, the rank of the matrix $2(g - \lambda_2)A - \lambda_3 gE$ may be equal to either 3, or 2, or 0. If the rank is 3, then no solutions exist (since $R \neq 0$). If the rank is 2, then the solutions of (14.49) are six points (14.48). Finally, if the rank is 0, then

$$\lambda_1 = -\frac{g^2}{2}, \quad \lambda_2 = g, \quad \lambda_3 = 0. \tag{14.50}$$

In particular, if $\lambda_2 \neq 0$, then $g \neq 0$.

Thus, in the case $\lambda_2 \neq 0$, the corresponding solutions of system (14.49) fill the two-dimensional sphere

$$\{(S, R) : S = gR, \langle R, R \rangle = 1\} \tag{14.51}$$

in the manifold $M^4_{1,g}$. Computing the values of the functions H and K at the points of this sphere, we obtain

$$h = \frac{g^2}{2}\left(\frac{R_1^2}{A_1} + \frac{R_2^2}{A_2} + \frac{R_3^2}{A_3}\right), \qquad k = g^2.$$

Hence, the image of the sphere (14.51) under the momentum mapping is the segment of the horizontal straight line $\{k = g^2\}$ on the plane $\mathbb{R}^2(h, k)$. Moreover, the preimages of the points lying on this segment are level lines of H (on the sphere (14.51)), which are given as the intersection of the sphere $\{\langle R, R \rangle = 1\}$ and the ellipsoid $\{g^2\langle A^{-1}R, R \rangle = 2h\}$.

Thus, the part of the bifurcation diagram for $\lambda_2 \neq 0$ is the segment

$$\tau_0 = \left\{k = g^2, \ \frac{g^2}{2A_1} \leq h \leq \frac{g^2}{2A_3}\right\}. \tag{14.52}$$

The preimage of the points $h = \dfrac{g^2}{2A_1}$, $h = \dfrac{g^2}{2A_2}$, $h = \dfrac{g^2}{2A_3}$ from this segment τ_0 contains critical points of H. For any other point $y \in \tau_0$, its preimage $\mathcal{F}^{-1}(y)$ consists of two critical circles.

Now consider the case $\lambda_2 = 0$.

Then, from the second equation of (14.49) we obtain $\lambda_1 = 0$. Moreover, the first equation of (14.49) implies that two of three coordinates S_1, S_2, S_3 are equal to zero. Hence we obtain three series of solutions:

$$\left(\frac{g}{R_1}, 0, 0, R_1, R_2, R_3\right), \qquad \lambda_3 = 2A_1, \qquad h = \frac{g^2}{2A_1 R_1^2}, \qquad k = \frac{g^2}{R_1^2}, \qquad (14.53_1)$$

$$\left(0, \frac{g}{R_2}, 0, R_1, R_2, R_3\right), \qquad \lambda_3 = 2A_2, \qquad h = \frac{g^2}{2A_2 R_2^2}, \qquad k = \frac{g^2}{R_2^2}, \qquad (14.53_2)$$

$$\left(0, 0, \frac{g}{R_3}, R_1, R_2, R_3\right), \qquad \lambda_3 = 2A_3, \qquad h = \frac{g^2}{2A_3 R_3^2}, \qquad k = \frac{g^2}{R_3^2}, \qquad (14.53_3)$$

where $R_1^2 + R_2^2 + R_3^2 = 1$.

For each of these series, the critical points of $\mathcal{F}$ fill two discs (two hemispheres $\{\langle R, R \rangle = 1, R_i \neq 0\}$), which are mapped to some subsets of straight lines $\{k = 2A_i h\}$. Furthermore, it is easy to see that, on each of the three subsets (14.53_i), the function K has a minimum equal to g^2, and its regular level lines consist of two circles.

Thus, the part of the bifurcation diagram corresponding to $\lambda_2 = 0$ consists of three rays

$$\tau_1 = \{k = 2A_1 h, \ k \geq g^2\},$$
$$\tau_2 = \{k = 2A_2 h, \ k \geq g^2\}, \qquad\qquad (14.54)$$
$$\tau_3 = \{k = 2A_3 h, \ k \geq g^2\},$$

whose origins lie on the segment τ_0 (this segment shrinks to a point for $g = 0$).

Taking the union of segment (14.52) and rays (14.54), we obtain the bifurcation diagram of the momentum mapping for the Euler case. It is presented in Fig. 14.21(a) for $g = 0$ and in Fig. 14.21(b) for $g \neq 0$.

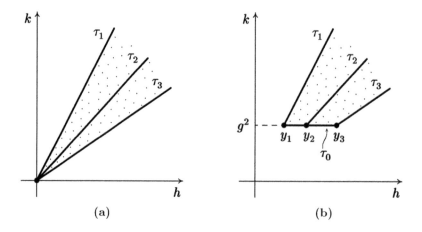

Figure 14.21

We now prove that K is a Bott function on each non-singular isoenergy surface $Q_{g,h}^3 = \{f_1 = 1, f_2 = g, H = h\}$, where $h \neq \dfrac{g^2}{2A_i}$. In other words, we need to verify that the critical circles of K (as a function on $Q_{g,h}^3$) are all non-degenerate. For this, according to the scheme discussed in Section 14.1 (Lemma 14.1), at each singular point, we must restrict the bilinear form with the matrix

$$G = G_K - \lambda_1 G_1 - \lambda_2 G_2 - \lambda_3 G_H \qquad (14.55)$$

to the tangent space to $Q_{g,h}^3$ and verify that its rank (after restriction) is equal to 2. Here G_1, G_2, G_H, G_K are the matrices composed from the second derivatives of the functions f_1, f_2, H, K respectively; and the numbers $\lambda_1, \lambda_2, \lambda_3$ are the coefficients of the linear combination of gradients (14.47).

The critical circles of the function K restricted to $Q_{g,h}^3$ belong to the preimage of the points of intersection of the bifurcation diagram and straight line $\{h = \text{const}\}$ in $\mathbb{R}^2(h,k)$. It is convenient to consider each part of the bifurcation diagram (i.e., the segment τ_0 and rays τ_1, τ_2, τ_3) separately.

Consider the ray τ_1. In this case, $\lambda_1 = \lambda_2 = 0$, and $\lambda_3 = 2A_1$. Therefore, matrix (14.55) takes the form

$$G = G_K - 2A_1 G_H = \begin{pmatrix} 0 & 0 & 0 & 0 & 0 & 0 \\ 0 & 2 - \dfrac{2A_1}{A_2} & 0 & 0 & 0 & 0 \\ 0 & 0 & 2 - \dfrac{2A_1}{A_3} & 0 & 0 & 0 \\ 0 & 0 & 0 & 0 & 0 & 0 \\ 0 & 0 & 0 & 0 & 0 & 0 \\ 0 & 0 & 0 & 0 & 0 & 0 \end{pmatrix}.$$

At the critical points corresponding to τ_1 (i.e., at points (14.53_1)), the gradients of f_1, f_2, and H are

$$\operatorname{grad} f_1 = \begin{pmatrix} 0 & , & 0 & , & 0 & , & 2R_1 & , & 2R_2 & , & 2R_3 \end{pmatrix},$$

$$\operatorname{grad} f_2 = \begin{pmatrix} R_1 & , & R_2 & , & R_3 & , & \dfrac{g}{R_1} & , & 0 & , & 0 \end{pmatrix},$$

$$\operatorname{grad} H = \begin{pmatrix} \dfrac{g}{A_1 R_1} & , & 0 & , & 0 & , & 0 & , & 0 & , & 0 \end{pmatrix}.$$

Therefore, the basis in the tangent space to $Q_{g,h}^3$ can be given, for example, in the following way:

$$e_1 = \begin{pmatrix} 0 & , & 0 & , & 0 & , & 0 & , & -R_3 & , & R_2 \end{pmatrix},$$

$$e_2 = \begin{pmatrix} 0 & , & 1 & , & 0 & , & -\dfrac{R_1 R_2}{g} & , & \dfrac{R_1^2}{g} & , & 0 \end{pmatrix},$$

$$e_3 = \begin{pmatrix} 0 & , & 0 & , & 1 & , & -\dfrac{R_1 R_3}{g} & , & 0 & , & \dfrac{R_1^2}{g} \end{pmatrix}.$$

Obviously, the restriction of the form G to the tangent space to $Q^3_{g,h}$ in terms of the basis e_1, e_2, e_3 has the matrix

$$\begin{pmatrix} 0 & 0 & 0 \\ 0 & 2-\dfrac{2A_1}{A_2} & 0 \\ 0 & 0 & 2-\dfrac{2A_1}{A_3} \end{pmatrix}.$$

Since A_1, A_2, A_3 are pairwise different, the rank of this matrix is 2. Therefore, the critical circles of the integral K lying in the preimage of τ_1 are non-degenerate critical submanifolds. Moreover, since we assume that $A_1 > A_2 > A_3$, the index of these critical circles is equal to 2, i.e., the function $K|_{Q^3_{g,h}}$ has a maximum on them.

The proof of non-degeneracy of critical circles and the computation of their indices can similarly be carried out for the rays τ_2, τ_3 and segment τ_0. As a result, we shall see that the critical circles related to τ_2 are non-degenerate and of saddle type, whereas the critical circles related to τ_3 and τ_0 are non-degenerate and of elliptic type (more precisely, K takes a minimal value on them).

It is convenient to distinguish three zones (intervals) for energy values on the h-axis. We denote them by I, II, III. Namely, we set

$$\mathrm{I} = \left(\frac{g^2}{2A_1}, \frac{g^2}{2A_2} \right), \qquad \mathrm{II} = \left(\frac{g^2}{2A_2}, \frac{g^2}{2A_3} \right), \qquad \mathrm{III} = \left(\frac{g^2}{2A_3}, \infty \right).$$

These three zones correspond to different types of isoenergy manifolds Q^3. More specifically (see Section 14.2.2),

$$Q^3 \simeq S^3 \text{ in zone I}, \quad Q^3 \simeq S^1 \times S^2 \text{ in zone II}, \quad Q^3 \simeq \mathbb{R}P^3 \text{ in zone III}.$$

We now describe the marked molecules W^* corresponding to each type of Q^3.

Theorem 14.2.

a) *Let the area constant g be different from zero. Then the marked molecules $\mathcal{E}_1^*, \mathcal{E}_2^*, \mathcal{E}_3^*$ of systems in the Euler case have the form presented in Fig. 14.22(a,b,c). These three different molecules correspond to the three energy zones I, II, III.*

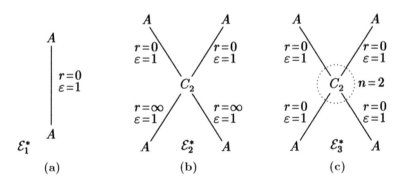

Figure 14.22

b) *If $g = 0$, then the topology of the Liouville foliation does not depend on the energy level and is described by the marked molecule $\mathcal{E}_3^*$ shown in Fig. 14.22(c).*

Proof. We first assume that $g \neq 0$. Let us apply the loop molecules method described in Sections 10.3–10.6 to the Euler case. There are three singular points y_1, y_2, y_3 on the bifurcation diagram (Fig. 14.21). We now prove that y_1 and y_3 correspond to a center–center singularity, and y_2 corresponds to a singularity of saddle–center type.

Consider the point y_1 (for y_3, the arguments are similar). Its preimage in $M^4_{1,g}$ under the momentum mapping $\mathcal{F}$ consists of two points given in terms of coordinates by

$$(R_1, R_2, R_3, S_1, S_2, S_3) = (\ 1,\ \ 0,\ \ 0,\ \ g,\ \ 0,\ \ 0),$$
$$(R_1, R_2, R_3, S_1, S_2, S_3) = (-1,\ \ 0,\ \ 0, -g,\ \ 0,\ \ 0).$$

Consider the first of them. As local coordinates in a neighborhood of this point in $M^4_{1,g}$, we take R_2, R_3, S_2, S_3. Our aim is to describe the functions H and K in terms of these coordinates and to verify that their Hessians satisfy the non-degeneracy condition (Definition 1.22). Instead of H, it is more convenient to consider the linear combination of H and K of the form $\widetilde{H} = H - K \cdot (2A_1)^{-1}$.

Then, up to the third order, $\widetilde{H}$ and K can be written in terms of the chosen local coordinates as follows:

$$\widetilde{H} = S_2^2 \left(\frac{1}{2A_2} - \frac{1}{2A_1} \right) + S_3^2 \left(\frac{1}{2A_3} - \frac{1}{2A_1} \right) + \dots,$$
$$K = (S_2 - gR_2)^2 + (S_3 - gR_3)^2 + \dots.$$

In the coordinates R_2, R_3, S_2, S_3, the matrix of Poisson brackets does not have a canonical form. That is why we make another coordinate change by the formula

$$p_1 = S_3, \quad q_1 = S_2, \quad p_2 = S_3 - gR_3, \quad q_2 = S_2 - gR_2.$$

In these new coordinates, the matrix of the Poisson structure has the canonical form (up to multiplication by the area constant g). In terms of these coordinates, the functions $\widetilde{H}$ and K becomes

$$\widetilde{H} = q_1^2 \left(\frac{1}{2A_2} - \frac{1}{2A_1} \right) + p_1^2 \left(\frac{1}{2A_3} - \frac{1}{2A_1} \right) + \dots,$$
$$K = q_2^2 + p_2^2 + \dots.$$

Thus, the pair of functions at the point in question has a non-degenerate singularity of center–center type. Similar arguments work for the points y_2 and y_3. The only difference is that, in the final formula for $\widetilde{H}$, we must cyclically permute the indices. As a result, the coefficients $\left(\frac{1}{2A_2} - \frac{1}{2A_1} \right)$ and $\left(\frac{1}{2A_3} - \frac{1}{2A_1} \right)$ (which were positive for y_1) change signs. Namely, both of them become negative at the point y_3; the type of the singularity remains, consequently, the same (i.e., center–center). At the point y_2, their signs become opposite, i.e., y_2 has saddle-canter type. Thus, we have determined the types of all the singular points of the bifurcation diagram.

Now consider the loop molecules corresponding to the singular points y_1, y_2, y_3 (Fig. 14.23). In this figure, by fat lines we indicate the arcs of the circles of small radius which lie inside the image of the momentum mapping.

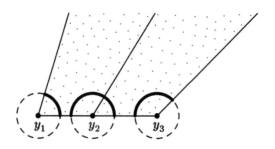

Figure 14.23

For the points y_1 and y_3 of center–center type, the loop molecule has the form $A\!-\!A$, and the mark r is equal to zero (Theorem 9.1). Therefore, by deforming the circle (centered at y_1) into a vertical segment, we do not change the molecule, but, on the other hand, we obtain the desired molecule for the isoenergy manifold Q^3 corresponding to the first energy zone I (i.e., for small energy levels). Notice that, in this case, the mark ε depends on the orientation of Q^3. Without loss of generality, we may assume that $\varepsilon = 1$.

Thus, Theorem 14.2 is proved for the zone I. Note that the complete molecule in fact consists of two copies of $A\!-\!A$, since the preimage of y_1 is disconnected and consists of two isolated center–center points (the same is true for y_3).

We now pass to the second zone II. We begin with computing the loop molecule corresponding to y_2. According to Theorem 9.2, it suffices to determine the l-type of the singularity. Recall that the l-type has the form (V, sA). Therefore, we only need to find the atom V corresponding to the preimage of the horizontal segment τ_0. The analysis of explicit expression for the points lying in the preimage $X = \mathcal{F}^{-1}(\tau_0)$ shows that X is a two-dimensional symplectic submanifold in $M_{1,g}^4$ diffeomorphic to S^2. Namely, X is defined as follows:

$$X = \{(S_1, S_2, S_3) = g(R_1, R_2, R_3), \ R_1^2 + R_2^2 + R_3^2 = 1\}.$$

Restricting the Hamiltonian H on X, we obtain

$$H|_X = g^2 \left(\frac{R_1^2}{2A_1} + \frac{R_2^2}{2A_2} + \frac{R_3^2}{2A_3} \right).$$

Note that X can be regarded as a surface in $\mathbb{R}^3(R)$ given by $R_1^2 + R_2^2 + R_3^2 = 1$. We are interested in the topological type of the saddle singularity of H restricted onto X. It is easy to see that the qualitative picture of level lines of H on this sphere has the form shown in Fig. 14.24. Therefore, the saddle singularity has type C_2. Then, according to Theorem 9.2, the loop molecule corresponding to the singularity y_2 is of the form presented in Fig. 14.25.

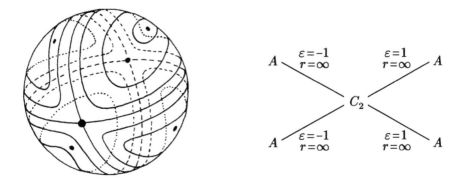

Figure 14.24

Figure 14.25

Thus, we have described all three loop molecules in the Euler case. Now we wish to compute the (isoenergy) molecules W^* corresponding to zones II and III.

It is easily seen from Fig. 14.26(a) that the desired molecule for the second zone II is obtained by gluing the loop molecules corresponding to the singularities y_1 and y_2 (see Section 10.3). As a result, we obtain the molecule shown in Fig. 14.26(a). Here we have glued two edges incident to atoms A. One of these edges is endowed with the mark $r = 0$, the other is endowed with the mark $r = \infty$. As a result, the atoms A have disappeared, and the new edge has obtained the mark $r = 0$ (see the rule for summing r-marks in Section 10.3).

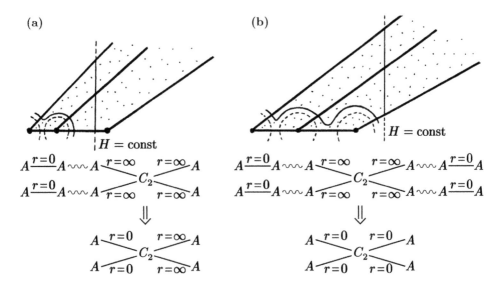

Figure 14.26

It remains to compute the molecule W^* for the third zone III. The method is the same, but here we glue the desired molecule from three loop molecules. This procedure and the result are shown in Fig. 14.26(b).

Thus, we have described the molecules for each energy zone and indicated the r-mark on each edge. Finally, let us compute the ε- and n-marks. Recall that, in some cases, ε depends on the choice of orientation on Q^3. In particular, this happens on the edges of type C_2 — A with the infinite r-mark. In this case, we shall assume (by changing orientation, if necessary) that $\varepsilon = 1$. If the r-mark on the edge C_2 — A is equal to zero, then ε does not depend on the choice of orientation and has a natural topological sense. To explain this, consider the singular trajectories γ_1 and γ_2 of the Hamiltonian flow that correspond to the atoms A and C_2 joined by the edge under consideration. Each of them has the natural orientation defined by the direction of the flow. On the other hand, since $r = 0$, these trajectories are homologous without orientations (see Fig. 14.27). Therefore, we can compare their orientations. If these orientations coincide, then $\varepsilon = 1$; if they are opposite, then $\varepsilon = -1$. This is just what we must verify.

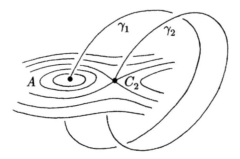

Figure 14.27

For this, it is convenient to use the fact that the trajectories of the Hamiltonian flow given by the integral K are all closed. In particular, γ_1 and γ_2 are trajectories of sgrad K. Clearly, the orientations given by sgrad K on these trajectories are the same. That is why we may just compare the orientations given on γ_i by the flows sgrad H and sgrad K. If we know the bifurcation diagram, we can easily do so. The trajectory γ_1 corresponds to some point lying on the ray τ_1. In view of Proposition 1.16, on γ_1 we have the relation $dH = (2A_1)^{-1}dK$. Similarly, on γ_2 we obtain $dH = (2A_2)^{-1}dK$. Since $A_1 > 0$ and $A_2 > 0$, it follows that the flows sgrad H and sgrad K have the same directions both on γ_1 and γ_2. Hence $\varepsilon = +1$ on the edge A — C_2 (in the case $r = 0$).

For the edge between τ_2 and τ_3, the argument is just the same.

The n-mark appears in the case of high energies only, i.e., in the zone III. It can be easily computed if the topological type of the isoenergy manifold Q^3 is known in advance. In our case, $Q^3 \simeq \mathbb{R}P^3$. The only possibility is that $n = \pm 2$. This follows from the comment to Proposition 4.4. Changing, if necessary, the orientation of Q^3, we may assume without loss of generality that $n = 2$. Thus, part (a) of Theorem 14.2 is proved.

Consider now the case $g = 0$. The bifurcation diagram has the form shown in Fig. 14.21(a). It is obtained from the bifurcation diagram for $g \neq 0$ by passage to the limit as g tends to zero: the segment τ_0 moves toward the origin and shrinks into a point. If we take some level $\{H = \text{const}\}$ from the zone III and set $g \to 0$, then the transformation of the bifurcation diagram happens at a distance from the chosen energy value. Therefore, the qualitative picture and topology of the Liouville foliation remain the same. Thus, the limit molecule for $g = 0$ coincides with the molecule corresponding to the third zone III. This completes the proof of part (b).   □

In conclusion, we shall show how the information on the topology of the Liouville foliation in the Euler case allows us to explain the following mechanical experiment. Take a book as an example of a rigid body. (Instead of a book we can take anything else; the only property we need is that the body has three natural symmetry planes, and its principal moments of inertia are quite different.) Let us dispose it in the horizontal plane, as shown in Fig. 14.28(a), and toss it up rotating simultaneously about the horizontal symmetry axis. Then we catch the book and see in what position it came back.

It turns out that the result essentially depends on how exactly the book was located before tossing. Namely, the book has three natural mutually orthogonal symmetry axes. Thus, we can toss it up with rotating about any of these axes. If the rotation is about the axis related to the least moment of inertia, then the book returns in the same position as it was before tossing. We shall see the same effect in the case of rotation about the axis related to the maximal moment of inertia: after several flips the book returns to the initial position (see Fig. 14.28(a)).

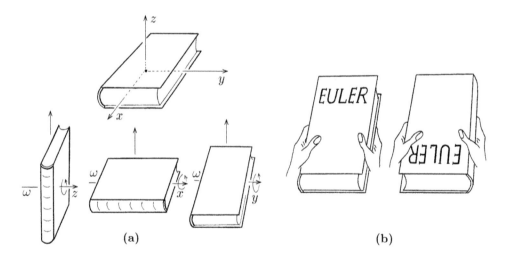

(a)   (b)

Figure 14.28

The situation will be quite different if we toss up the book with rotating about the middle axis of inertia. In this case, the book turns over. More precisely, if before tossing the back of the book was in your left hand, then having caught the book you observe that the back has turned in your right hand (Fig. 14.28(b)).

This curious fact can be easily explained from the viewpoint of the topology of the Liouville foliation. Indeed, the motion of the book is a good model of the Euler case in rigid body dynamics. We only need to forget about the motion of the center of mass and consider the "pure rotation" about the center of mass. Moreover, we may assume that the area constant is zero. The point is that we rotate the book about one of the axes of inertia. Therefore, the angular momentum vector $\vec{K}_0$ (in the initial position) is proportional to the angular velocity vector $\vec{\omega}_0$. The unit vertical vector $\vec{\gamma}_0$ is obviously orthogonal to ω_0 and, consequently, to $\vec{K}_0$. Since the area constant is defined to be the product $\langle \vec{K}, \vec{\gamma} \rangle$, and $\langle \vec{K}_0, \vec{\gamma}_0 \rangle = 0$ in the initial position, in our experiment this constant is zero. Thus we can interpret the motion of the book as a trajectory of the system in the Euler case on a fixed isoenergy surface Q^3. We have already calculated the molecule W^* that describes the foliation of Q^3 into invariant Liouville tori (Fig. 14.22(c)). Three types of motion correspond to three types of integral trajectories.

The first type corresponds to the stable periodic trajectories related to two "upper" atoms A in the molecule. This is the rotation about the minimal axis of the ellipsoid of inertia. The motion is stable, and it is clear that the book returns to the initial position (when you catch it).

The second type corresponds to the stable periodic trajectories related to two "lower" atoms A. This is the rotation about the maximal axis of the ellipsoid of inertia. This motion is also stable and this fact can easily be observed in our experiment.

The third type is the most interesting one. It corresponds to two hyperbolic periodic trajectories γ and γ' related to the saddle atom C_2. The real motion of the book is given by a certain integral trajectory which starts near one of two periodic hyperbolic solutions, say γ. Theoretically, we could rotate and toss up the book in such a way that the corresponding point would have stayed on γ all the time. But it is rather hard to do this on practice. A small perturbation of initial data makes the book move along a trajectory $\alpha(t)$ which is close to γ only for small t, but then moves away and, after some time, approaches the other periodic solution γ'. Such a behaviour is just the interpretation of the topology of C_2 from the mechanical point of view. In our experiment we see this as the turning over effect.

Let us emphasize once more that the global topology of the atom C_2 is very important here. If the saddle bifurcation had had another topological type, we would have seen totally different behavior of the book. If, for instance, the atom had been of type B, the book would have had to return to the initial saddle periodic solution γ. Therefore, the back of the book would not have turned over, but would have stayed in the left hand.

14.4. LIOUVILLE CLASSIFICATION OF SYSTEMS IN THE LAGRANGE CASE

The classical Lagrange case is a system with Hamiltonian (14.13).   Under a coordinate change in $\mathbb{R}^6 (S, R)$, this Hamiltonian can be reduced to the form

$$ H = \frac{1}{2} \left(S_1^2 + S_2^2 + \frac{S_3^2}{\beta} \right) + R_3 . $$

There are several generalizations of the Lagrange case. For example, one can consider the Hamiltonian with quadratic potential:

$$H = \frac{1}{2}\left(S_1^2 + S_2^2 + \frac{S_3^2}{\beta}\right) + R_3^2 \,;$$

or add the gyrostatic momentum:

$$H = \frac{1}{2}\left(S_1^2 + S_2^2 + \frac{S_3^2}{\beta}\right) + \lambda S_3 + R_3 \,.$$

We shall consider the Hamiltonian

$$H = \frac{1}{2}\left(S_1^2 + S_2^2 + \frac{S_3^2}{\beta}\right) + V(R_3) \,, \tag{14.56}$$

where $V(x)$ is a certain smooth function on the segment $[-1, 1]$. For Hamiltonian (14.56), as well as for the classical Lagrange case, the additional integral has the form $K = S_3$.

Consider the momentum mapping $\mathcal{F} = H \times K : TS^2 \to \mathbb{R}^2(h, k)$, where H is Hamiltonian (14.56), and describe the critical points of $\mathcal{F}$. Fixing the value $S_3 = k$, we look for critical points of the function $H_{g,k} = H|_{P^3}$, where $P^3 = \{f_1 = 1, f_2 = g, K = k\}$. The manifold P^3 is non-singular for $k \neq \pm g$ (for a while, we exclude the case $k = \pm g$ from our examination). The critical points of $H_{g,k}$ can be found from the condition

$$\operatorname{grad} H = \lambda_1 \operatorname{grad} f_1 + \lambda_2 \operatorname{grad} f_2 + \lambda_3 \operatorname{grad} K \,,$$
$$f_1 = 1, \qquad f_2 = g, \qquad K = k \,. \tag{14.57}$$

Denoting R_3 by x, we express all the unknowns of system (14.57):

$$R_1 = \sqrt{1 - x^2}\,\cos t, \qquad R_2 = \sqrt{1 - x^2}\,\sin t, \qquad R_3 = x,$$

$$S_1 = \frac{g - kx}{\sqrt{1 - x^2}}\,\cos t, \qquad S_2 = \frac{g - kx}{\sqrt{1 - x^2}}\,\sin t, \qquad S_3 = k,$$

$$\lambda_1 = -\frac{(g - kx)^2}{2(1 - x^2)^2}, \qquad \lambda_2 = \frac{g - kx}{1 - x^2}, \qquad \lambda_3 = \frac{k}{\beta} - x\frac{g - kx}{1 - x^2}, \tag{14.58}$$

where t is some parameter, and x is determined from the condition

$$V'(x) + \frac{(g - kx)(gx - k)}{(1 - x^2)^2} = 0, \qquad |x| < 1, \tag{14.59}$$

where $V'(x)$ denotes the derivative of $V(x)$ at the point x.

For every fixed point x satisfying (14.59), conditions (14.58) determine exactly one critical circle of the function $H_{g,k} = H|_{P^3}$. The value of $H_{g,k}$ on this circle is

$$W_{g,k}(x) = \frac{(g - kx)^2}{2(1 - x^2)} + \frac{k^2}{2\beta} + V(x),$$

and condition (14.59) can be rewritten as

$$W'(x) = 0, \qquad |x| < 1. \tag{14.60}$$

Thus, the critical circles of $H_{g,k}$ are parametrized by critical points of the function $W_{g,k}(x)$. This function $W_{g,k}$ is an analog of the reduced potential, and the projections of Liouville tori on the Poisson sphere are determined in this case by the condition $W_{g,k}(R_3) \leq h$.

Following the algorithm described in Lemma 14.1, we can find the indices of the critical circles (14.58) of the function $H_{g,k}$. In order to do this, we consider the matrix $G = G_H - \lambda_1 G_1 - \lambda_2 G_2 - \lambda_3 G_K$, where G_H, G_1, G_2, G_K are the Hesse matrices of the functions H, f_1, f_2, K, and then restrict the form determined by this matrix to the space orthogonal to $\operatorname{grad} f_1, \operatorname{grad} f_2, \operatorname{grad} K$. In our case, the matrix G is of the following form:

$$\begin{pmatrix} 1 & 0 & 0 & -\lambda_2 & 0 & 0 \\ 0 & 1 & 0 & 0 & -\lambda_2 & 0 \\ 0 & 0 & 1/\beta & 0 & 0 & -\lambda_2 \\ -\lambda_2 & 0 & 0 & -2\lambda_1 & 0 & 0 \\ 0 & -\lambda_2 & 0 & 0 & -2\lambda_1 & 0 \\ 0 & 0 & -\lambda_2 & 0 & 0 & V''(x) - 2\lambda_1 \end{pmatrix},$$

where $\lambda_1 = -\dfrac{(g - kx)^2}{2(1 - x^2)^2}$, $\lambda_2 = \dfrac{g - kx}{1 - x^2}$, and $V''(x)$ is the second derivative of $V(x)$. The gradients of f_1, f_2, K at the critical points (14.58) are equal to

$$\operatorname{grad} f_1 = \left(0,\, 0,\, 0,\, 2\sqrt{1-x^2}\cos t,\, 2\sqrt{1-x^2}\sin t,\, 2x\right),$$

$$\operatorname{grad} f_2 = \left(\sqrt{1-x^2}\cos t,\, \sqrt{1-x^2}\sin t,\, x,\, \frac{g-kx}{\sqrt{1-x^2}}\cos t,\, \frac{g-kx}{\sqrt{1-x^2}}\sin t,\, k\right),$$

$$\operatorname{grad} K = (0,\, 0,\, 1,\, 0,\, 0,\, 0).$$

$$\tag{14.61}$$

The basis in the space orthogonal to these gradients can be chosen, for example, in the form

$$e_1 = (\sin t,\, -\cos t,\, 0,\, 0,\, 0,\, 0),$$

$$e_2 = (0,\, 0,\, 0,\, \sin t,\, -\cos t,\, 0),$$

$$e_3 = \left(\frac{k-gx}{1-x^2}\cos t,\, \frac{k-gx}{1-x^2}\sin t,\, 0,\, x\cos t,\, x\sin t,\, -\sqrt{1-x^2}\right).$$

Calculating $G(e_i, e_j)$, we obtain the following matrix:

$$\tilde{G} = \begin{pmatrix} 1 & \dfrac{kx-g}{1-x^2} & 0 \\ \dfrac{kx-g}{1-x^2} & \dfrac{(kx-g)^2}{(1-x^2)^2} & 0 \\ 0 & 0 & (1-x^2)W_{g,k}''(x) \end{pmatrix}.$$

Its eigenvalues are

$$\mu_1 = 0, \qquad \mu_2 = 1 + \frac{(g - kx)^2}{(1 - x^2)^2}, \qquad \mu_3 = (1 - x^2)W_{g,k}''(x).$$

The zero eigenvalue μ_1 corresponds to the fact that the Hessian of $H_{g,k}$ is degenerate along the direction tangent to the critical circle. The eigenvalue μ_2 is always positive; and the sign of μ_3 coincides with the sign of $W_{g,k}''(x)$ (since $|x| < 1$). Thus, the critical circles of $H_{g,k}$ may be either saddle or minimal. Since critical circles are determined by (14.60), local maxima correspond to saddle circles, and local minima of $W_{g,k}(x)$ on the interval $(-1, 1)$ correspond to minimal circles. This allows us to describe the bifurcation diagrams for an arbitrary Hamiltonian of the form (14.56). We confine ourselves with the description of those Hamiltonians for which the molecule W has the simplest form.

Proposition 14.3. *Let H be a Hamiltonian of the form (14.56), where $V(x)$ is a smooth function on the segment $[-1, 1]$. Suppose that, for any $x \in (-1, 1)$, we have*

$$V''(x) \geq 0. \tag{14.62}$$

Then the additional integral $K = S_3$ is a Bott function on every non-singular isoenergy 3-manifold $Q = \{f_1 = 1, f_2 = g, H = h\}$. Moreover, K has either a local minimum or a local maximum on each critical circle in Q (i.e., there are no saddle critical circles).

COMMENT. However, the condition $V''(x) \geq 0$ should not be considered as a necessary condition for K to be a Bott function. The only difference that occurs in the case $V''(x) < 0$ is that some saddle critical circles may appear. The integral K will remain a Bott function on almost all isoenergy manifolds Q_h.

Proof. We have to prove that, on each isoenergy three-dimensional manifold $Q = \{f_1 = 1, f_2 = g, H = h\}$, the critical points of the function $K|_Q$ form non-degenerate critical circles on which K has either a minimum or a maximum. It is easy to see that these critical circles are, at the same time, critical for the functions $H_{g,k}$, i.e., for the Hamiltonian H restricted to the 3-manifolds $\{K = \text{const}\}$. Moreover, to verify the non-degeneracy condition and define the type of such a circle (saddle or center), we may choose any of the functions $H|_{\{K=\text{const}\}}$ and $K|_{\{H=\text{const}\}}$. In other words, instead of K we may consider the function $H_{g,k}$. But we have already carried out the calculation for this function. In particular, the eigenvalues of the Hessian are

$$\mu_1 = 0, \qquad \mu_2 = 1 + \frac{(g - kx)^2}{(1 - x^2)^2}, \qquad \mu_3 = (1 - x^2)W_{g,k}''(x).$$

Hence, μ_2 is always positive, and the sign of μ_3 is defined by the sign of $W_{g,k}''(x)$. To compute the sign of $W_{g,k}''(x)$, we introduce the following notation: $p = \dfrac{g + k}{2}$, $q = \dfrac{g - k}{2}$. Then

$$W_{g,k} = \frac{p^2}{1 + x} + \frac{q^2}{1 - x} + V(x) + \frac{k^2(1 - \beta)}{2\beta}.$$

Hence

$$W'_{g,k}(x) = -\frac{p^2}{(1+x)^2} + \frac{q^2}{(1-x)^2} + V'(x),$$

$$W''_{g,k}(x) = \frac{2p^2}{(1+x)^3} + \frac{2q^2}{(1-x)^3} + V''(x).$$

Since $x \in (-1,1)$ and $V''(x) \geq 0$, we obtain $W''_{g,k}(x) > 0$. Then the eigenvalues μ_2 and μ_3 are both positive. Therefore, the integral K is a Bott function. Moreover, its critical circles are non-degenerate, and K have either a minimum or a maximum on them. $\square$

The Hamiltonian of the classical Lagrange case is a particular case of (14.56), for which $V(x)$ is a linear function. Then $V''(x) = 0$, i.e., condition (14.62) is fulfilled, and, therefore, Proposition 14.3 holds. Hence, a transformation of the Liouville foliation on Q_h may happen only if we pass through a critical value h_0 of the Hamiltonian (i.e., the isoenergy 3-manifold Q_h changes the topological type itself). Thus, inside each region shown in Fig. 14.4, the molecule W is the same, i.e., the Liouville foliation has the same topological sense.

In Fig. 14.29, we present several possible bifurcation diagrams of the momentum mapping $\mathcal{F} = H \times K$ (for various values of g and β), where

$$H = \frac{1}{2}\left(S_1^2 + S_2^2 + \frac{S_3^2}{\beta}\right) + R_3$$

is the Hamiltonian of the classical Lagrange case. The images of the singular points $(0,0,\pm g,0,0,\pm 1)$ are denoted in Fig. 14.29 by bolded points. They are the points, where $dH|_{M_{1,g}} = 0$ and $dK|_{M_{1,g}} = 0$ simultaneously.

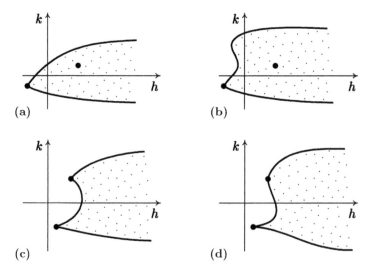

(a)

(b)

(c)

(d)

Figure 14.29

Let us note one useful and important property of the bifurcation diagrams in the Lagrange case. It turns out that each bifurcation diagram is bijectively projected onto the k-axis (with the only exception for the isolated point of the bifurcation diagram; see Fig. 14.29(a,b)). Indeed, formulas (14.58) show that the bifurcation diagram is given by the equation $h = W_{g,k}(x)$, where x is the solution of the equation $W'_{g,k}(x) = 0$. Since $W''_{g,k}(x) > 0$, the function $W'_{g,k}(x)$ is monotonically increasing. Besides, the function $W'_{g,k}(x)$ takes all values from $-\infty$ to $+\infty$. Therefore, the solution x always exists and is unique.

Theorem 14.3. *In the classical Lagrange case, the molecule W has the form A — A for any connected component of Q_h^3. These components can be of the following types: the sphere S^3, the projective space $\mathbb{R}P^3$, and the direct product $S^1 \times S^2$. The mark r on the edge of the molecule W^* depends on the topological type of Q only. Namely, the marked molecule has the following form:*

a) *A — A with $r = 0$ for S^3 (we denote this molecule by $\mathcal{L}_1^*$),*

b) *A — A with $r = 1/2$ for $\mathbb{R}P^3$ (we denote this molecule by $\mathcal{L}_2^*$),*

c) *A — A with $r = \infty$ and $\varepsilon = +1$ for $S^1 \times S^2$ (we denote this molecule by $\mathcal{L}_3^*$).*

COMMENT. In the case of S^3 and $\mathbb{R}P^3$, the mark ε depends on the choice of orientation on Q. That is why we may assume without loss of generality that $\varepsilon = +1$. Note that there is no mark n here. Thus, we obtain the complete Liouville classification of integrable systems of the classical Lagrange case (see also Table 14.1).

Table 14.1. Lagrange case

No	MOLECULE	Q^3
1	$r = 0$ $\varepsilon = 1$ A ———————— A	S^3
2	$r = \infty$ $\varepsilon = 1$ A ———————— A	$S^1 \times S^2$
3	$r = 1/2$ $\varepsilon = 1$ A ———————— A	$\mathbb{R}P^3$

Proof (of Theorem 14.3).   Since, according to Proposition 14.3, there are no saddle critical circles, the molecule W has the form A — A for each connected component of the isoenergy 3-manifold.   The mark r is calculated by using Proposition 4.3.   It remains to find the mark ε in the case, when Q is diffeomorphic to $S^1 \times S^2$.   In such a case, Q is glued from two solid tori whose axes are critical circles of the integral K.   Note that these circles are naturally included in a two-parameter family, which can be considered as a trivial S^1-fibration on $Q^3 = S^1 \times S^2$.   Moreover, this fibration is compatible with the Liouville foliation on Q^3 in the sense that each circle lies on a certain Liouville torus.   In fact, its fibers can be defined as integral curves of the Hamiltonian vector field $\operatorname{sgrad} K$ on Q (it is easy to see that all of them are closed).   Using this natural trivial S^1-fibration, we can choose admissible coordinate systems on the solid tori in such a way that the gluing matrix takes the form $C = \begin{pmatrix} \pm 1 & 0 \\ 0 & \mp 1 \end{pmatrix}$.   Recall (see Section 4.1) that, in this case, the first basis cycle is the contractible cycle on the solid torus, and the second basis cycle is the fiber of the trivial S^1-fibration just described.   The latter has the natural orientation induced by $\operatorname{sgrad} H$.   We now wish to show that the gluing matrix is in fact $C = \begin{pmatrix} +1 & 0 \\ 0 & -1 \end{pmatrix}$ or, equivalently, the orientations defined by $\operatorname{sgrad} H$ on the axes of the solid tori (i.e., second basis cycles) are opposite.

To this end, we need to compare the directions of $\operatorname{sgrad} H$ and $\operatorname{sgrad} K$ on two critical circles (i.e., on the axes of the solid tori).

Since $\operatorname{sgrad} H$ and $\operatorname{sgrad} K$ are linearly dependent on these circles, we have $\operatorname{sgrad} H = c_i \operatorname{sgrad} K$, where $i = 1, 2$. Let us show that c_1 and c_2 have opposite signs.   To see this, it suffices to use the fact that the bifurcation diagram Σ is bijectively projected to the k-axis and, consequently, can be presented as a graph of a certain continuous function $h = h(k)$. Now recall that c_1 and c_2 have a very natural interpretation (Proposition 1.16): they are the derivatives h'_k of the function $h = h(k)$ at the points where the vertical segment $\{h = \text{const}\}$ intersects

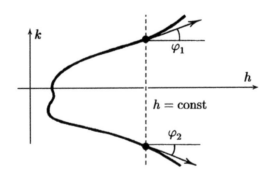

Figure 14.30

the bifurcation diagram Σ. It is easily seen from Fig. 14.30 that the signs of these derivatives are opposite, as required. Thus, $C = \begin{pmatrix} +1 & 0 \\ 0 & -1 \end{pmatrix}$ and, consequently, $\varepsilon = +1$. $\square$

We also can consider the Hamiltonian for the generalized Lagrange case:

$$H = \frac{1}{2}\left(S_1^2 + S_2^2 + \frac{S_3^2}{\beta} \right) + \lambda S_3 + V(R_3). \tag{14.63}$$

Corollary. *The statement of Theorem 14.3 is true for Hamiltonians* (14.63).

Proof. Notice that Hamiltonian (14.63) is a linear combination of Hamiltonian (14.56) and the integral $K = S_3$. Therefore, the momentum mapping $H \times K$ for (14.63) is a composition of the similar mapping $H \times K$ for (14.56) and the linear transformation of the plane $\mathbb{R}^2 (h, k)$ onto itself given by the matrix

$$\begin{pmatrix} 1 & \lambda \\ 0 & 1 \end{pmatrix}.$$

Since a non-degenerate linear transformation does not change indices of critical circles, the molecules W^* will have the same form as in Theorem 14.3. The arguments for computing the ε-mark also remain the same, since the main property of Σ that we have used in the proof of Theorem 14.3 is preserves under this linear transformation. $\square$

Theorem 14.4. *The isolated point of the bifurcation diagram for the Lagrange case (in those cases when such a point exists, i.e., for $g^2 < 4$) corresponds to the single singularity of focus–focus type in $M_{1,g}^4$ with coordinates $(0,0,g,0,0,1)$. Its loop molecule W^* is a circle (without any atoms) endowed with the monodromy matrix*

$$\begin{pmatrix} 1 & 0 \\ 1 & 1 \end{pmatrix}.$$

Proof. We only need to prove that the singularity corresponding to the isolated singular point of the bifurcation diagram is non-degenerate. In the non-degenerate case, such a singularity is necessarily of the focus–focus type. In addition, we need to compute the number of focus–focus points of the corresponding singular leaf. It turns out that such a point is unique. Indeed, the two singular points of the form $(0,0,\pm g,0,0,\pm 1)$ belongs to different levels of the integral $K = S_3$ (because $K = S_3 = \pm g$). Therefore, the level $\{K = S_3 = g\}$ contains just one singular point, as required. We now prove that this point has focus–focus type.

It suffices to verify that the eigenvalues of the linearized system at this singular point are complex numbers of the form

$$a + ib, \quad a - ib, \quad -a + ib, \quad -a - ib,$$

where a and b both differ from zero.

Consider the Euler–Poisson equations for the Hamiltonian

$$H = \frac{1}{2}\left(S_1^2 + S_2^2 + \frac{S_3^2}{\beta} \right) + R_3.$$

It has the form

$$\dot{S}_1 = S_2 S_3(1 - \beta^{-1}) - R_2, \qquad \dot{R}_1 = S_2 R_3 - S_3 R_2 \beta^{-1},$$
$$\dot{S}_2 = -S_1 S_3(1 - \beta^{-1}) + R_1, \qquad \dot{R}_2 = -S_1 R_3 + S_3 R_1 \beta^{-1},$$
$$\dot{S}_3 = 0, \qquad\qquad\qquad \dot{R}_3 = S_1 R_2 - S_2 R_1.$$

Consider the linearization of this system at the point $(0, 0, g, 0, 0, 1)$. As local co-ordinates on $M_{1,g}^4$ in a neighborhood of this point, we take variables S_1, S_2, R_1, R_2. Then the matrix of the linearized system becomes

$$\begin{pmatrix} 0 & g - \dfrac{g}{\beta} & 0 & -1 \\ -g + \dfrac{g}{\beta} & 0 & 1 & 0 \\ 0 & 1 & 0 & -\dfrac{g}{\beta} \\ -1 & 0 & \dfrac{g}{\beta} & 0 \end{pmatrix}.$$

Its eigenvalues are

$$a + ib, \quad a - ib, \quad -a + ib, \quad -a - ib,$$

where $a = \dfrac{\sqrt{4 - g^2}}{2}$, $b = g\left(\dfrac{1}{2} - \dfrac{1}{\beta}\right)$ (see [272] for details). If $g^2 < 4$, then a and b are real numbers, and $a \neq 0$.

However, if $g = 0$ or $\beta = 2$, then $b = 0$. In fact, this does not change the type of the singularity. The point is that we may replace the Hamiltonian H by a linear combination $\tilde{H} = H + \lambda K$. Clearly, such a procedure has no influence on the foliation, but the corresponding eigenvalues become different.

It remains to find the monodromy matrix. Since the singular leaf contains only one singular focus–focus point, it follows from Theorem 9.11 that this matrix has the desired form.   □

14.5. LIOUVILLE CLASSIFICATION OF SYSTEMS IN THE KOVALEVSKAYA CASE

Hamiltonian (14.14) in the Kovalevskaya case can be reduced to the form (14.25), i.e.,

$$H = \frac{1}{2}(S_1^2 + S_2^2 + 2S_3^2) + R_1,$$

by a linear transformation in $\mathbb{R}^6(S, R)$ which preserves the bracket (14.6). Then the additional integral will be of the form

$$K = \left(\frac{S_1^2 - S_2^2}{2} - R_1\right)^2 + (S_1 S_2 - R_2)^2. \tag{14.64}$$

Curves which separate domains on the plane $\mathbb{R}^2 (g, h)$ with different topological types of Q^3 for Hamiltonian (14.25) are described in Section 14.2 (Fig. 14.11). In order to describe all separating curves in the Kovalevskaya case, one must add to the curves in Fig. 14.11 those curves which separate domains with different molecules W^*. Thus, we need to determine how the molecule W^* corresponding to the isoenergy 3-manifold

$$Q_h^3 = \{f_1 = 1, f_2 = g, H = h\}$$

changes when h varies and g is fixed. This can be done by examining the bifurcation diagrams for the momentum mapping

$$K \times H \colon M_{1,g} \simeq TS^2 \to \mathbb{R}^2 (k, h) \,,$$

where H is Hamiltonian (14.25), and K is integral (14.64), both defined on $M_{1,g} \simeq TS^2 = \{f_1 = 1, f_2 = g\}$.

The bifurcation diagrams of this mapping are constructed in the book by M. P. Kharlamov [178]. The form of the bifurcation diagram depends on the value g. Qualitatively different diagrams are obtained in the following cases:

a) $g = 0$,

b) $0 < |g| < 1$,

c) $1 < |g| < (4/3)^{3/4}$,

d) $(4/3)^{3/4} < |g| < \sqrt{2}$,

e) $|g| > \sqrt{2}$.

They are presented in Fig. 14.31.

The bifurcation diagrams consist of the ray

$$\{k = 0, \ h > g^2\} \,, \tag{14.65}$$

part of the parabola

$$\left\{ k = (h - g^2)^2, \ \frac{g^2}{2} - 1 \le h \le g^2 + \frac{1}{2g^2} \right\} \,,$$

and the curve given in parametric form by

$$k = 1 + tg + \frac{t^4}{4} \,, \qquad h = \frac{t^2}{2} - \frac{g}{t} \,, \tag{14.66}$$

where $t \in (-\infty, 0) \cup (g, +\infty)$.

The cusp of curve (14.66), when $|g| \le (4/3)^{3/4}$, has the coordinates

$$\left(1 - \frac{3g^{4/3}}{4}, \frac{3g^{2/3}}{2} \right) \in \mathbb{R}^2 (k, h) \,. \tag{14.67}$$

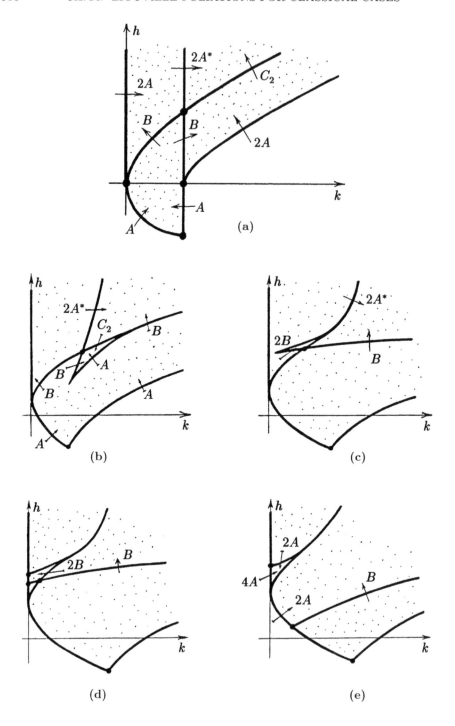

Figure 14.31

The point where the curve is tangent to the parabola has the coordinates

$$\left(\frac{1}{4g^4}, g^2 + \frac{1}{2g^2} \right) \in \mathbb{R}^2 (k, h) . \tag{14.68}$$

The bifurcations of Liouville tori at the critical values of the momentum mapping $K \times H$ are also given in Fig. 14.31. If the point moving along the plane $\mathbb{R}^2 (k, h)$ crosses the corresponding branch of the bifurcation diagram in the direction marked by the arrow, then the bifurcation of Liouville tori in the preimage of this point is described by the atom indicated near the corresponding arrow. The bifurcation diagrams (a), (b), (c), (d), (e) in Fig. 14.31 transform one into another when the parameter g changes continuously. Those parts of the bifurcation diagram which transform one into another (under variation of g) determine the same bifurcations of Liouville tori.

If we know the bifurcation of Liouville tori for all points of the bifurcation diagram, we can describe the molecules W of the isoenergy 3-manifold Q for all fixed g and h. To this end, we must examine how Liouville tori bifurcate at the preimage of the point moving along the line $\{h = c\}$ in $\mathbb{R}^2 (k, h)$. Let us look at what happens when we change the parameter c. From the explicit form of the bifurcation diagram, one can easily understand for what values of c the molecule will change. This happens in the following cases:

1) when c is a critical value of the function $H|_{M_1, g}$ (in this case, the topological type of Q^3 also changes);

2) when the line $\{h = c\}$ passes either through the cusp of curve (14.66), or through the point where the curve is tangent to the parabola, or through the origin of ray (14.65).

The images of critical points of the function $H|_{M_1, g}$ under the momentum mapping $K \times H$ are shown in Fig. 14.31 by bolded points. The separating curves corresponding to them have been already constructed (see Fig. 14.11). Taking into account (14.65), (14.67), (14.68), we obtain the equations of the remaining separating curves on the plane $\mathbb{R}^2 (g, h)$:

$$h = g^2 ,$$

$$h = \frac{3g^{2/3}}{2} , \qquad |g| \leq \left(\frac{4}{3} \right)^{3/4} , \tag{14.69}$$

$$h = g^2 + \frac{1}{2g^2} .$$

Having combined curves (14.69) with those shown in Fig. 14.11, we obtain a complete collection of separating curves for the Kovalevskaya case.

Let us note that the method suggested by M. P. Kharlamov in [178] does not give an answer to the question whether the additional integral K is a Bott function on Q^3 in the Kovalevskaya case. This can be verified by computing the indices of critical circles for the function $K|_{Q_h}$ and using Lemma 14.1.

The next theorem summarizes the results on the global topology of the Kovalevskaya case obtained by different methods in a series of papers [22], [44], [107], [178], [277], [344]. In the form presented below, the theorem was proved in [60].

Theorem 14.5. *For the system with Hamiltonian (14.14) (i.e., for the Ko-valevskaya case), the separating curves in the plane* $\mathbb{R}^2(g, h)$ *have the form shown in Fig. 14.32. They divide the plane into* 10 *domains of different types. In each domain, we indicate the pair* $(Q, \mathcal{K}^*)$, *i.e., the isoenergy 3-manifold* Q *and the corresponding loop molecule* $\mathcal{K}^*$. *As a result, we obtain the complete list which consists of* 10 *pairs:*

$$(S^3, \mathcal{K}_1^*), \quad (S^3, \mathcal{K}_2^*), \quad (S^3, \mathcal{K}_3^*), \quad (S^3, \mathcal{K}_4^*), \quad (S^1 \times S^2, \mathcal{K}_5^*), \quad (S^1 \times S^2, \mathcal{K}_6^*),$$
$$(K^3, \mathcal{K}_7^*), \quad (\mathbb{R}P^3, \mathcal{K}_8^*), \quad (\mathbb{R}P^3, \mathcal{K}_9^*), \quad (\mathbb{R}P^3, \mathcal{K}_{10}^*).$$

The marked molecules $\mathcal{K}_i^*$ *are listed in Table 14.2. The index* i *is related to numbering in Table 14.2. For all the points* (g, h) *that do not belong to the separating curves, the Kovalevskaya integral is a Bott function on the isoenergy surface* $Q = \{f_1 = 1, \ f_2 = g, \ H = h\}$. *Here* K^3 *denotes* $(S^1 \times S^2) \# (S^1 \times S^2)$. *Thus, we obtain the complete classification of the integrable Kovalevskaya systems up to Liouville equivalence.*

REMARK. In order to simplify notation, in Fig. 14.32 (as well as in some others), we simply write Q–i instead of the pair $(Q, \mathcal{K}_i^*)$.

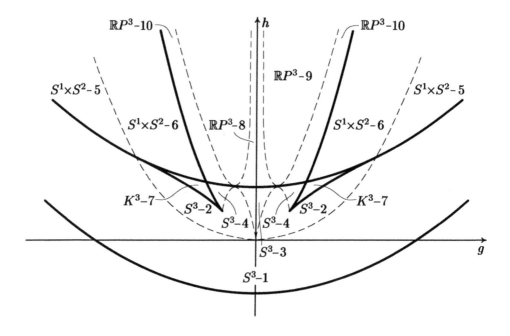

Figure 14.32

In Fig. 14.33, we present the bifurcation diagrams for the Kovalevskaya case together with horizontal dotted segments each of which corresponds to a certain energy level. Different segments correspond to qualitatively different energy levels in the sense that the topological invariants $\mathcal{K}^*$ related to these levels are all different. There is a natural correspondence between the energy levels shown in Fig. 14.33 and 10 regions shown in Fig. 14.32. This correspondence is as follows:

$$a_1 \to 1, \quad a_2 \to 3, \quad a_3 \to 8,$$
$$b_1 \to 1, \quad b_2 \to 2, \quad b_3 \to 3, \quad b_4 \to 8, \quad b_5 \to 9,$$
$$c_1 \to 1, \quad c_2 \to 2, \quad c_3 \to 4, \quad c_4 \to 10, \quad c_5 \to 9,$$
$$d_1 \to 1, \quad d_2 \to 2, \quad d_3 \to 7, \quad d_4 \to 6, \quad d_5 \to 4, \quad d_6 \to 10, \quad d_7 \to 9,$$
$$e_1 \to 1, \quad e_2 \to 5, \quad e_3 \to 6, \quad e_4 \to 10, \quad e_5 \to 9,$$

where Latin letters with subscripts denote the energy levels in Fig. 14.33, and the integers indicate the numbers of regions in Fig. 14.32 and in Table 14.2.

Now we are able to give a complete list of singularities of the momentum mapping $\mathcal{F} = (H, K) \colon M_{1,g}^4 \to \mathbb{R}^2$ that appear in the Kovalevskaya case. First, in the Kovalevskaya case, there are four different generic bifurcations. In our terms, they are atoms A, A^*, B, C_2. These singularities correspond to smooth regular pieces of the bifurcation diagram. Second, the bifurcation diagram has several singular points such as cusps, intersection points, and points of tangency. The types of these points can also be described. All of them are listed below.

The singular points $y_1, y_2, \ldots, y_{13}$ of the bifurcation diagram are indicated in Fig. 14.33. Singular points denoted by the same number correspond to singularities of the same topological type. We assume here that the area constant g differs from several exceptional values equal to 1, $(4/3)^{3/4}$, $2^{1/2}$ (at which the bifurcation diagram changes the type).

Theorem 14.6 [44].

a) *The singular points* y_1, y_3, y_7, y_{10}, y_{11}, y_{12} *correspond to non-degenerate singularities of the momentum mapping* $\mathcal{F} = (H, K) \colon M_{1,g}^4 \to \mathbb{R}^2$. *More precisely,* y_1 *and* y_{10} *correspond to singularities of center–center type,* y_{11} *and* y_{12} *correspond to singularities of center–saddle type,* y_3 *and* y_7 *correspond to singularities of saddle–saddle type. In the Kovalevskaya case, there are no singularities of focus–focus type.*

b) *The singular points* y_2, y_4, y_5, y_6, y_8, y_9, y_{13} *correspond to degenerate one-dimensional orbits of the action of* $\mathbb{R}^2$ *generated by the Hamiltonian* H *and integral* K *on* $M_{1,g}^4$.

c) *The loop molecules of these singular points are listed in Table* 14.3.

It is easy to see that Table 14.3 contains eight different loop molecules. Thus, in the Kovalevskaya case, there are singularities of eight different topological types defined by the listed loop molecules. Besides, there are four generic singularities, namely the atoms A, A^*, B, C_2 that describe the bifurcations on regular curves of the bifurcation diagram.

Table 14.2. Kovalevskaya case

No	MOLECULE	Q^3
1	$A \underset{\substack{}}{\overset{\substack{r=0 \\ \varepsilon=1}}{\rule{6cm}{0.4pt}}} A$	S^3
2	A ($r=0$, $\varepsilon=1$), A ($\varepsilon=1$, $r=0$) — B ($n=0$) — A ($r=1/2$, $\varepsilon=1$)	S^3
3	A ($r=0$, $\varepsilon=1$), A ($\varepsilon=1$, $r=0$) — B ($n=1$) — ($r=0$, $\varepsilon=1$) — B ($n=2$) — A ($r=0$, $\varepsilon=1$), A ($\varepsilon=1$, $r=0$)	S^3
4	A ($r=\infty$, $\varepsilon=1$), A ($r=0$, $\varepsilon=1$) — B — ($r=0$, $\varepsilon=1$); A ($r=\infty$, $\varepsilon=1$), A ($r=0$, $\varepsilon=1$) — B ($r=0$, $\varepsilon=1$) — B ($n=0$) — A ($r=1/2$, $\varepsilon=1$)	S^3
5	A ($r=\infty$, $\varepsilon=1$), A ($\varepsilon=1$, $r=\infty$) — B — A ($r=0$, $\varepsilon=1$)	$S^1 \times S^2$
6	A ($r=0$, $\varepsilon=1$), A ($r=0$, $\varepsilon=1$) — B ($n=0$) ($r=0$, $\varepsilon=-1$); A ($r=0$, $\varepsilon=1$), A ($r=0$, $\varepsilon=1$) — B ($n=0$) ($\varepsilon=-1$, $r=0$) — B ($n=-2$) — A ($r=0$, $\varepsilon=1$)	$S^1 \times S^2$

Table 14.2. Kovalevskaya case (continued)

No	MOLECULE	Q^3
7		$(S^1 \times S^2) \# (S^1 \times S^2)$
8		$\mathbb{R}P^3$
9		$\mathbb{R}P^3$
10		$\mathbb{R}P^3$

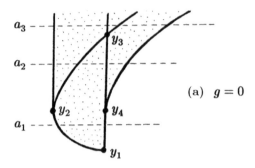

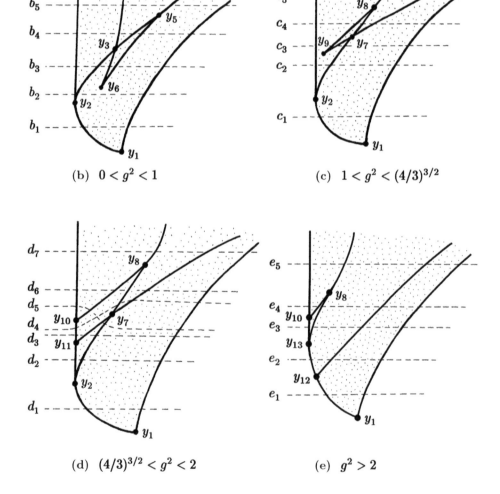

Figure 14.33

Table 14.3. Loop molecules for the Kovalevskaya case

14.6. LIOUVILLE CLASSIFICATION OF SYSTEMS IN THE GORYACHEV–CHAPLYGIN–SRETENSKIĬ CASE

Now we pass to the Goryachev–Chaplygin–Sretenskiĭ case. Hamiltonian (14.19) is reduced to the form (14.35) by a linear transformation which preserves the bracket (14.6) in $\mathbb{R}^6(S,R)$. Then the additional integral K takes the form

$$K = (S_3 + 2\lambda)(S_1^2 + S_2^2) - S_1 R_3. \tag{14.70}$$

The Goryachev–Chaplygin case is obtained from the Sretenskiĭ case when $\lambda = 0$.

The bifurcation diagrams of the momentum mapping $H \times K\colon M_{1,0} \to \mathbb{R}^2(h,k)$ are constructed in [178] by M. P. Kharlamov. Here H is Hamiltonian (14.35), K is integral (14.70) both defined on $M_{1,0} \simeq TS^2 = \{f_1 = 1, f_2 = 0\} \subset \mathbb{R}^6(S,R)$.

The bifurcation diagrams for different values of λ are shown in Fig. 14.34:

a) $\lambda = 0$,
b) $0 < \lambda < \dfrac{1}{\sqrt{3}}$,
c) $\dfrac{1}{\sqrt{3}} < \lambda < 1$,
d) $\lambda > 1$.

When λ is replaced by $-\lambda$, the bifurcation diagrams reflect in the line $\{k = 0\}$.

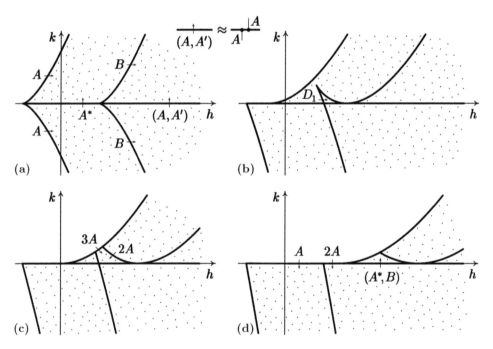

Figure 14.34

The bifurcation diagrams consist of the ray $\{k = 0, h \geq -1\}$ and of the following curves given in parametric form:

$$h = \frac{3}{2}t^2 + 4\lambda t + 2\lambda^2 \pm 1, \qquad k = t^3 + 2\lambda t^2.$$

The bifurcations of Liouville tori are shown in Fig. 14.34 in the same way as in the Kovalevskaya case. The notation (A^*, B) means that two bifurcations of Liouville tori corresponding to the atoms A^* and B occur simultaneously. The notation (A, A') means that there are one minimal and one maximal circles of the integral K in the preimage of these points of the bifurcation diagram.

As in the Kovalevskaya case, having found the projections of cusps and tangency points on the h-axis, one obtains equations for the separating curves on the plane $\mathbb{R}^2(\lambda, h)$:

$$h = 1 - \frac{2\lambda^2}{3} \quad \text{and} \quad h = 2\lambda^2 \pm 1.$$

Combining them with the curves shown in Fig. 14.17, we get the answer for the Goryachev–Chaplygin–Sretenskiĭ case (see Fig. 14.35). The Bott property for the additional integral K can be verified by straightforward calculation of indices of the critical circles.

Theorem 14.7 (A. A. Oshemkov, P. Topalov, O. E. Orel). *The separating curves in the plane $\mathbb{R}^2(\lambda, h)$ for the Goryachev–Chaplygin–Sretenskiĭ case are presented in Fig. 14.35. The additional Goryachev–Chaplygin–Sretenskiĭ integral (14.70) is a Bott function on all isoenergy 3-surfaces corresponding to points (λ, h) which do not belong to the separating curves. The complete list of invariants for the Goryachev–Chaplygin–Sretenskiĭ case consists of 8 pairs listed in Table 14.4:*

$$(S^3, \mathcal{G}_1^*), \quad (S^3, \mathcal{G}_2^*), \quad (S^3, \mathcal{G}_3^*), \quad (S^1 \times S^2, \mathcal{G}_4^*), \quad (S^1 \times S^2, \mathcal{G}_5^*),$$
$$(\mathbb{R}P^3, \mathcal{G}_6^*), \quad (\mathbb{R}P^3, \mathcal{G}_7^*), \quad (K^3, \mathcal{G}_8^*),$$

where K^3 denotes the connected sum $(S^1 \times S^2) \# (S^1 \times S^2)$.

In particular, for $\lambda = 0$, i.e., in the Goryachev–Chaplygin case, the additional integral is a Bott function on all regular isoenergy surfaces. The complete list of invariants in this case consists of two pairs: $(S^3, \mathcal{G}_1^*)$ and $(\mathbb{R}P^3, \mathcal{G}_8^*)$.

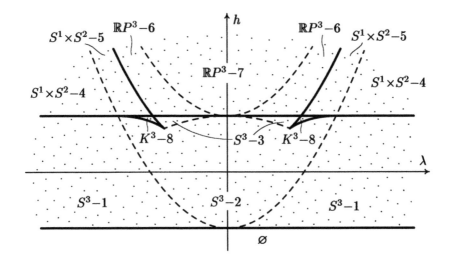

Figure 14.35

Table 14.4. Goryachev–Chaplygin–Sretenskiĭ case

No	MOLECULE	Q^3
1	$A \xrightarrow[\substack{}]{\substack{r=0 \\ \varepsilon=1}} A$	S^3
2	$A \xrightarrow[\substack{}]{\substack{r=0 \\ \varepsilon=1}} \underset{n=0}{A^*} \xrightarrow[\substack{}]{\substack{r=0 \\ \varepsilon=1}} A$	S^3
3	$A \xrightarrow[\substack{}]{\substack{r=0 \\ \varepsilon=1}} \underset{n=0}{A^*} \xrightarrow[\substack{}]{\substack{r=0 \\ \varepsilon=1}} D_1$ branching to $\substack{\varepsilon=1 \\ r=0}\ A$, $\substack{r=\infty \\ \varepsilon=1}\ A$, $\substack{r=0 \\ \varepsilon=1}\ A$	S^3
4	$A \xrightarrow[\substack{}]{\substack{r=0 \\ \varepsilon=1}} B$ branching to $\substack{r=\infty \\ \varepsilon=1}\ A$, $\substack{\varepsilon=1 \\ r=\infty}\ A$	$S^1 \times S^2$
5	$A \xrightarrow[\substack{}]{\substack{r=0 \\ \varepsilon=1}} \underset{n=-1}{B}$ branching: $\substack{r=1/2 \\ \varepsilon=-1}\ \underset{n=-1}{A^*} \xrightarrow[\substack{}]{\substack{r=0 \\ \varepsilon=1}} A$, $\substack{r=0 \\ \varepsilon=1}\ A$; $\substack{r=0 \\ \varepsilon=-1}\ \underset{n=0}{B}$ with $\substack{r=0 \\ \varepsilon=1}\ A$	$S^1 \times S^2$
6	$A \xrightarrow[\substack{}]{\substack{r=0 \\ \varepsilon=1}} \underset{n=-1}{B}$ branching: $\substack{r=1/2 \\ \varepsilon=-1}\ \underset{n=-1}{A^*} \xrightarrow[\substack{}]{\substack{r=0 \\ \varepsilon=1}} A$, $\substack{r=0 \\ \varepsilon=1}\ A$; $\substack{r=0 \\ \varepsilon=-1}\ \underset{n=-2}{B}$ with $\substack{r=0 \\ \varepsilon=1}\ A$	$\mathbb{R}P^3$

Table 14.4. Goryachev–Chaplygin–Sretenskiĭ case (continued)

No	MOLECULE	Q^3
7		$\mathbb{R}P^3$
8		$(S^1 \times S^2) \# (S^1 \times S^2)$

Thus, we obtain the complete description of Liouville foliations that occur in the Goryachev–Chaplygin–Sretenskiĭ case.

The bifurcation diagrams for the Goryachev–Chaplygin–Sretenskiĭ case are shown in Fig. 14.36 together with vertical dotted segments each of which corresponds to a certain energy level. Different segments (on the same bifurcation diagram) correspond to qualitative different isoenergy surfaces Q^3 in the sense that the topological invariants $\mathcal{G}^*$ related to these Q^3's are mutually distinct.

There is a natural correspondence between the energy levels shown in Fig. 14.36 and eight regions shown in Fig. 14.35. This correspondence is as follows:

$a_1 \to 2$, $a_2 \to 7$,

$b_1 \to 1$, $b_2 \to 2$, $b_3 \to 3$, $b_4 \to 6$, $b_5 \to 7$,

$c_1 \to 1$, $c_2 \to 2$, $c_3 \to 8$, $c_4 \to 3$, $c_5 \to 5$, $c_6 \to 6$, $c_7 \to 7$,

$d_1 \to 1$, $d_2 \to 4$, $d_3 \to 5$, $d_4 \to 6$, $d_5 \to 7$,

where the Latin letters with indices denote the energy levels in Fig. 14.36, and the integers corresponding to them indicate the numbers of regions in Fig. 14.35 and in Table 14.4.

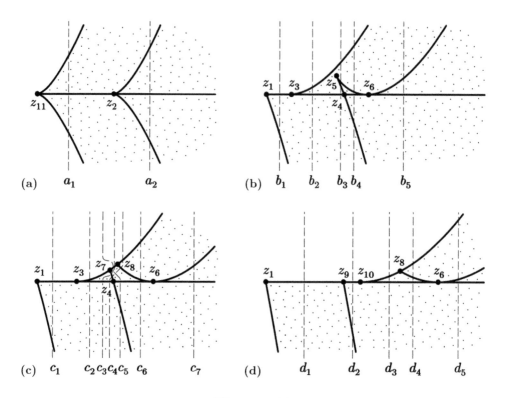

Figure 14.36

Thus, for each bifurcation diagram, we can easily see how the molecule of the system changes under variation of the energy level.

COMMENT. In Fig. 14.36(c), there are two intersecting dotted lines c_4 and c_5. For each particular value of λ, only one of these cases is realized depending on the position of the singular point z_8. The projection of z_8 to the horizontal axis may be either on the right or on the left of the point z_4. In the first case, we have c_5; in the second case, we have c_4.

Now we can collect a complete list of singularities of the momentum mapping $\mathcal{F} = (H, K) \colon M_{1,0}^4 \to \mathbb{R}^2$ that occur in the Goryachev–Chaplygin–Sretenskiĭ case. First, in this case, there are four atoms corresponding to generic bifurcations (singularities). They are A, A^*, B, D_1. As usual, these bifurcations correspond to smooth regular pieces of the bifurcation diagram. Second, there are 11 singular points of the bifurcation diagram listed below (cusps, tangency, and intersection points). The topological types of the singularities corresponding to these points can also be described (see Theorem 14.8 below).

The singular points $z_1, \ldots, z_{11}$ of the bifurcation diagram are indicated in Fig. 14.36. Singular points denoted by the same numbers correspond to singularities of the same topological type. We assume here that λ differs from several exceptional values equal to $\pm 1/\sqrt{3}, \pm 1$, at which the bifurcation diagram changes its type.

Theorem 14.8.

a) *The singular points z_1, z_4, z_7, z_8, z_9 correspond to non-degenerate singularities of the momentum mapping $\mathcal{F} = (H, K) \colon M_{1,0}^4 \to \mathbb{R}^2$ (Fig. 14.36). More precisely, z_1 and z_8 correspond to singularities of center–center type, z_7 and z_9 correspond to singularities of center–saddle type, and z_4 corresponds to a singularity of saddle–saddle type. There are no singularities of focus–focus type in the Goryachev–Chaplygin–Sretenskiĭ case.*

b) *The singular points z_3, z_5, z_6, z_{10} (Fig. 14.36) correspond to degenerate one-dimensional orbits of the action of $\mathbb{R}^2$ generated by the Hamiltonian H and integral K on $M_{1,0}^4$.*

c) *The singular points z_{11} and z_2 correspond to degenerate zero-dimensional orbits of the action of $\mathbb{R}^2$ (i.e., degenerate equilibrium points in $M_{1,0}^4$).*

d) *The loop molecules of the above singular points for the Goryachev–Chaplygin–Sretenskiĭ case are presented in Table 14.5.*

Table 14.5. Loop molecules for the Goryachev–Chaplygin–Sretenskiĭ case

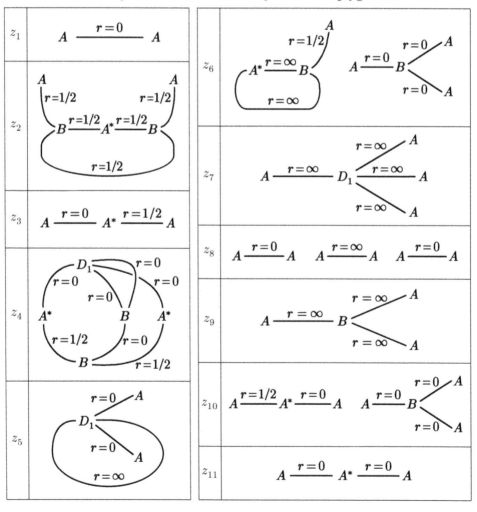

REMARK. This list in Table 14.5 contains exactly 10 different marked molecules. Thus, in the Goryachev–Chaplygin–Sretenskiĭ case, there are singularities of 10 different types that correspond to the listed molecules. Besides the singularities $z_1, \ldots, z_{11}$, there are four generic singularities corresponding to the atoms A, A^*, B, D_1. They describe the bifurcations of the Liouville foliation that correspond to regular curves of the bifurcation diagram.

14.7. LIOUVILLE CLASSIFICATION OF SYSTEMS IN THE ZHUKOVSKIĬ CASE

The Hamiltonian in the Zhukovskiĭ case is

$$H = \frac{(S_1 + \lambda_1)^2}{2A_1} + \frac{(S_2 + \lambda_2)^2}{2A_2} + \frac{(S_3 + \lambda_3)^2}{2A_3}.$$

The additional integral has the same form as in the Euler case:

$$K = S_1^2 + S_2^2 + S_3^2.$$

The bifurcation diagrams for the momentum mapping $K \times H \colon M_{1,g} \to \mathbb{R}^2$, where

$$M_{1,g} \simeq TS^2 = \{f_1 = 1, f_2 = g\} \subset \mathbb{R}^6 (S, R),$$

were constructed by M. P. Kharlamov in [178]. They are presented in Fig. 14.37 and can be given by explicit formulas in the following way.

Set

$$a_1 = A_1^{-1}, \quad a_2 = A_2^{-1}, \quad a_3 = A_3^{-1}$$

and suppose that $a_1 > a_2 > a_3 > 0$. Then the bifurcation diagram is given as a parametrized curve $(h(t), k(t))$ on the plane $\mathbb{R}^2 (h, k)$, where

$$h(t) = \frac{t^2}{2} \left(\frac{a_1 \lambda_1^2}{(a_1 - t)^2} + \frac{a_2 \lambda_2^2}{(a_2 - t)^2} + \frac{a_3 \lambda_3^2}{(a_3 - t)^2} \right),$$

$$k(t) = \frac{a_1^2 \lambda_1^2}{(a_1 - t)^2} + \frac{a_2^2 \lambda_2^2}{(a_2 - t)^2} + \frac{a_3^2 \lambda_3^2}{(a_3 - t)^2}.$$

This curve consists of three connected components that correspond to three intervals, where the parameter t can vary: (a_3, a_2), (a_2, a_1), and $(a_1, +\infty) \cup (-\infty, a_3)$. The connected components corresponding to intervals (a_3, a_2), (a_2, a_1) have one cusp each. The third component is a smooth curve.

More precisely, we should consider not the whole curve but part of it that belongs to the half-plane defined by $g^2 \le k$. Besides, the bifurcation diagram contains the vertical segment lying on the line $\{g^2 = k\}$ whose ends are the intersection points of this line with the third component of the curve, as shown in Fig. 14.37.

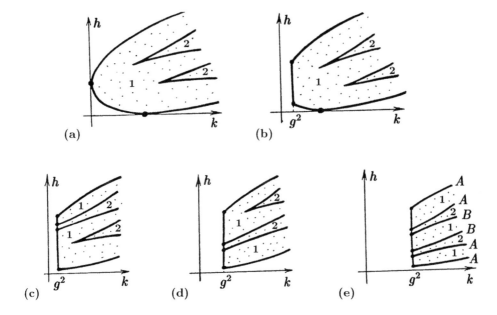

Figure 14.37

When $\lambda_1, \lambda_2, \lambda_3$ are all different from zero, the bifurcation diagrams Σ are shown in Fig. 14.37. For $g = 0$, they are presented in Fig. 14.37(a); and for $g \neq 0$, they are presented in Fig. 14.37(b–e). Under variation of g, the bifurcation diagram changes in the following way. The segment $\{k = g^2\}$ moves to the right cutting pieces of Σ, on which the cusp points are located. These pieces may be cut in different order. This depends on the parameters $A_1, A_2, A_3, \lambda_1, \lambda_2, \lambda_3$ of the Hamiltonian. Therefore, when g is increasing, we obtain one of the following sequences of diagrams: either (a)–(b)–(c)–(e) or (a)–(b)–(d)–(e) (see Fig. 14.37). For those regions on $\mathbb{R}^2(k, h)$ whose preimage under $K \times H$ is not empty, we also indicated in Fig. 14.37 the number of Liouville tori at the preimage of each point from the corresponding region. The bifurcation diagram is tangent (when $g = 0$) to the vertical axis $\{k = 0\}$ at the point

$$h_0 = \frac{\lambda_1^2}{2A_1} + \frac{\lambda_2^2}{2A_2} + \frac{\lambda_3^2}{2A_3},$$

and (when $g^2 < \lambda_1^2 + \lambda_2^2 + \lambda_3^2$) to the horizontal axis $\{h = 0\}$ at the point $k_0 = \lambda_1^2 + \lambda_2^2 + \lambda_3^2$. Six branches of this curve have the following asymptotics at infinity:

$$h \approx \frac{k}{2A_i} \pm \frac{\lambda_i \sqrt{k}}{A_i}. \qquad (14.71)$$

If the point moving along the plane $\mathbb{R}^2(k, h)$ crosses the bifurcation diagram, then the Liouville tori which lie in the preimage of this point undergo a certain bifurcation. The types of these bifurcations for the Zhukovskiĭ case is defined in [178]. They can be described as follows. Let us examine the line $\{h = c\}$, where c is sufficiently large. Then, in view of (14.71), this line crosses branches of the bifurcation diagram in some definite order. Let us enumerate them in this order: $x_1, x_2, x_3, x_4, x_5, x_6$ (see Fig. 14.38(a)).

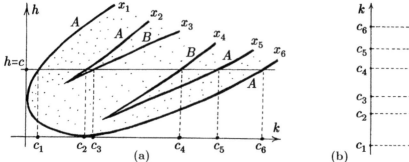
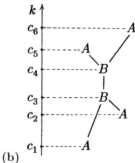

Figure 14.38

Consider the function $\widetilde{K} = K|_{Q_c}$, where $Q_c = \{f_1 = 1,\ f_2 = g,\ H = c\}$, and denote its critical values by $c_1, c_2, c_3, c_4, c_5, c_6$. The molecule W that demonstrates how the Liouville tori bifurcate along the line $\{h = c\}$ is shown in Fig. 14.38(b). The atoms A correspond to elliptic critical circles of $\widetilde{K}$, and the atoms B correspond to saddle critical circles. Thus, for all points of the curve, the bifurcations of the Liouville tori are described. Only minimal critical circles (i.e., those on which $\widetilde{K}$ has a minimum) lie in the preimage of the segment belonging to the line $\{k = g^2\}$.

In order to describe all possible types of molecules W, we need to determine how the line $\{h = c\}$ intersects the bifurcation diagram for different c. First of all, let us examine the case $g = 0$. Suppose the cusp point which divides the branches x_4 and x_5 has coordinates (k_1, h_1) on the plane $\mathbb{R}^2(k, h)$, and the cusp which lies between the branches x_2 and x_3 has coordinates (k_2, h_2). It is easy to see that the bifurcation of the Liouville foliation happens for $c = \min(h_1, h_2)$ and $c = \max(h_1, h_2)$ (Fig. 14.39). Besides, at the point $c = h_0$, the topological type of the isoenergy 3-manifold Q changes. The molecule W (without marks) remains the same but the numerical marks change, i.e., the marked molecule W^* transforms.

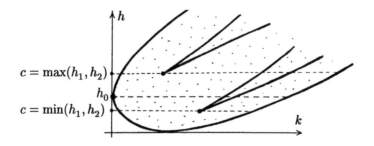

Figure 14.39

The case $g \neq 0$ can be examined in a similar way. The point is that the coordinates (k_1, h_1) and (k_2, h_2) of the cusp points do not change under variation of g. Therefore, in the case $g \neq 0$, the bifurcational values $c = \min(h_1, h_2)$ and $c = \max(h_1, h_2)$ remain the same. However, they disappear when the cusp point leaves the image of the momentum mapping. As a result, we see that two horizontal segments ending in cusps must be added to the curves that separate domains with different topological types of Liouville foliations on Q (Fig. 14.16). Then the topological type of Q and marked molecule W^* will coincide for all points of one domain. Different dispositions of these segments on the plane $\mathbb{R}^2(g, h)$ are determined by the disposition of the points h_1, h_2, h_0 on the h-axis (see also Fig. 14.40(a–e)).

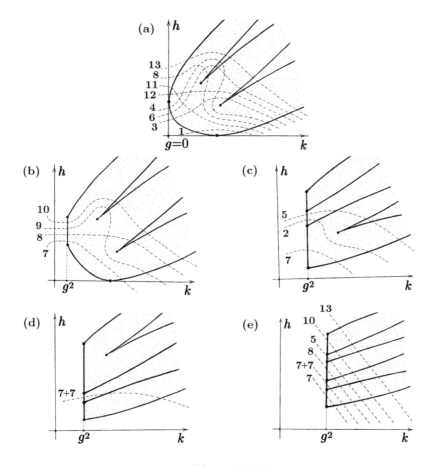

Figure 14.40

Let us draw curves separating domains with different topological type of Q by solid lines, and curves which divide domain with different marked molecules W^* by dotted lines. In each domain, we indicate the pair (the topological type of Q, the molecule W_i^* for Q).

Theorem 14.9 (A. A. Oshemkov, P. Topalov). *For the system with Hamiltonian* (14.16) *with* $\lambda_1, \lambda_2, \lambda_3 \neq 0$ *(i.e., the Zhukovskiĭ case), the separating curves on the plane* $\mathbb{R}^2(g,h)$ *for different values of parameters are of the form shown in Fig.* 14.41(a–f). *For each domain in Fig.* 14.41, *the topological type of the isoenergy 3-manifold* Q *and the corresponding marked molecule* W_i^* *are indicated. The complete list (for different parameters of the Hamiltonian) consists of 13 pairs presented in Table 14.6 and Fig.* 14.40(a–e):

$$(S^1 \times S^2, \mathcal{Z}_1^*), (S^1 \times S^2, \mathcal{Z}_2^*), (S^1 \times S^2, \mathcal{Z}_3^*), (S^1 \times S^2, \mathcal{Z}_4^*), (S^1 \times S^2, \mathcal{Z}_5^*), (S^1 \times S^2, \mathcal{Z}_6^*),$$
$$(S^3, \mathcal{Z}_7^*), (S^3, \mathcal{Z}_8^*), (S^3, \mathcal{Z}_9^*), (S^3, \mathcal{Z}_{10}^*), (\mathbb{R}P^3, \mathcal{Z}_{11}^*), (\mathbb{R}P^3, \mathcal{Z}_{12}^*), (\mathbb{R}P^3, \mathcal{Z}_{13}^*).$$

The additional integral K *is a Bott function on all the isoenergy 3-manifolds* $Q^3 = \{f_1 = 1, f_2 = g, H = h\}$ *for which the point* (g,h) *does not belong to the separating curves shown in Fig.* 14.41.

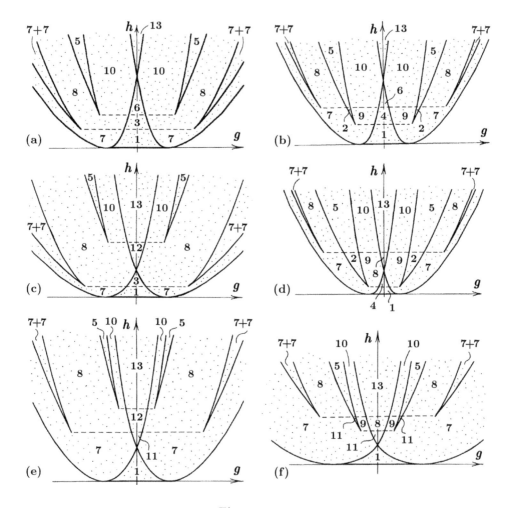

Figure 14.41

COMMENT. The numbers near dotted lines in Fig. 14.40 correspond to those in Table 14.6. The energy levels $\{H = \text{const}\}$ indicated in Fig. 14.40 by dotted lines are shown as smooth curves. In fact, these curves must be horizontal straight lines (in the plane $\mathbb{R}^2(k, h)$), but we have drawn them curved in order to present all possible situations on the same figure. Otherwise, we would have to draw too many different bifurcation diagrams for the momentum mapping $K \times H \colon M_{1,g} \to \mathbb{R}^2$ in order to include all the cases of mutual dispositions of the cusps (k_1, h_1), (k_2, h_2) and the point of tangency $(0, h_0)$.

Recall that Hamiltonian (14.12) corresponding to the Euler case is obtained from Hamiltonian (14.16) of the Zhukovskiĭ case when $\lambda_1 = \lambda_2 = \lambda_3 = 0$. The bifurcation diagram of the momentum mapping $K \times H$ in the Euler case is obtained from that in the Zhukovskiĭ case by passing to the limit as $\lambda \to 0$. The branches of the curve shown in Fig. 14.37 are combined pairwise (when $\lambda \to 0$): x_1 with x_2, x_3 with x_4, and x_5 with x_6. As a result, we obtain the bifurcation diagram shown in Fig. 14.42.

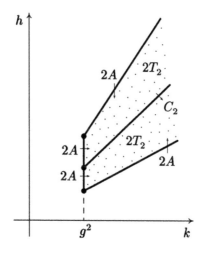

Figure 14.42

It consists of three rays

$$\left\{ h = \frac{k}{2A_i}, \ k \geq g^2 \right\}$$

and the segment

$$\left\{ k = g^2, \ \frac{g^2}{2A_3} \leq h \leq \frac{g^2}{2A_1} \right\}.$$

When the branches combine, the corresponding critical circles move to the same level of the additional integral. As a result, two new molecules $\mathcal{E}_2^*, \mathcal{E}_3^*$ appear in the Euler case (see Table 14.7).

Table 14.6. Zhukovskiĭ case

No	MOLECULE	Q^3
1	$A \overset{\begin{array}{c}r=\infty\\ \varepsilon=1\end{array}}{\rule{3cm}{0.4pt}} A$	$S^1 \times S^2$
2	$A \overset{r=0}{\underset{\varepsilon=1}{\rule{2cm}{0.4pt}}} B \big< \begin{array}{l} \overset{r=\infty}{\underset{\varepsilon=1}{}}\ A \\ \underset{r=\infty}{\overset{\varepsilon=1}{}}\ A \end{array}$	$S^1 \times S^2$
3	$A \overset{r=0}{\underset{\varepsilon=-1}{\rule{2cm}{0.4pt}}} \underset{n=0}{B} \big< \begin{array}{l} \overset{r=0}{\underset{\varepsilon=1}{}}\ A \\ \underset{r=0}{\overset{\varepsilon=1}{}}\ A \end{array}$	$S^1 \times S^2$
4	$A \overset{r=0}{\underset{\varepsilon=1}{\rule{2cm}{0.4pt}}} \underset{n=0}{B} \big< \begin{array}{l} \overset{r=0}{\underset{\varepsilon=1}{}}\ A \\ \underset{r=0}{\overset{\varepsilon=-1}{}}\ A \end{array}$	$S^1 \times S^2$
5	$\begin{array}{l} A \\ \quad \\ A \end{array} \big> \overset{r=\infty}{\underset{\varepsilon=1}{}} B \overset{r=\infty}{\underset{\varepsilon=1}{\rule{1.5cm}{0.4pt}}} B \big< \begin{array}{l} \overset{r=0}{\underset{\varepsilon=1}{}}\ A \\ \underset{r=0}{\overset{\varepsilon=1}{}}\ A \end{array}$ with $\varepsilon=1,\ \varepsilon=1$	$S^1 \times S^2$
6	$\begin{array}{l} A \\ \quad \\ A \end{array} \big> \overset{r=0}{\underset{\varepsilon=-1}{}} B \overset{r=\infty}{\underset{\underset{n=0}{\varepsilon=1}}{\rule{1.5cm}{0.4pt}}} B \big< \begin{array}{l} \overset{r=0}{\underset{\varepsilon=1}{}}\ A \\ \underset{r=0}{\overset{\varepsilon=1}{}}\ A \end{array}$ with $\varepsilon=1$	$S^1 \times S^2$
7	$A \overset{\begin{array}{c}r=0\\ \varepsilon=1\end{array}}{\rule{3cm}{0.4pt}} A$	S^3
8	$A \overset{r=\infty}{\underset{\varepsilon=1}{\rule{2cm}{0.4pt}}} B \big< \begin{array}{l} \overset{r=0}{\underset{\varepsilon=1}{}}\ A \\ \underset{r=0}{\overset{\varepsilon=1}{}}\ A \end{array}$	S^3
9	$A \overset{r=0}{\underset{\varepsilon=1}{\rule{2cm}{0.4pt}}} B \big< \begin{array}{l} \overset{r=0}{\underset{\varepsilon=1}{}}\ A \\ \underset{r=\infty}{\overset{\varepsilon=1}{}}\ A \end{array}$	S^3

Table 14.6. Zhukovskiĭ case (continued)

No	MOLECULE	Q^3
10	A $r=0$, $\varepsilon=1$ — B — $r=\infty$, $\varepsilon=1$ — B — $r=0$, $\varepsilon=1$ A; A $\varepsilon=1$, $r=\infty$ — B; B — $\varepsilon=1$, $r=0$ A	S^3
11	A ——— $\begin{array}{c} r=1/2 \\ \varepsilon=1 \end{array}$ ——— A	$\mathbb{R}P^3$
12	A — $\begin{array}{c} r=0 \\ \varepsilon=1 \end{array}$ — B $(n=2)$ $\begin{array}{c} r=0, \varepsilon=1 \\ \varepsilon=1, r=0 \end{array}$ A, A	$\mathbb{R}P^3$
13	A $r=0$, $\varepsilon=1$ — B — $r=\infty$, $\varepsilon=1$ — B — $r=0$, $\varepsilon=1$ A; A $\varepsilon=1$, $r=0$ — B $(n=2)$ — B — $\varepsilon=1$, $r=0$ A	$\mathbb{R}P^3$

Table 14.7. Euler case

No	MOLECULE	Q^3
1	A ——— $\begin{array}{c} r=0 \\ \varepsilon=1 \end{array}$ ——— A	S^3
2	A $r=\infty$, $\varepsilon=1$ — C_2 — $r=0$, $\varepsilon=1$ A; A $\varepsilon=1$, $r=\infty$ — C_2 — $\varepsilon=1$, $r=0$ A	$S^1 \times S^2$
3	A $r=0$, $\varepsilon=1$ — C_2 $(n=2)$ — $r=0$, $\varepsilon=1$ A; A $\varepsilon=1$, $r=0$ — C_2 — $\varepsilon=1$, $r=0$ A	$\mathbb{R}P^3$

14.8. ROUGH LIOUVILLE CLASSIFICATION OF SYSTEMS IN THE CLEBSCH CASE

Hamiltonian (14.20) in the Clebsch case

$$H = \frac{S_1^2}{2A_1} + \frac{S_2^2}{2A_2} + \frac{S_3^2}{2A_3} + \frac{\varepsilon}{2}(A_1 R_1^2 + A_2 R_2^2 + A_3 R_3^2)$$

includes four parameters A_1, A_2, A_3, and ε. Setting $A_i' = \sqrt{|\varepsilon|} A_i$ and dividing the Hamiltonian by $\sqrt{|\varepsilon|}$, we obtain Hamiltonian (14.20) with parameters A_1', A_2', A_3', and $\varepsilon = \pm 1$. Thus, we can examine only the Hamiltonians with $\varepsilon = \pm 1$, for which the curves on the plane $\mathbb{R}^2 (g, h)$ separating domains with different topological types of Q^3 have already been constructed in Section 14.2.7 (see Figs. 14.18 and 14.19).

The Clebsch integral has the form

$$K = \frac{1}{2}(S_1^2 + S_2^2 + S_3^2) - \frac{\varepsilon}{2}(A_2 A_3 R_1^2 + A_3 A_1 R_2^2 + A_1 A_2 R_3^2).$$

As has already been noted, if we know the bifurcation diagram of the momentum mapping $H \times K : M_{1,g} \simeq TS^2 \to \mathbb{R}^2 (h, k)$, we can easily obtain the bifurcation diagram for the system with the Hamiltonian $H' = \alpha H + \beta K$, where α and β are constants. It is obtained from the original diagram by a certain linear non-degenerate transformation of the plane $\mathbb{R}^2 (h, k)$. Since we are only interested in the bifurcations of Liouville tori that happen when the point moves along the line $\{h = \text{const}\}$ in $\mathbb{R}^2 (h, k)$, the molecule W for the Hamiltonian H' can be described by defining how the line $\{\alpha h + \beta k = \text{const}\}$ intersects the bifurcation diagram of the mapping $H \times K$.

Let us note the following important fact. It turns out that all the Hamiltonians in the Clebsch case can be obtained, for example, in the form of a linear combination $\alpha H_0 + \beta K_0$ of the commuting functions

$$H_0 = (S_1^2 + S_2^2 + S_3^2) + (c_1 R_1^2 + c_2 R_2^2 + c_3 R_3^2),$$
$$K_0 = (c_1 S_1^2 + c_2 S_2^2 + c_3 S_3^2) - (c_1^2 R_1^2 + c_2^2 R_2^2 + c_3^2 R_3^2),$$

where $c_1 + c_2 + c_3 = 0$. Thus, the general four-parametric Hamiltonian in the Clebsch case is obtained by taking a linear combination of the simple functions H_0 and K_0, which depend on two parameters only.

In Fig. 14.43, we show which Hamiltonians in the Clebsch case are obtained for different values of α and β. If the line $\{\alpha h + \beta k = 0\}$ lies in the domain I, then $\alpha H_0 + \beta K_0$ is the Hamiltonian of the form (14.20) with $\varepsilon = 1$; if it lies in the domain II, then $\alpha H_0 + \beta K_0$ is Hamiltonian (14.20) with $\varepsilon = -1$. If the line lies in the shaded domain, then there are non-compact manifolds among the isoenergy surfaces Q^3. Such linear combinations are not examined here.

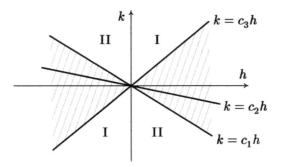

Figure 14.43

The bifurcation diagrams for the Clebsch case were constructed and studied by T. I. Pogosyan in [293], [294], [295]. They are shown in Fig. 14.44 for the mapping

$$H_0 \times K_0 : TS^2 \to \mathbb{R}^2 (h, k).$$

Three qualitatively different cases are possible:

a) $g^2 > p_2$,

b) $p_1 < g^2 < p_2$,

c) $g^2 < p_1$.

where p_1 and p_2 are certain constants depending on the parameters c_1, c_2, c_3 of the Hamiltonian H_0. When $f_2 = g$ and $f_2 = -g$, the diagrams completely coincide.

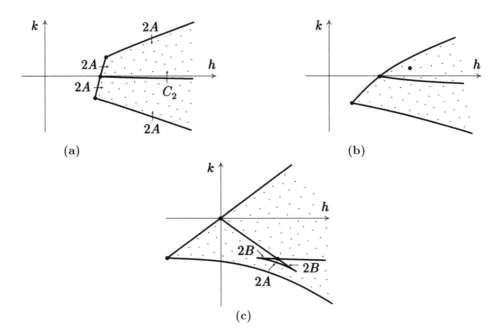

Figure 14.44

It is convenient to write the equation for the bifurcation curve in a parametric form, where h and k depend on two parameters x and y which are connected, in turn, by some relation. Namely,

$$h = -2x - \frac{g}{y}(c_1 c_2 + c_2 c_3 + c_3 c_1 + 3x^2),$$

$$k = c_1 c_2 + c_2 c_3 + c_3 c_1 - x^2 - \frac{g}{y}(x^3 - x(c_1 c_2 + c_2 c_3 + c_3 c_1) + 2c_1 c_2 c_3),$$

$$\text{where} \quad y^2 = (x - c_1)(x - c_2)(x - c_3).$$

The bifurcation diagram is actually part of this curve shown in Fig. 14.44.

The three lines

$$\{k = c_1 h + c_2 c_3\}, \quad \{k = c_2 h + c_3 c_1\}, \quad \{k = c_3 h + c_1 c_2\}$$

are asymptotes of the bifurcation curve. The three heavily drawn points in Fig. 14.44 are the images of points at which both grad H_0 and grad K_0 vanish on $M_{1,g} \simeq TS^2$. Their coordinates are $(g^2 + c_i, c_i(g^2 - c_i)) \in \mathbb{R}^2(h, k)$.

The bifurcations of Liouville tori are shown in Fig. 14.44. Notice that, for suitable values of parameters, non-degenerate singularities of all possible types occur (namely, center–center, center–saddle, saddle–saddle, and focus–focus).

Examining different lines $\{\alpha h + \beta k = c\}$, one can define the molecules W for the Hamiltonians $\alpha H_0 + \beta K_0$ for different c. The molecule W may change either when c is a critical value of the function $\alpha H_0 + \beta K_0$, or when the line $\{\alpha h + \beta k = c\}$ passes through a cusp of the curve. Having constructed the separating curve corresponding to cusps of the bifurcation diagram, we obtain a description of regions with different molecules W^* in the Clebsch case.

We now describe these separating curves on the plane $\mathbb{R}^2(g, h)$ in detail. First of all, among these curves, there are three parabolas given by the equations

$$h = \frac{g^2}{2A_i} + \varepsilon \cdot \frac{A_i}{2}, \quad \text{where} \quad i = 1, 2, 3, \varepsilon = \pm 1.$$

If we cross each of these parabolas, then the topological type of Q^3 changes.

We also must add to them another algebraic curve $(g(t), h(t))$ shown in Fig. 14.45 by a dotted line. Under crossing this curve the topology of Q remains the same, but the molecule W^* changes. The parametric equation of this curve can be taken in the form

$$g(t) = \frac{4P(t)^3 t \sigma_3}{Q(t)},$$

$$h(t) = \frac{\sigma_3(-10t^6 + 15\sigma_1 t^5 - 4(\sigma_1^2 + 2\sigma_2)t^4 + (4\sigma_1\sigma_2 + \sigma_3)t^3 + 2\sigma_1\sigma_3 t^2 - 4\sigma_2\sigma_3 t + 3\sigma_3^2)}{2t^2 Q(t)},$$

where

$$P(t) = \sqrt{\frac{(A_1 - t)(A_2 - t)(A_3 - t)}{t}}, \quad Q(t) = 2t^6 - 3\sigma_1 t^5 + 3\sigma_3 t^3 - 2\sigma_1\sigma_3 t^2 + \sigma_3^2,$$

$$\sigma_1 = A_1 + A_2 + A_3, \quad \sigma_2 = A_1 A_2 + A_2 A_3 + A_1 A_3, \quad \sigma_3 = A_1 A_2 A_3.$$

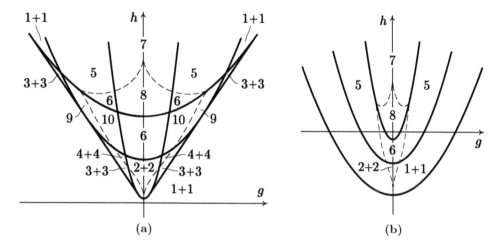

Figure 14.45

One of the possible arrangements of the parabolas $h = \dfrac{g^2}{2A_i} + \varepsilon \cdot \dfrac{A_i}{2}$, $i = 1, 2, 3$, and the curve $(g(t), h(t))$ on the plane $\mathbb{R}^2 (g, h)$ is presented in Fig. 14.45(a) for $\varepsilon = +1$, and in Fig. 14.45(b) for $\varepsilon = -1$. The whole picture depends on the parameters A_1, A_2, A_3. When they vary, the mutual disposition of the parabolas and the curve may change.

Theorem 14.10 (A. A. Oshemkov).   *The separating curves on the plane* $\mathbb{R}^2 (g, h)$ *for Hamiltonians* (14.20) *(i.e., in the general Clebsch case) are given by the above formulas. The additional integral* K *is a Bott function on each isoenergy 3-manifold* $Q = \{f_1 = 1,\ f_2 = g,\ H = h\}$ *provided the point* (g, h) *does not belongs to the separating curves. The list of all possible pairs of the form*

$$(Q, W) = (\text{isoenergy surface, rough molecule})$$

for the Clebsch case consists of 10 *pairs listed in Table* 14.8:

$$(S^3, \mathcal{C}_1),\ (S^3, \mathcal{C}_2),\ (S^1 \times S^2, \mathcal{C}_3),\ (S^1 \times S^2, \mathcal{C}_4),\ (S^1 \times S^2, \mathcal{C}_5),\ (S^1 \times S^2, \mathcal{C}_6),$$
$$(\mathbb{R}P^3, \mathcal{C}_7),\ (\mathbb{R}P^3, \mathcal{C}_8),\ (N^3, \mathcal{C}_9),\ (N^3, \mathcal{C}_{10}),$$

where $N^3 = (S^1 \times S^2) \# (S^1 \times S^2) \# (S^1 \times S^2)$.

COMMENT.   Let us explain how to find the molecule W for specific values of parameters A_1, A_2, A_3, g, h.   To this end, we must first substitute A_1, A_2, A_3 in the equations of the algebraic curve $(g(t), h(t))$ and three parabolas. We obtain a picture presented in Fig. 14.45. After this, we take the point (g, h) and determine what region this point belongs to.   Then we look at the number of this region in Fig. 14.45 and find the desired answer in Table 14.8.

Table 14.8. Clebsch case

No	MOLECULE	Q^3
1	$A \longrightarrow A$	S^3
2	$A \longrightarrow B <^A_A$	S^3
3	$A \longrightarrow A$	$S^1 \times S^2$
4	$A \longrightarrow B <^A_A$	$S^1 \times S^2$
5	$^A_A > C_2 <^A_A$	$S^1 \times S^2$
6	$^A_A > C_2 <^{B<^A_A}_{B<^A_A}$	$S^1 \times S^2$
7	$^A_A > C_2 <^A_A$	$\mathbb{R}P^3$
8	$^A_A > C_2 <^{B<^A_A}_{B<^A_A}$	$\mathbb{R}P^3$
9	$^A_A > C_2 <^A_A$	$(S^1 \times S^2)\# (S^1 \times S^2)\# (S^1 \times S^2)$
10	$^A_A > C_2 <^{B<^A_A}_{B<^A_A}$	$(S^1 \times S^2)\# (S^1 \times S^2)\# (S^1 \times S^2)$

14.9. ROUGH LIOUVILLE CLASSIFICATION OF SYSTEMS IN THE STEKLOV CASE

Hamiltonian (14.21) in the Steklov case includes four parameters: A_1, A_2, A_3, and ε. Let us consider the following cases:

$$H_0 = a_1 S_1^2 + a_2 S_2^2 + a_3 S_3^2 + 2(a_1^2 S_1 R_1 + a_2^2 S_2 R_2 + a_3^2 S_3 R_3) + a_1^3 R_1^2 + a_2^3 R_2^2 + a_3^3 R_3^2,$$
$$K_0 = S_1^2 + S_2^2 + S_3^2 - 2(a_1 S_1 R_1 + a_2 S_2 R_2 + a_3 S_3 R_3) - 3(a_1^2 R_1^2 + a_2^2 R_2^2 + a_3^2 R_3^2),$$

$$(14.72)$$

where $a_1 + a_2 + a_3 = 0$. As in the Clebsch case, an arbitrary Hamiltonian H in the Steklov case can be represented in the form

$$H = \alpha H_0 + \beta K_0 + \gamma f_1 + \delta f_2, \qquad (14.73)$$

where the parameters a_1, a_2, a_3 and the coefficients $\alpha, \beta, \gamma, \delta$ are

$$a_1 = \frac{\varepsilon}{3}(2A_2 A_3 - A_3 A_1 - A_1 A_2), \qquad a_2 = \frac{\varepsilon}{3}(2A_3 A_1 - A_1 A_2 - A_2 A_3),$$

$$a_3 = \frac{\varepsilon}{3}(2A_1 A_2 - A_2 A_3 - A_3 A_1),$$

$$\alpha = \frac{1}{2\varepsilon A_1 A_2 A_3}, \qquad \delta = \frac{5\varepsilon}{9}(A_1 + A_2 + A_3) - \frac{2\varepsilon}{9}\left(\frac{A_2 A_3}{A_1} + \frac{A_3 A_1}{A_2} + \frac{A_1 A_2}{A_3}\right),$$

$$\beta = \frac{1}{6A_1} + \frac{1}{6A_2} + \frac{1}{6A_3}, \qquad \gamma = \frac{2\varepsilon^2(A_1 A_2 + A_2 A_3 + A_3 A_1)^3}{27 A_1 A_2 A_3} - \varepsilon^2 A_1 A_2 A_3.$$

Following the above described algorithm, we can construct the bifurcation diagrams for the momentum mapping

$$K_0 \times H_0 \colon M_{1,g} \to \mathbb{R}^2(k, h), \qquad (14.74)$$

where H_0 and K_0 are both defined on $M_{1,g} \simeq TS^2 = \{f_1 = 1, \ f_2 = g\} \subset \mathbb{R}^6(S, R)$.

If we know the bifurcation diagrams of mapping (14.74), then we can construct the molecules W for Hamiltonian (14.73) by determining how the Liouville tori bifurcate in the preimage of the point moving along the line $\{\alpha h + \beta k = \text{const}\}$. Suppose that the parameters a_1, a_2, a_3 in expression (14.72) satisfy the condition

$$a_1 < 0 \leq a_2 < a_3.$$

The general case can be reduced to this one by a coordinate transformation.

We now describe the bifurcation diagrams of mapping (14.74) constructed by A. A. Oshemkov in [277] and [280]. For all g, this diagram is the union of the curve given in a parametric form

$$k = -8\mu g - 12\mu^2, \quad h = -4\mu^2 g - 8\mu^3, \quad -(g + a_3) \leq 2\mu \leq -(g + a_1), \qquad (14.75)$$

and three rays lying on the straight lines $\{h = a_i k + 4a_i^3 + 4ga_i^2\}$ (where $i = 1, 2, 3$).

The rays are defined as follows:

$$\{h = a_i k + 4a_i^3 + 4ga_i^2, \ k \geq T - 12a_i^2 - 8ga_i\} \subset \mathbb{R}^2(k, h), \tag{14.76}$$

where T is non-negative.

Curve (14.75) is of the form given in Fig. 14.46:

a) $g < -3a_3$,
b) $-3a_3 < g < -3a_2$,
c) $-3a_2 < g < -3a_1$,
d) $g > -3a_1$.

This curve is defined in a parametric form with parameter μ. For the point of the curve marked in Fig. 14.46 by the digit i (where $i = 1, 2, 3$), the value of μ is equal to $-\dfrac{1}{2}(g + a_i)$. This is the intersection point of the curve and the line

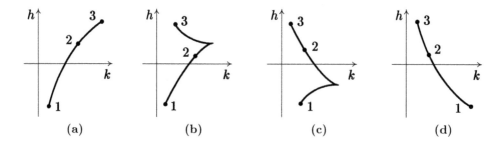

Figure 14.46

$\{h = a_i k + 4a_i^3 + 4ga_i^2\}$. For the cusp of the curve, we have $\mu = -g/3$, where $-3a_3 < g < -3a_1$. Besides, the parameter μ is equal to a_i at the point where the ray is tangent to the curve.

Summarizing all the above, we obtain the bifurcation diagrams shown in Fig. 14.47. In order to simplify the figures, the coordinate axes on the plane $\mathbb{R}^2(k, h)$ are not shown. The bifurcations of Liouville tori indicated in Fig. 14.47 will be defined below.

For each smooth piece of the bifurcation diagram, it is possible to compute the index of critical circles from the corresponding family. It has been done by A. A. Oshemkov in [277] and [280].

It turns out that all of these circles are non-degenerate with the only exception being those which belong to the preimage of the cusps or points of tangency. Having found the indices of the critical circles, one can describe the bifurcations of Liouville tori for critical values of (14.74). The bifurcation encoded by the atom A obviously corresponds to those parts of the bifurcation diagram for which the index is either 0 or 2. From this, it is easy to determine the number of Liouville tori in the preimage of every point which does not lie on the bifurcation diagram. It remains to determine the type of bifurcations arising from saddle circles.

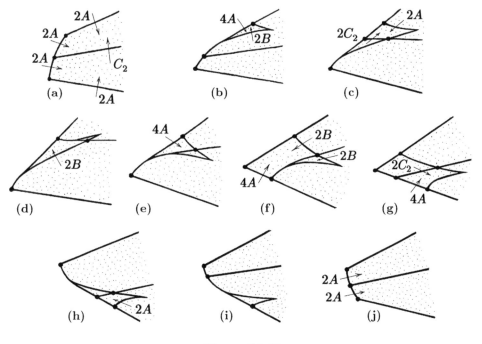

Figure 14.47

Let us examine, for example, the case shown in Fig. 14.47(g). This bifurcation diagram is shown in more detail in Fig. 14.48. The digits in Fig. 14.48 indicate the number of Liouville tori, the number of saddle circles (for example, $4s$) and the number of critical circles of elliptic type (for example, $2m$). Bifurcations along the arrows I and II are bifurcations of two Liouville tori into four tori via two saddle circles. There is only one bifurcation with such a property (see the list of atoms in Chapter 3). This bifurcation corresponds to two atoms B.

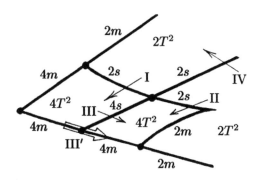

Figure 14.48

The bifurcation of Liouville tori along the arrow III has the same form as the bifurcation of critical circles in the preimage of a point which moves along the arrow III' (this follows from Theorem 9.2). It can be verified (see [277] and [280] for details) that the arrow III' corresponds to two copies of the atom C_2. Therefore, the bifurcation of Liouville tori along III has the same type and is described by two atoms C_2. It remains to determine the type of the bifurcation along the arrow IV. Under variation of g, the bifurcation diagram is deformed; for sufficiently large g, it has the form shown in Fig. 14.47(j). Besides, the rays are transformed into corresponding rays. Therefore, the bifurcation along IV has the same type as the bifurcation which occurs at the intersection of the middle ray in Fig. 14.47(j). It has type C_2 (this can be verified by the same method as we used for the bifurcation along III).

Examining all the other cases (a)–(j) presented in Fig. 14.47, one can determine all bifurcations of Liouville tori. The final result is shown in Fig. 14.47.

Theorem 14.11 (A. A. Oshemkov). *Let Hamiltonian (14.21) (Steklov case) be represented in the form (14.73). Then the bifurcation diagrams of the momentum mapping $H \times K: TS^2 \to \mathbb{R}^2$ are obtained from the bifurcation diagrams shown in Fig. 14.47 by a non-degenerate linear transformation of the plane $\mathbb{R}^2(k,h)$. The additional integral K is a Bott function on each non-singular isoenergy 3-manifold $Q_c = \{f_1 = 1,\ f_2 = g,\ H = c\}$ with the only exception of those for which the line $\{\alpha h + \beta k = c - \gamma - \delta g\}$ on the plane $\mathbb{R}^2(k,h)$ passes through the point where the ray is tangent to the curve or through the cusp (for the diagram shown in Fig. 14.47). The bifurcations of Liouville tori at critical values of the momentum mapping are shown in Fig. 14.47. The list of all possible molecules W in the Steklov case (for different α, β, g, h) consists of 6 molecules presented in Table 14.9.*

Table 14.9. Steklov case

No	MOLECULE	No	MOLECULE
1	A——————A	4	(molecule diagram)
2	(molecule diagram)	5	(molecule diagram)
3	(molecule diagram)	6	(molecule diagram)

14.10. ROUGH LIOUVILLE CLASSIFICATION OF INTEGRABLE FOUR-DIMENSIONAL RIGID BODY SYSTEMS

As was shown in Section 14.1, different generalizations of the classical problem of rigid body motion are described by the Euler equations on the Lie algebra e(3). Analogous equations may also be examined for other Lie algebras. In this section, we consider one integrable case of Euler equations on the Lie algebra so(4).

Elements of the Lie algebra so(4) are represented as skew-symmetric matrices with the ordinary commutator

$$[X, Y] = XY - YX. \tag{14.77}$$

Let skew-symmetric matrices be of the form

$$\begin{pmatrix} 0 & -M_3 & M_2 & p_1 \\ M_3 & 0 & -M_1 & p_2 \\ -M_2 & M_1 & 0 & p_3 \\ -p_1 & -p_2 & -p_3 & 0 \end{pmatrix}.$$

Then the Poisson bracket on so(4)* corresponding to commutator (14.77) is defined by the following relations:

$$\{M_i, M_j\} = \varepsilon_{ijk} M_k, \qquad \{M_i, p_j\} = \varepsilon_{ijk} p_k, \qquad \{p_i, p_j\} = \varepsilon_{ijk} M_k. \tag{14.78}$$

The Hamiltonian system for the Lie algebra so(4) is written in the form of the Euler equations:

$$\dot{M}_i = \{M_i, H\}, \quad \dot{p}_i = \{p_i, H\}, \tag{14.79}$$

where $H(M, p)$ is the Hamiltonian.

Bracket (14.78) in $\mathbb{R}^6 (M, p)$ is degenerate. The invariants of the Lie algebra so(4) are

$$\begin{aligned} f_1 &= M_1^2 + M_2^2 + M_3^2 + p_1^2 + p_2^2 + p_3^2, \\ f_2 &= M_1 p_1 + M_2 p_2 + M_3 p_3. \end{aligned} \tag{14.80}$$

They are Casimir functions of bracket (14.78) and, therefore, commute with any function $f(M, p)$. The common level surfaces of f_1 and f_2 are the orbits of the coadjoint representation of the Lie group SO(4):

$$O(d_1, d_2) = \{f_1 = d_1, \ f_2 = d_2\} \subset \mathbb{R}^6 (M, p). \tag{14.81}$$

The restriction of Poisson bracket (14.78) to these orbits is non-degenerate. The orbits are non-singular for $d_1 > 2|d_2|$. In this case, all of them are homeomorphic to $S^2 \times S^2$. If $d_1 = 2|d_2|$, then we obtain singular orbits homeomorphic to S^2. If $d_1 < 2|d_2|$, then $O(d_1, d_2) = \varnothing$. System (14.79) defines a Hamiltonian system with two degrees of freedom on generic orbits. Complete integrability of (14.79) is equivalent to the existence of one additional integral K, which is functionally independent with the Hamiltonian H on the orbits.

Let us consider the Hamiltonian of the form

$$H = a_1 M_1^2 + a_2 M_2^2 + a_3 M_3^2 + c_1 p_1^2 + c_2 p_2^2 + c_3 p_3^2. \tag{14.82}$$

If its parameters satisfy the relation

$$a_1 c_1 (a_2 + c_2 - a_3 - c_3) + a_2 c_2 (a_3 + c_3 - a_1 - c_1) + a_3 c_3 (a_1 + c_1 - a_2 - c_2) = 0, \tag{14.83}$$

then the corresponding Hamiltonian system is completely integrable. Equations (14.79) with Hamiltonian (14.82), (14.83) are sometimes said to be the equations of four-dimensional rigid body motion. This is an analog of the classical Euler case (see, for example, [124], [125], [245]).

Consider relation (14.83). It is easy to show that it is equivalent to one of the following two conditions:

$$a_1 + c_1 = a_2 + c_2 = a_3 + c_3 \tag{14.84}$$

or

$$
\begin{aligned}
a_1 c_1 &= q + r(a_1 + c_1), \\
a_2 c_2 &= q + r(a_2 + c_2), \\
a_3 c_3 &= q + r(a_3 + c_3),
\end{aligned} \tag{14.85}
$$

where q and r are certain constants.

If the first condition (14.84) is fulfilled, then Hamiltonian (14.82) takes the form

$$H_0 = b_1 M_1^2 + b_2 M_2^2 + b_3 M_3^2 - b_1 p_1^2 - b_2 p_2^2 - b_3 p_3^2. \tag{14.86}$$

The second condition (14.85) can be rewritten as

$$
\begin{aligned}
(r - a_1)(r - c_1) &= r^2 + q, \\
(r - a_2)(r - c_2) &= r^2 + q, \\
(r - a_3)(r - c_3) &= r^2 + q.
\end{aligned} \tag{14.87}
$$

If $r^2 + q = 0$, then at least three of the coefficients a_i, c_i $(i = 1, 2, 3)$ are equal to zero. Then a suitable linear coordinate transformation in $\mathbb{R}^6 (M, p)$ preserving bracket (14.78) reduces Hamiltonian (14.82) to one of the following:

$$H_1 = b_1 M_1^2 + b_2 M_2^2 + b_3 M_3^2, \tag{14.88}$$

$$H_2 = b_1 p_1^2 + b_2 p_2^2 + b_3 p_3^2. \tag{14.89}$$

Let $r^2 + q \neq 0$. Then Hamiltonian (14.82) can be written in the form

$$H = A H_3 + r f_1, \tag{14.90}$$

where

$$H_3 = b_1 M_1^2 + b_2 M_2^2 + b_3 M_3^2 + b_2 b_3 p_1^2 + b_3 b_1 p_2^2 + b_1 b_2 p_3^2,$$

$$b_1 = \frac{(c_2 - r)(c_3 - r)}{r^2 + q}, \quad b_2 = \frac{(c_3 - r)(c_1 - r)}{r^2 + q}, \quad b_3 = \frac{(c_1 - r)(c_2 - r)}{r^2 + q}, \tag{14.91}$$

$$A = \frac{(a_1 - r)(a_2 - r)(a_3 - r)}{r^2 + q}.$$

Since the addition of the invariant f_1 to the Hamiltonian of the form (14.82) does not change system (14.79), we see that Hamiltonian (14.82), (14.83) is equivalent to one of the Hamiltonians H_0, H_1, H_2, H_3, each of which depends on three parameters b_1, b_2, b_3 only.

Consider the Hamiltonian H_1. The integral in this case is $K_1 = M_1^2 + M_2^2 + M_3^2$. It is evident that Hamiltonian system (14.79) with the Hamiltonian H_1 (defined by (14.88)) determines exactly the same phase flow on $\mathbb{R}^6$ as in the ordinary Euler case (this can be easily observed after transformation $S = M$, $R = p$). The bifurcation diagram for the momentum mapping

$$K_1 \times H_1 : S^2 \times S^2 \to \mathbb{R}^2 (k_1, h_1)$$

is shown in Fig. 14.49. It consists of five segments lying on the lines $\{h_1 = b_i k_1\}$ $(i = 1, 2, 3)$, $2k_1 = d_1 \pm (d_1^2 - 4d_2^2)^{1/2}$, where d_1 and d_2 determine the orbit (14.81). The bifurcation diagram of the mapping

$$f_2 \times H_1 : S^5 \to \mathbb{R}^2 (d_2, h_1),$$

where $S^5 = \{f_1 = d_1\} \subset \mathbb{R}^6 (M, p)$, is shown in Fig. 14.50. It consists of three ellipses and two vertical segments $\{2d_2 = \pm d_1, \ b_1 d_1 < 2h < b_3 d_1\}$, which are tangent to all three ellipses (we assume that $0 < b_1 < b_2 < b_3$). This is the complete collection of separating curves for this case. The topological type of Q and the molecule W are indicated for each domain in Fig. 14.50. More precisely, the number associated with a region in Fig. 14.50 indicates the number of the corresponding molecule W in Table 14.10.

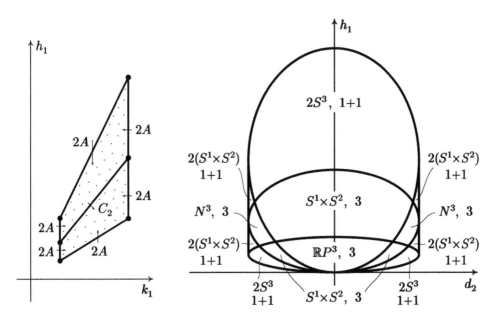

Figure 14.49   Figure 14.50

As we shall see below, the Hamiltonian H_1 is quite different from the other Hamiltonians of the form (14.82), (14.83). We now examine the remaining Hamiltonians H_0, H_2, H_3.

Consider the Hamiltonian H_0. The function

$$K_0 = (b_1+b_2)(b_1+b_3)p_1^2 + (b_2+b_3)(b_2+b_1)p_2^2 + (b_3+b_1)(b_3+b_2)p_3^2 \qquad (14.92)$$

can be taken as an integral which is functionally independent of H_0 on orbits (14.81).

It is easy to check that all Hamiltonians (14.82), (14.83) except for H_1 can be represented as a linear combination

$$H = \alpha H_0 + \beta K_0 + \gamma f_1 \,, \qquad (14.93)$$

where α, β, γ are some coefficients. Therefore, the bifurcation diagrams for an arbitrary Hamiltonian (14.82), (14.83) are obtained from those for the mapping

$$H_0 \times K_0 \colon S^2 \times S^2 \to \mathbb{R}^2 \, (h_0, k_0)$$

under a non-degenerate linear transformation of the plane $\mathbb{R}^2 \, (h_0, k_0)$ (see Sections 14.8 and 14.9).

The bifurcation diagrams of the mapping $H_0 \times K_0$ were constructed by A. A. Oshemkov in [278]. We now describe this construction. We shall assume that the coefficients of the Hamiltonian H_0 satisfy the condition

$$0 < b_1 < b_2 < b_3 \,. \qquad (14.94)$$

The cases with negative b_i's can be reduced to (14.94) by a linear transformation in $\mathbb{R}^6 \, (M, p)$ which preserves bracket (14.78).

Orbit (14.81) is determined by two parameters d_1 and d_2. When $2|d_2| < d_1$, the orbit is homeomorphic to $S^2 \times S^2$. The critical points of $\widetilde{H}_0 = H_0|_{S^2 \times S^2}$ can be found from the condition

$$\operatorname{grad} H_0 = \lambda_1 \operatorname{grad} f_1 + \lambda_2 \operatorname{grad} f_2 \,,$$
$$f_1(M, p) = d_1 \,, \qquad (14.95)$$
$$f_2(M, p) = d_2 \,.$$

This system has twelve solutions (i.e., critical points):

$$\begin{array}{lll}
(\pm A, 0, 0, \pm B, 0, 0), & (0, \pm A, 0, 0, \pm B, 0), & (0, 0, \pm A, 0, 0, \pm B), \\
(\pm B, 0, 0, \pm A, 0, 0), & (0, \pm B, 0, 0, \pm A, 0), & (0, 0, \pm B, 0, 0, \pm A),
\end{array} \qquad (14.96)$$

where $2A = \sqrt{d_1 + 2d_2} + \sqrt{d_1 - 2d_2}$ and $2B = \sqrt{d_1 + 2d_2} - \sqrt{d_1 - 2d_2}$.

Let us find the critical points of the function $\widetilde{K}_0 = K_0|_{Q_c^3}$, where $Q_c^3 = \{f_1 = d_1,\, f_2 = d_2,\, H_0 = c\}$, and c is a non-critical value of $\widetilde{H}_0$. As in the Steklov case (see Section 14.9), we write the condition

$$\operatorname{grad} K_0 = \mu_1 \operatorname{grad} f_1 + \mu_2 \operatorname{grad} f_2 + \mu \operatorname{grad} H_0$$

in the form

$$G_\mu \begin{pmatrix} M \\ p \end{pmatrix} = 0, \qquad f_1(M,p) = d_1, \qquad f_2(M,p) = d_2, \qquad (14.97)$$

where $G_\mu = G_{K_0} - \mu G_{H_0} - \mu_1 G_1 - \mu_2 G_2$ is the matrix obtained as the linear combination of the Hessians of f_1, f_2, H_0, K_0. From the explicit form of G_μ, it is easy to conclude that the rank of G_μ is equal to 3 or 5. In the case of rank 5, the solutions of (14.78) are points (14.96) only. Consider the case when rank $G_\mu = 3$. Let $d_2 \neq 0$ (the case $d_2 = 0$ will be treated separately). Then system (14.97) becomes

$$\mu_1 = -\mu^2 - \mu(b_1 + b_2 + b_3),$$
$$\mu_2^2 = 4\mu(\mu+b_1+b_2)(\mu+b_2+b_3)(\mu+b_3+b_1),$$
$$\mu_2 p_1 = 2\mu(\mu + b_2 + b_3)M_1,$$
$$\mu_2 p_2 = 2\mu(\mu + b_3 + b_1)M_2, \qquad (14.98)$$
$$\mu_2 p_3 = 2\mu(\mu + b_1 + b_2)M_3,$$
$$M_1^2 + M_2^2 + M_3^2 + p_1^2 + p_2^2 + p_3^2 = d_1,$$
$$M_1 p_1 + M_2 p_2 + M_3 p_3 = d_2.$$

While solving system (14.98), three qualitatively different cases appear:
a) $\varphi_1 < D < 1$,
b) $\varphi_2 < D < \varphi_1$,
c) $0 < D < \varphi_2$,
where $D = 2|d_2|/d_1$; φ_1 and φ_2 are some constants which depend only on the parameters b_1, b_2, b_3 of the Hamiltonian. The second equation from system (14.98) determines the curve on the plane (μ, μ_2) shown in Fig. 14.51. System (14.98) has solutions if and only if the point (μ, μ_2) lies on the heavily drawn segments of the curve in Fig. 14.51. Cases (a), (b), (c) in Fig. 14.51 correspond to different ranges of D indicated above.

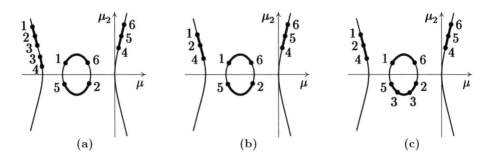

(a)   (b)   (c)

Figure 14.51

If the point (μ, μ_2) lies on the heavily drawn segments and does not coincide with any of bolded points denoted in Fig. 14.51 by numbers, then the solution of (14.98) consists of only two circles in $\mathbb{R}^6 (M, p)$. Under the mapping $H_0 \times K_0$, both these circles transform into the point

$$(h_0(\mu), k_0(\mu)) \in \mathbb{R}^2 (h_0, k_0),$$

where

$$h_0(\mu) = d_1 (2\mu + b_1 + b_2 + b_3) - d_2 \frac{d\mu_2}{d\mu},$$

$$k_0(\mu) = d_1 \mu^2 + d_2 \left(\mu_2 - \mu \frac{d\mu_2}{d\mu} \right).$$

(14.99)

Here $\mu_2(\mu)$ is the function which is defined by the second equation from (14.98). Thus we obtain mapping (14.99) of the heavily drawn segments of the curve into the plane $\mathbb{R}^2 (h_0, k_0)$. The image of this mapping is exactly the bifurcation diagram of the momentum mapping $H_0 \times K_0 \colon S^2 \times S^2 \to \mathbb{R}^2 (h_0, k_0)$.

The bifurcation diagrams are presented in Fig. 14.52. Cases (a), (b), and (c) in Fig. 14.52 correspond to cases (a), (b), and (c) in Fig. 14.51. The fact that the bifurcation diagrams are of the form shown in Fig. 14.52 can be proved in the following way.

Under mapping (14.99), the points marked in Fig. 14.51 by numbers transform into the points marked by the same numbers in Fig. 14.52. These points are the images of points (14.96) under the momentum mapping $H_0 \times K_0$. Their coordinates on the plane $\mathbb{R}^2 (h_0, k_0)$ are

$$\left(\pm b_i \sqrt{d^2_1 - 4d^2_2}, \quad (b_i + b_j)(b_i + b_k) \frac{d_1 \mp \sqrt{d^2_1 - 4d^2_2}}{2} \right),$$

(14.100)

$$\{i, j, k\} = \{1, 2, 3\}.$$

Thus, the bifurcation diagram is "glued" from segments of the curve shown in Fig. 14.51. According to Lemma 14.2, the relation

$$\frac{dk_0}{d\mu} = \mu \cdot \frac{dh_0}{d\mu}$$

is valid for functions $h_0(\mu)$ and $k_0(\mu)$ which define mapping (14.99). From this, it is easy to deduce the convexity of each segment of the bifurcation diagram. It remains to find out when the bifurcation diagram has cusps. The existence of a cusp on the bifurcation curve for some $\mu = \varepsilon$ is equivalent to the condition $\frac{dh_0}{d\mu}(\varepsilon) = 0$. Using explicit expression (14.99) for the function $h_0(\mu)$, one can show that cusps appear only in case (c) and are situated as shown in Fig. 14.52(c).

As in the Steklov case, one can determine the indices of critical circles and the bifurcations of Liouville tori.

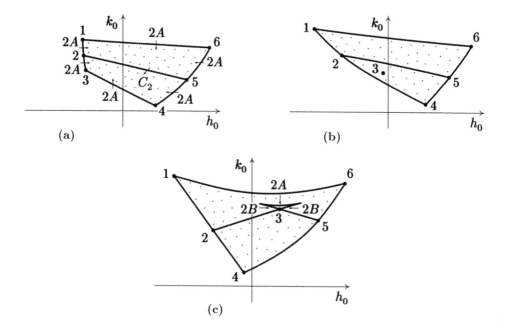

Figure 14.52

Theorem 14.12 (A. A. Oshemkov). *For a Hamiltonian system of the form (14.79) (i.e., for the equations of motion of a four-dimensional rigid body) with Hamiltonian (14.82), (14.83), the bifurcation diagram of the momentum mapping*

$$H \times K : S^2 \times S^2 \to \mathbb{R}^2 \qquad for \quad d_2 \neq 0$$

is obtained from diagrams shown in Fig. 14.52 under a suitable non-degenerate linear transformation of the plane $\mathbb{R}^2 (h_0, k_0)$. The additional integral is a Bott function on each non-singular isoenergy 3-manifold $Q_h^3 = \{f_1 = 1, f_2 = d_2, H = h\}$, provided the line $\{\alpha h_0 + \beta k_0 = h - \gamma d_1\}$ does not pass through a cusp of the bifurcation curve (here α, β, γ are the coefficients from (14.93)). Bifurcations of Liouville tori at critical values of the momentum mapping $H \times K$ are shown in Fig. 14.52.

While constructing bifurcation diagrams of the mapping $H_0 \times K_0$, we have assumed that $d_2 \neq 0$. Now let us examine the case when $d_2 = 0$. For the mapping $H_0 \times K_0 : \{f_1 = d_1, \ f_2 = 0\} \to \mathbb{R}^2 (h_0, k_0)$, the bifurcation diagram simplifies considerably. It consists of four segments and the part of parabola which is tangent to all these segments (see Fig. 14.53). The equations of lines on which these segments lie are

$$k_0 = (b_i + b_j)(b_k d_1 - h_0), \qquad \{i, j, k\} = \{1, 2, 3\}. \tag{14.101}$$

The equation of the parabola is $4d_1 k_0 = (h_0 - d_1(b_1 + b_2 + b_3))^2$. When $d_2 \to 0$, the segment of the bifurcation curve between the cups "approaches" the segment of the curve with end-points 1 and 6; and when $d_2 = 0$, it "combines" with it (see Fig. 14.52(c)).

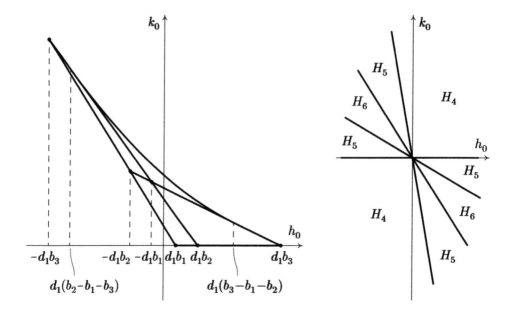

Figure 14.53   Figure 14.54

The bifurcation diagrams of the mapping $H_0 \times K_0$ (Fig. 14.52) allow us to classify isoenergy 3-manifolds for any Hamiltonian of the form (14.82), (14.83). This classification will be carried out in the same way as in Sections 14.3–14.9. Namely, we shall construct separating curves on the plane $\mathbb{R}^2(d_2, h)$ and indicate the molecule W for each region. As was remarked above, all Hamiltonians (14.82), (14.83) are linear combinations of H_0 and K_0 (except for the case (14.88) which has already been examined). However, for different linear combinations, separating curves may be qualitatively different from each other. We first distinguish the types of Hamiltonians that occur for different linear combinations of H_0 and K_0.

Let us transfer the lines which contain segments of the bifurcation diagram for $d_2 = 0$ (shown in Fig. 14.53) to the origin. Then we obtain the lines shown in Fig. 14.54. They divide the plane $\mathbb{R}^2(h_0, k_0)$ into domains. The type of Hamiltonian (14.82), (14.83) represented in the form (14.93) depends on the domain where the line $\{\alpha h_0 + \beta k_0 = 0\}$ is situated (here α and β are taken from (14.93)). Up to the sign, the following types of Hamiltonians correspond to the domains in Fig. 14.54:

$$H_4 = \frac{M_1^2}{A_1} + \frac{M_2^2}{A_2} + \frac{M_3^2}{A_3} + A_1 p_1^2 + A_2 p_2^2 + A_3 p_3^2,$$

$$H_5 = \frac{M_1^2}{A_1} + \frac{M_2^2}{A_2} + \frac{M_3^2}{A_3} - A_1 p_1^2 - A_2 p_2^2 - A_3 p_3^2, \qquad (14.102)$$

$$H_6 = \frac{M_1^2}{A_1} + \frac{M_2^2}{A_2} - \frac{M_3^2}{A_3} + A_1 p_1^2 + A_2 p_2^2 - A_3 p_3^2,$$

where $A_1, A_2, A_3 > 0$.

In fact, if the Hamiltonian H is represented in the form (14.93), then it can be written in the following way:

$$H = \alpha^2\beta^{-1}(y_1 M_1^2 + y_2 M_2^2 + y_3 M_3^2 + y_2 y_3 p_1^2 + y_3 y_1 p_2^2 + y_1 y_2 p_3^2) +$$
$$+ (\alpha(b_1 + b_2 + b_3) - \alpha^2\beta^{-1} + \gamma)f_1 , \tag{14.103}$$

where $y_1 = 1 - \beta\alpha^{-1}(b_2 + b_3)$, $y_2 = 1 - \beta\alpha^{-1}(b_3 + b_1)$, $y_3 = 1 - \beta\alpha^{-1}(b_1 + b_2)$, and b_1, b_2, b_3 are coefficients in H_0 and K_0. Therefore, if α and β differ from zero, then the Hamiltonian $H = \alpha H_0 + \beta K_0 + \gamma f_1$ is equivalent to a Hamiltonian of the form H_3 (see (14.91)). Suppose that y_1, y_2, y_3 are not equal to zero. Let us put

$$A_1 = \sqrt{\left|\frac{y_2 y_3}{y_1}\right|}, \quad A_2 = \sqrt{\left|\frac{y_3 y_1}{y_2}\right|}, \quad A_3 = \sqrt{\left|\frac{y_1 y_2}{y_3}\right|}. \tag{14.104}$$

Substituting expressions (14.104) into (14.103) and dividing by a constant, we see that Hamiltonian (14.103) is equivalent to

$$H = \varepsilon_1 \frac{M_1^2}{A_1} + \varepsilon_2 \frac{M_2^2}{A_2} + \varepsilon_3 \frac{M_3^2}{A_3} + \varepsilon_2\varepsilon_3 A_1 p_1^2 + \varepsilon_3\varepsilon_1 A_2 p_2^2 + \varepsilon_1\varepsilon_2 A_3 p_3^2 , \tag{14.105}$$

where $\varepsilon_i = \operatorname{sign} y_i$, $(i = 1, 2, 3)$. Hamiltonian (14.105) may be reduced to the form (14.102) by a coordinate change of the form

$$(M_1', M_2', M_3', p_1', p_2', p_3') = (M_1, p_2, p_3, p_1, M_2, M_3), \quad A_2' = A_2^{-1}, \quad A_3' = A_3^{-1},$$

which preserves bracket (14.78).

Comparing expressions (14.103) for y_1, y_2, y_3 and equations of lines (14.101) (and also taking into account the fact that $0 < b_1 < b_2 < b_3$), one obtains domains corresponding to the Hamiltonians H_4, H_5, H_6 shown in Fig. 14.54. If $\beta = 0$ in formula (14.93), then the Hamiltonian H is equivalent to the Hamiltonian H_0. In Fig. 14.54, the vertical line $\{h_0 = 0\}$ corresponds to this Hamiltonian. If either $\alpha = 0$ or one of the parameters y_i is equal to zero in expression (14.103), then the Hamiltonian H is equivalent to the Hamiltonian H_2 (see (14.89)). The four lines which separate domains in Fig. 14.54 correspond to these Hamiltonians.

Now let us describe the curves on the plane $\mathbb{R}^2(d_2, h)$ which separate domains corresponding to different topological types or Fomenko invariants of isoenergy surfaces $Q_h^3 = \{f_1 = d_1, f_2 = d_2, H = h\}$, where H is any Hamiltonian of the form H_0, H_2, H_4, H_5, H_6.

The curves which separate domains with different topological type of Q_h^3 are the images of critical points of the mapping

$$f_2 \times H : S^5 \to \mathbb{R}^2(d_2, h), \tag{14.106}$$

where $S^5 = \{f_1 = d_1\} \subset \mathbb{R}^6(M, p)$.

Calculating, one sees that, for any of the examined Hamiltonians, the critical points of mapping (14.106) fill two 2-spheres

$$\{p_1^2 + p_2^2 + p_3^2 = d_1/2, \ \ M_1 = \pm p_1, \ \ M_2 = \pm p_2, \ \ M_3 = \pm p_3\} \subset S^5 \qquad (14.107)$$

and three circles

$$\{M_2 = M_3 = p_2 = p_3 = 0, \ \ M_1^2 + p_1^2 = d_1\},$$
$$\{M_3 = M_1 = p_3 = p_1 = 0, \ \ M_2^2 + p_2^2 = d_1\}, \qquad (14.108)$$
$$\{M_1 = M_2 = p_1 = p_2 = 0, \ \ M_3^2 + p_3^2 = d_1\} \subset S^5 \subset \mathbb{R}^6 (M, p).$$

Besides, for the Hamiltonians H_4 and H_6, one obtains two more 2-spheres

$$\{(1 + A_1^2)p_1^2 + (1 + A_2^2)p_2^2 + (1 + A_3^2)p_3^2 = d_1,$$
$$M_1 = \pm A_1 p_1, \ \ M_2 = \pm A_2 p_2, \ \ M_3 = \pm \sigma A_3 p_3\} \subset S^5 \subset \mathbb{R}^6 (M, p), \qquad (14.109)$$

where $\sigma = 1$ for the Hamiltonian H_4, and $\sigma = -1$ for the Hamiltonian H_6.

Critical points (14.108) transform under mapping (14.106) into three ellipses for which the line $\{d_2 = 0\}$ is an axis of symmetry and the lines $\{2d_2 = \pm d_1\}$ are common tangents. Two-dimensional spheres (14.107) (they are singular orbits of the coadjoint representation) are mapped into two segments which lie on the lines $\{2d_2 = \pm d_1\}$. The spheres (14.109) are mapped into two other segments which lie on the lines $\{h = \pm 2d_2\}$, which are also common tangents to ellipses. As a result, one obtains separating curves shown in Fig. 14.55. The dotted line separates domains with different invariants W. This curve can be constructed if one determines coordinates of cusps of the bifurcation diagram shown in Fig. 14.52. As in the Clebsch case, it may intersect a different number of domains. The type of W is indicated by numbers in Fig. 14.55.

Theorem 14.13 (A. A. Oshemkov). *Any Hamiltonian of the form (14.82), (14.83) (a four-dimensional rigid body) is equivalent to one of the Hamiltonians of type $H_0, H_1, H_2, H_4, H_5, H_6$. Separating curves and the corresponding molecules W for these Hamiltonians are given*
  in Fig. 14.50   for H_1,
  in Fig. 14.55(a) for H_4,
  in Fig. 14.55(b) for H_6,
  in Fig. 14.55(c) for H_5,
  in Fig. 14.55(d) for H_2, where b_1, b_2, b_3 are of the same sign,
  in Fig. 14.55(e) for H_2, where b_1, b_2, b_3 have different signs,
  in Fig. 14.55(f) for H_0.
The additional integral in this case is a Bott function on each isoenergy surface $Q_h^3 = \{f_1 = d_1, \ f_2 = d_2, \ H = h\}$ provided the point (d_2, h) does not belong to the separating curve. The complete list of molecules for all Hamiltonians of the form (14.82), (14.83) consists of 9 molecules W presented in Table 14.10.

COMMENT. Under variation of parameters, the cusps of the dotted curve shown in Fig. 14.55 may move into another region. In such a case, the bifurcation diagrams will slightly differ from those in Fig. 14.55. But, any way, no new regions appear, and the list of molecules remains the same.

In conclusion, we discuss the relationship between the Clebsch case and the four-dimensional rigid body systems just examined. It is well known that the Lie algebra so(4) can be deformed into the Lie algebra e(3). Moreover, this deformation can be chosen so that the four-dimensional rigid body system transforms into the Clebsch case (see, for example, [252] and [253]). This deformation can be described as follows. Consider another bracket $\{\,,\}'$ in $\mathbb{R}^6\,(M,p)$:

$$\{M_i, M_j\}' = \varepsilon_{ijk} M_k\,, \quad \{M_i, p_j\}' = \varepsilon_{ijk} p_k\,, \quad \{p_i, p_j\}' = \frac{\varepsilon_{ijk} M_k}{N^2}\,. \tag{14.110}$$

Bracket (14.110) is obtained from (14.78) by multiplying each p_i by some constant N; this is equivalent to changing the basis in so(4). Hence we obtain that the kernel of bracket (14.110) is generated by the functions

$$f_1(N) = \frac{M_1^2 + M_2^2 + M_3^2}{N^2} + p_1^2 + p_2^2 + p_3^2\,, \tag{14.111}$$
$$f_2 = M_1 p_1 + M_2 p_2 + M_3 p_3\,.$$

Then (14.83) implies a sufficient integrability condition for Hamiltonian (14.82) with respect to bracket (14.110):

$$a_1 c_1 (a_2 - a_3) + a_2 c_2 (a_3 - a_1) + a_3 c_3 (a_1 - a_2)$$
$$+ \frac{a_1 c_1 (c_2 - c_3) + a_2 c_2 (c_3 - c_1) + a_3 c_3 (c_1 - c_2)}{N^2} = 0\,. \tag{14.112}$$

Obviously, as $N \to \infty$, bracket (14.110) transforms into bracket (14.6) on the coalgebra e(3)*, and functions (14.111) become the invariants of the Lie algebra e(3). Besides, as $N \to \infty$, relation (14.112) transforms into Clebsch's condition of integrability for the Hamiltonian

$$H = a_1 S_1^2 + a_2 S_2^2 + a_3 S_3^2 + c_1 R_1^2 + c_2 R_2^2 + c_3 R_3^2$$

on e(3)*.

Let us note that the Hamiltonians H_2, H_4, H_5, H_6 satisfy relation (14.112) for any N. Thus, the bifurcation diagrams for any N are obtained from the bifurcation diagrams shown in Fig. 14.52 by a non-degenerate linear transformation of the plane $\mathbb{R}^2 (h_0, k_0)$. This linear transformation depends on N. As $N \to \infty$, the points denoted by numbers $4, 5, 6$ in Fig. 14.52 "go to infinity", and in the limit one obtains the bifurcation diagrams for the Clebsch case shown in Fig. 14.44.

A similar situation occurs for separating curves. For bracket (14.110), the separating curves are also of the form shown in Fig. 14.55, but the vertical segments tangent to the ellipses lie on the lines $\{2d_2 = \pm d_1 N\}$. As $N \to \infty$, these segments also "go to infinity", and in the limit we obtain the separating curves for the Clebsch case (see Fig. 14.45). Cases (a) and (c) in Fig. 14.55 correspond to cases (a) and (b) in Fig. 14.45. There is no analog in the Clebsch case for Fig. 14.55(b), since some isoenergy 3-manifolds Q^3 become non-compact under the contraction.

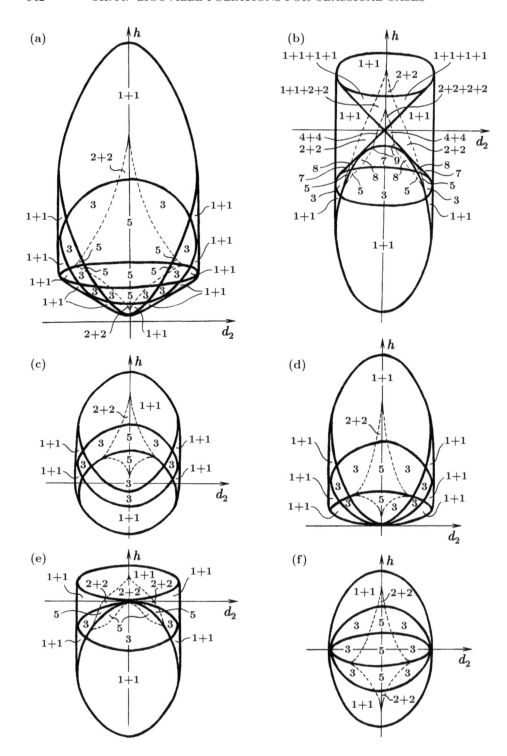

Figure 14.55

Table 14.10. Four-dimensional rigid body

No	MOLECULE
1	
2	
3	
4	
5	
6	
7	
8	
9	

14.11. THE COMPLETE LIST OF MOLECULES APPEARING IN INTEGRABLE CASES OF RIGID BODY DYNAMICS

The molecules W (without numerical marks) that occur in the main integrable cases in rigid dynamics are listed in Table 14.11. The total number of such molecules is 17. In Table 14.11, we also indicate the topological types of isoenergy surfaces on which these molecules are realized. Shaded boxes in Table 14.11 denote the cases where the answer is not described completely.

We tried to arrange the integrable cases in Table 14.11 according to their complexity. From this point of view, the Lagrange case turns out to be the simplest one, while the most complicated are the Steklov case and 4-dimensional rigid body.

Table 14.11. The list of molecules in integrable cases

No	MOLECULE	Lagrange	Euler	Zhukovskiĭ	Kovalevskaya	Sretenskiĭ	Goryachev–Chaplygin	Clebsch	Steklov	4-dim. body
1	A—A	S^3 $\mathbb{R}P^3$ $S^1{\times}S^2$	S^3	S^3 $\mathbb{R}P^3$ $S^1{\times}S^2$	S^3	S^3		S^3 $S^1{\times}S^2$	/////	/////
2	A—$B{<}^A_A$			S^3 $\mathbb{R}P^3$ $S^1{\times}S^2$	S^3 $S^1{\times}S^2$	$S^1{\times}S^2$		S^3 $S^1{\times}S^2$	/////	/////
3	A—A^{*}—A			$\mathbb{R}P^3$		S^3	S^3			
4	$^A_A{>}B$—$B{<}^A_A$			S^3 $S^1{\times}S^2$	S^3					
5	$^A_A{>}C_2{<}^A_A$	$\mathbb{R}P^3$ $S^1{\times}S^2$						N^3 $\mathbb{R}P^3$ $S^1{\times}S^2$	/////	/////
6	A—$B{<}^{B{<}^A_A}_{B{<}^A_A}$				S^3, K^3 $\mathbb{R}P^3$ $S^1{\times}S^2$					
7	A—$B{<}^{A^{*}-A}_{A^{*}-A}$				$\mathbb{R}P^3$					
8	$^A_A{>}C_2{<}^{A^{*}-A}_{A^{*}-A}$				$\mathbb{R}P^3$					

Table 14.11. The list of molecules in integrable cases (continued)

No	MOLECULE	Lagrange	Euler	Zhukovskii	Kovalev-skaya	Sretenskii	Goryachev–Chaplygin	Clebsch	Steklov	4-dim. body
9	$A\!-\!B\!<^{A}_{A}\!>\!B\!-\!A$					$\mathbb{R}P^3$	$\mathbb{R}P^3$			////
10	$A\!-\!A^*\!-\!D_1\!<^{A}_{A}\!\!<^{A}$					S^3 K^3				
11	$A\!-\!B\!<^{A^*\!-A}_{B<^A_A}$					$\mathbb{R}P^3$ $S^1\!\times\!S^2$				
12	$^{A}_{A}\!>\!C_2\!<^{B<^A_A}_{B<^A_A}$							N^3 $\mathbb{R}P^3$ $S^1\!\times\!S^2$	////	////
13	$^{A-B}_{A-B}\!\!>\!\!<^{B-A}_{B-A}$								////	////
14	$^{A}_{A}\!>\!C_2\!\supset\!C_2\!<^{A}_{A}$									////
15	$^{A}_{A}\!>\!C_2\!\supset\!C_2\!<^{B<^A_A}_{B<^A_A}$									////
16	$^{A-B}_{A-B}\!\!>\!\!<^{B-B<^A_A}_{B-B<^A_A}$									////
17	$^{A-B}_{A-B}\!\!>\!\!<^{C_2<^A_A}_{C_2<^A_A}$								////	

Here $K^3 = (S^1 \times S^2) \,\#\, (S^1 \times S^2)$, $N^3 = (S^1 \times S^2) \,\#\, (S^1 \times S^2) \,\#\, (S^1 \times S^2)$

Chapter 15

Maupertuis Principle and Geodesic Equivalence

The classical Maupertuis principle, which states a connection between the trajectories of a natural system and the geodesics of a certain Riemannian metric, is discussed in many papers devoted to the calculus of variations and mechanics [63], [101], [181], [196], [218], and [253].

In this chapter, we show how by using the Maupertuis principle one can construct new examples of integrable geodesic flows on the sphere. We describe this mechanism by using, as an example, classical integrable cases in rigid body dynamics. Moreover, we discuss one interesting generalization of the Maupertuis principle connected with the Dini theorem on geodesically equivalent metrics on two-dimensional surfaces.

15.1. GENERAL MAUPERTUIS PRINCIPLE

Let M^n be a compact smooth Riemannian manifold with metric $g_{ij}(x)$, and let T^*M be its cotangent bundle with standard coordinates x and p, where $x = (x^1, \ldots, x^n)$ are local coordinates of a point on M and $p = (p_1, \ldots, p_n)$ defines a covector from T_x^*M. Recall that T^*M is a smooth symplectic $2n$-manifold with the standard symplectic form $\omega = \sum dp_i \wedge dx^i$. Consider a natural Hamiltonian system on T^*M with the Hamiltonian

$$H = \sum g^{ij}(x)\, p_i p_j + V(x) ,$$

where g^{ij} denotes the inverse tensor to the metric, and $V(x)$ is a smooth potential given on M.

Consider the $(2n-1)$-dimensional isoenergy level $Q^{2n-1} = \{H(x,p) = h\}$ for a sufficiently large value of energy h such that $h > \max V(x)$. It is easy to observe that the submanifold Q^{2n-1} is the isoenergy level for another system given by the Hamiltonian

$$\widetilde{H} = \sum \frac{g^{ij}(x)}{h - V(x)}\, p_i p_j\,.$$

Indeed, $Q^{2n-1} = \{\widetilde{H}(x,p) = 1\}$. This system is the geodesic flow of the Riemannian metric $ds^2 = \widetilde{g}_{ij}\, dx^i dx^j$ on M, where

$$\widetilde{g}_{ij} = (h - V(x))\, g_{ij}(x)\,.$$

This immediately implies that the integral curves of the Hamiltonian systems $v = \operatorname{sgrad} H$ and $\widetilde{v} = \operatorname{sgrad} \widetilde{H}$ coincide (up to a reparametrization) on the level Q^{2n-1}. In particular, their projections on the configuration space M are also the same. This statement is usually called the Maupertuis principle.

Thus, we can speak about the Maupertuis map, which associates a certain Riemannian metric g to every natural system v restricted to an isoenergy level in such a way that the geodesics of g coincide with the integral curves of v. It is clear that the main properties of the natural system and the geodesic flow corresponding to it will be very similar. In particular, the fact of existence of first integrals is preserved "under the Maupertuis map".

Theorem 15.1. *Let $v = \operatorname{sgrad} H$ be a natural Hamiltonian system on T^*M, and let $\widetilde{v} = \operatorname{sgrad} \widetilde{H}$ be the corresponding (by virtue of the Maupertuis principle) geodesic flow.*

a) If v possesses a smooth integral $f(x,p)$ on the given isoenergy surface $Q = \{H = h\}$ (such integrals are usually called particular), then the geodesic flow $\widetilde{v}$ possesses a smooth integral $\widetilde{f}(x,p)$ (not particular, but general) on the whole cotangent bundle T^*M such that $f|_Q = \widetilde{f}|_Q$.

b) If v possesses an integral that is a polynomial of degree m in momenta, then the geodesic flow $\widetilde{v}$ possesses the integral that is a homogeneous polynomial of the same degree.

Proof. Let f be a particular integral of the natural system with the Hamiltonian $H = K + V$. It is clear that this function will be an integral of the geodesic flow related to the Hamiltonian $\widetilde{H}$ on the isoenergy surface $Q = \{H = h\} = \{\widetilde{H} = 1\}$. Using the homogeneity of the geodesic flow, one can extend f to the whole cotangent bundle (except, perhaps, the zero section) in such a way that f becomes a global integral of the geodesic flow:

$$\widetilde{f}(x,p) = f\left(x, \frac{p}{|p|}\right),$$

where the norm $|p|$ is related to the Riemannian metric $\widetilde{g}_{ij}$, i.e., $|p| = \sqrt{\widetilde{H}}$.

Here we use the fact that the vector field $\widetilde{v}$ is a geodesic flow, and therefore, its integral can be extended by homogeneity from a single isoenergy surface to the whole space. Some problems may appear on the zero section, but they can be easily avoided by multiplying $\widetilde{f}$ by an appropriate function of the form $\lambda(\widetilde{H})$.

Let us verify that $\widetilde{f}(x,p)$ is indeed an integral of the geodesic flow. Observe that $\widetilde{f}(x,p)$ coincides with the initial integral f on the initial isoenergy surface Q. It needs to show that $\widetilde{f}(x,p)$ is constant on the geodesics, i.e., $\widetilde{f}(x(t),p(t)) = \text{const}$. Clearly, the trajectory $\left(x(t), \dfrac{p(t)}{|p(t)|} \right)$ is a geodesic lying on the level $Q = \{\widetilde{H} = 1\}$. Hence we have

$$\widetilde{f}(x(t),p(t)) = f\left(x(t), \frac{p(t)}{|p(t)|} \right) = \text{const} ,$$

since f is constant on the geodesics lying on the given energy level Q.

It remains to show that the Maupertuis map preserves polynomial integrability. In other words, if the initial flow v has a polynomial integral of degree m, then the flow $\widetilde{v}$ also has a polynomial integral of the same degree m.

Lemma 15.1. *Let an integral f of the flow v have the form $X(p,x) + Y(p,x)$, where all monomials of $X(p,x)$ have even degrees, and all monomials of $Y(p,x)$ have odd degrees. Then each of polynomials $X(p,x)$ and $Y(p,x)$ is an integral of v.*

Proof. By computing the Poisson bracket of the integral f with the Hamiltonian $H = K + V$, where K is a quadratic form in p, and V is a potential, i.e., a smooth function which does not depend on p, we obtain

$$\{H,X\} + \{H,Y\} = 0 .$$

It is easy to see that $\{H,X\}$ and $\{H,Y\}$ are polynomials, which contain respectively only even and only odd degrees in momenta. If the identity $\{H,X\} + \{H,Y\} = 0$ holds on the whole cotangent bundle, then each term vanishes separately. In the case when f is an integral only on the isoenergy surface Q^{2n-1}, this remains still true. To verify this, consider the relation $\{H,X\} + \{H,Y\} = 0$ in each cotangent space $T_x^* M$ separately. The isoenergy hypersurface Q defines the ellipsoid in $T_x^* M$ given by the equation $H = h$. Since the polynomial $\{H,X\} + \{H,Y\}$ vanishes on this ellipsoid, it follows that this polynomial is divisible by $H - h$. Therefore,

$$\{H,X\} + \{H,Y\} = (H - h)Z .$$

It is important that this equality holds identically on the whole cotangent space $T_x^* M$, but not only on the ellipsoid. But in this case,

$$\{H,X\} + \{H,Y\} = (H - h)Z_{\text{even}} + (H - h)Z_{\text{odd}} ,$$

where $Z = Z_{\text{even}} + Z_{\text{odd}}$ is the decomposition of Z into the polynomials that contain only even and only odd degrees in momenta respectively. Since $H - h$ contains only terms of even degrees in momenta, it follows that $(H - h)Z_{\text{even}}$ is a polynomial with terms of even degrees only, and $(H - h)Z_{\text{odd}}$ is that with terms of odd degrees. Since the identity holds on the whole cotangent space, we have

$$\{H,X\} = (H - h)Z_{\text{even}} \quad \text{and} \quad \{H,Y\} = (H - h)Z_{\text{odd}} ,$$

i.e., each of these polynomials vanishes on the hypersurface $H - h = 0$. $\square$

This implies that, without loss of generality, we may always assume that the polynomial integral f consists of either even or odd degrees only. Therefore, f can be written as follows:

$$f = X_k(p, x) + X_{k-2}(p, x) + X_{k-4}(p, x) + \dots,$$

where X_s is a homogeneous polynomial of degree s in momenta with coefficients depending on x.

We now turn to the proof of item (b). As an integral $\tilde{f}$ of the flow $\tilde{v}$ we can take the following homogeneous polynomial of degree k:

$$\tilde{f} = X_k(p, x) + |p|^2 X_{k-2}(p, x) + |p|^4 X_{k-4}(p, x) + \dots.$$

It coincides with the integral f on the hypersurface Q. Hence $\{\tilde{f}, \tilde{H}\} = \{\tilde{f}, H\} = 0$ on the hypersurface Q. Due to homogeneity of $\tilde{f}$ this identity will hold on the whole cotangent bundle including the zero section. This completes the proof. $\square$

By using the Maupertuis principle, the classification of natural systems on two-dimensional surfaces that admit quadratic or linear integrals can be reduced to the above classification of integrable geodesic flows. As an example, we now formulate an analog of Theorem 11.7, which gives the local description of quadratically integrable geodesic flows.

Theorem 15.2. *Let the natural system with the Hamiltonian*

$$H = \sum g^{ij}(x)\, p_i p_j + V(x) = K + V$$

be given on a two-dimensional manifold M, and

$$F = \sum b^{ij}(x)\, p_i p_j + U(x) = B + U$$

be its quadratic integral.

Suppose that, at some point $x_0 \in M^2$, two quadratic forms K and B are not proportional. Then, in a neighborhood of this point x_0, there exist local regular coordinates (u, v) in which the Hamiltonian H and integral F have the following form:

$$H = \frac{p_u^2 + p_v^2}{f(u) + g(v)} + \frac{Z(u) + W(v)}{f(u) + g(v)},$$

$$F = \frac{g(v)p_u^2 - f(u)p_v^2}{f(u) + g(v)} + \frac{g(v)Z(u) - f(u)W(v)}{f(u) + g(v)},$$

where f, g, Z, and W are some smooth functions on M^2. In other words, if the natural system has a quadratic integral, then its variables can be separated.

Proof. Restricting the natural system to the energy level $\{H = h\}$ and using the Maupertuis principle, we obtain the geodesic flow with the Hamiltonian $\widetilde{H} = \dfrac{K}{h - V}$ and the quadratic integral $\widetilde{F} = B + U\widetilde{H}$. For the functions $\widetilde{H}$, $\widetilde{F}$ we can apply Theorem 11.7, which allows us to reduce the corresponding Riemannian metric to the Liouville form by choosing a suitable local coordinates u, v on M^2. As we can see from the proof of Theorem 11.7, these local coordinates do not depend on the choice of the constant h. Indeed, u and v are constructed from the holomorphic form R associated with the integral of the geodesic flow. The explicit formula for R implies that R is not changed under perturbations of h. More precisely, it is not changed if the integral of the flow is changed by adding a function proportional to the Hamiltonian. Therefore, in our case, the form R is completely determined by the quadratic form B and does not depend on h.

As a result, the functions $\widetilde{H}$, $\widetilde{K}$ obtained earlier from H and K by using the Maupertuis principle can be written in terms of u and v as follows:

$$\widetilde{H} = \frac{K}{h - V} = \frac{p_u^2 + p_v^2}{f_h(u) + g_h(v)},$$

$$\widetilde{F} = B + U\frac{K}{h - V} = \frac{g_h(v)p_u^2 - f_h(u)p_v^2}{f_h(u) + g_h(v)}.$$

Here the functions f_h, g_h depend on h as a parameter. But the coordinates u, v do not depend on h.

Hence, $K = \lambda^{-1}(p_u^2 + p_v^2)$ and $h - V = \lambda^{-1}(f_h(u) + g_h(v))$ for some smooth function $\lambda(u, v)$. On the other hand, the quadratic forms K and B commute with respect to the Poisson bracket, and, consequently, according to Theorem 11.7, they have the form

$$K = \frac{p_u^2 + p_v^2}{f(u) + g(v)}, \qquad B = \frac{g(v)p_u^2 - f(u)p_v^2}{f(u) + g(v)},$$

i.e., $\lambda(u, v) = f(u) + g(v)$. Hence we see that

$$V = \frac{f(u)h - f_h(u) + g(v)h - g_h(v)}{f(u) + g(v)}.$$

The function in the denominator does not depend on h. Therefore, without loss of generality, we can set

$$f(u)h - f_h(u) = Z(u), \qquad g(v)h - g_h(v) = W(v),$$

where Z and W do not depend on the parameter h.

Thus, the Hamiltonian H has the desired form. Now we turn to the integral F. We have

$$\widetilde{F} = B + U\frac{K}{h - V} = \frac{g_h(v)p_u^2 - f_h(u)p_v^2}{f(u) + g(v)}.$$

In this equality we know the functions B, K, V. Expressing the function U from them, we obtain

$$U = \frac{g(v)Z(u) - f(u)W(v)}{f(u) + g(v)} .$$

Thus,

$$H = \frac{p_u^2 + p_v^2}{f + g} + \frac{Z + W}{f + g} , \qquad F = \frac{gp_u^2 - fp_v^2}{f + g} + \frac{gZ - fW}{f + g} ,$$

as was to be proved. □

The separation of variables allows us to obtain the following explicit formulas for the trajectories.

Proposition 15.1.   *The trajectories of the natural system on M^2 with the Hamiltonian H and quadratic integral F (see formulas in Theorem 15.2) lying on the common level $\{H = h, F = a\}$ are given in separating variables (u, v) by the equations*

$$\int \frac{dx}{\sqrt{hf(x) - Z(x) + a}} \pm \int \frac{dy}{\sqrt{hg(y) - W(y) - a}} = c .$$

Proof.   Consider the Liouville torus that is one of connected components of the common level surface $\{H = h = \text{const}, F = a = \text{const}\}$. This torus is given by the following system of equations:

$$\frac{p_u^2 + p_v^2}{f + g} + \frac{Z + W}{f + g} = h , \qquad \frac{gp_u^2 - fp_v^2}{f + g} + \frac{gZ - fW}{f + g} = a .$$

We now use again the Maupertuis principle which states that the trajectories of our natural system coincide with geodesics of the Riemannian metric

$$ds^2 = ((hf - Z) + (hg - W))(du^2 + dv^2) .$$

But this metric has the Liouville form and we already know the explicit formulas for the geodesics, see Theorem 11.5. Finally, we obtain

$$\int \frac{dx}{\sqrt{hf - Z + a}} \pm \int \frac{dy}{\sqrt{hg - W - a}} = c . \qquad □$$

15.2. MAUPERTUIS PRINCIPLE
IN RIGID BODY DYNAMICS

We now apply the described construction to a specific case where M is the two-dimensional sphere S^2. First, we realize the cotangent bundle of the sphere as the following model. Consider the vector space $\mathbb{R}^6$ $(s_1, s_2, s_3, r_1, r_2, r_3)$ as the dual space to the Lie algebra $e(3) = so(3) + \mathbb{R}^3$ of the affine isometry group of the three-dimensional Euclidean space. The coordinates (s_1, s_2, s_3) correspond to the subgroup of rotations SO(3), and (r_1, r_2, r_3) correspond to the subgroup of translations. Then $\mathbb{R}^6 = e(3)^*$ is endowed with the natural Poisson bracket

$$\{s_i, s_j\} = \epsilon_{ijk} s_k, \quad \{s_i, r_j\} = \epsilon_{ijk} r_k, \quad \{r_i, r_j\} = 0.$$

In various versions of the rigid body dynamics, the variables r_i and s_i obtain specific mechanical meaning. For example, when studying the motion of a rigid body about a fixed point in the gravity field, the variables r_i denote the components of the unit vertical vector in the coordinate system that is fixed in the body, and s_i are the components of the angular momentum.

Consider the four-dimensional submanifold M_0^4 in $\mathbb{R}^6$, which is an orbit of the coadjoint representation, given by

$$r_1^2 + r_2^2 + r_3^2 = 1, \qquad r_1 s_1 + r_2 s_2 + r_3 s_3 = 0,$$

It is known that the Poisson bracket, being restricted to this submanifold, becomes non-degenerate, and moreover, the submanifold M_0^4 is symplectomorphic to the cotangent bundle of the sphere $T^* S^2$.

Let us consider the Hamiltonian system v on $\mathbb{R}^6 (s_1, s_2, s_3, r_1, r_2, r_3)$ with the Hamiltonian

$$H(r, s) = \langle B(r)s, s \rangle + V(r),$$

where $\langle \cdot, \cdot \rangle$ is the Euclidean inner product on $\mathbb{R}^3$, B is a symmetric positive definite matrix (depending, in general, on r), and $V(r)$ is a smooth potential. Note that the Hamiltonians of the equations that appear in rigid body dynamics have just this form.

Restricting v to the submanifold $M_0^4 = T^* S^2$, we obtain a certain natural system. Moreover, any natural system on the sphere can be obtained in this way.

Applying the Maupertuis principle to the system v, we obtain a new system $\widetilde{v}$ on $M_0^4 = T^* S^2$ with the Hamiltonian $\widetilde{H} = \dfrac{1}{h - V(r)} \langle B(r)s, s \rangle$. It is easy to see that the trajectories of v and $\widetilde{v}$ on the three-dimensional level $Q^3 = M_0^4 \cap \{H = h\}$ coincide up to a reparametrization. In particular, this implies the following result.

Theorem 15.3. *Let v be integrable on $M_0^4 \subset \mathbb{R}^6$, and let its integral $f(r, s)$ be a polynomial of degree n in the variable s with coefficients depending on r. Then the system $\widetilde{v}$ also admits an integral $\widetilde{f}$ that is a homogeneous polynomial (in s) of the same degree. In particular, $\widetilde{v}$ is integrable on M_0^4.*

Proof. According to the Maupertuis principle, the new vector field $\widetilde{v}$ is given by the Hamiltonian $\widetilde{H}$ of the form

$$\widetilde{H} = \frac{1}{h - V(r)} \langle B(r)s, s \rangle.$$

The relationship between the polynomials f and $\widetilde{f}$ can be explicitly indicated. Let $f = \sum P_i(s)$, where $P_i(s)$ is a homogeneous polynomial in s of degree i (with coefficients depending on r). Without loss of generality, we can assume that in this decomposition, all i are simultaneously either even or odd. If it is not the case, then, dividing f into the odd and even parts, one can easily verify that each of them is an integral of v independently. Then the integral $\tilde{f}$ of $\tilde{v}$ can be defined as follows:

$$\tilde{f} = \sum_{i=1}^{m} P_i(s)(\tilde{H}(s,r))^{\frac{m-i}{2}} \,.$$

Here $\dfrac{m-i}{2}$ is an integer; that is why $\tilde{f}$ is in fact a homogeneous polynomial of degree m.

The same arguments as in the previous section show that the function $\tilde{f}$ is a polynomial in s and is an integral of $\tilde{v}$. $\square$

Thus, according to the Maupertuis principle, to the initial system v on $\mathrm{e}(3)^*$ we assign a new system $\tilde{v}$ on the cotangent bundle T^*S^2. Its Hamiltonian $\tilde{H}$ is a positive definite quadratic form in the variables s, and therefore, describes the geodesic flow of some Riemannian metric on the sphere. We now indicate explicit formulas, which express this metric in terms of the Hamiltonian $\tilde{H}$.

First of all, we need explicit formulas for the symplectomorphism between the cotangent bundle T^*S^2 with the standard symplectic structure and the orbit $M_0^4 \subset \mathbb{R}^6 (r, s)$.

Let us realize the cotangent bundle of the sphere as a symplectic submanifold T^*S^2 in $T^*\mathbb{R}^3$. Let u_1, u_2, u_3 be Euclidean coordinates in $\mathbb{R}^3$, and let p_1, p_2, p_3 be the corresponding momenta. Identifying the tangent vectors with the cotangent ones by means of the Euclidean inner product, we can define the cotangent bundle T^*S^2 by the relations

$$u_1^2 + u_2^2 + u_3^2 = 1\,, \qquad u_1 p_1 + u_2 p_2 + u_3 p_3 = 0\,.$$

Consider now the mapping $\mu \colon T^*S^2 \to \mathrm{e}(3)^*$ given by

$$r = u\,, \qquad s = [u, p]\,,$$

where $[\cdot, \cdot]$ denotes the vector product in the Euclidean space.

Lemma 15.2.

a) *Under the mapping μ, the image of the cotangent bundle T^*S^2 in $\mathrm{e}(3)^*$ coincides with the orbit M_0^4.*

b) *The mapping $\mu \colon T^*S^2 \to \mu(T^*S^2) = M_0^4$ is a symplectomorphism.*

Proof. a) Clearly, the equation $u_1^2 + u_2^2 + u_3^2 = 1$ transforms into $r_1^2 + r_2^2 + r_3^2 = 1$ under the embedding μ. Further, the vector v orthogonal to u is mapped to the vector product of u and v. Clearly, this product $[u, v]$ is also orthogonal to u. Therefore, the vector $s = [u, v]$ satisfies the linear equation

$$r_1 s_1 + r_2 s_2 + r_3 s_3 = 0\,,$$

which is the "image" of the orthogonality relation $u_1 p_1 + u_2 p_2 + u_3 p_3 = 0$.

Moreover, it is easy to see that the mapping $\mu \colon T^*S^2 \to \mu(T^*S^2) = M_0^4$ is a diffeomorphism.

b) It remains to verify that the standard Poisson bracket in $\mathbb{R}^6\,(u,p)$

$$\{u_i,p_j\} = \delta_{ij}\,, \quad \{u_i,u_j\} = 0\,, \quad \{p_i,p_j\} = 0$$

transforms into the standard Lie–Poisson bracket on the coalgebra $e(3)^* = \mathbb{R}^6\,(r,s)$

$$\{s_1,s_2\} = s_3\,, \quad \{s_2,s_3\} = s_1\,, \quad \{s_3,s_1\} = s_2\,,$$
$$\{s_1,r_2\} = r_3\,, \quad \{s_2,r_3\} = r_1\,, \quad \{s_3,r_1\} = r_2\,,$$
$$\{r_1,s_2\} = r_3\,, \quad \{r_2,s_3\} = r_1\,, \quad \{r_3,s_1\} = r_2\,,$$
$$\{s_i,r_i\} = 0\,, \qquad \{r_i,r_j\} = 0\,.$$

This fact is verified by a straightforward calculation. Thus, we have proved that $T^*\mathbb{R}^3\,(u,v) \to e(3)^*(r,s)$ is a Poisson map. Taking into account that the embedding

$$T^*S^2 \simeq TS^2 \to T\mathbb{R}^3 \simeq T^*\mathbb{R}^3$$

is symplectic, we obtain that the mapping $\mu\colon T^*S^2 \to \mu(T^*S^2) = M_0^4$ is the desired symplectomorphism. □

Using μ, we can write explicit formulas for the metric g_{ij} that corresponds to the homogeneous quadratic Hamiltonian $\widetilde{H} = \langle B(s),s\rangle$ and the inverse formulas that express B through g_{ij}.

Theorem 15.4. *The Hamiltonian system on the orbit M_0^4 with the Hamiltonian $\widetilde{H} = \langle B(s),s\rangle = \sum B_{ij}(r)s_is_j$ describes the geodesic flow of the metric $ds^2 = \sum \overline{B}_{ij}(u)\,du_i\,du_j$ restricted to the standard sphere $S^2 = \{u_1^2 + u_2^2 + u_3^2 = 1\}$, where*

$$\overline{B}_{ij} = B_{ij}\lambda^{-1}\,,$$

and λ is the determinant of the form $B(u)$ restricted to the two-dimensional plane that is orthogonal to the radius vector u (here the determinant is computed in an orthogonal basis).

Thus, we have described some correspondence $B \to \overline{B}$ between quadratic forms in the three-dimensional space. It is easy to see that this correspondence is actually an involution. In particular, the following statement holds.

Theorem 15.5. *Under the assumptions of the previous theorem, the form B that determines the Hamiltonian $\widetilde{H}$ can be reconstructed from the metric $\overline{B}$ by the same procedure, namely,*

$$B = \overline{\overline{B}}\,.$$

Proof (of Theorems 15.4 and 15.5).

Consider the Hamiltonian $K(u,v)$ of the geodesic flow of the metric g_{ij} on the sphere S^2. Recall that (u,v) belongs to the cotangent bundle of the sphere. By the definition of the Hamiltonian K, its value $K(u,v)$ on the pair (u,v) is the square of v in the sense of the metric g^{-1} at the point $u \in S^2$. Here we consider g^{-1} as a quadratic form (metric) on vectors by lowering indices by means of the Euclidean inner product. On the other hand, the explicit formulas of the embedding of the cotangent bundle of the sphere onto the orbit in $e(3)^*$ imply that $K(u,v)$ coincides with the square of the vector $[u,v]$ in the sense of the quadratic form B. Thus, taking into account that $[u,v]$ is the vector in the tangent plane $T_u S^2$ obtained from v by rotating it through $\pi/2$, we get the following statement.

Lemma 15.3.   *The restriction of B onto the tangent plane $T_u S^2$ can be described as follows. To compute the inner product of two tangent vectors a and b with respect to the form B, we must rotate each of these vectors through $\pi/2$ and then take their inner product with respect to the form g^{-1}.*

Using this lemma, we can now compare the matrices of the two forms B and g on the tangent plane to the sphere. It is well known that in the tangent plane there is an orthonormal basis in terms of which the form g is presented as the diagonal matrix

$$g = \begin{pmatrix} c & 0 \\ 0 & d \end{pmatrix}.$$

Compute the matrix of B in the same basis. We obtain

$$B = \begin{pmatrix} 1/d & 0 \\ 0 & 1/c \end{pmatrix} = \frac{1}{cd} \begin{pmatrix} c & 0 \\ 0 & d \end{pmatrix} = \frac{g}{\det g}.$$

This completes the proof of Theorem 15.4.   □

It is seen from the formula obtained that the operation "bar" is an involution on the tangent plane to the unit sphere S^2. This implies the statement of Theorem 15.5.   □

REMARK.   If we consider the operation "bar" as an involution on the set of Riemannian metrics on the unit sphere embedded into $\mathbb{R}^3$ in the standard way (defined as explained in Lemma 15.3), then we can introduce pairs of Riemannian metrics which are dual one to the other. It is an interesting fact that in the sense of this duality, the metric on the ellipsoid is dual to the metric on the Poisson sphere.

15.3. CLASSICAL CASES OF INTEGRABILITY IN RIGID BODY DYNAMICS AND RELATED INTEGRABLE GEODESIC FLOWS ON THE SPHERE

It is well known that many equations in rigid body dynamics can be represented as Hamiltonian systems on $e(3)^* = \mathbb{R}^6(s,r)$. In the most general case, the corresponding Hamiltonian is

$$H = I_1 s_1^2 + I_2 s_2^2 + I_3 s_3^2 + L(r,s) + V(r),$$

where I_1, I_2, I_3 are constants (these are moments of inertia of the body), L is a linear function of s, and V is a smooth potential.

Consider a particular case where the Hamiltonian H does not contain linear (in momenta) terms, but includes some potential $V(r)$. Applying the Maupertuis principle in this situation, we can interpret this Hamiltonian system as the geodesic flow of a certain Riemannian metric. What kind of metrics do we obtain as a result? The answer easily follows from Theorem 15.4.

Theorem 15.6. *Let* $H = I_1 s_1^2 + I_2 s_2^2 + I_3 s_3^2 + V(r)$. *Then the trajectories of the corresponding Hamiltonian system on the orbit* M_0^4 *lying on the level* $\{H = h = \text{const}\}$ *coincide with the geodesics of the Riemannian metric*

$$ds^2 = \frac{h - V(u)}{I_1 I_2 I_3} \cdot \frac{I_1\, du_1^2 + I_2\, du_2^2 + I_3\, du_3^2}{\frac{u_1^2}{I_1} + \frac{u_2^2}{I_2} + \frac{u_3^2}{I_3}}$$

restricted from $\mathbb{R}^3$ *to the standard two-dimensional sphere* $S^2 = \{u_1^2 + u_2^2 + u_3^2 = 1\}$.

The most interesting for us are integrable cases in the rigid body dynamics, from which we select the cases associated with the names of Euler [111], [112], Lagrange [209], Kovalevskaya [191], Goryachev and Chaplygin [81], [146], and Clebsch [84]. All of them have Hamiltonians of the form described in Theorem 15.6 (i.e., do not contain terms linear in momenta). Applying to them the Maupertuis principle, we obtain a number of integrable geodesic flows.

A discussion of other cases of integrability in rigid body dynamics can be found in [18], [22], [36], [82], [89], [98], [106], [110], [144], [145], [147], [151], [180], [216], [266], [268], [289], [305], [322], [323], [324], [325], [326], [327], [361], [365], [366], and [367].

15.3.1. *Euler Case and the Poisson Sphere*

The Hamiltonian of the Euler case is

$$H = A s_1^2 + B s_2^2 + C s_3^2.$$

According to the Maupertuis principle (see Theorem 15.6), this case is associated with the following metric on S^2:

$$ds^2 = \frac{h}{ABC} \cdot \frac{A\, du_1^2 + B\, du_2^2 + C\, du_3^2}{\frac{u_1^2}{A} + \frac{u_2^2}{B} + \frac{u_3^2}{C}}.$$

where h is a fixed energy level. Here we suppose the above metric to be restricted to the standard sphere S^2 in $\mathbb{R}^3$ (u_1, u_2, u_3). As was pointed out, this metric is called the metric on the Poisson sphere. Some of its properties were discussed in Chapter 8.

It is an interesting question if it is possible to realize the Poisson sphere as a smooth sphere embedded (or immersed) into $\mathbb{R}^3$.

Theorem 15.7. *The Hamiltonian system of the Euler case (in the rigid body dynamics) coincides with the geodesic flow on the Poisson sphere.*

Recall that the Euler system is considered on the four-dimensional manifold M_0^4 imbedded into $\mathbb{R}^6 \simeq e(3)^*$ as a coadjoint orbit. From the viewpoint of rigid body dynamics, this means that the area constant is assumed to be zero. The same assumption is supposed to be held in all the cases discussed below.

15.3.2. Lagrange Case and Metrics of Revolution

The Hamiltonian in the Lagrange case is

$$H = As_1^2 + As_2^2 + Cs_3^2 + V(r_3).$$

Here the ellipsoid of inertia is a surface of revolution $(A = B)$. The corresponding metric on the sphere has the form

$$ds^2 = \frac{h - V(u_3)}{AAC} \cdot \frac{A\,du_1^2 + A\,du_2^2 + C\,du_3^2}{\frac{u_1^2}{A} + \frac{u_2^2}{A} + \frac{u_3^2}{C}}.$$

This metric is obviously invariant with respect to rotations about the u_3-axis. In this sense, it is similar to a metric on a sphere of revolution. In particular, its geodesic flow admits a linear integral. Note that in the classical Lagrange case related to the motion in the gravity field, the potential V has the form $V(r_3) = r_3$.

Theorem 15.8.

a) *The Hamiltonian system of the Lagrange case (on a fixed energy level) is smoothly orbitally equivalent to the geodesic flow of a metric on the two-dimensional sphere that is invariant under a smooth S^1-action (such metrics are usually called metrics of revolution). The explicit form of this metric is indicated above.*

b) *Conversely, any smooth metric of revolution on the sphere can be written in the above form under an appropriate choice of the potential $V(r_3)$. In particular, the geodesic flow of such a metric is smoothly orbitally equivalent to the Lagrange case with an appropriate potential $V(r_3)$.*

15.3.3. Clebsch Case and Geodesic Flow on the Ellipsoid

One of the most remarkable consequences of the Maupertuis principle is the smooth orbital equivalence between the integrable Clebsch case and the geodesic flow on the ellipsoid. This result was found by H. Minkowski and V. V. Kozlov in different time and in different ways.

Let us consider the integrable Hamiltonian system v in $e(3)^* = \mathbb{R}^6(s, r)$ that describes the motion of a rigid body in an ideal fluid in the classical Clebsch case. The corresponding Hamiltonian H and the additional integral f of this system are

$$H = as_1^2 + bs_2^2 + cs_3^2 - \frac{r_1^2}{a} - \frac{r_2^2}{b} - \frac{r_3^2}{c},$$

$$f = s_1^2 + s_2^2 + s_3^2 + \frac{r_1^2}{bc} + \frac{r_2^2}{ca} + \frac{r_3^2}{ab}.$$

Theorem 15.9 [105]. *The Hamiltonian system v of the Clebsch case restricted to the three-dimensional energy level $\{H = 0\}$ is smoothly orbitally equivalent to the geodesic flow $\tilde{v}$ on the ellipsoid*

$$\frac{x^2}{a} + \frac{y^2}{b} + \frac{z^2}{c} = 1.$$

Proof. Consider the three-dimensional level $Q = \{H = 0\}$. According to the Maupertuis principle, on this level, the trajectories of v coincide with the trajectories of the system $\tilde{v}$ with the Hamiltonian

$$\tilde{H} = \frac{as_1^2 + bs_2^2 + cs_3^2}{\frac{r_1^2}{a} + \frac{r_2^2}{b} + \frac{r_3^2}{c}}.$$

The Hamiltonian $\tilde{H}$ defines some Riemannian metric on S^2. Using Theorem 15.6, we see that this metric coincides with the metric

$$ds^2 = a\, du_1^2 + b\, du_2^2 + c\, du_3^2$$

restricted to the sphere embedded into $\mathbb{R}^3$ in the standard way. But this metric is obviously isometric to that on the indicated ellipsoid as was required. $\square$

The Maupertuis principle allows us to point out an interesting connection between the ellipsoid and the Poisson sphere.

Consider two Riemannian metrics

$$ds_0^2 = a\, du_1^2 + b\, du_2^2 + c\, du_3^2\,,$$

$$ds_1^2 = \frac{a\, du_1^2 + b\, du_2^2 + c\, du_3^2}{\frac{u_1^2}{a} + \frac{u_2^2}{b} + \frac{u_3^2}{c}}$$

in the three-dimensional Euclidean space.

After being restricted to the standard sphere, the first metric yields the ellipsoid metric, and the second one yields the Poisson sphere metric. We now consider a family of Riemannian metrics

$$ds_\alpha^2 = (1 - \alpha)\, ds_0^2 + \alpha\, ds_1^2\,,$$

where $0 \le \alpha \le 1$.

In other words, we consider a linear deformation of one metric into the other. It turns out that all these metrics ds_α^2 are quadratically integrable on the sphere. One can verify this, for example, by using the Maupertuis principle again. The point is that the metric ds_α^2 corresponds (under the "Maupertuis map") to the quadratically integrable system with the Hamiltonian of the Clebsch case

$$H = as_1^2 + bs_2^2 + cs_3^2 - \frac{r_1^2}{a} - \frac{r_2^2}{b} - \frac{r_3^2}{c}\,,$$

where the system should be restricted not to the level $\{H = 0\}$ as before, but to the level $\{H = h\}$. Here h and α are connected by the relation $\alpha = h(h+1)^{-1}$. Thus, varying the energy of the Clebsch system from zero to infinity (after applying the "Maupertuis map"), we deform the ellipsoid metric to the metric on the Poisson sphere.

The bifurcation diagram for the Clebsch case has been constructed by T. I. Pogosyan [294], [295] and A. A. Oshemkov [277] (see also Chapter 14). Varying the energy from zero to infinity, we move the straight line $\{H = h = \mathrm{const}\}$ from the left to the right. From the explicit form of the diagram, we can see that the moving line does not pass through any singular points of the bifurcation diagram. Therefore, the Liouville foliation does not change its topological structure under this deformation. As a result, we obtain another proof of the fact that the Jacobi problem (i.e., geodesic flow on the ellipsoid) and the Euler case (i.e., geodesic flow on the Poisson sphere) are Liouville equivalent.

15.3.4. Goryachev–Chaplygin Case and the Corresponding Integrable Geodesic Flow on the Sphere

We now apply the formula of Theorem 15.6 to the integrable Goryachev–Chaplygin case. Here the Hamiltonian and integral have the form

$$H = s_1^2 + s_2^2 + 4s_3^2 + r_1\,,$$
$$f = s_3(s_1^2 + s_2^2) - \frac{r_3 s_1}{2}\,.$$

According to the Maupertuis principle, we produce a new Hamiltonian $\widetilde{H}$ on the orbit M_0^4:

$$\widetilde{H} = \frac{s_1^2 + s_2^2 + 4s_3^2}{h - r_1}\,.$$

Here we set $h > 1$. The integral f is transformed to the integral $\widetilde{f}$, which is a homogeneous polynomial (in s) of degree 3, namely,

$$\widetilde{f} = s_3(s_1^2 + s_2^2) - \frac{r_3 s_1}{2(h - r_1)}(s_1^2 + s_2^2 + 4s_3^2)\,.$$

The Riemannian metric of the corresponding geodesic flow on the sphere is

$$ds^2 = \frac{h - u_1}{4} \cdot \frac{du_1^2 + du_2^2 + 4du_3^2}{u_1^2 + u_2^2 + u_3^2/4}\,.$$

Theorem 15.10 [57], [63]. *The integrable Goryachev–Chaplygin case generates (under the Maupertuis map) the geodesic flow on the two-dimensional sphere, which is integrable by means of the third degree integral indicated above. This integral is reduced neither to a linear nor to a quadratic one.*

Proof. Consider the (rough) molecule W for the geodesic flow of the Goryachev–Chaplygin metric on the sphere. As we know, this flow is orbitally equivalent to the Goryachev–Chaplygin case in rigid body dynamics, and consequently, has the same Liouville foliation. Therefore, the molecule W coincides with that for the Goryachev–Chaplygin case, which was calculated by A. A. Oshemkov [277] (see also Chapter 14). It is presented in Fig. 15.1. We continue the proof by assuming the contrary, i.e., that the integral $\widetilde{f}$ of the flow on the sphere can be reduced to a quadratic (or linear) one. In this case, we can use the results obtained in Chapter 12. There we have calculated the molecules W^* of all geodesic flows on the sphere, which are integrable by means of quadratic or linear integrals.

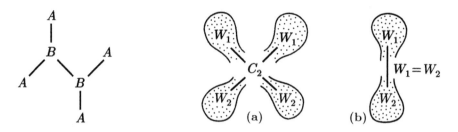

Figure 15.1   **Figure 15.2**

Using this description, we see that the molecule of the Goryachev–Chaplygin flow should have one of two possible forms shown in Fig. 15.2. The molecule in Fig. 15.2(a) corresponds to the case where $\tilde{f}$ is reducible to a quadratic integral, and the molecule in Fig. 15.2(b) corresponds to the case of a linear integral. Here W_1 is a tree whose branches are all directed upward, and W_2 is a tree whose branches are directed downward. Moreover, the vertices of the tree-graphs W_1 and W_2 should have a special type, in particular, have no star-vertices. Comparing these graphs with that in Fig. 15.1, we see that the graph W shown in Fig. 15.1 does not have the required structure. Since W is an invariant of the integrable system, we come to a contradiction; this completes the proof. □

15.3.5. *Kovalevskaya Case and the Corresponding Integrable Geodesic Flow on the Sphere*

The Hamiltonian H in the Kovalevskaya case is

$$H = \frac{1}{2}(s_1^2 + s_2^2 + 2s_3^2) + r_1 \, .$$

The Kovalevskaya integral is then expressed as

$$f = \left(\frac{s_1^2}{2} - \frac{s_2^2}{2} - r_1\right)^2 + (s_1 s_2 - r_2)^2 \, .$$

According to the Maupertuis principle, by using H, we produce a new Hamiltonian $\tilde{H}$ on the cotangent bundle of the sphere:

$$\tilde{H} = \frac{s_1^2 + s_2^2 + 2s_3^2}{h - r_1} \, ,$$

where $h > 1$. The integral f transforms into $\tilde{f}$, which has the form

$$\tilde{f} = \left(\frac{s_1^2}{2} - \frac{s_2^2}{2} - r_1 \frac{s_1^2 + s_2^2 + 2s_3^2}{h - r_1}\right)^2 + \left(s_1 s_2 - r_2 \frac{s_1^2 + s_2^2 + 2s_3^2}{h - r_1}\right)^2 \, .$$

Then the Riemannian metric of the corresponding geodesic flow on the sphere is given by

$$ds^2 = \frac{h - u_1}{2} \cdot \frac{du_1^2 + du_2^2 + 2du_3^2}{u_1^2 + u_2^2 + u_3^2/2} \, .$$

Theorem 15.11 [57], [63]. *The integrable Kovalevskaya case generates (under the Maupertuis map) the geodesic flow on the two-dimensional sphere, which is integrable by means of the fourth degree integral indicated above. This integral is reduced neither to a linear nor to quadratic one.*

Proof. The arguments repeat the proof of the previous theorem. We only need to compare the molecule W of the Kovalevskaya case (Fig. 15.3) calculated by A. A. Oshemkov in [277] (see also Chapter 14) with the molecules of quadratically and linearly integrable geodesic flows on the sphere (Fig. 15.2). The molecules are

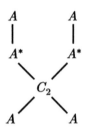

Figure 15.3

different, since W contains two atoms A^*, which are forbidden for the molecules illustrated in Fig. 15.2. Therefore, the Kovalevskaya metric does not belong to the class of metrics with linearly or quadratically integrable geodesic flows. The theorem is proved. □

Note that, in fact, the cases of Goryachev–Chaplygin and Kovalevskaya generate a one-parameter family of Riemannian metrics on the sphere. As one can see from the formulas, the coefficients of these metrics include the parameter $h > 1$, which can arbitrarily vary. Thus, we obtain two one-parameter families of integrable geodesic flows whose integrals have degree 3 and 4, respectively, and cannot be reduced to quadratic ones.

S. A. Chaplygin [82] and D. N. Goryachev [147] discovered a 4-parameter family of potentials on the sphere, that gives us integrable natural systems with an integral of degree 4. This family includes the Kovalevskaya top as a particular case. In the above notation, the corresponding Hamiltonian and integral have the following form:

$$H = \frac{1}{2}(s_1^2 + s_2^2 + 2s_3^2) - \left(\frac{a}{r_3^2} + 2b_1 r_1 r_2 + b_2(r_2^2 - r_1^2) + c_1 r_1 + c_2 r_2 \right),$$

$$F = 4\left(s_1 s_2 + 2a\frac{r_1 r_2}{r_3^2} - b_1 r_3^2 + c_1 r_2 + c_2 r_1 \right)^2$$

$$+ \left(s_1^2 - s_2^2 + \frac{2a(r_2^2 - r_1^2)}{r_3^2} + 2b_2 r_3^2 + 2c_1 r_1 - 2c_2 r_2 \right)^2,$$

where a, b_1, b_2, c_1, c_2 are arbitrary constants. In fact, to construct an integrable geodesic flow on the sphere, one has to set $a = 0$ in order for the potential to have no singularities. That is why we speak about four parameters only. Here, as in the Goryachev–Chaplygin case, for integrability we have to assume the area integral to be zero. By using the Maupertuis principle, we obtain a family of integrable geodesic flows on the sphere with an additional integral of the fourth degree. It is an interesting problem to analyze the topology of these flows depending on the choice of parameters.

Thus, there exist metrics on the sphere whose geodesic flows are integrable by means of integrals of degree 1, 2, 3, and 4.

15.4. CONJECTURE ON GEODESIC FLOWS WITH INTEGRALS OF HIGH DEGREE

In Chapter 11, we have completely classified the Riemannian metrics on the two-dimensional surfaces whose geodesic flows are integrable by means of linear and quadratic integrals. An analogous question on the description (and classification) of metrics whose geodesic flows are integrable by means of polynomial integrals of degree more than 2 still remains open. (Of course, here we mean that the degree of these integrals cannot be reduced.) Moreover, the above metrics (the Goryachev–Chaplygin metric and the Kovalevskaya metric) on the two-dimensional sphere remain, in essence, the only examples of metrics with polynomial integrals of degree more than 2. On the torus, no such examples are known. Moreover, numerous attempts to construct such metrics on the torus yet failed.

Conjecture A. On the two-dimensional torus, there are no Riemannian metrics whose geodesic flows admit polynomial integrals of degree $n > 2$ and do not admit any linear and quadratic integrals. In other words, the list of integrable geodesic flows on the torus from Chapter 11 is complete, i.e., gives us a complete classification (up to an isometry) of geodesic flows with polynomial in momenta integrals.

Conjecture B. On the two-dimensional sphere, there are no Riemannian metrics whose geodesic flows are integrable by means of an integral of degree $n > 4$ and do not admit integrals of degree ≤ 4.

Conjectures A and B were formulated by V. V. Kozlov and A. T. Fomenko.

There are a number of arguments in favor of Conjecture A. For example, see papers by V. V. Kozlov, N. V. Denisova [198], [199], M. L. Bialyĭ [76], V. V. Kozlov, D. V. Treshchev [201]. See also the survey by A. V. Bolsinov, A. T. Fomenko, V. V. Kozlov [63] and the book by V. V. Kozlov [196].

It is worth pointing out the global character of Conjectures A and B. The question is about the properties of metrics given on the whole sphere and the whole torus (i.e., globally). If one restricts oneself to the local aspect of the problem, then the situation will be immediately clarified. Namely, the following result by V. V. Kozlov holds.

Theorem 15.12. *In the domain $D_x \times \mathbb{R}^2_p$ (where D is a disc on the plane x_1, x_2), there exist integrable systems with the Hamiltonian*

$$H = \frac{p_1^2 + p_2^2}{2\Lambda(x_1, x_2)}$$

that admit a polynomial in momenta integral (independent of H) of arbitrary degree n, but do not admit any polynomial integral (independent of H) of degree less than n.

See the proof in the paper by V. Ten [336].[1]

Here we also point out two well-known integrable cases with integrals of degree 3 and 4. These are Toda lattices [37], [202], [340] and Calogero–Moser systems [77], [78], [249]. Using the Maupertuis principle, we can construct from them integrable geodesic flows on a disc with integrals of degree 3 and 4.

We first indicate the integrable cases of the Toda lattices with two degrees of freedom by listing their Hamiltonians H, together with the corresponding integrals F.

Case 1.

$$H = \frac{1}{2}(p_1^2 + p_2^2) + v_1 e^{\sqrt{3}x_1 + x_2} + v_2 e^{-\sqrt{3}x_1 + x_2} + v_3 e^{-2x_2},$$

$$F = \frac{1}{3}p_1^3 - p_1 p_2^2 + v_2(p_1 + \sqrt{3}p_2)e^{-\sqrt{3}x_1 + x_2} - 2v_3 p_1 e^{-2x_2}$$
$$+ v_1(p_1 - \sqrt{3}p_2)e^{\sqrt{3}x_1 + x_2}.$$

Case 2.

$$H = \frac{1}{2}(p_1^2 + p_2^2) + v_1 e^{x_1} + v_2 e^{x_2} + v_3 e^{-x_1 - x_2} + v_4 e^{-\frac{1}{2}(x_1 + x_2)},$$

$$F = p_1^2 p_2^2 + 2v_2 p_1^2 e^{x_2} + 2p_1 p_2\left(v_3 e^{-x_1 - x_2} + v_4 e^{-\frac{1}{2}(x_1 + x_2)}\right) + 2v_1 p_2^2 e^{x_1}$$
$$+ 2v_2 v_3 e^{-x_1} + 2v_1 v_3 e^{-x_2} + 4v_1 v_2 e^{x_1 + x_2} + \left(v_3 e^{-x_1 - x_2} + v_4 e^{-\frac{1}{2}(x_1 + x_2)}\right)^2.$$

Case 3.

$$H = \frac{1}{2}(p_1^2 + p_2^2) + e^{-x_1 - x_2} + \gamma_1 e^{x_1} + \frac{1}{2}\beta_1 e^{2x_1} + \gamma_2 e^{x_2} + \frac{1}{2}\beta_2 e^{2x_2},$$

$$F = p_1^2 p_2^2 + p_1^2(2\gamma_2 e^{x_2} + \beta_2 e^{2x_2}) + 2p_1 p_2 e^{-x_1 - x_2} + p_2^2(2\gamma_1 e^{x_1} + \beta_1 e^{2x_1})$$
$$+ e^{-2(x_1 + x_2)} + \beta_1 \beta_2 e^{2(x_1 + x_2)} + 2\beta_1 \gamma_2 e^{2x_1 + x_2} + 2\beta_2 \gamma_1 e^{x_1 + 2x_2}$$
$$+ 4\gamma_1 \gamma_2 e^{x_1 + x_2} + 2\gamma_1 e^{-x_2} + 2\gamma_2 e^{-x_1}.$$

We recall now the description of integrable cases in Calogero–Moser systems with two degrees of freedom. Consider n particles with unit mass located on a straight

---

[1] Just recently we have found out that K. Kiyohara has disproved the Conjecture B (Kiyohara K., *Math. Ann.*, **320** (2001), P. 487–505). He has constructed integrable geodesic flows on the 2-sphere with integrals of arbitrary degree n.

line at points $x_1, x_2, \ldots, x_n$ and interacting with a mutual interaction potential f. Then their dynamics is described by the Hamiltonian

$$H = \frac{1}{2} \sum_{i=1}^{n} p_i^2 + \sum_{i<j} f(x_i - x_j).$$

There are four types of potentials that give us integrable cases for n particles. The corresponding systems are usually called *Calogero–Moser systems*. These potentials are

1) $f(x) = \dfrac{1}{x^2}$,

2) $f(x) = \dfrac{1}{\sin^2 x}$,

3) $f(x) = \dfrac{1}{\sinh^2 x}$,

4) $f(x) = \dfrac{1}{\text{sn}^2 x}$,

where $\text{sn}(x) = \text{sn}(x|m)$ is the elliptic sine (or sine amplitude). Recall its definition. Consider the following elliptic integral:

$$u(\varphi, m) = \int_0^\varphi \frac{d\theta}{\sqrt{1 - m^2 \sin^2 \theta}}.$$

Here $0 \le m \le 1$. Then φ can be considered as a function $\varphi = \varphi(u, m)$. That is, φ can be thought as the inverse function to u. It is called the *amplitude* of the elliptic integral u and is denoted by $\text{am}(u)$. Then, by definition, $\text{sn}(u) = \sin \text{am}(u)$.

At the same time, the potential $f(x)$ can be rewritten in terms of the Weierstrass function $\wp(x)$. This function is connected with the elliptic sine by the following simple relation:

$$\wp(x) = e_3 + \frac{e_1 - e_3}{\text{sn}^2 \left(\sqrt{e_1 - e_3}\, x \right)}.$$

Here $\wp(x)$ is the Weierstrass function related to the orthogonal lattice with arbitrary periods $w_1 \in \mathbb{R}$ and $w_2 \in i\mathbb{R}$. Further, e_1, e_2, and e_3 denote the values of $\wp(x)$ at the half-periods, i.e.,

$$e_1 = \wp\left(\frac{w_1}{2}\right), \quad e_2 = \wp\left(\frac{w_1 + w_2}{2}\right), \quad e_3 = \wp\left(\frac{w_2}{2}\right).$$

In this case, the parameter m of the elliptic sine is expressed as

$$m^2 = \frac{e_2 - e_3}{e_1 - e_3}.$$

Note that the first three potentials $f(x)$

$$\frac{1}{x^2}, \quad \frac{1}{\sin^2 x}, \quad \frac{1}{\sinh^2 x}$$

can actually be considered as degenerations of $\wp$ as its periods (one or both) tend to infinity.

To produce an integrable geodesic flow (on a 2-disc) from these Hamiltonians, we must take a system with three particles. Then, the system always has a linear integral $F_1 = p_1 + p_2 + p_3$ and a cubic integral F. To construct a system with two (but not three) degrees of freedom, we need to make the reduction with respect to the linear integral F_1. The obtained system with two degrees of freedom will still have a cubic integral.

Let us explain this in more detail. The Hamiltonian H and cubic integral F are initially as follows:

$$H = \frac{1}{2}(p_1^2 + p_2^2 + p_3^2) - (f(x_1 - x_2) + f(x_1 - x_3) + f(x_2 - x_3)),$$
$$F = p_1 p_2 p_3 + p_1 f(x_2 - x_3) + p_2 f(x_1 - x_3) + p_3 f(x_1 - x_2).$$

By making the reduction with respect to $p_1 + p_2 + p_3$, we obtain an integrable system with two degrees of freedom with the following Hamiltonian and integral of degree three. To simplify the below formulas, we have done some scaling. The new coordinates (after the reduction) are denoted by p, y. We have

$$H = \frac{1}{2}(p_1^2 + p_2^2) - f(2y_2) - f\left(\sqrt{3}y_1 + y_2\right) - f\left(-\sqrt{3}y_1 + y_2\right),$$
$$F = \frac{1}{3}p_1^3 - p_1 p_2^2 - \left(p_1 + \sqrt{3}p_2\right) f\left(-\sqrt{3}y_1 + y_2\right)$$
$$- 2p_1 f(-2y_2) - \left(p_1 - \sqrt{3}p_2\right) f\left(\sqrt{3}y_1 + y_2\right).$$

Here f denotes any one of the above potentials. Note that in the first case, where $f = x^{-2}$, the expressions for H and F can be simplified. Namely,

$$H = \frac{1}{2}(p_1^2 + p_2^2) - \left(\frac{3(y_1^2 + y_2^2)}{2y_2(y_2^2 - 3y_1^2)}\right)^2,$$
$$F = p_1^3 - 3p_1 p_2^2 + \frac{9(3y_1^4 - 6y_1^2 y_2^2 - y_2^4)}{2y_2^2(y_2^2 - 3y_1^2)^2}p_1 - 36\frac{y_1 y_2}{(y_2^2 - 3y_1^2)^2}p_2.$$

Let us mention a paper by L. S. Hall [154], in which he made an attempt to examine natural systems that admit third degree integrals from the general viewpoint.

Let us notice that besides the Goryachev–Chaplygin and Kovalevskaya metrics, there exist other metrics on the sphere with integrals of degree 3 and 4. Such examples for integrals of degree 3 and 4 have been recently constructed by E. N. Selivanova (see [153] and [313]). It would be extremely interesting to obtain a classification of such metrics, as well as explicit formulas for them.

We now indicate two more examples of Riemannian metrics on a 2-disc (or the half-plane) whose geodesic flows are integrable by means of third degree integrals.

The first of them was found by C. Holt. It is

$$ds^2 = \left(\alpha - y^{-2/3}\left(\delta - \frac{3}{4}by^2 - bx^2 - cx\right)\right)(dx^2 + dy^2).$$

Here α, δ, b, and c are arbitrary constants. This metric is obtained by applying the Maupertuis principle to the natural system with the Hamiltonian

$$H = \frac{1}{2}(p_1^2 + p_2^2) + U(x, y),$$

where

$$U(x, y) = y^{-2/3}\left(\delta - \frac{3}{4}by^2 - bx^2 - cx\right).$$

Its integral has degree 3:

$$F = 2p_1^3 + 3p_1 p_2^2 + 3y^{-2/3}(2\delta - 2bx^2 - 2cx + 3by^2)p_1 - y^{1/3}(18bx + 9c)p_2.$$

The next example of a local metric whose geodesic flow admits a third degree integral is obtained from the natural system that was described by A. Fokas and P. Lagerstrom. Its Hamiltonian and integral are

$$H = \frac{1}{2}(p_1^2 + p_2^2) + (-y^2 + x^2)^{-2/3},$$
$$F = (xp_2 - yp_1)(p_1^2 - p_2^2) - 4(xp_2 + yp_1)(-y^2 + x^2)^{-2/3}.$$

In conclusion, we want to demonstrate how the degree of the first integral is connected with the topology of the Liouville foliation in the case of integrable geodesic flows on closed two-dimensional surfaces.

We have above described these foliations in terms of the so-called marked molecules. The molecules have been explicitly listed for each set of parameters (like f, g, L) (see Chapter 12). We now collect these results in the form of Table 15.1, where we list these molecules in the simplest cases when the Liouville foliation has the least possible number of singularities.

Comments about Table 15.1.

1. The functions f and g that are parameters of the metrics have the least possible number of critical points and these points are all non-degenerate.

2. The edges of the molecules must be endowed with numerical marks n, r, ε. These marks are completely indicated in Chapters 12, 14. Some of them are omitted in Table 15.1. However, we leave these marks in those cases where they are required to distinguish non-equivalent Liouville foliations. The point is that in some cases (as is seen from the table) the molecules without marks coincide, although the corresponding Liouville foliations are different. As an example, it is worth paying attention to the fact that the molecule related to the Kovalevskaya metric and that related to the geodesic flow on the Klein bottle coincide if we do not take the marks into account. This means that, from the local point of view, the corresponding Liouville foliations have the same structure. But the distinction of the marks implies that their global structures are different.

3. The signs "?" in Table 15.1 mean that the examples of the corresponding Riemannian metrics are unknown. (Recall that according to the conjecture formulated above, no such examples exist.)

Table 15.1. Molecules of integrable geodesic flows

deg	S^2	$\mathbb{R}P^2$	T^2	K^2
	f-metric	f-metric	(q,t,L)-metric	(L,g)-metric
1	A |$\;r=\frac{1}{2}$ A	A |$\;r=\frac{1}{4}$ A	A | B $()$ B | A	A | A^* | A^* | A
2	(L,f,g)-metric $A\quad A$ $\searrow\!\nearrow$ C_2 $\nearrow\!\searrow$ $A\quad A$ $n=2$	(L,f,g)-metric $A\quad A$ $\searrow\!\nearrow$ C_2 $\nearrow\!\searrow$ $A\quad A$ $n=4$	$(L,f,g,k/m)$-metric $A\quad\qquad A$ $\searrow\; r=\frac{k}{m}\;\nearrow$ $B\!-\!\!-\!B$ $r=-\frac{k}{m}\|\qquad\|r=-\frac{k}{m}$ $B\!-\!\!-\!B$ $\nearrow\; r=\frac{k}{m}\;\searrow$ $A\qquad\quad A$<hr>$(L,f,g,0)$-metric (global Liouville) $A\qquad\quad A$ $\searrow\; r=0\;\nearrow$ $B\!-\!\!-\!B$ $r=0\|\qquad\|r=0$ $B\!-\!\!-\!B$ $\nearrow\; r=0\;\searrow$ $A\qquad\quad A$	$(L,f,0)$-metric (quasi-linear case) $A\quad\quad A$ $\searrow\!\nearrow$ C_2 $\nearrow\!\searrow$ $K\quad\quad K$<hr>(L,f,g)-metric $A\qquad A$ $r=\frac{1}{2}\|\qquad\|r=\frac{1}{2}$ $A^*\quad A^*$ $\searrow\!\nearrow$ C_2 $\nearrow\!\searrow$ $A\qquad A$
3	Goryachev–Chaplygin metric A | $B\quad A$ $\nearrow\;\searrow\;\nearrow$ $A\quad B$ | A	**?**	**?**	**?**
4	Kovalevskaya metric $A\qquad A$ $r=0\|\qquad\|r=0$ $A^*\quad A^*$ $\searrow\!\nearrow$ C_2 $\nearrow\!\searrow$ $A\quad\;A$	**?**	**?**	metric $ds^2_{m,n,f}$ (quasi-quadratic case) A | $B\qquad K$ $\nearrow\;\searrow\;\nearrow$ $K\quad B$ | A

15.5. DINI THEOREM AND THE GEODESIC EQUIVALENCE OF RIEMANNIAN METRICS

In this section, we discuss the classical Dini theorem, which gives us a local description of Riemannian metrics on two-dimensional surfaces that admit non-trivial geodesic equivalencies [97]. In particular, this theorem gives us a possibility to construct examples of orbitally equivalent geodesic flows.

Definition 15.1. Two Riemannian metrics $G = (g_{ij})$ and $\widehat{G} = (\widehat{g}_{ij})$ given on a manifold M are called *geodesically equivalent* if the geodesic lines of the metric G coincide (as sets, i.e., without taking into account their parametrization) with those of the metric $\widehat{G}$.

The simplest example of geodesically equivalent metrics are metrics that differ only by a constant multiplier. In what follows, we will say that two metrics G and $\widehat{G}$ are *non-trivially geodesically equivalent* if they are geodesically equivalent, but are not proportional, i.e., cannot be obtained from each other via the multiplication by a constant.

Let G and $\widehat{G}$ be geodesically equivalent metrics. It is easy to see that their geodesic flows are then smoothly orbitally equivalent. Here we consider the flows as dynamical systems on the cotangent bundle or on the isoenergy surface. The orbital diffeomorphism is given by

$$(x, p) \rightarrow \left(x, \frac{|p|_G}{|\widehat{G}G^{-1}p|_{\widehat{G}}} \widehat{G}G^{-1}p \right).$$

Here (x, p) are usual coordinates on the cotangent bundle, x is a point on the manifold M, and p is a covector from $T_x^* M$. By $|p|_{\widehat{G}}$ and $|p|_G$, we denote the norms of p in the sense of $\widehat{G}$ and G. The same mapping written not in momenta but in velocities (i.e., in coordinates $(x, \dot{x})$) becomes simpler, that is,

$$(x, \dot{x}) \rightarrow \left(x, \frac{|\dot{x}|_G}{|\dot{x}|_{\widehat{G}}} \dot{x} \right).$$

Here $|\dot{x}|_{\widehat{G}}$ and $|\dot{x}|_G$ are the lengths of the velocity with respect to $\widehat{G}$ and G.

Notice, however, that not every orbital isomorphism between two geodesic flows is induced by a geodesic equivalence. The necessary condition is that this isomorphism commutes with the natural projection $T^*M \rightarrow M$ given by $(x, p) \rightarrow x$.

It turns out that the Dini theorem is a reflection of the following more general fact found by V. S. Matveev and P. I. Topalov [228].

Consider two Hamiltonian systems v and $\widehat{v}$ with two degrees of freedom. Restrict them to regular isoenergy 3-surfaces Q^3 and $\widehat{Q}^3$. Suppose these restrictions are smoothly orbitally equivalent. Recall that, by the smooth orbital equivalence, we mean the existence of a diffeomorphism $\xi: Q^3 \rightarrow \widehat{Q}^3$ that sends (oriented) trajectories of the first system to those of the second one. Then, using such a diffeomorphism ξ, we can canonically construct additional integrals of both

Hamiltonian systems [345]. The idea of this construction is the following. Consider the restrictions of the symplectic forms ω and $\widehat{\omega}$ to the isoenergy 3-surfaces Q and $\widehat{Q}$ and take the pull-back of $\widehat{\omega}$ from $\widehat{Q}$ to Q. Obviously, the obtained form $\xi^*\widehat{\omega}$ on Q has the following properties:

1) it is closed,

2) its kernel is generated by the Hamiltonian vector field v on Q.

Hence it is easy to see that v preserves $\xi^*\widehat{\omega}$ on Q. Indeed, by calculating the Lie derivative of $\xi^*\widehat{\omega}$ along v, we obtain

$$L_v\xi^*\widehat{\omega} = v\lrcorner\, d(\xi^*\widehat{\omega}) + d(v\lrcorner\,\xi^*\widehat{\omega}) = 0\,,$$

since $\xi^*\widehat{\omega}$ is closed (this annihilates the first term) and the kernel of $\xi^*\widehat{\omega}$ coincides with v (this annihilates the second term).

On the other hand, $L_v\omega = 0$ by virtue of the fact that v is a Hamiltonian vector field. Then the forms ω and $\xi^*\widehat{\omega}$ have the same kernel on the three-dimensional isoenergy manifold Q and, therefore, are proportional: $\xi^*\widehat{\omega} = f\omega$, where f is a smooth function on Q. Since v preserves both $f\omega = \xi^*\widehat{\omega}$ and ω, it preserves the function f. This means exactly that f is an integral of the flow v on the isoenergy surface Q.

Of course, this integral may turn out to be constant. This happens, for example, if the orbital equivalence of two flows is obtained by using the Maupertuis principle. But in some cases (in particular, in the Dini theorem), from the existence of an orbital equivalence, one can deduce the existence of a non-trivial integral.

Theorem 15.13 (U. Dini).

a) *Let Riemannian metrics G and $\widehat{G}$ on a two-dimensional surface M be non-trivially geodesically equivalent. Then the geodesic flows of both metrics are quadratically integrable. Moreover, the non-trivial quadratic integral F of the geodesic flow v of the metric G (as a function on the cotangent bundle) has the form*

$$F(x,p) = \frac{(\det G)^{2/3}}{(\det \widehat{G})^{2/3}}\,\langle G^{-1}\widehat{G}G^{-1}p, p\rangle\,.$$

b) *Conversely, let the geodesic flow of the metric G on the two-dimensional surface M be integrable by means of a quadratic integral $F(x,p) = \langle Fp, p\rangle$. Suppose that this integral is positive definite (it is always possible to achieve this by adding the Hamiltonian H with some coefficient to F). Then the initial metric G is geodesically equivalent to a new metric $\widehat{G}$ that is given on the tangent bundle of M by the formula*

$$\widehat{G} = (\det G)^{-2}(\det F)^{-2}GFG\,.$$

REMARK. Here G is the matrix of the initial metric (regarded as a quadratic form on the tangent bundle), F is the matrix of the quadratic integral (regarded as a quadratic form on the cotangent bundle), then GFG is a quadratic form on the tangent bundle (which is obtained from F by lowering its indices by means of G).

Proof (of the Dini theorem).   Consider the isoenergy 3-surfaces Q and $\widehat{Q}$. We shall assume them to be given as the Hamiltonian levels $Q = \{H = 1\}$ and $\widehat{Q} = \{\widehat{H} = 1\}$. As remarked above, the geodesic equivalence induces the orbital isomorphism $\xi \colon Q \to \widehat{Q}$ given by the formula

$$(x, p) \to \left(x, \frac{|p|_G}{|\widehat{G}G^{-1}p|_{\widehat{G}}} \widehat{G}G^{-1}p \right).$$

Since this formula is written in the explicit form, we can compute the form $\xi^*\widehat{\omega}$ and find the proportion coefficient for $\xi^*\widehat{\omega}$ and ω. On the isoenergy surfaces Q and $\widehat{Q}$ we consider three smooth vector fields D_0, D_1, D_2 and, respectively, $\widehat{D}_0, \widehat{D}_1, \widehat{D}_2$. We describe them in the case of Q. Here D_0 is a tangent vector to the ellipse given in the cotangent plane by the equation $H = 1$. We shall assume $|D_0| = 1$ so that the period of the field D_0 along each ellipse $\{H = 1\}$ is equal to 2π. The field D_1 is the initial Hamiltonian field v on Q. Finally, $D_2 = [D_0, D_1]$ is the standard commutator of the vector fields D_0 and D_1. Since D_1 belongs to the kernel of the forms ω and $\xi^*\widehat{\omega}$ on Q, it suffices to compute the values of these forms on the vectors D_0 and D_2 only, that is,

$$\omega(D_0, D_2), \qquad \xi^*\widehat{\omega}(D_0, D_2).$$

The ratio of these quantities is, as we saw above, an integral f of the Hamiltonian vector field $v = D_1$. Let us compute both expressions. To this end we need the following statement.

Lemma 15.4. *The following relation holds:*

$$[D_0, D_2] = -D_1.$$

REMARK.   In fact, two more interesting relations hold:

$$[D_1, D_2] = kD_0, \qquad [D_0, D_1] = D_2,$$

where k is the Gaussian curvature of the metric (lifted from M to $Q \subset T^*M$).

Proof.   It is convenient to use conformal coordinates, in terms of which the metric G has the form $ds^2 = \lambda(x_1, x_2)(dx_1^2 + dx_2^2)$. Then the Hamiltonian H becomes

$$H = \frac{p_1^2 + p_2^2}{2\lambda}.$$

Further, in terms of coordinates (x_1, x_2, p_1, p_2) the vector fields D_0 and D_1 take the form

$$D_0 = (0, 0, p_2, -p_1),$$

$$D_1 = \left(\frac{p_1}{\lambda}, \frac{p_2}{\lambda}, \frac{\partial \lambda}{\partial x_1} \cdot \frac{p_1^2 + p_2^2}{2\lambda^2}, \frac{\partial \lambda}{\partial x_2} \cdot \frac{p_1^2 + p_2^2}{2\lambda^2} \right).$$

By computing the commutator $[D_0, D_2] = [D_0, [D_0, D_1]]$ explicitly using the standard explicit formula, we obtain the desired expression. $\square$

Lemma 15.5. *The following identity holds:*

$$\omega(D_0, D_2) = 1/2\,.$$

Proof. Recall that $\omega = d\varkappa$, where the action 1-form $\varkappa$ has the form: $\varkappa = p\,dx = p_1\,dx_1 + p_2\,dx_2$. Then

$$d\varkappa(D_0, D_2) = D_0\,\varkappa(D_2) - D_2\,\varkappa(D_0) - \varkappa([D_0, D_2])\,.$$

Note that $\varkappa(D_0) = 0$, because D_0 goes to zero under the natural projection of Q onto the base M. Moreover, $\varkappa(D_2) = 0$. To see this, we compute $\omega(D_0, D_1)$. On the one hand, this expression is equal to zero, since D_1 lies in the kernel of ω. One the other hand, by using the above formula for the case $d\varkappa(D_0, D_1)$, we obtain

$$d\varkappa(D_0, D_1) = D_0\,\varkappa(D_1) - D_1\,\varkappa(D_0) - \varkappa([D_0, D_1])\,.$$

Here $D_1\,\varkappa(D_0) = 0$, because $\varkappa(D_0) = 0$. The first term is also equal to zero, because $\varkappa(D_1) = \varkappa(\mathrm{sgrad}\,H) = p(\dot x) = |\dot x|^2 = 1/2$ and, consequently, $D_0\,\varkappa(D_1) = D_0(1/2) = 0$. Therefore,

$$\omega(D_0, D_1) = d\varkappa(D_0, D_1) = -\varkappa([D_0, D_1]) = -\varkappa(D_2) = 0\,.$$

Thus,

$$d\varkappa(D_0, D_2) = -\varkappa([D_0, D_2]) = (\text{according to Lemma 15.5}) = \varkappa(D_1) = 1/2\,. \quad \square$$

Just in the same way we prove that $\widehat\omega(\widehat D_0, \widehat D_2) = 1/2$.

We now compute $\xi^*\widehat\omega(D_0, D_2) = \widehat\omega(d\xi(D_0), d\xi(D_2))$. First note that the vector field $d\xi(D_0)$ is proportional to $\widehat D_0$, since Q is mapped onto $\widehat Q$, and ξ is a fiber diffeomorphism. Here we consider Q and $\widehat Q$ as S^1-fiber bundles over M so that D_0 and $\widehat D_0$ are exactly tangent vectors to the S^1-fibers. Therefore, we have

$$d\xi(D_0) = \alpha \widehat D_0\,,$$

where $\alpha(x, p)$ is a certain scalar function. Besides, the vector field $d\xi(D_1)$ is proportional to $\widehat D_1$. This follows from the fact that ξ establishes an orbital isomorphism between the systems v and $\widehat v$. Thus,

$$d\xi(D_1) = \beta \widehat D_1\,,$$

where $\beta(x, y)$ is a certain scalar function.

Further, we have

$$
\begin{aligned}
\widehat\omega(d\xi(D_0), d\xi(D_2)) &= \widehat\omega(d\xi(D_0), d\xi([D_0, D_1])) \\
&= \widehat\omega(d\xi(D_0), [d\xi(D_0), d\xi(D_1)]) = \widehat\omega(\alpha\widehat D_0, [\alpha\widehat D_0, \beta\widehat D_1]) \\
&= \widehat\omega(\alpha\widehat D_0, \alpha\beta[\widehat D_0, \widehat D_1] + \alpha\widehat D_0(\beta)\widehat D_1 - \beta\widehat D_1(\alpha)\widehat D_0) \\
&= \alpha^2\beta\widehat\omega(\widehat D_0, [\widehat D_0, \widehat D_1]) = \alpha^2\beta\widehat\omega(\widehat D_0, \widehat D_2) = \alpha^2\beta/2\,.
\end{aligned}
$$

Thus, the proportion coefficient between ω and $\xi^*\widehat{\omega}$ is equal to $\alpha^2\beta$. In other words, $f = \alpha^2\beta$. It remains to find the coefficients, i.e., the functions α and β.

Lemma 15.6. *The following identity holds:*

$$\beta = |\widehat{G}G^{-1}p|_{\widehat{G}}|p|_G^{-1} \, .$$

Proof. First observe that the formula

$$\xi\colon (x,p) \to (x, |\widehat{G}G^{-1}p|_{\widehat{G}}^{-1}|p|_G \widehat{G}G^{-1}p)$$

can be rewritten in the tangent bundle as follows:

$$\xi(x,\dot{x}) = (x, \lambda\dot{x}) \, ,$$

where $\lambda = |\dot{x}|_G|\dot{x}|_{\widehat{G}}^{-1} = |\widehat{G}G^{-1}p|_{\widehat{G}}^{-1}|p|_G$. In other words, $\widehat{\dot{x}} = \lambda\dot{x}$, $\widehat{x} = x$, and hence $d\xi(v) = \lambda^{-1}\widehat{v}$. Simply speaking, the renormalization of the velocity vector happens. Thus, $\beta = \lambda^{-1}$. $\quad\square$

Lemma 15.7. *The following equality holds:*

$$\alpha = \frac{|p|_G^2}{|\widehat{G}G^{-1}p|_{\widehat{G}}^2} \cdot \frac{\sqrt{\det \widehat{G}}}{\sqrt{\det G}} \, .$$

Proof. Since $\xi\colon Q \to \widehat{Q}$ is a fiber diffeomorphism, its action can be considered on each fiber separately. That is why we may express this as follows:

$$\widehat{p} = \xi(p) = \lambda(p)A(p) \, ,$$

where $\lambda(p) = |p|_G|\widehat{G}G^{-1}p|_{\widehat{G}}^{-1}$ is a scalar function, and $A(p)$ is the linear operator defined by $A(p) = \widehat{G}G^{-1}p$. Consider the mapping ξ not only on the fiber (i.e., on the ellipse $Q \cap T_xM$), but also on the whole tangent plane T_xM. We want to compute the determinant of the differential $d\xi$ in two different ways.

For each tangent vector a, we have

$$d\xi(a) = (d\lambda(a))A(p) + \lambda(p)A(a) \, .$$

It is easy to see that, setting $a = p$, we get $d\lambda(p) = 0$. Therefore, $d\xi(p) = \lambda(p)A(p)$. Consider a basis a, p in the tangent plane T_xM and look how the linear operator $d\xi$ acts on it. It is not difficult to show that this linear operator can be presented in the following form:

$$d\xi = \lambda(p)A + \mu \,(\text{projection onto } A(p) \text{ along } p) \, ,$$

where μ is a certain scalar coefficient. Hence, the second term has no influence on the determinant of $d\xi$, and everything is defined by the first term only. Thus, we obtain

$$\det(d\xi) = \det(\lambda(p)A) = \lambda^2 \det A \, .$$

Now we compute the same determinant $\det(d\xi)$ in another way. In Fig. 15.4 we show two fibers (i.e., two ellipses in the tangent plane) which are connected by the mapping ξ. Consider the basis D_0, p at the point p (see Fig. 15.4).

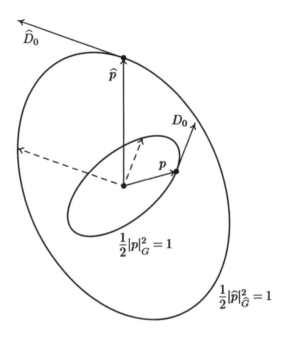

Figure 15.4

The vector D_0 is tangent to the ellipse and has the direction conjugated to p. After the parallel transport to the origin, the vector D_0 becomes on the same ellipse as p. Consider the images of these vectors under $d\xi$. We know that $d\xi(p) = \hat{p}$ and $d\xi(D_0) = \alpha \hat{D}_0$. Hence

$$\det(d\xi) = \frac{\text{the area of the parallelogram spanned on } d\xi(p) \text{ and } d\xi(D_0)}{\text{the area of the parallelogram spanned on } p \text{ and } D_0}$$

$$= \alpha \cdot \frac{\text{the area of the parallelogram spanned on } \hat{p} \text{ and} \hat{D}_0}{\text{the area of the parallelogram spanned on } p \text{ and } D_0}$$

$$= \alpha \cdot \frac{\sqrt{\det \hat{G}}}{\sqrt{\det G}}.$$

By equating the obtained expressions for $\det(d\xi)$, we find α, namely,

$$\lambda^2 \det A = \lambda^2 \det(\hat{G} G^{-1}) = \alpha \frac{\sqrt{\det \hat{G}}}{\sqrt{\det G}}.$$

Thus, $\alpha = \lambda^2 \dfrac{\sqrt{\det \hat{G}}}{\sqrt{\det G}}$. $\square$

By substituting the obtained expressions for the coefficients α and β into $f = \alpha^2 \beta$, we obtain

$$f = \frac{|p|_G^4}{|\widehat{G}G^{-1}p|_{\widehat{G}}^4} \cdot \frac{\det \widehat{G}}{\det G} \cdot \frac{|\widehat{G}G^{-1}p|_{\widehat{G}}}{|p|_G} = \frac{|p|_G^3}{|\widehat{G}G^{-1}p|_{\widehat{G}}^3} \cdot \frac{\det \widehat{G}}{\det G}.$$

This function is not quadratic in momenta, but we can produce from it the desired quadratic integral F. To this end, we need to raise f to the power $(-2/3)$ and multiply by the doubled Hamiltonian of the system v, which has the form $H = \frac{1}{2}|p|_G^2$. As a result, we get

$$F = 2f^{-2/3}H = |\widehat{G}G^{-1}p|_{\widehat{G}}^2 \frac{(\det G)^{2/3}}{(\det \widehat{G})^{2/3}}.$$

Since $\langle G^{-1}\widehat{G}G^{-1}p, p \rangle = |\widehat{G}G^{-1}p|_{\widehat{G}}^2$, we finally obtain the desired formula:

$$F = \frac{(\det G)^{2/3}}{(\det \widehat{G})^{2/3}} \langle G^{-1}\widehat{G}G^{-1}p, p \rangle.$$

The fact that F is non-trivial easily follows from the condition that the metrics G and $\widehat{G}$ are not proportional.

This completes the proof of the first part of the Dini theorem.

We now prove the second part of Theorem 15.13.

In some sense, the above arguments are invertible. Roughly speaking, the desired formula $\widehat{G} = (\det G)^{-2}(\det F)^{-2}GFG$ can be obtained from the relation just obtained

$$F = \frac{(\det G)^{2/3}}{(\det \widehat{G})^{2/3}}G^{-1}\widehat{G}G^{-1}$$

by expressing $\widehat{G}$ from G and F.

However, we can do the same in a different way. Since the geodesic flow of G is quadratically integrable, this metric admits Liouville coordinates x_1, x_2 (Theorem 11.7). In terms of these coordinates, we have

$$G = (f(x_1) + g(x_2))(dx_1^2 + dx_2^2).$$

Moreover, the quadratic integral F has the form

$$F = \frac{(c + g(x_2))p_1^2 + (c - f(x_1))p_2^2}{f(x_1) + g(x_2)}.$$

Then, for the metric $\widehat{G}$, we obtain the following formula:

$$\widehat{G} = (\det G)^{-2}(\det F)^{-2}GFG = \left(\frac{1}{c-f} + \frac{1}{c+g}\right)\left(\frac{dx_1^2}{c-f} + \frac{dx_2^2}{c+g}\right).$$

Thus, we have two Riemannian metrics given explicitly. Moreover, G is a Liouville metric, and $\widehat{G}$ is an almost Liouville metric. In both cases, we know the explicit formulas for the geodesics. By substituting G and $\widehat{G}$ into these formulas, one can easily make sure that the resulting equation for the geodesics just coincide. But this means that the metrics G and $\widehat{G}$ are geodesically equivalent.

This completes the proof of item (b). $\quad\square$

Corollary. *A Riemannian metric (in the two-dimensional case) admits a non-trivial geodesic equivalence if and only if its geodesic flow is quadratically integrable.*

COMMENT. In fact, by the Dini theorem, one usually means the statement that, under the above assumptions, the metric G can be reduced to a Liouville form. This is equivalent to the first part of Theorem 15.13, since the separation of variables follows immediately from the existence of a quadratic integral (see Chapter 6).

Note that the classical Dini theorem is usually understood in local sense. But we have formulated a global result, which is related also to closed two-dimensional surfaces. Its globality consists in the fact that the formula for the quadratic integral F can be written in a canonical way as a whole, that is, on the entire two-dimensional surface M. This leads to several interesting global consequences.

Corollary. *There are no geodesically equivalent metrics on two-dimensional closed surfaces with negative Euler characteristic (that is, different from the sphere, the projective plane, the torus, and the Klein bottle).*

The proof follows from the Dini theorem and the fact that there are no quadratically integrable geodesic flows on such surfaces (see Chapter 11).

Note that, on a two-dimensional surface M (including those of genus $g > 1$), there can exist *local Liouville metrics* with non-integrable geodesic flows. Here, by a local Liouville metric we mean a metric for which there exist local Liouville coordinates in a neighborhood of every point $x \in M$. In particular, in such a neighborhood, its geodesic flow possesses a quadratic integral. However, it cannot be possible, in general, to glue those (local) integrals together into a global quadratic integral. As an example, one can take a constant negative curvature metric on a sphere with handles, which is a local Liouville metric, but its geodesic flow is not (globally) integrable.

The second part of the global Dini theorem makes it possible to construct non-trivial examples of geodesically equivalent metrics on the sphere and torus. Moreover, all such metrics can be classified by using the uniqueness of a quadratic integral F. The only exception is the constant curvature metric on the sphere, for which there exist two independent quadratic integrals. In this case, the following statement holds.

Corollary. *Any Riemannian metric that is geodesically equivalent to a constant curvature metric on the sphere is a constant curvature metric itself.*

REMARK. At first glance it seems that we obtain a possibility to construct many new metrics that are geodesically equivalent to a given metric G (by applying the Dini theorem step by step). However, in fact, the "Dini map" $G \to \widehat{G}$ is, in some sense, an involution. Roughly speaking, after the second iteration, we return to the initial metric G. More precisely, the obtained family of geodesically equivalent metrics of type $\widehat{G}$ has only one parameter, i.e., is homeomorphic to an interval. By choosing an appropriate quadratic integral F, we can jump (by using the "Dini map" $G \to \widehat{G}$) from one point of this interval to any other.

15.6. GENERALIZED DINI–MAUPERTUIS PRINCIPLE

We have already discussed two well-known theorems, which give a method to construct examples of orbitally equivalent Hamiltonian systems. The first theorem is the Maupertuis principle, which states that a natural system restricted to its isoenergy surface is orbitally equivalent to the geodesic flow of some metric, which is explicitly described. The second statement is the Dini theorem, which says that the geodesic flow of a Riemannian metric on a two-dimensional surface possesses a non-trivial quadratic integral if and only if there exists another metric with the same geodesics. Moreover, from each such metric, we can canonically construct a quadratic integral, and vice versa, from each quadratic integral, we can uniquely construct a metric with the same geodesics. Note that the geodesic equivalence implies the orbital equivalence on isoenergy surfaces.

We now formulate the statement, which generalizes and combines these two theorems (in the case of dimension 2).

For a quadratic integral F, we use the notation $F = B - U$, where B is the quadratic (in momenta) part of F, and U is a smooth function on M which does not depend on momenta.

Theorem 15.14. *Consider the natural system with the Hamiltonian $H = K - V$ on a two-dimensional manifold M, where K is the kinetic energy, and $-V$ is the potential (the sign "minus" is taken for convenience). Let F be a quadratic (in momenta) integral of the system having the form $F = B - U$, where B is a positive definite quadratic form (in momenta), and the function U is positive everywhere on M. (These conditions can always be attained by taking an appropriate linear combination with the Hamiltonian H.) Consider the restriction of the initial Hamiltonian system to the invariant three-dimensional level $Q = \{F = 0\}$ of the integral F. Then, on the level Q, this system is smoothly orbitally equivalent to the geodesic flow given by the Hamiltonian*

$$\widetilde{H} = \frac{\det B^2}{\det K^2} \cdot \frac{K B^{-1} K}{U} .$$

Moreover, the orbital equivalence of these two systems takes place in a strong sense, i.e., it commutes with the natural projection onto the base M. This means that the geodesics of the metric and the trajectories of the initial natural system on M just coincide.

COMMENT. Thus, to construct orbitally equivalent systems, we can restrict a Hamiltonian system not only to a level of the Hamiltonian $\{H = \text{const}\}$ (as in the classical Maupertuis principle), but also to a level of any quadratic integral $\{F = \text{const}\}$.

Corollary. *The classical Maupertuis principle is a particular case of Theorem 15.14.*

Proof. Consider the function $F = H - h_0$ as a quadratic integral and apply the formula of Theorem 15.14.   □

Corollary. *The second part of the Dini theorem is a particular case of Theorem 15.14.*

Proof. In the case of geodesic flows, the potential V is absent, so we have $H = K$ and $F = B$. Restricting the geodesic flow to the level $\{F - 1 = 0\}$ and applying Theorem 15.14, we obtain

$$\widetilde{H} = \frac{\det B^2}{\det K^2} \cdot \frac{KB^{-1}K}{1}.$$

The formula obtained coincides with that from the second part of the Dini theorem (see Theorem 15.13 (b)).   □

Proof (of Theorem 15.13). We shall use the existence of Liouville coordinates for quadratically integrable natural systems on two-dimensional surfaces. According to Theorem 15.2, in terms of such coordinates u, v, the Hamiltonian H and the integral F have the following form:

$$H = \frac{p_u^2 + p_v^2}{f + g} + \frac{Z + W}{f + g},$$

$$F = \frac{gp_u^2 - fp_v^2}{f + g} + \frac{gZ - fW}{f + g}.$$

Let us compute $\widetilde{H}$ in terms of u and v. It is clear that

$$B = \frac{gp_u^2 - fp_v^2}{f + g}, \qquad K = \frac{p_u^2 + p_v^2}{f + g}, \qquad U = -\frac{gZ - fW}{f + g}.$$

By substituting this into the formula for $\widetilde{H}$, we obtain

$$\widetilde{H} = \frac{-fp_u^2 + gp_v^2}{Zf^{-1} - Wg^{-1}}.$$

The positive definiteness of F implies $f < 0$ and $Zf^{-1} - Wg^{-1} > 0$. Thus, $\widetilde{H}$ is of the almost Liouville type. In this case, we know the explicit formulas for geodesics (see Remark after Theorem 11.5). The result is as follows:

$$\int \frac{du}{\sqrt{-Z - af}} \pm \int \frac{dv}{\sqrt{-W - ag}} = c.$$

On the other hand, by using Proposition 15.1, we can obtain the explicit expressions for the trajectories of the initial natural system on the level $\{F = 0\}$. We get

$$\int \frac{du}{\sqrt{-Z + hf}} \pm \int \frac{dv}{\sqrt{-W + hg}} = c.$$

In both equations obtained, a and h are parameters. Setting $h = -a$, we can evidently identify the above equations. Therefore, the trajectories of these systems coincide, as required. $\square$

COMMENT. It is an interesting question what a higher-dimensional analog of the generalized Maupertuis–Dini principle is.

15.7. ORBITAL EQUIVALENCE OF THE NEUMANN PROBLEM AND THE JACOBI PROBLEM

A particular case of Theorem 15.14 is the statement on the orbital equivalence of the Neumann problem (the motion of a point on the standard sphere in a quadratic potential) and the Jacobi problem (geodesics on the ellipsoid). This fact has been observed by H. Knörrer [184]. In this section, we obtain this theorem as a direct and simple consequence of the generalized Maupertuis–Dini principle.

Consider the motion of a point on the unit sphere $S^2 = \{x^2 + y^2 + z^2 = 1\}$ in $\mathbb{R}^3 (x, y, z)$ with a quadratic potential (Neumann problem). Let us write this system in the sphere-conical coordinates ν_1, ν_2, ν_3 (see Chapter 8). Then the metric on the standard 2-sphere, being written in the coordinates ν_2, ν_3, takes the following form:

$$ds^2 = \frac{1}{4}(\nu_2 - \nu_3)\left(-\frac{d\nu_2^2}{P(\nu_2)} + \frac{d\nu_3^2}{P(\nu_3)}\right).$$

where $P(\nu) = (a + \nu)(b + \nu)(c + \nu)$. The quadratic potential

$$U = ax^2 + by^2 + cz^2$$

in sphere-conical coordinates becomes

$$U = a + b + c + (\nu_2 + \nu_3).$$

Thus, the Hamiltonian of the Neumann problem is

$$H = 2 \cdot \frac{-P(\nu_2)p_2^2 + P(\nu_3)p_3^2}{\nu_2 - \nu_3} + (a + b + c + \nu_2 + \nu_3).$$

We see that in these coordinates, the variables are separated; so the Neumann system has the quadratic integral

$$F = 2 \cdot \frac{\nu_3 P(\nu_2) p_2^2 - \nu_2 P(\nu_3) p_3^2}{\nu_2 - \nu_3} - \nu_2 \nu_3 \,.$$

This integral satisfies the conditions of Theorem 15.14 (generalized Maupertuis–Dini principle). Therefore, the trajectories of the Neumann problem lying on the level $\{F = 0\}$ coincide with the geodesics of the metric $\widetilde{G}$ related to the following Hamiltonian:

$$\widetilde{H} = \frac{\nu_2 P(\nu_2) p_2^2 - \nu_3 P(\nu_3) p_3^2}{\nu_2 - \nu_3} \,.$$

The metric $\widetilde{G}$ is not a metric on the ellipsoid yet, however, it is geodesically equivalent to the ellipsoid metric. To verify this, we apply the Dini theorem and construct a new (geodesically equivalent) metric $\widehat{G}$ from the metric $\widetilde{G}$. The metric $\widehat{G}$ is defined by the formula

$$\widehat{G} = (\det \widetilde{G})^{-2} (\det \widetilde{F})^{-2} \widetilde{G} \widetilde{F} \widetilde{G} \,.$$

Here the additional integral is taken as follows:

$$\widetilde{F} = \frac{\nu_2 \nu_3}{\nu_2 - \nu_3} \left(-P(\nu_2) p_2^2 + P(\nu_3) p_3^2 \right) \,.$$

A direct calculation yields

$$\widehat{G} = \left(\frac{1}{\nu_3} - \frac{1}{\nu_2} \right) \left(-\frac{d\nu_2^2}{\nu_2^2 P(\nu_2)} + \frac{d\nu_3^2}{\nu_3^2 P(\nu_3)} \right) \,.$$

It turns out that metric $\widehat{G}$ obtained is (isometric to) the metric on the ellipsoid. To see this, one just needs to make the following transformation:

$$\lambda_2 = \nu_3^{-1}, \qquad \lambda_3 = \nu_2^{-1} \,.$$

As a result, the metric is reduced to the form

$$ds^2 = (\lambda_2 - \lambda_3) \left(\frac{\lambda_2 \, d\lambda_2^2}{\widehat{P}(\lambda_2)} - \frac{\lambda_3 \, d\lambda_3^2}{\widehat{P}(\lambda_3)} \right) \,,$$

where

$$\widehat{P}(\lambda) = abc \left(\lambda + \frac{1}{a} \right) \left(\lambda + \frac{1}{b} \right) \left(\lambda + \frac{1}{c} \right) \,.$$

That is actually the metric on the ellipsoid $ax^2 + by^2 + cz^2 = 1$, which is written in elliptic coordinates.

In fact, we have constructed here a smooth mapping ξ from the ellipsoid into the unit sphere under which geodesics on the ellipsoid are sent to trajectories of the Neumann problem on the sphere: a point of the ellipsoid with elliptic coordinates λ_2, λ_3 is mapped to the point on the sphere with sphere-conical coordinates $\nu_2 = \lambda_3^{-1}$, $\nu_3 = \lambda_2^{-1}$. This mapping has a very natural geometric meaning: ξ is just the Gauss map from the ellipsoid to the sphere (i.e., $\xi(x) = \vec{n}(x)$, where $\vec{n}(x)$ is the unit normal to the ellipsoid at the point x).

Theorem 15.15 (Knörrer [184]). *Under the Gauss map of the ellipsoid, its geodesics turn to the trajectories of the Neumann problem on the sphere. In particular, the Neumann problem (the motion of a point on the standard sphere in the quadratic potential) on the zero level surface $\{F = 0\}$ of the quadratic integral F indicated above is orbitally equivalent to the Jacobi problem (the geodesic flow on the triaxial ellipsoid).*

Just in the same way, one can prove the generalization of this theorem obtained by A. P. Veselov [353]. He has shown that the trajectories of the generalized Jacobi problem (the motion of a point on the ellipsoid in the spherical potential; see the next section) are transformed to the trajectories of the Neumann problem under the Gauss map. Moreover, this mapping preserves the foliation of the (co)tangent bundle into three-dimensional level surfaces of the additional quadratic integrals.

15.8. EXPLICIT FORMS OF SOME REMARKABLE HAMILTONIANS AND THEIR INTEGRALS IN SEPARATING VARIABLES

In this section, we present the Hamiltonians and integrals of some well-known integrable systems in the elliptic and sphere-conical coordinates. These formulas can be useful for different kinds of calculations, in particular, for an explicit integration of systems, since the variables are separated in these coordinates.

Recall that by $\lambda_1, \lambda_2, \lambda_3$ and ν_1, ν_2, ν_3 we denote the elliptic coordinates and the sphere-conical coordinates in $\mathbb{R}^3$, respectively. These coordinates are connected with the usual Cartesian coordinates x, y, z in $\mathbb{R}^3$ by the formulas indicated in Chapter 13.

We consider here the following integrable Hamiltonian systems: geodesic flows on the sphere, on the ellipsoid, and on the Poisson sphere; the Neumann system that describes the motion of a point on the standard sphere in the quadratic potential $U = ax^2 + by^2 + cz^2$; the generalized Jacobi problem that describes the motion of a point on the ellipsoid in the spherical potential $U = x^2 + y^2 + z^2$; and finally, the Clebsch case in rigid body dynamics in an ideal fluid (with zero area constant). We give below a list of the Hamiltonians and integrals of these systems.

1. The geodesic flow on the standard sphere:

$$H = \frac{2}{\nu_2 - \nu_3}\left(-P(\nu_2)p_2^2 + P(\nu_3)p_3^2\right),$$

$$F = \frac{2}{\nu_3^{-1} - \nu_2^{-1}}\left(\frac{P(\nu_2)p_2^2}{\nu_2} - \frac{P(\nu_3)p_2^3}{\nu_3}\right).$$

2. The geodesic flow on the triaxial ellipsoid (Jacobi problem):

$$H = \frac{2}{\lambda_2 - \lambda_3}\left(\frac{P(\lambda_2)p_2^2}{\lambda_2} - \frac{P(\lambda_3)p_3^2}{\lambda_3}\right),$$

$$F = \frac{2}{\lambda_3^{-1} - \lambda_2^{-1}}\left(\frac{P(\lambda_2)p_2^2}{\lambda_2^2} - \frac{P(\lambda_3)p_3^2}{\lambda_3^2}\right).$$

3. The geodesic flow on the Poisson sphere (Euler case):

$$H = \frac{2}{\nu_3^{-1} - \nu_2^{-1}}\left(\frac{P(\nu_2)p_2^2}{\nu_2} - \frac{P(\nu_3)p_2^3}{\nu_3}\right),$$

$$F = \frac{2}{\nu_2 - \nu_3}\left(-P(\nu_2)p_2^2 + P(\nu_3)p_3^2\right).$$

4. The motion of a point on the sphere in a quadratic potential (Neumann problem):

$$H = \frac{2}{\nu_2 - \nu_3}\left(-P(\nu_2)p_2^2 + P(\nu_3)p_3^2\right) + \nu_2 + \nu_3,$$

$$F = \frac{2}{\nu_3^{-1} - \nu_2^{-1}}\left(\frac{P(\nu_2)p_2^2}{\nu_2} - \frac{P(\nu_3)p_2^3}{\nu_3}\right) - \nu_2\nu_3.$$

5. The motion of a point on the ellipsoid in the spherical potential (generalized Jacobi problem):

$$H = \frac{2}{\lambda_2 - \lambda_3}\left(\frac{P(\lambda_2)p_2^2}{\lambda_2} - \frac{P(\lambda_3)p_3^2}{\lambda_3}\right) + \lambda_2 + \lambda_3,$$

$$F = \frac{2}{\lambda_3^{-1} - \lambda_2^{-1}}\left(\frac{P(\lambda_2)p_2^2}{\lambda_2^2} - \frac{P(\lambda_3)p_3^2}{\lambda_3^2}\right) + \lambda_2\lambda_3.$$

6. The Clebsch case (with zero area constant):

$$H = \frac{2}{\nu_3^{-1} - \nu_2^{-1}}\left(\frac{P(\nu_2)p_2^2}{\nu_2} - \frac{P(\nu_3)p_2^3}{\nu_3}\right) - \nu_2\nu_3,$$

$$F = \frac{2}{\nu_2 - \nu_3}\left(-P(\nu_2)p_2^2 + P(\nu_3)p_3^2\right) + \nu_2 + \nu_3.$$

In all these formulas, $P(\lambda) = (\lambda + a)(\lambda + b)(\lambda + c)$.

COMMENT. Let us notice the following interesting fact. The Hamiltonian H and the integral F of the geodesic flow on the Poisson sphere (see item 3) are in fact obtained by interchanging the Hamiltonian and integral of the geodesic flow on the standard sphere (see item 1). This fact, however, can be observed without any formulas, since the additional integral in the Euler case is the scalar square of the kinetic momentum, that is, in fact, the Euclidean metric restricted to the sphere. So, we see here some kind of duality between two integrable systems. One of them is the geodesic flow on the sphere, the other is that on the Poisson sphere.

COMMENT. Items 4 and 6 are connected by an analogous duality. In other words, by interchanging the Hamiltonian H and the integral F of the Neumann problem, we obtain the Hamiltonian and the integral of the Clebsch case. Some kind of explanation of this duality consists in the fact that there are not so many integrable cases with simple potentials. It is clear that by interchanging H and F in the Neumann problem, we obtain some integrable natural system on the Poisson sphere with a quite simple potential. So, it is not surprising that this system coincides with the Clebsch case, which can also be thought of as a natural system on the Poisson sphere with a simple potential.

COMMENT. Recall that by the Clebsch case we mean a special case of Kirchhoff equations that describe the motion of a rigid body in an ideal fluid. These equations can naturally be written as a Hamiltonian system on the six-dimensional Lie coalgebra $e(3)^*$ of the orthogonal affine group $E(3)$. However, as is seen above, this system can be restricted to a special four-dimensional submanifold $M_0^4 \subset e(3)^*$, which is a coadjoint orbit given in $e(3)^* = \mathbb{R}^6\,(r, s)$ by

$$r_1^2 + r_2^2 + r_3^2 = 1\,,$$
$$r_1 s_1 + r_2 s_2 + r_3 s_3 = 0\,.$$

As a result, we obtain a natural system on the sphere (more precisely, on the Poisson sphere). Just this system is meant in item 6 (by the name of the "Clebsch case").

COMMENT. Many of the above results remain true for higher-dimensional analogs of the listed integrable systems.

COMMENT. By comparing items 2 and 6, we observe that, after having applied the Maupertuis principle to the Clebsch case, we obtain the geodesic flow on the ellipsoid. This fact itself has already been known; however, here it follows immediately from the explicit formulas for the two problems.

COMMENT. The above list of quadratically integrable Hamiltonians is natural and, in some sense, complete. To explain this idea, consider the following problem: describe the potentials U which give quadratically integrable natural systems on the sphere, ellipsoid, and Poisson sphere. This question has been discussed in a number of papers (see, for example, [38], [100], [109], [197], and [354]). In particular, in a paper by O. I. Bogoyavlenskiĭ [38], such potentials have been also described for n-dimensional case. In Theorem 15.2, we have obtained, in fact, the general form of such potentials. If the metric is reduced to a Liouville form (it is always possible in our case), then the desired potentials in Liouville coordinates u, v become

$$U = \frac{Z(u) + W(v)}{f(u) + g(v)},$$

where f, g, Z, W are some smooth functions. In our case, the potentials U should have the following form:

$$U_{\text{on the sphere}} = \frac{Z(\nu_2) - W(\nu_3)}{\nu_2 - \nu_3},$$

$$U_{\text{on the ellipsoid}} = \frac{Z(\lambda_2) - W(\lambda_3)}{\lambda_2 - \lambda_3},$$

$$U_{\text{on the Poisson sphere}} = \frac{Z(\nu_2) - W(\nu_3)}{\nu_2^{-1} - \nu_3^{-1}}.$$

In these formulas, we only need to require the functions (potentials) U to be smooth. This condition is not trivial, since the elliptic and sphere-conical coordinates have singularities. For example, if Z and W are polynomials, then they should be connected by the relation

$$Z = -W.$$

Let us see what we obtain in the case of simple polynomials W and Z. In the cases of the sphere and the ellipsoid, the first non-trivial potentials U can be obtained if Z and W are polynomials of degree 2. Then the corresponding potentials take the form

$$U_{\text{on the sphere}} = \frac{\nu_2^2 - \nu_3^2}{\nu_2 - \nu_3} = \nu_2 + \nu_3,$$

$$U_{\text{on the ellipsoid}} = \frac{\lambda_2^2 - \lambda_3^2}{\lambda_2 - \lambda_3} = \lambda_2 + \lambda_3.$$

As a result, we just obtain the Neumann problem and the generalized Jacobi problem (see items 4 and 5).

Of course, one can take any polynomials (not necessarily quadratic ones). Then the potential U becomes

$$U = \frac{Q(\nu_2) - Q(\nu_3)}{\nu_2 - \nu_3}.$$

It is easy to see that, in this case, U is a symmetric polynomial in ν_2 and ν_3. Therefore, U can be expressed as a polynomial in $\nu_2 + \nu_3$ and $\nu_2 \nu_3$. The function H obtained will be smooth on the sphere, since

$$\nu_2 + \nu_3 = ax^2 + by^2 + cz^2 - (a + b + c),$$

$$\nu_2 \nu_3 = abc\left(\frac{x^2}{a} + \frac{y^2}{b} + \frac{z^2}{c}\right).$$

A disadvantage of such potentials (we mean $\deg Z, \deg W > 2$) is that, in Cartesian coordinates, they become polynomials of degree at least 4. Therefore, in this construction, the Neumann problem is the simplest case, i.e., the first in this series of potentials.

In the case of the ellipsoid, the situation is similar. We only need to change the sphere-conical coordinates by elliptic ones in our arguments.

The functions Z and W need not to be polynomials. It is possible, for example, in the case of the ellipsoid, to put $Z = \lambda^{-1}$, $W = -\lambda^{-1}$. Then we obtain the following integrable potential:

$$U = \frac{\lambda_2^{-1} - \lambda_3^{-1}}{\lambda_2 - \lambda_3} = -\frac{1}{\lambda_2 \lambda_3} ;$$

this can be rewritten in Cartesian coordinates as

$$U = \frac{1}{abc\left(\frac{x^2}{a^2} + \frac{y^2}{b^2} + \frac{z^2}{c^2}\right)} .$$

Another natural family of integrable potentials can be obtained by setting $Z = \frac{\text{const}}{a + \lambda}$, $W = -\frac{\text{const}}{a + \lambda}$. Substituting this into the expression for the potential, we obtain

$$U = \text{const} \, \frac{\frac{1}{\lambda_2 + a} - \frac{1}{\lambda_3 + a}}{\lambda_2 - \lambda_3} = -\frac{\text{const}}{(\lambda_2 + a)(\lambda_3 + a)} = -\frac{a \cdot \text{const}}{(a - b)(a - c)} \cdot \frac{1}{x^2} .$$

Since const denotes an arbitrary constant, after renaming, we obtain the integrable potential $U = \frac{\alpha}{x^2}$. Clearly, we can proceed in the same way for the other coordinates y and z. As a result, we get the family of integrable potentials on the ellipsoid described by V. V. Kozlov in [197]:

$$U = \frac{\alpha}{x^2} + \frac{\beta}{y^2} + \frac{\gamma}{z^2} .$$

One should, however, remark that this potential has a singularity on the ellipsoid.

To obtain the case of the sphere, it is sufficient to replace λ by ν. For example, one can obtain the following potential:

$$U = \frac{\nu_2^{-1} - \nu_3^{-1}}{\nu_2 - \nu_3} = \frac{1}{-\nu_2 \nu_3} ,$$

or in Cartesian coordinates x, y, z,

$$U = \frac{1}{abc\left(\frac{x^2}{a} + \frac{y^2}{b} + \frac{z^2}{c}\right)} .$$

On the Poisson sphere the potential U has the form

$$U_{\text{on the Poisson sphere}} = \frac{Z(\nu_2) + W(\nu_3)}{\nu_2^{-1} - \nu_3^{-1}}.$$

The simplest case is $Z(\nu) = -W(\nu) = \nu$. This leads us to the potential

$$U = \frac{\nu_2 - \nu_3}{\nu_2^{-1} - \nu_3^{-1}} = -\nu_2\nu_3$$

which is just the Clebsch case. As above, Z and W are not necessary polynomials. The only condition is that the corresponding potential U should be smooth. So, the above list (items 4, 5, and 6) contains just the simplest integrable potentials.

COMMENT. By introducing the potentials like $\lambda_2^{-1}\lambda_3^{-1}$, we observe some new isomorphisms. Indeed, by adding the potential $-\dfrac{1}{\lambda_2\lambda_3}$ to the Hamiltonian H in the Jacobi problem, we obtain a new Hamiltonian

$$H = \frac{2}{\lambda_2 - \lambda_3}\left(\frac{P(\lambda_2)p_2^2}{\lambda_2} - \frac{P(\lambda_3)p_3^2}{\lambda_3}\right) - \frac{1}{\lambda_2\lambda_3}.$$

The corresponding system describes the motion of a point on the ellipsoid in the potential

$$U = \frac{1}{abc\left(\frac{x^2}{a^2} + \frac{y^2}{b^2} + \frac{z^2}{c^2}\right)}.$$

Now, by applying the Maupertuis principle, it is easy to verify that this system is equivalent to the Euler case (see item 3).

Corollary. *The Euler case (on the level $\{H = 0\}$) is orbitally equivalent to the natural system that describes the motion of a point on the ellipsoid*

$$\frac{x^2}{a} + \frac{y^2}{b} + \frac{z^2}{c} = 1$$

in the potential

$$U = \frac{-1}{abc\left(\frac{x^2}{a^2} + \frac{y^2}{b^2} + \frac{z^2}{c^2}\right)}.$$

Chapter 16

Euler Case in Rigid Body Dynamics and Jacobi Problem about Geodesics on the Ellipsoid. Orbital Isomorphism

16.1. INTRODUCTION

We have already discussed the theory of the classification of integrable Hamiltonian systems with two degrees of freedom up to homeo- and diffeomorphisms preserving trajectories. The main idea of this theory can be briefly formulated in the following way. Consider two integrable Hamiltonian systems with two degrees of freedom v_1 and v_2 restricted to their regular compact isoenergy submanifolds Q_1 and Q_2. It is assumed that these systems satisfy some natural conditions. We shall not list them here, referring the reader to Chapters 3–8, and pointing out that most known integrable Hamiltonian systems satisfy these conditions. In [46], [53] we have described complete sets of invariants which allow one to compare the dynamical systems (v_1, Q_1) and (v_2, Q_2) up to orbital equivalence, i.e., to answer the question whether there exists a homeomorphism (diffeomorphism) $\xi \colon Q_1 \to Q_2$ sending trajectories of the first system to those of the second one.

After having constructed the general classification theory, we may ask ourselves whether and how the orbital invariants of integrable Hamiltonian systems can be calculated in specific problems. Shall this theory really work, if we actually want to compare two concrete systems and find out if they are equivalent? In this chapter, we would like to demonstrate that the answer to this question is positive. We discuss here the results obtained in [55], [58], [59]. The calculation of orbital invariants for some concrete integrable systems are also in [270], [275], [312].

Note that there exist also other methods which allow one to discover isomorphisms between different integrable systems. See, for example, [1], [2], [35], [49], [184], [193], [353]. Besides, in the previous chapter, we also discuss some methods for constructing orbitally equivalent integrable systems on the basis of the Maupertuis principle and its generalizations. However, in this chapter, we approach the problem of orbital isomorphisms in a quite different way, using the theory of invariants of integrable systems.

We consider here two famous integrable Hamiltonian systems: the Jacobi problem about geodesics on the ellipsoid [167] and the integrable Euler case in rigid body dynamics [111], [112]. In this chapter, we prove the existence of an orbital *homeomorphism* between them.

Then, using smooth orbital invariants, we show that from the smooth point of view the Euler case and the Jacobi problem are not orbitally equivalent [51]. It turns out there exists a smooth invariant which is different for the two systems.

Before passing to exact formulations, we recall briefly the nature of orbital invariants, using these two classical problems as a model example.

16.2. JACOBI PROBLEM AND EULER CASE

Consider an ellipsoid X in the three-dimensional Euclidean space given by

$$\frac{x^2}{a} + \frac{y^2}{b} + \frac{z^2}{c} = 1 \,,$$

where $a < b < c$.

The geodesic flow on the ellipsoid is a Hamiltonian system on the cotangent bundle T^*X with the standard symplectic structure. The Hamiltonian of this system is

$$H(q,p) = \frac{1}{2} \sum g^{ij}(q) p_i p_j = \frac{1}{2} |p|^2 \,,$$

where $g_{ij}(q)$ is the induced Riemannian metric on the ellipsoid X, and $(q,p) \in T^*X$, $q \in X$, $p \in T_q^*X$. The isoenergy surface $Q^3 = \{2H = |p|^2 = 1\}$ in this case is a S^1-fibration over X (unit covector bundle). The geodesic flow on the ellipsoid admits an additional integral

$$f_J = abc \left(\frac{x^2}{a^2} + \frac{y^2}{b^2} + \frac{z^2}{c^2} \right) \left(\frac{\dot{x}^2}{a} + \frac{\dot{y}^2}{b} + \frac{\dot{z}^2}{c} \right) .$$

Here $(\dot{x}, \dot{y}, \dot{z})$ is the tangent vector to a geodesic (we identify tangent and cotangent vectors in the usual way).

The second system (Euler case) is given by the standard Euler–Poisson equations and describes the motion of a rigid body fixed at its center of mass:

$$\frac{dK}{dt} = [K, \Omega] \,,$$

$$\frac{d\gamma}{dt} = [\gamma, \Omega] \,.$$

Here vector $K = (s_1, s_2, s_3)$ is the kinetic momentum vector of the body, $\Omega = (As_1, Bs_2, Cs_3)$ is its angular velocity vector, $\gamma = (r_1, r_2, r_3)$ is the unit vertical vector (the coordinates of these vectors are written in the orthonormal basis which is fixed in the body and whose axes coincide with the principal axes of inertia). The parameters A, B, C of the problem are the inverses of the principal moments of inertia of the rigid body. We suppose they are all different, and $A < B < C$.

It is well known (see Chapter 14) that this system of differential equations is Hamiltonian in the six-dimensional space $\mathbb{R}^6(s_1, s_2, s_3, r_1, r_2, r_3)$ considered as the dual space of the Lie algebra $e(3) = so(3) + \mathbb{R}^3$, where $s_i \in so(3)$, and $r_i \in \mathbb{R}^3$. Recall that the Poisson structure is given here by the following formulas:

$$\{s_i, s_j\} = \varepsilon_{ijk} s_k , \quad \{s_i, r_j\} = \varepsilon_{ijk} r_k , \quad \{r_i, r_j\} = 0 .$$

The Hamiltonian of the system is

$$H = \frac{1}{2}(A s_1^2 + B s_2^2 + C s_3^2) .$$

Recall that the system of Euler–Poisson equations always has two additional integrals (the Casimir functions of the Poisson structure):

$$f_0 = |\gamma|^2 = r_1^2 + r_2^2 + r_3^2 ,$$
$$g = (K, \gamma) = s_1 r_1 + s_2 r_2 + s_3 r_3 .$$

Consider the four-dimensional invariant submanifold $M^4 = \{f_0 = 1, g = 0\}$ and restrict the system under consideration on it. The Poisson bracket defines a symplectic structure ω on M^4. It is not hard to check (see Chapter 15) that the symplectic manifold (M^4, ω) obtained is symplectomorphic to the cotangent bundle of the two-dimensional sphere.

Thus, under the above assumptions the Euler-Poisson equations can be viewed as a Hamiltonian system with two degrees of freedom on T^*S^2. This system is Liouville integrable by means of the additional integral

$$f_E = s_1^2 + s_2^2 + s_3^2 .$$

The isoenergy surface $Q^3 = \{2H = 1\}$ in the Euler case has the same topological structure as the one in the Jacobi problem, being diffeomorphic to the unit (co)vector bundle over the sphere.

As a result, both the systems (Jacobi problem and Euler case) can be considered as Hamiltonian systems on the cotangent bundle of the sphere. Moreover, the Euler case (under above assumptions) can be thought of as the geodesic flow of a special metric on the sphere (see Chapter 15). The sphere with this metric is usually called the Poisson sphere. The Poisson sphere and the ellipsoid are not isometric.

By $v_J(a, b, c)$ and $v_E(A, B, C)$ we denote the restrictions of the Jacobi and Euler systems to their isoenergy surfaces $Q_J = \{2H_J = 1\}$ and $Q_E = \{2H_E = 1\}$ respectively, where H_J and H_E are the Hamiltonians of the Jacobi problem and the Euler case indicated above.

REMARK. Note that, due to the homogeneity of H_J and H_E, the orbital structure of the systems does not depend on the choice of energy level. In other words, when the energy level changes, each system remains orbitally equivalent to the initial one.

Thus, we have two dynamical systems $v_J(a, b, c)$ and $v_E(A, B, C)$ given on diffeomorphic isoenergy three-dimensional manifolds. We want to find out whether these systems are similar in some sense. In particular, are they orbitally equivalent? If yes, then topologically or smoothly?

16.3. LIOUVILLE FOLIATIONS

Following the general scheme of the orbital classification theory for integrable Hamiltonian systems, if we want to find out whether two given systems are equivalent, then we should begin with studying and comparing their Liouville foliations.

Every integrable Hamiltonian system defines the structure of a foliation with singularities on the symplectic manifold (as well as on the isoenergy surface). If H is the Hamiltonian of the system, and f is its first integral independent of H almost everywhere, then the Liouville foliation is defined as the decomposition of the manifold into connected components of common level surfaces of f and H. In the general case, a Liouville foliation may depend on the choice of an additional integral. However, for non-resonant systems (under some additional assumptions like non-degeneracy) the foliation is uniquely defined, since almost all leaves can be characterized as the closures of trajectories. Besides, the Liouville foliation can be defined independently of an additional integral. It can be done, for example, as follows. We say that points $x, y \in M$ are equivalent if $f(x) = f(y)$ for any smooth additional integral f of the Hamiltonian system. It is clear that it is really an equivalence relation. Then, by definition, the Liouville foliation on M is said to be its decomposition into the equivalence classes (a leaf is an equivalence class).

Theorem 16.1. *The Liouville foliation related to the Hamiltonian systems* $v_J(a, b, c)$ *and* $v_E(A, B, C)$ *on isoenergy surfaces are diffeomorphic. In other words, the Jacobi problem and the Euler case are Liouville equivalent.*

Proof. To prove this statement one may just compute the marked molecule of the two systems. It has been done above. In both cases the molecules are similar and have the form shown in Fig. 12.35(a). Another method of proof is to construct an explicit deformation of one system into the other, which does not change the topology of the foliation. Such a deformation has been indicated in Section 15.3.3.

We confine ourselves to the explicit description of the Liouville foliation structure by constructing a quite simple model. Consider the sphere $S^2 = \{x^2 + y^2 + z^2 = 1\}$ and the unit covector bundle over it $Q^3 \xrightarrow{S^1} S^2$ (Hopf fibration). Consider a smooth function $h(x, y, z) = \alpha x^2 + \beta y^2 + \gamma z^2$ (where $\alpha < \beta < \gamma$) on the sphere and lift it to Q^3 in the natural way by assuming it to be constant on each leaf. The function h foliates Q^3 into its level surfaces. Denote this foliation by $\mathcal{L}_h$. Notice that the topology of this foliation obviously does not depend on the choice of a metric on the sphere and on the choice of α, β, γ.

It is easy to see that the foliation consists of four one-parameter families of Liouville tori, four circles (into which the tori from the families are shrunk), and one singular leaf of type $K \times S^1$, where K is a graph consisting of two circles intersecting transversally at two points.

We assert that the Liouville foliations of systems $v_J(a,b,c)$ and $v_E(A,B,C)$ are isomorphic to $\mathcal{L}_h$. Let us show this.

First consider the Euler case. Here the foliation is given by the function (additional integral)

$$f_E = s_1^2 + s_2^2 + s_3^2$$

on the common level surface of three functions

$$f_0 = r_1^2 + r_2^2 + r_3^2 = 1\,,$$
$$g = s_1 r_1 + s_2 r_2 + s_3 r_3 = 0\,,$$
$$2H = A s_1^2 + B s_2^2 + C s_3^2 = 1\,.$$

The first two of these functions determine the manifold M_0^4 diffeomorphic to the (co)tangent bundle of the sphere, the third one (Hamiltonian) selects the set of unit (co)vectors in the cotangent bundle. Here $\gamma = (r_1, r_2, r_3)$ is a point on the sphere, $K = (s_1, s_2, s_3)$ is a (co)tangent vector at this point.

We now make the following simple transformations. The idea is to interchange "coordinates" and "momenta". Put

$$x = \sqrt{A}\, s_1\,, \quad y = \sqrt{B}\, s_2\,, \quad z = \sqrt{C}\, s_3\,,$$
$$p_x = \frac{r_1}{\sqrt{A}}\,, \quad p_y = \frac{r_2}{\sqrt{B}}\,, \quad p_z = \frac{r_3}{\sqrt{C}}\,.$$

After this change the functions become

$$2H = x^2 + y^2 + z^2 = 1\,,$$
$$g = x p_x + y p_y + z p_z = 0\,,$$
$$f_0 = A p_x^2 + B p_y^2 + C p_z^2 = 1\,.$$

Thus, we can interpret the same isoenergy surface in another way, considering it as the unit covector bundle over the sphere $\{2H = x^2 + y^2 + z^2 = 1\}$ (but not over the sphere $\{f_0 = 1\}$ as before). As a result, the additional integral can be written as a function on the base

$$f_E = \frac{x^2}{A} + \frac{y^2}{B} + \frac{z^2}{C}\,,$$

which immediately leads us to the above model $\mathcal{L}_h$.

The similar construction can be carried out for the Jacobi problem, but we shall proceed in a different way. As we remarked above, both systems can be viewed as geodesic flows on the sphere. We have shown in Chapter 15, that the corresponding metrics admit a very simple deformation one to the other in the class of metrics with integrable geodesic flows. Moreover, under this deformation the Liouville foliation will be changed without bifurcations. As a result, we obtain a smooth isotopy between the Liouville foliations related to our systems $v_J(a, b, c)$ and $v_E(A, B, C)$. □

Thus, we have shown that the systems in question have the same Liouville foliation and described this foliation by means of a model example. It is clear that this condition is necessary for the orbital equivalence of the systems. However, we can say nothing yet about the behavior of trajectories on Liouville tori (i.e., on the leaves of the Liouville foliation). The next step is to examine orbital invariants of the systems.

16.4. ROTATION FUNCTIONS

The main orbital invariant of a system is the rotation function. Recall that the rotation number of a Hamiltonian system on a Liouville torus is defined in the following way. According to the Liouville theorem, on the torus there exists a coordinate system (φ_1, φ_2) in which the Hamiltonian system straightens and takes the form

$$\frac{d\varphi_1}{dt} = \omega_1 ,$$
$$\frac{d\varphi_2}{dt} = \omega_2 .$$

The rotation number on the given torus is defined to be the ratio $\rho = \omega_1/\omega_2$. Clearly, this number depends on the Liouville torus. On the isoenergy surface these tori form one-parameter families, whose parameter is, for example, an additional integral f. As a result, on each one-parameter family of tori the rotation function $\rho(f)$ appears.

In essence, the rotation function is the main orbital invariant of a system. However, we should pay attention to the fact that the rotation function depends, firstly, on the choice of basis on a Liouville torus and, secondly, on the choice of an additional integral of the system.

The dependence on the choice of basis can be easily avoided, if for the systems to be compared we choose the basis cycles on Liouville tori in a common way (we use here the fact that the foliations are isomorphic). For example, in our case for the Liouville foliation (see the model example above) we can suggest the following simple rule for choosing basis cycles.

Recall that the isoenergy surface has the structure of a S^1-fibration, each of whose fibers lies on some Liouville torus. It is easy to see that this S^1-fibration is uniquely defined up to isotopy. Therefore, as the first *uniquely defined* basic cycle, we can take the fiber λ of this fibration.

Next, consider an arbitrary one-parameter family of Liouville tori. These tori are contracted into a singular fiber (stable periodic trajectory). Therefore, on each of the tori we can uniquely define a disappearing cycle μ, which shrinks into a point as the tori tend to the stable periodic trajectory. We take μ as the second basis cycle.

Thus, we have defined bases on Liouville tori in a canonical way. Now, after having fixed the bases, we can write down the explicit formulas for the rotation functions.

The parameter t on a family of Liouville tori in the Jacobi problem is taken as the value of the integral f_J, and in the Euler case the analogous parameter τ is taken as the inverse value of the additional integral f_E, i.e., $1/f_E$. It is not hard to check that the parameters t and τ vary in the segments $[a, b]$ and $[A, B]$ respectively. The values a, b, c and A, B, C are bifurcational: at these points the bifurcations of Liouville tori happen. Any other value from the above segments corresponds to a pair of regular Liouville tori on which the rotation numbers coincide by symmetry argument.

We shall use the formulas for the rotation functions from Chapter 8. In fact, there we have calculated them in a different basis on Liouville tori. In order to obtain the expressions in the basis (λ, μ) just described, we need to find the transition matrix between the old basis and the new one. The old basis was related to the elliptic coordinates λ_2, λ_3 on the ellipsoid and had the form $\{\lambda_2 = \text{const}\}$ and $\{\lambda_3 = \text{const}\}$. It is easy to check that for $t \in (b, c)$ the old and new basis cycles are connected by relations

$$\lambda = \{\lambda_2 = \text{const}\} + \{\lambda_3 = \text{const}\},$$
$$\mu = \{\lambda_2 = \text{const}\}.$$

For $t \in (a, b)$ the analogous relations hold, but λ_2 and λ_3 have to be interchanged:

$$\lambda = \{\lambda_2 = \text{const}\} + \{\lambda_3 = \text{const}\},$$
$$\mu = \{\lambda_3 = \text{const}\}.$$

Using these relations and formulas for the rotation function in the old basis (Section 13.3.1), we obtain the following expressions for it in the new basis (λ, μ).

Proposition 16.1. *The rotation function in the Jacobi problem, written in the basis (λ, μ), has the following form:*
1) *for $a < t < b$*

$$\rho_J(t) = \frac{\int\limits_{-t}^{-a} \Phi(u, t)\, du}{\int\limits_{0}^{+\infty} \Phi(u, t)\, du},$$

2) *for $b < t < c$*

$$\rho_J(t) = -\frac{\int\limits_{-c}^{-t} \Phi(u, t)\, du}{\int\limits_{0}^{+\infty} \Phi(u, t)\, du},$$

where

$$\Phi(u, t) = \sqrt{\frac{u}{(u + a)(u + b)(u + c)(u + t)}}.$$

COMMENT. To prove these formulas one should use the following well-known relations in the theory of hyperelliptic integrals:

$$\int_{-c}^{-t} \Phi(u,t)\, du = \int_{-b}^{-a} \Phi(u,t)\, du + \int_{0}^{+\infty} \Phi(u,t)\, du \qquad \text{if } t \in (b,c),$$

$$\int_{-c}^{-b} \Phi(u,t)\, du = \int_{-t}^{-a} \Phi(u,t)\, du + \int_{0}^{+\infty} \Phi(u,t)\, du \qquad \text{if } t \in (a,b).$$

Analogous expressions for the rotation functions can be written in the Euler case. See also [18], [194], and [307]. All the arguments and transition formulas from the old basis to the new one are quite similar.

Proposition 16.2. *The rotation function in the Euler case, written in the basis (λ, μ), has the following form:*

1) *for $A < \tau < B$*

$$\rho_E(\tau) = \frac{1}{\pi} \int_{A}^{\tau} \Psi(u,\tau)\, du,$$

2) *for $B < \tau < C$*

$$\rho_E(\tau) = -\frac{1}{\pi} \int_{\tau}^{C} \Psi(u,\tau)\, du,$$

where

$$\Psi(u,\tau) = \frac{u}{\sqrt{(u-\tau)(C-u)(B-u)(A-u)}}\,.$$

Thus, we have described explicit formulas for the rotation functions in the corresponding bases. As we see, they do not coincide. But we should not expect the coincidence, because the parameters t and τ on the one-parameter families of Liouville tori were chosen independently of each other. The necessary condition for the orbital equivalence of the system is, of course, not the coincidence of the rotation functions, but their conjugacy. In other words, there must exist a monotone (strictly increasing) change $t = t(\tau)$ such that $\rho_J(t(\tau)) = \rho_E(\tau)$ on every of the four one-parameter families of tori. If we are interested in topological orbital equivalence, then the change must be a continuous map from $[A, C]$ to $[a, c]$. In the smooth case this change must be smooth.

It turns out that in the continuous case such a change exists for appropriate parameters (A, B, C) and (a, b, c).

To check this, it is sufficient to examine the limits and monotonicity of these functions.

Let us introduce the following notation

$$k(a, b, c) = \lim_{t \to a} \rho_J(t),$$

$$l(a, b, c) = \lim_{t \to c} \rho_J(t),$$

$$K(A, B, C) = \lim_{t \to A} \rho_E(\tau),$$

$$L(A, B, C) = \lim_{t \to C} \rho_E(\tau).$$

It is easy to check that the explicit formulas for these limits are

$$k(a,b,c) = \frac{\pi\sqrt{\frac{a}{(b-a)(c-a)}}}{\int\limits_{0}^{\infty}\Phi(u,a)\,du}\,,$$

$$l(a,b,c) = -\frac{\pi\sqrt{\frac{c}{(c-a)(c-b)}}}{\int\limits_{0}^{\infty}\Phi(u,c)\,du}\,,$$

$$K(A,B,C) = \frac{A}{\sqrt{(B-A)(C-A)}}\,,$$

$$L(A,B,C) = -\frac{C}{\sqrt{(C-A)(C-B)}}\,.$$

The following statement describes the qualitative behaviour of the rotation functions $\rho_E(\tau)$ and $\rho_J(t)$.

Proposition 16.3.
1) *The function $\rho_E(\tau)$ strictly increases on the intervals (A,B) and (B,C).*
2) *The function $\rho_J(t)$ strictly increases on the intervals (a,b) and (b,c).*
3) $\lim\limits_{\tau\to B\pm 0}\rho_E(\tau) = \lim\limits_{t\to b\pm 0}\rho_J(t) = \mp\infty.$

The first statement can be easily checked by straightforward calculation of the rotation function. The second one is a little more complicated, but the idea of the proof is the same: to differentiate the rotation function and just to check that the derivative is positive (see [59] for details). The third statement follows easily from the explicit formulas for the rotation functions. Let us notice, however, that tending of ρ to infinity as a torus approaches a saddle atom is a general fact (provided the basis is chosen correctly).

Thus, the qualitative behavior of the rotation functions in the Jacobi problem and Euler case is quite similar. It is clear that for monotone functions the only condition for continuous conjugacy is the coincidence of their limits at the ends of the intervals they are defined on. Therefore, the necessary and sufficient condition for continuous conjugacy of the two rotation functions ρ_J and ρ_E are two equalities

$$l(a,b,c) = L(A,B,C)\,,$$
$$k(a,b,c) = K(A,B,C)\,.$$

It follows from the general theory of the topological orbital classification of integrable systems that, in addition to the above invariants (i.e., two limits of the rotation functions), in the given case there is only one more invariant, namely, the Λ-invariant. Recall that, for each saddle bifurcation, Λ is defined to be the set of multipliers of hyperbolic periodic trajectories lying on the singular leaf. These multipliers are considered up to proportionality: $(\lambda_1 : \lambda_2 : \ldots : \lambda_n)$. However, in both cases (Euler and Jacobi) this invariant is trivial, that is, $\Lambda = (1 : 1)$. The point is that in the molecule W there is the only saddle atom C_2 (in other

words, the only saddle bifurcation).   Besides, both systems admit a natural
symmetry that interchanges the vertices of C_2, and therefore, the multipliers related
to these vertices coincide. This means that $\Lambda = (\lambda : \lambda) = (1 : 1)$.

In fact, speaking about the vertices of the atom C_2, we just mean the closed
geodesic γ of hyperbolic type (such a geodesic exists and is unique), which we should
consider as two different geodesics $\gamma(t)$ and $\gamma(-t)$.

In fact, we have calculated the complete orbital invariant for each of the systems,
i.e., the t-molecule. Recall that the t-molecule is obtained from the usual marked
molecule W^* by adding the rotation vectors on all of its edges and the Λ-invariant
on atoms.   There are no other (topological) orbital invariants for the Euler and
Jacobi systems.

Theorem 16.2. *The explicit form of the t-molecules for the Euler and Jacobi
systems is presented in Fig. 16.1.*

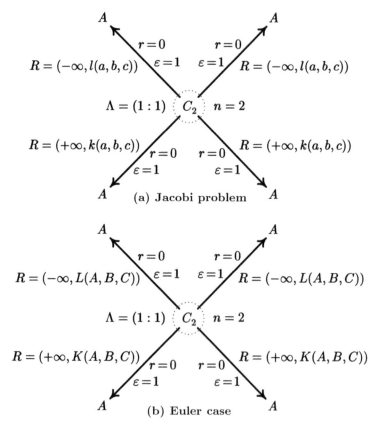

(a) Jacobi problem

(b) Euler case

Figure 16.1

Thus, to prove the equivalence of the Euler case and Jacobi problem we only
need to choose triples of parameters (a, b, c) and (A, B, C) for which the above
equalities hold.

16.5. THE MAIN THEOREM

Let us formulate the general idea that makes it possible to establish the existence of isomorphisms between different classes of systems by means of the theory of invariant, and then apply it in the case considered. Let us be given two classes of systems $\{v\}$ and $\{v'\}$, each of which depends on some parameters $(x_1, x_2, \ldots x_k) \in \mathcal{U}$ and $(X_1, X_2, \ldots, X_l) \in \mathcal{U}'$. Suppose that the systems considered have the same topological type of Liouville foliation, so we need to compare only a finite number of invariants which depend continuously on the parameters. Denote the space of these invariants by $\mathcal{I}$ (i.e., the space in which they take their values). As a result, we obtain two mappings $\phi : \mathcal{U} \to \mathcal{I}$ and $\phi' : \mathcal{U}' \to \mathcal{I}$, which send every set of parameters to the set of invariants for the corresponding dynamical system. In the space $\mathcal{I}$ two subsets (two "surfaces") $\phi(\mathcal{U})$ and $\phi'(\mathcal{U}')$ appear. The points of their intersections correspond to the pairs of equivalent systems. Using this natural and visual construction, one can also recognize what values of parameters and invariants correspond to equivalent systems, and conversely, which systems are not equivalent.

We now turn to the Euler case and Jacobi problem. Consider two mappings

$$\xi : (a, b, c) \to (k(a, b, c), l(a, b, c)) \in \mathbb{R}^2 ,$$
$$\Xi : (A, B, C) \to (K(A, B, C), L(A, B, C)) \in \mathbb{R}^2 ,$$

which assign the pair of orbital invariants to each ellipsoid and to each rigid body. The properties of these mappings and their images in the "space of invariants" turned out to be absolutely identical for the classes of systems $\{v_J(a, b, c)\}$ and $\{v_E(A, B, C)\}$.

Proposition 16.4 [58], [271].

a) $\xi(a, b, c) = \xi(a', b', c')$ if and only if the triples of parameters (a, b, c) and (a', b', c') are proportional, i.e., the corresponding ellipsoids are similar.

b) $\Xi(A, B, C) = \Xi(A', B', C')$ if and only if the triples of parameters (A, B, C) and (A', B', C') are proportional, i.e., the inertia ellipsoids of the corresponding rigid bodies are similar.

c) The images of ξ and Ξ coincide and have the following form on the two-dimensional plane $\mathbb{R}^2(x, y)$:

$$\{x > 0, \ y < -1\}.$$

This statement implies immediately the following main theorem.

Theorem 16.3. The Jacobi problem (geodesic flow on the ellipsoid) and the Euler case (in rigid body dynamics) are topologically orbitally equivalent in the following exact sense. For any rigid body there exists an ellipsoid (and vice versa, for any ellipsoid there exists "a rigid body") such that the corresponding systems $v_J(a, b, c)$ and $v_E(A, B, C)$ are topologically orbitally equivalent. The parameters a, b, c and A, B, C related to equivalent systems are uniquely defined up to proportionality.

REMARK. We put "a rigid body" in quotation marks, because in our theorem the parameters A, B, C are assumed to be arbitrary, whereas for a real rigid body they must satisfy the triangle inequality.

16.6. SMOOTH INVARIANTS

In conclusion, we discuss the question about the smooth orbital equivalence between the Jacobi problem and the Euler case. To answer it, we can use the smooth orbital classification theory for integrable Hamiltonian systems with two degrees of freedom. According to Chapter 8, the smooth orbital invariants are some power series (this is connected with the necessity to sew the derivatives of all orders). Since we are going, in fact, to prove the non-equivalence of the systems in the smooth sense, it is natural to begin with examination of first terms of these series to find at least one invariant which distinguishes them.

To that end it is sufficient to consider the rotation function again. This time we shall examine their asymptotic behavior as $t \to b$ and $\tau \to B$, i.e., as a Liouville torus approaches the saddle critical level of the additional integral. In our case, this level, from the topological point of view, is the direct product $K \times S^1$, where S^1 is the circle, and K is the planar graph presented as two circles intersecting at two points. This singular level contains two hyperbolic trajectories $\{x_1\} \times S^1$ and $\{x_2\} \times S^1$, where x_1 and x_2 are the vertices of K.

We begin with the following general remark. Let $\rho(t)$ be the rotation function of some integrable Hamiltonian system related to a fixed pair of basis cycles (λ, μ) on a one-parameter family $T^2(t)$ of Liouville tori. Here t is a parameter of the family. Let t_0 be a bifurcational value of t, i.e., the tori $T^2(t)$ tend to a singular leaf as $t \to t_0$. Suppose that this leaf contains a closed hyperbolic trajectory of the system, and consider the asymptotics of $\rho(t)$ as $t \to t_0$.

Proposition 16.5. *Let the first basis cycle λ be isotopic to the hyperbolic trajectory lying on the singular leaf. Then, as $t \to t_0$, we have*

$$\rho(t) = \Lambda \ln |t - t_0| + q(t),$$

where $q(t)$ is a function continuous at t_0. The coefficient Λ is the sum of the inverse values of the multipliers of all hyperbolic trajectories lying on the singular leaf and belonging to the closure of the family of tori.

It follows from this statement that Λ is a smooth orbital invariant of the system (even in the sense of C^1-smoothness), because the multipliers of hyperbolic trajectories are preserved under C^1-diffeomorphisms.

This fact, however, easily follows from the conjugacy condition for the rotation functions. Indeed, if we make a smooth change

$$t = t(\tau) = t_0 + a_0(\tau - \tau_0) + a_1(\tau - \tau_0)^2 + \ldots = t_0 + (\tau - \tau_0)g(\tau),$$

where $g(\tau)$ is a smooth function, $g(\tau_0) \neq 0$, and $t(\tau_0) = t_0$, then

$$\rho(t(\tau)) = \Lambda \ln |(\tau - \tau_0)g(\tau)| + q(t(\tau)) = \Lambda \ln |\tau - \tau_0| + \widetilde{q}(\tau),$$

where $\widetilde{q}(\tau)$ is continuous at τ_0.

Thus, the coefficient Λ in the asymptotics of $\rho(t)$ near the critical leaf which contains hyperbolic trajectories is a smooth orbital invariant of a Hamiltonian system. For the Euler case and the Jacobi problem we denote this invariant by $M(A, B, C)$ and $m(a, b, c)$, respectively.

Having known an explicit form of the rotation function, we can easily compute these numbers, using the following auxiliary lemma.

Lemma 16.1. *Let*

$$f(t) = \int_a^b \frac{g(u)\,du}{\sqrt{(u-b)(u-t)}}\,,$$

where $t > b$. *Then, as* t *tends to* b, *the following representation holds*:

$$f(t) = -g(b)\ln|t-b| + c(t)\,,$$

where $c(t)$ *is continuous at* $t = b$.

From this lemma we immediately obtain the following explicit formulas:

$$M(A,B,C) = -\frac{1}{\pi}\frac{B}{\sqrt{(C-B)(B-A)}}\,,$$

$$m(a,b,c) = -\frac{\sqrt{\frac{b}{(c-b)(b-a)}}}{\int_0^\infty \Phi(u,b)\,du}\,.$$

Now to the two above invariants we add one more new invariant and consider two two-dimensional surfaces in the three-dimensional space $\mathbb{R}^3\,(x,y,z)$, which we consider as the space of values of invariants:

$$\mathcal{E} = \left\{\begin{array}{l} x = K\,(A,B,C) \\ y = L\,(A,B,C) \\ z = M(A,B,C) \end{array}\right\}\,, \qquad \mathcal{J} = \left\{\begin{array}{l} x = k\,(a,b,c) \\ y = l\,(a,b,c) \\ z = m(a,b,c) \end{array}\right\}\,.$$

These surfaces are the images of the spaces of parameters in the space of invariants. We are speaking here about a "surface", taking into account the easily seen fact that the systems related to the proportional triples of parameters are smoothly orbitally equivalent and, therefore, are mapped to the same point in the space of invariants.

If we know the mutual location of these surfaces, we can draw some conclusions.

If the surfaces do not intersect, then there exists no pair (*rigid body, ellipsoid*) for which the corresponding dynamical systems are smoothly equivalent (even in the sense of C^1-smoothness). If the surfaces coincide, then this fact is a forcible argument for possible smooth equivalence of the systems considered, because such a coincidence may hardly occur by chance. Of course, the examination has to be continued in this case: we must compare all the other smooth invariants.

If the surfaces intersect along a curve, then this means that the corresponding systems, as a rule, are not equivalent. But there exist some exceptional pairs (*rigid body, ellipsoid*), for which the coincidence of at least three invariants happens. In such a situation we would have to compare the other smooth invariants for these exceptional pairs. However, in this case it would be natural to examine the question about C^1-equivalence. It can be shown that besides the above three invariants there is only one more C^1-invariant. So, to get the complete answer, it would be sufficient to analyze the mutual location of two two-dimensional surfaces in the four-dimensional space of invariants. Their intersection points would correspond to the pairs of C^1-equivalent systems. That is a possible scheme of analysis, which can be applied in a general situation.

In our case, because of the complication of explicit formulas, we have carried out computer analysis of the problem [51].

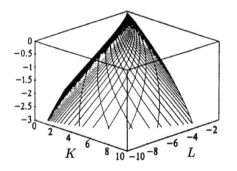

Figure 16.2. Euler case

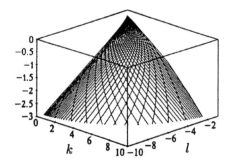

Figure 16.3. Jacobi problem

The surfaces $\mathcal{E}$ (Euler case) and $\mathcal{J}$ (Jacobi problem) are illustrated in Figs. 16.2 and 16.3. As we see, the qualitative behavior of these surfaces is very similar. They are both the graphs of some functions $z = z_{\mathcal{E}}(x, y)$ and $z = z_{\mathcal{J}}(x, y)$. This follows from Proposition 16.4.

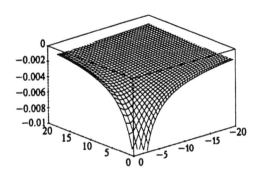

Figure 16.4. Difference between Euler case and Jacobi problem

In Fig. 16.4 we illustrate the surface

$$z = z_{\mathcal{E}}(x, y) - z_{\mathcal{J}}(x, y),$$

which shows the difference between the Euler and Jacobi cases. More precisely, this surface shows the difference between the third invariants M and m under the condition that the first two invariants coincide, i.e., $K(A, B, C) = k(a, b, c)$ and $L(A, B, C) = l(a, b, c)$.

Thus, the numeric examination shows that the surfaces $\mathcal{E}$ (Euler case) and $\mathcal{J}$ (Jacobi problem) do not intersect: the difference $z_{\mathcal{E}}(x,y) - z_{\mathcal{J}}(x,y)$ remains always negative, tending asymptotically to zero as the surfaces go to infinity.

Of course, this method is just a numerical experiment, and on these grounds one cannot formulate the result as a strict theorem. However, it is a strict result that these two surfaces do not coincide. This exactly means that continuously orbitally equivalent pairs (*rigid body, ellipsoid*), as a rule, are not smoothly equivalent (even in the sense of C^1-equivalence).

However, in our opinion, having carried out this numerical analysis, we can say with certainty that actually no smoothly equivalent pairs $v_J(a,b,c)$ and $v_E(A,B,C)$ exist.

16.7. TOPOLOGICAL NON-CONJUGACY OF THE JACOBI PROBLEM AND THE EULER CASE

Discussing the equivalence of the Euler and Jacobi problems, we can ask another question: can they be topologically conjugate for some values of their parameters? In other words, does there exist a homeomorphism between isoenergy surfaces which sends one Hamiltonian flow to the other and preserves parametrization on integral curves? The negative answer has been obtained by O. E. Orel [274].

Let us note, first of all, that, from the point of view of conjugacy, the Euler case and the Jacobi problem are three-parametric. When we spoke above about the orbital equivalence, we had, in essence, two parameters, because the (A,B,C) and (a,b,c) should be considered up to proportionality (the orbital type of the system did not change under homothety). Now the absolute values of the semi-axes of an ellipsoid and of the principal moments of inertia of a rigid body are essential. Note that, in the case $a = b = c$, the ellipsoid becomes the sphere. Analogously, for $A = B = C$, the Euler system describes the dynamics of the rigid ball. It is easy to see that in this case the systems just coincide, and we exclude this trivial case from the further consideration.

Theorem 16.4. *The geodesic flow on any ellipsoid (different from the sphere), restricted to its constant energy three-dimensional manifold, is not topologically conjugate to any system of the Euler case. In other words, for any values of parameters a,b,c and A,B,C (except for $a = b = c$ and $A = B = C$) the systems $v_J(a,b,c)$ and $v_E(A,B,C)$ are not topologically conjugate.*

Proof. We need again to calculate and compare some invariants of the systems in question.

As was shown above, the topological orbital type of an integrable system is completely determined in this case by two invariants: $k(a,b,c)$ and $l(a,b,c)$ for the Jacobi problem and, respectively, $K(A,B,C)$ and $L(A,B,C)$ for the Euler case. These invariants have a natural meaning. The point is that each of the systems to be compared has two periodic stable trajectories. The invariants k and l are the limits of the rotation numbers of the dynamical system as Liouville tori shrink into these trajectories. In the Jacobi problem these periodic trajectories correspond to two closed stable geodesics which are equatorial sections of the ellipsoid in the directions orthogonal to its largest and smallest semi-axes. In the Euler problem the analogous periodic trajectories correspond to the rotations of the rigid body around its maximal and minimal axes of inertia. Here the corresponding limits of the rotation number give the invariants K and L.

Note that the invariants k, l and K, L are functions of the parameters a, b, c and A, B, C respectively. The conditions, which are necessary and sufficient for the topological orbital equivalence of our systems, can be written as

$$k(a, b, c) = K(A, B, C) \quad \text{and} \quad l(a, b, c) = L(A, B, C).$$

Since we are interested now in the problem of comparing these two systems from the point of view of their conjugacy, then we should add at least three more new invariants to the two invariants mentioned above. These are the periods of three closed singular trajectories. Two of them have been just described. One needs to add one more periodic unstable trajectory to them, namely, the hyperbolic geodesic on the ellipsoid and, respectively, the unstable rotation of the rigid body around the middle axis of inertia. Denote these three additional invariants by t_1, t_2, t_3 for the Jacobi problem and by T_1, T_2, T_3 for the Euler case. As a result, the topological conjugacy class of $v_J(a, b, c)$ is defined by a set of invariants, which includes at least the following five numbers:

$$k, l, t_1, t_2, t_3 ;$$

and, respectively, for the Euler system $v_E(A, B, C)$:

$$K, L, T_1, T_2, T_3 .$$

It is useful to look at the explicit expressions for the periods of closed trajectories in the Euler and Jacobi problems. In the Jacobi problem the period of a closed geodesic is just equal to its length. Therefore, for the periods in the Jacobi problem we obtain

$$t_1 = \int_0^{2\pi} \sqrt{a \cos^2 t + b \sin^2 t} \, dt,$$

$$t_2 = \int_0^{2\pi} \sqrt{a \cos^2 t + c \sin^2 t} \, dt,$$

$$t_3 = \int_0^{2\pi} \sqrt{b \cos^2 t + c \sin^2 t} \, dt.$$

In the Euler case the periods of motion along the three closed trajectories are given by

$$T_1 = \pi\sqrt{2C}, \quad T_2 = \pi\sqrt{2B}, \quad T_3 = \pi\sqrt{2A}.$$

In both cases the energy level $\{H = h_0\}$ is fixed, and h_0 equals 1.

The Jacobi problem is three-parametric; therefore, by assigning the above *five* numbers to each triaxial ellipsoid, we obtain a smooth mapping from the *three-dimensional* space of triaxial ellipsoids into the *five-dimensional* Euclidean space. As a result, we obtain some 3-surface in $\mathbb{R}^5$. Denote it by J^3. Following the same scheme in the Euler case, we also obtain some 3-surface E^3 in the same five-dimensional space $\mathbb{R}^5$.

To check the topological non-conjugacy of the systems in question, it is sufficient to show that these two three-dimensional surfaces do not intersect in $\mathbb{R}^5$ (except for the only point corresponding to the case $a = b = c$ and $A = B = C$, i.e., the case of the standard sphere). This fact can be checked analytically (see [274]). $\square$

References

1. Adler M.P. and van Moerbeke P., Completely integrable systems, Euclidean Lie algebras and curves and linearization of Hamiltonian systems, Jacobi varieties and representation theory. *Adv. Math.*, **30** (1980), P. 267–379.

2. Adler M. P. and van Moerbeke P., The Kowalewski and Hénon–Heiles motions as Manakov geodesic flows on SO(4). A two-dimensional family of Lax pairs. *Comm. Math. Phys.*, **113** (1988), No. 4, P. 659–700.

3. Adler M.P. and van Moerbeke P., The complex geometry of the Kowalewski–Painlevé analysis. *Invent. Math.*, **97** (1989), P. 3–51.

4. Adler M.P. and van Moerbeke P., Kowalewski's asymptotic method, Kac–Moody algebras, and regularizations. *Comm. Math. Phys.*, **83** (1982), P. 83–106.

5. Albert C., Brouzet R., and Dufour J.-P. (Eds.), *Integrable Systems and Foliations*. Birkhäuser-Verlag, Basel–Boston–Berlin (1997).

6. Andronov A.A. and Pontryagin L.S., Rough systems. *Dokl. Akad. Nauk SSSR*, **14** (1937), No. 5, P. 247–250.

7. Andronov A.A., Leontovich E.A., Gordon I.I., and Maier A.G., *Qualitative Theory of Second-Order Dynamic Systems*. Wiley, New York (1973).

8. Anosov D.V., Rough systems. *Trudy Mat. Inst. Steklov*, **169** (1985), P. 59–93.

9. Anosov D.V., Aranson S.Kh., Bronstein I.U., and Grines V.Z., Smooth dynamical systems. In book: *Ordinary Differential Equations and Smooth Dynamical Systems* (*Encyclopaedia of Math. Sci., Vol. 1: Dynamical Systems I*). Springer-Verlag, Berlin–Heidelberg–New York (1988).

10. Anoshkina E.V., Topological classification of an integrable case of Goryachev–Chaplygin type with generalized potential in rigid body dynamics. *Uspekhi Mat. Nauk*, **47** (1992), No. 3, P. 149–150.

11. Anoshkina E.V., Topological classification of an integrable case of Goryachev–Chaplygin type with generalized potential in rigid body dynamics. *Trudy Mat. Inst. Steklov*, **205** (1994), P. 11–17.

12. Anoshkina E.V., On the topology of the integrable case of a gyrostat motion in a potential field of the Goryachev type. *Vestnik Moskov. Univ., Ser. Mat. Mekh.*, **53** (1998), No. 1, P. 23–29.

13. Apanasov B.N., *Geometry of Discrete Groups and Manifolds*. Nauka, Moscow (1991).

14. Aranson S.Kh. and Grines V.Z., Topological classification of flows on closed two-dimensional manifolds. *Uspekhi Mat. Nauk*, **41** (1986), No. 1, P. 149–169.

15. Arnol'd V.I., *Mathematical Methods of Classical Mechanics* (2nd edn.). Springer-Verlag, Berlin–New York (1989).

16. Arnol'd V.I. and Il'yashenko Yu.S., Ordinary differential equations. In book: *Ordinary Differential Equations and Smooth Dynamical Systems* (*Encyclopaedia of Math. Sci., Vol. 1: Dynamical Systems I*). Springer-Verlag, Berlin–Heidelberg–New York (1988).

17. Arnol'd V.I. and Givental' A.B., Symplectic geometry. In book: *Symplectic Geometry and its Applications* (*Encyclopaedia of Math. Sci., Vol. 4: Dynamical Systems IV*). Springer-Verlag, Berlin–Heidelberg–New York (1990).

18. Arkhangel'skiĭ Yu.A., *Analytical Dynamics of a Rigid Body*. Nauka, Moscow (1977).

19. Asimov D., Round handles and non-singular Morse–Smale flows. *Ann. Math.*, **102** (1975), No. 1, P. 41–54.

20. Atiyah M., Convexity and commuting Hamiltonians. *Bull. London Math. Soc.*, **14** (1982), P. 1–15.

21. Audin M., *The Topology of Torus Actions on Symplectic Manifolds*. Birkhäuser-Verlag, Basel–Boston–Berlin (1991).

22. Audin M., *Spinning Tops*. Cambridge University Press, Cambridge (1996).

23. Audin M., Courbes algébriques et systèmes intégrables: géodésiques des quadriques. *Expositiones Math.*, **12** (1994), P. 193–226.

24. Audin M. and Silhol R., Variétés abéliennes réelles et toupie de Kowalevski. *Compositio Math.*, **87** (1993), P. 153–229.

25. Babenko I.K., *Les flots géodésiques quadratiquement intégrables sur les surfaces fermées et les structures complexes correspondantes*. Preprint: Inst. de Recherche Math. Avancée (1996).

26. Babenko I.K. and Nekhoroshev N.N., On complex structures on two-dimensional tori, admitting metrics with non-trivial quadratic integral. *Matem. Zametki*, **58** (1995), No. 5, P. 643–652.

27. Bar-Natan D., On the Vassiliev knot invariants. *Topology*, **34** (1995), No. 2, P. 423–472.

28. Bates L., Monodromy in the champagne bottle. *J. of Appl. Math. and Phys.* (*ZAMP*), **42** (1991), P. 837–847.

29. Bautin N.N. and Leontovich E.A., *Methods and Means for a Qualitative Investigation of Dynamical Systems on the Plane* (2nd edn.). Nauka, Moscow (1990).

30. Berzin D.V., *Geometry of coadjoint orbits of special Lie groups*. Thesis, Moscow State Univ. (1997).

31. Besse A.L., *Manifolds All of Whose Geodesics are Closed*. Springer-Verlag, Berlin–Heidelberg–New York (1978).

32. Birkhoff G.D., *Dynamical Systems*. AMS, Providence (1927).

33. Birkhoff G.D., Sur le problème restreint des trois corps, I, II. *Ann. Scuola Norm. Super. Pisa*, **4** (1935), P. 267–306; **5** (1936), P. 1–72.

34. Blaschke W., *Einführung in die Differentialgeometrie*. Springer-Verlag, Berlin–Göttingen–Heidelberg (1950).

35. Bobenko A.I., Euler equations on so(4) and e(3). Isomorphism of integrable cases. *Funkts. Analiz i ego Prilozh.*, **20** (1986), No.1, P. 64–66.

36. Bogoyavlensky O.I., New integrable problem of classical mechanics. *Comm. Math. Phys.*, **94** (1984), P. 255–269.

37. Bogoyavlensky O.I., On perturbations of the periodic Toda lattices. *Comm. Math. Phys.*, **51** (1976), P. 201–209.

38. Bogoyavlensky O.I., Integrable cases in rigid body dynamics and integrable systems on S^n spheres. *Izvest. Akad. Nauk SSSR, Ser. Matem.*, **49** (1985), No.5, P. 899–915.

39. Bolotin S.V., Variational methods for constructing chaotic motions in rigid body dynamics. *Prikl. Mat. Mekh.*, **56** (1992), No.2, P. 230–239.

40. Bolotin S.V., Homoclinic orbits of geodesic flows on surfaces. *Russian J. Math. Phys.*, **1** (1993), No.3, P. 275–288.

41. Bolotin S.V. and Kozlov V.V., Symmetry fields of geodesic flows. *Russian J. Math. Phys.*, **3** (1995), No.3, P. 279–295.

42. Bolotin S.V. and Negrini P., A variational criterion for non-integrability. *Russian J. Math. Phys.*, **5** (1997), No.4, P. 415–436.

43. Bolsinov A.V., Commutative families of functions related to consistent Poisson brackets. *Acta Appl. Math.*, **24** (1991), P. 253–274.

44. Bolsinov A.V., Methods of calculation of the Fomenko–Zieschang invariant. In book: *Topological Classification of Integrable Systems (Advances in Soviet Mathematics, Vol. 6)*. AMS, Providence (1991), P. 147–183.

45. Bolsinov A.V., Compatible Poisson brackets on Lie algebras and completeness of families of functions in involution. *Izvest. Akad. Nauk SSSR, Ser. Matem.*, **55** (1991), No.1, P. 68–92.

46. Bolsinov A.V., A smooth trajectory classification of integrable Hamiltonian systems with two degrees of freedom. *Matem. Sbornik*, **186** (1995), No.1, P. 3–28.

47. Bolsinov A.V., Smooth trajectory classification of integrable Hamiltonian systems with two degrees of freedom. The case of systems with flat atoms. *Uspekhi Mat. Nauk*, **49** (1994), No.4, P. 173–174.

48. Bolsinov A.V., On the classification of Hamiltonian systems on two-dimensional surfaces. *Uspekhi Mat. Nauk*, **49** (1994), No.6, P. 195–196.

49. Bolsinov A.V., Multidimensional Euler and Clebsch cases and Lie pencils. In book: *Trudy Semin. po Vektor. i Tenzor. Analizu, Vol. 24*. Izdatel'stvo Moskovskogo Universiteta, Moscow (1991), P. 8–12. [English transl. in: *Tensor and Vector Analysis*. Gordon and Breach, Amsterdam (1998), P. 25–30.]

50. Bolsinov A.V., Fomenko invariants in the theory of integrable Hamiltonian systems. *Uspekhi Mat. Nauk*, **52** (1997), No.5, P. 113–132.

51. Bolsinov A.V. and Dullin H., On the Euler case in rigid body dynamics and the Jacobi problem. *Regular and Chaotic Dynam.*, **2** (1997), No.1, P. 13–25.

52. Bolsinov A.V. and Fomenko A.T., Trajectory classification of integrable Euler type systems in rigid body dynamics. *Uspekhi Mat. Nauk*, **48** (1993), No.5, P. 163–164.

53. Bolsinov A.V. and Fomenko A.T., Orbital equivalence of integrable Hamiltonian systems with two degrees of freedom. A classification theorem. I, II. *Matem. Sbornik*, **185** (1994), No.4, P. 27–80; No.5, P. 27–78.

54. Bolsinov A.V. and Fomenko A.T., Orbital classification of simple integrable Hamiltonian systems on three-dimensional constant-energy surfaces. *Dokl. Akad. Nauk SSSR*, **332** (1993), No.5, P. 553–555.

55. Bolsinov A.V. and Fomenko A.T., The geodesic flow of an ellipsoid is orbitally equivalent to the Euler integrable case in the dynamics of a rigid body. *Dokl. Akad. Nauk SSSR,* **339** (1994), No. 3, P. 293–296.

56. Bolsinov A.V. and Fomenko A.T., Application of classification theory for integrable Hamiltonian systems to geodesic flows on 2-sphere and 2-torus and to the description of the topological structure of momentum mapping near singular points. *J. of Math. Sci.,* **78** (1996), No. 5, P. 542–555.

57. Bolsinov A.V. and Fomenko A.T., Integrable geodesic flow on the sphere generated by Goryachev–Chaplygin and Kowalewski systems in the dynamics of a rigid body. *Matem. Zametki,* **56** (1994), No. 2, P. 139–142.

58. Bolsinov A.V. and Fomenko A.T., Trajectory invariants of integrable Hamiltonian systems. The case of simple systems. Trajectory classification of Euler-type systems in rigid body dynamics. *Izvest. Akad. Nauk SSSR, Ser. Matem.,* **59** (1995), No. 1, P. 65–102.

59. Bolsinov A.V. and Fomenko A.T., Orbital classification of the geodesic flows on two-dimensional ellipsoids. The Jacobi problem is orbitally equivalent to the integrable Euler case in rigid body dynamics. *Funkts. Analiz i ego Prilozh.,* **29** (1995), No. 3, P. 1–15.

60. Bolsinov A.V. and Fomenko A.T., Unsolved problems in the theory of topological classification of integrable systems. *Trudy Mat. Inst. Steklov,* **205** (1994), P. 18–31.

61. Bolsinov A.V. and Fomenko A.T., On dimension of the space of integrable Hamiltonian systems with two degrees of freedom. *Trudy Mat. Inst. Steklov,* **216** (1996), P. 45–69.

62. Bolsinov A.V. and Fomenko A.T., *Introduction to the Topology of Integrable Hamiltonian Systems.* Nauka, Moscow (1997).

63. Bolsinov A.V., Kozlov V.V., and Fomenko A.T., Maupertuis principle and the geodesic flows on the sphere appearing from integrable cases in rigid body dynamics. *Uspekhi Mat. Nauk,* **50** (1995), No. 3, P. 3–32.

64. Bolsinov A.V. and Matveev V.S., Integrable Hamiltonian systems: topological structure of saturated neighborhoods of non-degenerate singular points. In book: *Tensor and Vector Analysis.* Gordon and Breach, Amsterdam (1998), P. 31–56.

65. Bolsinov A.V., Matveev S.V., and Fomenko A.T., Topological classification of integrable Hamiltonian systems with two degrees of freedom. List of systems of small complexity. *Uspekhi Mat. Nauk,* **45** (1990), No. 2, P. 49–77.

66. Bolsinov A.V., Matveev V.S., and Fomenko A.T., Two-dimensional Riemannian metrics with integrable geodesic flows. Local and global geometry. *Matem. Sbornik,* **189** (1998), No. 10, P. 5–32.

67. Bolsinov A.V., Oshemkov A.A., and Sharko V.V., On classification of flows on manifolds. I. *Methods of Functional Analysis and Topology,* **2** (1996), No. 2, P. 190–204.

68. Borisov A.V. and Simakov N.N., The bifurcations of period doubling in rigid body dynamics. *Regular and Chaotic Dynam.,* **2** (1997), No. 1, P. 64–74.

69. Borisov A.V. and Tsygvintsev A.V., Kovalevskaya exponents and integrable systems of classical dynamics. *Regular and Chaotic Dynam.,* **1** (1996), No. 1, P. 15–37.

70. Bott R., Non-degenerate critical manifolds. *Ann. Math.,* **60** (1954), P. 249–261.

71. Bott R., On manifolds all of whose geodesics are closed. *Ann. Math.,* **60** (1954), No. 3, P. 375–382.

72. Brailov A.V., Some cases of complete integrability of Euler equations and their applications. *Dokl. Akad. Nauk SSSR*, **268** (1983), No. 5, P. 1043–1046.

73. Brailov A.V. and Fomenko A.T., The topology of integral submanifolds of completely integrable Hamiltonian systems. *Matem. Sbornik*, **133(175)** (1987), No. 3, P. 375–385.

74. Brailov Yu.A. and Kudryavtseva E.A., Stable topological non-conjugacy of Hamiltonian systems on two-dimensional surfaces. *Vestnik Moskov. Univ., Ser. Mat. Mekh.*, **54** (1999), No. 2, P. 20–27.

75. Busemann H., *The Geometry of Geodesics*. Academic Press, New York–London (1955).

76. Bialy M.L., First integrals that are polynomial in momenta for a mechanical system on the two-dimensional torus. *Funkts. Analiz i ego Prilozh.*, **21** (1987), No. 4, P. 64–65.

77. Calogero F., Exactly solvable one-dimensional many-body problems. *Lett. Nuovo Cimento*, **13** (1975), P. 411–416.

78. Calogero F., Solution of the one-dimensional N-body problems with quadratic and/or inversely quadratic pair potentials. *J. Math. Phys.*, **12** (1971), P. 419–436.

79. Casler B.G., An embedding theorem for connected 3-manifolds with boundary. *Proc. Amer. Math. Soc.*, **16** (1965), No. 4, P. 559–566.

80. Cavicchioli A., Repovš D., and Skopenkov A., An extension of the Bolsinov–Fomenko theorem on orbital classification of integrable Hamiltonian systems. *Rocky Mountain J. of Math.*, **30** (2000), No. 2, P. 447–476.

81. Chaplygin S.A., A new case of rotation of a heavy rigid body, supported at one point. In book: *Collected Works, Vol. 1*. Gostekhizdat, Moscow–Leningrad (1948), P. 118–124.

82. Chaplygin S.A., A new particular solution of the problem on a rigid body motion in liquid. In book: *Collected Works, Vol. 1*. Gostekhizdat, Moscow–Leningrad (1948), P. 337–346.

83. Chinburg T., Volumes of hyperbolic manifolds. *Diff. Geom.*, **18** (1983), P. 783–789.

84. Chinburg T., A small arithmetic hyperbolic 3-manifolds. *Proc. Amer. Math. Soc.*, **100** (1987), P. 140–144.

85. Clebsch A., Über die Bewegung eines Körpers in einer Flüssigkeit. *Math. Annalen*, **3** (1871), P. 238–262.

86. Colin de Verdière Y. and Vey J., Le lemme de Morse isochore. *Topology*, **18** (1979), P. 283–293.

87. Coxeter H.S.M. and Moser W.O.J., *Generators and Relations for Discrete Groups* (3rd edn.). Springer-Verlag, Berlin–Heidelberg–New York (1972).

88. Cushman R.H., Geometry of the bifurcations of the Hénon–Heiles family. *Proc. Royal Soc., London, Ser. A*, **382** (1982), P. 361–371.

89. Cushman R.H. and Bates L.M., *Global Aspects of Classical Integrable Systems*. Birkhäuser-Verlag, Basel–Boston–Berlin (1997).

90. Cushman R. and Knörrer H., The momentum mapping of the Lagrange top. In book: *Differential Geometric Methods in Physics* (*Lecture Notes in Math., Vol. 1139*). Springer-Verlag, Berlin–Heidelberg–New York (1985), P. 12–24.

91. Cushman R. and van de Meer J.-C., The Hamiltonian Hopf bifurcation in the Lagrange top. In book: *Géométrie Symplectique et Mécanique* (*Lecture Notes in Math., Vol. 1416*). Springer-Verlag, Berlin–Heidelberg–New York (1991), P. 26–38.

92. Darboux G., Sur le problème de Pfaff. *Bull. Sci. Math.*, **6** (1882), P. 14–36, 48–68.

93. Darboux G., *Leçons sur la théorie générale des surfaces et ses applications géométriques du calcul infinitésimal. Vol. 1–4.* Gauthier-Villars, Paris (1887–1896).

94. Denisova N.V., On the structure of the symmetry fields of geodesic flows on the two-dimensional torus. *Matem. Sbornik*, **188** (1997), No.7, P. 107–122.

95. Dimitrov I., Bifurcations of invariant manifolds in the Gel'fand–Dikiĭ system. *Phys. Lett. A*, **163** (1992), P. 286–292.

96. Dinaburg E.I., Connection between different entropy characteristics of dynamical systems. *Izvest. Akad. Nauk SSSR, Ser. Matem.*, **35** (1971), No.2, P. 324–366.

97. Dini U., Sopra un problema che si presenta nella teoria generale delle rapprese- tazioni geografice di una superficie su di unáltra. *Ann. di Math., Ser. 2*, **3** (1869), P. 269–293.

98. Dirac P., Generalized Hamiltonian dynamics. *Canadian J. of Math.*, **2** (1950), P. 129–148.

99. Donagi R. and Markman E., Spectral covers, algebraically completely integrable Hamiltonian systems, and moduli of bundles. In book: *Integrable Systems and Quantum Groups(Lecture Notes in Math., Vol. 1620)*. Springer-Verlag, Berlin– Heidelberg–New York (1996).

100. Dragovich V.I., On integrable potential perturbations of the Jacobi problem for the geodesics on the ellipsoid. *J. Phys. A: Math. Gen.*, **29** (1996), P. 317-321.

101. Dubrovin B.A., Fomenko A.T., and Novikov S.P., *Modern Geometry: Methods and Applications, Vol. 1–2.* Springer-Verlag, Berlin–Heidelberg–New York (1984–1985).

102. Dubrovin B.A., Krichever B.A., and Novikov S.P., Integrable systems, I. In book: *Symplectic Geometry and its Applications (Encyclopaedia of Math. Sci., Vol. 4: Dynamical Systems IV)*. Springer-Verlag, Berlin–Heidelberg–New York (1990).

103. Dufour J.-P., Molino P., and Toulet A., Classification des systèms intégrables en dimension 2 et invariants des modèles de Fomenko. *Compt. Rend. Acad. Sci. Paris*, **318** (1994), P. 942–952.

104. Duistermaat J.J., On global action-angle variables. *Comm. Pure Appl. Math.*, **33** (1980), P. 678–706.

105. Dullin H.R., Wittek A., Efficient calculation of actions. *J. Phys. A: Math. Gen.*, **27** (1994), P. 7461–7474.

106. Dullin H.R., Wittek A., Complete Poincaré sections and tangent sets. *J. Phys. A: Math. Gen.*, **28** (1995), P. 7157–7180.

107. Dullin H.R., Juhnke M., and Richter P., Action integrals and energy surfaces of the Kovalevskaya top. *Int. J. of Bifurcation and Chaos*, **4** (1994), No.6, P. 1535–1562.

108. Dullin H.R., Matveev V.S., and Topalov P.Ĭ., On integrals of the third degree in momenta. *Regular and Chaotic Dynam.*, **4** (1999), No.3, P. 35–44.

109. Eleonskiĭ V.M. and Kulagin N.E., On new cases of integrability of Landau–Liefshitz equations. *Zh. Eksp. Teor. Fiz.*, **83** (1983), No.2, P. 616–629.

110. Eliasson L.H., Normal forms for Hamiltonian systems with Poisson commuting integrals — elliptic case. *Comm. Math. Helv.*, **65** (1990), P. 4–35.

111. Euler L., *Mechanics.* GITTL, Moscow–Leningrad (1938).

112. Euler L., Du mouvement de rotation des corps solides autour d'un axe variable. *Mémoires Acad. Sci. Berlin*, **14** (1758), P. 154–193.

113. Farkas H.M. and Kra I., *Riemann Surfaces.* Springer-Verlag, Berlin–Heidelberg– New York (1980).

114. Flaschka H. and Ratiu T., A Morse theoretic proof of Poisson–Lie convexity. In book: *Integrable Systems and Foliations*. Birkhäuser-Verlag, Basel–Boston–Berlin (1997), P. 49–71.

115. Fleitas G., Classification of gradient-like flows on dimensions two and three. *Bol. Soc. Bras. Mat.*, **6** (1975), P. 155–183.

116. Fokas A., Lagerstrom P., Quadratic and cubic invariants in classical mechanics. *J. Math. Anal. Appl.*, **74** (1980), No. 2, P. 325–341.

117. Fomenko A.T., The topology of surfaces of constant energy in integrable Hamiltonian systems and obstructions to integrability. *Izvest. Akad. Nauk SSSR, Ser. Matem.*, **50** (1986), No. 6, P. 1276–1307.

118. Fomenko A.T., Morse theory of integrable Hamiltonian systems. *Dokl. Akad. Nauk SSSR*, **287** (1986), No. 5, P. 1071–1075.

119. Fomenko A.T., Bordism theory for integrable Hamiltonian non-degenerate systems with two degrees of freedom. New topological invariant of multidimensional integrable systems. *Izvest. Akad. Nauk SSSR, Ser. Matem.*, **55** (1991), P. 747–779.

120. Fomenko A.T., Topological invariants of Hamiltonian systems integrable in the sense of Liouville. *Funkts. Analiz i ego Prilozh.*, **22** (1988), No. 4, P. 38–51.

121. Fomenko A.T., Topological invariant roughly classifying integrable strictly non-degenerate Hamiltonians on four-dimensional symplectic manifolds. *Funkts. Analiz i ego Prilozh.*, **25** (1991), No. 4, P. 23–35.

122. Fomenko A.T., Symplectic topology of completely integrable Hamiltonian systems. *Uspekhi Mat. Nauk*, **44** (1989), No. 1, P. 145–173.

123. Fomenko A.T., Topological classification of all integrable Hamiltonian differential equations of general type with two degrees of freedom. In book: *The Geometry of Hamiltonian Systems (Proc. of Workshop Held June 5–16, 1989. Berkeley)*. Springer-Verlag (1991), P. 131–339.

124. Fomenko A.T., *Integrability and Non-integrability in Geometry and Mechanics*. Kluwer Academic Publishers, Amsterdam (1988).

125. Fomenko A.T., *Symplectic Geometry* (2nd edn.). Gordon and Breach, Amsterdam (1995).

126. Fomenko A.T., The theory of invariants of multidimensional integrable Hamiltonian systems (with arbitrary many degrees of freedom). Molecular table of all integrable systems with two degrees of freedom. In book: *Topological Classification of Integrable Systems (Advances in Soviet Mathematics, Vol. 6)*. AMS, Providence (1991), P. 1–36.

127. Fomenko A.T., Theory of rough classification of integrable non-degenerate Hamiltonian differential equations on four-dimensional manifolds. Application to classical mechanics. In book: *Topological Classification of Integrable Systems (Advances in Soviet Mathematics, Vol. 6)*. AMS, Providence (1991), P. 302–344.

128. Fomenko A.T. and Fuks D.B., *A Course of Homotopic Topology*. Nauka, Moscow (1989).

129. Fomenko A.T. and Matveev S.V., Hyperbolic geometry and topology. In book: *Fomenko A.T., Kunii T.L., Topological Modeling for Visualization*. Springer-Verlag, Tokyo (1997), P. 289–312.

130. Fomenko A.T. and Nguyen T.Z., Topological classification of integrable non-degenerate Hamiltonians on the isoenergy three-dimensional sphere. In book:

Topological Classification of Integrable Systems (*Advances in Soviet Mathematics*, Vol. 6). AMS, Providence (1991), P. 267–269.

131. Fomenko A.T. and Sharko V.V., Exact round Morse functions, inequalities of the Morse type and integrals of Hamiltonian systems. *Ukrainsk. Matem. Journal*, **41** (1989), No.6, P. 723–732.

132. Fomenko A.T. and Trofimov V.V., *Integrable Systems on Lie Algebras and Symmetric Spaces*. Gordon and Breach, Amsterdam (1988).

133. Fomenko A.T. and Zieschang H., On the topology of three-dimensional manifolds arising in Hamiltonian mechanics. *Dokl. Akad. Nauk SSSR*, **294** (1987), No.2, P. 283–287.

134. Fomenko A.T. and Zieschang H., On typical topological properties of integrable Hamiltonian systems. *Izvest. Akad. Nauk SSSR, Ser. Matem.*, **52** (1988), No.2, P. 378–407.

135. Fomenko A.T. and Zieschang H., A topological invariant and a criterion for the equivalence of integrable Hamiltonian systems with two degrees of freedom. *Izvest. Akad. Nauk SSSR, Ser. Matem.*, **54** (1990), No.3, P. 546–575.

136. Forster O., *Lectures on Riemann Surfaces*. Springer-Verlag, Berlin–Heidelberg–New York (1977).

137. Franks J., The periodic structure of non-singular Morse–Smale flows. *Comm. Math. Helv.*, **53** (1978), No.2, P. 279–294.

138. Funk P., Über Flächen mit lauter geschlossenen geodätischen Linien. *Math. Annalen*, **74** (1913), P. 278–300.

139. Gaidukov E.V., Asymptotic geodesics on a Riemannian manifold non-homeomorphic to the sphere. *Dokl. Akad. Nauk SSSR*, **169** (1966), No.5, P. 999–1001.

140. Gavrilov L., Explicit solutions of the Goryachev–Chaplygin top. *Compt. Rend. Acad. Bulg. Sci.*, **40** (1987), No.4, P. 19–22.

141. Gavrilov L., *The complex geometry of Lagrange top*. Preprint: No.61 du Laboratorie de Mathematiques Emile Picard, Université Toulouse III (1995).

142. Gavrilov L., Bifurcations of invariant manifolds in the generalized Hénon–Heiles system. *Physica D*, **34** (1989), P. 223–239.

143. Gavrilov L., Ouazzani-Jamil M., and Caboz R., Bifurcation diagrams and Fomenko's surgery on Liouville tori of the Kolossoff potential $U = \varphi + (1/\varphi) - k \cos \varphi$. *Ann. Sci. École Norm. Sup., Série 4*, **26** (1993), P. 545–564.

144. Goldstein H., *Classical Mechanics* (2nd edn.). Addison-Wesley, Reading, MA (1980).

145. Golubev V.V., *Lectures on the integration of equations of motion of a heavy rigid body about a fixed point*. Gostekhizdat, Moscow–Leningrad (1953).

146. Goryachev D.N., On the motion of a rigid body around a fixed point in the case $A = B = 4C$. *Matem. Sbornik*, **21** (1900), No.3, P. 431-438.

147. Goryachev D.N., New cases of integrability of dynamical Euler equations. *Warsaw Univ. Izv.*, **3** (1916), P. 1–15.

148. Gromoll D., Klingenberg W., and Meyer W., *Riemannsche Geometrie im Grossen*. Springer-Verlag, Berlin–Heidelberg–New York (1968).

149. Guillemin V., *The Radon Transform on Zoll Surfaces*. Preprint: Cambridge (1976).

150. Guillemin V. and Sternberg S., *Symplectic Techniques in Physics*. Cambridge University Press, Cambridge (1984).

151. Hadamard J., Sur la précession dans le mouvement d'un corps pesant de révolution fixé par un point de son axe. *Bull. Sci. Math.*, **19** (1895), P. 228–230.

152. Haefliger A. and Reeb G., Variétés (non separées) a une dimension et structures feuilletées du plan. *Enseign. Math.*, **3** (1957), P. 107–126.

153. Hadeler K.P. and Selivanova E.N., On the case of Kovalevskaya and new examples of integrable conservative systems on S^2. *Regular and Chaotic Dynam.*, **4** (1999), No. 3, P. 45–52.

154. Hall L.S., A theory of exact and approximate configuration invariants. *Physica D*, **8** (1983), P. 90–116.

155. Hamilton W.R., On a general method in dynamics, by which the study of the motions of all free systems attracting or repelling points is reduced to the search and differentiation of one central relation or characteristic function. *Philos. Trans. of Royal Soc.*, (1834), P. 247–308; Second essay on a general method in dynamics. *Philos. Trans. of Royal Soc.*, (1835), P. 95–144.

156. Helgason S., *Differential Geometry and Symmetric Spaces.* Academic Press, New York–London (1962).

157. Hilbert D. and Cohn-Vossen S., *Anschauliche Geometrie.* Springer-Verlag, Berlin (1932).

158. Hopf H., Über die Abbildungen der dreidimensional Sphären auf der Kugelfläche. *Math. Annalen*, **104** (1931), P. 637–665.

159. Horozov E., Perturbations of the spherical pendulum and Abelian integrals. *J. Reine and Angew. Math.*, **408** (1990), P. 114–135.

160. Hurwitz A. and Courant R., *Funktionentheorie* (3rd edn.). Springer-Verlag, Berlin (1929).

161. Huygens C., *Three Memoirs on Mechanics.* Izdat. Akad. Nauk SSSR, Moscow (1951).

162. Huygens C., L'horloge à Pendule. In book: *Oeuvres Complètes, Vol. 18.* Paris (1673), P. 69–368.

163. Iliev I.P. and Semerdjiev Kh.I., On holonomic mechanical systems admitting quadratic integrals. *Izvest. Vuzov*, **2** (1972), P. 51–53.

164. Ilyukhin A.A., *Spatial Problems of Non-linear Theory of Elastic Rods.* Naukova Dumka, Kiev (1979).

165. Ivanov A.O. and Tuzhilin A.A., The geometry of minimal nets and one-dimensional Plateau problem. *Uspekhi Mat. Nauk*, **47** (1992), No. 2, P. 53–115.

166. Ito H., Convergence of Birkhoff normal forms for integrable systems. *Comm. Math. Helv.*, **64** (1989), P. 412–461.

167. Jacobi C.G.J., *Vorlesungen über Dynamik* (2nd edn.). G. Reimer, Berlin (1884).

168. Kalashnikov V.V., Description of the structure of Fomenko invariants on the boundary and inside Q-domains, estimates of their number on the lower boundary for the manifolds S^3, $\mathbb{R}P^3$, $S^1 \times S^2$, and T^3. In book: *Topological Classification of Integrable Systems (Advances in Soviet Mathematics, Vol. 6).* AMS, Providence (1991), P. 297–304.

169. Kalashnikov V.V., *A class of generic integrable Hamiltonian systems with two degrees of freedom.* Preprint: Preprint No. 907, University Utrecht, Dept. of Math. (1995).

170. Kalashnikov V.V., The Bott property and generic position properties of integrable Hamiltonian systems. *Uspekhi Mat. Nauk*, **48** (1993), No. 6, P. 151–152.

171. Kalashnikov V.V., On the typicalness of Bott integrable Hamiltonian systems. *Matem. Sbornik*, **185** (1994), No. 1, P. 107–120.

172. Kalashnikov V.V., *Singularities of integrable Hamiltonian systems*. Thesis, Moscow State Univ. (1998).

173. Kalashnikov V.V., Typical integrable Hamiltonian systems on a four-dimensional symplectic manifold. *Izvest. Akad. Nauk SSSR, Ser. Matem.*, **62** (1998), No. 2, P. 49–74.

174. Kalashnikov V.V., A geometrical description of minimax Fomenko invariants of integrable Hamiltonian systems on S^3, $\mathbb{R}P^3$, $S^1 \times S^2$, T^3. *Uspekhi Mat. Nauk*, **46** (1991), No. 4, P. 151–152.

175. Kalashnikov V.V., Topological classification of quadratically integrable geodesic flows on the two-dimensional torus. *Uspekhi Mat. Nauk*, **50** (1995), No. 1, P. 201–202.

176. Kalashnikov V.V., On the topological structure of integrable Hamiltonian systems similar to a given one. *Regular and Chaotic Dynam.*, **2** (1997), No. 2, P. 98–112.

177. Katok S.B., Bifurcational sets and integral manifolds in rigid body problem. *Uspekhi Mat. Nauk*, **27** (1972), No. 2, P. 126–132.

178. Kharlamov M.P., *Topological Analysis of Integrable Problems in Rigid Body Dynamics*. Leningrad University, Leningrad (1988).

179. Kharlamov M.P., Topological analysis of classical integrable cases in the dynamics of a rigid body. *Dokl. Akad. Nauk SSSR*, **273** (1983), No. 6, P. 1322–1325.

180. Kharlamov P.V., *Lectures on Rigid Body Dynamics*. Novosibirsk State University, Novosibirsk (1965).

181. Kiyohara K., *Two classes of Riemannian manifolds whose geodesic flows are integrable* (*Memoirs of the AMS, Vol. 130, No. 619*). AMS, Providence (1997).

182. Kiyohara K., Compact Liouville surfaces. *J. Math. Soc. Japan*, **43** (1991), P. 555–591.

183. Klingenberg W., *Lectures on Closed Geodesics*. Springer-Verlag, Berlin–Heidelberg–New York (1978).

184. Knörrer H., Geodesics on quadrics and a mechanical problem of C. Neumann. *J. Reine and Angew. Math.*, **334** (1982), P. 69–78.

185. Knörrer H., Geodesics of the ellipsoid. *Invent. Math.*, **39** (1980), P. 119–143.

186. Knörrer H., Singular fibers of the momentum mapping for integrable Hamiltonian systems. *J. Reine and Angew. Math.*, **355** (1985), P. 67–107.

187. Kolokol'tsov V.N., Geodesic flows on two-dimensional manifolds with an additional first integral that is polynomial in the velocities. *Izvest. Akad. Nauk SSSR, Ser. Matem.*, **46** (1982), No. 5, P. 994–1010.

188. Kolokol'tsov V.N., New examples of manifolds with closed geodesics. *Vestnik Moskov. Univ., Ser. Mat. Mekh.*, **39** (1984), No. 4, P. 80–82.

189. Kolokol'tsov V.N., *Polynomial integrals of geodesic flows on compact surfaces*. Thesis, Moscow State Univ. (1984).

190. Korovina N.V., Totally symmetric bifurcations of Morse functions on two-dimensional surfaces. *Vestnik Moskov. Univ., Ser. Mat. Mekh.*, **54** (1999), No. 2, P. 13–19.

191. Kowalewski S., Sur le problème de la rotation d'un corps solide autour d'un point fixe. *Acta Math.*, **12** (1889), P. 177–232.

192. Kozlov V.V., Integrable and non-integrable Hamiltonian systems. *Sov. Sci. Rev. C. Math. Phys.*, **8** (1988), P. 1–81.

193. Kozlov V.V., Two integrable problems in classical dynamics. *Vestnik Moskov. Univ., Ser. Mat. Mekh.*, **36** (1981), No. 4, P. 80–83.

194. Kozlov V.V., *Methods of Qualitative Analysis in Rigid Body Dynamics.* Izdatel'stvo Moskovskogo Universiteta, Moscow (1980).

195. Kozlov V.V., Topological obstacles to integrability of natural mechanical systems. *Dokl. Akad. Nauk SSSR*, **249** (1979), No. 6, P. 1299–1302.

196. Kozlov V.V., *Symmetries, Topology and Resonances in Hamiltonian Mechanics.* Izdatel'stvo Udmurtskogo Gosudarstvennogo Universiteta, Izhevsk (1995).

197. Kozlov V.V., Some integrable generalizations of the Jacobi problem about geodesic on the ellipsoid. *Prikl. Mat. Mekh.*, **59** (1995), No. 1, P. 3–9.

198. Kozlov V.V. and Denisova N.V., Polynomial integrals of geodesic flows on the two-dimensional torus. *Matem. Sbornik*, **185** (1994), No. 12, P. 49–64.

199. Kozlov V.V. and Denisova N.V., Symmetries and topology of dynamical systems with two degrees of freedom. *Matem. Sbornik*, **184** (1993), No. 9, P. 125–148.

200. Kozlov V.V. and Kolesnikov N.N., On the integrability of Hamiltonian systems. *Vestnik Moskov. Univ., Ser. Mat. Mekh.*, **34** (1979), No. 6, P. 88–91.

201. Kozlov V.V. and Treshchev D.V., On integrability of Hamiltonian systems with a torical configuration space. *Matem. Sbornik*, **135(177)** (1988), No. 1, P. 119–138.

202. Kozlov V.V. and Treshchev D.V., Polynomial integrals of Hamiltonian systems with exponential interaction. *Izvest. Akad. Nauk SSSR, Ser. Matem.*, **53** (1989), No. 3, P. 537–556.

203. Kruglikov B.S., On the continuation of the symplectic form and a pair of functions in involution from $S^1 \times I \times T^2$. *Trudy Mat. Inst. Steklov*, **205** (1994), P. 98–108.

204. Kruglikov B.S., The existence of a pair of additional Bott integrals for a resonance Hamiltonian system with two degrees of freedom. *Trudy Mat. Inst. Steklov*, **205** (1994), P. 109–112.

205. Kruglikov B.S., The exact smooth classification of Hamiltonian vector fields on two-dimensional manifolds. *Matem. Zametki*, **61** (1997), No. 2, P. 179–200.

206. Kudryavtseva E.A., Realization of smooth functions on surfaces as height functions. *Matem. Sbornik*, **190** (1999), No. 3, P. 29–88.

207. Kudryavtseva E.A., Stable topological invariants and smooth invariants of conjugation for Hamiltonian systems on surfaces. In book: *Topological Methods in the Theory of Hamiltonian Systems.* Faktorial, Moscow (1998), P. 147–202.

208. Kudrayvtseva E.A., Reduction of Morse functions on surfaces to canonical form by smooth deformation. *Regular and Chaotic Dynam.*, **4** (1999), No. 3, P. 53–60.

209. Lagrange J.L., *Mécanique analytique, Vol. 1, 2.* Paris (1788).

210. Leontovich E.A. and Maier A.G., On trajectories defining the qualitative structure of the sphere's division into trajectories. *Dokl. Akad. Nauk SSSR*, **14** (1937), No. 5, P. 251–257.

211. Leontovich E.A. and Maier A.G., On the scheme defining the topological structure of division into trajectories. *Dokl. Akad. Nauk SSSR*, **103** (1955), No. 4, P. 557–560.

212. Lerman L.M. and Umanskiĭ Ya.L., Classification of four-dimensional integrable systems and the Poisson action of $\mathbb{R}^2$ in extended neighborhoods of simple singular points. I, II, III. *Matem. Sbornik*, **183** (1992), No. 12, P. 141–176; **184** (1993), No. 4, P. 103–138; **186** (1995), No. 10, P. 89–102.

213. Lerman L.M. and Umanskiĭ Ya.L., Structure of the Poisson action of $\mathbb{R}^2$ on a four-dimensional symplectic manifold. I, II. *Selecta Math. Sov.*, **6** (1987), P. 365–396; **7** (1988), P. 39–48.

214. Liouville J., Note sur l'intégration des équations différentielles de la dynamique, présentée au bureau des longitudes le 29 juin 1853. *J. Math. Pures et Appl.*, **20** (1855), P. 137–138.

215. Lützen L., *Joseph Liouville*. Springer-Verlag, New York (1990).

216. Lyapunov A.M., A new case of integrability of the equation of motion of rigid body in a fluid. In book: *Collection of Works, Vol. 1*. Akad. Nauk SSSR, Moscow (1954), P. 320–324.

217. Maier A.G., On trajectories on oriented surfaces. *Matem. Sbornik*, **12 (54)** (1943), No. 1, P. 71–84.

218. Marsden J. and Ratiu T., *Introduction to Mechanics and Symmetry*. Springer-Verlag, New York (1994).

219. Marsden J. and Weinstein A., Reduction of symplectic manifolds with symmetry. *Reports Math. Phys.*, **5** (1974), P. 121–130.

220. Matveev V.S., Calculation of values of the Fomenko invariant for a saddle–saddle type point of an integrable Hamiltonian system. In book: *Trudy Semin. po Vektor. i Tenzor. Analizu, Vol. 25, Part 1*. Izdatel'stvo Moskovskogo Universiteta, Moscow (1993), P. 75–104.

221. Matveev V.S., Integrable Hamiltonian systems with two degrees of freedom. The topological structure of saturated neighborhoods of saddle–saddle and focus–focus type points. *Matem. Sbornik*, **187** (1996), No. 4, P. 29–58.

222. Matveev V.S., An example of a geodesic flow on the Klein bottle that is integrable by means of a polynomial of degree four in momenta. *Vestnik Moskov. Univ., Ser. Mat. Mekh.*, **52** (1997), No. 4, P. 47–48.

223. Matveev V.S., Quadratically integrable geodesic flows on the torus and the Klein bottle. *Regular and Chaotic Dynam.*, **2** (1997), No. 1, P. 96–102.

224. Matveev V.S., *Singularities of a momentum mapping and the topological structure of integrable geodesic flows*. Thesis, Moscow State Univ. (1996).

225. Matveev V.S. and Oshemkov A.A., Algorithmic classification of invariant neighborhoods for points of the saddle–saddle type. *Vestnik Moskov. Univ., Ser. Mat. Mekh.*, **54** (1999), No. 2, P. 62–65.

226. Matveev V.S. and Topalov P.Ĭ., Jacobi fields of integrable geodesic flows. *Regular and Chaotic Dynam.*, **2** (1997), No. 1, P. 103–116.

227. Matveev V.S. and Topalov P.Ĭ., Conjugate points of hyperbolic geodesics of quadratically integrable geodesic flows on closed surfaces. *Vestnik Moskov. Univ., Ser. Mat. Mekh.*, **53** (1998), No. 1, P. 60–62.

228. Matveev V.S. and Topalov P.Ĭ., Geodesic equivalence of metrics on surfaces and their integrability. *Dokl. Akad. Nauk SSSR*, **367** (1999), No. 6, P. 736–738.

229. Matveev S.V., Universal deformations of special polyhedra. *Uspekhi Mat. Nauk*, **42** (1987), No. 3, P. 193–194.

230. Matveev S.V., Special spines of piecewise linear manifolds. *Matem. Sbornik*, **92(134)** (1973), No. 2, P. 282–293.

231. Matveev S.V., Special spine transformations and the Zeeman hypothesis. *Izvest. Akad. Nauk SSSR, Ser. Matem.*, **51** (1987), No. 5, P. 1104–1115.

232. Matveev S.V., The complexities of 3-manifolds and their list in increasing order of complexity. *Dokl. Akad. Nauk SSSR*, **301** (1988), No. 2, P. 280–283.

233. Matveev S.V., A method of representation of 3-manifolds. *Vestnik Moskov. Univ., Ser. Mat. Mekh.*, **30** (1975), No. 3, P. 11–20.

234. Matveev S.V., Classification of sufficiently large three-dimensional manifolds. *Uspekhi Mat. Nauk*, **52** (1997), No. 5, P. 147–174.

235. Matveev S.V. and Fomenko A.T., Constant energy surfaces of Hamiltonian systems, enumeration of three-dimensional manifolds in increasing order of complexity, and computation of volumes of closed hyperbolic manifolds. *Uspekhi Mat. Nauk*, **43** (1988), No. 1, P. 5–22.

236. Matveev S.V. and Fomenko A.T., Morse type theory for integrable Hamiltonian systems with tame integrals. *Matem. Zametki*, **43** (1988), No. 5, P. 663–671.

237. Fomenko A.T. and Matveev S.V., *Algorithmic and Computer Methods for Three-Manifolds*. Kluwer Academic Publishers, Dordrecht–Boston–London (1997).

238. Matveev S.V., Fomenko A.T., and Sharko V.V., Round Morse functions and isoenergetic surfaces of integrable Hamiltonian systems. *Matem. Sbornik*, **135(177)** (1988), No. 3, P. 325–345.

239. Matveev S.V. and Savvateev V.V., Three-dimensional manifolds with simple special spines. *Colloq. Math.*, **32** (1974), No. 2, P. 83–97.

240. Meyer K.R., Energy functions for Morse–Smale systems. *Amer. J. Math.*, **90** (1968), No. 4, P. 1031–1040.

241. Milnor J., *Morse Theory*. Princeton University Press, Princeton (1963).

242. Milnor J., *Lectures on the h-Cobordism Theorem*. Princeton University Press, Princeton (1965).

243. Mishchenko A.S., Integrating of geodesic flows on symmetric spaces. *Matem. Zametki*, **31** (1982), No. 2, P. 257–262.

244. Mishchenko A.S., Integrating of geodesic flows on symmetric spaces. In book: *Trudy Semin. po Vektor. i Tenzor. Analizu, Vol. 21*. Izdatel'stvo Moskovskogo Universiteta, Moscow (1983), P. 13–22.

245. Mishchenko A.S. and Fomenko A.T., Euler equations on finite-dimensional Lie groups. *Izvest. Akad. Nauk SSSR, Ser. Matem.*, **42** (1978), No. 2, P. 396–415.

246. Morgan J.W., Non-singular Morse–Smale flows on 3-dimensional manifolds. *Topology*, **18** (1979), No. 1, P. 41–53.

247. Moser J.K., On the volume elements on a manifold. *Trans. Amer. Math. Soc.*, **120** (1965), No. 2, P. 286–294.

248. Moser J.K., Various aspects of integrable Hamiltonian systems. In book: *Dynamical Systems* (*Proceedings, CIME, Bressanone, Italy, June 1978*). Birkhäuser, Boston (1980), P. 233–289.

249. Moser J.K., Three integrable Hamiltonian systems connected with isospectral deformations. *Adv. Math.*, **16** (1975), P. 197–220.

250. Moser J.K., *Lectures on Hamiltonian Systems* (*Memoirs of the AMS, No. 81*). AMS, Providence (1968).

251. Nekhoroshev N.N., Action-angle variables and their generalizations. *Trudy Mosk. Mat. Obsch.*, **26** (1972), No. 1, P. 181–198.

252. Novikov S.P., The Hamiltonian formalism and many-valued analogue of Morse theory. *Uspekhi Mat. Nauk*, **37** (1982), No. 5, P. 3–49.

253. Novikov S.P. and Shmeltser I., Periodic solutions of Kirchhoff's equations for the free motion of a rigid body in a fluid and the extended Lyusternik–Shnirel'man–Morse theory. *Funkts. Analiz i ego Prilozh.*, **15** (1982), No. 3, P. 54–66.

254. Nguyen T.Z., On a general position property of simple Bott integrals. *Uspekhi Mat. Nauk*, **45** (1990), No. 4, P. 161–162.

255. Nguyen T.Z., Topological invariants of integrable geodesic flows on the multidimensional torus and sphere. *Trudy Mat. Inst. Steklov*, **205** (1994), P. 73–91.

256. Nguyen T.Z., On the complexity of integrable Hamiltonian systems on three-dimensional isoenergy submanifolds. In book: *Topological Classification of Integrable Systems (Advances in Soviet Mathematics, Vol. 6)*. AMS, Providence (1991), P. 229–255.

257. Nguyen T.Z., *Contact 3-Manifolds. Integrable Hamiltonian Systems, and Exotic Symplectic Structures in* $\mathbb{R}^4$. Preprint: Intern. Centre for Theor. Phys., Trieste (1992).

258. Nguyen T.Z., Symplectic topology of integrable Hamiltonian systems, I.: Arnold–Liouville with singularities. *Compositio Math.*, **101** (1996), P. 179–215.

259. Nguyen T.Z., *The symplectic topology of integrable Hamiltonian systems*. Thèse, Université de Strasbourg (1994).

260. Nguyen T.Z., Singularities of integrable geodesic flows on multidimensional torus and sphere. *J. Geometry and Physics*, **18** (1996), P. 147–162.

261. Nguyen T.Z., Decomposition of non-degenerate singularities of integrable Hamiltonian systems. *Lett. Math. Phys.*, **33** (1995), P. 187–193.

262. Nguyen T.Z. and Fomenko A.T., Topological classification of integrable non-degenerate Hamiltonians on the isoenergy three-dimensional sphere. *Uspekhi Mat. Nauk*, **45** (1990), No. 6, P. 91–111.

263. Nguyen T.Z., Polyakova L.S., and Selivanova E.N., Topological classification of integrable geodesic flows with an additional quadratic or linear in momenta integral on two-dimensional orientable Riemannian manifolds. *Funkts. Analiz i ego Prilozh.*, **27** (1993), No. 3, P. 42–56.

264. Nguyen T.Z. and Polyakova L.S., A topological classification of integrable geodesic flows of the two-dimensional sphere with quadratic in momenta additional integral. *J. Nonlin. Sci.*, **6** (1992), P. 85–108.

265. Okuneva G.G., Some geometrical properties of the reduced configuration space in rigid body dynamics. *Vestnik Moskov. Univ., Ser. Mat. Mekh.*, **41** (1986), No. 4, P. 55–59.

266. Okuneva G.G., Integrable Hamiltonian systems in analytic dynamics and mathematical physics. In book: *Topological Classification of Integrable Systems (Advances in Soviet Mathematics, Vol. 6)*. AMS, Providence (1991), P. 33–66.

267. Olshanetsky M.A. and Perelomov A.M., Completely integrable Hamiltonian systems connected with semisimple Lie algebras. *Invent. Math.*, **37** (1976), P. 93–108.

268. Reyman A.G. and Semenov-Tian-Shansky M.A., Group-theoretical methods in the theory of finite-dimensional integrable systems. In book: *Integrable Systems. Nonholonomic Dynamical Systems (Encyclopaedia of Math. Sci., Vol. 16: Dynamical Systems VII)*. Springer-Verlag, Berlin–Heidelberg–New York (1994).

269. Orel O.E., An analysis of the neighborhood of a degenerate one-dimensional orbit of Poisson action of $\mathbb{R}^2$ on M^4. *Trudy Mat. Inst. Steklov*, **205** (1994), P. 113–130.

270. Orel O.E., The rotation function for integrable problems that are reducible to Abel equations. Trajectory classification of Goryachev–Chaplygin systems. *Matem. Sbornik*, **186** (1995), No. 2, P. 105–128.

271. Orel O.E., The rotation functions in the problem of trajectory classification of geodesic flows on the ellipsoid and the Euler problem in rigid body dynamics. *Vestnik Moskov. Univ., Ser. Mat. Mekh.*, **51** (1996), No. 1, P. 24–32.

272. Orel O.E., A criterion of trajectory equivalence of integrable Hamiltonian systems in neighborhoods of elliptic orbits. A trajectory invariant of the Lagrange problem. *Matem. Sbornik*, **188** (1997), No. 7, P. 139–160.

273. Orel O.E., Algebro-geometric Poisson brackets in the problem of exact integration. *Regular and Chaotic Dynam.*, **2** (1997), No. 2, P. 90–97.

274. Orel O.E., On non-conjugacy of the Euler case in rigid body dynamics and the Jacobi problem about the geodesics on the ellipsoid. *Matem. Zametki*, **61** (1997), No. 2, P. 252–258.

275. Orel O.E. and Takahashi S., Trajectory classification of integrable Lagrange and Goryachev–Chaplygin problems by computer analysis methods. *Matem. Sbornik*, **187** (1996), No. 1, P. 95–112.

276. Orlik P., *Seifert Manifolds*. Springer-Verlag, Berlin–Heidelberg–New York (1972).

277. Oshemkov A.A., Fomenko invariants for the main integrable cases of the rigid body motion equations. In book: *Topological Classification of Integrable Systems (Advances in Soviet Mathematics, Vol. 6)*. AMS, Providence (1991), P. 67–146.

278. Oshemkov A.A., Topology of isoenergy surfaces and bifurcation diagrams for integrable cases of rigid body dynamics on so(4). *Uspekhi Mat. Nauk*, **42** (1987), No. 6, P. 199–200.

279. Oshemkov A.A., Morse functions on two-dimensional surfaces. Coding of singularities. *Trudy Mat. Inst. Steklov*, **205** (1994), P. 131–140.

280. Oshemkov A.A., Description of isoenergetic surfaces of some integrable Hamiltonian systems with two degrees of freedom. In book: *Trudy Semin. po Vektor. i Tenzor. Analizu, Vol. 23*. Izdatel'stvo Moskovskogo Universiteta, Moscow (1988), P. 122–132.

281. Oshemkov A.A. and Sharko V.V., Classification of Morse–Smale flows on two-dimensional manifolds. *Matem. Sbornik*, **189** (1998), No. 8, P. 93–140.

282. Palis J. and de Melo W., *Geometric Theory of Dynamical Systems. An Introduction*. Springer-Verlag, New York–Heidelberg–Berlin (1982).

283. Paternain G.P., On the topology of manifolds with completely integrable geodesic flows. *Ergod. Theory and Dynam. Systems*, **12** (1992), P. 109–121.

284. Paternain G.P. and Spatzier R.J., New examples of manifolds with completely integrable geodesic flows. *Adv. Math.*, **108** (1994), No. 2, P. 346–366.

285. Paternain G.P., Entropy and completely integrable Hamiltonian systems. *Proc. Amer. Math. Soc.*, **113** (1991), No. 3, P. 871–873.

286. Peixoto M.M., On the classification of flows of 2-manifolds. In book: *Dynamical Systems (Proc. Symp. Univ. of Bahia, Salvador, Brasil, 1971)*. Academic Press, New York–London (1973), P. 389–419.

287. Peixoto M.M., Structural stability on two-dimensional manifolds. I, II. *Topology*, **1** (1962), No. 2, P. 101–120; **2** (1963), No. 2, P. 179–180.

288. Peixoto M.C. and Peixoto M.M., Structural stability in the plane with enlarged boundary conditions. *Anais Acad. Brasil. Ciências*, **31** (1959), No. 2, P. 135–160.

289. Perelomov A.M., *Integrable Systems in Classical Mechanics and Lie Algebras*. Nauka, Moscow (1990).

290. Pidkuiko S.I. and Stepin A.M., Polynomial integrals of Hamiltonian systems. *Dokl. Akad. Nauk SSSR*, **239** (1978), No. 1, P. 50–51.

291. Piovan L., Cyclic coverings of Abelian varieties and the Goryachev–Chaplygin top. *Math. Annalen*, **294** (1992), P. 755–764.

292. Pogorelov A.V., *External Geometry of Convex Surfaces*. Nauka, Moscow (1969).

293. Pogosyan T.I., Construction of the bifurcational set in one of the rigid body dynamics problems. *Mekhanika Tverdogo Tela*, **12** (1980), P. 9–16.

294. Pogosyan T.I., Domains of possible motion in the Clebsch problem. Critical case. *Mekhanika Tverdogo Tela*, **15** (1983), P. 3–23.

295. Pogosyan T.I., Critical integral surfaces in the Clebsch problem. *Mekhanika Tverdogo Tela*, **16** (1984), P. 19–24.

296. Pogosyan T.I. and Kharlamov M.P., Bifurcation set and integral manifolds in the problem of motion of a rigid body in a linear force field. *Prikl. Mat. Mekh.*, **43** (1979), No. 3, P. 419–428.

297. Poincaré H., Sur les lignes géodésiques des surfaces convexes. *Trans. Amer. Math. Soc.*, **6** (1905), P. 237–274.

298. Poincaré H., *Les Méthodes Nouvelles de la Mécanique Céleste, T. 1–3*. Gauthier-Villars, Paris (1892/1893/1899).

299. Poisson S., Mémoire sur la variation des constantes arbitraires dans les questions de la méchanique. *J. de l'Ecole Polytechnique*, **8** (1809), No. 15, P. 266–344.

300. Polyakova L.S., Invariants of integrable Euler and Lagrange cases. *Uspekhi Mat. Nauk*, **44** (1989), No. 3, P. 171–172.

301. Polyakova L.S., Topological invariants for some algebraic analogs of the Toda lattice. In book: *Topological Classification of Integrable Systems (Advances in Soviet Mathematics, Vol. 6)*. AMS, Providence (1991), P. 185–208.

302. Prasolov V.V. and Soloviyov Yu.P., *Elliptic Functions and Algebraic Equations*. Faktorial, Moscow (1997).

303. Raffy M.L., Détermination des éléments linéaires doublement harmoniques. *J. de Math.*, **4** (1894), No. 10, P. 331–390.

304. Reeb G., Sur les points singuliers d'une forme de Pfaff complètement intégrable ou d'une founction numérique. *Compt. Rend. Acad. Sci. Paris*, **22** (1946), P. 847–849.

305. Reyman A.G. and Semenov-Tian-Shansky M.A., A new integrable case of the motion of the 4-dimensional rigid body. *Comm. Math. Phys.*, **105** (1986), P. 461–472.

306. Russmann H., Über das Verhalten analytischer Hamiltonscher Differentialgleichungen in der Nahe einer Gleichwichtslosung. *Math. Annalen*, **154** (1964), P. 284–300.

307. Sadov Yu.A., Action-angle variables in the Euler–Poinsot problem. *Prikl. Mat. Mekh.*, **34** (1970), No. 5, P. 962–964.

308. Schlichenmaier M., *An Introduction to Riemann Surfaces, Algebraic Curves and Moduli Spaces*. Springer-Verlag, Berlin–Heidelberg–New York (1989).

309. Scott P., The geometries of 3-manifolds. *Bull. London Math. Soc.*, **15** (1983), No. 56, P. 401–487.

310. Selivanova E.N., Topological classification of integrable Bott geodesic flows on the two-dimensional torus. In book: *Topological Classification of Integrable Systems (Advances in Soviet Mathematics, Vol. 6)*. AMS, Providence (1991), P. 209–228.

311. Selivanova E.N., Classification of geodesic flows of Liouville metrics on the two-dimensional torus up to topological equivalence. *Matem. Sbornik*, **183** (1992), No. 4, P. 69–86.

312. Selivanova E.N., Trajectory isomorphisms of Liouville systems on the two-dimensional torus. *Matem. Sbornik*, **186** (1995), No. 10, P. 141–160.

313. Selivanova E.N., New examples of integrable conservative systems on S^2 and the case of Goryachev–Chaplygin. *Comm. Math. Phys.*, **207** (1999), No. 3, P. 641–663.

314. Selivanova E.N. and Stepin A.M., On the dynamical properties of geodesic flows of Liouville metrics on the two-dimensional torus. *Trudy Mat. Inst. Steklov*, **216** (1996), P. 158–175.

315. Siegel C.L., On the integrals of canonical systems. *Ann. Math.*, **42** (1941), No. 3, P. 806–822.

316. Smale S., Topology and mechanics, I, II. *Invent. Math.*, **10** (1970), P. 305–331; **11** (1970), P. 45–64.

317. Smale S., On gradient dynamical systems. *Ann. Math.*, **74** (1961), No. 1, P. 199–206.

318. Smale S., Morse inequalities for dynamical systems. *Coll. Math. Transl.*, **11** (1967), No. 4, P. 79–87.

319. Smale S., Differentiable dynamical systems. *Bull. Amer. Math. Soc.*, **73** (1967), P. 747–817.

320. Smolentsev N.K., On the Maupertuis principle. *Sibir. Matem. Zh.*, **20** (1979), No. 5, P. 1092–1098.

321. Spatzier R.J., Riemannian manifolds with completely integrable geodesic flows. *Proc. Symp. in Pure Math. (Amer. Math. Soc.)*, **54** (1993), No. 3, P. 599–608.

322. Sretenskiĭ L.N., Motion of the Goryachev–Chaplygin gyroscope. *Izvest. Akad. Nauk SSSR, Ser. Matem.*, **17** (1953), No. 1, P. 109–119.

323. Sretenskiĭ L.N., On some cases of motion of a heavy rigid body with a gyrostat. *Vestnik Moskov. Univ., Ser. Mat. Mekh.*, **18** (1963), No. 3, P. 60–71.

324. Sretenskiĭ L.N., On some cases of integrability of the equations of the gyrostat motion. *Dokl. Akad. Nauk SSSR*, **149** (1963), No. 2, P. 292–294.

325. Steklov V.A., Remarque sur un problème de Clebsch sur le mouvement d'un corps solide dans un liquide indéfini et sur le problème de M. de Brun. *Compt. Rend. Acad. Sci. Paris*, **135** (1902), P. 526–528.

326. Steklov V.A., *On the Motion of a Rigid Body in a Fluid.* Har'khov (1893).

327. Stepin A.M., Integrable Hamiltonian systems. I, II. In book: *Qualitative Methods in the Study of Non-linear Differential Equations and Non-linear Oscillations.* Inst. Matem. Acad. Nauk USSR, Kiev (1981), P. 116–170.

328. Sternberg S., *Lectures on Differential Geometry.* Prentice Hall, Inc. Englewood Cliffs, N.J. (1964).

329. Taimanov I.A., Topological obstructions to integrability of geodesic flows on non-simply connected manifolds. *Izvest. Akad. Nauk SSSR, Ser. Matem.*, **51** (1987), No. 2, P. 429–435.

330. Taimanov I.A., The topology of Riemannian manifolds with integrable geodesic flows. *Trudy Mat. Inst. Steklov*, **205** (1994), P. 150–163.

331. Taimanov I.A., On the topological properties of integrable geodesic flows. *Matem. Zametki*, **44** (1988), No. 2, P. 283–284.

332. Takahashi S., Orbital invariant of integrable Hamiltonian systems. In book: *Fomenko A.T., Kunii T.L., Topological Modeling for Visualization.* Springer-Verlag, Tokyo (1997), P. 349–374.

333. Tatarinov Ya.V., On the study of the phase topology of compact configurations with symmetry. *Vestnik Moskov. Univ., Ser. Mat. Mekh.*, **28** (1973), No. 5, P. 70–77.

334. Tatarinov Ya.V., Portraits of classical integrals in the problem of motion of a rigid body with fixed point. *Vestnik Moskov. Univ., Ser. Mat. Mekh.*, **29** (1974), No. 6, P. 99–105.

335. Tatarinov Ya.V., *Lectures on Classical Dynamics.* Izdatel'stvo Moskovskogo Universiteta, Moscow (1981).

336. Ten V.V., Local integrals of geodesic flows. *Regular and Chaotic Dynam.*, **2** (1997), No. 2, P. 87–89.

337. Thimm A., Integrable geodesic flows on homogeneous spaces. *Ergod. Theory and Dynam. Systems*, **1** (1981), No. 4, P. 495–517.

338. Thurston W.P., *Three-Dimensional Geometry and Topology.* Princeton University Press, Princeton (1997).

339. Thurston W.P., Three-dimensional manifolds, Kleinian groups and hyperbolic geometry. *Bull. Amer. Math. Soc.*, **6** (1982), No. 3, P. 357–381.

340. Toda M., *Theory of Nonlinear Lattices.* Springer-Verlag, Berlin–Heidelberg–New York (1981).

341. Topalov P.Ĭ, An action variable and the Poincaré Hamiltonian in a neighborhood of a critical circle. *Uspekhi Mat. Nauk*, **50** (1995), No. 1, P. 213–214.

342. Topalov P.Ĭ, The homological properties of Fomenko–Zieschang invariant marks. *Trudy Mat. Inst. Steklov*, **205** (1994), P. 164–171.

343. Topalov P.Ĭ, Inclusion of the Klein bottles to the topological classification theory of Hamiltonian systems. *Uspekhi Mat. Nauk*, **49** (1994), No. 1, P. 227–228.

344. Topalov P.Ĭ, Calculation of the fine Fomenko–Zieschang invariant for the main integrable cases in rigid body motion. *Matem. Sbornik*, **187** (1996), No. 3, P. 143–160.

345. Topalov P.Ĭ, Tensor invariants of natural mechanical systems on compact manifolds and the corresponding integrals. *Matem. Sbornik*, **188** (1997), No. 2, P. 137–157.

346. Topalov P.Ĭ, The critical points of the rotation function of an integrable Hamiltonian system. *Uspekhi Mat. Nauk*, **51** (1996), No. 4, P. 147–148.

347. Topalov P.Ĭ, *Calculation of topological invariants of integrable Hamiltonian systems.* Thesis, Moscow State Univ. (1996).

348. Topalov P.Ĭ., The Poincaré map in regular neighborhoods of Liouville critical leaves of an integrable Hamiltonian system. *Regular and Chaotic Dynam.*, **2** (1997), No. 2, P. 79–86.

349. Trofimov V.V., Generalized Maslov classes on the path space of a symplectic manifold. *Trudy Mat. Inst. Steklov*, **205** (1994), P. 172–199.

350. Umanskiĭ Ya.L., The scheme of a three-dimensional Morse–Smale dynamical system without closed trajectories. *Dokl. Akad. Nauk SSSR*, **230** (1976), No. 6, P. 1286–1289.

351. Vershik A.M. and Gershkovich V.Ya., Non-holonomic dynamical systems, geometry of distributions and variational problems. In book: *Integrable Systems. Non-holonomic Dynamical Systems (Encyclopaedia of Math. Sci., Vol. 16: Dynamical Systems VII).* Springer-Verlag, Berlin–Heidelberg–New York (1994).

352. Veselov A.P., Finite-zone potentials and an integrable system on the sphere with a quadratic potential. *Funkts. Analiz i ego Prilozh.*, **14** (1980), No. 1, P. 48–50.

353. Veselov A.P., Two remarks about the connection of Jacobi and Neumann integrable systems. *Math. Zeitschrift*, **216** (1994), P. 337–345.

354. Veselov A.P., Landau–Liefshitz equation and integrable systems in classical mechanics. *Dokl. Akad. Nauk SSSR*, **270** (1983), No. 5, P. 1094–1097.

355. Vey J., Sur certain systèmes dynamiques séparables. *Amer. J. Math.*, **100** (1978), P. 591–614.

356. Vinberg E.B., Hyperbolic reflection groups. *Uspekhi Mat. Nauk*, **40** (1985), No. 1, P. 29–66.

357. Waldhausen F., Eine Klasse von 3-dimensionalen Mannigfaltigkeiten. I, II. *Invent. Math.*, **3** (1967), No. 4, P. 308–333; **4** (1967), P. 88–117.

358. Wang X., The C^*-algebras of Morse–Smale flows on two-manifolds. *Ergod. Theory and Dynam. Systems*, **10** (1990), P. 565–597.

359. Weeks J., *Hyperbolic structures of three-manifolds*. PhD Thesis, Princeton Univ. (1985).

360. Weinstein A., Symplectic geometry. *Bull. Amer. Math. Soc.*, **5** (1981), P. 1–13.

361. Whittaker E.T., *A Treatise on the Analytical Dynamics of Particles and Rigid Bodies*. Cambridge University Press, Cambridge (1927).

362. Williamson J., On the algebraic problem concerning the normal forms of linear dynamical systems. *Amer. J. Math.*, **58** (1936), No. 1, P. 141–163.

363. Williamson J., On the normal forms of linear canonical transformations in dynamics. *Amer. J. Math.*, **59** (1937), P. 599–617.

364. Yamato K., A class of Riemannian manifolds with integrable geodesic flows. *J. Math. Soc. Japan*, **47** (1995), No. 4, P. 719–733.

365. Yehia H.M., New integrable cases in dynamics of rigid bodies. *Mech. Res. Com.*, **13** (1986), No. 3, P. 169–172.

366. Yehia H.M., New integrable cases in the gyrostat motion problem. *Vestnik Moskov. Univ., Ser. Mat. Mekh.*, **42** (1987), No. 4, P. 88–90.

367. Zhukovskiĭ N.E., On the motion of a rigid body having cavities filled with homogeneous liquid. *Zh. Russk. Fiz-Khim. Obsch.*, **17** (1885), No. 6, P. 81–113; No. 7, P. 145–149; No. 8, P. 231–280.

368. Zoll O., Über Flächen mit Schären geschlossener geodätischen Linien. *Math. Annalen*, **57** (1903), P. 108–133.

Subject Index